Telecommunications Technology Handbook

Second Edition

For a listing of recent titles in the *Artech House Telecommunications Library,* turn to the back of this book.

Telecommunications Technology Handbook

Second Edition

Daniel Minoli

Artech House
Boston • London
www.artechhouse.com

Library of Congress Cataloging-in-Publication Data
Minoli, Daniel, 1952–
   Telecommunications technology handbook / Daniel Minoli.—2nd ed.
      p.   cm.— (Artech House telecommunications library)
      Includes bibliographical reference and index.
      ISBN 1-58053-528-3 (alk. paper)
      1. Telecommunication.   2. Data transmission systems.   I. Title.   II. Series.

   TK5101.M565   2003
   621.382—dc22   2003057716

British Library Cataloguing in Publication Data
Minoli, Daniel, 1952–
   Telecommunications technology handbook.—2nd ed.—(Artech House telecommunications library)
   1. Telecommunication systems   2. Telecommunication
   I. Title
   621.3'82

   ISBN 1-58053-528-3

Cover design by Igor Valdman

i) Figures 1.5 and 1.8–1.13 have been reproduced with the prior authorization of the Union
as copyright holder;
ii) the sole responsibility for selecting extracts for reproduction lies with the beneficiary of
this authorization alone and can in no way be attributed to the ITU;
iii) the complete volume(s) of the ITU material, from which the texts reproduced are extracted, can be obtained from:

International Telecommunication Union
Sales and Marketing Division
Place des Nations - CH-1211 GENEVA 20 (Switzerland)
Telephone: + 41 22 730 61 41 (English)
+41 22 730 61 42 (French) / +41 22 730 61 43 (Spanish)
Telex: 421 000 uit ch / Fax: +41 22 730 51 94
E-mail: sales@itu.int / http://www.itu.int/publications

Section 8.3.5 reprinted with permission from IEEE Std 802.16©, "A Technical Overview
of the WirelessMAN—Air Interface for Broadband Wireless Access" by Carl Eklund,
Nokia Research Center, Roger B. Marks, National Institute of Standards and Technology,
Kenneth L. Stanwood and Stanley Wang, Ensemble Communications Inc., *IEEE Communications Magazine*, June 2002; Copyright 2002©, by IEEE. The IEEE disclaims any responsibility or liability resulting from the placement and use in the described manner.

International Standard Book Number: 1-58053-528-3
Library of Congress Catalog Card Number: 2003057716

10 9 8 7 6 5 4 3 2 1

For Gino & Angela, Anna, Emmanuelle, Emile, and Gabrielle.

Contents

Preface

This book follows in the tradition of the first edition of the *Telecommunications Technology Handbook*. This well-received text was written in the late 1980s and looked toward the technologies that were expected to be important in the coming decade. The main theme then was the introduction of *digital* transmission in enterprise networks. This included ISDN and BISDN (ATM) and T1/T3/SONET transmission links. Wireless was given early importance covering satellite/VSAT and terrestrial wireless data. Fiber-optic communication was recognized as important, as were the precursors of SANs (our channel extension chapter), standards (a long chapter), higher speed LANs and MANs (FDDI and so on). Security and network management were highlighted. All of these were important topics that made the book popular in the 1990s. This second edition is intended to have the same scope, reach, impact, and vision. It looks toward the technologies that are expected to be commercially important in the coming decade.

In the early 2000s, the telecommunications industry finds itself facing a challenge. A series of process-level shortfalls in the late 1990s, for example, in terms of accurate demand forecasting, conservative cost analyses for carriers' operations, realistic promises to investors regarding gross margins and profits, depth and innovation of service offerings, emphasis on technology rather than services, seduction into a "planned obsolescence of plant" mind-set that deprecated (and attempted to depreciate) a brand new plant in 2 years, among other shortfalls, all contributed to a sense of malaise from which the industry needs to rescue itself.

The good news is that telecommunications as an industry will continue to grow in the future; the fact, however, is that it will do so a slower real rate. It is critical to understand what is important and what is not. It is important to determine what will get the industry humming again and what will not.

The main goal of the industry in the next 3 to 7 years (and, hence, the theme of this second edition) will be the introduction of *broadband/Ethernet-based/wireless* transmission in enterprise networks. Technologies that are expected to be important include optical networking (all-optical and switched lambdas, also known as intelligent optical networking), Gigabit Ethernet, 10 Gigabit Ethernet, and 10 Gigabit Ethernet in the WAN (including VLANs and RPR), and new customer-practical connectivity options (such as multilink frame relay, transparent LAN services, and MPLS). VPNs will achieve increased penetration both as stand–alone services and as ways to support secure communications. Quality of service is important both for VPN support as well as for packet voice and digital video/multimedia applications.

A majority of industry observers were hoping for a recovery of the industry, starting in 2004–5 and continuing thereafter. This book aims to provide guidance to various industry principals as to which segments and areas are likely to be the most important, as the revitalization process for the industry starts to take shape. As noted the postcrash focus of the industry is expected to be on new applications, securely delivered on broadband/Ethernet-based/IP/wireless enterprise (and carriers') networks.

Acknowledgments

The author thanks Jyothi Chennu, Scott Hannah, Roy Rosner, Pushpendra Mohta, Ben Occhiogrosso, John Amoss, Lucille Bass, Israel Gitman, Tom Oser, Kazem Sohraby, and Jean Francois Rousseau.

The author also wishes to thank Joon Choi and Danny Lahav. Some portions of Chapter 4 are based on the white paper "Optical Transport Network—Solution to Network Scalability and Manageability" by OptiX Networks. OptiX Networks Inc. is a privately held U.S.-based fabless communications semiconductor company backed by strategic investors including TranSwitch Corporation. OptiX designs and supplies to network systems equipment providers, on an OEM basis, very-high-speed silicon semiconductor integrated circuit devices operating at 40 and 10 Gbps, serving the SONET/SDH and OTN market.

The author wishes to thank Mike Henderson, division director, CTO Organization, Transport Switching Systems (TSS) Division of Mindspeed Technologies, Inc., for contributing an excellent Chapter 6 covering SONET. Mindspeed Technologies designs, develops, and sells a complete portfolio of semiconductor networking solutions that facilitates the aggregation, transmission, and switching of data, video, and voice from the edge of the Internet to linked metropolitan area networks and long-haul networks. Since joining Mindspeed Technologies, Mr. Henderson's focus has been on high-speed access to the home. He initiated the splitterless DSL development effort within Mindspeed (then known as Rockwell Semiconductor Systems) and is the lead inventor on a fundamental patent for splitterless DSL technology used in ITU ADSL specification G.992.2. Mr. Henderson has had a number of articles published in the technical press and has made presentations and participated in panels at major trade shows, including Comdex and Networld Interop.

Material for Chapter 7 is based on a quality white paper by PMC-Sierra, Inc., which is thanked for its timely contribution. The May 2002 white paper, authored by Steve Gorshe, is titled "Transparent Generic Framing Procedure (GFP) Technology." PMC-Sierra develops high-speed broadband communications semiconductors and MIPS-based processors for access, metropolitan transport, and optical transport network equipment. Mr. Gorshe is a principal engineer in the Product Research Group and oversees ICs for SONET, optical transmission, and access systems. Mr. Gorshe is a senior member of the IEEE and coeditor for the

IEEE *Communications* magazine's Broadband Access Series. He is the chief editor for the ANSI T1X1 subcommittee and is responsible for SONET and optical network interface standards. Mr. Gorshe has also been a technical editor for T1.105, T1.105.01, T1.105.02, and T1.105.07 within the SONET standard series as well as the ITU–T SG15 G.7041 (GFP) recommendation. He has 24 patents issued or pending and has published several papers. The second section of Chapter 7 is based on a white paper by Mimi Dannhardt, "Ethernet over SONET Technology," also from PMC–Sierra, Issue 1, February 2002. Mimi Dannhardt is a technical advisor in the Product Research Group at PMC–Sierra and oversees ICs for SONET, optical transmission, and access systems. PMC–Sierra, in addition to its broadband semiconductors mentioned above, provides next-generation solutions for core and edge routers, multiservice switches, multiservice provisioning platforms, optical ADMs and cross–connects; 3G wireless base station controllers, and media gateways.

CHAPTER 1

Introduction and Overview

1.1 Introduction

Globally, telecommunications was a $1.2 trillion industry in 2002. During the 1990s, the growth rates of network deployment were reported to be 40% to 100% per year, depending on the industry segment. However, a significant slowdown occurred at the turn of the decade, which observers have variously called "a brutal telecommunications meltdown [1], "a depression," "a crumbling," "a gloom" [2], "woes" [3], "an abysmal predicament," "a telechasm" [4], "an implosion," "a catastrophic decline" [5], "the most devastating meltdown in history" [6], "a tumult" [7], "collapse in demand" [8], "coming to grips with an awful reality" [9], a "telecom slump" [10], "*annus horribilis* … [that] saw the near implosion of the telecommunications industry" [11], "beleaguered industry"[12], "utter crisis" (FCC Chief Michael Powell [13]), "worsening tech-stock debacle" [14], "seemingly endless contraction in telecom spending" [15], "market that continues to deteriorate" [16], and other similar characterizations. A typical quote from press-time financial wires states that the "blight is expected to claim more victims and further sicken the survivors before a healthier and markedly different industry emerges" [17]; another typical quote is "… we continue to see the telecom sector recovery lagging behind…" [18]. President George W. Bush said the United States will "go after" executives who mislead investors, and he called some telecom accounting practices "outrageous."

The good news is that telecommunications as an industry will continue to grow in the future; the sobering fact, however, is that it will do so at a slower real rate, maybe in the 5% to 6% per year level, maybe slightly less in the short term. Perhaps the earlier growth rates were overstated or overcalculated [19–21], considering the dozens of financial restatements that have been issued and the accounting probes and discoveries of "self-dealings" and "bandwidth swaps." Often self-serving constituencies confuse the issue by quoting growth in tertiary factors, such as bytes transported, Web pages published, domain names locked up, and so on, rather than quoting the ultimate true measure of growth, namely, revenue. For example, worldwide telecommunications revenues grew from $450 billion in 1992 to $1.2 trillion in 2002; it is well worth noting that this is only a 10.3%

compounded annual growth rate (CAGR) for that decade. Where is the 40%, 80%, 160%, or 320% yearly growth quoted by some?

Although major positive events occurred during the 1990s, for example, the rollout of the commercialized Internet, the deployment of advanced optics, and the penetration of wireless technologies, there were also negatives. Specifically, there was a near complete abandonment by telecom management of any analytical methodology for assessing demand within a stated statistical confidence level, along with an all-too-frequent abandonment of basic Finance 101 principles [22]. New management people entering the industry had no hard schooling in statistics and forecasting, network design, traffic engineering, economic engineering, or even basic common-sense economics. Demand was promulgated by this management by hearsay, not proper research, and after any sort of real demand cooled down, demand was artificially generated by undertaking "bandwidth swaps" [23–25] among a number of then-existing carriers, as well as massive use of "capitalized labor," to show a more positive bottom line. Consequently, "telecom companies have struggled with hefty debt loads, falling revenues, and loss of billions of dollars of market value" [26, 27]. Probably no other industry could survive while employing the recent business practices of many of the telecom managers in terms of the massive and often unjustified debt build-out, poor inventory management often with 95% to 97% of the products (bandwidth) "sitting on the shelf," unscientific or altogether missing forecasting, lack of valid market research (listening to consultants at trade shows rather listening to the enterprise end users), poor customer communication, and self-inflicted short-fuse plant/platform obsolescence. It is worth noting that nearly all industries try to keep their plant in use for the longest possible time, rather than declaring, by the pen of self-serving agents, that a new high-end plant is completely useless after 3 years. One typical postmortem is as follows:

> The malaise of the past few years can be laid at the feet of those companies that followed flawed business strategies based on false assumptions about the magnitude and nature of the changes within the industry. In addition, the inability to tie the actual growth of capacity to the actual revenue potential from that capacity led to mistaken calculations of actual market sizes. The effects of competition were misjudged, since there was no analytical method to determine the exact capacity that each carrier would have or bring on-line over the period. Revenue projections became vastly more difficult, in part, due to limited factual reporting of actual network usage by the major carriers. The hype surrounding the Internet did not help either [28].

While the disappearance of $3 trillion in investments in the late 1990s and early 2000s [29, 30] ($2 trillion in equity and $1 trillion in bonds) certainly deserves to be seriously studied by way of books, articles, business case analyses, business school dissertations, and establishment of new industry business practices, so as to avoid similarly self-inflicted "meltdowns" in the future, this author resists the temptation to perform said analysis herewith and offer his business guidance for how the industry can proceed at a healthier pace, but reserves the opportunity to make a contribution to that most needed analysis for another venue [31, 32].

In looking at new services and company valuations, one needs to consider the fact that major portions of the $1.2 trillion of telecom yearly revenue is comprised of landline voice, cellular voice, point-to-point data, switched data, and Internet. Consequently, a new service could have a relatively small market potential in absolute dollars, perhaps a "few" billion dollars (say, $5 billion, as an example). Valuations are generally one to three times forward revenue [33]; for the Internet as a whole this places the current valuation at $50 billion to $150 billion. Therefore, the valuation of a new service (and companies providing it) is typically one to three times this "few billions." As indicated earlier, the historical growth of telecom as a whole has been around 10% a year and will likely remain at that rate (or even diminish) in the future. Even considering the growth of the Internet by itself, and placing its "revenue" at $1 billion in 1992 and $50 billion in 2002, the CAGR is a "relatively modest" 48%. (For comparison, mobile services went from an estimated $10 billion in 1992 to an estimated $400 billion in 2002, or a CAGR of 45%, but with a much higher absolute value.) Most individuals as well as corporations have an upper bound of how much of their top line gets spent in telecom. Studies show this figure to be from 2% to 4% of the revenue (business) or income (residential) line. Furthermore, most individuals also have limited time to spend on some potential new service or concept; these individuals have to allocate their time to working, traveling, eating, socializing, family interactions, exercising, thinking, answering the telephone and cell phone, training, reading a book, watching TV, reading the paper, reading e-mails, surfing the Internet, watching a DVD movie, listening to music, and so on (although some of these tasks can be done simultaneously). And, while the growth of a specific industry segment could be large (with growth being a ratio), the absolute value of that market segment, and, hence, of the value of the company equity, is usually small in the short to intermediate term. This important paragraph sets a crucial watermark that should be kept in mind throughout the reading of the rest of the book.

By design, the focus of the discussion in this book is on telecommunications infrastructure. This telecommunications technologies handbook covers major technologies that are likely to impact the rest of the decade. In spite of the slowdown, the main technological and service themes of the industry for the rest of the decade (and, therefore, the scope of this book) are (1) the introduction of *broadband* [34] *Ethernet-based* data transmission in enterprise networks in the local area arena and possibly in the metropolitan area arena (this is via transparent local area network services), (2) the introduction of *wireless* data transmission in enterprise networks and in public (nomadic) wide area networks [35], and (3) the identification and introduction of new practical applications beyond access and transport for the public switched data network we call the Internet. We see these as the three pillars of this industry at this time. In addition to advances in enterprise networks, eventually we will also see generally available *higher speed data services for residential users*. Carriers will need to deploy these technologies in a rational, analytical, and cost-effective manner to support these major trends.

Looking out over the next 3 to 5 years for the rest of the decade, one can assume that there are not going to be any major communications-related *breakthroughs*: The laser is not poised to be invented, the microprocessor is not poised to be introduced, the Internet is not poised to be commercialized, more fiber is not poised to be deployed in conjunction with a capacity-driven need for higher and higher transmission speeds, and cellular telephony is not poised to be launched. As a long shot, there could theoretically be a regulatory "breakthrough," for example, someone might finally decide that indeed we should have a viable competitive environment at the local level. This could only be accomplished by limiting entrants to, say, three per market, and then supporting these three entrants with all sorts of tax incentives for, say, 15 years, which is likely what it takes to establish a well-rooted carrier. Short of technical and regulatory breakthrough, therefore, the industry focus will need to be on *applications* and on *profit-based use of the deployed inventory of long-haul and transoceanic capacity*, which, according to published reports, was 97% idle at the time of this writing. At the same time, the focus will be on maximizing the use of the already-deployed plant.

Drilling down a notch from the three themes listed above, technologies that are expected to be important for the rest of the decade include optical networking (all-optical and switched lambdas, also known as intelligent optical networking); *Gigabit Ethernet* (GbE) and *10-Gigabit Ethernet* (10GbE) in the *local area network* (LAN) environment as well as in the *wide area network* (WAN) environment, including *virtual local area networks* (VLANs) and possibly *resilient packet ring* (RPR); and new customer-practical connectivity options, such as multilink frame relay, *transparent LAN services* (TLS), and *multiprotocol label switching* (MPLS). *Virtual private networks* (VPNs) may achieve increased penetration both as stand-alone intranet capability and as a way to support secure communications; increased use of Internet-based services beyond just public information access (VPN services being just one example) is expected during the decade. Quality of service (QoS) is important both for VPN support as well as for packet voice and digital video/multimedia applications. Secure communications also in support of homeland security are likely to see increased penetration.

Also, what is important for carriers is the introduction of new revenue-generating services and graduation beyond a transmission-capacity-size-based pricing model, just as these carriers have graduated to a large degree from a distance-based pricing model for long-haul communications that had been in place at least from the 1930s through the mid-1990s. In some circles, people feel that unless the telecom industry can learn to exist in an environment where either the transmission bandwidth provided by the carrier to the user doubles every 18 months, while keeping the same price, or where the carrier is able to accept a halving of the price for the same transmission bandwidth every 18 months due to competitive market efficiency (a model similar to the one that has existed and has been successfully dealt with in the chip/computer/consumer-electronics space), then the ravages of continued Chapter 11 events and customer inconveniences and possible eventual abandonment will be an ever-present predicament.

Furthermore, for the foreseeable future, there has to be a realization that while *Internet Protocol* (IP)-based applications consume 20 times more bandwidth globally than voice (and it is always easy to introduce new bandwidth-intensive applications of any sort), voice services (averaged across all carriers) will continue to represent about 80% of the carriers' income; enterprise data services (also averaged across all carriers) will continue to represent about 15% of the carriers' income; and Internet services will represent, on the average, 5% of the carriers' income. Although single-digit shifts in percentages may occur during the next few years, the price erosion is such that the net impacts may well wash out, and the revenue ratios remain in relative (but not identical) proximity to the ones cited.

From April 2000 to at least the middle of 2003, we have seen a constant stream of "negative" articles in the financial press about the telecom industry [36]. On a daily basis, from one to a half-a-dozen negative articles in *just one newspaper,* such as the *Wall Street Journal,* have appeared for the cited period; with an average of three to four articles a day, or about *2,000 negative articles in 2 years.* Why such disappointment by the financial industry? We believe the answer to be simple: Not only is telecom management responsible for the issues listed at the beginning of this chapter impacting business execution (e.g., unscientific or altogether missing forecasting, and so on), but they promulgated what amounted to "get-rich-overnight" schemes, implying that one could achieve sustainable multibillion dollar market capitalizations overnight (the reality is that this may take 7 to 10 years). This management also promulgated that gross margins for companies such as *competitive local exchange carriers* (CLECs) and *DSL local exchange carriers* (DLECs) could be 80% (the reality being that the gross margin is more like 25% to 35%). It was not made clear that the telecom industry is like a utility from an investment point of view: Payback is more in the 7- to 10-year period, assuming though that management is not seduced into either replacing the infrastructure after 2 years or over-building the network (euphemistically labeled as "deploying a scalable network") on day 1, so that utilization of that plant remains in the 5% range for years and the profit is nonexistent (as a result of having to maintain a costly, overbuilt plant).

The fundamental business questions that constantly need to be asked for every project, every service, every idea, and every technology are as follows:

- Who is the paying customer for this service, idea, technology, and so on?
- How much is the customer willing to pay for this service?
- How much is the customer willing or able to spend?
- Will customers settle for a risky new entrant, or will they prefer to deal with a well-established, reliable, well-capitalized incumbent?
- Are there existing services that are very similar to the one in question? In such a case, is the present service at least 25% to 35% cheaper than the *present mode of operation* (PMO)? How long will the arbitrage last or will the original service's price also fall within a short period of time?

Consider an example of the last point above: An individual may be willing to pay $5 for a weekly periodical, but not $20 for that periodical. Furthermore, the individual may be willing to spend $15 a week at most on weekly periodicals—purchasing three different ones—but not $100 a week in periodicals. These questions must be routinely and satisfactorily answered for a new concept, network, deployment, or investment, if there has to be light (and positive bottom line) at the end of the tunnel.

One of the problems is that all carriers offer virtually the same services using the same exact platform from one of the top five equipment manufacturers in the world. There is no service differentiation whatsoever, in which case, the carriers must compete on price only. This, in turn, drags prices down, which, in turn, makes carrier operations nonsustainable. Differentiated services, developed by creative planners and validated via market research while keeping in mind that 80% gross margins will not be achievable with these either, are needed. This de-emphasizes the price competition and fosters sustainability. Also, no carrier can afford to have a single service, a single architecture, a single technology, and a single QoS, such as the *Ethernet competitive local exchanges* (ELECs) had.

1.1.1 Segmentation

This book uses the organization of the Internet architecture to discuss the set of technologies alluded to above. The model is based on five layers as follows:

1. Physical;
2. Network interfaces for transmission-specific media [e.g., Media Access Control (MAC) for LAN technologies];
3. Internet (IP, associated routing protocols);

4. Transport (TCP and UDP);

5. Applications (FTP, Web, SNMP, and associated presentation and session control applications).

1.1.1.1 Physical

An assessment of the field can start with a technology assessment of optics, since this technology is now the workhorse of communications. On the Layer 0 mechanism one can place an all-optical (Layer 1) network that is photonic from end to end. An optical network runs more efficiently if lambdas (high capacity optical user's streams) can be switched and provisioned; this is accomplished with a control mechanism such as the *generalized MPLS* (GMPLS). While waiting for an all-optical network, service providers can employ *dense* (or *coarse*) *wavelength-division multiplexing* (DWDM/CWDM) capabilities in support of broadband. Because of the major embedded base of the Synchronous Optical Network/Synchronous Digital Hierarchy (SONET/SDH), these technologies are here to stay; however, carriers and suppliers may contemplate the introduction of next generation SONET to better serve broadband customers. Wireless technologies, including *wireless personal networks* (WPANs), *wireless LANs* (WLANs), *wireless WANs* (WWANs), fixed wireless, and *free-space optics* (FSO) (e.g., operating at 10 microns), are other Layer 1 bearers that will continue to acquire importance in coming years.

1.1.1.2 Network Interfaces for Transmission-Specific Media

Network interfaces for transmission-specific media include the MAC for LAN technologies. The assessment of the field can then focus on key services at Layer 2 that are expected to play a role in the present decade: *frame relay* (FR), *asynchronous transfer mode* (ATM), and MPLS. Although these technologies were developed and deployed last decade, they will continue to be important for the foreseeable future. The fact remains, however, that many end users are still tied to T1-based (1.544-Mbps) technologies because other connectivity services are still being priced expensively. Carriers need to look at the practical, affordability level to determine how to move beyond T1 in order to open up new revenue possibilities. This entails, among other things, looking at multilink FR and other approaches. One can proceed with the Layer 2 assessment by looking at advances in LAN technologies, particularly GbE and 10GbE. Considering the advantages of an Ethernet-based approach to communications, considerable interest has been evinced in the past few years on another Layer 2 service that we mentioned earlier, namely, TLS (also known as Ethernet Private Line Service). New technologies such as GbE, *Rapid Spanning Tree Protocol* (RSTP), and MPLS, in fact, facilitate the deployment of TLS. Also, recently we have seen a flurry of activities in deploying new Layer 2 metropolitan Ethernet architectures (VLANs, GbE, RPR).

1.1.1.3 Internet and Transport

Layer 3/4 technologies such as IPv4, IPv6, *Stream Control Protocol* (SCTP), *Transmission Control Protocol* (TCP), and *Real Time Protocol* (RTP) are important. Security technologies are important in their own merit, but also in support of VPNs. As noted, Layer 3 (and also Layer 2) VPNs are expected to see increased usage over the decade, as well as Internet usage in general. QoS in packet networks is critical both for the support of VPNs just named, but also to support packetized voice and multimedia.

1.1.1.4 Applications

New applications are critical. During the past 25 years, it has been new applications that have fueled growth in the industry. Generally existing applications (things that have been around for 5–10 years or more) tend to grow from a few percentage points a year (e.g., voice) to perhaps 10 to 20 percentage points a year (e.g., private line services). New technologies supporting old services only serve an arbitrage function in terms of pricing and vendor mix ("out with the old and in with the new"). At each point in time, new applications (e.g., mobile voice/cellular, Internet access, Web hosting, and wireless data) have enjoyed high growth rates (50% to 100% a year or even more). Therefore, the focus of stakeholders has to be on ground-breaking new applications, since, as noted, there will not be ground-breaking new technologies in the short term. Carriers also want to look at *voice over packet* (VoP), particularly with an eye to new services such as advanced call centers using IP-PBX technology. Carriers may also want to pursue digital video, but again with an eye to new applications.

Figure 1.1 depicts the service, service infrastructure, and transport infrastructure model implicit in the previous discussion. Figure 1.2 illustrates at a high level a general topological taxonomy of networks, including various user-level networks, metropolitan access networks, metropolitan

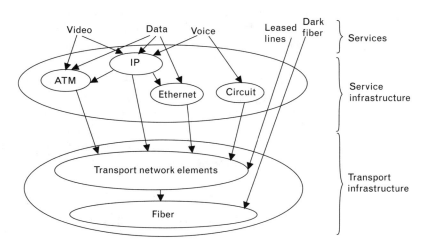

FIGURE 1.1
Typical communications services and service infrastructure.

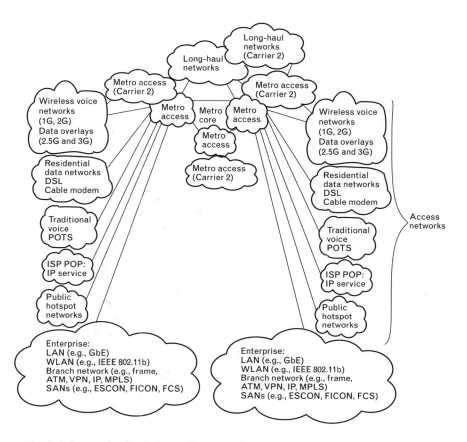

FIGURE 1.2   *Topological example of typical press-time networks.*

core networks, long–haul networks, and global networks (this one implicit in the figure). Also see Figure 1.3. Figure 1.4 provides a proper perspective on revenues, to which we have already alluded.

The industry has major potential on a going-forward basis if some directional imperatives (discussed later) are afforded adequate attention. Figure 1.5 depicts the subscriber base for three key telecom constituencies. As noted later, just the 12 U.S. companies have yearly revenue of about $370 billion and an early 2003 market capitalization of about $392 billion.

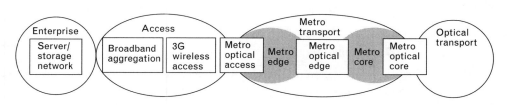

FIGURE 1.3   *Another view of the network taxonomy. (Courtesy PMC-Sierra.)*

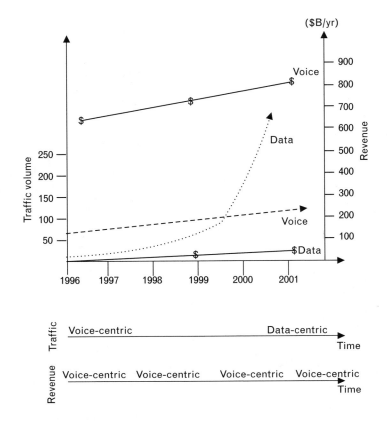

FIGURE 1.4   *Revenue example of typical press-time networks.*

FIGURE 1.5
*Subscriber base for
three key telecom
constituencies. (Source:
ITU World
Telecommunication
Indicators Database
and World
Telecommunication
Regulatory Database,
2001. Reprinted with
permission.)*

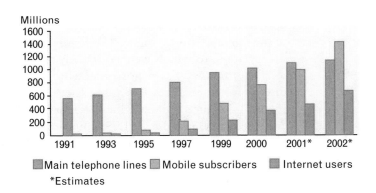

1.1.2   Our Investigation

Given the challenging state of the telecom industry planners, practitioners,
venture capitalists, and readers are interested in the answers to questions
such as these: What are the directions in telecom? What's "hot"? What's
"out"? The short introduction above hints at the fact that several new

business principles have to be put in play for this industry to regain a solid foothold. While it is always fashionable to blame regulations and regulators, we avoid addressing this topic and its merits or lack thereof, except to say that *recent cries by some equipment vendors* are insincere in our opinion, insofar as equipment vendors are simply looking to move equipment. These vendors made no such cries in the late 1990s, only in the slower paced 2000s. They did so as a matter of convenience for the purpose of getting the *incumbent local exchange companies* (ILECs) to purchase their equipment when the emerging carriers all but disappeared; only then did these vendors make self-serving reference to the "need to reform regulation." Hence, sidestepping the regulation question (except for including one published view in the Appendix to this chapter), here are some directional imperatives:

- The industry needs to focus on services, not on impetuously short-fuse life cycle technologies.
- The industry needs a focused set of solutions to technical issues, not a panoply of multiple redundant technical approaches to the same identical opportunity.
- The industry needs to improve the business case in such a manner as to deploy sustainable operations, along with a true understanding of costs, margins, inventory, right-sized solutions, and strategic directions.
- The industry needs to reduce costs for carriers while understanding that operations and administration are typically 50% or more of the total carrier expenditure. At the same time, carriers need to learn to do more with less. This means realigning staff and skill repertoires and hiring new management.
- The industry needs to rethink the pricing model, so that it will be based on perceived value not on intrinsic tie-ins with amounts of "raw materials" (specifically, amount of bandwidth) that comprise the service. Some maintain the position that we have not figured out how to make the Internet pay [37].
- The industry needs to make better use of the deployed assets, rather than a constant injection of new assets that end up having low or insignificant utilization. This also helps to keep the debt/equity ratio in appropriate safe zones.
- The industry needs to conduct primary market research, with direct end-user input and analytical forecasting and design techniques.
- The industry needs objective, conflict-of-interest-free, and knowledgeable analysts.

These, we believe, are the imperative industry directions for the near term. From a technology point of view, broadband communication for local and long-haul and wireless communications are likely to be important technologies for the decade. We are also seeing the deployment of DSL-based Internet services on the part of the ILECs.

For the present investigation, the approach of the book will be to work from the physical layer upward to the application layer. We start in Chapter 2 with a technology discussion of optics. As discussed in Chapter 3, one can deploy an all-optical network on the physical layer apparatus. Optical networks may ultimately be designed in such a manner that lambdas can be switched and provisioned; this can be accomplished with a mechanism such as GMPLS (Chapter 4). DWDM and CWDM capabilities to support broadband services are discussed in Chapter 5. At this time, the major embedded base is SONET/SDH; consequently, carriers and suppliers are contemplating the continued use of this technology topic, which is addressed in Chapter 6, as well as the introduction of next generation SONET services, as addressed in Chapter 7, rather than more "revolutionary" systems. Wireless technologies (WPAN, WLAN, WWAN, fixed wireless, and FSO) are other Layer 1 transmission vehicles that will continue to acquire importance. Chapter 8 reviews this topic.

The book then moves up the architecture model by looking at communication services and network interfaces. Chapter 9 discusses key services at this level that are expected to play a role for the foreseeable future: FR, ATM, and MPLS. The fact remains that many end users are still tied to T1-based technologies; consequently, Chapter 10 looks at the practical and affordability level of how to move beyond T1. This entails looking at multilink FR and other approaches. Chapter 11 continues the communication services and network interfaces discussion by looking at advances in LAN technologies, such as GbE and 10GbE. As already noted, new technologies facilitate the deployment of TLS; Chapter 12 covers the TLS topic in detail. Finally, Chapter 13 looks at one possible application that could stimulate traffic on future networks: digital video. Much more needs to be done at the application level, but that is a topic for another book.

1.2 Protocol Model Baseline

This section reviews the protocol baseline of the Internet that forms the backdrop for the networking architectures under discussion.

1.2.1 Protocol Framework

Table 1.1 contains a snapshot of the state of standardization of protocols used in the Internet, as determined by the *Internet Engineering Task Force* (IETF). It is an October 2001 snapshot of the current official protocol standards list and the best current practices list [38].

TABLE 1.1   STANDARD PROTOCOLS ORDERED BY STD

ACRONYM	TITLE	RFC#	STD#
	Internet Official Protocol Standards	3000	1
	Assigned Numbers	1700	2
	Requirements for Internet Hosts—Communication Layers	1122	3
	Requirements for Internet Hosts—Application and Support	1123	3
	[Reserved for Router Requirements. See RFC 1812.]		4
IP	Internet Protocol	791	5
ICMP	Internet Control Message Protocol	792	5
	Broadcasting Internet datagrams	919	5
	Broadcasting Internet datagrams in the presence of subnets	922	5
	Internet Standard Subnetting Procedure	950	5
IGMP	Host extensions for IP multicasting	1112	5
UDP	User Datagram Protocol	768	6
TCP	Transmission Control Protocol	793	7
TELNET	Telnet Protocol Specification	854	8
TELNET	Telnet Option Specifications	855	8
FTP	File Transfer Protocol	959	9
SMTP	Simple Mail Transfer Protocol	821	10
SMTP-SIZE	SMTP Service Extension for Message Size Declaration	1870	10
MAIL	Standard for the format of ARPA Internet text messages	822	11
NTP	[Reserved for Network Time Protocol (NTP). See RFC 1305.]		12
DOMAIN	Domain names—concepts and facilities	1034	13
DOMAIN	Domain names—implementation and specification	1035	13
	[Was Mail Routing and the Domain System. Now historic.]		14
SNMP	Simple Network Management Protocol	1157	15
SMI	Structure and identification of management information for TCP/IP-based internets	1155	16
Concise-MI	Concise MIB definitions	1212	16
MIB-II	Management Information Base for Network Management of TCP/IP-based internets: MIB-II	1213	17
EGP	[Was Exterior Gateway Protocol (RFC 904). Now historic.]		18
NETBIOS	Protocol standard for a NetBIOS service on a TCP/UDP transport: concepts and methods	1001	19
NETBIOS	Protocol standard for a NetBIOS service on a TCP/UDP transport: detailed specifications	1002	19
ECHO	Echo Protocol	862	20
DISCARD	Discard Protocol	863	21

TABLE 1.1   STANDARD PROTOCOLS ORDERED BY STD (CONTINUED)

ACRONYM	TITLE	RFC#	STD#
CHARGEN	Character Generator Protocol	864	22
QUOTE	Quote of the Day Protocol	865	23
USERS	Active users	866	24
DAYTIME	Daytime Protocol	867	25
TIME	Time Protocol	868	26
TOPT-BIN	Telnet Binary Transmission	856	27
TOPT-ECHO	Telnet Echo Option	857	28
TOPT-SUPP	Telnet Suppress Go Ahead Option	858	29
TOPT-STAT	Telnet Status Option	859	30
TOPT-TIM	Telnet Timing Mark Option	860	31
TOPT-EXTOP	Telnet Extended Options: List Option	861	32
TFTP	TFTP Protocol (Revision 2)	1350	33
RIP1	[Was Routing Information Protocol (RIP). Replaced by STD 56.]		34
TP-TCP	ISO transport services on top of the TCP: Version 3	1006	35
IP-FDDI	Transmission of IP and ARP over FDDI networks	1390	36
ARP	Ethernet Address Resolution Protocol, or converting network protocol addresses to 48-bit Ethernet address for transmission on Ethernet hardware	826	37
RARP	Reverse Address Resolution Protocol	903	38
IP-ARPA	[Was BBN Report 1822 (IMP/Host Interface). Now historic.]		39
IP-WB	Host Access Protocol specification	907	40
IP-E	Standard for the transmission of IP datagrams over Ethernet networks	894	41
IP-EE	Standard for the transmission of IP datagrams over experimental Ethernet networks	895	42
IP-IEEE	Standard for the transmission of IP datagrams over IEEE 802 networks	1042	43
IP-DC	DCN local network protocols	891	44
IP-HC	Internet Protocol on Network System's HYPER channel: Protocol specification	1044	45
IP-ARC	Transmitting IP traffic over ARCNET networks	1201	46
IP-SLIP	Nonstandard for transmission of IP datagrams over serial lines	1055	47
IP-NETBIOS	Standard for the transmission of IP datagrams over NetBIOS networks	1088	48
IP-IPX	Standard for the transmission of 802.2 packets over IPX networks	1132	49
ETHER-MIB	Definitions of Managed Objects for the Ethernet-like interface types	1643	50
PPP	Point-to-Point Protocol (PPP)	1661	51
PPP-HDLC	PPP in HDLC-like Framing	1662	51
IP-SMDS	Transmission of IP datagrams over SMDS	1209	52

TABLE 1.1   STANDARD PROTOCOLS ORDERED BY STD (CONTINUED)

ACRONYM	TITLE	RFC#	STD#
POP3	Post Office Protocol Version 3	1939	53
OSPF2	OSPF Version 2	2328	54
IP–FR	Multiprotocol Interconnect over Frame Relay	2427	55
RIP2	RIP Version 2	2453	56
RIP2–APP	RIP Version 2 Protocol Applicability Statement	1722	57
SMIv2	Structure of Management Information Version 2	2578	58
CONV-MIB	Textual conventions for SMIv2	2579	58
CONF-MIB	Conformance statements for SMIv2	2580	58
RMON–MIB	Remote Network Monitoring Management Information Base	2819	59
SMTP–Pipe	SMTP Service Extension for Command Pipelining	2920	60
ONE–PASS	One-Time Password System	2289	61

1.2.2   Some Notable Trends

Recent trends in the standardization process in IETF are focused on VoIP, wireless, QoS, MPLS, PPP/SONET, and a number of subsidiary topics. These trends highlight areas of current and near-term emphasis. Table 1.2 gives a sense of the recent focuses in the industry, reinforcing the topics addressed in this book.

1.3   Some Modeling Considerations

This section provides some quantitative information on the telecom industry that can serve as a baseline to the discussion.

1.3.1   Size of U.S. Industry

Table 1.3 illustrates both the positive and negative elements of the industry in the early 2000s. Figure 1.6 provides an example of stock/capitalization performance and Table 1.4 shows typical 2002 quotes from Bloomberg, and Table 1.5 provides a perspective. The positives are exemplified by the sheer market power of the industry, which was well in excess of $370 billion of revenues just for the top 12 U.S./North American telecom companies. That is approximately $1,200 a year for every man, woman, and child in the United States. For comparison, the food industry in the United States is about $500 billion a year, or approximately $1,800 a year per person.

TABLE 1.2   RECENT TOPICS OF INTEREST

AREA	RFC	TOPIC
VoIP		
	2871	Framework for telephony routing over IP
	3149	MGCP Business Phone Packages
	3087	Control of Service Context using SIP Request-URI
	3054	Megaco IP Phone Media Gateway Application Profile
	3050	Common Gateway Interface for SIP
	3047	RTP Payload Format for ITU-T Recommendation G.722.1
	3015	Megaco Protocol Version 1.0
	2833	RTP Payload for DTMF Digits, Telephony Tones, and Telephony Signals
	2543	Session Initiation Protocol (SIP)
Wireless/Mobile		
	3141	CDMA2000 Wireless Data Requirements for AAA
	3115	Mobile IP Vendor/Organization-Specific Extensions
	3024	Reverse Tunneling for Mobile IP, revised
	2977	Mobile IP Authentication, Authorization, and Accounting Requirements
	2501	Mobile Ad Hoc Networking (MANET): Routing Protocol Performance Issues and Evaluation Considerations
QOS		
	3140	Per-hop behavior identification codes
	3086	Definition of differentiated services per domain behaviors and rules for their specification
	2998	Framework for integrated services operation over Diffserv networks
	2990	Next steps for the IP QoS architecture
	2963	Rate adaptive shaper for differentiated services
MPLS ... Tunneling		
	3070	Layer 2 Tunneling Protocol (L2TP) over FR
	3035	MPLS using LDP and ATM VC switching
	3034	Use of label switching on FR networks specifications
	3032	MPLS Label Stack Encoding
	3031	Multiprotocol Label Switching Architecture
	2917	Core MPLS IP VPN architecture
	2661	Layer 2 Tunneling Protocol
	2637	Point-to-Point Tunneling Protocol
PPP/SONET		
	3153	PPP Multiplexing
	2823	PPP over SDL using SONET/SDH with ATM-like framing
	2615	PPP over SONET/SDH
Compression		
	3173	IP Payload Compression Protocol (IPComp)
	3096	Requirements for robust IP/UDP/RTP header compression
	3051	IP Payload Compression using ITU-T V.44 Packet Method
	2509	IP header compression over PPP
	2508	Compressing IP/UDP/RTP headers for low-speed serial links
	2507	IP header compression
Supportive Standards		
	3022	Traditional IP Network Address Translator (Traditional NAT)

TABLE 1.2   RECENT TOPICS OF INTEREST (CONTINUED)

AREA	RFC	TOPIC
Supportive Standards	3016	RTP Payload Format for MPEG-4 audiovisual streams
	2989	Criteria for evaluating AAA protocols for network access
	2903	Generic AAA Architecture
	2892	Cisco SRP MAC Layer Protocol
	2869	RADIUS extensions
	2865	Remote Authentication Dial In User Service (RADIUS)
	2960	Stream Control Transmission Protocol
	2748	COPS (Common Open Policy Service) Protocol
Other		
	3130	Notes from the state of the technology: DNSSEC
	3126	Electronic signature formats for long-term electronic signatures
	3125	Electronic signature policies
	3033	Assignment of the information field and protocol identifier in the Q.2941 Generic Identifier and Q.2957 User-to-User Signaling for the Internet Protocol
	2979	Behavior of and requirements for Internet firewalls
	2947	Telnet encryption: DES3 64-bit cipher feedback
	2939	Procedures and IANA guidelines for definition of new DHCP options and message types
	2929	Domain Name System (DNS) IANA considerations
	2916	E.164 number and DNS
	2914	Congestion control principles
	2906	AAA authorization requirements
	2816	Framework for integrated services over shared and switched IEEE 802 LAN technologies
	2815	Integrated service mappings on IEEE 802 networks
	2764	Framework for IP-based VPNs
	2757	Long thin networks
	2729	Taxonomy of communication requirements for large-scale, multicast applications
	2728	Transmission of IP over the vertical blanking interval of a television signal
	2725	Routing Policy System Security
	2719	Framework Architecture for Signaling Transport
	2703	Protocol-Independent Content Negotiation Framework
	2669	DOCSIS MIB Cable Device Management Information Base for DOCSIS-compliant cable modems and cable modem termination systems
	2663	IP Network Address Translator (NAT) terminology and considerations
	2542	Terminology and goals for Internet fax
	2541	DNS security operational considerations

The excesses of the late 1990s were clearly felt in the early 2000s. The ravages of not undertaking statistically valid planning, forecasting, and risk analysis are very clear in the depressed stock value and the market capitalization of the telecom companies, as exemplified by a snapshot of the financial markets at the time of this writing, exemplified in Table 1.3.

TABLE 1.3   A SNAPSHOT OF THE INDUSTRY AS OF FEBRUARY 6, 2003

SYMBOL	MARKET	MARKET CAP (FEBRUARY 6, 2003)	REVENUES (2001)	GROSS INCOME (2001)	NET INCOME (2001)
T	NYSE	$13,938,076,400	$52,550,000,000	$26,454,000,000	$6,983,000,000
WCOEQ	Other OTC	$453,284,685	$35,179,000,000	$20,440,000,000	$1,384,000,000
FON	NYSE	$10,904,662,020	$26,071,000,000	$13,076,000,000	($1,408,000,000)
AWE	NYSE	$16,281,390,500	$13,610,000,000	$7,582,000,000	($963,000,000)
PCS	NYSE	$3,997,986,040	$26,071,000,000	$13,076,000,000	($1,408,000,000)
LU	NYSE	$5,906,255,340	$21,294,000,000	$2,058,000,000	($16,226,000,000)
NT	NYSE	$8,894,156,040	$17,511,000,000	$3,344,000,000	($27,302,000,000)
CSCO	Nasdaq-NM	$95,516,258,260	$22,293,000,000	$11,072,000,000	($1,014,000,000)
VZ	NYSE	$103,548,250,000	$67,190,000,000	NA	$389,000,000
BLS	NYSE	$42,774,850,460	$24,130,000,000	NA	$2,570,000,000
SBC	NYSE	$82,374,236,430	$45,908,000,000	NA	$7,242,000,000
Q	NYSE	$7,476,524,370	$19,695,000,000	$12,584,000	($4,023,000,000)
Total		$392,065,930,545	$371,502,000,000	97,114,584,000	(-33,776,000,000)

Note: AT&T's revenues are presplit.

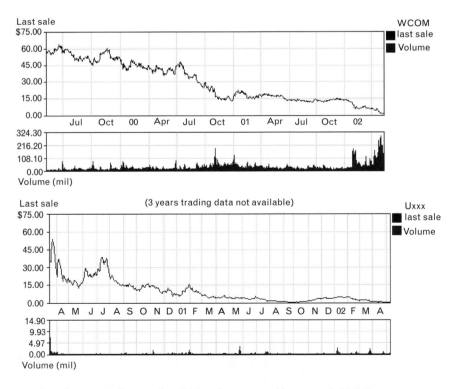

FIGURE 1.6   *Stock performance/valuation of an IXC and a questionably managed CLEC.*

TABLE 1.4   A MEDLEY OF BLOOMBERG HEADLINES FROM MID-2002 AND MID-2003

05/10/2002 17:02 PM. U.S. Stocks Fall, Led by Computer Shares; Microsoft Declines

…from the three-month daily average. Almost two stocks declined for every one that gained on the Big Board.

Telephone Shares Tumble

All 13 members of the S&P index of phone service companies fell. The group lost 3.7 percent, extending the decline this year to 32 percent. Investors are concerned repaying billions of dollars in debt issued to fuel acquisitions and build networks may get more difficult because of overcapacity and falling prices.

SBC Communications lost 90 cents to $30.45 and Verizon Communications Inc. dipped 80 cents to $39.35.

Qwest Communications Inc. dropped 86 cents to $5.04 after saying it will take longer to pay sales commissions to lower expenses. The fourth-biggest local-phone company has a $3.39 billion bank loan due at year's end.

WorldCom Inc. tumbled 43 cents to $1.58 a day after $32 billion of its debt was downgraded to junk by Moody's Investors Service. The second-biggest long distance phone company has plunged 89% this year.

Western Wireless Corp. plunged $2.33 to $3.35 after the provider of mobile-telephone service in 19 western U.S. states said its first-quarter loss widened as U.S. sales fell.…

05/14/2002 18:48. WorldCom Sets One-Day Trading Record for U.S. Stock

Clinton, Mississippi, May 14 (Bloomberg) – WorldCom Inc. fell in the most active day for a U.S. stock, with 670.5 million shares traded, after Standard & Poor's said it will remove the second-largest U.S. long-distance phone company from its main index.

The company eclipsed the record set by Enron Corp. on Nov. 28 after Dynegy Inc. scuttled plans to buy the now-bankrupt energy trader. WorldCom shares fell 20 cents, or 14 percent, to $1.24. More than 150 million shares were exchanged in the last half-hour of trading alone.

WorldCom's credit ratings were cut to junk last week. The company last month forced Chief Executive Bernard Ebbers to resign after profit tumbled and debt increased to $30 billion. Shares of companies that are removed from the S&P 500 index often fall because portfolio managers must sell their shares.

"They're not a player anymore," said Brian Bruce, who manages $8 billion for PanAgora Asset Management in Boston and owns WorldCom shares. "It's very likely that they'll be reorganized in some form," including a possible bankruptcy filing, he said.

05/17/2002 17:21. BellSouth to Cut 4,000–5,000 Jobs After Sales Fall

Atlanta, May 17 (Bloomberg) – BellSouth Corp., the biggest local-telephone company in nine southeastern U.S. states, will eliminate 4,000 to 5,000 jobs after sales fell.

… Job-placement firm Challenger, Gray & Christmas Inc. has said communications companies announced plans to shed 120,698 jobs in the first four months of 2002, 31 percent more than in the same period last year. Investors have said thousands more jobs in the industry likely will be cut to bring costs in line with lower spending on calls, other services and equipment.

The company also trimmed its estimate for capital spending, excluding Cingular, to $4.2 billion to $4.4 billion, from a previous estimate of as much as $5 billion. Last year, as it eliminated 4,200 jobs, BellSouth spent $6 billion.

SBC, BellSouth's partner in Cingular, on Tuesday announced that it was eliminating 5,000 jobs this quarter after reporting its first loss in almost five years in the first period.

BellSouth today cited heightened competition and "regulatory pricing pressures" for the cutbacks.

06/26/2002 07:03. Alcatel Cuts Forecasts, Sees Full-Year Operating Loss

Paris, June 26 (Bloomberg) – Alcatel SA, Europe's biggest maker of telecommunications equipment, abandoned a profit forecast for this year as clients cut spending. The stock plunged to its lowest level in at least thirteen years.

The company expects an operating loss for 2002, said Chief Financial Officer Jean-Pascal Beaufret. Alcatel no longer predicts sales will rebound from the second quarter and said costs for closing factories and firing workers will double to 1.2 billion euros ($1.2 billion) this year. The company plans to eliminate another 10,000 jobs by the end of next year.

TABLE 1.4   A MEDLEY OF BLOOMBERG HEADLINES FROM MID-2002 AND MID-2003 (CONTINUED)

Sales at phone-equipment makers have plunged as companies such as Deutsche Telekom AG and Vodafone Group Plc slashed investments and network operators like Global Crossing Ltd. went bust. WorldCom Inc., an Alcatel client, yesterday forecast losses for 2001 and the first quarter of 2002 after discovering it had inflated profit.

"We're at panic level," said Boris Boehm, who helps manage about $5.8 billion at Nordinvest and sold some of his Alcatel stock today. "This is not the end of the slide for Alcatel."

Alcatel shares fell as much as 2.52 euros, or 27 percent, to 6.83. Today's drop is the steepest since September 1998, when the stock shed 38 percent after Chief Executive Officer Serge Tchuruk said the company would miss full-year profit. Rivals such as Ericsson AB and Nokia Oyj also plunged.

"Evaporating Demand"

Tchuruk yesterday said he still believes in the long-term growth potential of the industry even as demand "evaporates." "When the operators reduce their spending, we take it full in the face," he said at a conference in Paris. "But I don't think they can break capital spending as much for very long."

Nokia cut its sales growth forecast last week. Lucent Technologies Inc., another rival, said this month that sales would decline for a fifth straight quarter. The company is paring another 6,000 jobs by September, after halving its workforce in the past 18 months.

At Alcatel, sales in the second quarter will probably be "about flat" compared with the first, Beaufret said. Alcatel had previously said it expected revenue growth to pick up in the second quarter.

"This industry is a long way from the bottom," said Susan Anthony, an analyst at Credit Lyonnais Securities who recommends investors sell Alcatel shares.

Alcatel is also trimming jobs in order to reduce costs by about 12 percent by the end of next year. It was already reducing its workforce by a third and selling factories to lower costs. Now, the company aims to cut its workforce to about 70,000 by the end of next year, down from about 80,000 at the end of this year and 99,000 at the end of 2001.

New measures are needed "in the environment equipment makers are facing today," Beaufret said.

Alcatel cut 10,000 jobs last year to be able to break even on quarterly sales of 5 billion euros rather than the 6 billion euros needed in 2000. This year, the break-even point stands at 4.5 billion euros, the company said. In 2003, Alcatel aims to break even on sales of 4 billion euros.

Ericsson to Cut as Many as 13,000 More Jobs on Losses

Stockholm, 4/29/2003—Ericsson AB, the world's largest maker of wireless networks, will cut as many as 13,000 jobs, or a fifth of the workforce, after its eighth straight quarterly loss. Ericsson's headcount will fall to 47,000 in 2004, or back to 1968 levels, Chief Executive Officer Carl-Henric Svanberg said at a press meeting. Ericsson earlier said it would slash its workforce to below 60,000 from 105,000 at the end of 2000. The first-quarter loss was 4.3 billion kronor ($520 million).

"Cuts as rough as these show they're really reacting to the situation," said Marko Alaraatikka, who helps manage the equivalent of $2.7 billion at Evli Investment Management, including Ericsson shares. Svanberg, who took over as CEO this month, is stepping up cost reductions as sales are sliding for a third year. Predecessor Kurt Hellstroem already slashed more than 40,000 jobs. Sweden's Ericsson and rivals including Alcatel SA and Nortel Networks Corp. have shed more than 200,000 jobs since 2000 as phone companies reined in spending on base stations and switches. The 127-year-old company's stock, one of Sweden's most widely held, has lost 96 percent since the March 2000 peak, wiping out about $280 billion in market value.

Sales for the quarter fell 30 percent to 25.9 billion kronor from 37 billion kronor. Ericsson said it expects second-quarter sales to rise "slightly" from the previous three months. Orders dropped 35 percent to 27.1 billion kronor. "Operators are holding back on already tight budgets," Svanberg told reporters. "We must ensure we return to profit even if demand continues to fall." Industry sales of wireless networks, which account for two-thirds of Ericsson's revenue, will drop by more than 10 percent this year, the company said. Ericsson has more than a quarter of the market for cellular-phone networks.

Ericsson's dependence on sales of wireless networks, a market some analysts say will shrink 20 percent this year, makes it more vulnerable than Alcatel. The French company's offerings range from satellites to high-speed

Table 1.4 A Medley of Bloomberg Headlines from Mid-2002 and Mid-2003 (continued)

> Internet equipment, providing some protection against the slowdown. Alcatel has also so far come farther than Ericsson in slashing costs, investors said. Alcatel CEO Serge Tchuruk today said he expects demand from phone carriers this year to contract "a few points" more than the 15 percent drop he predicted earlier in 2003.
>
> Measures announced before today would have cut Ericsson's headcount to 54,000 by year-end, beating its original target, Ericsson said. The company had 61,000 employees on March 31, 2003. About half the job cuts will be in Sweden, Svanberg said. "The name of the game is still to cut costs," said Stuart O'Gorman, who holds Alcatel shares but not Ericsson stock in the $700 million he helps manage at Henderson Global Investors. "The encouraging thing is that finally they're admitting that revenue won't massively rebound back to bubble levels," Henderson's O'Gorman said. Ericsson's CEO "seems to be saying that the company should be profitable even if sales don't rebound. That's encouraging and puts them back in line" with competitors.

Considering the business proposition comprised of the deployed facilities, the customer base, the barriers to entry [e.g., rights of way (ROW), capital, physical assets, technical expertise, and computer support systems and databases], the established reputation, and so on, a telecom company should be valued at at least three to five times annual revenues, maybe even more. This, of course, depends on the debt load of the company, which should always be subtracted from the prima facie value of the company, just like liabilities are subtracted from assets when calculating the net worth of an individual. The snapshot included in Table 1.3 shows a 1-time actual valuation, which is clearly very low (note, as an example, that AT&T's value is about 0.25-times yearly revenue). This says, in effect, that investors (at least temporarily) lost faith in the industry and the current management at these companies. The top three U.S. long-distance companies have piled up about $80.5 billion of debt building communication networks to reach

Table 1.5 Fifty Major Telecom Bankruptcies

Major Service Provider Bankruptcies

Long-Haul and Fiber Network Wholesalers:
Aleron, 360networks, Digital Teleport, Ebone/GTS, Enron Broadband, FLAG Telecom, Global Crossing, GST, Impsat, KPNQwest, Sigma Networks, Sphera, Storm Telecommunications, Teleglobe, Telergy, Velocita, Viatel, Williams Communications

CLEC/DLEC and Cable Companies:
Adelphia Communications, Adelphia Business Solutions, Broadband Office, Convergent Communications, Covad Communications, e.spire, FastComm, ICG Communications, Metromedia Fiber Network (MFN), McLeodUSA, Mpower, Network Plus, New Global Telecom, NorthPoint, Rhythms NetConnections, OnSite Access, Spectrotel, XO Communications, WINfirst, Yipes Communications, Zephion

Wireless/Satellite:
Advanced Radio Telecom (ART), OmniSky, Metricom/Ricochet, Nextwave Telecom, Motient, Teligent, WinStar, GlobalStar, Iridium, StarBand

ISP:
PSINet, Ardent Communications, colo.com, Excite@Home, Exodus, iBeam, NetRail

Source: Converge! Network Digest, retrieved from http://www.convergedigest.com/Mergers/bankruptcy.asp, June 25, 2002.

more customers. Investors, concerned that the once-lucrative long-distance business may not recover, are therefore demanding a premium to the buy their bonds. The bankruptcy of Worldcom and the $100 billion loss of AOL Time Warner are other data points in this continuum of financial/planning mishaps. It follows that service providers are "squeezed between cut-rate competitors and capital markets that will not finance growth ahead of profits" [37]. Hence, "telecom companies are in such a bad financial state that they are not spending money on software or other phone-equipment products" [39], sending a chill through the entire high-tech industry. While some may have been lulled into thinking that this was a only a 2002–2003 issue, this spending behavior and its ensuing consequences for the supplier-industry at large are apt to repeat themselves if the following systemic issues are not addressed head-on: mindless overbuilding of networks, overstating potential returns, ignoring end users, avoiding primary market research, developing technologies in search of a problem, ostracizing analytical methods for forecasting and planning, and relying on novices in the telecom field, or hiring people who do not understand the true issues.

Worldwide telecom revenues were predicted to grow from under $1.2 trillion in 2002 to almost $1.9 trillion in 2007 [28]. While the overall CAGR is 10.0%, *North America* (NA) has the slowest growth rate at 8.7% annually, but maintains its position as the region with the largest telecommunications services revenue. As the most mature market overall with respect to telecom services, its growth is most dependent on new services as opposed to subscriber growth. *Europe/Middle East/Africa* (EMEA) exhibits somewhat stronger revenue growth at 9.3% per year, due to growth in the wireless market and growth from less developed subregions of Eastern Europe, the Middle East, and Africa. The faster growing regions are *Latin America and the Caribbean* (LAC) and *Asia/Pacific* (AP). LAC is dominated by the fast-growing economy of Mexico, and though the large regional economies in Argentina and Brazil are presently under considerable pressure, the forecast is that continuing pent-up demand for telecommunications service—much of which is satisfied by wireless services—will account for its high CAGR relative to the worldwide composite. The AP region is experiencing the highest growth overall. The size of its underserved populations and its generally higher GDP growth rates, combined with several countries whose economies rely heavily on high-tech industries, is a formula for growth.

A glimmer of hope for new carriers appeared in May 13, 2002, when the U.S. Supreme Court upheld *Federal Communications Commission* (FCC) rules forcing the ILECs to provide wholesale prices to competitive carriers. The FCC had set rates that the ILECs can charge competitors based on hypothetical costs the competitors would incur building a new telephone network from the ground up using the best technology available, not the traditional costs the ILECs incurred in actually building their network. The traditional costs are much higher. The ILECs argued that the FCC formula makes it impossible for them to recover the money they spent to build and upgrade their networks. In addition, the Justice Department stated that the FCC has the authority to force the ILECs to bundle portions of their network together at the request of a competitor instead of leasing each component to a competitor separately. Some view this as a victory for consumer and business customers and the competitive carriers [40], while others view the decision as "too little too late" to reverse a sharp fall in the number of CLECs, which has been estimated, from information provided by the Association of Local Telecommunications Services, at 180 in 1998, 330 in 1999, 190 in 2000, and 80 in 2001.

However, soon thereafter, ". . . in a big win for the Baby Bells, a federal appeals court ordered the FCC to overhaul regulations aimed at introducing competition into the markets for local telephone service and high-speed Internet access. The U.S. Court of Appeals for the District of Columbia Circuit ordered the FCC to reconsider rules that now require the Bells to 'unbundle' and rent individual parts of their local-phone networks to their smaller competitors at heavily discounted rates... the Court sided with the Bells, ruling that the FCC had gone too far with its unbundling rules..." [41]. In early 2003, the FCC signaled an (apparent) desire to curtail or eliminate the wholesale prices available to CLECs under the *Unbundled Network Element Platform* (UNE-P) arrangement.

1.3.2   Growth of the Internet During the Late 1990s into the Early 2000s

This section looks at the growth of the Internet, based on materials published by the ITU [42]. Back in 1988, only seven countries were connected to the U.S. National Science Foundation Internet backbone. Access to the Internet grew to 200 countries a decade later. The total number of Internet users in 2000 was estimated to be 315 million. Around 5% of the world is now on-line (see Figure 1.7). However, we may be approaching saturation. In many developed markets, those who want to be on-line already are. The growth rate in the number of users declined to its lowest level ever (35%) in 2000. This is partly due to the downturn in the Internet economy, reflected by the sharp fall in dot-com stock prices. Although almost a third of people in developed countries are on-line, this figure is less than 2% in developing countries. It is no surprise that the birthplace of the Internet, the United States, is the world's largest Internet market with almost 100 million regular users at the beginning of 2001. The next largest is Japan with some 39 million users, of which over half can also access the Internet from mobile phones.

There were 100 million Internet hosts (accessible computer nodes) at the beginning of 2001. The host computer growth rate dropped in 2000, as a reflection of the downturn in the dot-com economy. The most popular host suffixes are ".com" with 33 million and ".net" with 23 million; together they account for 60% of all hosts. The country suffix with the

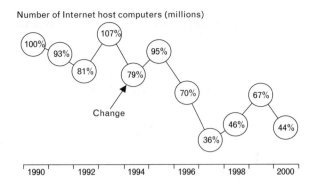

FIGURE I.7
*Growth of the Internet
in the 1990s.
(Source: ITU, 2001.
Reprinted with
permission.)*

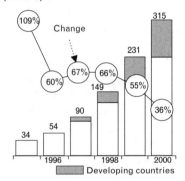

most hosts is Japan (.jp), which had 3.4 million in 2000. On a per capita basis, the nation with the most hosts using its country code is the tiny Pacific Island of Niue (.nu) with almost five hosts per person, though few of these are located in that territory. The growth of Web servers was significant in the early days—rising from 75,000 at the end of 1995 to more than 25 million by the end of 2000. Growth in 2000 was 158%. Web servers accounted for 25% of all Internet hosts at the time of this writing. Web servers terminating with the sought-after ".com" suffix number 15 million and account for 57% of the total. The country domain suffix with the most Web servers is the United Kingdom (.uk) at 1.7 million, or 7% of the total.

The point at which Internet capacity exceeded international telephone circuit capacity for the first time happened in 2000. Worldwide international Internet capacity was close to 300 Gbps, almost five times greater than in 1999. One noticeable aspect of the geography of the international Internet is the preeminence of the United States. For historical reasons, many countries route Internet traffic to the United States (see Figure 1.8). Unlike international telephone cables, where countries on both sides of the link pay for half the cost, countries wishing to connect to the United States must pay for the full cost of the circuit. In October 2000, a new ITU-T Recommendation (D.50) called on companies to negotiate with each other more equitable ways of sharing the cost of international Internet circuits.

FIGURE 1.8
*Internet traffic
patterns.
(Source: ITU,
adapted from
TeleGeography.)*

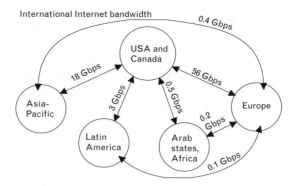

Though there were approximately 15,000 *Internet service providers* (ISPs) around the world in the early 2000s, there has been a high degree of market concentration with the top 20 serving approximately 45% of the market. The world's largest ISP at press time was America Online of the United States with more than three times more subscribers than the second largest, Germany's T-Online. Market concentration is expected to increase with bigger ISPs buying up smaller ones and larger ISPs merging.

Figure 1.9 (ITU sources) depicts the percentage of population using the Internet in various countries. At the beginning of the year 2000, there were 120 million Internet *subscribers* around the world, or just under a third of the estimated number of *users*. The ITU notes that geographic isolation combined with an intense interest in all things new, good knowledge of English, cold weather, and high-income may explain Iceland's lead [42]. Like Iceland, all the other Nordic countries have high levels of Internet penetration. Canada, ranked fourth, has a higher penetration than its southern neighbor, the United States. Two Asian countries make the top 10, the Republic of Korea and Singapore. The indicator *percentage of*

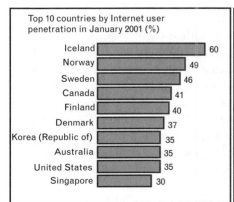

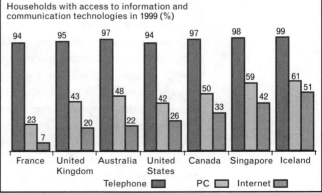

FIGURE 1.9   *Demographics of Internet use. (Source: ITU, 2001. Reprinted with permission.)*

households with a telephone is used by policy makers to measure the development of universal telephone service. The *percentage of households with Internet access* is a measure of universal Internet service. Among countries that collect this statistic, Iceland ranks first, with over half its households having access to the Internet (at the end of 1998). It is worth comparing the number of households with Internet access to those with personal computers and telephone service, because these are prerequisites for dial-up Internet access. In Singapore, which has the second highest home Internet access level in the world, the incumbent telephone operator provides all of its telephone subscribers with an Internet account. In reality, only the 59% of the households in Singapore with a personal computer can make use of this. According to one survey of 18 economies, in October 2000, people in the Hong Kong *special administrative region* were the most intensive users of the Internet, averaging more than 10 hours per month. The least-intensive users were the Irish; they logged on an average of just 4 hours per month.

It is interesting that statistics about the Internet are always in terms of users, hosts, Web pages, follow-through clicks, megabytes transmitted, and transactions sent—rarely if ever are the statistics given in terms of actual revenue. This is because, in spite of all the hoopla, the figure is relatively low. The best estimate of the worldwide Internet revenue (on the service side) is around $50 billion a year in 2002 [43] or 5% of the total—the cacophony of hype [44] would, however, lead one to believe that the numbers are different. This equates to about 220 million accounts each paying $20 per month for access. If the population of accounts were to double and there was no price erosion, the cumulative revenue would double to $100 billion a year. Considering the recent slowdown in Internet population growth to, say, 20% a year, and about 5 years would be needed to reach this figure.

It is critical that providers understand the economics of Internet services. One hears all too often about how the amount of traffic on the Internet exceeds that of voice. We have already noted that, although IP-based applications consume 20 times more bandwidth globally than voice (and it is always easy to introduce new bandwidth-intensive applications of any sort), voice services (averaged across all carriers) will continue to represent about 80% of the carriers' income. Table 1.6 shows that the revenue (economic wealth creation) for 1 gigabit of voice traffic transported (domestically) is now $15.63, while the revenue for 1 gigabit of Internet traffic is $0.14 (and this in spite of the fact that by 2002 long-distance calling rates dropped to about 7 cents a minute—less than half the 15-cent average of 5 years earlier). So, why all the fuss? Because the population of voice users has traditionally grown at 2% to 3% per year, while the population of Internet users has been growing at the 100% (mid-1990s) and 30% (early 2000s) per year range. However, it appears self-evident that when the number of Internet users reaches saturation, revenue generated simply by the transport side of the value proposition (namely, Internet access, backbone transport

TABLE 1.6   REVENUE PER BIT OF VOICE AND INTERNET TRAFFIC

YEAR	COST/MINUTE	BPS	REVENUE PER BIT	REVENUE FOR 1 GB	REVENUE FOR 10 GB
1997	0.12	64,000	$0.000000031250	$31.25	$312.50
2002	0.06	64,000	$0.000000015625	$15.63	$156.25

YEAR	MONTHLY COST (200 HOURS)	BPS	REVENUE PER BIT		
1997	19.99	56,000	$0.000000000496	$0.50	$4.96
2002	39.99	384,000	$0.000000000145	$0.14	$1.45

system, and so on) will be low and capped. The industry, therefore, desperately needs truly new applications, for example, fees-for-service content and e-commerce, to keep the revenue engine going.

The economic-generation engine is clearly moving in directions other than toward simple Internet access/transport and population growth, since both of these factors have asymptotic limitations. Internet pricing is an important determinant of access. There are two components to a traditional dial-up Internet price. The first is the ISP price. This can be a flat fee for unlimited use or a certain number of hours or it can vary with time. The second component of the price is usage charges for local telephone service. In some countries local calls are not charged; a flat rate is applied. Countries without local telephone charging, such as Canada, tend to have the lowest tariffs and high levels of Internet penetration. At the other extreme, countries such as Belgium with so-called "free" Internet access—meaning there is no separate ISP charge—may actually have very high tariffs if the local call charge is high. The average price for 30 hours of dial-up access among the 30-odd member countries of the *Organisation for Economic Cooperation and Development* (OECD) was $56 in June 2000. As stated, revenue from other than access and transit (transport) is needed.

Telephone companies in many countries are major beneficiaries of dial-up Internet traffic since they earn revenue from telephone usage charges. Of the top 10 ISPs in the world, half emerged from telephone companies. In many European countries, Internet dial-up traffic accounts for around one-third of local telephone traffic. In most countries, Internet dial-up traffic far exceeds international telephone calling. Demand is growing for faster Internet access speeds. The initial response for meeting demand for higher bandwidth was *Integrated Services Digital Network* (ISDN) lines. ISDN provides only marginally faster speed (64 Kbps) than conventional dial-up access. By the beginning of 2000, there were 24 million ISDN subscribers in the world, primarily in North America, Western Europe, and Japan, accounting for 7% of all telephone lines in service then. Less than 3% of Internet subscribers had broadband local access offerings such as *asynchronous digital subscriber line* (ADSL) or cable modem at the end of 1999 [42].

However, we are seeing clear signs of retrenchment to a more sustainable level growth. After all, the amount of time that consumers can allocate between newspapers, TV, cable TV, Internet, DVDs, music, IPODs, e-mail, education, and family, social, and job lives is clearly limited. A typical press-time quote recognizes this by stating as follows:

> It isn't just a matter of slower expansion.... Sweden's Ericsson said the market for mobile-network equipment would shrink 15% in 2002.... Makers of semiconductor production gear forecast that their business would contract 19%.... There is no optimism in the market.... Offices and households in industrialized countries have acquired PCs, cell-phones, and other basic productivity tools, and see little pressing need to upgrade. Some corporations also binged on technology to prepare for the year 2000... with that crisis passed, and with the Web boom maturing, there are simply too many employees at too many companies in large segments of the industry. Add to those factors a gloomy global economy and a catastrophic surplus of telecommunications capacity ... and the widening accounting scandals. . . [45].

1.3.3   Growth of Wireless During the Late 1990s and Early 2000s

In the early 1990s, there were about 10 million mobile cellular telephone subscribers around the world. This figure had grown by almost 70 times to more than 725 million, or one mobile phone for every eight inhabitants, at the beginning of 2001. Growth has been steady at an average of 50% per year since 1996. However, the mobile phone boom is also beginning to show signs of slowing down, although there were 234 million new mobile subscribers in the year 2000. A press-time article states: "The gloom over prospects for mobile-phone business has become so thick that even the industry's strongest players are losing credibility" [2]. The introduction of second generation networks such as the *Global System for Mobile Communications* (GSM) European standard sparked an increase in mobile growth [46]. Now operators aim at bolstering revenues by persuading customers to use new data services, such as picture messaging, wireless Internet access, and e-mail. However, those were the original promises of personal communication services (PCS), which have not yet materialized in 15 years.

At current growth rates, the number of mobile subscribers will surpass that of fixed telephones in the middle of this decade (see Figure 1.10). As of 2000, there were 35 markets—both developed and developing—where this transition has already taken place (e.g., Finland, Austria, Hong-Kong, Israel, Portugal, and Venezuela). More than 70% of the population in Europe already owned a mobile phone at press time; however, this has saturation implications. The mobile telephone is becoming a way of life for many, transcending the limitations of fixed telephones. One phenomenon

FIGURE 1.10 *Fixed and mobile telephone subscribers (in millions). (Source: ITU. Reprinted with permission.)*

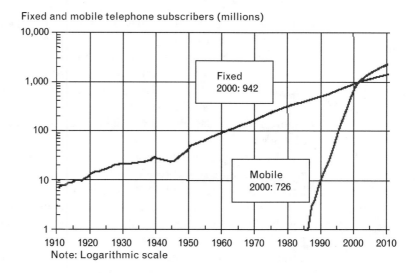

Fixed and mobile telephone subscribers (millions)

Fixed
2000: 942

Mobile
2000: 726

Note: Logarithmic scale

is that fixed household telephone penetration is holding steady or even dropping as users opt for mobiles. In Finland, the percentage of households with a fixed telephone has dropped from 94% to 83% during the last 10 years, while that of mobile phones has increased from 7% to 60%. In developing countries, competition and prepaid cards are proving a successful combination for driving mobile growth. The rise of mobile in developing countries is perhaps most powerfully conveyed by the fact that, based on current growth, China was expected to surpass the United States and emerge as the world's largest cellular market sometime in 2002. Unlike fixed telephone penetration, which generally peaks at around one telephone for every two inhabitants, mobile penetration has not yet reached an upper limit. The highest mobile penetration is found in Taiwan-China (see Figure 1.11). Ten years ago it had less than 100,000 mobile phones; today

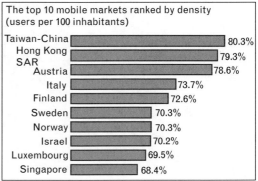

FIGURE 1.11   *Top mobile economies in 2000. (Source: ITU. Reprinted with permission.)*

four out of every five Taiwanese has one. Mobile penetration increased by 25 points in 2000 among the top 10 most wireless economies. At that rate, most adults will soon have at least one mobile phone [46].

The value of mobile in the overall telecommunication market is growing. In 2000, mobile revenues were $273 billion, accounting for almost one-third of the total worldwide telecommunication revenues (see Figure 1.12). On average, a mobile cellular subscriber generates revenue of $39 per month.

Two of the top 10 national mobile operators in 2000 are new (Verizon and Cingular) and result from the merger of North American cellular companies (see Table 1.7). European cellular companies lost market capitalization in 2002 out of concern that the companies would not be able to boost revenue as subscriber growth slows in Europe. These European companies have been counting on technology allowing faster access to the Internet to boost revenue from existing customers; however, infrastructure costs to support these services will be a major retardant factor [47].

1.3.4 Wireless Internet

Wireless Internet is an example of new applications being contemplated. The questions posed in Section 1.1, however, need to be answered. Observers make the statement that "...to believe in the [industry's] long-term projections is to accept on faith that the industry will come up with a new killer application. For now investors are only too willing to conclude that any such breakthrough is a long way off and revenue growth over the next couple of years [2002–2004] will be dismal" [48].

FIGURE 1.12
*Mobile service
revenues (in billions of
dollars).
(Source: ITU.
Reprinted with
permission.)*

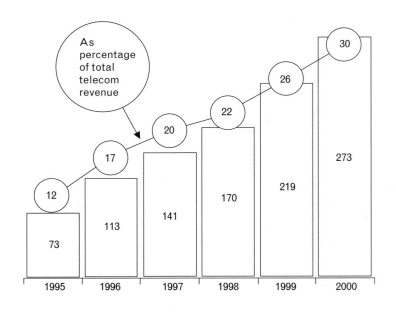

TABLE 1.7   TOP MOBILE OPERATORS, RANKED BY
SUBSCRIBERS (2000)

OPERATOR (COUNTRY)	SUBSCRIBERS		REVENUE	
	Total (000s)	Change 1999–2000 (%)	(USD millions)	Change 1999–2000 (%)
1. China Mobile (China)	65,260	97	13,481	56
2. NTT DoCoMo (Japan)★	36,030	17	37,093	26
3. Verizon (United States)	27,505	88	14,236	86
4. TIM (Italy)	21,600	17	7,466	6
5. Cingular (United States)	19,681	19	11,612	16
6. Mannesmann (Germany)	19,245	103	7,403	54
7. Deutsche Telekom (Germany)	19,141	111	6,102	30
8. AT&T (United States)	15,716	29	10,448	37
9. Omnitel (Italy)	14,920	43	4,992	39
10 France Télécom (France)	13,941	42	5,023	38
Top 10	253,039	53.8	111,857	35

★ Year beginning April 1.

A variety of Internet-like applications are available for today's mobile phones. The second generation GSM mobile phone introduced a simple but popular application called *Short Message Service* (SMS) that allows text messages to be sent from one telephone to another. This e-mail for telephones has proven popular for market segments ranging from youngsters to the hard of hearing. According to the GSM Association, in the month of January 2001, approximately 15 billion messages were sent around the world, for an average of 30 per GSM subscriber; this is almost quadruple the 4 billion sent in January 2000 [49] (again, transaction data, but no revenue numbers).

The *Wireless Application Protocol* (WAP), commercially launched in 2000, allows mobile phones to browse the Internet. Users access Web sites specially adapted to fit the screen size of a mobile phone. By the end of 2000, more than 100 mobile operators had launched WAP. According to the WAP Forum, there are about 10,000 WAP sites, and although around

50 million WAP phones are in circulation, there were fewer than 5 million users at the end of 2000. Acceptance of WAP has been tepid since it has been impacted by a number of problems including a shortage of handsets, slow speed, and a lack of applications.

Japan was the first country in the world to launch mobile Internet services when NTT DoCoMo started its *i-mode* service in 2000. By March 2001, 22 million people were using i-mode. There are two other mobile Internet services in Japan, *EZWeb* and *J-Sky*. The three services had 35 million subscribers between them in March 2001, with almost 70% of all Japanese Internet surfers logging in from a mobile. One of the attractions of i-mode is that, unlike WAP, HTML-based Web sites can be easily adapted. There were 1,600 sites on NTT DoCoMo's i-mode portal. In addition, there were 40,000 i-mode-compatible sites. Another i-mode success factor is that it is "always on" and priced according to information retrieved, not usage time.

The industry is now contemplating so-called *third generation* (3G) mobile networks. On paper 3G networks are significant for two reasons. Firstly, it marks the first time that there will be a global standard for mobile networks; up until now, mobile systems have been based on a variety of national and regional standards with about a dozen different ones in operation around the world. This has had a detrimental effect on compatibility and prices. Secondly, the new 3G networks will support broadband Internet access. The realities, however, are daunting. Considering the major market capitalization and direct bond/debt losses (in the several-hundred-billion-dollar range) that have occurred at the beginning of the decade, who is going to be willing to invest the hundreds of billions of dollars required to bring out continentwide 3G services without even knowing what the actual services will be, what the end-user penetration will be, and what the revenue/billing model will be? It typically takes around $1,000 or more of capital per active user link. Hence, to support 200 million users globally, $200 billion will be required; to support 400 million users, $400 billion; and to support 800 million users, $800 billion. Investors will need patience. The recovery of the $1,000 initial costs is typically achievable at $20 per month; another $30 per month (for a total of $50/month) is attributable to the service proper, be it long-distance voice, Internet transit, video content/royalties, and so on. The maximum price point the user may be willing to accept is $50/month. This means that it will take about 4 years just to recover the initial investment, and any net income will be at a point past that. Also, most services that have found consumer appeal first won over the business community (e.g., voice mail, e-mail, cell phones, and PCs). Therefore, 3G may first need to make a compelling penetration in the business market.

The 3G standard goes by the ITU name IMT-2000. Adopted by the world telecommunication community in November 1999, IMT-2000 will support high bandwidth including a minimum speed of 144 Kbps and a target speed of 2 Mbps under low–mobility environments. The launch of IMT-2000 networks, if they are eventually deployed, will drive toward the globalization of the mobile industry. This is because, for the first time, advantages will include a world standard, enhancing roaming, and interoperability. The European Union was an early advocate of 3G networks and had called on its members to rapidly license the system and launch networks by a target date of January 2002. Most Western European countries awarded licenses in 2000 (see Table 1.8). In Asia, NTT DoCoMo intended to be the first operator in the world to launch 3G services and was granted an IMT-2000 license in June 2000. Although the operator widely publicized plans to be the first to start a service in May 2001, it instead started with a test launch with commercial operation postponed to a point past press time (2002) [46]. In general, a pronounced slowdown has been observed in the early 2000s with regard to rolling out 3G.

TABLE 1.8   WESTERN EUROPE 3G IMT-2000 (UMTS) LICENSES

COUNTRY	NUMBER OF LICENCES	STATUS	PRICE (USD MILLIONS)
Austria	6	Awarded in Nov. 2000	714
Belgium	3	Awarded in Mar. 2001	419
Demark	Probably 4	Auction to start in Sept. 2001	—
Finland	4	Awarded in March 1999	—
France	2	Two bids to be awarded by 31 May 2001	9,340
Germany	6	Awarded in Aug. 2000	46,214
Greece	Probably 4	Auction in the second half of 2001	—
Ireland	Probably 4	To be awarded in May 2001	—
Italy	5	Awarded in Oct. 2000	10,084
Luxembourg	Probably 4		—
Netherlands	5	Awarded in July 2000	2,515
Portugal	4	Awarded in Dec. 2000	357
Spain	4	Awarded in March 2000	—
Sweden	4	Awarded in Dec. 2000	—
United Kingdom	5	Awarded in April 2000	35,411
European Union	48		105,054
Norway	4	Awarded in Nov. 2000	43
Switzerland	4	Awarded in Dec. 2000	120
Total	56		150,216

Proponents say that another indicator of the emergence of mobile is that there are now more mobile subscribers than personal computers or Internet users in the world. With the advent of mobile Internet, mobile operators (who in Europe alone spend $105 billion on acquiring spectrum licenses) hope that every mobile subscriber will also be an Internet user.

1.4   Rest of the Book

The rest of this book will focus on a subset of the plethora of technologies and technical approaches that are available or emerging, specifically for those technologies identified earlier in the chapter that are expected to have relevance in the near-term future. Some of these technologies and services could get the industry humming again.

ENDNOTES

[1]   *Wall Street Journal,* April 30, 2002, p. A9.

[2]   *Wall Street Journal,* May 5, 2002, p. A3.

[3]   *Wall Street Journal,* May 7, 2002, p. D1.

[4]   "Spitzer's Telecom Meltdown," Wall Street Journal, April 29, 2002, p. A18.

[5]   *Wall Street Journal,* May 1, 2002 p. B11.

[6]   *Wall Street Journal,* May 1, 2002, p. A8.

[7]   *Wall Street Journal,* May 1, 2002, p. A8.

[8]   *Wall Street Journal,* May 21, 2002, p. B6.

[9]   Baer, J., "Nortel to Cut 3,500 Jobs, May Sell Optical Parts Unit," *Bloomberg,* May 29, 2002.

[10] Heinzl, M., *Wall Street Journal,* May 30, 2002, p. B5.

[11] Kariya, S., "Where the Jobs Are," *IEEE Spectrum,* January 2002, p. 29.

[12] Berman, D., "Velocita Files for Bankruptcy in Another Telecom Setback," *Wall Street Journal,* June 4, 2002, p. B15.

[13] Dreazen, Y., "FCC's Powell Says Telecom 'Crisis' May Allow a Bell to Buy World-Com," *Wall Street Journal,* July 15, 2002, p. A1.

[14] *Wall Street Journal,* July 15, 2002, p. C1.

[15] Berman, D. K., "Lucent Posts a Wider Net Loss and Plans to Slash Work Force," *Wall Street Journal,* July 24, 2002, p. B9.

[16] Schuetz, M., "Alcatel Has EU1.44Bln Loss, Say No Rebound in Sight," *Bloomberg,* July 25, 2002.

[17] Nuzum, C., "Telecom Meltdown Persists and Spreads, 1st Quarter Results Show," *Dow Jones Business News,* April 29, 2002.

[18] Drucker, J., "AT&T Posts a Loss of $12.7 Billion," *Wall Street Journal,* July 24, 2002, p. B9.

[19] "…in growing the business we, along with others in the industry, outpaced demand, and as a result, are overbuilt…," *Wall Street Journal,* May 21, 2002, p. B7.

[20] "Phone companies raced to expand their long-distance networks in the late 1990s and 2000 … many carriers ran out of money, overbuilt their network, or failed to generate enough sales to justify their expansion plans," Baer, J., "Nortel to Cut 3,500 Jobs, May Sell Optical Parts Unit," *Bloomberg,* May 29, 2002.

[21] " 'The WorldCom disclosures confirm that accounting improprieties of unprecedented magnitude have been committed in the public markets,' the SEC said in an e-mailed statement," Giles, T., "WorldCom Says It Misreported $3.9 Billion in Expenses," *Bloomberg,* June 26, 2002.

[22] Consider this typical news story (*Wall Street Journal*, May 21, 2002, p. B7): "…internal systems and controls (are too) flawed to review the results in keeping with normal accounting procedures … preliminary ranges of restated results … show wider losses and lower revenues that previously stated."

[23] A May 1, 2002, newswire story like the following is typical (company name omitted): "DALLAS, Texas, May 1, 2002. Shore Deary, LLP: notice is hereby given that on April 25, 2002, a class action suit was filed in the Lufkin Division of the United States District Court for the Eastern District of Texas on behalf of all purchasers of the common stock of xxxx between May 10, 2001, and March 22, 2002, against xxxx and certain of its officers and directors seeking remedies under the Securities Act of 1934. The Complaint charges xxxx and certain of its officers and directors with issuing a series of material misrepresentations to the market during the class period, failing to adequately disclose a change in the company's business model and the risks involved in that change, and issuing financial statements that violated Generally Accepted Accounting Principles ("GAAP"), thereby artificially inflating the price of Universal Access's publicly traded securities. The alleged GAAP violations include recording revenue for contingent contracts and for "capacity swaps" with other telecommunications companies. As alleged in the complaint, these capacity swaps were merely a trading of services, which had no real business purpose other than to artificially inflate the revenues of the participating companies. The complaint alleges that these actions violated sections 10(b) and 20(a) of the Securities Exchange Act of 1934. Plaintiff seeks to recover damages on behalf of himself and the class members and is represented by the law firms of Shore Deary LLP; Nix, Patterson & Roach, LLP; and Patton, Haltom, Roberts, McWilliams & Greer, LLP."

[24] A quote from *Bloomberg* reads "… companies including Qwest Communications International Inc. and Global Crossing Ltd. are being investigated for similar transactions in which they simultaneously sold and bought back capacity on their networks with the same customer. Some investors questioned whether the transactions had any business purpose other than to boost reported sales for both companies…," Cimilluca, D., "Broadwing Stemmed a Sales Drop by Changing Accounting," *Bloomberg,* May 20, 2002, 18:30.

[25] A typical news story (*Wall Street Journal,* May 20, 2002, p. C1) reads "…were the deals illegal 'round-tripping,' in which companies exchange equal amounts of cash but bend accounting rules to show revenue increases? Or was there an underlying business purpose to the deals, which went wrong because optimistic assumptions about the need for more capacity turned out to be badly flawed—but not fraudulent?"

[26] *Wall Street Journal,* April 30, 2002, p. A9.

[27] "The American equity market has taken on the flavor of a casino to many investors," said Timothy O'Brien, who manages $265 million in the Evergreen Utility and Telecommunications Fund. "The cleanup process is not going to be pretty. This is already the ugliest I can remember," Giles, T., *Bloomberg,* "WorldCom Accused by SEC of Fraud, Bankruptcy May Loom," June 26, 19:28.

[28]   These comments based on Insight Research Corporation, *2003 Telecommunications Industry Review: An Anthology of Market Facts and Figures,* retrieved from http://www.insight-corp.com/review03.html, January 2003.

[29]   "Spitzer's Telecom Meltdown," *Wall Street Journal,* April 29, 2002, p. A18.

[30]   "All told, shareholders have lost an estimated $2 trillion from the 60% drop in market capitalization as a result of the telecom meltdown," *Wall Street Journal,* April 30, 2002, p. A9.

[31]   The telecom industry and the investors in general (which the capital-intensive telecom industry obligatorily requires), desperately need unbiased analysts knowledgeable about the field. Analysts should not have the dubious distinction of being personal friends of the CEOs they cover or "friends of the corporation" or private advisors of the CEOs of the companies they recommend. And they should not stand to make considerable money with their own insider-based recommendations, as some analysts described in *Wall Street Journal* articles at press time were alleged to do.

[32]   This observation is at least a beginning: "Investors have to watch for over leveraging, adequate corporate governance and management misrepresentations," St.Onge, J., "Adelphia Communications Seeks Bankruptcy Protection," *Bloomberg,* June 25, 2002.We would add as a bare minimum the following: Look for executive officers and senior management who utilize a full gamut of technical and scientific methodologies in decision making, and avoid hiring executives that only have superficial credibility based *not* on hard accomplishments but only (as many job advertisements for CEO positions call for) "presence" and "charisma."

[33]   The price P of a security would then be $P = V/O = 3R/O$ where R = revenue, V = valuation, and O = outstanding shares. If we assumed that the P/E ratio has to be around 20, then we get that $P = 2R/O$, when we assume that $E = I/O$, with I = net income and $I = R/10$ (on the assumption that the management of the company could deliver a 10% bottom line to the investors).The derivation is $P/E = 20$; or $P/(I/O) = 20$; or $P/[(R/10)/O] = 20$. If the management only delivered a 5% bottom line $(R/20)$, then the same equations resolve as $P = R/O$.

[34]   Do not interpret the word *broadband* to mean "cable-TV based." The word always had and should always have a more encompassing meaning.

[35]   Unless implied otherwise by the context, our use of the word *wireless* refers to mobile systems (for at least one end of the link), not fixed wireless systems.

[36]   For example, see A. Lataur, D. Berman, and S. Thurman, "A Wrong Number for Telecom," *WSJ,* April 28, 2003, p. C1.

[37]   McQuillan, J., NGN Ventures 2001, San Francisco, April 2001.

[38]   Reynolds, J. K., et al., *Internet Official Protocol Standards,* RFC 3000, 2001.

[39]   "European Stocks Decline, Led by Logica, Vodafone, Nokia, CMG," *Bloomberg,* May 10, 2002.

[40]   Dreazen, Y., "High Court Rules Against Baby Bells," *Wall Street Journal,* May 14, 2002, p. A2.

[41]   Dreazen, Y., "Appeals Court Sides with Bells in Dispute over Network Sharing," *Wall Street Journal,* May 28, 2002, p. B4.

[42]   ITU Telecommunication Indicators Update, March 2001.

[43]   The number was $45 billion in 1999 or $30 per subscriber per month.

[44]   Interestingly, the term *hype* is thrown around a lot with regard to various technologies (of the 1990s), but it hasn't as of yet been applied to the Internet.

[45]   Boudette, N. E., et al., "Global Tech Sector Expects Grim 2nd Half," *Wall Street Journal,* July 25, 2002, p. B6.

[46]   ITU Telecommunication Indicators Update, June 2001.

[47]   *Bloomberg News,* May 3, 2002.

[48]   *Wall Street Journal,* May 6, 2002, p. A14.

[49]   This section is based on [45].

Appendix 1A: One View on Regulation

This appendix contains one view on the regulatory issue. We do not necessarily agree or disagree with the positions. The purpose is simply to document a press-time snapshot of the issue, so that looking back over time one can determine if the issues raised were correct and passed the test of time.

H. J. Jenkins Jr., "Telecom's Curse: Too Many Lobbyists," *Wall Street Journal,* May 8, 2002, p. A19.

It hardly seems like a mere five years ago that WorldCom was plopping into the middle of the business world's consciousness with a $30 billion bid for long-distance carrier MCI.

Bernie Ebbers, he of the cowboy boots, was offering MCI shareholders WorldCom stock, then priced at a multiple of nearly 100. Investors at the time were enthralled by the idea of the senile old phone companies being overrun by the Internet. His rival for MCI's affections, GTE's Charles Lee, was one of those old telephone guys but he had a different idea. He dared to offer a lower price for MCI but in Plain Old Cash.

Mr. Ebbers won the battle but it's clear now who won the war. WorldCom teeters toward bankruptcy[1] as its former chief licks his financial wounds on the sidelines. Meanwhile, GTE went on to marry itself with Verizon, the super-monopolist that dominates the Northeast and looks like one of the few survivors of the spending frenzy of recent years.

We will not weep for Mr. Ebbers, but such is the outcome when billions are thrown at opportunities designed by Washington rather than the market. Looking back, regulatory policy had become insane. Local Bells were being allowed to merge with Bells on the ground that, hey, they were already monopolies so what difference did it make if they got bigger? Yet regulators stood fast against a sensible coalescence of the local and long-distance businesses to form a single, undifferentiated business called service.

This was half the telecom meltdown right here. Out of every dollar paid for long distance, 93 cents goes to the overhead, billing and marketing costs of maintaining a separate long-distance industry. Any business built on such a non sequitur can't be long for the world. Long distance needed an orderly burial, but regulatory politics and telecom's lobbying wars got in the way.

1. Of course, WorldCom did eventually file for Chapter 11.

The other half of the disaster has its roots in Washington's regulatory fumbling of broadband. Mike Armstrong has been working valiantly to dig AT&T out of the dying long-distance business by buying cable TV networks and converting them to two-way broadband. But it hasn't been a fun ride for AT&T shareholders. No sooner did he begin investing billions to build a digital pipe into millions of homes than competitors were demanding "open access" from regulators and rivals over his attempt to merge his cable systems with those of Comcast.

But the cable guys have hardly been innocents themselves, constantly badgering Congress and the FCC to hog-tie the Baby Bells' own broadband effort. State legislators in Oklahoma finally got fed up and just deregulated Southwestern Bell's broadband business. In turn, the company announced plans last month to roll out service to 37 more towns.

Sadly, not even Intel and a host of Silicon Valley worthies who recently marched on Washington have managed to break a similar impasses there. They just keep bouncing off Ernest Hollings, the irascible Democrat who chairs a key Senate panel and likes being the scourge of whatever big business his staff points him at.

History will want to know how we came to throw away $1 trillion of bondholders' money while knocking $2 trillion off the collective wealth of telecom shareholders in the 1990s. Here's how:

Much of the wasted spending was actually predicted on a funny notion embodied in the 1996 telecom law—or at least the law as interpreted by Reed Hundt, the Al Gore protégé who served as FCC chairman. His vision, dreamed up out of the whisperings of countless lobbyists, was that regulation should hold the incumbent Bells in place while start-up competitors cherry picked their businesses to death.

This notion had a certain charm but was hardly necessary to keep the world moving forward. Wireless, satellite, cable and various other challenges were coming along anyway. Better that this investment had been made without illusion and without pathetic reliance on regulators, which turned out to be one of the worst gambles of all time. In the end the 1996 act was successful mainly for the permanent gusher of campaign contributions it created.

Mr. Armstrong at AT&T believes the 1996 model can still work, and he's had some luck with state regulators lately. But a business plan built on expecting the Bells to cooperate in their own overthrow would seem a recipe for frustration. "Economies always win out. Crazy, convoluted schemes never work," says one telecom CEO.

In comparison, WorldCom's Mr. Ebbers did not seem to have any road map at all beyond continuing to make deals and grab telecom properties while telecom was a scarce resource.

Hard to believe, but there was a moment when it seemed it might run out. AOL and other Internet hookup providers had gone to flat-rate, all-you-can eat pricing. The standard three-to-five minute phone call the old system was designed to handle had suddenly become a call that lasted all week. Old time Bell engineers were worried the system would bog down under the surge in demand.

While it lasted, Mr. Ebbers was earning astronomical margins from Internet backbone. Indeed, even with all the lousy policy out of Washington, telecom might still be levitating today if the courts had upheld Napster and Hollywood was busting a gut to offer a digital business model of its own.

Too bad. Now the industry's biggest millstone is its own lobbying armies. Telecom players need to do deals—lots of them—to clean up and consolidate, but trench warfare between the factions in Washington has made getting anything almost impossible.

We have got a wireless industry with too many players, a consumer long-distance industry that needs to be absorbed by the Bells, and needless regulatory obstacles to broadband. The only way anything is going to get fixed is if everyone puts down their battle-axes and agrees to let technological rather than regulatory competition drive the future.

CHAPTER 2

Optics Technologies

This chapter looks at the telecom transport mechanism of choice for the present and the foreseeable future: *fiber optics*. The first edition of this book recognized the importance of fiber optics and a substantive chapter was dedicated to the topic. Fiber optics supports what will be one of the key three pillars of the telecom industry for the rest of the decade in terms of importance and relevance: broadband communications. Major advances in fiber-optic technology have been made during the past few years, particularly in the context of multiplexing multiple information streams onto a single fiber pair. In fiber optics *frequency-division multiplexing* (FDM) is known parochially as *wavelength-division multiplexing* (WDM) [1, 2]. These advances in multiplexing have resulted in the ability to carry increased information over a single fiber pair: One can now routinely support 10 Gbps on a single beam of "light" using *time-division multiplexing* (TDM) methods and work is under way to develop technology to support 40 Gbps, also with TDM. WDM allows the carriage of numerous (8–160) beams (frequencies) over a single fiber pair or even a single fiber, providing near-terabit per second speeds.

In this chapter we look at the fundamental science of fiber optics, fiber types including extended band fibers, DWDM principles, amplifiers, optical add/drop multiplexers, optical cross-connect systems, and optical switches. LAN applications are also discussed. Figure 2.1 depicts the positioning of optical technology in the wider context of delivering broadband communications and multimedia services.

2.1 Transmission Systems

Figure 2.2 depicts the portion of the electromagnetic spectrum that is of interest in fiber-optic communication. The key elements of a fiber-optic transmission system are as follows:

- *Transmitter:* a *laser diode* (LD) or a *light-emitting diode* (LED);
- *Fiber-optic waveguide:* a single-mode fiber in long-haul applications or a multimode fiber in LAN applications;
- *Receiver:* a photodetector diode.

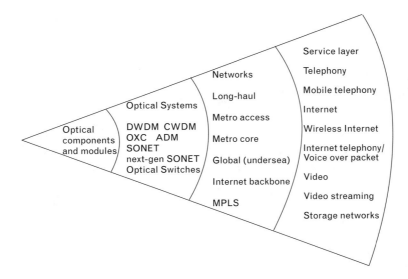

FIGURE 2.1   *Fiber technology underpinning for communication.*

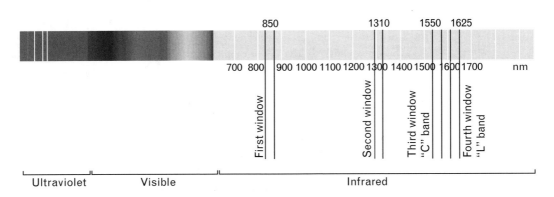

FIGURE 2.2   *Optical bands of interest.*

Figure 2.3 depicts a basic transmission system that also embodies multiplexing, specifically WDM. Typically, for long–haul and metropolitan applications one has termination TDM electronic-level equipment at the edges of the link to support tributary lower speed signals, such as a SONET terminal. This SONET transmission/multiplex equipment aggregates electrical signals that the fiber-optic transmitter converts to a specific-wavelength optical signal for transmission over the fiber. The receiver converts the optical signal back to electrical signals. As implied, several of these TDM signals can be "stacked" onto a single physical medium. The FDM technique that enables carriage of multiple (stacked) signals involves aggregating various *amplitude shift keying* (ASK)-modulated "light" beams operating at different frequencies/wavelengths.

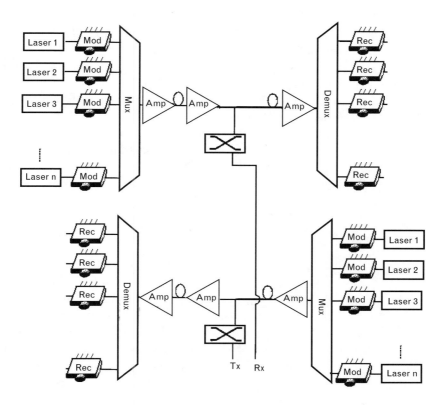

FIGURE 2.3   *Components of an optical transmission system.*

Unrepeated transmission distances for fiber-optic systems have increased significantly over the years. This is advantageous for applications such as oceanic systems between the United States and Europe and the United States and Asia. However, typical land-based systems deployed today, particularly in multiplexed applications, utilize amplifiers or repeaters every several-dozen miles. This use of repeaters increases the source-to-destination distances supported by the fiber-optic link to thousand of miles. Typically carriers install a fiber infrastructure based on cables with a relatively high fiber count of the order of several hundred strands; this deployment strategy provides bandwidth opportunities, particularly in the long-haul portion of the network.

Two approaches are taken to optical transmission: *noncoherent transmission* and *coherent transmission*. Noncoherent methods are the norm for commercial systems today and involve on–off keying and direct detection of signal. Noncoherent transmission is a variant of amplitude modulation; specifically, it is a unipolar ASK technique in which the amount of information sent is based on the source/reception clocking rate. Optical signals are typically in the form of *nonreturn-to-zero* (NRZ) or *return-to-zero* (RZ) or pulses. For the "off" state, the light from an LED source (for LAN applications) is turned off; on the other hand, lasers (for long-haul applications) are only dimmed (utilizing NRZ) to about 10 dB below the "on" level. Line signals are also used for clock recovery and are typically scrambled to provide a balance between 0's and 1's.

Extending this basic method, several noncoherent optical channels can be operated over the low-loss/low-dispersion regions of an optical fiber, as noted earlier. This WDM approach requires filters to separate the optical channels at the receiving end. Initially, such multiplexers were designed to support two channels in the 1,300-nm wavelength region and three channels in the 1,500-nm region. Commercial products now support several dozen channels and systems in development support several hundred channels of 2.4- to 10-Gbps (hence, the term *dense WDM*). This beam density is accomplished by utilizing narrow-spectrum lasers and direct detection. Each wavelength in a WDM system can be from a traditional link, for example, an OC-48 link; hence, one does not have to make most previously deployed equipment obsolete. One merely needs laser transmitters chosen for wavelengths that match the WDM demultiplexer to make sure each channel is properly decoded at the receiving end. If one uses an OC-48 SONET input, one can have 4×2.5 Gbps = 10 Gbps up to 32×2.5 Gbps = 80 Gbps. Whereas 32 channels were typical in the early 2000s, systems deployed by the mid-2000s are expected to offer 80 to 128 channels. Furthermore, one is not limited to SONET/SDH; one can support GbE, or mix and match SONET and GbE or any other digital signals [3].

Even DWDM, however, does not utilize the fiber medium to its maximum potential. The wavelength regions with low loss (from 1,270–1,350 nm and from 1,480–1,600 nm) provide a total wavelength band of 200 nm; this corresponds to a frequency bandwidth of about 29 THz; yet, it is used to support only relatively few channels—100 channels at 10 Gbps equate to 1 terabit per second, or about 1/300th of the theoretical capacity. Newer fibers extend the usable frequency range even further.

2.2   Optical Fibers

Optical fiber acts as a low-attenuation (less than 0.5 dB/km or so) waveguide for photonic signals. Refraction, which is the bending of a light beam, is one of the most important phenomena in optical communication. Refraction occurs because light moves faster through some materials than through others. The index of refraction is defined in reference to two media and represents the ratio of the speeds of light propagation in the two media.

The difference in index of refraction gives the optical fiber its transmission (waveguide) properties. Optical fiber is manufactured with a cylindrical core of a higher index of refraction than the surrounding cladding material. The fiber's core is made of transparent glass with a relatively high index of refraction; cladding is made of glass with a relatively lower index (see Figure 2.4). The fiber (core and surrounding cladding) acts as a waveguide for the optical signal. Because there is a difference of speed in the two optical elements of the media, the light is reflected and refracted at the boundary (since, by design, the index of refraction changes "abruptly" at the core–cladding boundary, the resulting fiber is called a *step-index* fiber [4]). Optical fibers used in telecommunications at the present time are based on silica. Silica has a low optical signal loss in the infrared region of the spectrum ranging from 800 to 1,600 nm. The index of refraction of either the core or the cladding (or both) is altered from the values of pure silica by the inclusion of small amounts of materials such as germanium or phosphorus, in a process called *doping*. Figure 2.5 depicts the attenuation performance of various fiber types.

FIGURE 2.4
Light transmission in a fiber waveguide.

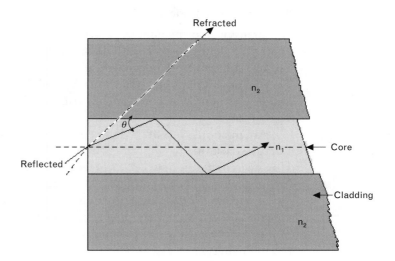

FIGURE 2.5
Attenuation of silica fiber.

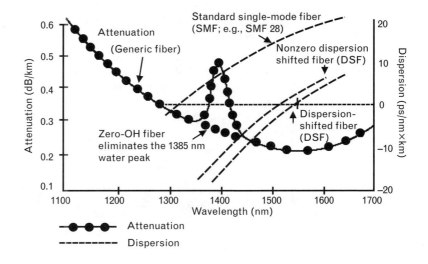

Two families of step-index fiber exist: multimode and single-mode fibers. Multimode fibers have a core-to-cladding diameter of 62.5/125 μm (although other ratios are also available). In a multimode fiber, the core carries multiple "modes" or rays of light. Multimode fibers are more typical of LANs or data center installations. When the core dimension is only a few times larger than the wavelength of the signal to be transmitted, only one of the originating rays ("modes") will be transmitted; this results beneficially in low loss. For wavelengths of 1,300 to 1,550 nm, the core needs to be less than 10 μm in diameter for single-mode transmission properties to hold. With a single-mode fiber, the information-carrying capacity is much higher; however, because the core is small, more expensive light sources are required.

2.2.1 Fiber-Optic Medium Performance

Any transmission medium degrades signal quality. Typically one experiences (1) attenuation of a signal or waveform and (2) distortion of its shape. For fiber, however, the degradation is much less than that of other media. Because of this reduced attenuation, optical fiber can carry a signal over a longer distance or at much higher bandwidth than, say, twisted pair or coaxial cable.

Attenuation is loss of the light energy as it travels along the waveguide: Energy is lost due to impurities, imperfections, and physical phenomena. However, light not only becomes weaker (attenuated) as it goes through a long segment of fiber, but individual pulses of light may become broadened or experience overlapping. Either effect makes the signal more difficult to recover at the remote end. The term that describes the waveform distortion is *dispersion*. Other impairments affecting optical transmission include the following:

- *Rayleigh scattering.* This is caused by the microscopic nonuniformities of glass and its refractive index. These develop due to small variations in the density of glass as it cools during the manufacturing process. A ray of light is partially scattered into many directions, therefore, some of the light energy is lost (see Figure 2.6). The attenuation due to scattering decreases with wavelength because the structure of glass is much finer than the wavelength. Rayleigh scattering lessens as wavelengths grow longer; scattering affects short wavelengths more than long wavelengths and limits the use of wavelengths below 800 nm.

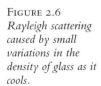
FIGURE 2.6
Rayleigh scattering caused by small variations in the density of glass as it cools.

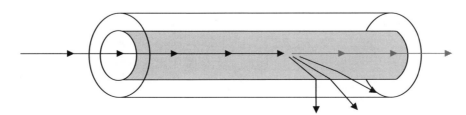

- *Absorption.* This is caused by interactions of the signal with impurities in the glass, causing unwanted radiation of the signal. For silica, loss peaks at a wavelength of about 1,390 nm caused by OH–radical impurity. Losses for high-quality fiber are typically less than 0.40 and 0.22 dB/km at 1,310 and 1,550 nm, respectively. As noted earlier, new fibers have been developed that eliminate the 1,385-nm water peak.

- *Bending, specifically macrobending and microbending.* Fiber bends result from optical fiber coating, cabling, packaging, installation, and aging. Macrobending is the loss that is due to macroscopic deviations of the axis from a straight line. Microbending loss occurs due to the sharp curvatures involving local axis displacements of a few micrometers and spatial wavelengths of a few millimeters.

- *Polarization-mode dispersion (PMD).* PMD is caused by slight fiber asymmetry: During fabrication or during the drawing process, a core or cladding that is not totally circular may result. PMD is the result of the accumulation of weak birefringence along the various fiber spans. (A link of any length exceeding 2 or 4 km is typically comprised of spliced fibers that originate from different manufacturing events.) An input light pulse of an initial polarization will decompose into two pulses with two different polarizations separated in time. The time separation leads to a decreased system (signal) margin or even outages. PMD also depends on the temperature of the fiber, sometimes making transmission problematic on a time-of-year basis. This issue becomes more pronounced and problematic at higher speeds (i.e., OC–192 and OC–768).

For single-mode-based fiber-optic transmission systems, the link length, absent regeneration, is limited by attenuation rather than by dispersion for data rates up to 0.5 Gbps. At higher data rates, however, the link length is limited by dispersion. Length issues are generally more of a concern for long-haul systems than for metropolitan or LAN applications. To address these impairments, links are extended by adding regeneration.

2.2.2   Dispersion Compensation Approaches

Earlier we noted that the channel bandwidth in a fiber-optic system is limited by the amount of pulse spreading; this is because pulse spreading results in a higher *bit error rate* (BER) (see Figure 2.7). The total dispersion (pulse spreading or broadening) and the resulting fiber bandwidth are characterized by two basic effects: *modal dispersion* for multimode transmission system and *chromatic dispersion* for single-mode transmission systems [5]. Dispersion is measured either in (1) bandwidth in megahertz per kilometer or (2) in time spread in picoseconds per kilometer (picoseconds of pulse-spreading per kilometer of fiber per nanometer of source spectral width [ps/km–nm]).

Figure 2.7
Dispersion.

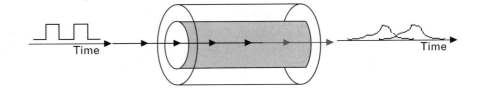

2.2.2.1   Intermodal Delay Distortion

Because of the physical size of the core, multiple rays (modes) enter and are carried by a multimode fiber. Due to its large uniform core, rays of light in a multimode fiber propagate through the core at different angles. Rays that go straight through the core have a shorter path (traverse the fiber faster) than those at a large entrance angle that bounce off the inside wall of the core. This causes intermodal delay distortion, which in turn limits the bandwidth capacity of the fiber. The problem can be reduced by using a multimode, graded-index fiber, for which the index of refraction changes gradually across the cross section (diameter) of the fiber. Here high-angle rays, with a longer physical distance to travel, spend more time in regions of the silica/glass with a lower index of refraction where the speed of light is faster. This compensating effect results in giving all rays almost the same transit time; the result is that graded-index fiber has a higher information capacity than step-index fiber. The compensation, however, is never perfect, and, consequently, this fiber still cannot support large data rates for more than a few kilometers. Furthermore, it is more difficult to manufacture this type of fiber. The net outcome is that graded-index fiber has had relatively little if any commercial deployment. The bottom line is that multimode fiber has limitations and is, therefore, used only in small-diameter networks such as LANs and campus environments.

2.2.2.2   Chromatic Dispersion

When a pulse of light is injected into a fiber, the optical energy does not all reach the far end at the same time; this causes the exit pulse to be broadened (dispersed). This phenomenon is called *chromatic dispersion*. Chromatic dispersion is manifested in two ways: (1) as *material dispersion* resulting from slightly different indices of refraction that the fiber presents to different wavelengths of light and (2) as *waveguide dispersion* resulting from longer in-fiber paths taken by longer wavelengths; this is caused by the fact that, unfortunately, the light source (LD or LED) does not generate a sharp single-wavelength signal, but a signal comprised of a number of light components. The light pulse is usually composed of a small spectrum of wavelengths, and different wavelengths travel at different speeds in fiber. Due to manufacturing variations and aging, laser operating wavelengths span a range of operating wavelengths. LDs are much tighter than LEDs in terms of the spectrum they generate. Chromatic dispersion limits the bandwidth that can be achieved over a fiber-optic link. Note that when using WDMs, multiple wavelengths travel down a fiber by design.

Because the optical sources used produce signals over a range of wavelengths and because each wavelength component of the pulse travels at a slightly different velocity, the pulse spreads in time as it travels down the fiber. In the visible spectrum of light, material dispersion causes longer wavelengths to travel faster than shorter ones; in the near-infrared region (1,100–1,300 nm) the opposite happens: Longer wavelengths travel slower. There is also a temperature dependence to chromatic dispersion; although this dependence can generally be ignored for speeds up to 10 Gbps, it becomes an issue for higher speeds—its impact is 16 times more severe at 40 Gbps than at 10 Gbps. Negative values of chromatic dispersion have the same effect as positive values, but the sign indicates if longer or shorter wavelengths travel faster (see Figure 2.8). Three approaches are used to reduce the effect of chromatic dispersion:

1. Use a light source that contains only one wavelength. This approach, however, is not implementable in complete terms, because, as noted, every source has some spectral spread. Typical lasers produce light at a number of different wavelengths spread over 5 to 10 nm or more. More sophisticated designs can reduce the spectrum; devices implementing these designs, however, are expensive.

2. Select a transmitter center wavelength and fiber design in such a manner so that the effect is minimized (this is the most common approach to the issue.) The optimal center wavelength for the transmitter (on a non-WDM-used fiber) is the point where chromatic dispersion is zero. This wavelength is called λ_0 or *zero-dispersion point* (ZDP). As seen earlier in Figure 2.3, silica-based optical fiber has low loss at 1,300 nm, and even lower loss at 1,550 nm. In the

FIGURE 2.8
Dispersion as a function of wavelength.

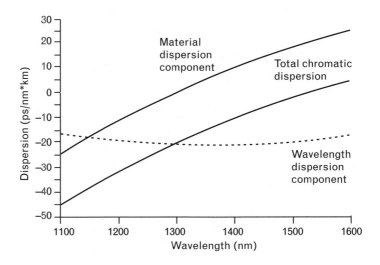

1,300-nm region, two signals 10 nm apart in wavelengths (a typical laser line width) will have practically little difference in velocity. However, a laser pulse with a 1-nm spectral width centered at 1,550 nm will be broadened by 16 ps after traveling 1 km. (If the spectral width is 10 nm and fiber length is 10 km, the pulse will be broadened by 1,600 ps.) This necessitates the use of either *dispersion-shifted fiber* (DSF) or *dispersion compensation fiber* (DCF).

The ZDP varies from fiber to fiber and is an important parameter for fiber manufacturers to control. ZDP occurs naturally in pure silica glass at 1,270 nm. Because, in principle, single-mode fibers work with a single wavelength, one way to exploit ZDP is to find a laser that emits light at 1,270 nm. However, the design entails other factors. For long-haul applications, it is desirable to operate at 1,550 nm because of lower attenuation compared to 1,300 nm; however, single-mode fiber shows a large chromatic dispersion at 1,550 nm. It is possible to design more sophisticated index profiles so that the point where chromatic dispersion is zero falls near 1,550 nm instead of 1,300 nm. This optical fiber is referred to as DSF. DSF supports longer unrepeatered distances at higher information transmission rates than are supported with unshifted fiber. As noted earlier, glass waveguides incur losses due to Rayleigh scattering. To take advantage of this, dopant materials can be added to glass until its ZDP is shifted into the range between 1,300 and 1,600 nm. Ideally (in non-WDM situations), the laser would operate at the ZDP wavelength, so that the bandwidth would be reasonably high; however, as noted, laser operating wavelengths typically span a range of wavelengths.

3. Utilize a certain amount of DCF at the receiving end. By properly selecting a terminating fiber that has the opposite effect of the transmission fiber in terms of dispersion, one can reduce the impact of dispersion. The add-on fiber takes the rays that traveled faster and slows them down more than the rays that traveled slower. Unfortunately a rather long spool of fiber is typically required: It could take several kilometers of fiber. The compensating fiber is placed in a "pizza box" and these "pizza boxes" have to be selected based on the length of the transmission link under consideration. Another problem is that this add-on fiber further impacts attenuation since the additional fiber must be taken into account in the power budget. A relatively new approach is to use tunable dispersion compensators.

In summary, chromatic dispersion can be managed, although this imposes a cost and an attenuation premium. At higher speeds the issue becomes more pressing (particularly at 40 Gbps on non-DSF).

2.2.3   Fiber Products

The common single-mode fiber types are standard fiber (*non-DSF* [NDSF], e.g., Corning SMF-28), DSF, and the various types of *nonzero dispersion-shifted fiber* (NZDSF, e.g., Corning LEAF or LS or Lucent True-wave). The three principal types of fiber used in telecommunication links and their ITU-T specifications are as follows:

1. NDSF, ITU-T G.652;
2. DSF, ITU-T G.653;
3. NZDSF, ITU-T G.655.

The major types of single-mode fibers and their application are summarized in Table 2.1. A new fiber type has recently been introduced that supports acceptable performance across the entire operating wavelength range, including an attenuation at 1,383 nm of less than or equal to 0.31 dB/km with a hydrogen-induced attenuation increase of less than or equal to 0.01 dB/km. This fiber belongs to the family of *extended wavelength band* (EWB) fibers defined in ITU G.652.C. Table 2.2 identifies industry standards for single-mode fiber as well as extended wavelength fiber (as seen in Figure 2.9). (e.g., see [6]).

Conventional single-mode fibers are classified as either matched-cladding or depressed-cladding designs. In the former type, the cladding has the same index of refraction from the core–cladding interface to the outer surface of the fiber. In the latter type, there are two layers of cladding material, and the index of refraction of the inner cladding, which is adjacent to the fiber core, is lower than that of the outer cladding (typically undoped silica).

TABLE 2.1   FIBER TYPES AND APPLICABILITY

FIBER TYPE	APPLICATION/SUITABILITY
NDSF (standard SM fiber)	Workhorse of telecom; accounts for more than 95% of deployed plant. Fiber is suitable for SONET/SDH/TDM use in the 1,310-nm region or DWDM use in the 1,550-nm region (with dispersion compensators). Fiber can also support the 10GbE standard at distances of more than 300m.
DSF	Suitable for SONET/SDH/TDM use in the 1,550-nm region, but not suitable for DWDM in this region.
NZDSF	Usable for both SONET/SDH/TDM and DWDM use in the 1,550-nm region.
Newer generation fibers	Examples include EWB/dispersion-flattened fibers that allow the utilization of wavelengths farther from the optimum wavelength without pulse spreading; and fiber that allows the energy to travel further into the cladding, creating a small amount of dispersion to counter four-wave mixing.

Note: Four-wave mixing (FWM) occurs when two or more frequencies of light propagate through an optical fiber together. As long as a condition known as *phase matching* is satisfied, light is generated at new frequencies using optical power from the original signals. Generation of light through four-wave mixing has serious implications for DWDM. The generation of new frequencies from two or three input signals is shown in Figure 2.10. http://www.npl.co.uk/photonics/nonlinear/four_wave_mixing.html.

TABLE 2.2   STANDARDS FOR BASIC FIBER TYPES

CATEGORY	STD SMF	EWB SMF
IEC 6073-2-50	B1.1	B1.3 (extended wavelength band)
ITU G.652	G.652.B	G.652.C (extended wavelength band)
TIA 492CAAB	IVa	IVa (dispersion-unshifted with low water peak)

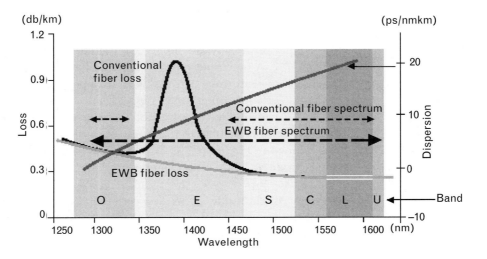

FIGURE 2.9   *Optical bands and parameters within bands.*

FIGURE 2.10
Four-wave mixing.
Additional frequencies
generated through
FWM in the
(a) partially degenerate
and (b) nondegenerate
case. (Source:
National Physical
Laboratory of U.K.)

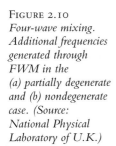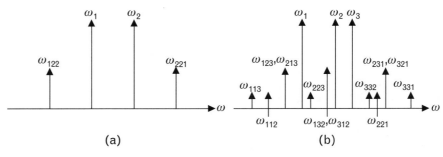

As noted, single-mode fibers are designed in such a fashion that only one signal mode propagates. In this environment, the information transmission capacity is limited by the phenomenon of chromatic dispersion, which, as discussed earlier, is a result of the wavelength-dependent velocities of propagation in the fiber material. Conventional single-mode fiber (e.g., SMF-28) has its minimum chromatic dispersion (that is to say, its maximum transmission capacity) near 1,310 nm. We discussed DSF as being designed so that the wavelength of minimum dispersion is "shifted" to 1,550 nm. At this wavelength, the attenuation can be significantly lower than at 1,310 nm. Single-mode fibers can also be designated as dispersion-flattened fibers. Dispersion-flattened single-mode fibers are attractive because the dispersion is low in both the 1,310-nm window and in the 1,550-nm window.

Single-mode fibers are characterized by the *mode field diameter* (MFD) and by the cutoff wavelength. The MFD is a measure of the width of the guided optical power's intensity distribution in the core and cladding. Typical values for the MFD are in the 10.0-μm range. The cutoff wavelength is the wavelength below which a single-mode fiber propagates more than one mode and above which only a single (fundamental) mode can operate; typical values are 1,200 nm.

Multimode fibers carry dozens of propagating modes, each of which travels at a slightly different velocity. These fibers tend to be used in GbE and 10GbE LAN applications, although these also can utilize single-mode fiber. The core diameter of the multimode fiber normally used is 62.5 μm, but other special designs of multimode fibers exist. Multimode fiber can be of the step-index or graded-index type, although most fiber deployed is of the step-index kind. Roughly speaking, 850- and 1,310-nm multimode systems can operate at distances from a few hundred meters to a couple of kilometers; see Table 2.3 for examples [7].

2.3 Fiber-Optic Elements

This section surveys some of the basic elements that comprise a fiber-optic system. These include optical transmitters, receivers, amplifiers, multiplexing components, and switching elements.

2.3.1 Optical Transmitters

Optical transmitters convert digital electrical signals into pulsed optical signals, and they couple these signals into the fiber waveguide for transmission. Transmitters need to be small in size, enjoy conversion and coupling

TABLE 2.3 DISTANCES SUPPORTED BY MULTIMODE FIBER

APPLICATION	WAVELENGTH	RANGE (M)
10BASE-FL	850-nm multimode	2,000
100BASE-FX	1,310-nm multimode	2,000
100BASE-SX	850-nm multimode	300
1000BASE-SX	850-nm multimode	220/550★
1000BASE-LX	1,310-nm single mode	550

★ 62.5-μm/50-μm fiber.

efficiency, support high pulsing speed, have environmental resistance, and be reliable. The two major kinds of semiconductor transmitter in use are LEDs and LDs. LDs are typically used with single-mode fibers, whereas LEDs are typically used with multimode fibers (however, other combinations are possible). Requirements for LDs include precise wavelength generation, narrow spectrum width, sufficient power, and control of chirp (the change in frequency of a signal over time). Semiconductor lasers satisfy the first three requirements; chirp, however, can be impacted by the techniques used to modulate the signal.

2.3.1.1 LEDs

There are two types of LEDs technologies: *surface emitters* and *edge emitters*. The surface-type photonic emission pattern is typically 120° × 120°. The peak power coupled into a multimode fiber is about −10 to −20 dBm (the power coupled into single-mode fiber is about −27 to −37 dBm). The spectral width of the emission is about 35 to 50 nm for short wavelengths and about 80 to 120 nm for long wavelengths.

The edge-emitting diode has a more "focused" emission pattern, typically 120° × 30°. It follows that the power coupled into the fiber is as much as 10 dB greater than for surface emitters. The spectral width is about 30 to 90 nm over the band of interest. However, the spectral width of the LED source and the dispersion characteristic of the fiber limit the bandwidth that can be used. Also, because the output optical power varies with dc voltage and with temperature, compensation of the LED is required: A thermoelectric cooler can be used to stabilize the LED temperature over a range of temperature differences between the LED and the ambient.

2.3.1.2 LDs

LDs have a typical emission pattern of 30° × 30°; this provides for higher coupling efficiency than for LEDs, particularly for single-mode fibers. LDs give a peak power of −10 to +10 dBm, typically near 0 dBm, into either multimode or single-mode fibers. The spectra of LDs are much narrower than for LEDs, ranging from 3 to 5 nm for typical multilongitudinal-mode lasers and 1 nm or less for single-longitudinal-mode lasers.

With the advent of DWDM, the need arises for tunable lasers that can be used in lieu of a large inventory of fixed-wavelength line cards. Tunable lasers reduce system maintenance and troubleshooting activities, thereby reducing system operating and maintenance costs. Today's DFB are unique to a given wavelength. In case of an outage on a system, the carrier must find the exact replacement laser. As is probably obvious, a DWDM system with 100 channels presents a challenge to the service provider in terms of sparing and maintenance. A tunable laser is a single replacement for all static sources and sparing is done with a single, common, configurable card. Also, these devices support wavelength conversion, wavelength switching, and wavelength add/drop. Tunable laser technologies include: DFB, *distributed Bragg reflector* (DBR), *vertical cavity surface emitting laser* (VCSEL), and *external cavity diode laser* (ECDL).

The least expensive fiber-optic transmitters, the LEDs, are used for multimode transmission at 850 or 1,310 nm, but are limited in speed. *Fabry-Perot* (FP) laser diodes are relatively inexpensive and are typically used for single-mode transmission at 1,310 nm. DFB laser diodes are most expensive and are used for single-mode transmission at 1,550 nm. VCSEL is now popular for GbE operation at 850 nm. The wholesale cost of a LED transmitter/receiver (transceiver) component is about $20; a FP laser transceiver, $100; a DFB laser, $600; and a gigabit VCSEL, $70.

Tunable lasers are also at the basis of bandwidth-(lambda)-on-demand and *intelligent optical networking* (OIN). The OIN is an architecture that is based on all-optical communication along with a control mechanism for both the *user-network interface* (UNI) as well as for the *network node interface* (NNI). Using tunable lasers, the transport nodes can match available capacity to real-time demand: Routes can be set up across the network for an end-to-end "lambda" circuit; this also allows dynamic setup and teardown of lambdas based on downstream requirements.

2.3.2 Fiber-Optic Receivers

Optical receivers convert on–off light pulses (photons) by photodetection into electrical signals. The electrical signals are processed by electronic components to provide the output data stream. Quality receivers are able to detect light levels down to hundreds of nanowatts. There are two kinds of photodetector diode, the *positive intrinsic negative* (PIN) type, with a composition of InGaAs, and the *avalanche photodetector diode* (APD) type, with a composition of Ge. Typically, the APD type can detect a lower optical signal power than the PIN type by about 10 dB (at 10^{-9} BER). An ideal receiver should offer good sensitivity (low received power for 10^{-9} BER), wide dynamic range, relatively low cost to manufacture and operate, reliability, low dependence on temperature, and so on. The APD has the highest sensitivity and good dynamic range but is difficult to produce; InGaAs photoconductors have poorer sensitivity but are simpler and have the potential to be integrated with low-noise amplifiers. Over time, advances are expected to provide improved receivers with integrated optic and electronic circuits.

2.3.3 WDM Elements

Going back at least half-a-century, network designers have looked for simple, straightforward methods for migration of networks to higher performance levels. Throughout the 1950s and 1960s, analog and then digital carrier systems of ever-increasing capacity were being developed. A carrier system is a multiplexed transmission system supporting multiple channels.

Fiber-optic (lightwave) systems have been deployed at least since 1979 (one such system was deployed in Trumbull, Connecticut, in October 1979 operating at DS3 rates). The 90-Mbps systems (FT3C) were deployed as early as 1983. In the mid- to late 1980s systems supporting 400 to 565 Mbps emerged, followed by systems operating at 800 to 1,100 Mbps. In the early 1990s, systems operating at 1,600 to 2,200 Mbps (1.6–2.2 Gbps) emerged. Finally, SONET/SDH systems emerged supporting up to 2.4 Gbps (OC-48); 10-Gbps (OC-192) systems have been introduced in the past couple of years, and 40-Gbps systems are under development. These advances are based on TDM technologies, in which the clock driving the on–off *intensity modulation* (IM) light process is driven at a higher pulse rate. As one approaches the limit as to how short the light pulse can be made, at least two other methods are used to increase the carrying capacity on the fiber waveguide: (1) coherent transmission, a variant of full FDM; and (2) WDM, which is in effect a FDM method in which distinct light beams are each modulated with IM methods. WDM has become the de facto technology to extend link throughput beyond 10 Gbps. Several elements support WDM systems [8].

2.3.3.1 Thin-Film Filters

Thin-film filters can be used to demultiplex wavelengths in a WDM system. Typically, several wavelengths are passed through the filter on a single beam of light; as the beam of light encounters the variable filter, the demultiplexing function occurs. The process can also be designed to perform the multiplexing functions. Thin-film filters are a cost-effective solution for low to medium channel counts. Thin-film technology is the primary method used for networks with 16 to 40 channels, with 400- or 200-GHz spacing. The goal is to find economically feasible methods to produce 100- and 50-GHz channel spacing or channel counts. A solution in support of the goal of narrower channel spacing and higher channel counts is *interleaving technology*. Interleaving combines two separate multiplexing and demultiplexing devices; here, one multiplexer or demultiplexer "interleaves" the odd channels, while the second "interleaves" the even channels. For example, 200-GHz spacing is "interleaved" to support 50-GHz applications. This approach adds channels in groups of four instead of groups of two.

2.3.3.2 Fiber Bragg Grating

Fiber Bragg grating (FBG) technology is employed to separate and combine wavelengths. This goal is achieved by changing the structural core of standard single-mode fiber. At manufacturing time, the strand of fiber is exposed to a high-power ultraviolet light, which creates a periodic change in the refractive index along the core by creating ridges on a short section of a fiber. The periodic change in the core of the fiber enables narrowband wavelength selection. Each fiber grating isolates a single wavelength. The limitation of this technology is that it performs better at moderate channel counts (less than 16 channels), and it is difficult to scale for the higher channel count-systems.

2.3.3.3 Arrayed Waveguides Gratings

Arrayed waveguide gratings (AWGs) are built by glass layers of carefully engineered composition deposited on a silica or silicon substrate. To guide the light pulses through the AWG, the layers of glass are patterned and etched. The input signal (arriving from a distant destination over a single fiber) is split by the origination coupler and passed on to multiple curved waveguides. Each waveguide is a different length; this enables the light signals to travel at different speeds across the AWG. This difference in speed gives rise to interference and diffraction when the signals enter the receiving coupler. Diffraction separates the wavelengths and allows the individual light signals to exit the AWG at the same time, but on different fibers. These fibers are then connected to receive-SONET terminals. The opposite process occurs to map the signal from multiple transmit-SONET terminals onto a single fiber. Using AWGs as an alternative to thin-film filters and FBGs reduces the overall cost and insertion loss of the network. AWGs have become the technology of choice for high-channel-count DWDM systems. DWDM technology is covered in more detail in Chapter 5.

2.3.4 Regenerators and Optical Amplifiers

We have already noted that as the signal travels through the fiber waveguide it undergoes attenuation. This requires that the signal be regenerated as it travels down the waveguide, particularly for long-haul applications. Preferably, one wants to be able to handle the amplification function in an optical domain, without the need for routine *optical–electrical–optical* (OEO) conversions. Recently, optical amplification has been introduced in long-haul transmission links, obviating the need for routine *retiming, reshaping, and regenerating* (3R, or RRR).

The *erbium-doped fiber amplifier* (EDFA) is a well-known technology particularly useful for DWDM applications (see Figure 2.11; also see

FIGURE 2.11
EDFA.

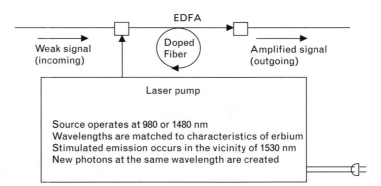

Figure 2.12 for a specific example). A basic EDFA consists of isolators, a pump laser, an erbium-doped fiber, and a coupler. The isolators are used at the front- and back-ends of the amplifier to prevent pump-laser emission and to suppress light reflections coming from the erbium-doped fiber. The pump laser "pumps" power into the fiber carrying the light signal. The coupler is used to connect the pump laser directly to the amplifier. In a simple EDFA, the amplification takes place in 50m or so of coiled erbium fiber inside the amplifier; the fiber is doped with a chemical, erbium, which allows for maximum amplification at 1,550 nm. The use of a *gain flattening filter* (GFF) can extend the range of operation of an EDFA amplifier.

Unfortunately, EDFAs can generate noise in the fiber-optic network. As a result, electrical regeneration is still needed at various points in the network to remove the noise. The drawback is that electrical regeneration is expensive and limits the amount of bandwidth that can be supported. In the traditional way to regenerate an optical signal, the wavelength must be

FIGURE 2.12
*Example of an
EDFA package.
(Courtesy BaySpec,
Inc.)*

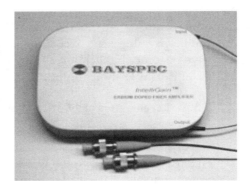

Specifications		
Parameters	IntensiGain™ L-Band in-line amplifier	Notes
Wavelength range	1570 – 1603 nm	
Saturated output power	≥ 14 dBm	Pin = – 3 dBm
Small signal gain	≥ 24 dB	Pin = – 20 dBm
Noise figure	≤ 6.0 dB	Pin = – 10 dBm
Polarization sensitivity	< 0.3 dB	
Return loss (input and output)	> 35 dB	Pin = –10 dBm at pump off
Gain flatness (peak-to-peak)	≤ 1.0 dB	Pin = –10 dBm with GFF
Operating current	< 2.5 A	
Operating voltage	5.0 V DC	
Power consumption	< 13 W	
Operating temperature	–5 to 65 °C	
Dimension	120 mm × 95 mm × 13 mm	

converted from an optical signal to an electronic signal and back to an optical signal. Hence, in a typical DWDM network, each optical wavelength traveling through a fiber has to be routed through a SONET regenerator that performs the 3R functions. Although performance varies by manufacturer, wavelengths traveling at higher signal rates (e.g., OC-192) could need several bays of SONET equipment to regenerate just a few wavelengths.

A new technology for next generation DWDM systems mitigates the costly regeneration process: Raman amplification. Whereas EDFA amplification takes place in the 50m of fiber coiled inside the box, Raman amplification takes place outside the box, along the regular fiber that is laid in the ground. This improves the performance of amplification and reduces the number of regenerators required in the network. The noise figure-of-merit (*signal-to-noise ratio* or SNR) for a Raman system is lower (better) than for an EDFA system. Raman amplification is a phenomenon in which a fiber pumped at a certain wavelength exhibits gain 100 nm away. Because they do not need a specially doped fiber, Raman amplifiers can de designed to pump the fiber in the transmission link (in the ground or on a pole); they act as distributed amplifiers compensating for attenuation along the fiber. Raman amplifiers have good gain flatness and dynamic gain shaping; they also support low signal power (−10 dBm being typical). *Forward error correction* (FEC) techniques provide another 5 to 10 dB of (additional) margin.

2.3.5 Optical Network Elements in Common Use

This section lists key optical network elements in common use.

2.3.5.1 SONET Terminals

SONET terminals per se are not considered optical network elements; however, they provide very critical edge multiplexing functions and are widely deployed. Since the early 1990s, SONET/SDH has been the major technology for fiber-optic networks throughout the world. SONET is the synchronous hierarchy defined by standards produced by the T1 Committee in the United States; the international equivalent is SDH, which is defined by the ITU-T Recommendations. Although the lower rates within the DS1/DS3 digital hierarchy in the United States and the E1-based hierarchy in Europe used different rates, a single set of rates at the higher levels has been agreed to [9]. Besides transport, SONET/SDH supports performance monitoring, fault isolation, protection switching, signal interleaving, and scaling capabilities; furthermore, it enjoys explicit standards-based interoperability features. Using TDM techniques that are intrinsic in SONET, data are now routinely transmitted at 2.5 Gbps (OC-48) and, increasingly, at 10 Gbps (OC-192); recent advances have resulted in speeds of 40 Gbps (OC-768).

Figure 2.13 depicts the basic reference model of a SONET link. Typical functions supported by SONET include TDM channelization (also

FIGURE 2.13
SONET reference model.

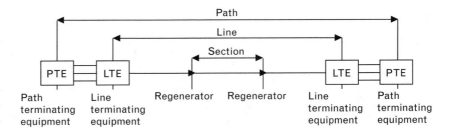

supporting lower speed "virtual tributaries"), ring, mesh, and linear topologies, 50-ms protection switching, midspan meets, data communication/*operation support system* (OSS) channel, and bandwidth manageability (with migration paths from OC-48 to OC-192 to OC-768). Table 2.4 identifies the rates defined and the terms used to identify them. Although standardization and interoperability are seen in SONET terminals, one can still find single-vendor SONET networks elements on any given ring due to a level of vendor-specific design regarding provisioning and maintenance aspects of the various products.

2.3.5.2   Optical Add/Drop Modules

Optical add/drop multiplexers (OADMs) are used to deliver signals to their intended destinations. OADM can add or drop an individual wavelength or a set of wavelengths without electronic conversion. OADMs allow certain wavelengths to pass through the node uninterrupted; OADMs also have broadcast capabilities that enable information on several channels (typically four) to be dropped and simultaneously continue as "express" channels.

OADMs can be of a preset fixed wavelength or can be dynamically configurable for one or more wavelengths. The filtering effect is achieved by using a number of technologies, for example, thin-film and FBG technology. One of the key design factors associated with OADMs, however,

TABLE 2.4   HIERARCHICAL RATES
FOR SONET/SDH

RATE (MBPS)	SONET	SDH
51.84	STS-1/OC-1	
155.52	STS-3/OC-3	STM-1
622.08	STS-12/OC-12	STM-4
1,244.16	STS-24/OC-24	
2,488.32	STS-48/OC-48	STM-16
9,953.28	STS-192/OC-192	STM-64
3,9813.12	STS-768/OC-768	STM-256

is loss (attenuation). A circulator is used to access or transport (add or drop) the signals of different wavelengths traveling on the fiber; circulators are passive devices that guide wavelengths from port to port, in one direction only. The purpose of circulators is to guide the wavelengths that are filtered in and out of the OADM without interrupting any of the light signals. As an example, the input light signal is carrying multiple $(\lambda_1, \lambda_2, \lambda_3, ..., \lambda_x)$ channels; as it moves through the OADM, FBG technology and the circulators enable channel λ_x to be "dropped" and channel λ_y to be "added" (see Figure 2.14).

OADMs can be placed between two end terminals along any route and be used in place of an optical amplifier. Commercially available OADMs allow carriers to drop or add up to four STM–16/OC–48 channels between DWDM terminals. OADMs are well suited for meshed or branched network configurations, as well as for ring architectures; ring architectures are used to enhance survivability. The OADM technology allows planners to move closer to an all-optical network. OADMs are useful *network elements* (NEs) in the context of metropolitan core/metropolitan access networks. OADMs are reasonably inexpensive, and, in particular, are cheaper than electronic ADMs. Fixed-lambda systems are mature and are widely deployed; configurable-lambda systems are becoming available; configurable-lambdas systems are planned to be used in the ION.

2.3.5.3 Optical Cross-Connect Systems

Many current generation optical networks use *optical cross-connects* (OXCs). These NEs are capable of interconnecting multiple DWDM systems and switching optical channels in and out of various DWDMs without having to convert the signal into a lower speed electrical signal. First generation OXCs had optical inputs and optical outputs but use an electronic switching matrix in between. In this environment, in order to switch the wavelengths, the cross-connect converts optical signals to electrical signals, and then converts them back to optical signals. This conversion makes the switching process relatively inefficient and expensive. Second generation

FIGURE 2.14
Equipment savings with OADMs. (a) All traffic needs to be regenerated. (b) Pass-through traffic is all optical.

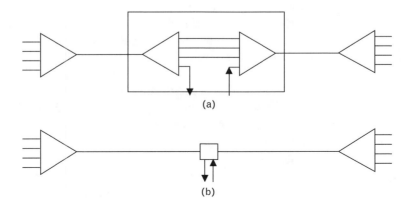

(a)

(b)

OXCs have the capability of switching wavelengths without an electrical conversion, thereby supporting an optical-to-optical (O–O) operation. OXCs also support interconnection between the DWDM systems and IP routers, SONET/SDH multiplexers, and ATM switches.

The deployment of OXCs is advantageous to all-optical networks because they support higher bandwidth and also because they enable flexible network designs. For example, OXCs can perform traditional SONET/SDH functions, including restoration, signaling, performance monitoring, fault management, and multiplexing. By including these SONET/SDH functions in the optical layer, carriers are able to reconfigure their networks more flexibly, while perhaps also lowering deployment and operation costs. OXCs enable the optical core to manage high-speed traffic, while the traditional SONET/SDH equipment is moved to the edge to manage the lower speed interfaces and mixtures of protocols. Also, non-SONET/SDH equipment (e.g., GbE switches, 10GbE switches, RPR devices, and so on) can be supported by the optical core.

Three families of technologies have been developed to build all-optical cross-connects: microelectromechanical systems, bubble technology, and liquid crystal technology. (Keep in mind that an OXC is really a "small" optical switch.)

Microelectromechanical Systems

Microelectromechanical systems (MEMS) are used in wavelength switching and optical cross-connects. MEMS is comprised of an array of microscopic tilting mirrors sculpted from semiconductor materials such as silicon. MEMS are based on the mechanical positioning of microscopic mirrors (silicon wafers with a matrix of movable mirrors) that direct wavelength signals between input and output ports. When switching output ports, the mirror at the intersection of output 1 and output 2 rises and redirects the light signal coming from input 1, while the other mirrors on the matrix remain flat. MEMS devices are attractive because of their miniaturization and potential mass production. The main drawback of MEMS devices is the issue of movable parts and reliability. In 3D arrays the mirrors can be tilted in any direction. 3D MEMS arrays can be used for large-scale optical cross-connects, particularly if groups of wavelengths need to be switched from one fiber to another. In 2D arrays the mirrors can only flap up and down, incorporating a single axis. 2D MEMS arrays are used in smaller, first generation, all-optical switches because they are further along in the development process. Because MEMS is based on mechanical movement, it offers the best isolation performance available. For the same reason MEMS cannot be used for drop-and-continue applications.

Bubble Technology

With bubble technology an input signal enters the device with the goal of exiting at a specific output, say, output 1. When an input signal and output signal intersect in the device, a vapor bubble is formed on command. This point of intersection houses an index-matching liquid that enables the formation of the bubble. In the presence of a bubble, the input signal is reflected toward output 2 (without the formation of a bubble, the light signal passes straight through the device toward output 1.) The "bubble" device is of interest because of its connection with a proven technology and lack of moving parts. However, scaling the system can become a problem: Each time the service provider adds a block, more insertion loss is generated and accumulated across the system.

Liquid Crystal Technology

LCDs (also known as LXT) consist of active and passive elements. The active element is the liquid crystal cell and the passive element is the polarization beam splitter. The switching action is dependent on whether or not an electrical field is present. When there is no electrical field in the device, the input signal reflects off the liquid crystal pixel and rotates 90 degrees. When an electrical field is present, the light signal travels straight through the active cell without changing direction. Two advantages of LCD technology are (1) the lack of movable parts, making it more reliable than MEMS technology, and (2) the manufacturing process (the same manufacturing process as flat-panel displays), a technology that is mature, so that materials, fabrication, infrastructure, and the processes are already in place to supply the large quantities of the optical device. The drawbacks of liquid crystal technology are thermal sensitivity, slower switching speeds, and limited scalability. The thermal sensitivity issue can be overcome by temperature controllers, improved design, or improved liquid crystal fluid; however, scalability remains an issue because of cost.

2.3.5.4 Optical Switching

High-beam count optical switching is the enabling technology that will make ION possible. This technology allows direct switching of constituent wavelengths (either intrinsic to the network or customer originated). Optical switching eliminates optical-electrical-optical (OEO) conversions: Whereas a traditional DCS takes electrical signals in and sends electrical signals out, the optical switch takes optical signals in and sends optical signals out. An optical switch can be classified into two categories. The first category changes the optical signal into electrical and performs a switching function before changing it back to optical. This is known as an *OEO switch* or *opaque switch*. The second category is a pure optical switch where OEO conversion is not necessary; this category is also called a *photonic switch,* because it switches photons, or lights, rather than electrons in the opaque switches [10, 11].

Optical switching matrices need optical characteristics such as low insertion loss, high isolation, and low back reflection. In addition, they need to be reliable, fast, compact, low power consuming, and scalable.

Techniques that are suitable for optical switching matrices include but are not restricted to MEMS, some MEMS-like technologies (thermo-capillary, piezoelectric), liquid crystals, bubbles, thermo- and acousto-optics, and holograms. These technologies offer different characteristics that are summarized in Table 2.5 [12]. Figure 2.15 depicts an example of an optical-optical-optical (OOO, also known as optical) switching architecture.

MEMS were discussed earlier, so we turn now to LXT devices. In LXT, light is passed through liquid crystals that alter their transmission characteristics when an electric current is applied. Most LXT devices depend on the polarization properties of the light to accomplish switching. LXT devices can throttle the amount of light that passes through them. They can, therefore, be used as variable attenuators or power splitters with configurable splitting ratio. The latter enables various drop-and-continue applications. LXT switches have good scalability; however, the technology will first be used for modestly sized wavelength-selective devices. LXT can be combined with electroholographic technology for increasing switching speeds. LXT offers low-power consumption.

TABLE 2.5   SWITCHING TECHNOLOGIES

	ACOUSTO-OPTIC	BUBBLES	HOLOGRAMS	LXT	MEMS	THERMO-OPTIC
Scalability	Medium	Medium	High	Medium	Very high	Medium
Switching speed	Fast	Medium	Very fast	Medium	Medium	Medium
Reliability	Good	Good	Good	Good		Good
Power consumption	Unknown	Unknown	Medium, but requires high voltages	Very low	Medium	Vey high when based on silica

Source: [12].

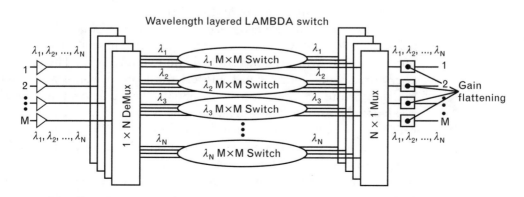

FIGURE 2.15   *Example of OOO switching. (Courtesy of Corvis.)*

As discussed previously, bubbles formed by printer ink pens have surfaces that act like mirrors. Bubbles are currently employed in optical switches by reusing parts of desk jet printer technology. The technology is intended for use in add/drop multiplexers and scalable switches.

Thermo-optics consists of passive splitters. When these filters are heated, their refractive index is changed to alter the way in which they divide wavelengths between either of two output ports. They are wavelength sensitive and need temperature control. AWG can be integrated with arrays of thermo-optic switches to create an add/drop multiplexer. Polymers are probably the best material because AWGs are temperature sensitive and polymers keep the heat localized at the switch. As a waveguide approach, thermo-optics eliminates beam alignment issues created by free-space approaches but are susceptible to cross-talk, polarization dependence, and loss issues.

Acousto-optics uses sound waves to deflect light that passes through crystals or fused couplers. Acousto-optics will be used for wavelength-selective switches.

With holograms, electrically energized Bragg gratings are created inside a crystal. When voltage is applied, the grating deflects the light to an output port. When no voltage is present, the light passes straight through. Holograms compete with 3D MEMS on scalability but are better suited to switching individual wavelengths rather than groups of wavelengths. Their high switching speed (nanosecond range) means that holograms can be used for optical packet-by-packet routing. Like LXT devices, holograms can support drop-and-continue applications.

Opaque (OEO) switches support switching of signals from STS-1 to STS-768; however, their cost per port increases as bandwidth increases. The ability to manage across all of these signals allows for cost-effective grooming and multiplexing for sublambda signals. Full visibility to all payload and overhead is required to effectively administer service policies in the network. OEO switches can, however, become expensive. Termination of the optical line results in significant capital costs, as well as real estate footprint costs and power requirements. In addition, scalability for OEO switches can be a challenge when dealing with the switching requirement of hundreds of terabits per second in the future. Photonic [13] switches allow switching at the photonic layer to be done at a relatively low cost: Photonic switches have a fixed cost per port regardless of the amount of bandwidth through each port, because they switch light and do not discriminate between 1 lambda or 100 lambdas in the light. In addition, photonic switches are bit-rate independent. For this reason, at very high bandwidths, the cost of switching photonically is less expensive compared to opaque switches. A single switch can allow scalability into the hundreds of terabits per second. The drawbacks of photonic switches are their inability to perform many of the functions achievable with OEO switches. For example, grooming of bandwidth below the "light path" can only be accomplished with opaque switches. Also, the inability to monitor electrical payload and overhead makes service policy management by the switch itself less attractive. Nevertheless, the cost and scalability benefits of photonic switches can still be attractive, especially in the applications where there is no need for sublambda bandwidth management, and the pass-through traffic dominates [10, 11].

Until recently, optical networks were mostly static: Connections take a long time to provision. Furthermore, networks have been susceptible to manual errors. The International Telecommunications Union (ITU) has published an architectural standard for *automatically switched transport networks* (G.astn, now known as G.8070) that provides dynamic optical connectivity though automatic routing and switching of transport bandwidth. With optical switches and G.astn, it is possible to bring intelligence to the optical layer. This topic is discussed in Chapter 4.

2.3.6 Applications

As will be discussed in greater detail in Chapters 3 through 6, telecom companies' long-range desires are to migrate to all-optical networks. In such a network, an intelligent control apparatus (a signaling mechanism) would be desireable for setup and teardown of optical connections across the network to support bandwidth-on-demand requirements. Bandwidth-on-demand technology is postulated on the assumption that bandwidth is scarce and therefore cannot be "always on" for the user. This assumption may or may not be valid. Note that switched virtual circuits have not been as commercially successful as packet switching, FR, or cell relay. The complexity of the capacity management may outweigh the benefits of allocating bandwidth somewhat more efficiently. A related assumption is that there are "spots" in the network where the bandwidth is put in place only when the demand is there.

All-optical network architectures are driven by the fact that electrical processing is expensive and the electronics require much power, space, and *operation, administration, maintenance, and provisioning* (OAM&P). In particular, a move toward an all-optical network simplifies the protocol stack for devices such as routers that wish to make use of the network bandwidth. Originally a broadband connection entailed IP running on ATM, which in turn ran on SONET/SDH over an optical infrastructure. One migration possibility would be to run IP on ATM directly on optical. Another migration would be to run IP on SONET/SDH, which in turn runs on the optical fiber. The target architecture could be to run IP directly over optical. However, one must keep in mind that certain "services" are required of the lower layer, such as restoration, switching, and framing. Figure 2.16 depicts the advantages of utilizing optical technologies, just from an equipment point of view.

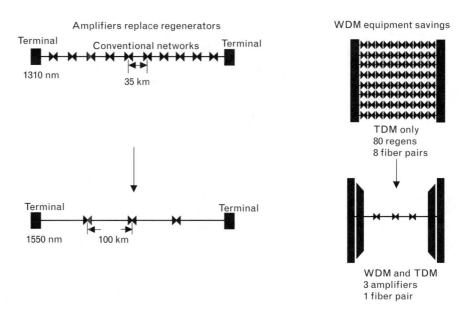

FIGURE 2.16
*Savings achievable
with optical
components.*

With an eye to the increasing deployment of DWDM equipment in the wide area backbone and also in metropolitan area networks, the American National Standards Institute (ANSI) and ITU-T started standards activity in the late 1990s and early 2000s to define a common method for managing multiple wavelength systems—the *optical transport network* (OTN). SONET/SDH-based networks have played a major role over the years in addressing the network capacity demands, and it is expected that SONET systems will continue to do so for the foreseeable future. Considering the changing network requirements in terms of data services support, future networks may require the capabilities of OTN; this assumes that data services will have the distinction of not just requiring 20 times the bandwidth of voice (at the macro level), but also bringing in a much higher share of the carrier's revenue.

The growth of data services and the Internet have led to an evolution in the optical networking infrastructure. This has resulted in requirements for networks capable of transporting a variety of *heterogeneous types of signals* directly across multiple wavelengths on optical backbones [14]. Network infrastructure must also permit flexible and transparent management of each optical link and each wavelength as a discrete optical channel, each with its own OAM&P information. To enable development of this next generation OTN technology, a network architecture, information structures, and bit rates, developed through ITU-T Study Group 15 standardization activities, are required. OTN is based on DWDM technology and will enable the evolution of the plant to a data-optimized, multiservice network infrastructure. OTNs and related topics are discussed in Chapter 3. Signaling in OTN environments is discussed in Chapter 4.

ENDNOTES

[1]   Remember that frequency and wavelength are related by the speed of light, namely $f = c/\lambda$.

[2]   When a large number of beams are multiplexed, this technique is known as DWDM.

[3]   Retrieved from http://www.optronics.gr/Tutorials/dwdm.htm.

[4]   One fiber type (with commercial deployment) offers a gradual change in the index across the cross section of the fiber; this is called *graded-index* fiber.

[5]   *Modal dispersion* is also called *intermode* or *multimode dispersion. Chromatic dispersion* is also called *intramode* or *spectral dispersion.*

[6]   George, J., "Ethernet in the First Mile Optical Architectures and Fibers," IEEE EFM Meeting Minutes, Hilton Head, SC, March 2001, www.ieee802.org/3/efm/public.

[7]   Bird, R., *Fiber Optics 101: A Primer,* retrieved from http://certcities.com/editorial/features/story.asp?EditorialsID=25.

[8]   Some material in Sections 2.3.3, 2.3.4, and 2.3.5 is based on the report "Optical Players Investment Opportunities, First Union Securities Inc.," June 2000.

[9]   Retrieved from http://www-comm.itsi.disa.mil/on/on_itu.html.

[10]   Wu, B., "Optical Switch Variations: How Will They Play?" retrieved from http://www.xchangemag.com/articles/161solutions5.html, 2002.

[11]   Nortel Networks, Optical Internet Business, http://www.nortelnetworks.com.

[12]   Adva Optical promotional material, retrieved from http://www.advaoptical.com. The rest of the section is from the same source.

[13]   These are also called *photonic cross-connects* (PXCs).

[14]   Intel Corporation, information retrieved from ftp://download.intel.com/design/network/products/optical/appbref/24950501.pdf.

CHAPTER 3

All-Optical Networks

....

3.1   Introduction

This chapter discusses advances in all-optical networks, in particular the Optical Transport Network (OTN). The SONET/SDH specifications of the late 1980s and early 1990s were designed for supporting the transmission of a single wavelength (also known as λ, beam, or lightpath) per fiber; typically a send-and-receive wavelength pair operates over two fibers. OTN, on the other hand, is designed for the transmission of multiple wavelengths per fiber, which is characteristic of DWDM systems. At the same time, a need has emerged of late for networks capable of transporting a variety of heterogeneous types of signals directly over wavelengths carried on the same optical backbone. This need is in fact addressed by the OTN. Vendors consider the "next generation WDM network" to be the OTN.

As is the case for SONET/SDH, OTN is standards based, so that interoperability among various equipment manufacturers and interfaces is ensured. In the early 2000s, the ITU reached agreement on several new standards for next generation optical networks capable of transporting transparent wavelength services, SONET/SDH services, and data streams (Ethernet, Fibre Channel, ATM, FR, and IP). The new standards provide the ability to combine multiple client signals within a wavelength, to facilitate optimal utilization of transport capacity, and to realize improved cost effectiveness of transport capacity, while allowing switching at service rates of 2.5, 10, and 40 Gbps. The new standards also specify optical equipment functions such as performance monitoring, fault isolation, and alarming[1]. In this chapter we look at standards related to the OTN along with architectural considerations. Motivations and positioning of all-optical networks are also discussed. The signaling aspects of the OTN are discussed in greater detail in Chapter 4.

ITU Recommendations G.709 and G.872 define an OTN consisting of optical channels within an optical multiplex section layer within an optical transmission section layer network; the optical channels are the individual lightpaths (beams). The OTN promises to deliver network scalability and manageability and to address the requirements of next generation networks that have a goal of efficiently transporting data-oriented traffic [2]. Service providers' networks need to have the ability to scale *when future demand materializes once again*; furthermore, these networks need to be manageable in a cost-effective manner. As noted in earlier chapters, industry growth is now somewhere in the 10% to 15% per year range, although there were hundreds of articles at press time making reference to the then-existing bandwidth glut [3, 4]. Because of the carnage that took place in the early 2000s when supply greatly exceeded demand, three key points related to proper positioning are noted here before we proceed with our discussion of OTN technology itself:

1. Developers need to make much more sophisticated statements than superficial assertions such as the typical boiler plate "Fueled by the Internet usage, network capacity demands are increasing at an explosive rate … scalable networks are needed." Much better justification is required to provide some degree of credibility to a proposed new technology. In recent years, "scalability" has achieved a self-serving agenda for equipment providers, obfuscating issues of real demand, and selling to operators the idea of building networks of such size that would only be fully utilized several years down the road, often as much as 10 years later. At the same time, these vendors have preached a message of near-immediate obsolescence: "If your equipment is more that 3 years old, then it is out-of-date." Clearly this dynamic does not add up, and is self-defeating: How can one be pressured to build a network with a given generation of equipment that one will not fill for 10 years and, at the same time, be subjected to dogmatic proclamations of 3-year obsolescence? There should be no surprise that many stocks for telecom equipment vendors hit bottom at around $1 in 2002. Good financial paybacks are difficult for any technology being deployed in less than 4 or 5 years under any reasonable economic metric; therefore, any equipment that is deployed *must* have a life cycle of at least 6 years, particularly when considering the planning, engineering, and test-and-turnup expense associated with deploying NEs or an entire network. Scalability by itself is a void concept that can lead to very distorted network designs, particularly when the concept is ill understood and ill applied. Scalability is side A of a coin; side B has "modularity" as its key concept. Modularity implies that one starts small by choice (not the opposite, that is, "start big" in the name of scalability), and, *as resources are needed*—that is, as they are driven by demand—they are added gradually, transparently, and cost effectively.

2. A key takeaway regarding the recent industry situation is appropriate. When *some* demand for cross-country private lines at

DS-3, OC-3, OC-12, and, eventually, at OC-48 rates arose in the 1997–1999 time frame under the thrust of the evolving commercialized Internet, the carriers of the day were *slow to provision the circuits*. It took these carriers 12 to 15 months to internalize the fact that there was a demand for high-speed facilities, when they were (are) almost completely focused on T1 lines. This *delay in provisioning was misinterpreted* by some as a shortcoming on the supply side of the equation. This led people to think that there was a major need for new construction, and it led to the launch of many new parallel carriers, which ultimately did lead to an oversupply, at least in the near term and in the long-haul portion of the network. All that was needed, instead, was for the existing carriers to promptly internalize the fact that they were seeing a demand for high-speed facilities (rather than being almost completely focused on T1 lines) and should place a telephone call to an equipment provider and replace the SONET terminals on the existing fiber with DWDM systems that could double, quadruple, multiply by 8, multiply by 16, multiply by 32, multiply by 64, or multiply by 128 the amount of usable bandwidth that could be secured on the same existing plant. The problem, hence, was slow market response and appreciation for customers' needs, not an actual shortcoming or gap in supply [5].

3. Networks should be designed by networking professionals. Carriers have a requirement to deploy infrastructures that are manageable, cost effective, sustainable, and produce a financial bottom line. Over the years people outside the carrier space have tried to tell carriers how to design their networks. These efforts have either died on "the vine" or led to major investments followed by bankruptcies on discovery that the approach was not financially sustainable in the medium to long term. An example of the former was the effort to secure ubiquitous unspecified bit rate (UBR)-like service in ATM, where users wanted to just get "any spare capacity" that was laying around. And they wanted that capacity without having to pay for it—without having to pay for the infrastructure costs needed to provide a financially self-sustaining ATM service to start with, which is necessary to make any capacity at all available and would include, as a by-product, UBR capacity. An example of the latter was the delivery of DSL services by companies that neither controlled (owned) the fundamental resource (the local loop) nor the engineering, provisioning, or monitoring process of said resource. Users of electricity do not tell the power company how to design the power network. Users of water do not tell the water utility how to design the main distribution system.

Cable–TV users do not tell the cable company how to design the cable distribution system or where to deploy the satellite systems. Users of bandwidth cannot tell the carrier how to design the network, although users can and should certainly articulate their requirements, but those requirements should not be cycled through the agenda of the equipment vendors. A trade-magazine-selling debate was active at press time as to whether the carriers will deploy a next-target (some say next generation) overlay network based on OTN with end users signaling their needs to the network, or whether the next-target network will be a peer-to-peer arrangement where the end user in effect controls the network [6] (see Figure 3.1 [7]). The latter model will never prevail in an environment where the providers of capital (to build these next-target networks) expect a financial dividend for their investment. Not only will the network of the future be either an OTN network or some OTN equivalent, but in our estimation, the signaled-λ will not turn out to be a popular data solution in the short term any more than other *switched virtual circuit* (SVC) technologies that have appeared in the past 35 years have gained popularity. Fundamentally, the lack of commercial interest on switched paths has to do with the fact that the data sources and sink points for enterprise users are almost always fixed at specific locations (at least for those flows requiring multigigabit links), and the overhead of establishing (setup and teardown) a "circuit" to send a few connectionless packets is never truly practical in most circumstances. Enterprise network connections are like trunks—trunks between layer 3 switches (routers), similar to trunks between voice switches. However, over the years, considerable standardization energies have gone into developing X.25 SVC, ISDN X.25 support, FR SVCs, and ATM SVCs. In this book we afford the reader an exposure to these optical signaling issues and we will let the market decide the commercial success of these services; however, it is worth pondering on previous history.

These observations are not intended to take away from the OTN effort. Just that appropriate context of rationalization, growth, "scalability," deployment timetables, and capability-set development has to be kept in mind. The new ITU recommendations *allow* for scalability. This is good. But it does not mean that carriers need to build a huge inventory on Day 1 and oversupply the inventory for 10 years to come. For example, in building a new highway, it is good to allocate land around it to eventually *allow* future expansion of the highway from two lanes to five lanes, but it would not generally be wise in most instances to build five lanes in each direction on Day 1.

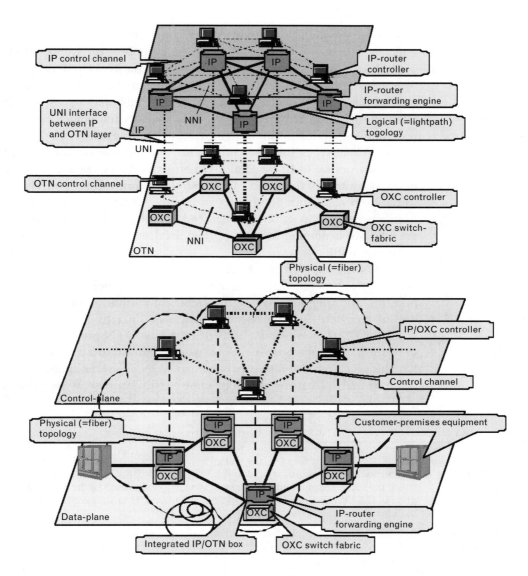

FIGURE 3.1   *Comparison between peer-to-peer and overlay model for OTN. (Source: Telecom Lab Italia.)*

3.2 Motivations, Goals, and Approaches

Existing TDM SONET/SDH networks were originally designed for voice and leased-line services. According to proponents, the growing trend of data traffic is now posing technical challenges not only in terms of volumes but also related to the burst and asymmetrical nature of such traffic [8], at least in the outermost edges of the network. (In the core of a large data/IP network like the Internet or a corporate intranet backbone, data traffic is neither bursty nor asymmetric.) At the same time, DWDM technology, which was initially introduced to increase transport capacity, is currently available for implementing advanced optical networking functionality. The existing transport networks are expected to evolve to next generation optical networks. The expectation is that these networks should fulfill new emerging requirements such as fast and automatic end-to-end provisioning, optical rerouting and restoration, support of multiple clients and client types, deployment of *optical virtual private networks* (OVPN), and interworking of IP-based and OTN networks. According to proponents, indicators of this trend can be extracted from various projects' consortia, from standardization bodies around the world (e.g., see Table 3.1), from emerging product vendor lines, and from the initiatives of some carriers who are beginning to offer in their services portfolio flexible optical connections [8].

OTN is composed of a set of *optical network elements* (ONEs) interconnected by optical fiber links that are able to provide functionality of transport, multiplexing, routing, management, supervision, and survivability of optical channels carrying client signals, according to the requirements given in ITU Recommendation G.872 [9]. The OTN provides the ability to route wavelengths and, therefore, when deployed in a ring topology, it has the same survivability capabilities as SONET/SDH rings. Furthermore, the OTN has the ability to improve survivability by reducing the

TABLE 3.1   KEY STANDARDIZATION ENTITIES DOING WORK IN OPTICS (PARTIAL LIST)

ITU Telecommunications Standardization Sector (ITU-T)	ITU-T is a subunit of the ITU, the United Nations body chartered with establishing telecom and radio standards around the globe. The ITU-T has been instrumental in laying the standardized foundation for optical networking (including the basic "ITU-grid" of DWDM channels). A key relevant standard is SDH used in networks outside of the United States and Japan.
Institute of Electrical and Electronics Engineers (IEEE)	The IEEE promotes standards related to electrical and information technologies. The 802 Committee is the developer of all LAN standards to date, including the Ethernet family of specifications.
Internet Engineering Task Force (IETF)	IETF is an informal body that is responsible for the evolution of the Internet architecture.
Committee T1 and T1 X1 Subcommittee	Committee T1, sponsored by the *Alliance for Telecommunications Industry Solutions* (ATIS) and accredited by the ANSI, develops technical standards and reports on the interconnection and interoperability of telecommunications networks in the United States. T1X1 is a subcommittee of Committee T1; it has developed SONET standards over the years.
Optical Internetworking Forum (OIF)	OIF is an industry-bred organization consisting of several hundred equipment manufacturers, telecom service providers, and users whose goal is to foster interoperability among optical equipment. It developed the optical UNI 1.0.
European Telecommunications Standards Institute (ETSI)	ETSI is a not-for-profit organization whose mission is to produce the telecommunications standards that will be used for decades to come throughout Europe and beyond. ETSI unites about 1,000 members from more than 50 countries inside and outside Europe, and represents administrations, network operators, manufacturers, service providers, research bodies, and users.

number of electro-optical network elements; these elements may, in principle, be prone to failures.

According to developers, the OTN provides a cost-effective, high-capacity, survivable, and flexible transport infrastructure. The elimination of multiple service network overlays and the elimination of fine-granularity sublayers imply a reduction in the number and types of network elements and a concomitant reduction in capital and operating costs for the network provider. With OTN, fine-granularity transport interfaces that are specific to protocol and bit rates are no longer required, in turn reducing the network provider's OAM&P costs. At the same time, the data-optimized service infrastructure is best positioned for evolving data-oriented applications [10]. Figure 3.2 depicts the consolidation that is possible, and Table 3.2 lists some of the benefits of the OTN. The optical transport functions include the following:

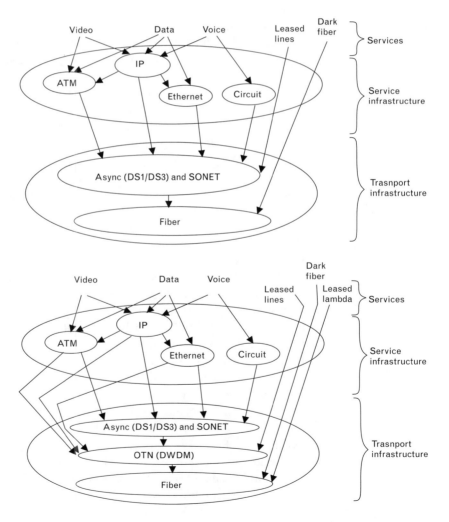

FIGURE 3.2
Pre-OTN and post-OTN service support.

TABLE 3.2   OTN (G.709) BENEFITS

Maintenance signals per wavelength

Fault-isolation capabilities

Small overhead

FEC (better performance)

Protocol agnostic

- Multiplexing function;
- Cross-connect function, including grooming and configuration;
- Management functions;
- Physical media functions.

The *Generic Framing Protocol* (GFP) provides a means for packing non-voice traffic into a SONET or OTN frame. A virtual concatenation mechanism called the *link capacity adjustment scheme* (LCAS) provides additional flexibility. With conforming implementation, interworking between equipment from different vendors (e.g., the vendor manufacturing a router plug-in and a vendor manufacturing the network equipment to support the connection) is achievable.

As described, a distinguishing characteristic of the OTN is its ability to transport any digital signal, independent of client-specific aspects (protocol agnostic). The flexibility of OTN is built on the protocol and bit-rate independence of the information-carrying optical beams in the fiber waveguide. This transparency enables the OTN to carry many different types of traffic over an optical channel regardless of the protocol [GbE, 10GbE, ATM, SONET, asynchronous fiber optic transmission system (FOTS), and so on] or bit rate (155 Mbps, 1.25 Gbps, 2.5 Gbps, and so on). According to the general functional modeling described in ITU Recommendation G.805, the OTN boundary is placed across the optical channel/client adaptation in such a way as to include the provider-specific (server-specific) processes yet leave out the client-specific processes [9]. Some claim that SONET/SDH has been optimized to serve voice-based traffic with a TDM framing that does not allow efficient use of bandwidth in bursty data traffic environments. As noted, proponents make the pitch that with the emergence of ever-growing Internet and intranet data-oriented traffic on the network, a new network mechanism is needed to meet the network demands for scalability and manageability. This is fine, except to note for the record, once again, that the traffic at the core of a network is rarely if ever bursty, since Probability 101 principles clearly demonstrate that the sum of 15 to 20 independently and identically distributed random variables leads to a distribution that is Gaussian in nature; furthermore, when the link has a service time that is small compared to the arrival rate (e.g., a 1,000-person company being run an a gateway router with a single T1 uplink), the distribution is actually D/D/1. Also, with bandwidth being a commodity, a strict focus on "efficient use" may be largely anachronistic at this time. With the "glut" of bandwidth (existing, or easily caused at points in the future by the rapid deployment of DWDM equipment) and the low cost of bandwidth, a rationalization for the introduction of an entirely new packet-oriented technology, with a price tag of billions of dollars, *just* for the sake of achieving a 10%, 20%, or 50% bandwidth improvement is, prima facie, rather weak.

3.3   OTN Standards Support

Today's global communications world has many definitions for optical networks and many technologies that support them. This has resulted in a number of different *study groups* (SGs) within the ITU-T, for example, SG4, SG11, SG13, and SG15 developing recommendations related to optical transport. Recognizing that without a coordinated effort there is the danger of duplication of work as well as the development of incompatible and noninteroperable standards, the ITU WTSC 2000 [11] designated SG15 as lead study group on optical technology, with the mandate to define and maintain an overall standards framework in collaboration with other SGs and standards bodies [9, 12]. The OTN consists of the required networking capabilities and the technology required to support them.

ITU-T Recommendations G.872, G.709, and G.959.1, the initial set of standards in the OTN series approved in 1999 and in 2001, address the OTN architecture, interface frame format, and physical layer interfaces, respectively. This work is carried out by SG15 of the ITU. The ITU-T G.709, *Network Node Interface for the Optical Transport Network*, provides the basic standard specification for OTN. OTN elements' transport of a client signal over an optical channel is based on the digital signal wrapping technique defined in ITU-T Recommendation G.709. The wrapper provides transmission protection using FEC. The OTN architecture is further defined under ITU-T G.872, *Architecture of Optical Transport Networks*. By complying with ITU-T G.709, network nodes from various vendors can interoperate, although in actual practice carriers rarely if ever mix transport equipment from two vendors. OTN provides carrier-class robustness.

In addition to basic transport, there is also interest in real-time provisioning of lambdas, the details of which are discussed in Chapter 4. Working Group 15 of the ITU-T has grouped these new signaling architectures into two protocol-independent framework models: the general *automatic switched transport network* (ASTN) and the more specific *automatic switched optical network* (ASON). Furthermore, the OIF has taken the ITU-T ASTN and ASON models and extended the signaling capabilities of the IETF GMPLS and incorporated it into its OIF *optical-user-to-network interface* (O-UNI) 1.0 documentation [13].

We now provide a more detailed survey of key recommendations. OTN-related recommendations developed by SG15 toward the end of 2001 [14] include the following (also see Table 3.3):

- G.709, *Network Node Interface for the Optical Transport Network (OTN)*, February 2000; Amendment 1, November 2001. This recommendation is also referred to as Y.1331, and is being renumbered G.7090. It provides the requirements for the *optical transport hierarchy* (OTH) signals at the network node interface, in terms of definition of an *optical transport module of order n* (OTM-n), structures for an OTM-n, formats for mapping and multiplexing client signals, and functionality of the overheads. The initial version covers OTM-0 (single channel,

TABLE 3.3   KEY OTN-RELATED ITU-T STANDARDS

G.691	*Optical Interfaces for Single Channel SDH Systems with Optical Amplifiers, and STM-64 Systems*
G.692	*Optical Interfaces for Multi-Channel Systems with Optical Amplifiers*
G.693	*Optical Interfaces for intra-office systems*
G.7041/Y.1303	*Generic Framing Procedure (GFP)* specifies interface mapping and equipment functions for carrying packet-oriented payloads
G.709	*Network Node Interface for the OTN*
G.7710/Y.1701	*Common Equipment Management Function Requirements* is a generic equipment management recommendation
G.7712/Y.1703	*Architecture and Specification for Telecommunications Management Networks (TMN)*
G.7713/Y.1704	*Distributed Call and Connection Management*
G.7714/Y.1705	*Generalized Automatic Discovery Techniques*
G.798	*Characteristics of OTN Hierarchy Equipment Functional Blocks*
G.8080/Y.1304	*Architecture for the Automatic Switched Optical Network (ASON)*
G.8251	*The Control of Jitter and Wander Within the OTN*
G.871/Y.1301	*Framework of Optical Transport Network Recommendations*
G.872	*Architecture of the OTN*
G.874	*Management Aspects of the OTN Element*
G.874.1	*OTN Protocol-Neutral Management Information Model for the Network Element View*
G.957	*Optical Interfaces for Equipment and Systems Relating to the SDH*
G.958	*Digital Line Systems* based on the synchronous digital hierarchy for use on optical fiber cables
G.959.1	*OTN Physical Layer Interfaces*

reduced functionality) and OTM-16r (16 channels with reduced functionality, i.e., no overhead signal). An amendment to Recommendation G.709, entitled *Interfaces for Optical Transport Networks,* describes the mappings for TDM multiplexed signals in the OTN, as well as extensions to allow even higher bit-rate signals to be carried using virtual concatenation.

- ITU-T Recommendation G.7041/Y.1303, *Generic Framing Procedure (GFP),* specifies interface mapping and equipment functions for carrying packet-oriented payloads including IP/PPP, Ethernet, Fibre Channel, and *Enterprise Systems Connection/Fiber Connection* (ESCON/FICON) over optical and other transport networks. This recommendation, together with ITU-T Recommendation G.709 on *Interfaces for Optical Transport Networks,* provides the full set of mappings necessary to carry IP traffic over DWDM systems.

- ITU-T Recommendation G.7710/Y.1701, *Common Equipment Management Function Requirements,* is a generic equipment management recommendation and was derived from knowledge gained through the development of SONET and SDH equipment. This recommendation provides the basis for management of equipment for new transport network technologies including the optical transport network.

- New ITU-T Recommendation G.7712/Y.1703, *Architecture and Specification of Data Communication Network,* extends capabilities originally built for *Telecommunications Management Networks* (TMNs). It allows the use of IP protocols as well as *Open Systems Interconnection* (OSI) protocols, and supports new services such as ASTN through communication among the transport plane, the control plane, and the management plane.

- G.871, *Framework of Optical Transport Network Recommendations,* October 2000. This recommendation provides a framework for coordination among the various activities in ITU-T on OTN to ensure that recommendations covering the various aspects of OTN are developed in a consistent manner. As such, this recommendation provides references for definitions of high-level characteristics of OTN, along with a description of the relevant ITU-T recommendations that were expected to be developed, together with the time frame for their development [15]. This is also numbered Y.1301.

- G.872, *Architecture of Optical Transport Network,* February 1999; revision, November 2001. This recommendation describes the functional architecture of OTNs using the modeling methodology described in ITU-T G.805. The OTN functionality is described from a network-level viewpoint, taking into account an optical network layered structure, client characteristic information, client/server layer associations, networking topology, and layer network functionality providing optical signal transmission, multiplexing, routing, supervision, performance assessment, and network survivability. This recommendation is limited to the functional description of OTNs that support digital signals. The support of analog or mixed digital/analog signals is not included [15]. A revision includes (among other features), the ability to support TDM signals over the OTN.

- ITU-T Recommendation G.798, *Characteristics of Optical Transport Network Hierarchy Equipment functional Blocks*, specifies the charac-teristics of OTN equipment, including supervision, information flow, processes, and functions to be performed by this equipment. See Table 3.4 for an illustrative question/charter document on G.798.

- ITU-T Recommendation G.8251, *The Control of Jitter and Wander Within the Optical Transport Network*, includes the network limits for jitter and wander, as well as jitter and wander tolerances required of equipment for OTNs. These parameters relate to the variability of the bit rate of signals carried over the optical transport network.

- ITU-T Recommendation G.874, *Management Aspects of the Optical Transport Network Element*, specifies applications, functions, and requirements for managing optical networking equipment utilizing OSS. It covers the areas of configuration management, fault management, and performance management for client optical network elements.

- ITU-T Recommendation G.874.1, *Optical Transport Network Protocol-Neutral Management Information Model for the Network Element View*, provides a means to ensure consistency among the models for OTN equipment for specific management protocols, including *Common Management Information Service Element* (CMISE), *Common Object Request Broker Architecture* (CORBA), and *Simple Network Management Protocol* (SNMP).

Related standards include the following [15]:

- G.692, *Optical Interfaces for Multichannel Systems with Optical Amplifiers,* October 1998. This recommendation specifies multichannel optical line system interfaces for the purpose of providing future transverse compatibility among such systems. It defines interface parameters for systems of four and eight channels operating at bit rates of up to STM-16 on fibers, as described in G.652, G.653, and G.655 with nominal span lengths of 80, 120, and 160 km and target distances between regenerators of up to 640 km. A frequency grid anchored at 193.1 THz with interchannel spacings at integral multiples of 100 GHz is specified as the basis for selecting channel central frequencies. *Note:* During drafting, this recommendation was referred to as "G.mcs."

TABLE 3.4 STUDY GROUP 15 QUESTION 9/15: TRANSPORT EQUIPMENT AND NETWORK PROTECTION/ RESTORATION

1. Background and justification

Two key technologies are driving the development of transport equipment and standards:

The explosive growth of the Internet and other packet-based traffic carried over networks;

The development of OTN, which provides the ability to dramatically increase the bandwidth that can be carried over networks.

Responsibilities under this Question include the following areas of standardization related to these new technologies:

Specification of all equipment functions and supervision processes related to the OTN and SDH layer networks;

Specification of adaptation functions and supervision processes for transport of packet traffic (e.g., IP, ATM, MPLS) via the SDH or OTN layer networks;

Specification of survivability capabilities and development of a strategy for multilayer survivability interactions;

Examining the impact on synchronization layer functions, which may differ for transport of different client layers (e.g., IP traffic).

Recommendations related to transport technologies (e.g., SDH, PDH, OTN) used in the access environment and not covered by other Questions of ITU-T SG 15, such as flexible multiplexers, are also covered by this Question. Responsibility under this Question includes the following draft or published recommendations:

G.681: Functional characteristics of interoffice and long-haul line systems using optical amplifiers, including optical multiplexing

G.705: *Characteristics of Plesiochronous Digital Hierarchy (PDH) Equipment Functional Blocks*

G.781: *Synchronization Layer Functions*

G.783: *Characteristics of Synchronous Digital Hierarchy (SDH) Equipment Functional Blocks*

G.798: *Characteristics of Optical Transport Network (OTN) Equipment Functional Blocks*

G.806: *Characteristics of Transport Equipment—Description Methodology and Generic Functionality*

G.841: *Types and Characteristics of SDH Network Protection Architectures*

G.842: *Interworking of SDH Network Protection Architectures*

2. Question

What equipment functions must be specified to enable compatible transport equipment in interoffice and long-distance networks, including evolution to the optical transport network?

What additional characteristics of transport equipment should be recommended to provide enhanced survivability capabilities and a cohesive strategy for multilayer survivability interactions?

What considerations are necessary for equipment optimized for transport of IP-type traffic?

3. Study Items

Study items to be considered include the following:

Completion of equipment recommendations for the OTN (OTN equipment relates to equipment functions and the assembly of optical components and subsystems).

Equipment Recommendations to provide enhanced survivability capabilities and a cohesive strategy for multilayer survivability interactions. This includes revisions needed to Recommendations G.841 and G.842. These recommendations will cover SDH and OTN layer protection, as well as multilayer survivability, including interactions with protection at packet layers.

Specifications of equipment functions necessary to implement an ASON.

Enhancements to SDH equipment characteristics or network protection protocols to support paths with longer delay times, such as satellite systems, submarine cable systems, radio relay systems, or systems with numerous regenerators.

Enhancements to the transport equipment recommendations in order to meet the needs of the access network.

Enhancements required to the transport equipment recommendations in order to meet the needs of the GII and transport of Internet and other packet-based traffic. This includes examining the impact, if any, on synchronization layer functions.

Incorporating the impact of recent technological developments into equipment recommendations.

Clarification and resolution of technical issues in current and draft recommendations.

4. Specific Tasks

Revise Recommendations G.781, G.841, G.842 as required.

Complete Recommendation G.798 by 2001.

Develop additional recommendations from progress on the above study points.

- G.693, *Optical Interfaces for Intra-Office Systems,* November 2001. This recommendation provides parameters and values for optical interfaces of single-channel intraoffice systems of nominal 10- and 40-Gbps aggregate bit rates. Applications are described with target distances of 0.6 and 2 km.

- G.957, *Optical Interfaces for Equipment and Systems Relating to the Synchronous Digital Hierarchy,* December 1990; revision 1, March 1993; revision 2, July 1999. This recommendation specifies optical interface parameters for equipment and systems based on the synchronous digital hierarchy to enable transverse compatibility.

- G.958, *Digital Line Systems Based on the Synchronous Digital Hierarchy for Use on Optical Fiber Cables,* December 1990; revision, November 1994; deleted, January 2002. This recommendation provides requirements for digital synchronous line systems based on SDH specified in G.707, G.708, and G.709. The content of this recommendation has been included in the new G.783 and Corrigendum 1 to G.798.

- G.959.1, *Optical Transport Network Physical Layer Interfaces,* February 2001. This recommendation focuses on optical parameter values for pre-OTN single-channel and multichannel interdomain interfaces and provides a framework for OTN physical interfaces.

- G.8070 (2001—formerly G.astn). This recommendation specifies requirements for the ASTN. The ASTN provides a set of control functions for the purpose of setting up and releasing connections across a transport network. The requirements contained in this recommendation are technology independent. The architecture of switched transport networks meeting the requirements in this recommendation and the technical details required to implement these networks for particular transport technologies are found in other recommendations [16]. This standard is discussed further in Chapter 4.

· G.8080 (formerly G.ason). This recommendation describes the reference architecture for the control plane of the ASON that supports the requirements identified in Recommendation G.8070, which is a client/server model of optical networking. The reference architecture is described in terms of the key functional components and the interactions among them. This recommendation describes the set of control plane components used to manipulate transport network resources in order to provide the functionality of setting up, maintaining, and releasing connections. The use of components allows for the separation of call control from connection control and the separation of routing and signaling. G.8080 takes path-level and call-level views of an optical connection and applies a distributed call model to the operation of these connections. With a call model in place, a network operator may now bill for calls and, based on the parameters of the class of service requested for the call, select the connections with the type of protection or restoration required to meet the class of service [16]. It was expected that the first issue of G.ason would be based on a SONET/SDH transport network. This standard is further discussed in Chapter 4.

OTN specifications were not exhaustively completed by press time. For example, G.709 still required considerable work in defining the multiplexing and transport overheads. G.872, the OTN architecture document, was undergoing revision and expansion. G.798, which defines the OTN functions that are needed for midspan meets, had not been completed yet. G.841 and G.842, which deal with OTN protection, had yet to be started. Finally, G.874 and G.875, which provide the network management information model and functional requirements, are also incomplete. In the area of optical signaling, the ITU–T has been involved in the creation of a standardized architecture for optically signaled networks. However, as of press time the ITU had not yet settled on a signaling mechanism: It was considering a number of different approaches, with GMPLS being a contender; two other contenders included a signaling scheme based on ATM PNNI, and a proposal for an entirely new set of protocols. GMPLS was the front runner at a point in time because of the desire by some developers to leverage control plane work brought forth with MPLS; the final disposition is unknown, and, if MPLS falls out of favor, the ITU may select a completely new scheme [16]. However, as noted earlier, 35 years of history have shown that neither X.25 SVCs, nor FR SVCs, nor ATM SVCs have experienced commercial success; we venture to guess that in the short term (through 2007), switched lightpaths are not going to see major commercial success either for end-user enterprise applications.

3.4 Introductory Concepts

3.4.1 Evolutionary Approach

Does the OTN need to be evolutionary or revolutionary? Enterprise users are in need of affordable and reliable broadband bandwidth. End users would settle for a DS1 line, a DS3 line, an OC-3 line, an OC-12 line, or an OC-48 line between their data sources and data sinks, without many questions asked, if this bandwidth could be readily provisionable when needed (say, in 10 business days), be inexpensive, and be reliable. If an enterprise user got an OC-12 between several sites of interest to them at an affordable price, the user would not care that the overhead on the line was 2.29%; or that by introducing some new "virtual packing," the overhead could be reduced to 2.22%; or that by using packet technology the provider could overbook 10 logical lines onto 8 physical lines, and so on. Carriers own the "mine" when it comes to bandwidth; bandwidth is a carrier's staple, their currency. Carriers can afford an overhead of 2.29% instead of spending millions (billions?) of dollars to revamp their network to achieve an overhead of 2.22%. On the other hand, equipment vendors tend to make noise regarding the need for bandwidth efficiency, so that carriers, instead of throwing bandwidth at the problem, of which they have plenty, are sold on the idea of purchasing new equipment (which they could otherwise do without) to "squeeze" some efficiency out of the "pipe" they already own in abundance. Trade press statements such as those that follow are indicative of this mindset: "The enterprise folks do not like the cost and complexity of a SONET frame. Changes in the voice and data traffic mix are prompting carriers to find ways of enabling clients to request services from the underlying optical infrastructure." We venture to guess that most users are not aware of what the SONET frames looks like or what the overhead is. Note that 27 (3 × 9) bytes of overhead for every 810 (9 × 90) bytes of enveloped data is not inefficient, being only 3.33% overhead. For simple comparison, all IP people live with an overhead of 14 bytes for Ethernet framing on the 1,518 payload, plus 20 bytes for the IP header inside the Ethernet frame, plus 20 bytes for the TCP header inside the IP packet, for a total of 54 bytes or 3.55% overhead. We are not advocating intrinsic inefficiency, but rather putting a context to the discussion.

According to the trade press, data users would "like to see Gigabit Ethernet or 10 Gigabit Ethernet (GbE/10GbE) in the metro/WAN for the near- to mid-term, with long-term solutions possibly centering on Resilient Packet Ring (RPR)." These just-named technologies were tried in the early 2000s but abandoned as a carrier infrastructure because they are designed for local, not long distance; they are designed for one administration (company), not the multiple domains (companies) intrinsic to a carrier environment; they are not secure; they do not have traffic management capabilities; they do not support QoS either in terms of guaranteed bandwidth or in terms of delay and jitter; they are not as reliable as traditional systems; they take 5 to 50 sec to reconfigure around an outage, whereas SONET/SDH takes 50 ms; Ethernet does not integrate effectively with existing SONET networks, even when implementing the WAN PHY within the 10GbE standard (because of timing issues); and they were not financially cost effective and sustainable when used in a carrier application, with the idea of also turning a bottom-line profit.

Despite rumors of its demise, SONET-based architecture and derivative services are alive and well [17]. The overwhelming majority of carriers implement and use the protocol, and voice supporters are now further adapting SONET to the new requirements. Carriers are planning to deploy a transport network that can carry a plethora of traffic types over it. OTN is the high-speed successor to SONET/SDH, and it is planned to be delivered over an ASON. On the other hand, some data folk are advocating an IP-centric approach based on GMPLS [6].

As discussed earlier in the chapter, there are two basic approaches to a next-target network: the overlay model and the peer-to-peer model. The overlay model calls for maintaining two discrete networks: a layer 1 optical network and the client network. The optical network is provisioned separately from the client network. Routers, telephone switches, and other WAN equipment that comprise the client's network establish connections between devices with nominal information about the underlying network. To request capacity from the underlying network, clients access the optical network through UNIs, which shield the complexity from those on the outside. Devices within the optical network rely on NNIs to access network information (see Figure 3.3 [18]). Traditional voice operators tend to back the overlay model. The peer-to-peer folks argue for a single network

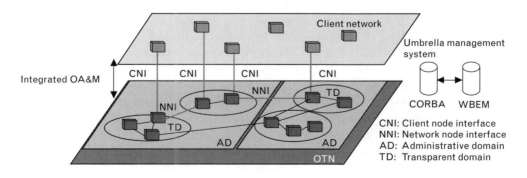

FIGURE 3.3   *OTN model. (Source: Telecom Lab Italia [18].)*

in which the equipment at the network's edge (largely data equipment) decides how bandwidth is allocated within the network's core. The distinctions between the data and the underlying optical networks become blurred, as the peer-to-peer model merges the requisite signaling protocols for establishing and tearing down connections. Not surprisingly, this is a model largely promoted by the data vendors that comprise the IETF [6].

3.4.2   Framework

OTN [19] sees connections as three components: optical channels, optical multiplex sections, and optical transmission sections. Optical channels (connections, or trails) are conceptually similar to the SONET path: the *optical channel layer* (OCh) transports client signals between two endpoints on the OTN. *Optical multiplex sections* (OMSs) describe the WDM portions that support these optical channels. OMSs are similar to SONET lines, but account for multiple wavelengths (SONET does not); hence, OMS data streams consist of many optical channels aggregated together. At the lowest level of the OTN link is the *optical transmission section* (OTS). OTSs enable transmission of signals over individual fiber spans, as is the case in SONET. The OTS defines a physical interface that details optical parameters such as wavelength, power level, and SNR.

As is the case in SONET, OTN defines the OTH network hierarchy. Although SONET's STS-1s are concatenated to support higher speeds, OTH's base unit, the *optical transport module* (OTM), supports higher speeds by bundling together wavelengths. OTMs can span multiple wavelengths of different carrying capacities. To indicate this difference, OTMs utilize two suffixes: OTM-n.m, where *n* refers to the maximum number of wavelengths supported at the lowest bit rate on the wavelength while *m* indicates the bit rate supported on the interface. Three throughput rates are supported: 2.5 Gbps, indicated by a 1; 10 Gbps, indicated by a 2; and 40 Gbps, indicated by a 3. An interface might support some combination of these, namely, a 2.5- and a 10-Gbps combination (1 and 2), a 10- and a 40-Gbps combination (2 and 3), or a combination of all three speeds (1, 2, and 3). Hence, an OTM-3.2 indicates an OTM that spans three wavelengths, each operating at at least 10 Gbps. Similarly, an OTM-5.12 indicates a channel that spans five wavelengths and can operate at either 2.5 or 10 Gbps. When OTMs are sent on the network, they are sent as 64-byte (512-bit) frames. The frame has four regions. The frame alignment area is for network operational purposes. Then there are three overhead areas: an *optical channel transport unit* (OTU) specific overhead area, an *optical channel payload unit* (OPU) specific overhead area, and an *optical channel data unit* (ODU) specific overhead area. These all describe various facets of the optical channel.

Bandwidth on demand (also called "instant provisioning" by some) requires a signaling protocol to set up the paths or connections. Work on

SONET and OTN signaling is being defined in the ASTN and ASON specifications. In the ITU-T (overlay) model the network consists of clients. These clients might be multiplexers, WDM systems, an Ethernet switch, or some other devices implementing GFP. These clients connect into the network through one of three different types of network interfaces: UNIs, *external- network-to-network interfaces* (E-NNIs), and *internal-network-to-network interfaces* (I-NNIs). The UNI defines how customers can access their providers' networks. Only the basic information is provided, namely, the name and address of the endpoint, authentication and admission control of the client, and connection service messages. Carriers, however, need to share more information among themselves and within their domains using their own network devices. E-NNIs provide reachability or summarized network address information, along with authentication and admission control and connection service messages. E-NNIs can also be used between two business units belonging to the same provider, or they can reduce the amount of topology information exchanged between networks. With E-NNIs, the exact paths are not known to the partner, but the available clients that can be "called" are known. The I-NNIs enable devices to get the topology or routing information for the carrier's network, as well as connection service messages and information necessary to optionally control network resources.

Through these interfaces, clients can request three different types of circuits: provisioned, signaled, and hybrid. *Provisioned circuits*, also called hard-permanent circuits, are what we commonly characterize as a leased line. Either through a network management station or manual intervention, each individual network element along a particular path is configured with the required information to establish a connection between two endpoints. *Signaled circuits* are established dynamically by the endpoint that requests connectivity and bandwidth. These types of connections require network-addressing information to establish a connection with an endpoint. *Hybrid connections* are a new type of circuit. As the name suggests, they are a hybrid between provisioned and signaled circuits. Hybrid circuits have provisioned connections into the ASTN, but then rely on switched connections within the ASTN to connect with other nodes. Because two types of connections are combined, hybrid connections are also known as *soft provisioned connections* (SPCs). To the end node, an SPC and a regular permanent circuit appear the same.

3.5 Basic OTN Technical Concepts

OTN [20] evolved from the concept of Digital Wrapper. OTN adds an optical channel layer whereby each wavelength is wrapped in a digital envelope. The digital envelope consists of the header in which overhead bytes are carried and the trailer that performs FEC functions. Between the header and FEC is the payload section that allows for all existing network communications protocols to be mapped with no disruption to the protocol. It follows that the OTN is protocol independent. By virtue of the overheads, each OTN node can carry and transmit management and control information throughout the network, making performance monitoring and other network management possible on a per λ basis. The FEC portion performs detecting and correcting transmission errors encountered across an optical link. Unlike *bit interleaved parity* (BIP)-8 error monitoring as found in SONET/SDH, FEC has the capability to correct the errors—a feature that can allow service providers to offer and support different levels of service agreements. By minimizing errors, FEC plays an additional crucial role in extending the reach of the fiber span, deemed especially critical at high transmission rates.

The OTN frame structure is defined for three OTU bit rates: OTUk, where $k = 1, 2, 3$ corresponds to 2.5, 10, and 40 Gbps, respectively. The OTN G.709 client mapping and frame structure are shown in Figure 3.4. The OTUk frame is composed of the following entities:

- *OPUk: optical channel payload nit order of k.* The OPUk includes payload and overhead. The payload includes the client information with its specific mapping technique and the overhead includes the information to support the adaptation of the specific client. Each type of client has its specific overhead structure.

- *OTUk: optical channel transport unit order of k.* The OTUk includes FEC and overhead for management and performance monitoring (*section monitoring* [SM]). The FEC is based on Reed-Solomon coding as specified in ITU-T G.975 standard.

- *ODUk: optical channel data unit order of k.* The ODUk is composed of several overheads for path *performance monitoring* (PM), *tandem connection monitoring* (TCM), communication channels, and protection control (APS/PCC).

FIGURE 3.4
*OTN ITU-T G.709
client mapping.*

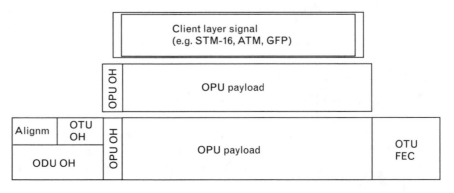

In the next generation DWDM OTNs, each wavelength can be managed, configured, and monitored as a single entity. The OTN frame structure allows end-to-end *operations, administration, and maintenance* (OAM) across the network, among multiple vendors' equipment. The OAM layer at the OTN frame overhead is minimal (especially when compared to that of SONET/SDH at high bit rates) and efficient, consuming less than 0.4% of the frame. The OTUk overheads include the SM and *general communication channel* (GCC) fields. The OTUk, ODUk, and OPUk overheads are shown in Figure 3.5. The SM fields are used for the trail trace identifier (TTI), BIP-8, backward defect indication (BDI), *backward error indication and backward incoming alignment error* (BEI/BIAE), and *incoming alignment error* (IAE). The ODUk overhead includes six levels of tandem connection monitoring (TCM1–6), PM, communication channels (GCC and EXP), protection control (APS/PCC), TCM activation, and *fault localization* (FTFL). The PM overhead is composed of TTI, BIP-8, BDI, BEI, and a status bits indicating the presence of a maintenance signal (STAT). The TCM overhead is composed of TTI, BIP-8, BDI, BEI/BIAE, and STAT.

For testing purposes, the OTN includes pseudorandom binary sequence (PRBS) and NULL signal mapping into the OPUk for performance monitoring measurements. Other OTN maintenance signals are available at all levels. The OTUk includes the OTU *generic AIS* (GAIS), BDI, IAE, BIAE, and BEI. The ODUk includes maintenance indications

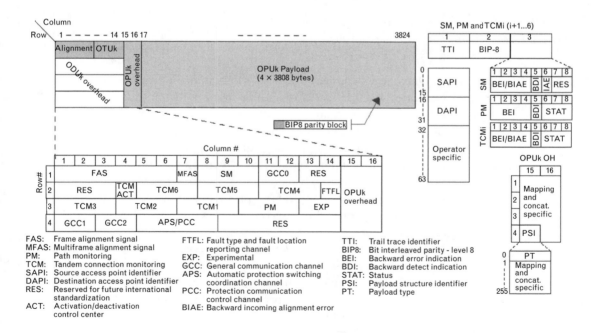

FIGURE 3.5 *OTN overhead bytes. (Source: Optix Networks.)*

at the PM (BEI and BDI) and TCM (BEI, BIAE, and BDI). Additional OTM maintenance signals are *open connection indication* (ODU-OCI) used for optical cross-connects and switching fabrics to indicate when an output port is not connected to input port. Locked (ODU-LCK) is used to indicate that a connection is administratively locked and to prevent test patterns on this connection. An *alarm indication signal* (ODU-AIS) is used to indicate a failed signal. All of these maintenance signals are used as *forward defect indication* (FDI) to signal the receiving equipment of different failures. The FTFL field is used in conjunction with section and tandem connection endpoints on detection of an SF or SD condition to automatically locate fault/degradation in a specific network operator domain.

When migrating to higher rate optical links, FEC mechanisms are a necessity. FEC is a signal coding/decoding technique that adds redundancy data to a signal. This redundancy data identifies and corrects corrupted data and thus reduces the BER. The OTN uses the Reed–Solomon RS(255,239) error correction code defined in the ITU-T G.975 standard. This code utilizes about 7% of the OTN bandwidth. The *net electrical coding gain* (NECG) describes the performance of the FEC code. The addition of the FEC improves the coding gain by approximately 6.2 dB (compared to uncoded data) for an output BER of 10^{-15}.

The use of FEC adds performance monitoring and early warning for optical link performance degradation. The early warning allows controlled protection before a failed signal is detected. Another use of BER monitoring capabilities such as inverted 1's and 0's provides feedback and control to the optical modules' operating conditions while the link continues to operate with lower errors and without active alarms. Using FEC can significantly improve the BER performance of the client signals and expand the optical span.

One of the major advantages of OTN is that it provides client mapping capabilities and backward capability support for existing SONET/SDH protocols without requiring changes to the format, bit rate, or timing. The OTN allows transmission of different data packet types using the new GFP mapping supporting the ever-growing data-driven needs. The new GFP mapping reduces the layers between the fiber and the IP layer and thus results in very efficient use of bandwidth. The mapping capability advantages make OTN a protocol agnostic carrier allowing service transparency for SDH/SONET, Ethernet, ATM, IP, MPLS, and other protocols/clients.

The mapping of *constant bit rate* (CBR)-type clients is handled by a justification mechanism. Two mapping types are defined: synchronous and asynchronous. In synchronous mapping, justification is not required due to the fact that the OTU rate is proportional to the client rate. In asynchronous mapping, the OTU and client rates are independent and use a *negative justification opportunity* (NJO) and *positive justification opportunity* (PJO) for rate adaptation. If the client rate is higher than the OTU rate, the NJO byte is used to accommodate the additional rate. For the case of a slower client rate, the PJO byte is used to compensate the "missing" rate. The justification is indicated by three *justification control* (JC) bytes reporting the justification operation as shown in Figure 3.6.

The OTN TDM multiplexing is defined as asynchronous mapping for the following options: 4 × ODU1 into ODU3, 4 × ODU2 into ODU3, and 16 × ODU1 into ODU3. A multiplexing example of four ODU1 signals into an ODU2 is shown in Figure 3.7. The ODU1 signals, including

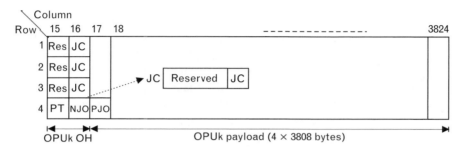

FIGURE 3.6   *OTN justification mechanism.*

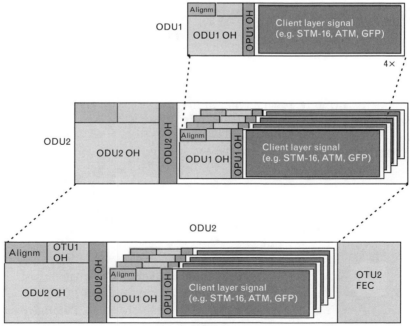

ITU-T G.709 OTN multiplexing

FIGURE 3.7   *OTN multiplexing.*

the frame alignment overhead and an all-0's pattern in the OTUk overhead locations, are adapted to the ODU2 clock via justification (asynchronous mapping). These adapted ODU1 signals are byte interleaved into the OPU2 payload area, and their justification control and opportunity signals (JC, NJO) are frame interleaved into the OPU2 overhead area. ODU2 overhead is added, after which the ODU2 is mapped into the OTU2. OTU2 overhead and frame alignment overhead are added to complete the signal for transport via an OTM signal.

In summary, based on requirements to make the network more scalable and manageable to address the ever-changing bandwidth requirements, OTN provides the optimal solution for the next generation transport systems. Necessary overhead bytes consume minimal overhead while providing ample OAM information. The embedded FEC mechanism ensures a longer reach and preserves signal integrity. The client information mapping allows for support of various protocols including SONET/SDH. Network system requirements mandate the support and implementation of OTN in all next generation transport systems. The new recommendations support efficient transport of popular data protocols, such as Ethernet and Fibre Channel, as well as other wideband and broadband services like SDH/SONET, ATM, FR, audio/video, and IP-based services. They also specify detailed equipment functions to support performance monitoring, fault isolation, and alarming, including support for optically transparent subnetworks.

3.6   Implementations

Currently, several types of networks infrastructures are in place:

- *Transport:* async DS1/DS3 systems, SONET-based systems, and vendor-based DWDM systems;
- *Layer 2 packet:* frame networks overlaid on ATM and ATM overlaid on async or sync optical transport;
- *Layer 3 packet (data):* private lines overlaid on async or sync optical transport, IP overlaid on ATM, and IP overlaid on async or sync optical transport.

At face value it is desirable to migrate to a single infrastructure in order to reduce OAM&P costs. As an amplification of Figure 3.2, Figure 3.8 illustrates a next-target environment consisting of a single, data-optimized, multiservice infrastructure interconnected over the OTN. The move to this next-target environment, however, will not happen overnight, particularly if the raw demand for bandwidth remains depressed;

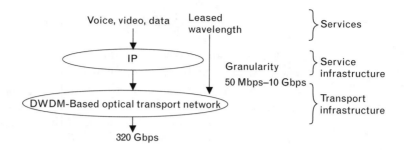

FIGURE 3.8
Next-target network.

nonetheless, a DWDM-based OTN provides the common, enabling link between the architecture of today and the next-stage network [10]. Researchers are developing high-speed lightwave transport systems designed to transmit signals at 40 Gbps or more per wavelength. As the bit rate per channel is increased from the current 2.5 to 10 Gbps to 40 Gbps, large capacities can be achieved with fewer wavelength channels. This will allow the network configuration to be simplified. Field experiments with OTN supporting over 1 Tbps (43 Gbps × 25 wavelengths) were under way at press time [21].

The need for higher capacity over time is self-evident. The questions, however, are how much and how soon? High capacity is inherent in a DWDM-based solution: Each wavelength can support up to 10 Gbps or more; 32 such wavelengths can be multiplexed onto a single fiber (pair). The resulting aggregate capacities of 320 Gbps represent a significant pool of transmission capacity. Previous studies by the author [22] indicate that at the present time corporations do well with 6 Mbps per 100 office workers in 16,000 square feet of office space (with about 60 Mbps per 100 users about 5 years out). Hence, 75,000,000 office workers evenly distributed in 100 U.S. metropolitan areas (for argument's sake) would require 7,500 × 6 Mbps or 45 Gbps per city, at this time. One DWDM/OTN ring per city would be more than sufficient to support this traffic. If one assumed that there were five carriers per city, the equivalent of five OTN systems would suffice. It would take a traffic level of around 45 Mbps per 100 users to saturate one DWDM/OTN system. Notice that 100 users all having a 10-Mbps segment to their desks results in 1 Gbps of total local bandwidth. The assumption can be made that 90% of the traffic remains local and that, of the 10% that leaves the building, a 15-for-1 concentration (due to boundary-source burstyness) can be achieved, leading to the 6-Mbps bandwidth need referenced above—once such users migrated to 100-Mbps Ethernet segments, the aggregate bandwidth leaving a 100-person building would be 60 Mbps.

Therefore, the deployment of OTN has a business driver associated with it (read "a problem in search of a solution, rather than a solution in search of a problem"). However, our estimate is that one would only need 5 × 100 such systems a year in the United States, spread over 5 years (that is, 100 systems a year, or at $2 million to $4 million each, a market of $200 million to $400 million). The rest of the world is typically 1.5 times the U.S. baseline, bringing this market to about $1 billion a year.

There is interest in validating the concept of OTN before deployment can start on a commercial scale. The *Layers Interworking in Optical Networks* (LION) *Program* [23] goals of the European Information Society provide a crisp description of the issues at play. The goal of the LION project is to design and test a resilient and managed transport network realized by an OTN carrying different clients (e.g., SDH, ATM, IP based) with interworking and interconnection between layer networks and domains. The identified requirements need to be validated in a testbed where IP routers and SDH equipment are integrated over an OTN infrastructure. The following objectives are important: Define the interworking and interconnection requirements between client/server layer networks and domains; define the functional requirements of an IP-based transport network; enhance the functional architecture of an OTN to account for new emerging features (e.g., digital optical container); implement, integrate, and test NNIs between transparent domains and *client node interfaces* (CNIs), both based on digital optical container; implement, integrate, and test an "umbrella" management system over a testbed; develop and test strategies for integrated resilience controlled by an overall OAM and a management system adopting QoS demanding applications; and make techno–economic evaluations of an IP-based network over an OTN. Network operators expect the OTN to be a client-independent network supporting different layer networks. Fulfilling these expectations requires studies on interoperability, interworking, and interconnection of an OTN carrying different client transport networks. The target of the LION project is to develop and test OTN architectures, providing a support for the convergence of different transport networks carrying services and in multiple-domain environments. The LION project defines the requirements for interworking and interconnection between layer networks and domains. One of the major achievements was expected to be the implementation, integration, and testing of NNI between transparent domains and CNI, both based on digital optical containers. Integrated strategies for resilience in a multilayer transport network will be investigated and tested over the testbed. Furthermore an integrated multilayer transport network requires appropriate and practical OAM and an integrated management system to guarantee effective network interoperability; in particular, proper signaling between client/server layer networks and between domains is required. In this context, the LION project will design and test an "umbrella" management architecture that enables integration of TMN, *Web-Based Enterprise Management* (WBEM) [24], and SNMP on the network-level management. In particular, to investigate the interoperability of different management technologies, two peer-to-peer "umbrella" open management systems will be implemented, one with CORBA and another with WBEM. To complete the design of the multiple-client optical transport infrastructure, techno-economical evaluations will be carried out that will also consider, as an input parameter, the QoS requirements for network operators. The activities carried out in the LION project will allow it to contribute to a strong European position in support of standardization activities (OIF, ITU, ETSI, IETF) and in the telecommunications market in the areas of optical network interfaces, interworking, network management, OAM, and optimized architectures.

While waiting for deployment of the OTN, enterprise users will continue to have access to SONET-based services. It is not clear when OTN-based technology will be deployed in the United States on a broad scale, but a 2005-or-beyond expectation would not be totally unreasonable. While carrying data traffic across an access ring, a SONET link is not bandwidth efficient on that initial ring; however, it is all relative. For example, a 10-Mbps Ethernet connection is typically transported across an STS-1 (51.84-Mbps) link. But this is no different than when a user gets a T1 tail (1.544 Mbps) to support a 384-Kbps fractional T1 line or a FR circuit with a 64-Kbps CIR. This is why *grooming* is used, in which outer-edge traffic is repacked into multiplexed links of higher payload concentration. Carriers build networks using hierarchical sets of rings, not one giant ring that spans a metropolitan area. Ignoring the fact that bandwidth may be a commodity, virtual concatenation mechanisms have been developed of late to enable channels in SONET to be combined to support improved efficiencies. At a grooming point in the network, Ethernet streams, for example, could be multiplexed and carried across VT1.5-6v (10.368-Mbps) links instead of an STS-1 link; similarly, 100-Mbps Ethernet links could be multiplexed and carried across an STS-2c (103.68-Mbps) link instead of an STS-3c (155.520-Mbps) link.

3.7 Other Developments

In 2001 the ITU published a new global standard for very short reach optical interfaces intended for intraoffice applications; the goal is to make optical technology economically effective in this context. The expectation is that the number of intraoffice links will increase as a consequence of the total growth of network capacity and the evolution toward the ASON, resulting in deployment of optical switches and transmission terminals in the same offices. The ITU-T G.693 agreement enables manufacturers to support this migration with interoperable equipment designs. The G.693

short reach optical interface addresses the urgent need for low-cost, high-speed optical links to interconnect colocated equipment such as routers, DWDM terminals, and SDH add/drop multiplexers.

Interface specifications in Recommendation ITU-T G.693 focus on link distances up to 2 km. (Other existing ITU-T standards on optical interfaces address longer link distances.) This standard has been agreed under the fast-track approval procedure called AAP. Under this procedure, a comment period is initiated when a study group gives consent to approve the draft text of a recommendation that it considers mature.

Recommendation G.693 provides interface specifications for target distances of 0.6 and 2 km on single-mode fibers as specified in Recommendations G.652, G.653, and G.655. Applications using both O-band (nominally 1,310-nm) and C-band (nominally 1,550-nm) wavelengths are specified. Each interface specification has one of four discrete values of maximum attenuation. These values are 4, 6, 12, and 16 dB. The interface specifications support maximum data rates of approximately 10 or 40 Gbps.

ENDNOTES

[1]   Retrieved from http://www.convergedigest.com/standards/standardsarticle.asp? ID=1272.

[2]   Choi, J., and D. Lahav, *Optical Transport Network—Solution to Network Scalability and Manageability,* retrieved from http://www.optixnetworks.com/images/innerpgs/w_p.pdf.

[3]   For example, consider this quote: "…but like the rest of the telecom industry, Velocita encountered a glut of service from rival carriers and prices continued to drop … the company joins a host of telecom companies—including Global Crossing Ltd., Williams Communications Group Inc., KPNQwest NV [360Networks, MFN, and Yipes], that are seeking bankruptcy court protection…," Berman, D., "Velocita Files for Bankruptcy in Another Telecom Setback," *Wall Street Journal,* June 4, 2002, p. B15.

[4]   "Reckless optimism created excess capacity in industries such as telecommunications, experts say," St. Onge, J., "Bankruptcy Filings Continue to Surge Even as Economy Improves," *Bloomberg,* July 3, 2002.

[5]   Just because a restaurant owner—not realizing that a yearly convention is in town and that he will need 50 pounds of baked pasta on a given day—fails to cook the pasta (and keeps the hundred of pounds of unbaked goods in the restaurant's storage area), it does not mean that there is a shortage of pasta and that eight new pasta factories are needed. It merely means that this businessperson needs to be more responsive and more market savvy.

[6]   "Networkers will no longer need to purchase monthly or yearly contracts for one-time capacity—instead, carriers will be able to sell lines with capacity for durations of an hour or days. Those lines might be configured much like a frame relay or ATM circuit, in which you pay for a certain amount of basic transport, but then can get higher capacities as necessary. Or perhaps networkers will pay for capacity as if it were a smart phone line: as long as the bits flow, the meter runs; stop transmitting, and the tab stops there." Greenfield, D., "Optical Standards: A

Blueprint for the Future, Ever Feel Lost in the Current Plethora of Optical Standards? Here's the Map that Puts It All Together," *Network Magazine,* October 5, 2001, p. 15 ff.

[7] Cavazzoni, C., et al., "Envisaging Next-Generation Data-Centric Optical Networks," *DRCN 2001: Third Int. Workshop on Design of Reliable Communication Networks,* Budapest, October 2001, Cod. Doc. DPP 2001.02515.

[8] Maracich, G., and A. Manzalini, "Road Map Towards Next Generation Optical Networks," *Workshop of the IST Project LION "Layers Interworking in Optical Networks,* retrieved from http://www.telecomitalialab.com/pdf/wshop3.pdf.

[9] ITU Work Groups, "Optical Transport Networks & Technologies," retrieved from http://www.itu.int/itudoc/itu-t/com15/otn/79506.pdf.

[10] Allen, B., and J. Rouse, "Optical Transport Networks—Evolution, Not Revolution," OPTera Metro Solutions White Paper, Nortel Networks, 2000.

[11] World Telecommunication Standardization Conference 2000 (WTSC-2000), Montreal, Canada, October 2000.

[12] To maintain differentiation from the standardized OTN based on Recommendation G.872, this lead study group activity is titled *Optical Transport Networks & Technology* (OTNT).

[13] Shahane, D., "Building Optical Control Planes: Challenges and Solutions," *Communication Systems Design,* January 7, 2002.

[14] ITU Public Relations, "Recommendations Adopted by Study Group 15 in October 2001," retrieved from http://www.itu.int/newsroom/Recs/SG15Recs. html.

[15] Retrieved from http://www-comm.itsi.disa.mil/on/on_itu.html.

[16] *Specs and Standards: ITU,* retrieved from http://www.lightreading.com/document. asp?doc_id=7098&page_number=15, January 8, 2002.

[17] Minoli, D., P. Johnson, and E. Minoli, *SONET-Based Metropolitan Area Networks,* New York: McGraw-Hill, 2002.

[18] Manzalini, A., "IST Project LION (Layers Interworking in Optical Networks)," *Workshop on Efficient Transport Networks,* March 22, 2000, retrieved from http://www.eurescom.de/~pub/seminars/past/2000/IP/06Manzalini/06aManzalini/06Manzalini.pdf.

[19] This subsection is based on the reference listed in [6].

[20] Section 3.5 is based in its entirety on [2].

[21] NTT, Network Innovation Laboratories, Photonic Transport Network Laboratory, retrieved from http://www.onlab.ntt.co.jp/en/pt/40G.

[22] Minoli, D., P. Johnson, and E. Minoli, *SONET-Based Metropolitan Area Networks,* New York: McGraw-Hill, 2002.

[23] Retrieved from http://www.ibcn.intec.rug.ac.be/projects/IST/LION.

[24] WBEM is a system for unified administration of network, systems, and software resources that was proposed by Microsoft, Intel, Compaq, Cisco, BMC Software, and others. It allows users to manage distributed systems using any Web browser. In June 1998, WBEM was turned over to the *Desktop Management Task Force* (DMTF) standards body (http://www.dmtf.org). It incorporates a new *Common Information Model* (CIM) protocol that defines the objects to be managed. CIM v2.0 is the current version. WBEM originally specified a new *Hypermedia Management Protocol* (HMMP) for transporting management information, but now *eXtensible Markup Language* (XML) will be used instead. Original plans to standardize an *object manager* (OM) that collects management data and acts as an interface to supporting applications have been dropped; vendors must now define this component themselves. HTTP is the access

mechanism for WBEM. Compaq, HP, and Dell are planning to support WBEM. Tivoli supports WBEM in its NetView 5.1 product that runs on Windows NT and Unix. Cisco supports WBEM in its CiscoWorks 2000 product. Microsoft will support WBEM in Windows 98 and Windows 2000. (See *The Must-Have Reference for IP and Next Generation Networking*, Anritsu Company, http://www.us.anritsu.com/downloads/files/musthave.pdf.)

CHAPTER 4

Intelligent Optical Networks and GMPLS

The previous chapter discussed the standards baseline for a new generation of optical networks, the OTN. In planning for a new network, it would, in principle, be useful to have the ability for the end user to set up optical connections from and to various points in the network in real time and as needed. This chapter amplifies the discussion of Chapter 3 and looks at the issue of signaling in more detail. The ASTN/ ASON mechanism discussed in this chapter aims at providing the OTN with an intelligent optical control plane for dynamic network provisioning. Its subtending capabilities include network survivability, protection, and restoration. ASTN specifications were under development and review by the ITU at press time, and were being positioned as the global framework for ION.

We should note at the outset that, although switched-lambda services are within our technical grasp in the core of the network, these services are a *real challenge* in practical end-to-end applications. For example, in the United States, there are approximately 4.5 million commercial buildings, with more than 735,000 of those providing office space. Yet, only approximately 20,000 buildings had fiber loops as of 2002. (The number was forecast to reach about 30,000 by 2007, but the telecom slowdown of the early decade may well impact that goal.) Of the 20,000 buildings cited, many are "carrier hotels," central-office-like real estate housing the telecommunications equipment of competitive carriers. That leaves the number of actual end-user-penetrated buildings even smaller. The issue is that it generally takes about a quarter of a million dollars to penetrate a building, and often even more.

Although it is fashionable to assume that a large number of customers will require gigabit-level connectivity in a building, the stark realities are quite different. As a quick anecdotal calculation, one can assume that—at least for the foreseeable future—only the Fortune 5000 companies [1] are candidates for switched-lambda services. Assuming that these companies had on an average 10 locations that are tier 1 network locations requiring gigabit-level connectivity, that would be 50,000 data sources/sinks. If we assume that these companies had all chosen the "intelligent building" locations that are on fiber (20,000 cited above), this would translate to

2.5 customers per building. Also see Figure 4.1 [2], which shows the expected revenue as a function of communication link speed. This author has for many years used a value of 3 as the number of potential broadband customers in a building [3, 4]. Without even considering the cost of the network, the cost of operations, and the sustaining/required net income that needs to be realized, a 36-month amortization for the two ends of the circuit would result in a per-circuit *monthly recurring charge* (MRC) of $500,000/3/36 = $4,629; including the other items, this will easily reach an MRC of $10,000 to $15,000. This figure is rather large; many Fortune 5000 companies currently rely on FR technology, spending $200 to $1,000 per month per link.

Network features of interest to enterprise users almost invariably have to be end to end to be of value: Reliability has to be end to end; QoS has to be end to end; security has to be end to end; and, similarly, provisionability has to be end to end. It is immaterial to the user that the core can be provisioned in 2 sec, when the user has to wait 9 months to get fiber installed and deployed in a building. Today we can already provision an OC-3 in seconds using DCS technology, but if a building on the outskirts of town does not have a fiber loop, it could take months to obtain an end-to-end OC-3. Herewith the assumption is made that these issues just highlighted are well understood by the developers of next-target (next generation) optical networks and, hence, that the features discussed in the chapter are not just "pie-in-the-sky" for most potential enterprise users.

4.1   ION Overview

4.1.1   Framework

With today's DWDM technology the capacity of a single fiber can be increased 160-fold. DWDM, however, does not hold the ultimate solution to address bandwidth demand. As covered in the previous chapter,

FIGURE 4.1
Forecast U.S. market for broadband services.

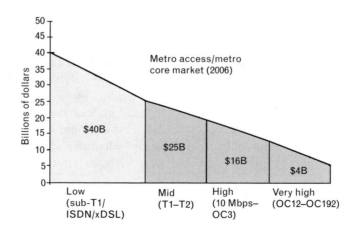

carrier-grade standards–supported work for a next-target (next generation) intelligent optical network has been undertaken under the ASTN and ASON specifications of the ITU-T. The ASTN/ASON model is based on mesh network architectures; mesh topologies are now, in fact, being deployed by service providers. Figure 4.2 provides a pictorial view of ASTN/ASON. Several industry entities were already advocating signaling concepts regarding IONs in the late 1990s and early 2000s. For example, *Optical Domain Service Interconnect* (ODSI) was a coalition of service providers and equipment vendors who wanted to enable user devices (router, ATM switch, and so on) to be able to dynamically request bandwidth from the optical network core. The coalition was formed in January 2000 and terminated its activities in December 2000 after having successfully shown a demonstration of the concept. Two typical examples of advocacy for ASTN/ASON by proponents are as follows:

> While DWDM and SONET have assisted in meeting the bandwidth and capacity requirements of Internet traffic, they cannot enable carriers to deliver new revenue generating services, such as web hosting and VPNs. Service providers are dealing with the need to blend the reliability of their existing equipment with the need to incorporate new technologies—enabling them to rapidly and flexibly provision bandwidth, plus develop new value-added service offerings based on a dynamically provisioned network. This demands a new model, the key to which is the role of the control plane, which will form the bridge between the management plane (a superset of today's capabilities) and the OTN itself [5].

> While DWDM significantly increases the capacity of the network, there is still a need to more rapidly and cost effectively harness that raw capacity and deliver new optical services—dynamically. Intelligent

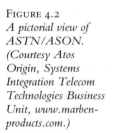

FIGURE 4.2
A pictorial view of ASTN/ASON. (Courtesy Atos Origin, Systems Integration Telecom Technologies Business Unit, www.marben-products.com.)

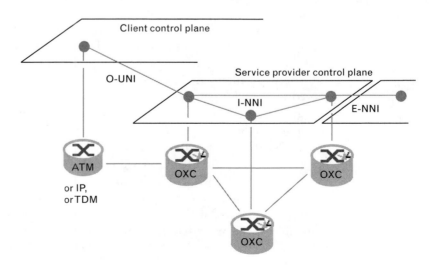

optical switches are that answer. By providing cohesive unity between DWDM and SONET/SDH architectures, these switches combine hardware advances and sophisticated software intelligence to implement innovative features based on emerging routing and signaling standards. Intelligent optical switches are the key component to evolve the network to support dynamic provisioning, sophisticated high bandwidth applications, remote performance monitoring and a reduction of duplicate restoration capacity. Intelligent optical switches support both ring and mesh architectures, allowing service providers to evolve their existing infrastructures while immediately cutting both capital and operating costs. Whereas ring architectures mandate the reservation of 100% excess capacity, mesh architectures leave the choice of protection to the service providers themselves, reducing costs by as much as 70% without compromising critical restoration times [6].

What subset of this panorama will eventually be realized in the intermediate term remains to be seen.

The G.astn (G.8070) architectural standard design assures (at least in theory) that service providers can deliver global, end-to-end lambda services. Whether lambda services are needed on a broad scale between continents (at least in the near to midterm) remains to be established. ASTN also presents service providers with an evolution from early vendor products that were sold as proprietary intelligent switching platforms. Utilizing standards, interoperability across a multiple-vendor network environment and across multiple service provider networks can be expected—this in turn ensures future-proofing the carrier and end-user investment. To realize the ASTN/ASON, signaling standards are needed. GMPLS and OIF O-UNI were being considered by the ITU-T in the early part of the decade. Both GMPLS and the OIF O-UNI can be mapped to ASTN/ASON models; hence, these predecessor protocols were being considered for such implementations. The ITU-T, however, was also considering alternative proposals to these, including a completely new protocol or a modified form of the signaling and routing mechanisms used by ATM's private network-to-network interface (PNNI). Table 4.1 recaps the key standards associated with ASTN/ASON, which we first introduced in Chapter 3; also see Figure 4.3. Standards activities have included key documents such as *ODSI Functional Specification; ITU-T Automatic Switched Transport Network* (ASTN/SG15) and *Automatic Switched Optical Network* (ASON/SG13); IETF GMPLS; and OIF UNI 1.0 [7].

The ASTN/ASON model focuses on providing the OTN with an intelligent optical control plane incorporating dynamic network provisioning combined with network survivability, protection, and restoration. In effect, a hierarchy has been created with fibers at the bottom, followed by groups of λ's (wavebands), individual λ's, SONET/SDH tributaries, and

TABLE 4.1   ASON/ ASTN Standards as of Early 2003

G.7712 (G.dcn), *Architecture and Specification of the Data Communications Network (DCN),* February 2001

G.807, *Requirements for ASTN,* July 2001

G.7041 (G.gfp), *Generic Framing Procedure,* June 2001

G.7042 (G.LCAS), *Link Capacity Adjustment Scheme for Virtual Concatenated Signals,* October 2003 (expected)

G.808 (G.gps), *Generic Line Protection Switching,* October 2003 (expected)

G873.1 (G.otnprot), *Optical Transport Network (OTN) Linear Protection,* October 2003 (expected)

G8080 (G.ason), *Architecture for ASON,* July 2001

G.dsn, *Optical System Design and Engineering Considerations,* June 2001

G.7713 (G.dcm), *Distributed Connection Management,* February 2001

G.7714 (G.sdisc), *Automatic Service Discovery in ASON,* February 2001

G.ndisc, *Automatic Neighbor Discover in ASON,* March 2001

G.7710/Y.1701 (G.cemr), *Common Equipment Management,* October 2001

G.7715 (G.rtg), *ASON Routing (NNI),* May 2004 (expected)

packet switch capable (PSC) connections at the top (see Figure 4.4). As connections are established at each level of the hierarchy, they need to be propagated to other nodes in the network so they can, if desired, use these paths to establish higher level connections [5]. Dynamic routing necessitates direct information exchange between neighboring nodes, in order to discover topology information. To achieve this, the out-of-band control plane is augmented with a link layer node-to-node protocol. The ASTN/ASON specifications are framework models and, therefore, are protocol independent. (For example, as noted, both GMPLS and O-UNI protocols can be mapped to ASTN/ASON models, and they may become accepted implementations, although as noted, the ITU-T is also considering alternative protocols.) The long-term benefits from having ASTN/ASON assets include the ability to support circuit provisioning in minutes; dynamic restoration and resiliency; flexible service selection, dynamic resource allocation; reduced carrier software development; and interdomain, intercarrier QoS [8]. In many ways, the ASTN/ASON model is relatively traditional (evolutionary) and incorporates aspects of other network technologies, such as the PSTN and ATM, although the current implementations for O-UNI and GMPLS are based on IP specifications [5].

GMPLS was getting a lot of attention in the early 2000s. It largely consists of [9] (also see Figure 4.5) the following:

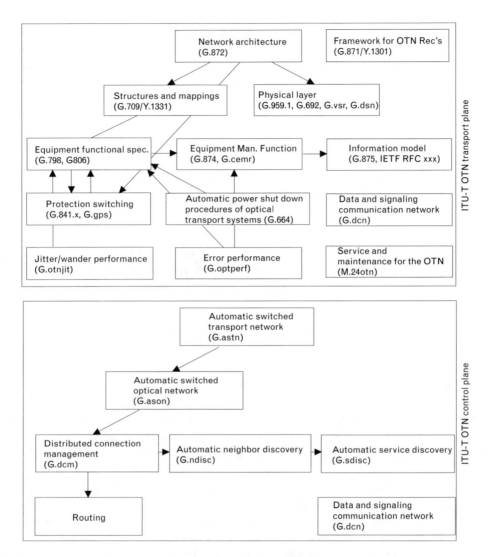

FIGURE 4.3   *Transport plane and control plane protocol apparatus. (Source: Vissers, M., "Optical Transport Network & Optical Transport Module: Digital Wrapper," presented at Beyond SONET/SDH Conference, April 2001.)*

- Extensions to the dynamic routing protocols *Intermediate System–Intermediate System* (IS-IS) and *Open Shortest Path First* (OSPF) in order to provide management applications with the topology of the optical network, optical NE capabilities, resource availability, and so on;

- Extensions to the signaling protocols *Constraint-Based Routing Label Distribution Protocol* (CR–LDP) and *Reservation Protocol with Traffic Engineering* (RSVP-TE) in order to establish, maintain, and tear down optical paths;

FIGURE 4.4
OTN/ASON
/ASTN environment
example.

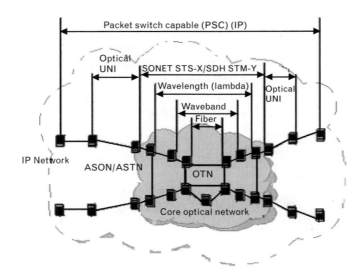

FIGURE 4.5
Example of a
GMPLS-based stack.
(Courtesy Atos
Origin, Systems
Integration Telecom
Technologies Business
Unit.)

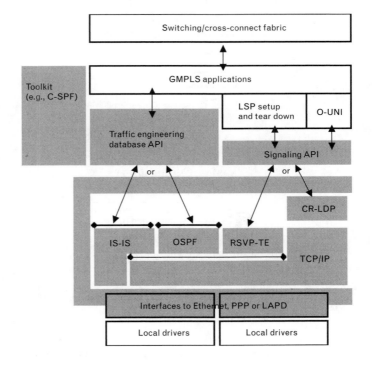

- A new *Link Management Protocol* (LMP) for control channel management and link information exchange between neighboring optical NEs.

(With regard to IETF's GMPLS, note that as of the end of 2002, no Internet drafts had been advanced to RFC status.)

User client equipment (e.g., a router) connects into the network via the UNI. The UNI defines how customers can access the provider's network. Clients can request three different types of circuits over the UNI using a signaling protocol: provisioned, signaled, and hybrid. The signaling mechanism was still under discussion at press time. Interfaces supported in ASTN/ASON are (see Figure 4.6) as follows:

- UNI (interface between service control planes of service requester and provider);

- NNI (interface between two networks; both can be optical, or one can be IP and the other optical): I-NNI (interior): trusted relationship (same or different administrative domain) or E-NNI (exterior): untrusted relationship (different administrative domain).

The dynamic aspects of the ASTN/ASON requirements (e.g., provisioning and restoration) require complex interactions between the *optical control channels* (OCCs) and the transport plane (this being done over a *connection control interface* [CCI] as shown earlier in Figure 4.3). In turn, this implies an interaction between the signaling and routing protocols, not

FIGURE 4.6
Interfaces defined in ASTN/ASON.

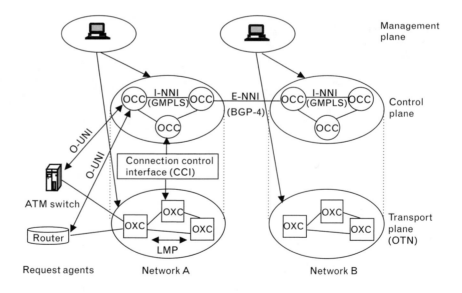

UNI: An interface between the control planes of the service requester and service provider; uses bidirectional signaling
(Example: OIF's O-UNI)
NNI: An interface between two networks or subnetworks
(both networks can be optical, or one network can be IP and the other optical)
    Interior NNI (I-NNI): An interior interface that has a trusted relationship and can belong to the same or different administrative domains (Example IETF's GMPLS)
    Exterior NNI (E-NNI): An exterior interface that has an un-trusted relationship and usually belongs to different administrative domains

unlike the interaction between the UNI and PNNI protocols in ATM. The result is an out-of-band control mechanism in which the signaling and data paths can flow differently through the network.

As an example, as noted earlier, the IETF has defined LMP to support link management and to exchange information between neighboring nodes. LMP is designed to support the delivery of necessary information to the routing protocol; it also supports traditional link layer OAM&P functions (e.g., those performed by the SONET *data communication channel* [DCC]). A modified *Constrained Shortest Path First* (CSPF) engine is used to calculate a path through the network that meets required QoS criteria such as bandwidth, delay, jitter, the characteristics of fibers, route diversity, and security. The *traffic engineering* (TE)-specific parameters are stored at each network node (as attributes of a link–state database), and this information is used by CSPF to determine appropriate routing. Once generated by a CSPF engine, this constrained route information is used by a signaling protocol, such as GMPLS, to establish the connection with each node, verifying that the resources are available as the signaling message transits the network [5]. Extensions to some of the existing protocols may be required. For example, with the hierarchy and multiple connections between two OXCs, each connection may carry its own or multiple control plane channels. This has the effect of possibly increasing the control plane state and computational burden, which in turn can impact the cost and overall performance. A solution is to bundle lightpaths so that a single control plane channel can support multiple physical connections.

Some consider the ASTN/ASON technology to be not as much for support of future lambda services, but for improved carrier-internal OAM&P functionality. The administration of traditional SDH or SONET and DWDM optical transport networks relies on a static architecture (e.g., TMN) that uses management protocols such as TL1 or CMISE to manage individual network elements. However, such a management plane may no longer be adequate to support the administration requirements of evolving optical networks. Such requirements include faster, interoperable, and dynamic provisioning of optical paths; a better integration with data networks; a more efficient use of network resources by providing protection through meshed topologies rather than rings; better support of transparent technologies; and the ability to let service providers offer new services such as on-demand provisioning of optical paths. ASTN specifies how some administration functions that were traditionally handled by the management plane, in particular, the provisioning of optical paths, should now be handled by a control plane that relies on dynamic routing and signaling protocols; such functions make up the NNI [9]. Because carrier operations (OAM&P) represent a major portion of the carrier's expense budget, any technology or advancement that truly addresses OAM&P is of interest.

New investments are required by carriers in order to realize the OTN/ASTN/ASON. The long-term benefits include rapid circuit provisioning and the other capabilities listed earlier. Many of these features, however, are available in one form or another in today's SONET/SDH–based networks. The reality of the fact is that today very few planners inter-mix equipment from various vendors within a planned network, unless they are forced to do so by a merger or acquisition or by external circumstances or so on. Planners do not by design typically use SONET equipment from Nortel and Lucent on the same ring; or routers from Cisco and Juniper on the same peer-level backbone, FR switches from Lucent and Cisco on the same tier 1 network, and so on. Hence, the idea of "interworking" cannot be oversold. Also, the instantaneous provisionability frankly is only a chimera: It will not occur until a majority of buildings of interest are on fiber facilities. Today in the United States, only 5% of the office buildings are equipped with fiber—for instantaneous provisionability to be a reality, perhaps 50%, 65%, 80%, or 95% of the buildings would have to have fiber networks.

In December 2001 under the banner "The 40 Gigabit per Second Phone Call: Global Standards for Automatically Switched Optical Networks Enable New Market Services," the ITU announced that new global standards for ASONs and their control mechanisms had been adopted as a result of an aggressive work program initiated by ITU to support bandwidth-on-demand applications. The standards, which add switching capability to the installed optical fiber infrastructure, were developed and agreed on in less than a year. The ASON family of standards from ITU-T SG15 builds on OTN standards completed earlier and moves the industry, according to the ITU, toward the *optical Internet*. The ITU states that these standards can create tremendous business opportunities for network operators and service providers, giving them the means to deliver end-to-end, managed bandwidth services efficiently, expediently, and at reduced operational cost. ASON standards can also be implemented to add dynamic capabilities to new optical networks or established SDH networks. The expected business benefits, according to the ITU, include the following:

- Increased revenue-generating capabilities through fast turnup and rapid provisioning, as well as wavelength-on-demand services to increase capacity and flexibility;

- Increased return on capital from cost-effective and survivable architectures that help protect current and future network investments from forecast uncertainties;

- Reduced operations costs as a result of more accurate inventory and topology information, resource optimization, and automated processes that eliminate manual steps.

ASON control mechanisms provide support for both switched wavelength and subwavelength connection services in transport networks to provide bandwidth on demand. Wavelength connection services make use of an entire wavelength of light, whereas subwavelength services use a channel within a wavelength. The ASON control mechanisms also enable fast optical restoration. Traditionally, transport networks have used protection rather than restoration to provide reliability for connections. With protection, connections are moved to dedicated or shared routes in the event of a failure of fiber or network equipment. With restoration, the endpoints can "redial" to reestablish the connection through an alternative route as soon as a loss of the original connection is detected. Restoration is a defined advantage for carriers because it makes better use of the network capacity and, with this new standard, it can be performed faster than with most proprietary restoration systems available today. These recommendations are an important step, in the view of the ITU, toward the completion of the ASON series of standards. Work ahead includes the addition of detailed protocol specifications and the expansion of features for interoperable network restoration.

ITU-T Recommendation G.807, the first standard in the ASTN series approved in 2001, addressed the network-level architecture and requirements for the control plane of ASTN independent of specific transport technologies. Agreement was reached on three new standards in the ASON:

- New ITU-T Recommendation G.8080/Y.1304, *Architecture for the Automatically Switched Optical Network (ASON)*, specifies the architecture and requirements for the automatic switched transport network as applicable to SDH transport networks, defined in Recommendation G.803, and OTNs, defined in Recommendation G.872. This new recommendation is based on requirements specified in Recommendation G.807.

- New ITU-T Recommendation G.7713/Y.1704, *Distributed Call and Connection Management*, gives the requirements for *distributed connection management* (DCM) for both the UNI and the NNI. The requirements in this recommendation specify the signaling communications between functional components to perform automated connection operations, such as setup and release of connections. It describes DCM messages, attributes, and state transitions in a protocol neutral fashion.

- New ITU-T Recommendation G.7714/Y.1705, *Generalized Automatic Discovery Techniques*, describes automatic discovery processes that support distribution connection management. Applications of automatic discovery addressed include neighbor discovery and adjacency discovery. The requirements, attributes, and discovery methods are described in a protocol neutral fashion.

An agreement was also reached on Recommendation G.7712/Y.1703, *Architecture and Specification of Data Communication Network*. The initiative "Realizing the Next Generation Optical Network" is applicable to ASON in that it specifies the architecture and requirements for a *data communications network* (DCN) to support the exchange of ASON messages in addition to the traditional TMN communication. These communications take place among the transport plane, control plane, and management plane for ASON signaling and network management.

Under the banner "Realizing the Next Generation Optical Network, New Standards Support Ultra-High Capacity Transport," the ITU had previously announced that its Telecommunication Standardization Sector had reached agreement on new global standards for equipment and management of the next generation optical network, namely, the OTN. The new standards support carrier-grade ultrahigh-capacity transport networks capable of supporting fully transparent wavelength services. As covered in Chapter 3, these new SG15 recommendations provide telecommunications equipment manufacturers with the necessary tools to produce interoperable products, allowing carriers to build and manage ultrahigh-capacity optical networks.

In particular, these standards support efficient transport of popular data protocols such as Ethernet and Fibre Channel, together with other wideband and broadband services including SDH/SONET, ATM, FR, audio/video, and IP-based services. They also specify detailed equipment functions to support performance monitoring, fault isolation, and alarming, including support for optically transparent subnetworks. The OTN series of standards facilitate end-to-end connectivity between optical transport elements in a global network, as we covered in Chapter 3.

An important new feature provided by these standards is the ability to combine multiple client signals within a wavelength to allow maximum utilization and cost effectiveness of transport capacity while still allowing switching at the OTN service rates of 2.5, 10, and 40 Gbps. These new standards complement the OTN interface standards agreed on in early 2001 by providing the equipment and management specifications necessary for vendors to provide interoperable solutions. Optical networking continues to be a multibillion dollar business sector satisfying the unceasing demand for more transport capacity for data, video, and voice traffic. According to the ITU, as applications have become more complex and demand increases, the standards for ultrahigh-capacity transport networks will be in place to allow the development of future services that could increase the demand for network bandwidth.

4.1.2   Positioning of GMPLS in ASTN/ASON

This section discusses the advocacy for GMPLS by some constituencies and provides a short overview of GMPLS and O-UNI. These topics are then expanded in Section 4.2.

The interconnection of the IP clients (routers) and the optical control plane can be designed as being either loosely or tightly coupled: "loose coupling" refers to an overlay model, and "tight coupling" refers to a peer model. (There is also a hybrid model, which is related to the peer model but shares attributes of both models [10].) The overlay signaling mechanism enables the client (the service requester) to add, modify, or delete connections over the carrier's network, without providing the client with visibility to the carrier's network topology. Therefore, the overlay model (although it may use the same IP layer protocols as utilized in the client network for its routing and signaling), maintains a separation at the client-to-network interface by keeping the IP client routing, signaling protocols, topology distribution, and addressing scheme independent from the ones used by the optical layer of the carrier network [5]. This approach is likely to be adopted by carriers because they want to be able to control the resources of the transport network. This has traditionally been their view going back to at least the mid-1970s (with X.25 network services), if not earlier.

GMPLS is a proposed peer-to-peer signaling protocol that had a strong constituency at press time. GMPLS extends MPLS with the necessary mechanisms to control routers, DWDM systems, ADMs, photonic cross-connects, and possibly other optical NEs. GMPLS closely follows the peer-to-peer network model of IP technology. In a GMPLS environment, every participating *label-switching router* (LSR) sees the entire network state. Originally, there was no definition of a UNI or NNI in GMPLS, but the OIF started the process of defining a UNI for the GMPLS specification. The OIF has enhanced MPLS/GMPLS to broaden the scope of the IETF-developed capabilities, particularly in the area of addressing. (GMPLS extends the addressing mechanism to nodes that may not be IP capable, including support for *network service access point* [NSAP] mechanisms.) GMPLS does add link bundling and hierarchy capabilities; it also supports a bidirectional view of optical connections, as opposed to the unidirectional *labeled switch paths* (LSPs) that are established in MPLS.

The UNI specifies the mechanism by which a *user network interface client* (UNI-C) is able to invoke transport network services with a *user network interface network* (UNI-N), which is an NE in the carrier's network. In this environment, edge (customer) routers can make their route calculations based on information collected through a GMPLS-enhanced routing protocol, such as OSPF or IS-IS. When the route is calculated, GMPLS utilizes RSVP-TE or CR-LDP to set up the LSPs. While MPLS deals only with what GMPLS identifies as PSC interfaces, GMPLS adds four other types of interfaces: *Layer 2 switch capable* (L2SC) interfaces recognize frames and cells. *TDM capable* (or TDM) interfaces forward data based on a data's time slot. *Lambda switch capable* (LSC) interfaces such as photonic cross-connects work on individual wavelengths. A *fiber switch cable* (FSC) interfaces work on individual or multiple fibers.

The GMPLS UNI functionality (at the logical level) is limited to three activities: connection creation, connection deletion, and connection status inquiry. Connection creation allows a connection of specified attributes to be activated between two points; these connections may be subject to policies that the operator defines, such as user group restrictions or security procedures. Connection deletion takes down an established connection. Connection status inquiry allows nodes to retrieve certain connection parameters by querying the network. Development work remains to be done before the OIF UNI can be widely adopted. The connections that can be established today are limited to SONET or SDH. Interdomain management functions also need to be developed [11].

This peer model supports the I–NNI interface of the ASTN/ASON model and assumes that all devices in the network have a complete topological view and that they participate in routing. This would be supported, for example, in the collection of OCCs for a single network, especially if this network is privately owned by the client (rather than being a carrier's network). In principle, in a hierarchical network comprised of optical switches, SONET/SDH ADMs, and IP layer 2/layer 3 devices, the entire optical core can be visible to an edge router using the same *interior gateway protocol* (IGP) routing instance over the network, such as OSPF or IS–IS [5]. As we noted, this approach is unlikely to be adopted by carriers because they do not want to lose control of their network by disseminating critical OTN information (such as network bandwidth, capacity, and topology).

4.2 ION Concepts

As covered in the previous section, the IETF has been active in tracking ASON work. The architectural choices for the interaction between IP and optical network layers, particularly the routing and signaling aspects, are key to the successful deployment of next generation networks. Of interest is the March 2002 IPO WG Internet draft (draft-ietf-ipo-ason-02.txt) by O. Aboul-Magd, B. Jamoussi, S. Shew, Gert Grammel, Sergio Belotti, and Dimitri Papadimitriou titled *Automatic Switched Optical Network Architecture and Its Related Protocols*, March 2002. This draft [12] describes the main architectural principles of the automatic switched optical networks work that have recently been approved by the ITU–T [13, 14]. ASON architecture defines a set of reference points (interfaces) that allows ASON clients to request network services across those reference points. The protocols that run over ASON interfaces are not specified in [13, 14]. IP-based protocols such as GMPLS [15] can be considered such that the ASON/ASTN work can benefit from the protocols design work progressing at the IETF. To cross-fertilize the discussion, the basic concepts are described next.

4.2.1 Introduction

The existing transport networks provide SONET/SDH and WDM services whose connections are provisioned via network management. This process is both slow (weeks to months) relative to the switching speed and costly to the network providers. An ASON is an OTN that has dynamic connection capability. It encompasses SONET/SDH, wavelength, and potentially fiber connection services in both OEO and all-optical networks. Such capability provides a number of added values:

- *Traffic engineering of optical channels.* In such a case, bandwidth assignment is based on actual demand patterns.

- *Mesh network topologies and restoration.* Mesh network topologies can in general be engineered for better utilization for a given demand matrix. Ring topologies might not be as efficient due to the asymmetry of traffic patterns.

- *Managed bandwidth to core IP network connectivity.* A switched optical network can provide bandwidth and connectivity to an IP network in a dynamic manner compared to the relatively static service available today.

- *Introduction of new optical services.* The availability of switched optical networks will facilitate the introduction of new services at the optical layer. Those services include bandwidth on demand and OVPNs.

This section describes the main ASON architecture principles. This work has recently been approved by the ITU-T.

4.2.2 ASON Architecture Principles

This section gives a quick summary of ASON architecture principles as defined in [14]. The interested reader may refer to [14] and the references therein. ASON defines a control plan architecture that allows the set up and tear down of calls (and the connections that support a call) as a result of a user request. To achieve global coverage and the support of multiple client types, the architecture is described in terms of components, and a set of reference points and rules must be applied at the interface points between clients and the network and between networks.

4.2.2.1 ASON Reference Points

In ASON architecture there is the recognition that the optical network control plan will be subdivided into domains that match the administrative domains of the network. The transport plane is also partitioned to match the administrative domains. Within an administrative domain, the control plane may be further subdivided, for example, by actions from the management plane. This allows the separation of resources into, for example, domains for geographic regions, which can be further divided into domains that contain different types of equipment. Within each domain, the control plane may be further subdivided into routing areas for scalability, which may also be further subdivided into sets of control components. The transport plane resources used by ASON will be partitioned to match the subdivisions created within the control plane.

The interconnection between domains, routing areas, and, where required, sets of control components is described in terms of reference points. The exchange of information across these reference points is described by the multiple abstract interfaces between control components. The physical interconnection is provided by one or more of these interfaces. A physical interface is provided by mapping an abstract interface to a protocol. The reference point between an administrative domain and an end user is the UNI. The reference point between domains is the E-NNI. The reference point within a domain between routing areas and, where required, between sets of control components within routing areas is the I-NNI. Figure 4.7 shows a possible domain subdivision and the reference points between them.

The difference between the I-NNI and the E-NNI is significant. I-NNI is applied in a single routing area where all equipment supports the same routing protocol, and detailed routing information can be exchanged between the different nodes. On the other hand, E-NNI is mainly concerned with reachability between the domains that employ different routing and protection methodologies.

4.2.2.2   Call and Connection Control Separation

Call and connection control are treated separately in the ASON architecture. Call control is a signaling association between one or more user applications and the network to control the setup, release, modification, and maintenance of sets of connections. Call control is used to maintain the association between parties, and a call may embody any number of underlying connections, including zero, at any instance of time.

Call control is provided at the ingress/egress of the network or at domain boundaries. Call control is applicable at the E-NNI and UNI reference points. Call and connection control separation allows intermediate (relay) network elements to support only procedures needed for the support of switching connections. Access to call information at domain boundaries allows domains that use different protection or restoration

FIGURE 4.7
ASON/ASTN
global architecture.

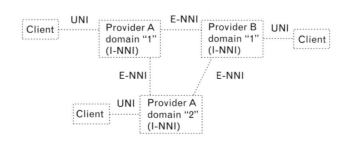

mechanisms to interwork [e.g., a metropolitan network using a unidirectional path switched ring (UPSR) with a backbone network using mesh restoration] without the need for all domains to understand all of the possible protection/restoration schemes.

With call and connection control separation, a single call may embody a number of connections (more than one) among user applications. This allows for the introduction of enhanced services where a single call is composed of more than one application, for example, voice and video. In other situations, this separation between call and connection control is beneficial to the service provider, especially in the areas of restoration and maintenance. In those situations, it is cost saving to maintain the call state while restoration actions are under way.

4.2.2.3 Policy and Security

According to the ASON architecture, policy is defined as the set of rules applied at a system boundary and implemented by port controller components. System boundaries may be nested to allow for correct modeling of shared policies with any scope. A system is defined as any (arbitrary) collection of components. In general, a system boundary will coincide with a domain boundary; this allows the application of a common policy for all interfaces that cross the domain boundary. The nesting of system boundaries allows the application of additional (more stringent) policies if the domain boundaries are between cost centers within a single network (administration) or between different networks (administrations). Policy is applied at individual interfaces crossing the reference points described earlier.

4.2.2.4 Federation

Connection control across multiple domains requires cooperation between controllers in the different domains. A federation is defined as the community of domains that cooperate for the purpose of connection management. Two types of federations are defined: the joint federation model, in which one connection controller has authority over connection controllers that reside in different domains, and the cooperative model, in which there is no concept of a parent connection controller.

4.2.3 ASON Control Plane Requirements

A well-designed control plane architecture should give service providers better control of their network, while providing faster and improved accuracy of circuit setup. The control plane itself should be reliable, scalable, and efficient. It should also be sufficiently generic to support different technologies and differing business needs and different partitions of functions by vendors (i.e., different packaging of the control plane components). In summary, the control plane architecture should:

- Be applicable to a variety of transport network technologies (e.g., SONET/SDH, OTN, PXC). To achieve this goal, it is essential for the architecture to isolate technology-dependent aspects from technology-independent aspects and address them separately.

- Be sufficiently flexible to accommodate a range of different network scenarios. This goal may be achieved by partitioning the control plane into distinct components. This allows vendors and service providers to decide the location of these components, and also allows the service provider to decide the security and policy control of these components.

The control plane should support either *switched connections* (SCs) or SPCs of basic connection capability in transport networks. These connection capability types are (1) unidirectional point-to-point connections, (2) bidirectional point-to-point connections, and (3) unidirectional point-to-multipoint connections. The control of connectivity is essential to the operation of a transport network. The transport network itself can be described as a set of layer networks, each acting as a connecting function whereby associations are created and removed between the inputs and outputs of the function. These associations are referred to as *connections*. Three types of connection establishment are defined: provisioned, signaled, and hybrid.

Establishment of *provisioned connection* is triggered by a management system and is referred to as a hard permanent connection. A *signaled connection* is established on demand by the communicating endpoints using a dynamic protocol message exchange in the form of signaling messages. In a *hybrid connection,* a network provides a permanent connection at the edge of the network and utilizes a switched connection within the network to provide end-to-end connections between the permanent connections at the network edges.

The most significant difference between the three methods just mentioned is seen in the party that sets up the connection. In the case of provisioning, connection setup is the responsibility of the network operator, whereas in the signaled case, connection setup may also be the responsibility of the end user. Additionally, third-party signaling should be supported across a UNI.

4.2.4 ASON Functional Architecture

The components of the control plane architecture are as follows:

1. *Connection controller function (CC).* The connection controller is responsible for coordination among the link resource manager, routing controller, for the purpose of the management and supervision of connection setup, release, and modification.

2. *Routing controller (RC).* The role of the RC is to respond to requests from CC for route information needed to setup a connection, and to respond to requests for topology information for network management purposes.

3. *Link resource management (LRM).* The LRM components are responsible for the management of subnetwork links, including the allocation and deallocation of resources, providing topology and status information.

4. *Traffic policing (TP).* The role of the TP is to check that the incoming user connection is sending traffic according to agreed-on parameters.

5. *Call controller:* The two types of call controllers are a calling/called party call controller and a network call controller. The role of the call control is the generation and processing of call requests.

6. *Protocol controller (PC).* The PC provides the function mapping of the parameters of the abstract interfaces of the control components into messages that are carried by a protocol to support interconnection via an interface.

4.2.5 ASON Reference Points and GMPLS Protocols

The ASON CP defines a set of interfaces or reference points:

- *UNI.* UNI runs between the optical client and the network.
- *I-NNI.* I-NNI defines the interface between the signaling network elements within the same domain or between routing areas.
- *E-NNI.* E-NNI defines the interface between ASON control planes belonging to different domains.

The different ASON interfaces are described in the next few sections. Candidate GMPLS-based protocols for use at the different interfaces are also discussed.

4.2.5.1 ASON UNI

The ASON UNI allows ASON clients to perform a number of functions including these:

- *Connection Create.* Allows the clients to signal to the network to create a new connection with specified attributes. Those attributes might include bandwidth, protection, restoration, and diversity.
- *Connection Delete.* Allows ASON clients to signal to the network the need to delete an already existing connection.

- *Connection Modify.* Allows ASON clients to signal to the network the need to modify one or more attribute for an already existing connection.

- *Status Enquiry.* Allows ASON clients to inquire about the status of an already existing connection.

Other functions that might be performed at the ASON UNI include client registration, address resolution, and neighbor and service discovery. Those functions could be automated or manually configured between the network and its clients.

Client registration and address resolution are tightly coupled to the optical network address scheme. Requirements for optical network addresses and client names are outlined in [16]. In general, the client name (or identification) domain and optical address domain are decoupled. The client ID should be globally unique to allow for the establishment of end-to-end connections that encompass multiple administration domains. For security, the nodal addresses used for routing within an optical domain are not allowed to cross network boundaries. The notion of closed user groups should also be included in ASON addressing to allow for the offering of OVPN services.

ASON UNI realization requires the implementation of a signaling protocol with sufficient capabilities to satisfy UNI functions. Both LDP [17] and RSVP-TE [18] have been extended for use with the signaling protocol across the ASON UNI. The extensions involve the definition of the necessary type-length-value (TLVs) or objects to be used for signaling connection attributes specific to the optical layer. New messages are also defined to allow for connection status enquiry. The OIF has adopted both protocols in its UNI 1.0 specification [19]. Note, however, that recent efforts within the IETF have been made to move CR-LDP off the standards track, with only RSVP-TE as the recommended solution. Also note that as of press time, RSVP-TE has enjoyed more field deployment.

4.2.5.2   ASON I-NNI

The I-NNI defines the interface between adjacent CCs in the same domain or between routing areas. The two main aspects of I-NNI are signaling and routing.

Path selection and setup through the optical network requires a signaling protocol. Transport networks typically utilize explicit routing, where path selection can be done either by operator or software scheduling tools in management systems. In ASON, end-to-end optical channels (connections) are requested with certain constraints. Path selection for a connection request should employ constrained routing algorithms that balance multiple objectives:

- Conform to constraints such as physical diversity;

• Balance the load of network traffic to achieve the best utilization of network resources;

• Follow policy decisions on routing such as preferred routes.

To facilitate the automation of the optical connection setup, nodes in the optical network must have an updated view of its adjacencies and of the utilization levels at the various links of the network. This updated view is sometime referred to as *state information*.

State information dissemination is defined as the manner in which local physical resource information is disseminated throughout the network. First the local physical resource map is summarized into logical link information according to link attributes. This information can then be distributed to the different nodes in the network using the control plane transport network IGP. ASON I-NNI could be based on two key protocols, IP and MPLS. Because MPLS employs the principle of separation between the control and the forward planes, its extension to support I-NNI signaling is feasible.

GMPLS [15] defines MPLS extensions to suit types of label switching other than the in–packet label. Those other types include time slot switching, wavelength and waveband switching, and position switching between fibers. Both CR-LDP [20] and RSVP-TE [21] have been extended to allow for the request and the binding of generalized labels. With generalized MPLS, an LSP is established with the appropriate encoding type (e.g., SONET, wavelength, and so on). LSP establishment takes into account specific characteristics that belong to a particular technology. MPLS traffic engineering requires the availability of routing protocols that are capable of summarizing link state information in their databases. Extensions to IP routing protocols, OSPF and IS–IS, in support of link state information for generalized MPLS are described in [22, 23].

4.2.5.3   ASON E-NNI

E-NNI is the external NNI between different domains. Those domains may belong to the same network administration or to different administrations. In some sense, E-NNI can be viewed as being similar to the UNI interface with some routing functions to allow for the exchange of reachability information between different domains.

BGP is the IP-based protocol that is commonly deployed between different domains. It could be used to summarize reachability information between different ASON domains in the same manner as it has been in use today for IP networks. BGP is rich in policy, which makes it a good candidate to satisfy service requirements such as diversity where policies could be used in choosing diverse routes. (Note, however, that while BGP is well suited for the transport of richly policy-labeled information between routing domains, it is not well optimized for rapid convergence times; this may have implications when considering restoration times versus initial path selection.)

4.2.6 ASON/ASTN CP Transport Network (Signaling Network)

In this section, we detail some architectural considerations for the makeup of the transport network that is used to transport the control plane information. For circuit-based networks, the ability to have an independent transport network for message transportation is an important requirement.

The control network represents the transport infrastructure for control traffic and can be either an inband or out-of-band network. An implication of this is that the control plane may be supported by a different physical topology from that of the underlying ASON. Control networks must satisfy certain fundamental requirements in order to ensure that control plane data can be transported in a reliable and efficient manner. In the event of control plane failure (for example, communications channel or control entity failure), although new connection operations will not be accepted, existing connections will not be dropped. Control network failure would still allow dissemination of the failure event to a management system for maintenance purposes. This implies a need for separate notifications and status codes for the control plane and ASON. Additional procedures may also be required for control plane failure recovery.

We recognize that the interworking of the control networks is the first step toward control plane interworking. To maintain a certain level of ease, it is desirable to have a common control network for different domains/subnetworks or types of network.

Typically, control plane and transport functions may coexist in a network element. However, this may not be true in the case of third-party control. This situation needs further study. Furthermore, issues in the control plane vis-à-vis the transport network also require further study. The ASON CP transport network requirements includes the following:

- Control plane message transport should be secure. This requirement stems from the fact that the information exchanged over the control plane is service-provider specific and security is of utmost importance.

- Control message transport reliability has to be guaranteed in almost all situations, even during what might be considered catastrophic failure scenarios of the controlled network.

- The control traffic transport performance affects connection management performance. Connection service performance largely depends on its message transport. Time-sensitive operations, such as protection switching, may need certain QoS guarantees. Furthermore, a certain level of survivability of the message transport should be provided in case of control network failure.

- The control network needs to be both upward and downward scalable in order for the control plane to be scalable. Downward scalability may be envisioned where the ASON network offers significant static connections, reducing the need for an extended control network.

- The control plane protocols should not assume that the signaling network topology is identical to that of the transport network. The control plane protocols *must* operate over a variety of signaling network topologies.

Given the above requirements, it is critical that the maintenance of the control network itself not pose a problem to service providers. As a corollary, this means that configuration-intensive operations should be avoided for the control network.

Common channel signaling links are associated with user channels in the following ways:

- Associated, whereby signaling messages related to traffic between two network elements are transferred over signaling links that directly connect the two network elements.

- Nonassociated, whereby signaling messages between two network elements A and B are routed over several signaling links, while traffic signals are routed directly between A and B. The signaling links used may vary with time and network conditions.

- Quasiassociated, whereby signaling messages between nodes A and B follow a predetermined routing path over several signaling links, while the traffic channels are routed directly between A and B.

Associated signaling may be used where the number of traffic channels between two network elements is large, thereby allowing a single signaling channel to be shared among a large number of traffic channels.

Quasiassociated signaling may be used to improve resiliency. For example, consider a signaling channel that has failure mechanisms independent of the traffic channels. Failure of the signaling channel will result in loss of signaling capability for all traffic channels even if all traffic channels are still functional. Quasiassociated signaling mitigates against this by employing alternative signaling routes. In other words, the signaling network must be designed such that failure of a signaling link should not affect the traffic channels associated with that signaling channel.

4.2.7 Transport Network Survivability and Protection

This section describes the strategies that can be used to maintain the integrity of an existing call in the event of failures within the transport network. The terms *protection* (replacement of a failed resource with a preassigned standby) and *restoration* (replacement of a failed resource by rerouting using spare capacity) are used to classify these techniques. In general, protection actions are completed in the tens of milliseconds range, whereas restoration actions normally complete in times ranging from hundreds of milliseconds to up to a few seconds.

The ASON control plane provides a network operator with the ability to offer a user calls with a selectable *class of service* (CoS), (e.g., availability, duration of interruptions, and errored seconds). Protection and restoration are mechanisms used by the network to support the CoS requested by the user. The selection of the survivability mechanism (protection, restoration, or none) for a particular connection that supports a call will be based on the policy of the network operator, the topology of the network, and the capability of the equipment deployed. Different survivability mechanisms may be used on the connections that are concatenated to provide a call. If a call transits the network of more than one operator, then each network should be responsible for the survivability of the transit connections. Connection requests at the UNI or E-NNI will contain only the requested CoS, not an explicit protection or restoration type.

The protection or restoration of a connection may be invoked or temporarily disabled by a command from the management plane. These commands may be used to allow scheduled maintenance activities to be performed. They may also be used to override the automatic operations under some exceptional failure conditions.

The protection or restoration mechanism should:

- Be independent of, and support any, client type (e.g., IP, ATM, SDH, Ethernet);

- Provide scalability to accommodate a catastrophic failure in a server layer, such as a fiber cable cut, which impacts a large number of client layer connections that need to be restored simultaneously and rapidly;

- Utilize a robust and efficient signaling mechanism, which remains functional even after a failure in the transport or signaling network;

- Not rely on functions that are not time critical to initiate protection or restoration actions. Therefore, consideration should be given to protection or restoration schemes that do not depend on fault localization.

4.2.8   Relationship to GMPLS Architecture

The relationship between the ASON/ASTN control plane architecture and GMPLS-based protocols was established in Section 4.2.4, where we showed how the different GMPLS protocol could be used for the realization of the different ASON/ASTN external interfaces.

Recently, a GMPLS architecture [24] has been introduced. It is important to note that there is no conflict between the GMPLS architecture and the network architecture presented here. ASON/ASTN provides a functional architecture of a control plane that allows the establishment of switched paths in optical networks. It provides the set of external interfaces necessary for the ASTN/ASON network to have a global reach. It does that, however, in a protocol-independent fashion that can be realized in different ways provided that its requirements are satisfied.

The GMPLS architecture focuses more on the applications of GMPLS-defined protocols, for example, CR-LDP for the setup of generalized LSP (GLSP) at the different interfaces of the network such as I-NNI and UNI. It does that in a more comprehensible way than is described in Section 4.2.4.

4.2.9   Other ASON/ASTN-Related ITU Activities

This section describes other activities that are currently under way at the ITU and are related to ASON/ASTN architecture.

4.2.9.1   Common Equipment Management (G.cemr)

The G.cemr [25] recommendation specifies those *equipment management function* (EMF) requirements that are common for SDH and OTN. EMF provides the means through which a network element level manager manages the *network element function* (NEF). These kinds of functions are not detailed in the current GMPLS work because of its focus on control plane-related aspects. Network management aspects are the subjects of other working groups in IETF such as OPS WG.

MPLS Solution   Element manager functions are not part of the control plane specifications in GMPLS.

Requirements

- Network management applications should perform *fault, configuration, accounting, performance and service management* (FCAPS).
- Path setup can be triggered by means of a network management system using control plane mechanisms.
- For path setup, control plane and network management systems should cooperate to allow path provisioning by network management as well as provisioning using the control plane.

4.2.9.2   Data Communications Network (G.7712/Y.1703)

In [26] the various functions constituting a telecommunications network can be classified into two broad functional groups. One is the transport functional group, which transfers any telecommunications information from one point to another point(s). The other is the control functional group, which realizes various ancillary services and operations and maintenance functions. The DCN provides transport for the applications associated with the control functional group. Examples of applications transported by the DCN are transport network operations/management applications, DCN operations/management applications, ASTN control plane applications, and voice communications. The IP-based DCN provides layer 1 (physical), layer 2 (data link), and layer 3 (network) functionality and consists of routing/switching functionality interconnected via links. These links can be implemented over various interfaces, including WAN interfaces, LAN interfaces, and ECCs. This recommendation provides the architecture requirements for an IP-based DCN, the requirements for interworking between an IP-based DCN and an OSI-based DCN, and the IP-based DCN interface specifications.

GMPLS Solution   Because in GMPLS the signaling and management plane are independent from each other, different kinds of networks can be used for both tasks. Today GMPLS itself can be managed by the use of GMPLS MIB (draft-nadeau-mpls-gmpls-te-mib-00.txt and references). In the view of GMPLS, each node is capable of processing signaling and routing messages whereby the topology of the transport network and the control plane network are the same.

Requirements

- The management and signaling functions should be decoupled from each other.
- The DCN should support in-fiber-inband, in-fiber-out-of-band, and out-of-fiber/out-of-band signaling for any kind of technology.
- DCN should be dimensioned to support fast restoration by providing fast transport of restoration messages.
- DCN should be IP based and support IP addressing.
- IP routing mechanisms (OSPF, IS-IS) should be used.

4.2.9.3   Distributed Connection Management (G.7713/Y.1704)

Recommendation G.7713/Y.1704 [27] covers the areas associated with the signaling aspects of automatic switched transport network, such as attribute specifications, the message sets, the interface requirements, the DCM state diagrams, and the interworking functions for the distributed connection management.

GMPLS Solution   In GMPLS a permanent communication between the network devices is established that is necessary to exchange reachability and

TE information (e.g., routing protocol provides reachability and TE attributes information). Link bundling plays a key role by augmenting the scalability of the routing protocol. A user device or a management station can optionally trigger a connection setup and initiates a control plane action:

1. In a first phase, the edge device (where the trail termination is located) has to determine the route (trail) either by a route calculation (e.g., explicit route computation through C-SPF) or by receiving a precalculated route from an external device, such as a traffic engineering tool.

2. Then it signals the trail request to the involved nodes across the network, reserving the bandwidth without allocating it.

3. When bandwidth reservation has been performed the trail is implemented.

Some optimizations have also been added to speed up the process of implementing the connection. The full procedure is explained in more detail in draft–ietf–mpls–generalized–signaling–05.txt.

Requirements   GMPLS control plane components could be applied to ASTN to achieve a distributed connection control taking into account draft–ietf–mpls–generalized–signaling–05.txt.

4.2.9.4   Generalized Automatic Discovery (G.7714/Y.1705)

Recommendation G.7714/Y.1705 [28] describes the specifications for automatic discovery (referred to as *autodiscovery*) to aid DCM and routing in the context of automatically switched transport networks (ASTN/ ASON). In this recommendation, three major instances of discovery are addressed: (1) adjacency discovery, (2) neighbor discovery, and (3) service discovery. In addition, the results of neighbor discovery are also used for establishing logical adjacencies between nodes at the control plane.

Adjacency Discovery
Adjacency discovery is described as the process of verifying physical connectivity between two ports on adjacent network elements over a specific physical layer. Depending on the physical packaging of the functions within a network element, two types of associations need to be discovered as part of adjacency discovery.

GMPLS Solution   The *Optical Link Interface* (OLI) concept and requirement are proposed in conjunction with the LMP-WDM protocol to cover the functions provided by adjacency discovery. From the GMPLS perspective, information exchange occurs between a "passive" and an "active" element, such as between a DWDM (OLS) system and an OXC. Ongoing work with ITU referred to as G.vbi (*virtual backplane interface*) will complete the picture.

Requirements

- Adjacency discovery should be provided through a simple protocol mechanism for reporting the health and properties of OLSs based on a well-defined set of parameters.

- It should be extensible so that we can start with a set of the most needed parameters initially and be able to extend later by adding new parameter types and new parameters within a type.

- The initial focus is on SONET and SDH equipment. However, the OLI must be extensible to support other types of equipment such as Ethernet and G.709 OTN.

- The adjacency must be reliable; we cannot assume a one-to-one relationship between OLS and client. That is, an OLS client will most likely be attached to multiple different OLSs, and a single OLS may have multiple different clients at a single location.

Neighbor Discovery

Recommendation G.7714/Y.1705 [28] provides the requirements and message sets for the automatic neighbor for the UNI, I-NNI, E-NNI, and *physical interface* (PI). The requirements in this recommendation specify the discovery process across these interfaces that aid automated connection management.

GMPLS Solution   MPLS is based on an IP-based control plane incorporating protocols defined for routing and neighbor discovery defined in OSPF and IS-IS. To achieve a single control plane across multiple technology layers, a single method for neighbor discovery and routing is mandatory. LMP extensions for neighbor discovery have solved the "potential" problem of the usage of a routing protocol at the UNI (when considering for instance OIF UNI 1.0 specification).

Requirements

- Neighbor discovery should be used to detect and maintain node adjacencies. For this, the mechanisms already defined at the IETF for OSPF and IS-IS should be used.

- Topology information and resource information should be decoupled. While topology information remains unchanged, resource utilization can change dynamically when setting up a new path. To support this concept, the links between two adjacent nodes should be bundled. In the case of any single link failure within the bundle, the topology information remains stable, although the capacity information may change.

- The control plane should detect changes in the resources and enable a timely reaction if established path or network resources are affected by this change.

Service Discovery

Recommendation G.7714/Y.1705 [28] provides the requirements and message sets for the automatic service discovery for the UNI, I-NNI, E-NNI, and PI. The requirements in this recommendation specify the discovery process across these interfaces that aids automated connection management.

GMPLS Solution   GMPLS focuses on intradomain implementation on which OIF based its UNI specification. An OIF–UNI GMPLS profile can be considered when discussing UNI implementations. Extensions of LMP enables the exchange of service discovery information at the UNI 1.0 specification.

Requirements

- Service discovery mechanisms should be aligned with the mechanisms provided by GMPLS such that a seamless integration of UNI and NNI can be supported.

4.2.9.5   OTN Routing (G.rtg)

A first outline of G.rtg (G.7715) has been presented to the ITU Q14/SG15 meeting. The recommendation is expected to be completed by mid-2004.

GMPLS Solution   GMPLS is supposed to be based on OSPF/IS-IS routing mechanisms and more explicitly on the traffic engineering extensions of these protocols, such as CSPF. See also draft-kompella-ospf-gmpls-extensions-01.txt, *OSPF Extensions in Support of Generalized MPLS*, and draft–ietf-isis-gmpls-extensions-02.txt, *IS-IS Extensions in Support of Generalized MPLS*. Today ongoing work related to GMPLS/BGP has started as well in order to cover interdomain routing specification for nonpacket-based networks (such as draft-parent-optical-bgp-00.txt). GMPLS also allows the use of explicit or implicit routing.

Requirements

- Support of explicit and implicit routing;
- Support of OSPF and ISIS routing protocols for intradomain and subsequently BGP for interdomain routing;

- Support of constrained based routing in order to conform, for example, to constraints such as physical diversity and to achieve traffic engineering objectives in the transport network. Examples are to adhere to operator policies on routing such as preferred routes or to conform to network specific constraints.

4.2.9.6   OTN Connection Admission Control (G.cac, now G.7717)

Connection admission control (CAC) is necessary for authentication of the user and controlling access to network resources. CAC should be provided as part of the control plane functionality. It is the role of the CAC function to determine if sufficient free resources are available to allow a new connection. If that is the case, the CAC may permit the connection request to proceed; alternatively, if enough resources are not available, it should notify the originator of the connection request that the request has been denied. Connections may be denied on the basis of available free capacity or alternatively on the basis of prioritization. CAC policies are outside the scope of standardization.

GMPLS Solution   CAC in the sense of authentication and access control is not explicitly addressed in GMPLS because a trusted relationship in a single-operator, multiple-vendor network is assumed. The work related to connection admission is performed, for example, in OIF. Related issues such as security of signaling protocols is already included in RSVP-TE and CR-LDP. However, if a given LSP cannot be established through the network (for reasons as diverse as resource unavailability, overbooking, control plane congestion, and so on), it is simply rejected.

Requirements   None.

4.2.9.7   OTN Link Management (G.lm)

Work was under way at press time.

GMPLS Solution   The LMP defined in draft-ietf-ccamp-lmp-00.txt is used to manage the resources available between two nodes and to check the connections. It is closely related to the unnumbered interface and bundling concepts described in draft-kompella-mpls-bundle-05.txt, *Link Bundling in MPLS Traffic Engineering*.

Requirements

- Link management should form a consistent network level resource view between adjacent nodes.
- The use of link management should decouple resource information from topology information that is bound to the bundling concept.
- LMP as defined in IETF (draft-ietf-ccamp-lmp-00.txt) should be considered for G.lm.

4.3 Deployment Status

In early 2003 work on the ASTN/ASON, GMPLS, and O–UNI standards was progressing at a steady pace with some vendors already claiming to have field deployed prestandard implementations. The challenges to full-scale adoption of these models lie in several areas: (1) finishing the remaining standards development work, (2) persuading carriers to adopt IP technology in the control plane, and (3) deriving the economic benefits carriers will accrue from deploying optical signaling technology [5].

With reduced capital expenditures in the early part of the decade, service providers have shown a degree of reluctance to invest in new technologies unless careful economic and rate-of-return analyses are conducted regarding what needs to be an increased revenue opportunity. There also has to be an assurance that the embedded base of hundreds of billions of dollars can continue to be utilized in a cohesive manner with the proposed new technology. Proponents make the pitch that the ability to more rapidly provision bandwidth enhances revenue by enabling service providers to support new value-added services and realize substantial cost savings. Provisioning capabilities, however, have to be end to end to be of real value. Once the capabilities are available end to end, or at least edge to edge, the dynamic bandwidth allocation that can be accomplished with optical signaling (particularly in the core of the network) can be leveraged to maximize the reuse of existing transmission facilities, to reduce the need to overprovision, and to optimize service reliability. It remains to be seen what the actual commercial deployment of ASTN/ASON will be in the near-term (2003–2006).

ENDNOTES

[1]   To provide a context, note that there are about 16,000 publicly traded companies in the United States; this is not to imply that only publicly traded companies need gigabit networks, but these companies tend to be the largest companies around.

[2]   Minoli, D., P. Johnson, and E. Minoli, *Ethernet-Based Metropolitan Area Networks*, New York: McGraw-Hill, 2002.

[3]   The author has more than 18 years experience in the area of broadband services for data/IP applications.

[4]   If our assumption about "intelligent buildings" is not correct and sources and sinks are spread out over a larger base, then the number of customers per building is even lower.

[5]   Shahane, D., "Building Optical Control Planes: Challenges and Solutions," *Communication Systems Design,* January 7, 2002.

[6]   "Transitioning the Optical Network," 2002 White Paper Materials, retrieved from http://www.tellium.com/optical/opticaloverview.html.

[7]   "Optical Networks and IP over DWDM," retrieved from http://www.cis.ohio-state.edu/~jain/refs/opt_refs.htm.

[8]   Retrieved from http://www-classes.usc.edu/engr/ee-s/650/Lecture%20note/lecture10.pdf.

[9]   Promotional material from Atos Origin, Systems Integration Telecom Technologies Business Unit, Paris, France, retrieved from http://www.marben-products.com.

[10]   The hybrid model, which was just beginning to be discussed at press time, provides a mechanism for limited information sharing, typically using *border gateway protocol version 4* (BGP-4) to transfer the reachability information between optical networks. The approach maintains the separation of the routing protocol engine between the IP and optical domains, but transfers the information from one routing protocol engine to the other (in the ASTN/ASON model, this is similar to the capabilities of E-NNI).

[11]   Greenfield, D., "Optical Standards: A Blueprint for the Future," *Network Magazine,* October 5, 2001.

[12]   This section is based on its entirety on O. Aboul-Magd et al., *Automatic Switched Optical Network (ASON) Architecture and Its Related Protocols,* IPO WG Internet draft, draft-ietf-ipo-ason-02.txt, March 2002. Copyright © The Internet Society. All Rights Reserved. This document and translations of it may be copied and furnished to others, and derivative works that comment on or otherwise explain it or assist in its implementation may be prepared, copied, published and distributed, in whole or in part, without restriction of any kind, provided that the above copyright notice and this paragraph are included on all such copies and derivative works.

[13]   Mayer, M. (Ed.), *Requirements for Automatic Switched Transport Networks (ASTN),* ITU G.8070/Y.1301, V1.0, International Telecommunications Union, May 2001.

[14]   Mayer, M. (Ed.), *Architecture for Automatic Switched Optical Networks (ASON),* ITU G.8080/Y1304, V1.0, International Telecommunications Union, October 2001.

[15]   Ashwood-Smith, P., et al., *Generalized MPLS—Signaling Functional Description,* draft-ietf-mpls-generalized-signaling-04.txt, work in progress, May 2001.

[16]   Lazar, M., et al., "Alternate Addressing Proposal," OIF2001.21, OIF Contribution, January 2001.

[17]   Aboul-Magd, O., et al., *LDP Extensions for Optical User Network Interface (O-UNI) Signaling,* draft-ietf-mpls-ldp-uni-optical-01.txt, work in progress, July 2001.

[18]   Yu, J., et al., *RSVP Extensions in Support of OIF Optical UNI Signaling,* draft-yu-mpls-rsvp-oif-uni-00.txt, work in progress, December 2000.

[19]   Rajagopalan, B. (Ed.), "User Network Interface (UNI) 1.0 Signaling Specifications," OIF2000.125.7, OIF Contribution, October 2001.

[20]   Ashwood-Smith, P., et al., *Generalized MPLS Signaling: CR-LDP Extensions,* draft-ietf-mpls-generalized-cr-ldp-03.txt, work in progress, May 2001.

[21]   Ashwood-Smith, P., et al., *Generalized MPLS Signaling: RSVP-TE Extensions,* draft-ietf-mpls-generalized-rsvp-te-03.txt, work in progress, May 2000.

[22]   Kompella, K., et al., *IS-IS Extensions in Support of Generalized MPLS,* draft-ietf-isis-gmpls-extensions-01.txt, work in progress, November 2000.

[23]   Kompella, K., et al., *OSPF Extensions in Support of Generalized MPLS,* draft-kompella-ospf-gmpls-extensions-01.txt, work in progress, November 2000.

[24]   Mannie, E. (Ed.), *Generalized Multi-Protocol Label Switching (GMPLS) Architecture,* draft-ietf-ccamp-gmpls-architecture-00.txt, work in progress, June 2001.

[25]   Draft New Recommendation G.cemr, *Common Equipment Management Function Requirements,* International Telecommunications Union, June 2001

[26] Daloia, C. (Ed.), *Architecture and Specification of Data Communication Network,* ITU Recommendation G.7712/Y.1703, October 2001.

[27] Lin, Z. (Ed.), *Distributed Call and Connection Management,* ITU Recommendation G.7713/Y.1704, October 2001.

[28] Sankaranarayanan, S. (Ed.), *Generalized Automatic Discovery Techniques,* ITU Recommendation G.7714/Y.1705, October 2001.

CHAPTER 5

Dense and Coarse WDM

The previous two chapters provided a view into the optical developments and deployments that are expected to be seen in the 2003–2007 time frame. This chapter looks at the constituent technology that is available at the present time, wavelength–division multiplexing. WDM can be utilized in an off-the-shelf manner to design long-haul networks, global networks, and metropolitan networks (when the technology is shown to be economically viable). WDM technology will continue to be important for the foreseeable future. Hence, our coverage of the topic herewith. Multiplexing is fundamental to communications: The concept goes back to the 1880s with the multiplexing of several telegraphic channels by Emile Baudot onto a single copper-based transmission medium.

WDM systems and high-capacity WDM systems, called *dense WDM* (DWDM), have seen major deployment in the United States (and in the rest of the world) since the late 1990s. The technology was developed, to a large degree, in the early and mid-1990s [1]. WDM systems provide high-capacity transmission of multiple tributary signals over a single optical fiber (pair). By using FDM techniques on a fiber transmission system, one can achieve *fiber-optic pair gain,* in which multiple signals ("lambdas") are "upconverted" ("transponded") to operate at different frequencies and missed (combined). Each of the constituent signals can carry OC-48 or OC-192 channels between endpoints [2]. As implied in the previous two chapters, WDM is viewed as operating at the optical layer; SONET/SDH is viewed as an overlay onto the WDM layer. DWDM is the workhorse of today's optical backbone networks. The goal is to eventually migrate DWDM systems into the OTN and the eventually into ASON/ASTN machinery discussed earlier in the book. The development of DWDM is the first step toward an all-optical network: Using DWDM techniques the carrier can reroute signals in the optical domain without necessarily converting these signals to electrical signals, and can add or drop signals by inserting or withdrawing wavelengths as needed.

The proliferation of bandwidth-intensive applications, such as the delivery of graphics-rich hypertext over the Internet and corporate intranet, has driven demand for broadband connectivity and transport in the recent past. WDM technologies significantly increase the traffic-carrying capacity of existing fiber spans by combining multiple optical signals of different wavelengths on a single fiber pair [3]. DWDM coupler devices perform the actual multiplexing/demultiplexing of the different optical wavelengths. From 2 to 1,000 channels have been demonstrated, with typical commercial systems in the field currently carrying 16 or 32 channels each operating at an OC-48 (2.5-Gbps) or OC-192 (10-Gbps) data rate; commercial-grade systems for large applications are now also available in the 64- and 128-channel configuration, but these are not as widely deployed as of yet. WDM allows the use of existing electronics and existing fibers, but shares fibers by transmitting different channels at different wavelengths ("colors") of light. Optical transport systems that already use fiber-optic amplifiers as repeaters also do not require upgrading for most WDM systems.

The observation pertinent to turn-of-the-decade DWDM technology is that it has a vendor-specific flavor, even though the ITU has defined a set of standard wavelengths for manufacturers to use in order to facilitate basic interworking between equipment. In particular, the provisioning and management plane are invariably vendor specific. The OTN/ASON/ASTN standards described in Chapters 3 and 4 are designed to address this issue and at the same time advance the features and capability space, as we discussed in these earlier chapters.

This chapter looks at key technologies that support DWDM as well as DWDM applications. Until now, this technology has been optimized for, and has seen major deployment in, long-haul networks, based on *traditional cost-of-transmission-versus-cost-of-multiplexing economic trade-off considerations*, although vendors have engaged in a very hard sell to convince carriers that there is a major need for this technology in the metro access/metro core space. While such metro applications certainly do exist, the total market is significantly bounded. *Coarse WDM* (CWDM) techniques may be more applicable to the metro (access) space from an economic and technological point of view. CWDM is characterized by wider channel spacing than DWDM and typically does not require cooled optics.

In long-haul applications, in addition to reducing the cost of a link (and, hence, the cost of bandwidth), DWDM systems simplify the deployment of network capacity. Deployment of new network capacity can be achieved by installing additional or higher bit-rate interfaces in the DWDM systems at either end of the fiber link (that is, increase the number of lambdas). In this upgrade path, the existing optical amplifiers amplify the new channel without additional regenerators, although, in specific instances, new regenerators could be required.

An analytical approach to the question of the use of WDM in the *metro segment of the network* will demonstrate that the application for this technology in this space has a rather limited market potential, at least in the near term. As noted earlier in the book, about 20,000 buildings in the United States each will have about three broadband (Gbps-like) customers in the near term. Averaging over the top 50 cities in the United States, this results in an average of 400 buildings, or 1,200 customers per city. If each customer needed 1 Gbps, this is an aggregate throughput of about 1.2 Tbps, or about three top-of-the-line DWDM systems per city per carrier. Assuming that there is a natural market dynamic for the support of three comparable competitors per city, this equates to an average of nine top-of-the-line DWDM systems per city. However, bandwidth is not the only consideration in metro environments, and the number of drops and the geography have to be taken into account. If we assume that a ring can support 16 to 32 drops, these kinds of average statistics would need 15 to 30 access rings and a couple of metro core rings/meshes per carrier per city.

With these general positioning observations behind us, we can proceed with an assessment of the technology.

5.1 Introduction to the Technology

Three approaches are used for expanding network capacity: (1) installing more cables, (2) increasing the system bit rate to multiplex more TDM tributary signals on existing systems, or (3) WDM [4].

Installing more cables is obviously fairly expensive. In metropolitan areas this approach may be required in order to bring more buildings onto the fiber network. Fortunately, fiber cable has become relatively inexpensive and installation methods have become somewhat more efficient (e.g., mass fusion splicing), although overall labor costs—possibly including unionized labor in various disciplines—remain significant. Hence, this appears to be the required course of action in the *metro access* portion of the network. Regarding the *metro core*, conduit space may not be available or major construction might be necessary; therefore, the approach of adding new cable may not be the most cost-effective approach. In the long-haul system, new construction can be equally expensive.

With regard to the second method, increasing the system data rate may not prove optimal from a financial point of view. Many systems are already running at SONET OC-48 rates (2.5 Gbps) and upgrading to OC-192 (10 Gbps) is relatively expensive, requiring the changing out of many or all of the electronics in a network. The circuitry that makes possible speeds of 10 and 40 Gbps is still complex and costly; OAM&P support is also relatively demanding. At the engineering level, significant technical issues remain that may restrict the deployment of this technology in specific instances. For example, as discussed in Chapter 2, signals transmitted at OC-192 speeds are 16 times more affected by chromatic dispersion than signals operating at OC-48 speed. Polarization-mode dispersion, which limits the distance a light pulse can travel without degradation, is also an issue at the higher speeds. Finally, the increased transmission power required by the higher bit rates also introduces nonlinear effects that can affect signal quality. The OC-48-to-OC-192 and OC-192-to-OC-768 migrations add four times the capacity. In some cases, this increase in bandwidth is more than would be necessary, while in other cases is not enough.

The third alternative, WDM, has proven more cost effective in many instances. Figures 5.1, 5.2, 5.3, and 5.4 highlight the key concepts associated with DWDM. Figure 5.1 depicts the basic concept of multiplexing. At the transmitting end, as noted, a DWDM multiplexer combines a multiplicity of wavelengths into a composite signal at a specified and/or standardized frequency interval; the combined signal is transmitted over a fiber–optic network on a single fiber (pair). At the receiving end, a DWDM demultiplexer separates the composite signal into their respective frequency (wavelength) components, which are subsequently directed to different ports for output. According to the principle that the optical path is reversible, a demultiplexer can be used as a multiplexer if the output ports of the demux device are used as the input ports and the input port is used as the output port, and vice versa [5]. Figure 5.2 shows an early implementation of the general concept of Figure 5.1. Figure 5.3 depicts pictorially the packing of multiple SONET/SDH links onto a single transport system, and Figure 5.4 shows a practical implementation. DWDM mux/demux technology should typically support the following features: athermal packaging, high

FIGURE 5.1
Basic multiplexing concept.
(Source: Optronix Limited.)

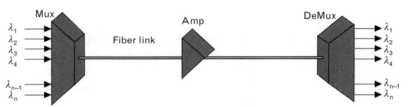

Lasers with precise, stable wavelengths

Optical fiber that exhibits low loss and transmission
performance in the relevant wavelength spectre

Flat-gain optical amplifiers to boost the signal on longer spans

Photodetectors and optical demultiplexers using
thin film filters or diffractive elements

Optical add/drop multiplexers and optical cross-connect components

FIGURE 5.2
Basic WDM concept.
(Source: Optronix Limited.)

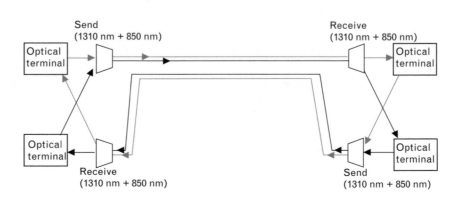

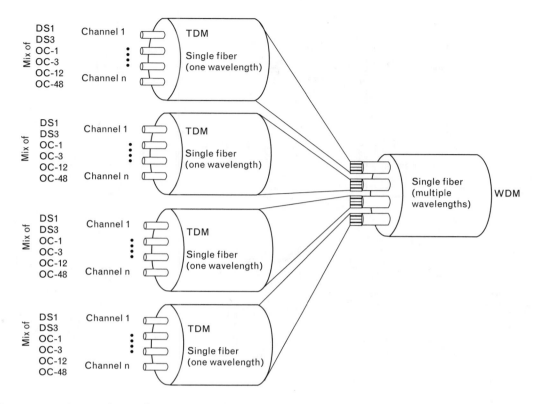

FIGURE 5.3 *A pictorial view of DWDM. (Source: Optronix Limited.)*

channel counts, wide passband, ultralow insertion loss, low channel cross-talk, polarization insensitivity, and high channel uniformity.

A DWDM system supports the following functions:

- *Interfacing.* Client-side physical and logical interfaces must be supported to receive the input signal (this function is performed by transponders) and to distribute the output signals at the remote end.

- *Generating the signal.* The source laser must provide a stable signal within a specific and narrow bandwidth that carries the digital data with IM.

- *Combining the signals.* Suitable multiplexers are needed to combine the optical signals.

- *Transmitting the signals.* The combined signal must be relayed reliably to the remote end. The signal may need to be optically amplified along the way. The effects of signal degradation, loss, and cross-talk must be taken into account in the design. These effects can be minimized by controlling variables such as channel spacings, wavelength tolerance, and laser power levels.

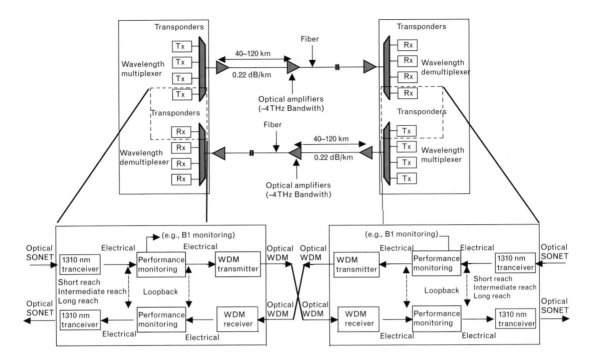

FIGURE 5.4 *Basic WDM implementation. (Source: Optronix Limited.)*

- *Separating the received signals.* At the receiving end, the combined signal must be separated into individual optical beams. This function is
generally technically difficult.

- *Receiving the signals.* The demultiplexed signal is received by an appropriate photodetector.

- *Supporting OAM&P functions.* This refers to the ability to administer
the device for end-to-end services.

Figure 5.5 depicts the injection of multiple optical channels in a fiber
and some of the key performance metrics that are of interest. Figure 5.6

FIGURE 5.5
*Injection of multiple
channels in a fiber and
key performance
metrics.
(Source: Optronix
Limited.)*

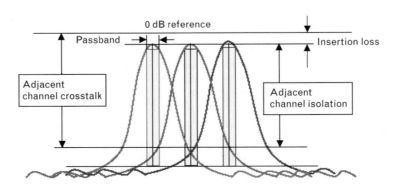

shows an example of a DWDM mux/demux element, to provide the reader with a visual picture of the basic components.

To support basic interoperability, the ITU has specified a family of specifications, notably, the ITU grid (see Table 5.1); other standards-related entities have also published specifications. ITU Recommendation G.692 defines a laser grid for point-to-point WDM systems based on 100-GHz wavelength spacings with a center wavelength of 1,553.52 nm (reference frequency of 193.1 THz). As noted, typical press-time systems can generally handle up to 40 DWDM channels per 100-GHz spacing application and 80 DWDM channels for 50-GHz spacing applications in a single C- or L-band (as defined in Chapter 2). The ITU grid defines a minimum channel spacing of 100 GHz in the frequency, domain corresponding to

TABLE 5.1   ITU GRID

FREQUENCY (THz)	WAVELENGTH (NM)	FREQUENCY (THz)	WAVELENGTH (NM)	FREQUENCY (THz)	WAVELENGTH (NM)
196.1	1,528.77	164.6	1,540.56	193.1	1,552.52
196.0	1,529.55	194.5	1,541.35	193.0	1,553.33
195.9	1,530.33	194.4	1,542.14	192.9	1,554.13
195.8	1531.12	194.3	1542.94	195.8	1,554.94
195.7	1,531.9	194.2	1,543.73	192.7	1,555.75
195.6	1,532.68	194.1	1,544.53	192.6	1,556.56
195.5	1,533.47	194.0	1,545.32	195.5	1,557.36
195.4	1,534.25	193.9	1546.12	192.4	1,558.17
195.3	1,535.04	193.8	1,546.92	192.3	1,558.98
195.2	1,,535.82	193.7	1,547.72	192.2	1,559.79
195.1	1,536.61	193.6	1,548.51	192.1	1,560.61
195.0	1,537.40	193.5	1,549.32	192.0	1,561.42
194.9	1,538.19	193.4	1,550.12	191.9	1,562.23
194.8	1,538.98	193.3	1,550.92	191.8	1,563.05
194.7	1,539.77	193.2	1,551.72	191.7	1,563.86

approximately 0.8 nm in the wavelength domain [1]. A channel plan describes a number of factors that need to be taken into account in an optical communications network including number of channels, channel spacings and width, and channel center wavelengths. Although this grid defines the standard, vendors are free to use wavelengths in arbitrary ways and to choose from any part of the spectrum. Vendors can also deviate from the grid by extending the upper or lower bounds or by spacing the wavelengths more closely, typically at 50 GHz, to double the number over channels. However, interworking will suffer if vendors utilize variants of the basic standard.

Acceptance and deployment of DWDM technology will drive the expansion of the optical layer and allow service operators to exploit the bandwidth capacity that is inherent in optical fiber. However, these DWDM systems are pre-OTN/ASTN/ASON, therefore, they are not necessarily interoperable with evolving technology. Some operators are taking a wait-and-see attitude regarding major new deployment until the existing bandwidth inventory is allocated to paying customers and/or a new generation of (OTN/ASTN/ASON) technology becomes available.

Vendors of DWDM systems and technology include Nortel, Cisco, Alcatel, Ciena, and Lucent, to list just a few. Many U.S. carriers have settled on DWDM at STM-16/OC-48 rates as their technology of choice for gaining more capacity at the long-haul level, but OC-192 systems and OC-768 systems are also emerging; 16-channel DWDM has been deployed throughout the carrier infrastructure, and 32-channel systems (and higher) are also entering the field. The widespread introduction of this technology, however, has at the same time led to a long-haul bandwidth glut and to price disruptions; in turn, this has set expectations for price points at the metro access/metro core that may or may not be achievable.

As telecommunications and data services have become more critical to business operations, and customers have concentrated more company traffic over a small set of telecommunication facilities, carriers have had to ensure that their networks are fault tolerant. To meet these service-continuity requirements, providers have had to deploy backup routes. Often carriers employ simple 1:1 redundancy in ring or point-to-point configurations (see Figure 5.7). This design approach can double the need for bandwidth on the network infrastructure. However, with bandwidth being relatively easily available (thanks to DWDM technology), carriers are perhaps best served by utilizing their own resources to meet user needs, rather than by purchasing equipment from external sources to conserve a resource such as bandwidth for which there may not, in fact, be immediate demand. Some proponents have advanced technologies such as RPR; however, the idea of conserving metropolitan bandwidth was already explored with rather similar technologies in the late 1980s (when in fact bandwidth was at an even higher premium) with limited commercial success (e.g., IEEE 802.6 MAN and Bellcore's SMDS).

FIGURE 5.7
*Backup facilities add
to the bandwidth
requirement.
(Source: Optronix
Limited.)*

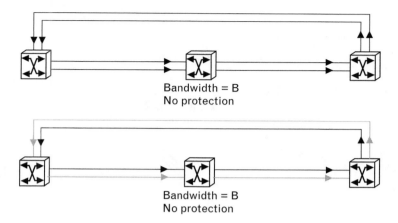

Bandwidth = B
No protection

Bandwidth = B
No protection

5.1.1 How Does WDM Work?

It is relatively easy to understand WDM. Consider the fact that one can see many different colors of light—red, green, yellow, blue, and so on—all at once. The colors are transmitted through the air together and may mix, but the colors can be easily separated using a simple device like a prism, just like we separate the "white" light from the sun into a spectrum of colors with the prism (Figure 5.8). This technique was first demonstrated with optical fiber in the early 1980s when telecom fiber-optic links still used multimode fiber. Light at 850 and 1,300 nm was injected into the fiber at one end using a simple fused coupler. At the far end of the fiber, another coupler split the light into two fibers, one sent to a silicon detector more sensitive to 850 nm and one to a germanium or InGaAs detector more sensitive to 1,300 nm. Filters removed the unwanted wavelengths, so each detector then was able to receive only the signal intended for it (see Figure 5.9 [6]).

By the late 1980s, all telecom links were single-mode fiber, and coupler manufactures learned how to make fused couplers that could separate 1,300- and 1,550-nm signals adequately to allow WDM with simple, inexpensive components. These early WDM systems use the two widely spaced wavelengths (in the 1,310- and 1,550-nm regions), and were therefore also called *wideband WDM*. However, these systems had limited usefulness, since fiber was designed differently for 1,300 and 1,550 nm, due to the dispersion characteristics of glass. Fiber optimized at 1,300 nm was

FIGURE 5.8
*Separating a beam of
light into its colors.
(Source: Optronix
Limited.)*

Prism

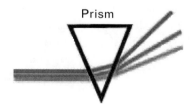

FIGURE 5.9
*WDM with couplers
and filters. (Source:
Optronix Limited
[6].)*

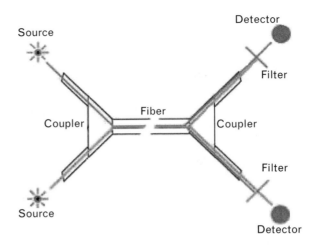

FIGURE 5.9
*WDM with couplers
and filters. (Source:
Optronix Limited
[6].)*

used for local loop links, while long-haul and submarine cables used dispersion-shifted fiber optimized for performance at 1,550 nm [6].

With the advent of fiber-optic amplifiers used for/in repeaters in the late 1980s, emphasis shifted to the 1,550-nm transmission band. WDM only made sense if the multiplexed wavelengths were in the region of the fiber amplifiers' operating range of 1,520 to 1,560 nm. It was not long before WDM equipment was able to put four signals into this band, with wavelengths about 10 nm apart. Hence, the early 1990s saw a second generation of WDM, sometimes called *narrowband WDM,* in which two to eight channels were used. These channels were now spaced at an interval of about 400 GHz in the 1,550-nm window. By the mid-1990s, DWDM systems were emerging with 16 to 40 channels and spacing from 100 to 200 GHz. By the late 1990s, DWDM systems had evolved to the point where they were capable of 64 to 160 parallel channels, densely packed at 50- or even 25-GHz intervals [7]. Systems with higher density are on the horizon. The specific limits of this technology are not yet known and have probably not been reached, though systems were available as of mid-year 2000 with a capacity of up to 160 lambdas on one fiber. Figure 5.10 provides a summary of the generations of products in this space.

Generally, the input end of a WDM system is relatively simple. It is a simple coupler that combines all of the inputs into one output fiber (see Figure 5.11). These couplers (multiplexers) have been available for many years, offering 2, 4, 8, 16, 32, or even 64 inputs. It is the demultiplexer that is the difficult component to manufacture. The demultiplexer takes the input fiber and collimates the light into a narrow, parallel beam of light. It shines on a grating (a mirror-like device that works like a prism) that separates the light into the different wavelengths by sending them off at different angles. An optical element captures each wavelength and focuses it into a fiber, creating separate outputs for each separate wavelength of light.

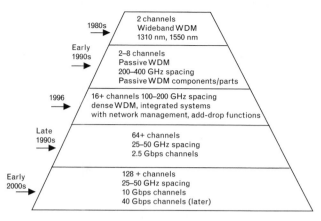

FIGURE 5.10   *WDM generations.*

System-level evolutions over time (1977–2007)

Phase 1 (1977–Present): Asynchronous TDM (optical DS3s etc.)
Phase 2 (1989–Present): Synchronous TDM (SONET/SDH)
Phase 3 (1990–Present): Wavelength division multiplexing
Phase 4 (1994–Present): Dense wavelength division multiplexing
Phase 5 (2001–Present): Intelligent Optical Networking (ION)/Optical Time Division Multiplexing (OTDM)

Evolution of all-optical systems over time (early 1990s–mid 2000s)

	Early 1990s	Mid 1990s	Late 1990	Early 2000s	Mid 2000
Mesh networks			100 GHz WDM 50 GHz WDM Static OADMs	Dynamic OADM	Dynamic optical switches Wavelength changers
Ring networks		200 GHz WDM	100 GHz WDM 50 GHz WDM Static OADMs	Dynamic OADM	Dynamic optical switches Wavelength changers
Linear networks	400 GHz WDM	200 GHz WDM	100 GHz WDM 50 GHz WDM Static OADMs	Dynamic OADM	Dynamic optical switches Wavelength changers

Evolution of optical technologies over time (mid 1990s–late 2000s)

Phase	Phase 1	Phase 2	Phase 3
Time frame	1996–2000	2001–2005	2006–2010
Major application	Voice	Voice and Data	Voice and data and optical applications
Major architecture	SONET	SONET and OTN	SONET and OTN/ASON/ASTN and optical packet
Characteristic	Static	Wavelength routing	Optical processing
Technologies	Thin film filters Fiber Bragg gratings Arrayed waveguides Optical amplifiers/EDFA Lasers Receivers Forward error correction (FEC)	Optical switching MEMS (2D, 3D) Bubble jet Tunable lasers Dynamic gain flattening Dynamic OADM Raman amplifiers Tunable dispersion compensation Some GMPLS deployment OC-768	Optical time division multiplexing Optical 3R Lambda conversion Optical FEC Optical switching 40 or 100 Gbps Ethernet Some FTTH

5.1.2   Enabling Technologies

This section looks at some key DWDM–enabling technologies in more detail.

5.1.2.1   Sources

As we saw in Chapter 2, two types of semiconductor lasers are widely used, monolithic Fabry-Perot lasers and *distributed feedback* (DFB) lasers. DFBs are particularly well suited for DWDM applications, since they emit a

FIGURE 5.11
WDM demultiplexer.

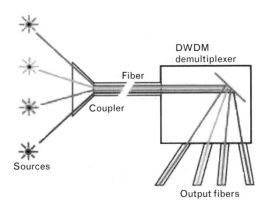

near-monochromatic light, are capable of high speeds, have a favorable SNR, and have good linearity. DFB lasers also have center frequencies in the region around 1,310 nm, and also from 1,520 to 1,565 nm. The latter wavelength band is compatible with EDFAs. Wide-spectrum tunable lasers are under development; these will be important in ASON/ASTN networks.

5.1.2.2   Transmission

A number of transmission-related technical issues have to be taken into account in the design of DWDM systems. Among others are chromatic dispersion, attenuation, PMD, and nonlinear effects. Linear-type factors, such as attenuation and dispersion, can be compensated; however, nonlinear effects are cumulative. PMD was discussed in Chapter 2 and is caused by ovality of the fiber-optic medium as a result of the manufacturing process or from external stressors; PMD becomes a problem at speeds at or above OC-192. Nonlinear effects manifest themselves when optical power is high, hence, they are important in long-haul DWDM systems. Pertinent nonlinear effects include the following:

- *Polarization mode dispersion.* PMD is caused by a combination of intrinsic effects (birefringence and mode coupling) and extrinsic effects (bends, twists) on cable. Studies show, though, that transmission up to 250 miles is possible at OC-192.

- *Stimulated Brillouin scattering.* SBS limits the power that can be injected into a single-mode fiber. As more and more OC-48/OC-192 systems are placed over a fiber, the output power of the EDFAs can cause SBS problems. Techniques have recently been developed to boost the SBS threshold.

- *Self-phase modulation.* SFM introduces chirping, which in turn interacts with fiber dispersion to cause pulse broadening or compression (depending on the dispersion profile of the fiber).

• *Four-wave mixing*. This limits multichannel transmission on dispersion-shifted fiber that has its zero dispersion in the EDFA bandwidth range. Monitoring of input power levels into the EDFA can address this problem when conventional fiber is used for transmission. New fiber designs (including NZ-DSF) have been introduced to address the issue.

A number of these effects were discussed in Chapter 2. Four-wave mixing is the most critical effect in DWDM. Four-wave mixing is caused by the nonlinear nature of the refractive index of the optical fiber; nonlinear interactions among different DWDM beams create sidebands that cause interchannel interference. Three frequencies interact to produce a fourth frequency, resulting in cross-talk and SNR degradation. The net effect of four-wave mixing is to limit the overall channel capacity. Four-wave mixing effects cannot be filtered out, either optically or electrically; the disruptive effect increases with the length of the fiber. A number of these issues also impact TDM-based data rates at 10 Gbps. The import of all of this is that due to its susceptibility to four-wave-mixing, DSF is unsuitable for WDM applications; NZ-DSF takes advantage of the fact that a small amount of chromatic dispersion can be used to mitigate four-wave mixing.

5.1.2.3 Amplification

Another technology that facilitates DWDM is the development of *optic amplifiers* (OAs), particularly EDFAs (see Figure 5.12), for use as repeaters. As we saw in Chapter 2, optical signals become attenuated as they travel through the fiber link and, therefore the signal must be periodically regenerated. Due to attenuation and other impairments, there are limits to how long a fiber segment can propagate a signal with integrity before it has to be fully regenerated. Prior to the availability of OAs, the system required a repeater for every signal transmitted. The OA has made it possible to amplify all of the wavelengths as an ensemble and without OEO conversion; this implies that full regeneration needs to be done less often.

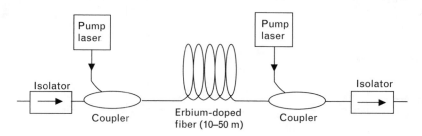

FIGURE 5.12 *WDM's EDFAs. When excited, erbium emits light around 1.54 micrometers. (This is the wavelength for optical fibers used in DWDM.) A weak signal enters the erbium-doped fiber, into which light at 980 or 1,480 nm is injected using a pump laser. This injected light stimulates the erbium atoms to release their stored energy as additional 1,550-nm emissions. As this process continues along the fiber, the signal grows stronger.*

Optical amplifiers can amplify numerous wavelengths of light simultaneously, as long as all are in the wavelength range of the fiber-optic amplifier. EDFAs are available for the C-band and the L-band. They work best in the range of 1,520 to 1,560 nm; hence, most DWDM systems are designed for that range. Now that fiber has been made with mitigated impact from the OH absorption bands at 1,400 and 1,600 nm, the possible range of DWDM has broadened considerably. Technology needs to be developed for wider range fiber amplifiers if we are to take advantage of the new fibers [6].

In SONET/SDH networks each separate fiber carrying a single optical signal (typically at 2.5 Gbps) required a separate electrical regenerator every 40 to 100 km. As additional fibers are "turned up" in this environment, the total cost of regenerators can become very significant. Costs include (1) the equipment costs of the regenerators themselves, (2) the facilities to house and power them, and (3) almost as critical, the OAM&P costs. Figure 5.13 depicts a pre-DWDM and a post-DWDM environment. DWDMs make use of optical amplifiers, as noted. As the figure shows, a single optical amplifier can reamplify all of the channels on a DWDM fiber without demultiplexing and processing them individually. Although optical amplifiers could be utilized in the SONET link case to extend the distance before having to regenerate the signal, the solution is impractical because we would still need an amplifier for each fiber. The bottom line is that because with DWDM multiple signals are transported on a single fiber pair, fewer pieces of equipment are required. Long-haul systems typically need large equipment cabinets for amplifiers and/or regenerators; this makes the systems expensive

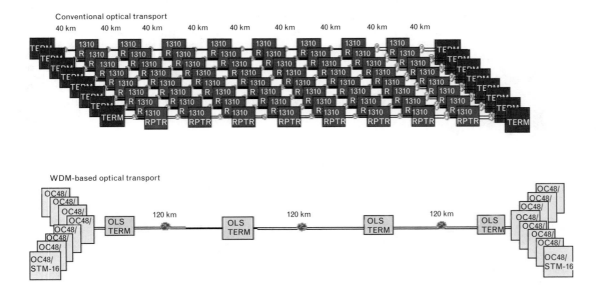

FIGURE 5.13   *Pre- and post-DWDM environments. (Source: Optronix Limited.)*

from an equipment, power, environment, space, maintenance, sparing, design, and troubleshooting point of view, which limits their applications in the metro access/metro core segment. Significantly reducing the expense of regenerators required for each fiber, particularly in long-haul and submarine applications, results in considerable savings. The OA amplifies the signals, but it does not reshape, retime, or retransmit them as a regenerator. This implies that the amplifier is less expensive than a regenerator. However, the optical signals still need to be regenerated periodically over long-distance links. (With newer technologies, signals can be transmitted hundreds to thousands of kilometers without requiring regeneration.)

Optical amplifiers maintain continuity by amplifying the multiple signals simultaneously as they travel down the fiber. The initial amplifier technology was EDFA. Now new extended-band amplifiers are being studied and developed. These amplifiers carry the signal to optical demultiplexers, where they are split into the original channel. Semiconductor optical amplifiers have been proposed for use in optical cross-connects for wavelength conversion in the optical domain and for amplification of a large number of ports. The DWDM wavelengths fall within two bands: a "blue" band between 1,527.5 and 1,542.5 nm and a "red" band between 1,547.5 and 1,561 nm. Each band is dedicated to a particular direction of transmission. EDFA allows for the following:

- Amplification in the optical domain (no OEO conversion required);
- Amplification of multiple signals simultaneously;
- Format independence (analog, digital, pulse, and so on);
- 20 to 30 dB gain;
- Bit-rate independence when designed for the highest bit rate;
- Current bandwidth: 30 nm in the 1,530- to 1,560-nm range (80 nm in the future possible);
- High bandwidth that can be extended with improved designs.

The key performance parameters of OAs are gain, gain flatness, noise level, and output power. EDFAs are typically capable of gains of 30 dB or more. Gain needs to be flat because all signals must be amplified uniformly. Because the signal gain provided with EDFAs is inherently wavelength dependent, gain flattening filters are often built into modern EDFAs. Low noise is a necessity because this effect is cumulative and cannot be filtered out. The SNR is the limiting factor in the number of amplifiers that can be concatenated and, therefore, the length of a fiber link before 3R functions (reshape, retime, retransmit) needs to be injected.

WDM amplifier technology has improved significantly in the past few years. It is now mature and is used in many networks. These amplifiers support transmission of multiple OC-48 (2.4-Gbps), OC-192 (10-Gbps), and evolving OC-768 [8] signals over conventional and newer NZ-DSF. Typical repeaters (amplifiers) enable 32 channels to be inserted and spaced 0.8 nm apart in field-deployable systems. One needs flat-gain performance in the bandwidth of the amplifier, as well as high power; high power is critical in applications where one needs to transmit over long distances. High-end DWDM systems are designed with extended reach optics support for the major fiber types in use today including NDSF, DSF, and NZ-DSF fibers (e.g., Lucent TrueWave™ Classic/Plus and Corning SMF-LS™/ LEAF™). (Fiber types were discussed in Chapter 2.) High capacity does not come for free: Long-haul applications require periodic regeneration to ensure sufficient signal strength and quality at the receiving end. This results in substantial repeater bays, which adds to the total transmission system cost. Although the cost of one regenerator may appear to be small, these costs can dominate when multiplied over the total length of a route, especially for heavy routes where several systems might be contained within the same fiber sheath.

5.1.2.4   Tributary Conversion

With *narrowband* or *DWDM*, the outputs of 2, 4, 8, 16, or more SONET terminals are optically multiplexed into one fiber pair. A transponder converts the tributary optical signal (e.g., SONET/SDH) back to an electrical signal and performs the 3R functions. This electrical signal is then used to drive the WDM laser. Each transponder within the system converts its client's signal to a slightly different wavelength. The wavelengths from the various transponders in the system are then optically multiplexed. In the receive direction of the DWDM system, the reverse process takes place. Individual wavelengths are filtered from the multiplexed fiber and delivered to individual transponders; these transponders convert the signal to electrical and drive a standard interface to the tributary system.

As noted earlier, the wavelengths used are all within the 1,550-nm range to allow for optical amplification using EDFA technology to maximize reach. The wavelength of the optical output from a SONET network element is roughly centered at 1,310 or 1,550 nm with an approximate tolerance of ±20 nm. Such wavelengths would have to be so widely spaced that after optical multiplexing only a few would be within the EDFA passband. Therefore, DWDM systems must translate the output wavelength from the SONET NE to a specific, stable, and narrow-width wavelength in the 1,550-nm range that can be multiplexed with other similarly constrained wavelengths. The device that does the translation function is the transponder.

5.1.3   Technical Details

This section [9] looks more closely at the underlying technology for DWDM systems. Table 5.2 identifies key concepts and terminology.

TABLE 5.2   KEY DWDM-RELATED PARAMETERS

0–dB Reference level	Defined at the straight-through power level when the devices to be tested are removed. Many parameters are defined below with respect to the 0–dB reference.
Adjacent channel cross-talk (dB)	The relative power level coming from an adjacent channel, referenced to the 0–dB power level. It is usually measured at the wavelengths of the passband boundaries. Because three pass bandwidths will be given, we prefer to measure the adjacent channel cross-talk at the center of a passband. Note that adjacent channel is interchangeably used, but it is referenced to the peak power of the given channel.
Athermal	The thermal stability of the devices. If the performance parameters, such as wavelength and insertion loss, are well below some defined critical values over the operating temperature, the device is said to be *athermal*. To claim a mux/demux device as athermal, its thermal wavelength stability should be better than 1.0 pm/°C and temperature-dependent insertion loss should be smaller than 0.015 dB/°C.
Center wavelength (nm)	The wavelength at which a particular signal channel is centered. The ITU has defined the standard optical frequency grid (channel center frequency) with 100-GHz spacing based on a reference frequency of 193.10 THz (1,552.52 nm), the so-called *ITU grid*. A channel center wavelength is chosen at the wavelength corresponding to the ITU grid, and should be distinguished from the actual center position of each passband of the device.
Center wavelength offset (pm)	A relative drift of the actual central wavelength of a particular channel with respect to the standard ITU grid. Note that the center wavelength offset is a cumulative value resulting from optical misalignment, aging, and temperature change over the operating temperature range.
Channel	A single signal channel consists of a frequency band that has a finite pass bandwidth and is centered at a given frequency such as one specified by the ITU grid. In DWDM, each channel corresponds to one particular wavelength and carries an individual data stream.
Channel pass bandwidth (nm)	A maximum wavelength (or frequency) range around the corresponding center wavelength (or frequency) at a given relative power level, for example, specified at a 0.5-dB down power level referenced to 0 dB.
Channel spacing (GHz)	The frequency interval between the center frequencies of any two neighboring channels in DWDM components or modules.
Channel uniformity (dB)	The maximum difference of insertion loss across all signal channels for all polarization states over the operating temperature range. Channel uniformity is a measure of how evenly power is distributed between the output ports of the devices.
Chromatic dispersion (ps/nm)	The worst case dispersion within any clear window for all polarization states of a device.
Clear window (nm)	Defined as a wavelength band around each ITU center wavelength. The worst case values are specified within the "clear window" of all channels and for all polarizations. This approach eliminates the need to calculate the effects of the filter line shape, center channel accuracy, and polarization. Typically, the clear window is defined as 25% of the channel spacing. For 100-GHz spacing, the clear window is defined as the ITU center wavelength ±12.5 GHz (±0.10 nm).
Cumulative cross-talk (dB)	The relative power level coming from all other channels including adjacent and nonadjacent channels, referenced to the 0–dB power level.
Directivity (dB)	Also called *near-end cross-talk*, this is the ratio of the optical power launched into an input port to the optical power returning to any other input port.
Flat-top passband (nm)	Specifies a class of DWDM mux/demux devices whose transmission spectrum profiles within the passband are relatively flat by comparison with the Gaussian profile.
Gaussian passband (nm)	Specifies a class of DWDM mux/demux devices whose transmission spectrum profiles within the passband are essentially Gaussian.

TABLE 5.2   KEY DWDM-RELATED PARAMETERS (CONTINUED)

Insertion loss (dB) of a device	The relative power level transmitted to the output end referenced to the 0-dB reference level when a device is inserted for all polarization states and over the operating temperature range. It usually measures the power difference between the lowest peak power among all channels and the 0pdB reference.
Insertion loss (dB) of a single channel	The relative power level transmitted to the output end referenced to the 0-dB reference level when a device is inserted for all polarization states and over the operating temperature range. It may be defined as (1) peak transmission within the passband or (2) the minimum transmission within any clear window.
Nonadjacent channel cross-talk (dB)	The relative power level coming from all other channels except the two adjacent channels, referenced to the 0-dB power level. Commonly, only the first two nonadjacent channels (left- and right-hand sides) are accounted for.
Operating temperature (°C)	The temperature range over which the device can be operated and maintain its specifications.
Polarization-dependent loss (PDL) (dB)	Maximum insertion loss variation among all polarization states over the operating temperature range. Two definitions of PDL are commonly adopted: (1) PDL at the ITU center wavelength and (2) worst case PDL across the entire passband.
Polarization mode dispersion (PMD) (ps)	The pulse spreading caused by a change of polarization properties.
Return loss (dB)	The relative power level reflected back to the input end in the backward direction referenced to the 0-dB reference level when a device is inserted.
Ripple (dB)	The power difference between the minimum and maximum insertion loss within the entire passband or clear window for all polarizations.
Storage temperature (°C)	The temperature range over which the device can be stored and cycled without damage, and can be operated properly according to its specifications over operating temperature.
Wavelength range (nm)	The spectral region over which the device is operated. All signal channels are confined within this wavelength range, for example, C-band or L-band.

Source: Based on materials from BaySpec, Inc.

5.1.3.1 Available Mechanisms

As alluded to in Chapter 2, a number of technologies can be utilized for wavelength separation in a demultiplexer. The basic physical techniques can be classified into three categories: (1) refraction, (2) interference, and (3) diffraction. Examples of technologies in each category include prism-based systems, *thin-film filters* (TFFs), AWGs, FBGs, and free-space diffraction gratings. Each technology has its own intrinsic trade-offs relating to technical complexity, implementation challenges, cost, performance, reliability, and production yields. Table 5.3 provides a comparison of these technologies. Figure 5.14 provides a pictorial view of DWDM multiplexing/demultiplexing technologies. Some inherent loss is associated with multiplexing and demultiplexing. This loss is dependent on the number of channels but can be addressed with optical amplifiers.

TABLE 5.3   PERFORMANCE COMPARISON OF
VARIOUS MUX/DEMUX DEVICES

	VPG	AWG	TFF	FBG
Athermal operation	Y	N	Y	Y
Denser channel spacing	Y	N	N	N
High channel count	Y	Y	N	N
High channel isolation	Y	N	Y	Y
High channel uniformity	Y	Y	N	N
High production yield	Y	N	N	N
High thermal stability	Y	N	Y	Y
Low chromatic dispersion	Y	Y	N	N
Low cost per channel	Y	N	N	N
Low insertion loss	Y	N	N	N
Low PDL	Y	N	Y	Y
Small packaging dimension	Y	N	N	N
Technology maturity	Y	N	Y	Y
Wide passband	Y	Y	Y	Y

Refraction technology, such as prism-based mux/demux, has been shown to be less than ideal. TFFs based on multiple beam interference or FBGs based on diffraction have been utilized in multichannel mux/demux devices; here discrete filters are cascaded in a serial manner. The insertion loss increases with the channel number, and the channel uniformity becomes problematic. Cost is a major consideration: In addition to the high cost of individual filters, the required combination with other technologies (e.g., optical interleavers for TFF devices and optical circulators for FBGs), also increases system costs. Furthermore, these two types of devices have high chromatic dispersion (± 50 to ± 200 ps/nm), making these technologies unsuitable for high data transmission rate applications (e.g., 40 Gbps). Hence, considering the cost and performance factors, these two types of DWDM mux/demux devices are not suitable for systems with a high channel count.

In the recent years, AWGs have appeared to be an attractive technology for high channel count mux/demux devices, enabling one to process optical signals in a parallel manner. The low chromatic dispersion (typically ± 5 to ± 10 ps/nm) of AWGs makes them usable for 40-Gbps systems. The drawback of this technology relates to the fact that manufacturing AWGs involves a series of complex production processes and requires bulky facilities. Another key consideration is that AWGs require active temperature

FIGURE 5.14
*Pictorial view of
DWDM technologies.*

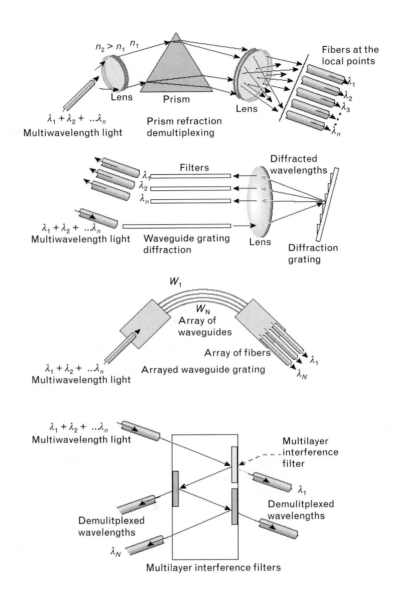

control in order to stabilize the thermal wavelength drift and temperature-dependent loss variations.

On the issue of component cost, phased-array WDM devices (also called arrayed waveguide gratings [10]) are emerging as important passive devices and may be the right kind of technology for metro access and possibly for some metro core applications in smaller diameter tier 2 or tier 3 cities. New technology is available that can produce a 1 × 8 WDM chip (8 channels) for around $450, a 1 × 16 WDM chip for around $700, and a 1 × 32 WDM chip for $1,200. Component vendors may avail themselves of a sol-gen-based method that reduces manufacturing costs by more than an order of magnitude compared to other methods. Sol-gen is a mature glass technology that has already been used in several fields. The *photonic hybrid active silica integrated circuit* (PHASIC) is a Canadian-developed method that combines features of organic polymers and glass to produce a sophisticated hybrid material for integrated optics in a cost-effective manner; companies can use advanced photomicrolithography to "print" optical circuits and devices directly into the hybrid glass.

Literature [11] documents the connection between molecular (chemical) hybridization, supramolecular hybridization, and functional hybridization (device and system-level constructs). This literature describes how low-temperature *hybrid sol-gel glasses* (HSSG) combine attractive features of organic polymers and inorganic glasses, giving rise to a new process for fabrication of WDMs, couplers, power splitters, and waveguides and gratings. The process combines chemical synthesis and sol-gel processing with straightforward photomask techniques. HSSGs are material systems whose properties can be predictably tuned, from properties that are closely related to organic polymers, all the way to properties that are pretty much like conventional inorganic glasses. This creates a development opportunity for the chemist and material scientist, through appropriate design, to realize compositions that can be adapted to a range of optical devices. In academic lingo, "In combining broad spectrum molecular chemistry with physical processing, HSGG recommends itself as the material locus for resolving antagonisms that arise out of conflicting technical objectives in photonics." Unlike conventional/monolithic approaches (that make use of semiconductor materials), hybrid architectures for glass optoelectronics enable the development of a range of passive glass components such as couplers, splitters, waveguides, WDMs, mirrors, gratings and lenses [12]. The five existing (conventional) glass fabrication methods all suffer drawbacks when compared with sol-gel techniques. These conventional methods are (1) sputtering, (2) thermal oxidation and nitridation, (3) *chemical vapor deposition* (CVD), (4) *plasma enhanced low temperature chemical vapor deposition* (PECVD), and (5) *flame hydrolysis deposition* (FHD). Some vendors (such as PIRI, Siemens, and Kymata) use FHD methods. Here a hydrogen–oxygen nozzle flame burns gases; the combustion of these gases produces a glass soot that is melted at ~1,000°C. This process requires more than 6 hours and is repeated three times. The second coating step is followed by a series of coating and vacuum etching steps. The photonic hybrid silica integrated circuit method, on the other hand, uses a hybrid sol-gen glass; components are made by spin- or dip-coating of fluids and at 100 to 200°C. Next the devices are created by photolithography directly in the HSSG, avoiding vacuum film deposition and repetitive steps. One issue with passive devices, however, is power: After a few nodes the original power is reduced appreciably; however, this issue is manageable in a metroaccess environment with relatively small-diameter rings. The phased-array technology is similar in concept to the *passive optical network* (PON) systems being discussed by the industry for residential use.

In summary and in practical terms, the manufacturing complexity leads to low yield, thereby increasing the per unit cost; poor performance leads to degraded signal quality and degraded system performance such as high insertion loss, high channel cross-talk, low channel uniformity, and high polarization-dependent loss.

Free-space diffraction gratings have emerged of late as a promising technology to overcome the cost, manufacturability, yield, practice, and technical complexity drawbacks encountered in TFF, FBG, and AWG-based DWDM mux/demux devices.

5.1.3.2   Technologies Based on Volume Phase Gratings

Free–Space Diffraction Gratings

A diffraction grating is a conventional optical device used to spatially separate the different wavelengths or colors contained in a beam of light. The device consists of a collection of diffracting elements (narrow parallel slits or grooves) separated by a distance comparable to the wavelength of light under study. These diffracting elements can be either reflective or transmitting, forming a reflection grating or transmission grating. An electromagnetic wave containing a multiplicity of wavelengths incident on a grating will, upon diffraction, have its electric field amplitude, or phase, or both modified and, as a result, a diffracting pattern is formed in space. Diffraction gratings can also be classified into two types of gratings: amplitude and phase according to the physical nature of diffracting elements. An amplitude grating is produced through mechanically ruling a thin metallic layer deposited on a glass substrate or photograph (lithography), whereas a phase grating consists of a periodic variation of the refractive index of the grating material. The gratings are termed *free-space* gratings because the phase difference among diffracted beams is generated in the free space, rather than in dispersion media-like waveguides [5].

Volume Phase Gratings

A *volume phase grating* (VPG) is also called a thick phase grating according to the well-known Q-parameter, defined as

$$Q = \frac{2\pi\lambda d}{n_B \Lambda^2 \cos\alpha}$$

where λ is wavelength, d is the thickness of the grating, n_g is the refractive index of the material, Λ is the grating period, and α is the incident angle. The phase grating is called "thin" for $Q < 1$ and "thick" for $Q > 10$. The parameters involved are defined in Figure 5.15.

It is desirable to have a thick transmission VPG that is designed and manufactured to provide the highest diffracting efficiency (up to 99%) and the largest angular dispersion for DWDM devices. The VPG is made from a diffractive element sandwiched between two substrates, each of which is

FIGURE 5.15
A transmission phase grating configuration.

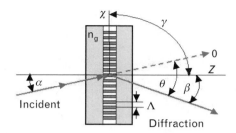

formed from low scattering glass whose external surface is coated with an antireflection coating to enhance the passage of radiation. The diffractive element is a volume hologram comprising a photosensitive medium with thickness ranging from a few to tens of micrometers, such as a layer of proprietary photopolymer materials. Through exposing an interference pattern coming from two mutually coherent laser beams to the photosensitive medium layer, a periodic modulation to the refractive index of the medium is formed, which typically has a sinusoidal profile. This is the volume phase grating. The fabrication of holographic elements for different purposes has been described in several references. The manufacturing cost of forming holographic elements is low because the work is basically a photographic process.

Diffraction by VPGs

The high diffraction efficiency and large angular dispersion capability of a VPG provides a proven technology to demultiplex equally spaced DWDM signals. For a thick grating, the diffraction must simultaneously satisfy the well-known grating equation and Bragg condition. From the former, the angular dispersion is found to be

$$\frac{d\theta}{d\lambda} = \frac{m}{\Lambda \cos \theta}$$

where m is the diffraction order, Λ is the grating constant, and θ is the diffraction angle. On the receiving plane, each channel covers a spatial range determined by

$$\Delta x_1 = \left(\frac{d\theta}{d\lambda}\right)\Delta\lambda_1 L \quad \Delta x_2 = \left(\frac{d\theta}{d\lambda}\right)\Delta\lambda_2 L, \quad \ldots$$

where L is the distance between the grating and the receiving plane, $\Delta\lambda_k$ is the wavelength range of the kth channel centered at the wavelength λ_k. Thus, the channel signals that are output can be obtained at the successive positions $x_1 = x_0 + \Delta x_1, x_2 = x_1 + \Delta x_2, x_3 = x_2 + \Delta x_3, \ldots$, where x_0 is the reference position. The geometry of a demultiplexer is illustrated in Figure 5.16.

FIGURE 5.16
*Demultiplexing a
composite signal
containing a plurality
of wavelengths using a
free-space transmission
VPG.*

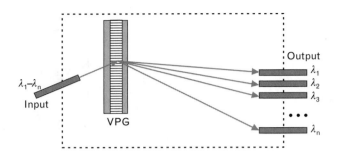

Additional Specifics on Free-Space VPGs

A free-space VPG-based demultiplexing device comprises an input port, a collimating lens, a diffraction grating element (VPG), a focusing lens, and a receiving fiber array (output ports). Figure 5.17(a) shows a schematic diagram illustrating a simplified configuration for a demultiplexer apparatus. The input port receives the composite signal containing a plurality of wavelengths from a DWDM network, and the output ports receive the demultiplexed channel signals. The divergent beam from the input fiber is collimated with the lens before the VPG. The parallel beam after the collimating lens is subsequently incident on the diffraction grating element, by which different wavelength components are diffracted to different directions. The dispersed beam is then focused by a focusing lens onto the fiber array for output that is made by a series of single-mode fibers. The output fiber assembly can be made by stacking one or more rows of closely spaced, end-flushed, antireflection-coated optical fibers, preferably well aligned in silicon V-grooves. Figure 5.17(b) illustrates the transmitted signal spectrum that is output from the fiber array. The passband profile is typically Gaussian.

Figure 5.18 shows a design diagram of an optical mux/demux device. When the depicted device operates as a demultiplexer, an incoming optical beam containing a plurality of wavelengths is transmitted to the device by one of the optical fibers (input), the end of which is located at the focal plane of a collimating lens. The beam containing multiplexed wavelengths thus is collimated by the lens and then impinged on the VPG unit, which

FIGURE 5.17   *(a)
Simplified optical con-
figuration of a VPG-
based DWDM
mux/demux. (b)
Transmission power
spectrum of the device.*

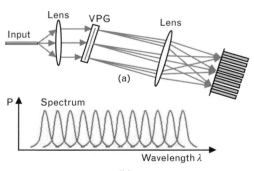

FIGURE 5.18
Double-pass configuration of a DWDM demultiplexer based on transmission VPG.

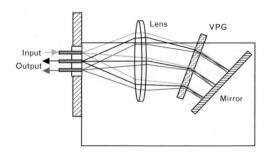

diffracts and angularly separates the beam into a number of individual colli-mated beams according to the number of wavelengths containing in the mixed beam. The spatially separated beams will be redirected back to the same grating by a 100% mirror, which provides further spatial separation of the individual wavelengths. The spatially dispersed collimated beams are then focused by the same focusing lens to its focal plane and received directly by a series of optical fibers (output).

Passband Profiles: Gaussian and Flat–Top

Two types of transmission spectra are commonly available with the use of free-space VPG-based demultiplexers, whose passband profiles are classified into *Gaussian* and *flat-top*. The Gaussian-type passband profile is inherent to the demultiplexer devices based on the diffraction grating technology. The underlying physics are described later. On one hand, the diffraction from a grating together with a focusing lens generates a dispersed spectrum on the receiving plane according to the wavelength. In the narrow wavelength range, such as in the C-band, the spectrum is approximately uniform. On the other hand, single-mode fibers used as output ports support the funda-mental fiber mode, which is approximately Gaussian distributed. The cou-pling between the uniform diffracted field (frequency + space domains) and the Gaussian mode field of fiber (space domain) results in a Gaussian-type spectrum (frequency domain), as seen in Figure 5.17(b). The spectrum across the passband region can be broadened by the use of subsidiary tech-niques, such as microlenses. This type of transmission profile exhibits very low chromatic dispersion and its spectrum is flat across the passband. The latter is an important feature for dispersion compensation.

An actual transmission spectrum obtained from a high-end [13] 100-GHz and 40-channel DWDM demultiplexer is shown in Figure 5.19, which appears to be broad Gaussian. Some customers prefer to use Gaus-sian passband mux/demux because their lasers in transmitters can be accu-rately locked at given frequencies.

In some optical communications systems where the multiplexing and demultiplexing components are used in a cascade way, a flat-top passband profile is desirable. The broad pass bandwidth is also required whenever exit wavelength drifts and thermal variations exist. This is because the

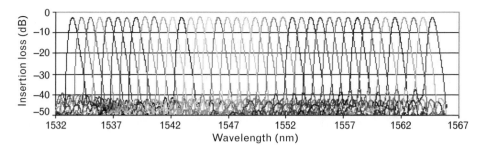

FIGURE 5.19   *Broad Gaussian transmission spectrum of a 100-GHz and 40-channel DWDM demultipexer.*

wider the pass window, the more tolerance on the laser offset specification, and thus the easier for the system to be operated. There are several approaches to flattening the passband spectrum of free-space grating-based mux/demux devices. They include the techniques of using a carefully designed microlens array to reshape the light field before coupling to the receiving fibers, of using a periodic window notch filter to counteract the Gaussian passband regions, and of using new class of grating design. It is on recently that flat-top mux/demux devices based on free-space diffraction gratings have been practically available in volume. They have started to replace the thin-film filter-based components.

Figure 5.20 shows an example of transmission spectrum obtained from a high-end [13] 100-GHz flat-top DWDM demultiplexer. The passband at the 0.5-dB down-power level is greater than 0.35 nm and insertion loss is less than 4.5 dB. Here, a 20-channel demultiplexing device is used in order to clearly show the flattened and broadened passband profile.

In summary, VPG is becoming the most effective technology to provide high-performance, low-cost DWDM mux/demux devices. It is superior to the other technologies. In particular, the state-of-the-art athermal and robust packaging makes the devices fully passive and allows them to work over the ambient temperature of −10 to 80°C. Ultralow insertion loss of ≤ 2 dB is a

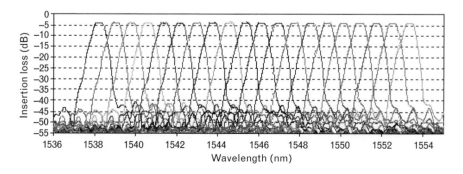

FIGURE 5.20   *Flat-top transmission spectrum of a 100-GHz and 20-channel DWDM demultiplexer.*

major advantage. Furthermore, it is predicable that the competitive price based on VPG technology can be reduced to ~$50 per channel in a few years. Table 5.4 depicts some typical parameters for this kind of technology.

5.2 Operation of a Transponder-Based DWDM System

Figure 5.21 shows the end-to-end operation of a unidirectional DWDM system [7]. The following steps describe the system shown in Figure 5.21:

TABLE 5.4   SPECIFICATION FOR 100 GHz 40-CHANNEL DWDM MUX/DEMUX DEVICES

PARAMETERS		UNIT	DATA	
Passband Profile			Broad Gaussian	Flat-top
Channel number			40	40
Channel spacing		GHz	100	100
Center wavelength range			ITU grid, C-Band or L-band	ITU grid, C-Band or L-band
Center wavelength offset to ITU grid		pm	$\leq \pm40$	$\leq \pm40$
Insertion loss		dB	≤ 3.5 (Typ. 3.0)	≤ 5.0 (Typ. 4.5)
	@ 0.5 dB down		≥ 0.18	≥ 0.30
Passband	@ 1.0 dB down	nm	≥ 0.24	≥ 0.36
	@ 3.0 dB down		≥ 0.40	≥ 0.51
Polarization-dependant loss		dB	< 0.3	< 0.3
Chromatic dispersion		ps/nm	$< \pm5.0$	$< \pm5.0$
Polarization-mode dispersion		ps	< 0.3	< 0.3
Adjacent channel cross-talk		dB	< -30	< -30
Nonadjacent channel crosstalk		dB	< -35	< -35
Cumulative cross-talk		dB	< -25	< -25
Channel loss uniformity		dB	≤ 5.0	≤ 5.0
Return loss		dB	> 45	> 45
Operating ambient temperature		°C	0–70	0–70
Storage temperature		°C	−40–85	−40–85
Mechanical package		mm	$100 \times 70 \times 16$	$150 \times 75 \times 16$
Fiber type			SMF-28, ribbon or fan-out	SMF-28, ribbon or fan-out
Connectors			Upon request and customer specified	Upon request and customer specified

Source: Courtesy BaySpec's Chromatica 100-GHz products.

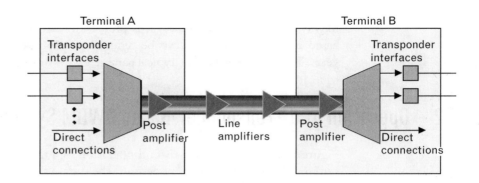

FIGURE 5.21
*Elements of a
unidirectional
DWDM system.*

1. The transponder accepts input in the form of standard single-mode or multimode laser. The input can come from different physical media and different protocols and traffic types.

2. The wavelength of each input signal is mapped to a DWDM wavelength.

3. DWDM wavelengths from the transponder are multiplexed into a single optical signal and launched into the fiber. The system might also include the ability to accept direct optical signals to the multiplexer; such signals could come, for example, from a satellite node.

4. A postamplifier boosts the strength of the optical signal as it leaves the system (optional).

5. Optical amplifiers are used along the fiber span as needed (optional).

6. A preamplifier boosts the signal before it enters the end system (optional).

7. The incoming signal is demultiplexed into individual DWDM lambdas (or wavelengths).

8. The individual DWDM lambdas are mapped to the required output type (for example, OC–48 single-mode fiber) and sent out through the transponder.

5.3   Network Design Evolutions

A DWDM optical infrastructure allows carriers the flexibility to expand capacity in any portion of their network, thus addressing specific problem areas that are congested due to high bandwidth demands.

DWDM provides a "grow as you go" infrastructure, that allows for a 10- to 100-fold capacity expansion as new services increase the demand for bandwidth. The "grow as you go" approach is implemented by adding the lambda plug-ins over time [14]. As noted, high-end commercially available

DWDMs can deliver up to 320 Gbps of capacity; these are 32-channel systems with each channel operating at 10 Gbps (OC-192 SONET/SDH). Optical technology under development at the commercial level can provide 160 lambdas at 10 Gbps each over a single fiber pair for long-distance networking services (more lambdas are achievable in laboratory systems). However, many observers now believe demand was overestimated in the late 1990s and early 2000s, at least in the long-haul and Internet segments of the market, and that those very large systems will not be needed in the near term.

The agnostic nature of DWDM technology allows multiple networks (ATM, IP, SONET) to share the same optical core, but possibly with some bandwidth inefficiencies (e.g., yes, one can map a GbE link directly over a lambda, but one will be dedicating a channel to support 1 Gbps where one could otherwise derive 10 Gbps/OC-192). DWDM provides a means for carriers to integrate the diverse technologies of their existing networks onto one physical infrastructure. OTN will enhance this capability further. DWDM systems are bit-rate and format independent and can accept any combination of interface rates and formats, although SONET tributaries are the most common.

5.3.1   Long-Haul Application Scope

In long-haul networks, the primary value provided by DWDM in combination with optical line amplifiers *is the cost-effective transmission of high aggregate bit rates over large distances on a single fiber pair*. The large distances in long-haul networks make deploying new fiber challenging. Long-haul carriers have traditionally been the first to embrace advances in transmission technologies to gain additional capacity while leveraging their existing equipment [15].

Because of its costs, at this time DWDM is more suited to longer reach applications rather than shorter. This situation could change in the future if developers were to begin to grasp what the real requirements are in the metro access/metro core space. For long-haul applications, DWDM offers a cost-effective complement to OC-48/OC-192 TDM technology that is embodied in SONET/SDH systems. Assuming a decent-quality fiber plant, WDM methods enable the planner to maximize network capacity and address service scalability needs. And because it defers—or eliminates—the capital outlays and long lead times associated with deploying new fiber cable, DWDM is a useful solution for high-growth routes that have an immediate need for increased bandwidth. Carriers that are building or expanding their long-haul networks could find DWDM to be an economical way to incrementally increase capacity, rapidly provide needed expansion, and "future-proof" their infrastructures against unforeseen bandwidth demand. Network wholesalers can take advantage of DWDM to lease capacity, rather than entire fibers, either to existing operators or to new market entrants. However, in aggregate, DWDM technology is not cheap. Systems can cost from $50,000 to $100,000 per "lambda," particularly when redundant and long-reach applications are considered; furthermore, these figures do not include the cost of the repeaters. Systems costing from half a million to $1 million or more are not unusual. Power consumption and floor space requirements are usually high, and unless tunable laser systems are available, complex design/engineering and sparing disciplines are involved, with plug-ins specific to given lambdas being required. Furthermore, "future-proofing" may in fact imply that after a route upgrade, there is no need for additional equipment for several years; this is, therefore, "bad news" for equipment suppliers.

DWDM is well suited for long-distance telecommunications operators that use either point-to-point or ring topologies. The availability of 16, 32, or more new transmission channels, where there used to be one, improves an operator's ability to expand capacity and simultaneously set aside backup bandwidth without installing new fiber. Proponents make the case that this large amount of capacity is critical to the development of self-healing rings. By deploying DWDM terminals, an operator can construct a protected 40-Gbps ring with 16 separate communication signals using only two fibers. However, unless a major underlying engine is continuously driving demand "through the roof," this kind of technology is a "one-time (in a long time) upgrade," with obvious market sizing implications.

Deployments of WDM in the early 1990s were based on *wideband* technology. Wideband WDM doubles the capacity of fiber plant by optically coupling the outputs of two terminals in a *fiber-optic transmission system* (FOTS); one terminal operates in the 1,310-nm range and the other in the 1,550-nm range. Although this is a cost-effective solution for applications with restricted reach (distance), wideband WDM systems, which tend to consist of little more than an optical coupler and splitter, suffer from the absence of maintenance capabilities and scalability for long-haul applications [15]. However, if they are priced well, these WWDM (or CWDM) solutions, perhaps with some added features, may be applicable to metro applications.

Even in long-haul applications, design considerations aimed at optimizing the cost profile are not always straightforward. In particular, TDM-only solutions that support increasing speed kept pace with a number of advances in the WDM technology during the mid- to late 1990s, at least for medium-size trunking applications (up to 10 Gbps). For example, a TDM-based solution has only one 10-Gbps SONET terminal. A WDM system that transports an aggregate capacity of 10 Gbps requires four 2.5-Gbps terminals in addition to a WDM terminal per end. Because TDM technology has typically quadrupled its capacity for a cost multiplier of 2.5, the 10-Gbps solution appears to be more cost effective. However, if the TDM system also requires four 2.5-Gbps terminals to provide the first stage of multiplexing, the 10-Gbps solution might actually be more costly.

(Note that if the 2.5-Gbps terminals are already in the network, they represent a sunk cost and might not be included in the cost analysis.)

Before OAs were developed and deployed, higher speed TDM-based systems were more cost effective than WDM because the TDM systems allowed multiple lower speed electronic regenerators at a point in the network to be replaced with a single higher speed regenerator at that point in the network; originally this was not the case with the WDM design. The introduction of OAs with the ability to amplify the entire ITU grid frequencies simultaneously allows multiple lower speed electronic regenerators at a site to be replaced with one optical amplifier, making WDM more cost effective in this context.

The original DWDM systems were optimized for long-haul interexchange applications; therefore, these WDM systems support mainly point-to-point or linear configurations. Although these systems provide fiber capacity relief, the management of add, drop, and pass-through traffic must be done manually. Current-generation DWDM products can perform linear ADM functions in long-haul or interexchange applications. These products also support ring topologies, which increase the reliability and the survivability of the user-level connection. Products that support ring and/or mesh configurations are more suited to metro applications. Tracking the technology evolution, OC-192 SONET systems have emerged in the recent past that can transport an OC-192 (10 Gbps) on one fiber pair. These can be deployed discretely (as a simple SONET/SDH system), or in conjunction with a WDM system.

The majority of embedded fiber in long-haul networks is standard, single-mode (G.562-defined) fiber with dispersion in the 1,550-nm window, which limits the distance for OC-192 transmission. The majority of the legacy fiber plant cannot support high-bit-rate TDM. Earlier vintage fiber has some attributes that lead to significant dispersion and would, therefore, be incompatible with high-bit-rate TDM.

State-of-the-art OC-192 SONET technology is a viable alternative for new fiber builds because the fiber parameters can be controlled through placement of the appropriate fiber type and because new fiber handling procedures can be accommodated. However, for existing fiber applications, the ability to install 10-Gbps TDM systems depends on fiber properties such as chromatic and polarization-mode dispersion, which may differ from fiber span to fiber span. In addition, high-rate TDM systems require specialized fiber handling and termination procedures when compared to lower rate OC-48 systems. In contrast, current DWDM systems can transport up to 32 wavelengths at OC-48 each, giving an aggregate capacity of 80 Gbps on one fiber pair. This means WDM technology surpasses TDM in terms of the aggregate capacity offered on a single fiber pair, while maintaining the same fiber handling procedures developed for OC-48 TDM systems [6]. SONET and WDM systems typically have built-in FEC features that provide a 10^{-15} BER. DWDM has a number of advantages even over the latest TDM option STM-64/OC-192. However, one needs to recognize that in some instances TDM may offer a better solution than DWDM. NZ-DSF, for example, is flexible enough for the latest TDM equipment, but it is expensive and may limit the ability of carriers to migrate to the greater bandwidth available through DWDM at STM-16/OC-48 rates.

In conclusion, it is not possible to offer generalized rules for design, but each specific situation must be examined individually, particularly taking into account the embedded base and/or existing plant.

5.3.2   Metro Access/Metro Core Application Scope

The pertinent question for the current discussion is whether DWDM has practical application to the metro access/metro core space. Any network planner knows that as one moves closer and closer to the tier 1 backbone of the network (the long haul), the need becomes more intensive and concentrated, while at the edges of the network (the metro access), the needs are less intensive and concentrated; therefore, the long-haul backbone is the more natural "habitat" for DWDM technology rather than the metro access and metro core space. There has been a lot of hype in the recent past about metro DWDM. The reality is that in its general form, DWDM has *some* (limited) applicability to metropolitan environments, probably for a handful of POP-to-POP rings [16]. If DWDM systems now on the market were redesigned to be optimized according to the requirements for metropolitan environments, increased applicability could result. If the systems were redesigned to meet specific price points (see later discussion), then their applicability would be enhanced.

When the long-haul industry saw major retrenchments at the turn of the decade, a number of optical vendors took the easy (but short-lived) course of relabeling the equipment that had been developed by them for long-haul applications by pasting a "metro DWDM" label onto the equipment while at the same time generating new marketing collaterals, rather than redeveloping, as would have been more appropriate, equipment that is optimized and right-sized for metro access/metro core applications from both a density and cost point of view. It appears that, at least for the next few years, the opportunity for metro DWDM is somewhat limited. As noted, this technology may see *some* penetration in a metro core application of POP-to-POP rings, but extremely limited application in the metro access segment (see later discussion). Analysis shows that said vendors have products that would cost from $190,000 per building all the way to $720,000 per building. This is totally untenable for metro access, because carriers need to keep their in-building costs to be no more than $20,000 or thereabouts. A price of $10,000 per dropped lambda ($20,000 for a redundant system with two lambdas) is desirable for metro access, particularly when the component cost of metro optics to build 8- or 16-channel

CWDM, WWDM, or even DWDM is actually about $2,000 per lambda or less. In particular, if systems are designed in such a manner that each building needs to incur the full cost even though only one or two lambdas are dropped off at that building, then the equipment will see very limited penetration at the metro access level.

However, current-generation systems (CWDM and DWDM) still lack the full flexibility and affordability that is required in metro applications. Because the wavelengths that are added and dropped are fixed, there is no mechanism to increase the add/drop capacity without physically reconfiguring the system. Networking in a TDM network is provided by multiplexers, such as traditional linear and ring ADMs and cross-connects. These NEs must convert the optical signal to an electrical signal before performing their specific function, then convert the result back to an optical signal for transmission on the next span; this back-to-back electro-optic conversion is a major cost component of these network elements.

The following observations from a key vendor are refreshing: "Although capacity per fiber is important in the local networks, the value equation is different because the topologies and traffic patterns are more meshed, resulting in shorter spans with less capacity per span. The capacity demand is still within the realm of TDM technology and the cost advantages of optical amplification cannot be fully exploited because of the shorter distances. As such, WDM technology must bring a richer set of values to meet the diverse challenges of the local network" [15].

In long-haul networks, the major portion of the network transport cost is due to the numerous regenerators along a link (typically every 35 to 70 miles). Normally, not only does one need to do the amplification, but one also needs an optical-to-electrical conversion, further followed by an electrical-to-optical conversion. Utilizing OAs reduces this cost by eliminating the back-to-back electro-optic conversion. As the cost of optical amplifier technology drops, WDM becomes increasingly more economical in long-haul networks. However, in local networks, where distances are shorter, the bulk of the *core* network transport cost is due to the multiplexers and cross-connects; in the *access* network, it is the cost of the fiber spur. Hence, reductions in the cost of these components will prove advantageous to the bottom line. The one major factor driving the potential success of DWDM in the metro access/metro core is the cost per unit (specifically per lambda) per building. The cost per lambda should be in the $2,000 to $10,000 range; the cost per building should be in the $10,000 to $20,000 range; and the system should carry 4, 8, or 16 protected lambdas (8, 16, or 32 distinct lambdas).

One key consideration must be how many broadband customers actually reside in one building. It is worth noting that a carrier that needs to connect 400 buildings (say, with 15 to 20 metro access rings of 16 buildings each) and "pick up" 5 to 10 broadband customers per building is advantaged financially compared to a follow-on carrier that needs to connect the same 400 buildings with an overlay of 20 new rings, and only "pick up" 1 to 2 broadband customers per building. If the building had 3 broadband customers, the economics begin to work, but only when the per-building nodal cost is low (e.g., in the case of midrange next generation SONET, rather than more expensive DWDM).

Hence, the deployment of high-channel capacity WDM systems in the metro access is potentially ineffective from a financial point of view and leads to overkill. Even if one assumes that there are 8 buildings on a ring and each building has 3 broadband customers per building that a carrier is able to bring on line, note that the typical WAN speed today is 10 to 100 Mbps; this equates to a maximum 2.4 Gbps per ring, which fits very well the profile of a next generation SONET solution. If by 2006 the bandwidth need grows fourfold, an OC-192 next generation SONET solution would still be adequate. However, again, it is important to understand that the issue is less technical and more financial: If a DWDM solution were to enjoy cost parity with a next generation solution, then the planner would have no problem selecting the former over the latter.

Metro access/metro core networks need to carry many different types of traffic over an optical channel, regardless of the protocol (Ethernet, ATM, SONET, and so on) or bit rate (10 Mbps, 2.5 Gbps, and so on). Native data interfaces, such as Ethernet at various speeds, can be connected directly to the transport platform without additional adaptation. As we discuss in the next chapter, next generation SONET targets precisely this segment. For the picture shown in Figure 5.22 to be more than just a picture, the cost per building for the DWDM needs to be around $10,000 to $20,000 for the GbE and 10GbE applications, and $5,000 to $10,000 for 10-Mbps and 100-Mbps TLS.

Nearly all enterprise data traffic now originates in Ethernet format (10/100/GbE) while the backbone is built on optical rates of 2.5 and 10 Gbps with STS1 granularity. This is why it is important for metro WDM devices to support the data-oriented interfaces; this would not be a requirement for long-haul DWDM. On the other hand, a traditional (telecom) metro optical network continues to focus primarily on legacy DS-1 and DS-3 granularity within the MAN and through to the backbone. A full-featured metro WDM must support both the traditional interfaces as well as the new data interfaces. Enterprise routers currently do support direct OC-3c, OC-12c, and OC-48c/192c WAN links. These private-line interfaces on end-user equipment (that is, on intranet routers) are common because of their ubiquitous deployment throughout metro optical networks (e.g., extensive deployment of SONET transport and cross-connects). However, this comes at a cost: High-end SONET plug-ins can cost upward of $50,000. Therefore, it is desirable to include these interfaces in the NE, because of the economies of scale that can be secured by network providers, which can then be passed on to the enterprise user.

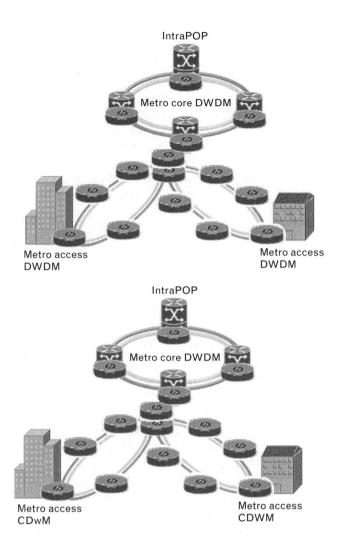

FIGURE 5.22
*Metro DWDM
applications example.*

IntraPOP

Metro core DWDM

Metro access
DWDM

Metro access
DWDM

IntraPOP

Metro core DWDM

Metro access
CDwM

Metro access
CDWM

Equipment vendors working on WDM are realizing that data-oriented services are now consuming more network capacity than voice, due to increasing numbers of users and escalating capacity demands from each user. The protocols, bit rates, and interfaces used by the enterprise networks that support these applications are different from those used in the carrier networks. The former are increasingly Ethernet based; the latter are typically OC-x based. Conversion between the two domains is needed. This conversion can be done by the user and would have to be done at multiple places where the enterprise network touches the carrier's network; or, better yet, it could be done by the carrier on behalf of the user. One example may suffice: Carriers have come to realize that it is better to carry voice over distances in digital format. However, the network does not impose that requirement to the user: All residential users generate analog voice over their telephone handsets and the network adapts the user's signal into a digital signal that is better suited for network handling. Similarly, because all enterprise networks are now based on Ethernet/IP/TCP technology, it would be preferable for the carrier's network to accept this handoff and then convert it, if necessary, to a SONET/WDM format. Because Ethernet interfaces are very cost effective for the end user compared to traditional WAN interfaces, TLS-based services (supported by Ethernet-ready edge NEs) would be an attractive offering to corporations. It also saves costs for the network provider by eliminating adaptation functions to ATM; while ATM can support speeds of tens of Mbps, it does not support equally well interfaces operating at 1 or 10 Gbps. Native data interfaces for other protocols on edge NEs, such as ESCON or FICON, would provide similar values depending on user-specific requirements.

At the end of the day, however, native data interface are not necessarily intrinsic to WDM only. Next generation SONET aims at supporting these same interfaces. Hence, the choice between WDM and next generation SONET is really (beyond initial cost) a traffic engineering consideration: How much growth can be anticipated in a metro access subnetwork?

During the 1990s, the metro network was primarily concerned with the transport of voice-based TDM circuits (DS1, DS3) for private-line services. In the recent past, the issues related to metro networks were about how to support the emergence of data services based on IP and lambdas, as well as how to support the continued growth of legacy TDM. Another related issue is to how to construct the metro network such that it can support the shift from legacy TDM to a data–centric network based on IP.

By deploying a WDM-based photonic network, the service provider gains an access or interoffice transport infrastructure that is flexible and scalable. However, planners need to ascertain that what could turn out to be a high NE cost does not become multiplicative *and all 8, 16, or 32 buildings on the ring are each forced to incur a high nodal cost*. Whereas a long–haul network may be a two-point link, or at most be a ring/mesh with a half dozen or so high–capacity points (say, in several states), metro access rings typically have a larger number of nodes on them, unless a nonredundant building access is utilized. A 16-building access ring using next generation SONET with a $30,000 device would cost the provider $480,000 in nodal equipment; a ring using DWDM with a $60,000 device would cost the provider $960,000.

There are certain key OAM&P implications for metro core WDM networks as well. At this time, long-haul applications do not require dynamic management of individual wavelengths (except with the new architectures such as ASON/ASTN discussed in Chapter 4). Therefore, one optical surveillance channel carried on a dedicated wavelength is sufficient to control and monitor all remote WDM network elements such as optical line amplifiers. Photonic networks in metropolitan access and interoffice applications manage individual wavelengths. Because individual wavelengths have different endpoints and take different paths across the network, one optical surveillance channel carried on a separate wavelength is not sufficient. Fault information is required for, and must accompany, each wavelength. Per-wavelength fault information is needed for fault detection and isolation, which are necessary to support survivability mechanisms and trigger fast maintenance and repair actions. In photonic networks, service providers can also verify the connectivity and monitor the performance of each connection [15].

Proponents make the case that with metro DWDM, user requests for increased bandwidth or different protocols can be filled quickly, so the network provider can realize increased revenue. This may be true in the metro core network, but is unlikely to be achieved in general in the metro access. New services such as optical-channel leased lines that provide end-to-end protocols and bit-rate independent connections can be offered, affording new revenue for the network provider. In WDM-based photonic networks, transport interfaces that are specific to protocol and bit rates are no longer required, minimizing the network provider's inventory and operating costs.

On the other hand, it is important to keep in mind that the market for traditional handoff from intranet routers will continue to exist for many years, and may in fact exceed the Ethernet-handoff market for the foreseeable future, as anecdotal informationtends to show [17]. The camp that supports native data interfaces asks the following rhetorical question: Given that WAN traffic from the enterprise into the metro network is primarily Ethernet data today and likely to continue to dominate in the future, should the conventional view that legacy private-line services are the most efficient means for metro optical networking be challenged? In the long term (5 to 7 years out), metro optical networking based on Ethernet native LAN interfaces could become prevalent. This is because Ethernet router vendors accrue additional costs when adapting native data interfaces into private-line interfaces (DS1, DS3, OC-N) using what is referred to as *telecom adapters*. The additional cost translates into private-line interfaces costing as much as three to four times more than native LAN interfaces (100BaseT, GbE). In addition, the cost of GbE interfaces is expected to drop rapidly as their deployment rises in Enterprise networks. Bandwidth growth has been limited to private-line-like bandwidth increments such as DS-1, DS-3, or OC-x. Because these private-line interfaces are provisioned as physical connections from the customer premises onto the metro network, incremental bandwidth needs require new physical connections to be established. However, as noted, the reality is that the legacy interfaces will be around for a number of years. Clearly most carriers have invested billions of dollars in SONET optical products, ATM switches, and DCS cross-connects; hence, they are highly motivated to price and design these services correctly, in order to achieve a return on their investment, particularly in the new telecom economics market. Eventually, carriers will migrate to new technologies; the challenge is how to make a graceful migration from the embedded network to a network based on data-oriented interfaces.

In summary, the issues in metro optical networks are significantly different than those faced for long-haul applications. Whereas long-haul networks are focused on lowering the cost per bit for point-to-point transport, metro networks are more concerned with cost-effective connectivity to a *multitude* of endpoints (that is, connectivity to a large set of buildings), service transparency, end-user protocol support, speed of provisioning, and end-to-end connectivity.

5.4   CWDM

In the component industry, a significant amount of effort is focused on reducing channel spacing to enable DWDM systems to support more wavelengths. Today, typical wavelength spacings using thin-film filters are at 200 GHz, moving to 100 GHz and even 50 GHz. The precision associated with this channel spacing results in challenging yields and higher component costs. However, as noted in the previous section, the metro market, especially metro access, will operate at a much lower channel density, allowing as much as several nanometers between channels (i.e., a four- or eight-channel system could operate in the neighborhood of 1,310, 1,480, 1,540, and 1,560 nm.) Meanwhile, the cost of producing less precise components (lasers, fibers, and so on) has fallen as yields have improved.

In 2002, the ITU set a global standard for metro optical fiber networks that was intended to expand the use of CWDM in metropolitan networks. According to the ITU-T this standard is necessary to meet the increasing demand of voice, data, and multimedia services for low-cost short-haul optical transport solutions, and it is expected to produce savings for telecommunications operators, that it is hoped will be passed on to consumers.

The worldwide optical metro network market was forecast to increase from $1.1 billion in 2001 to $4.3 billion by 2005, according to reports by Gartner DataQuest [18]. Bolstered by this agreed-on new standard, CWDM is poised to capture a sizeable share of this market. CWDM applications are especially good for coverage of up to 50 km. Where the distances are shorter and the need for capacity less, CWDM applications are able to use wider channel spacing and less expensive equipment, yet achieve the same quality standards of long-haul optical fiber systems, according to ITU-T SG15.

DWDM optical systems, which carry a large number of densely packed wavelengths, require a thermoelectric cooler to stabilize the wavelength emission and absorb the power dissipated by the laser. This consumes power while adding cost. However, for short transmission distances a "coarse" wavelength grid can reduce terminal costs by eliminating the temperature control and allowing the emitted wavelengths to drift with ambient temperature changes.

This agreed-on standard will be a stabilizing force for manufacturers of filters with wide channel spacing, for manufacturers of uncooled lasers with an expanded number of wavelengths, and for system manufacturers looking to offer low-cost short-haul optical transport solutions. Taken together, a well-defined grid will increase the rate of CWDM product and market development.

ITU-T Recommendation G.694.2 is the most recent in the series; it specifies physical layer attributes of optical interfaces. Recommendation G.694.2 provides a grid of wavelengths for target distances up to about 50 km on single-mode fibers as specified in Recommendations G.652, G.653, and G.655. The CWDM grid is made up of 18 wavelengths defined within the range of 1,270 to 1,610 nm spaced by 20 nm. ITU-T Recommendation G.694.2 has been agreed to under the fast-track approval procedure. Under this procedure, a comment period is initiated when a study group gives consent to approve the draft text of a recommendation that it considers to be mature. The combined announcement and comment period took just under 2 months. This standard was, therefore, expected to become effective before the end of 2002. Other recommendations in this series include the following:

- *ITU-T G.691 (2000)*. Optical interfaces for single-channel STM-64, STM-256 systems and other SDH systems with optical amplifiers;

- *ITU-T G.692 (1998)*. Optical interfaces for multichannel systems with optical amplifiers;

- *ITU-T G.693 (2001)*. Optical interfaces for intraoffice applications;

- *ITU-T G.694.1 (2002)*. Spectral grids for WDM applications: DWDM frequency grid;

- *ITU-T G.957 (1999)*. Optical interfaces for equipments and systems relating to SDH;

- *ITU-T G.959.1 (2001)*. Optical transport network physical layer interfaces.

The wavelength plan contained in the new ITU-T Recommendation G.694.2 has a 20-nm channel spacing to accommodate lasers that have high spectral width and/or large thermal drift. This wide channel spacing is based on economic considerations related to costs of lasers and filters, which vary with channel spacing. To accommodate multiple channels on each fiber, the agreed wavelength grid covers most of the recently approved bands of the single-mode optical fiber spectrum, from less than 1,300 nm to more than 1,600 nm.

CWDM systems support transmission distances up to 50 km. Within this distance, CWDM can support various topologies—hubbed rings, point-to-point, and PONs. CWDM is well suited for metro applications (e.g., CWDM local rings that connect central offices to major DWDM metro express rings) and for access applications (e.g., access rings, PONs).

CWDM systems can be used as an integrated platform for multiple clients, services, and protocols for enterprise users. The channels in CWDM can have different bit rates. CWDM also provides flexibility with regard to traffic demand changes through the ease of adding or dropping channels into or from systems.

CWDM related standardization activities in SG15 are continuing with work on a new draft recommendation (G.capp) specifying optical parameters and values for physical layer interfaces in CWDM applications.

ENDNOTES

[1]   Current systems offer from 4 to 128 channels of wavelengths. The higher numbers of wavelengths have led to the name DWDM. The technical requirement is only that (1) the lasers be of very specific wavelengths and the wavelengths be very stable and (2) the DWDM demultiplexers be capable of distinguishing each wavelength without unacceptable cross-talk.

[2]   One can also carry a non-SONET framed signal; however, the majority of systems now deployed are used as adjuncts to SONET, namely, they carry a SONET signal.

[3]   It is also possible to pack both a sending and a receiving signal onto a single strand, but most telecom systems utilize a fiber *pair*.

[4]   The deployment of unused strands in the same sheath that could be brought on as needed (by adding electronics) is viewed here as "installing new cables."

[5]   Yu, D., and W. Yang, "VPG-Based DWDM Mux/Demux Devices and Specifications," retrieved from BaySpec, Inc., http://www.bayspec.com/WP, February 10, 2002.

[6]   Retrieved from http://www.optronics.gr/Tutorials/dwdm.htm.

[7]   "Introduction to DWDM for Metropolitan Networks," Cisco Systems White Paper, retrieved from http://www.cisco.com, 2000.

[8]   Note that 40-Gbps transmission systems are now the "next frontier" in optical communication. At the transmitter end, appropriate source modulation is critical to deal with effects such as fiber dispersion and nonlinearity. Modulation requirements for efficient transmission at 40 Gbps impact chirp, drive voltages, insertion loss, extinct ratio, power handling capability, and wavelength dependence. Components such as LiNbO3, GaAs polymer, and integrated electroabsorption modulators may play a role in this context.

[9]   Portions of this section are based on promotional materials from BaySpec Inc. (http://www.bayspec.com). BaySpec designs, manufactures, and markets fiber-optic components and modules for the optical networking industry. Utilizing patented VPG technologies, BaySpec has developed a new generation of optical networking products that provides cost and performance benefits for enterprise, metro, and long-haul applications.

[10] Phased-array WDMs are imaging devices: They image the field of an input waveguide onto an array of output waveguides in a dispersive way. In phased-array-based devices, the imaging is provided by an array of waveguides, the length of which has been chosen so as to obtain the required optical/dispersive properties. Phased-array-based devices are simpler and more robust than other devices (such as a

grating-based devices, which require vertical etching to obtain a vertically etched reflection grating).

[11] For example, see Najafi, S. I., et al., "Passive and Active Sol Gel Materials and Devices," SPIE Press Vol. CR68, Bellingham, WA: SPIE, 1997, pp. 253–285; Najafi, S. I., et al., "Ultra-Violet Light Imprinted Sol-Gel Silica Glass Waveguide Devices," *Optics Communications,* Vol. 128, 1996; Najafi, S. I., et al., "Sol Gel Integrated Optics Coupler by Ultra-violet Light Imprinting," *Electronics Letters,* Vol. 31, 1995; Najafi, S. I., et al., "Ultra-Violet Light Imprinted Sol-Gel Silica Glass Low-Loss Waveguides for Use at 1550 Nanometers," *Optical Engineering,* Vol. 36, 1997; Najafi, S. I., et al., "Gratings Fabrication by Ultra-Violet Light Imprinting and Embossing in Sol-Gel Silica Glasses," *Proc. SPIE,* Vol. 2695, 1996; Andrews, M. P., et al., "Erbium-Doped Sol-Gel Glasses for Integrated Optics," *Proc. SPIE,* Vol. 2397, 1995; Najafi, S. I., et al., "Erbium in Photosensitive Hybrid Organoaluminosilicate Sol-Gel Glasses," *Proc. SPIE,* Vol. 2997, 1997; Najafi, S. I., et al., "Fabrication of Ridge Waveguides: A New Sol-Gel Route," *Applied Optics,* 1997; Najafi, S. I., et al., "UV-Light Imprinted Bragg Grating in Sol-Gen Ridge Waveguide with Almost 100% Reflectivity," *Electronics Letters,* 1997; Najafi, S. I., "Photoinduced Structural Relaxation and Densification in Sol-Gel-Derived Nanocomposite Thin Films: Implications for Integrated Optics Device Fabrication," *Canadian Journal of Chemistry,* Nov. 1998.

[12] Opportunities also exist, however, for combining these techniques with monolithic semiconductor-based optoelectronic multichip module techniques, to bring forth complex, high-performance, low-cost data and telecommunications modules.

[13] BaySpec's Chromatica products.

[14] For the purpose of sizing the market in the discussion here, however, we assumed that the entire system was deployed on day 1.

[15] Allen, B., and S. Wong, "Is WDM Ready for Local Networks?," Document 56005.25-05-99, Nortel White Paper, 1997.

[16] Excerpts from a Communications Industry Researchers' November 2001 report titled "Metro Optical Networking Market Opportunities" make these points: "Are we there yet? For close to five years, the metro DWDM market has failed to fulfill the vaunted promises of significant growth opportunities. CIR's research prognosticates that 2003 will be the first big year for substantial metro DWDM deployment to occur in the U.S., with the market reaching $500 million in 2005…. Simply put, metro optical networking market development will remain stuck in neutral until we see more widespread acceptance of DWDM ring technology by the RBOCs and other metro carriers. It has been the classic chicken and the egg scenario since early 1997, in which the industry has been waiting for a critical mass of wavelengths in the metro space to enable both the Optical Add/Drop multiplexer and optical switching markets to develop. But, until the vendors can provide a combination of lower cost and engineering-friendly DWDM products with true ring capability, the all-optical vision in the metro segment will not begin to be realized. CIR believes that the truth about the metro space reflects the economic realities of today, and not the market hype indicating that there are quick-fix elixirs. Most carriers will continue to consolidate spending and look for operational and cost efficiencies. Many of the emerging carriers that burst onto the scene with great fanfare promoting better, faster, and cheaper services are now on shaky financial footing. The overall economy's slowdown will translate into slowing demand for services, which means that incumbent carriers will strive to get by with as much of their legacy gear as they can as they work through the correction. CIR believes that companies such as Lucent, Nortel, Fujitsu and Cisco will most likely dominate the space based on their installed base and sheer size. Riverstone will lead Ethernet vendors, but by balancing Ethernet start-up providers with international accounts and OEM arrangements…," retrieved from http://www.cir-inc.com.

[17]   This based on the order flow seen at a metropolitan area service provider.

[18]   This material is based directly on ITU-T sources.

CHAPTER 6

SONET and SDH

Although evolving technologies are appearing on the deployment horizon, such as those described in the previous chapters, SONET and SDH are the de facto infrastructure technologies for communications services at the present time and for the near-term future. Given the large embedded-base of these technologies in the United States, Europe, and Asia, this chapter provides a detailed view of the topic. Effectively, all communication in the United States between DS-3 and OC-48 takes place over SONET systems. A portion of the DS1 traffic is also carried over virtual tributaries in SONET. Vendors are also bringing out data-aware "next generation" SONET platforms that support data interfaces directly on the SONET NE (when these NEs are intended to be placed in metro access/metro core buildings to support FE, GbE, 10GbE, of FCS/SAN applications); these next generation platforms discussed in Chapter 7.

This chapter[1] describes how SONET/SDH works, down to the octet level. SONET/SDH is not an easy subject. We "layer" the description, starting with the simpler, overview subjects and then moving on to the more complex subjects. When data are transmitted over a communications medium, a number of things must be provided on the link, including framing of the data, error checking, and the ability to manage the link, to name just a few. For optical communications these functions have been standardized by the ANSI T1X1.5 committee as SONET and by the ITU as SDH and are covered in detail in the chapter. Although SONET and SDH have

---

1. This chapter was contributed in its entirety by P. Michael Henderson. It is based on a white paper entitled *Fundamentals of SONET/SDH*, generated by Mr. Henderson at Mindspeed Technologies, Inc., February 14, 2001. Mindspeed Technologies, the Internet infrastructure business of Conexant Systems, Inc., designs, develops, and sells a complete portfolio of semiconductor networking solutions that facilitate the aggregation, transmission, and switching of data, video, and voice from the edge of the Internet to linked metropolitan area networks and long-haul networks. Mindspeed's products, ranging from physical layer devices to higher layer network processors, can be classified into two general categories: access products and wide area network (WAN) transport products. Access products include multiservice access gateway solutions, including voice over IP, and a broad family of multimegabit DSL products that are used in a variety of network access platforms such as remote access concentrators, voice gateways, digital loop carriers, DSL access multiplexers, and integrated access devices. WAN transport products, focused on packet-based optical networks, include T/E carrier, ATM and SONET/SDH transceivers, switch products, network processors, and software solutions. These products are used in various types of network equipment, including high-speed routers, ATM switches, optical switches, add/drop multiplexers, digital cross-connect systems, and DWDMs.

many similarities, there are some significant differences, especially in terminology. The discussion in this chapter is focused primarily on SONET and SONET terminology. This does not mean that SONET is more important than SDH—it is just easier to explain things from a SONET point of view because SONET is a subset of SDH. Once one understands SONET, it is easier to understand SDH; the Appendix at the end of the chapter addresses the SDH perspective.

SONET was developed in the United States through the ANSI T1X1.5 committee. ANSI work commenced in 1985 with the CCITT (now ITU) initiating a standardization effort in 1986. From the very beginning, conflict arose between the U.S. proposals and the ITU. The United States wanted a data rate close to 50 Mbps in order to carry DS-1 (1.544-Mbps) and DS-3 (44.736-Mbps) signals. The Europeans needed a specification that would carry their E1 (2.048-Mbps), E3 (34.368-Mbps), and 139.264-Mbps signals efficiently. The Europeans rejected the 50-Mbps proposal as bandwidth wasteful and demanded a base signal rate close to 150 Mbps. Eventually a compromise was reached that allowed the U.S. data rates to be a subset of the ITU specification, known formally as SDH.

6.1   Introduction to SONET/SDH

The end-to-end connection through a SONET/SDH network is always called the *path*. The connection between major nodes, such as between add/drop multiplexers, is called a *line*. And the link between an add/drop multiplexer and a regenerator, or between two regenerators, is called a *section*. Figure 6.1 shows these connections graphically. Keep in mind that back in the mid-1980s, the only kind of signal amplification available was electronic regeneration—optical amplifiers had not yet appeared.

FIGURE 6.1
*SONET reference
model.*

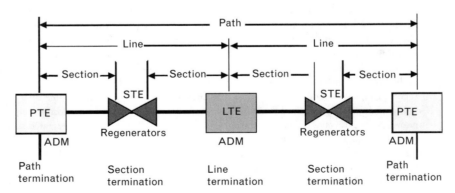

PTE: Path terminating equipment
LTE: Line terminating equipment
STE: Section terminating equipment

The basic SONET frame is set up as shown in Figure 6.2, as 9 rows of 90 octets. It is transmitted from left to right and top to bottom. That is, the octet in the upper left corner is transmitted first followed by the second octet, first row, and so on. When we get to octet 90, we come back and start with the first octet of the second row. (It is important to realize that this two-dimensional representation is just for convenience. The bits are simply transmitted one after another in a serial stream. We could also represent this SONET frame as a linear sequence of 810 octets. Every 90 octets we would have 3 overhead octets. The two-dimensional representation is more convenient, allowing the whole frame to be shown on a page, without making each octet tiny.)

Framing is accomplished by the first two octets, called the A1 and A2 octets. When the frame is transmitted, all octets except A1, A2, and J0 are scrambled to avoid the possibility that octets in the frame might duplicate the A1/A2 octets and cause an error in framing. The bit pattern in the A1/A2 octets is 1111 0110 0010 1000 (hex 0xf628). The receiver searches for this pattern in multiple consecutive frames [1], allowing the receiver to gain bit and octet synchronization. Once bit synchronization is gained, everything is done, from there on, on octet boundaries—SONET/SDH is octet synchronous, not bit synchronous.

The first three columns of a SONET frame are called the *transport overhead* (TOH). The 87 columns following the TOH are called the *synchronous payload envelope* (SPE). Within the SPE there is another column of overhead, called the *payload overhead* (POH), the location of which varies because of timing differences between networks. This leaves 86 columns by 9 rows for usable payload in this basic frame.

FIGURE 6.2
The basic SONET STS-1 frame.

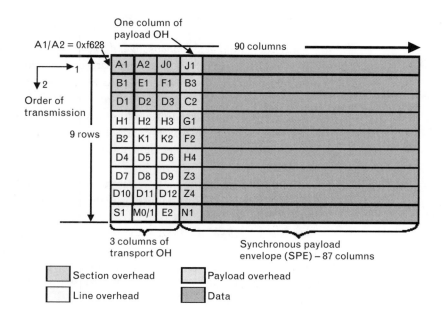

Later we see that the frame for the lowest SDH rate, STM-1, contains 270 columns by 9 rows, and that it contains 9 columns of transport overhead. If nothing were done, this would leave 261 columns for payload, including payload overhead. However, one form of the SDH payload, known as the *virtual container 3* (VC-3) has a structure very similar to the STS-1 87 columns by 9 rows of payload. When a VC-3 is combined with its associated H1, H2, and H3 pointers, it is known as an *administrative unit 3* (AU-3). Of course, there are three AU-3s in the payload area of an STM-1. SONET designers were interested in carrying their legacy plesiochronous traffic. Because of this, they tied everything to the existing voice traffic. And in the digital telephone network, voice is digitized according to ITU Recommendation G.711. That is, the granularity of the voice samples is one octet, and samples are taken 8,000 times per second, or every 125 μs. So this basic SONET frame is designed to carry voice conversation. This means that a SONET frame has a period of 125 μs. This is an important point. It turns out that every SONET frame repeats every 125 μs, no matter how fast the line speed gets. As the line rate goes up, the SONET frame gets bigger by some number of octets, just sufficient to keep the frame rate at 8,000 frames per second.

For this first-level basic SONET frame, this gives a data rate of 51.84 Mbps (90 columns times 9 rows, times 8,000 times per second, times 8 bits per octet). This signal is known as a *Synchronous Transport Signal–Level 1* (STS-1). Once the scrambler is applied to the signal, it is known as an *Optical Carrier–Level 1* signal or OC-1. STS-*N* rates are the same as the OC-*N* rates.

Because there are 86 nonoverhead columns of 9 rows, a SONET STS-1 frame has a usable payload rate of 49.536 Mbps, sufficient bandwidth to carry 774 simultaneous voice conversations [2]. This is in excess of the 672 simultaneous voice conversations carried in a DS-3, allowing one DS-3 to be easily mapped into a SONET STS-1 channel.

Lower rate *plesiochronous digital hierarchy* (PDH) traffic (DS-1, E1, DS-1C, DS-2, or DS-3) is encapsulated with additional framing octets, designed to allow the PDH traffic to be carried within a SONET/SDH channel. This is known as a *virtual tributary* (VT) in SONET and a *virtual container* (VC) in SDH. Multiple DS-1 circuits, for example, may be combined into a single SONET channel, up to 28 DS-1s in an STS-1. The techniques used to map plesiochronous traffic into SONET/SDH are covered later.

The ITU established a base rate close to 150 Mbps for SDH. Specifically, the rate the ITU established is three times the SONET STS-1 rate (three times 51.84 Mbps or 155.52 Mbps) and is called the *Synchronous Transport Module–Level 1* (STM-1) signal. SDH also uses a 9-row frame but an STM-1 signal has three times as many columns as the STS-1 signal (270 octets instead of 90 octets). This pattern repeats for the higher level of SONET/SDH—an STS-12 signal—has 9 rows but 1,080 columns (12 times 90 columns), and so on.

The overhead grows in the same proportion. In an STS-1 signal we have three columns of transport overhead. An STS-3/STM-1 signal has 9 columns of transport overhead. An STS-12/STM-4 signal has 36 columns of transport overhead. And an STS-768/STM-256 signal has 2,304 columns of transport overhead, which is quite significant.

Not all levels of SONET/SDH signals are used in the network. SONET goes from STS-1 to STS-3 because the STS-3 rate matches the lowest level of SDH, the STM-1. After that, both rates go up by factors of 4. There is nothing special about the factor of 4. Making changes to the network is expensive and difficult, and network providers only do it when it provides a significant gain or advantage. The basic concept for data rates has been "four times the data rate for twice the cost." The most common rates for SONET and SDH are listed in Table 6.1.

6.2   SONET/SDH Interleaving

An STS-3 can be thought of as three STS-1 bit streams transmitted in the same channel so that the resulting channel rate is three times the rate of an STS-1. And when multiple streams of STS-1 are transmitted in the same channel, the data are octet multiplexed [3]. For example, an STS-3 signal will transmit octet A1 of stream 1, then octet A1 of stream 2, then octet A1 of stream 3, then octet A2 of stream 1, octet A2 of stream 2, and so on (see Figure 6.3). This multiplexing is carried out for all levels of SONET and SDH, including STS-192 and STS-768. Because of this, SONET/SDH maintains a frame time of 125 μs.

Figure 6.4 shows an STS-3 frame created from the interleaving shown in Figure 6.3. It consists of 9 rows and 270 columns, of which 27 columns

TABLE 6.1   SONET/SDH DIGITAL HIERARCHY

SONET NAME	SDH NAME	LINE RATE (MBPS)	SYNCHRONOUS PAYLOAD ENVELOPE RATE (MBPS)	TRANSPORT OVERHEAD RATE* (MBPS)
STS-1	None	51.84	50.112	1.728
STS-3	STM-1	155.52	150.336	5.184
STS-12	STM-4	622.08	601.344	20.736
STS-48	STM-16	2,488.32	2,405.376	84.672
STS-192	STM-64	9,953.28	9,621.504	331.776
STS-768	STM-256	39,813.12	38,486.016	1,327.104

★ Overhead associated with the transport overhead columns only. Excludes overhead contributed by the POH.

FIGURE 6.3
Interleaving of three SONET STS-1 frames into an STS-3 frame.

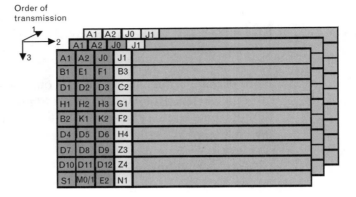

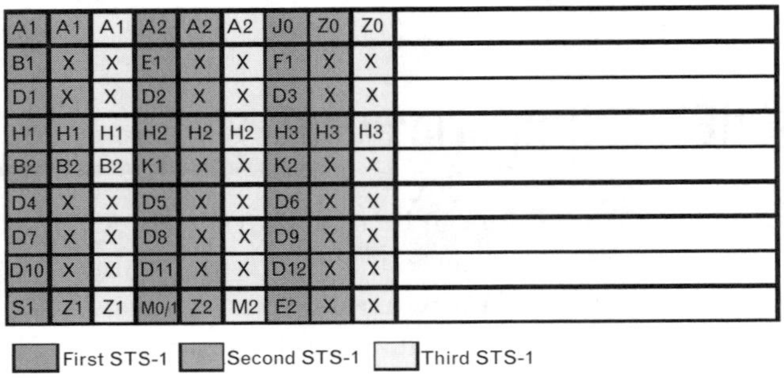

FIGURE 6.4
A SONET STS-3 frame showing how the STS-1s are interleaved.

are overhead. Focusing on the overhead, we see that the first three octets in the first row are all A1 octets. This is because we took three frames, like the one shown in Figure 6.2, and octet interleaved them. So we took the first octet from the first STS-1 frame (which is an A1 octet), then we took the first octet from the second STS-1 frame (which is an A1 octet), and then we took the first octet from the third STS-1 frame (which is also an A1 octet). Then we took the second octet from the first frame (which is an A2 octet), then the second octet from the second frame (which is an A2 octet), and finally the second octet from the third frame (which is also an A2 octet), and so on.

Another way to look at this interleaving is to take the first column from the first STS-1 frame and lay it in as the first column of the STS-3 frame. Then we take the first column from the second STS-1 frame and lay it next to the first column of the STS-3 frame that we are building. Then we take the first column from the third STS-1 frame and place it besides the existing two columns of the STS-3 frame. Then take the second column from the first STS-1 frame and put it into the fourth column slot of the STS-3 frame, and so on. Although this creates the proper STS-3 frame, keep in mind that the information is transmitted by row, octet by octet.

We have not talked about what all the octets mean yet, but look at the STS-3 frame. One can see that the first column of the STS-3 frame is the same as the first column of the STS-1 frame. And columns 4 and 7 of the STS-3 frame are the same as columns 2 and 3, respectively, of the first STS-1 frame. Because we interleaved these columns, this is to be expected. But look at columns 2, 3, 5, 6, 8, and 9 of the STS-3 frame. These are nothing like columns 1, 2, and 3 of the second and third STS-1 frames (although we see some similarity in row 1 and row 4). What is going on? It turns out that for many of the overhead octets, only one set is needed. This is not true for the overhead in row 1 or row 4, however. These are required in every column.

One type of payload is simply three of the individual STS-1 payloads (see Figure 6.5). The payloads would actually be inside of the frames but are shown separately to illustrate the fact that there are three of them.

We have not said anything yet about how to interpret these interleaved SONET/SDH streams. Suppose you needed to transmit data that exceeded an STS-1 data rate (51.84 Mbps). How would you do it? It turns out that SONET/SDH provides the ability to concatenate a number of STS-1 payloads to create a higher speed payload. This type of data stream is indicated with a lowercase "c" following the name (e.g., STS-3c or OC-3c). Concatenated channels have their payloads locked; that is, when payload adjustments are made (due to differences in clocks), they are done simultaneously for the entire frame and for N octets (this is explained later in the chapter). Thus, an STS-3c is a single payload (SPE) data stream of 150.336 Mbps. An STS-48c is a single payload (SPE) data stream of 2.405 Gbps.

Figure 6.6 shows the payload for a concatenated data channel. Figure 6.7 shows the relation of that concatenated payload to the transport overhead. Note that there is only one payload and that it is pointed to by the H1, H2,

FIGURE 6.5
Three individual STS-1 payloads carried in an STS-3 signal.

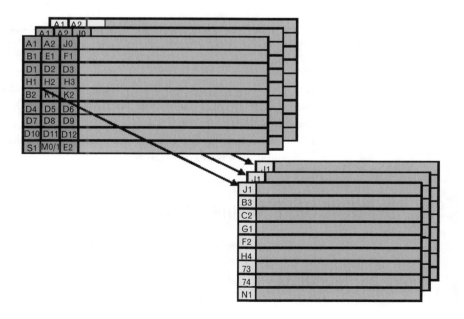

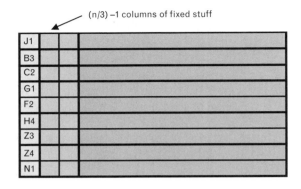

FIGURE 6.6
*A concatenated
payload. Note the
stuff columns after the
POH. These stuff
columns only appear
for SONET N > 3.*

FIGURE 6.7   *An
STS-3c concatenated
payload is pointed to
by the H1, H2, and
H3 pointers in the
first STS-1. Note
that no stuff columns
are shown because
they are only required
for N > 3.*

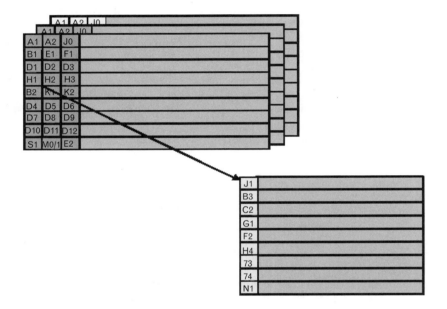

and H3 pointers in the first STS-1. How the pointers work is discussed in more detail later. Figure 6.6 is a bit tricky because it shows two "stuff" [4] columns after the POH. The note on the figure gives the equation for when these stuff columns are used: $(N/3) - 1$. For $N = 3$, this equation equals zero. So there are no stuff columns for an STS-3c, although there are stuff columns for all higher levels of concatenated SONET and SDH payloads.

6.3   Detail of the Transport Overhead

Now, let us examine each of the overhead octets in the transport overhead (the first three columns of an STS-1 frame). We examine the octets in the POH column later. Refer to Figure 6.8. The descriptions of the overhead octets are based on the ANSI T1.105 and ITU G.707 documents [5].

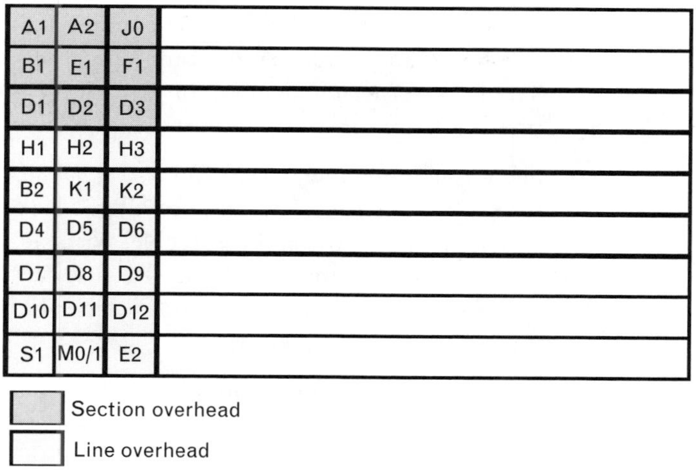

FIGURE 6.8
A SONET STS-1 frame showing detail of the transport overhead.

A1	A2	J0	
B1	E1	F1	
D1	D2	D3	
H1	H2	H3	
B2	K1	K2	
D4	D5	D6	
D7	D8	D9	
D10	D11	D12	
S1	M0/1	E2	

▨ Section overhead

☐ Line overhead

Framing octets (A1, A2)—*These octets allow the receiver to find the start of the SONET/SDH frame.* The A1 octet is 1111 0110 (hex 0xf6), whereas the A2 octet is 0010 1000 (hex 0x28). For SONET levels greater than STS-1 and less than or equal to STS-192, the A1 octet will be found in row 1, columns 1 to N (where N is the SONET level). The A2 octet will be found in row 1, columns $N + 1$ to $2N$. Framing for STS-768 uses the same A1, A2 values but limits the placement to columns 705 to 768 for A1 and columns 769 to 832 for A2 [6]. SDH uses the same values for the framing octets.

Section trace (J0)—*This octet allows two connected sections to verify that the connection is still alive and still connected to the right terminations.* The J0 octet is used to repetitively transmit either a 1-octet fixed length string or a 16-octet message so that a receiving terminal in a section can verify its continued connection to the intended transmitter. Only the first STS-1 carries the J0 value. This octet in the other STS-1s in a higher level SONET signal are reserved for future standardization. These reserved octets are referred to as Z0 octets. SDH uses this octet for the same purpose.

Parity (B1)—*The B1 octet is used by the receiver to estimate the bit error rate.* This octet is known as the *bit interleaved parity* (BIP-8) octet. Because the octet has 8 bits, 8 parities are computed, one for each bit of the octets of the frame. That is, you take the first bit of all of the octets in the frame and then set the first bit of the B1 octet so that the parity of these bits is even. Then you take the second bit of all of the octets in the frame, and set the second bit of the B1 octet so that it gives even parity, and so on. The parity represented by this octet is the parity of the *previous* frame. It is used to estimate the BER on the line. The B1 octet is only defined for the first STS-1 of an STS-N signal. SDH uses this octet for the same purpose.

Orderwire (E1)—*This octet is not important.* This octet was intended to be used for a voice channel between two technicians as they installed and tested an optical link. It is almost never used today. Technicians carry cellular phones and use these for communications when doing an installation. The E1 octet is only defined for the first STS-1 of an STS-*N* signal. SDH uses this octet for the same purpose.

Section user channel (F1)—*This octet is not important.* This octet is reserved for use by the network service provider. This octet is passed from section to section within a line and is readable, writable, or both at each section terminating equipment in that line. The use of this function is optional. The F1 octet is defined only for STS-1 number 1 of an STS-*N* signal. SDH uses this octet for the same purpose.

Section data communication channel (D1, D2, D3)—*These octets form a communication channel to send administrative messages.* These octets are allocated for section data communication and should be considered one 192-Kbps message-based channel for alarms, maintenance, control, monitor, administration, and other communication needs between section terminating equipment. This channel is available for internally generated, externally generated, and manufacturer-specific messages. These octets are defined only for STS-1 number 1 of an STS-*N* signal. SDH uses these octets for the same purposes.

Pointers and pointer action (H1, H2, H3)—*These octets are very important and are described in a later section.* These octets point to the payload (SPE), provide flags to indicate when the payload location changes, and provide a location for a data octet when a negative pointer adjustment is made. The operation of these pointers will be described in more detail in a later section. SDH handles pointers in the same way; however, the minimum SDH rate of STM-1 contains three H1 octets, three H2 octets, and three H3 octets.

Line parity (B2)—*The B2 octet is used by the receiver to estimate the bit error rate.* This octet operates in a fashion similar to the B1 octet, except that it excludes all of the section overhead octets and only applies to an STS-1. Because it only applies to an STS-1, there is a B2 octet in columns 1 to *N* of an STS-*N* signal. Note that this octet carries parity for the *previous* frame. SDH uses this octet for the same purpose.

Automatic protection switching (APS) channel (K1, K2)—*These octets are described in a later section.* These octets are used for APS signaling between line level entities. These octets are defined only for STS-1 number 1 of an STS-*N* signal. The operation and functionality of these octets are described in more detail in a later section of this chapter. SDH uses these octets for the same purpose.

Line data communications channel (D4–D12)—*These octets form a communication channel to send administrative messages.* These octets are allocated for line data communication and should be considered as one 576-Kbps message-based channel for alarms, maintenance, control, monitor, administration, and other communication needs between line-terminating entities. This is available for internally generated, externally generated, and manufacturer-specific messages. These octets are defined only for STS-1 number 1 of an STS-N signal. SDH uses this octet for the same purpose. SDH uses these octets for the same purpose but with additional codings.

Synchronization messaging (S1)—*This octet is not important to our discussion.* This octet is allocated for transporting synchronization status messages. S1 is defined only for STS-1 number 1 of an STS-N signal. Currently only bits 5–8 of S1 are used to transport synchronization status messages. These messages provide an indication of the quality level of the synchronization source of the SONET signal. SDH uses this octet for the same purpose.

STS-1 REI (M0)—*This octet sends the number of errors detected by the B octets back to the transmitter so it knows the line status as well as the receiver.* The octet in row 9, column $N + 1$ (where N is the value of the STS-N) can have two meanings. When the signal is an STS-1, it has the meaning of *remote error indicator* (REI). Currently only bits 5–8 of the M0 octet are to be used as a line REI function. These bits are used to convey the count of errors detected by the line BIP-8 (B2) octet back to the transmitter device. This count has 9 legal values, namely, 0 to 8. The remaining possible 7 values are to be interpreted as zero errors. Bits 1–4 of M0 are reserved for future use. Because there is no rate in SDH equivalent to STS-1, SDH does not define an M0 value for this octet (but see M1 later).

STS-N REI (M1)—*This octet sends the number of errors detected by the B octets back to the transmitter so it knows the line status as well as the receiver.* In a SONET signal at rates from STS-3 to STS-192, one octet, the M1 octet, is allocated for a line REI function. The M1 octet is located in the third STS-1 in order of appearance in the octet interleaved STS-N frame. The entire M1 octet is used to convey the count of errors detected by the line BIP-8 (B2) octet. This count has (8 times N) + 1 legal values, namely, 0 to 8N errors. For rates below STS-48, the remaining possible 255 − (8 times N) values are interpreted as zero errors. For the STS-48 and STS-192 rates, if the line BIP-8 detects greater than 255 errors, the line REI will relay a count of 255 errors. SDH uses this octet for the same purpose.

Orderwire (E2)—*This octet is not important.* This octet has the same purpose for line entities as the E1 octet has for section entities, and it is not used any more than the E1 octet. SDH uses this octet for the same purpose.

6.3.1   Payload Pointer Processing

So why do we have pointers in SONET/SDH? It all has to do with differences in clocks and accommodating those clocks. Now we look a bit closer at clocks, as used in communications systems. Suppose we have some type of network box that takes traffic from one side of the box, perhaps processes the traffic, and then sends the traffic out the other side. See Figure 6.9, for instance. Here, we are only going to look at one direction of the line, assuming that bits are flowing into the box from the left and leaving the box on the right.

Note in Figure 6.9 that the bit rate on the two sides is different. The bits are flowing in at 1 million bits per second but leaving at 1 million and 1 bits per second. Between the two lines is a FIFO or some kind of buffer that is used to minimize the problem we are going to discuss. Because the output line is faster, let's assume that we start the system off by allowing the line on the left to fill the FIFO before we start sending the bits out the right side.

Now, the right side line will send 1 bit more each second than the left side can put in the FIFO. Suppose we have an 8-bit FIFO. After 8 sec, the right side line will go to the FIFO to get a bit to transmit but there will not be any bits there. To fix this, the right side can send a "stuff" octet allowing the left side to fill the FIFO again. The important thing here is that the "stuff" octet be identifiable to the receiver connected to the right side line. That receiver can then throw the stuff octet away, leaving only valid data. Of course, throwing this stuff octet away may cause problems for that receiver, because it has no data to send for a short period of time. SONET/SDH actually uses this technique, and one to handle the case of a faster line feeding a slower line. The techniques for doing this are described in this section.

Technically, SONET/SDH are synchronous systems, meaning that all of the clocks are the same in the system (or so close as to be the same). However, even when all of the clocks are the same, jitter may exist that must be accommodated. And in reality, the clocks are not always the same in a SONET/SDH system.

Suppose that data are coming into a device slower (or faster) than they are being transmitted out the other side. Although buffers can be used to mitigate the effect of different clocks, eventually something has to be done to adjust for the difference between the receiving and transmitting clocks. This is where the pointer and pointer action octets (H1, H2, H3) come in.

The H1, H2 octets are the pointer octets, comprising 16 bits. The first 4 bits are the *new data flag* (NDF) bits and are set to 0110 during normal

FIGURE 6.9
A hypothetical network box that takes traffic from the left side and sends it out the right side.

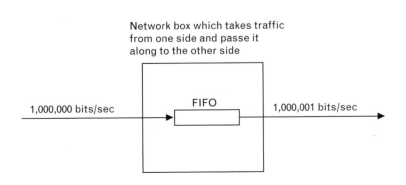

Network box which takes traffic
from one side and passe it
along to the other side

FIFO

1,000,000 bits/sec   1,000,001 bits/sec

operation. We will see that one way to introduce a new pointer value is by setting the NDF and including the new pointer. The next 2 bits have no meaning in SONET but are used in SDH [7].

The last 10 bits are the actual pointer and can vary from 0 to 782. A value of zero indicates that the payload (the SPE) starts at the first octet after the H3 octet. If the payload started at the second octet after the H3 octet, the pointer would have a value of one, and so on. Figure 6.10 shows the layout of the H1, H2 pointers, and Figure 6.11 shows the location of the

FIGURE 6.10
The usage of bits in the H1, H2 pointer octets. (Source: Draft Standard T1.105, October, 2000.)

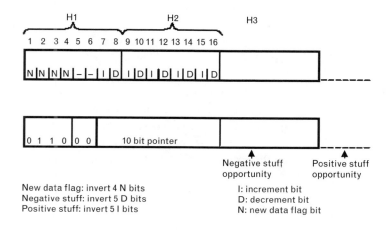

FIGURE 6.11 *Pointer values for an STS-1 SPE. The numbers indicate the value that would be carried in the last 10 bits of the H1, H2 octets to point at that specific octet. For example, the H1, H2 pointer would contain zero to point to the octet after the H3 octet, and would contain 782 to point to the last octet in row 3 of the next frame. Another common pointer is to point to the first payload octet of the next frame, value 522. (Source: Draft Standard T1.105, October 2000.)*

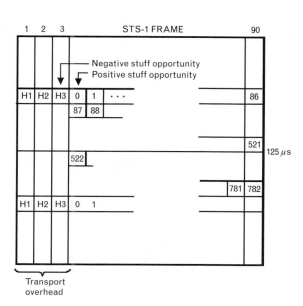

SPE for different values of the pointer. For the time being, ignore the "I" and "D" labels in Figure 6.10. The meanings of "I" and "D" are explained a little later in this section.

The H3 octet is used to carry a payload octet under certain conditions. Let's examine that. Suppose the incoming clock was faster than the outgoing clock. Eventually, we will accumulate an extra octet in our receiving buffer, over and above what we can transmit. How are we going to get this extra octet "caught up"?

The way we do it is to put that extra octet in the H3 location. So when we transmit one SONET frame of 810 octets, we actually transmit 784 octets of payload (86 columns times 9 rows, plus one H3 octet), rather than 783 octets of payload. So now we have "caught up" with that extra octet that accumulated on the receiving side (see Figure 6.12).

A similar problem occurs if the incoming clock is slow. Eventually, they will be an octet "deficit" in the receiving buffer that one needs to accommodate. This is done by putting a "stuff" octet in the location after the H3 octet. So when we transmit a SONET frame of 810 octets, we transmit only 782 octets of payload, rather than the 783 octets of payload. See Figure 6.13, which show this stuff octet adjustment.

But this affects the pointers. The SPE is still just 783 octets long so the SPE actually moved "backward" for a negative justification or "forward" for a positive justification in the SONET frame. Because the H1, H2 pointers point to the start of the SPE, these pointers have to be adjusted to account for this movement. Let's now discuss how this occurs.

Note in Figure 6.10 that the 10 bits of the pointer are labeled IDIDIDIDID. The "I" indicates "increment" and the "D" indicates decrement. When a new pointer value is to be introduced, the pointer (the last

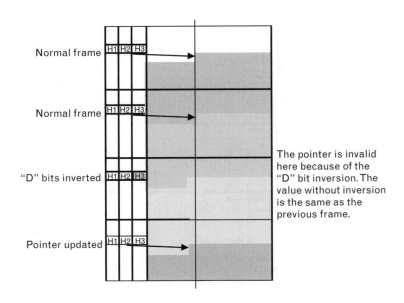

FIGURE 6.12
*Negative STS-1
pointer adjustment.
The "pointer" is the
last 10 bits of the H1,
H2 octets. The
meaning of the "D
bits" is explained in
the text.*

Normal frame   H1 H2 H3

Normal frame   H1 H2 H3

"D" bits inverted   H1 H2 H3

Pointer updated   H1 H2 H3

The pointer is invalid here because of the "D" bit inversion. The value without inversion is the same as the previous frame.

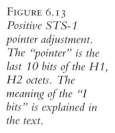

FIGURE 6.13
Positive STS-1 pointer adjustment. The "pointer" is the last 10 bits of the H1, H2 octets. The meaning of the "I bits" is explained in the text.

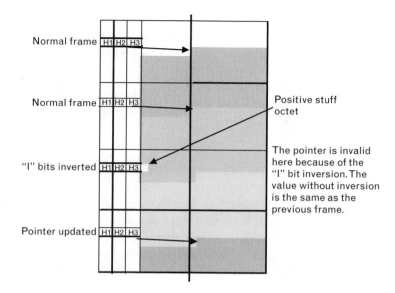

10 bits) is modified to indicate whether the adjustment is a positive or negative adjustment. If the adjustment is to be positive, the I bits are then inverted; if the adjustment is to be negative, the D bits are inverted.

The SONET equipment has its own register, which it uses to track the location of the SPE, and it compares the value in the last 10 bits of H1, H2 to that register to detect whether the adjustment is positive or negative (that is, which bits are inverted). Another way for the pointer to change is for the NDF to be set to indicate that a new pointer value is being introduced. We will cover that in more detail later.

Under normal operation, when the pointer is to be changed, the pointer will be modified, with the I bits inverted if the pointer is to be incremented and with the D bits inverted if the pointer is to be decremented. The SONET equipment compares the pointer value with the value in its register to determine if the action is an increment or a decrement. Let's look at this a bit closer. We start with the positive pointer adjustment. It may help if you also look at Figure 6.13 during the discussion on positive justification and Figure 6.12 during the discussion on negative justification.

Let's assume that the pointer is initially pointing at the 127th octet in the payload (I chose this number because it looks to be about where the pointer is pointing in Figures 6.12 and 6.13). Because the pointer value is one less (remember that the pointer value is zero to point at the first octet), the pointer will have a decimal value of 126, or 0x7e in hex. In terms of the 10 bits of the pointer, it will be 00 0111 1110.

Let's walk through a pointer change. The first row in Figure 6.14 indicates the bit values of H1, H2 for normal operational frames prior to the positive justification. On the frame where we do the positive justification,

FIGURE 6.14
The bit values for H1, H2 for a positive pointer adjustment using inverted I bits.

Positive justification using inverted "I" bits

Frame status	New data flag				Unused		I	D	I	D	I	D	I	D	I	D
Normal frame	0	1	1	0	X	X	0	0	0	1	1	1	1	1	1	0
Invert "I" bits	0	1	1	0	X	X	1	0	1	1	0	1	0	1	0	0
New ptr value	0	1	1	0	X	X	0	0	0	1	1	1	1	1	1	1
New ptr value	0	1	1	0	X	X	0	0	0	1	1	1	1	1	1	1
Normal frame	0	1	1	0	X	X	0	0	0	1	1	1	1	1	1	1

the H1, H2 octets will have the values indicated in the second row of Figure 6.14. Because we are doing a positive justification, all of the I bits will be inverted. The SONET equipment has a register where it keeps the pointer value so it can easily compare the value in the last 10 bits of H1, H2 to its register to see what is changed (and that the I bits are inverted). It is in this SONET frame that the octet after the H3 octet is stuffed with a non-data octet (see Figure 6.13). When a positive adjustment is done, the stuff octet after the H3 octet is ignored, so its content is meaningless—but it is usually all zeros.

Because the SONET equipment has been told that we are doing a positive justification, it can increment its register to point to the adjusted payload, even though the pointer in the H1, H2 octets has been "corrupted" by the I bit inversion.

On the next frame, the pointer has the new, incremented value. This is repeated for the next frame, and the fourth frame is considered a "normal" frame, available for another pointer adjustment.

Now, let's look at a negative adjustment. Assume the same starting conditions as above, payload at location 127, with a pointer value of 126. The first row in Figure 6.15 indicates the bit values of H1, H2 for normal operational frames prior to the negative justification. On the frame where we are going to do the negative justification, the H1, H2 octets will have the values indicated in the second row of Figure 6.15. Because we are doing a negative justification, all of the D bits will be inverted. The SONET equipment has a register where it keeps the pointer value so it can easily compare the value in the last 10 bits of H1, H2 to its register to see what has changed (and that the D bits are inverted). It is in this SONET frame that the H3 octet is used to carry a data octet.

FIGURE 6.15
The bit values for H1, H2 for a negative pointer adjustment using inverted D bits.

Negative justification using inverted "D" bits

Frame status	New data flag				Unused		I	D	I	D	I	D	I	D	I	D
Normal frame	0	1	1	0	X	X	0	0	0	1	1	1	1	1	1	0
Invert "D" bits	0	1	1	0	X	X	0	1	0	0	1	0	1	0	1	1
New ptr value	0	1	1	0	X	X	0	0	0	1	1	1	1	1	0	1
New ptr value	0	1	1	0	X	X	0	0	0	1	1	1	1	1	0	1
Normal frame	0	1	1	0	X	X	0	0	0	1	1	1	1	1	0	1

Because the SONET equipment has been told that we are doing a negative justification, it can decrement its register to point to the adjusted payload, even though the pointer in the H1, H2 octets has been "corrupted" by the D bit inversion.

On the next frame, the pointer has the new, decremented value. This is repeated for the next frame, and the fourth frame is considered a "normal" frame, available for another pointer adjustment.

When the H3 location is used for data, the actual payload octet placed in the H3 location is the octet that would normally have gone in the location just after the H3 octet. That is, the payload octets of row 4 slide left by one position. This leaves an "empty" octet in location 90 of row 4, which is filled by the octet that would have been in location 4 of row 5 (meaning that the payload of row 4 is 88 octets instead of 87 octets). So all of the payload octets from row 4 on are advanced one octet, including the next SPE.

In the next SONET frame, the 10 bits of pointer will point at the SPE (either one greater for a positive adjustment or one less for a negative adjustment). This adjustment can only occur every fourth SONET frame.

Pointer adjustments can also be signaled by the sender setting the NDF. Let's examine only a positive pointer adjustment and assume that the pointer is being adjusted by one octets for some administrative reason (see Figure 6.16).

The first line indicates a normal frame. In the next frame (the second row), the sending equipment has set the NDF and put a new pointer value in the last 10 bits of H1, H2. Note that the value is greater by one, indicating that the SPE has moved forward by one octet. Pointer values can change by much more than one octet with the use of the NDF. Suppose the sending equipment needed to make a large change in the location of the SPE for some reason. It could signal this change by the use of the NDF (this would likely cause errors to the customer, however).

Just like the pointer change indicated with the inverted I or D bits, pointer adjustments utilizing the NDF cannot occur more often than every fourth SONET frame. One reason for the limit of every fourth SONET frame is that another way a new pointer value can be set is if a new value appears for three consecutive frames, even if the I or D bits or the NDF is not set. After three consecutive frames with the same new pointer value, the receiver will accept the new value.

FIGURE 6.16
The bit values for H1, H2 for a one octet positive pointer adjustment using the new data flag.

One octet positive adjustment using the NDF

Frame status	New data flag				Unused		I	D	I	D	I	D	I	D	I	D
Normal frame	0	1	1	0	X	X	0	0	0	1	1	1	1	1	1	0
NDF indicator	1	0	0	1	X	X	0	0	0	1	1	1	1	1	1	1
New ptr value	0	1	1	0	X	X	0	0	0	1	1	1	1	1	1	1
New ptr value	0	1	1	0	X	X	0	0	0	1	1	1	1	1	1	1
Normal frame	0	1	1	0	X	X	0	0	0	1	1	1	1	1	1	1

Another reason for the limit of every fourth SONET frame is the positive pointer adjustment case where the pointer rolls over from 782 to 0. If one works out the situation in this case, one sees that the pointer in the frame following the frame with the NDF inverted will point to the same SPE as the (implied) pointer of the frame with the inverted NDF. This means that the system has to ignore the pointer for one SONET frame and the first valid pointer will be the third SONET frame. (The frame with the NDF inverted is one, the next frame is two, and the third frame has a valid pointer.)

A similar special case occurs when the pointer rolls over backward from 0 to 782. To begin, the zero value pointer is pointing at the octet after the H3 octet. The next frame has the NDF set and the H3 octet is used for data. The implied pointer is pointing to the H3 octet. The beginning of the next SPE is the last octet in row 3 of the SONET frame after the frame with the NDF inverted. However, the pointer in that frame cannot point backward in the frame, so the SONET equipment has to handle this situation as a special case. The pointer in that SONET frame points to the last octet in row 3 of the next (third) SONET frame. This can be confusing, but if one draws pictures similar to Figure 6.12 and Figure 6.17 one can see how it works.

This all works fine for an STS-1, which has only 783 octets in the payload. But what do we do when we have a concatenated payload, such as an STS-3c? Now we have 2,349 payload octets but the pointer is only 10 bits in length, giving the ability to point at only 1,024 locations (0 to 1,023). And much higher levels of concatenation are possible, such as an STS-192c. How do we get the pointers to work with these payloads? The answer is that we still use a value of 0 to 782 in the pointer but multiply it by N where N is the value in the STS-N. So for an STS-3c, we multiply the pointer value by 3 and adjustments take place three octets at a time. For STS-192c, we multiply the value in the pointer by 192 and adjustments take place 192 octets at a time.

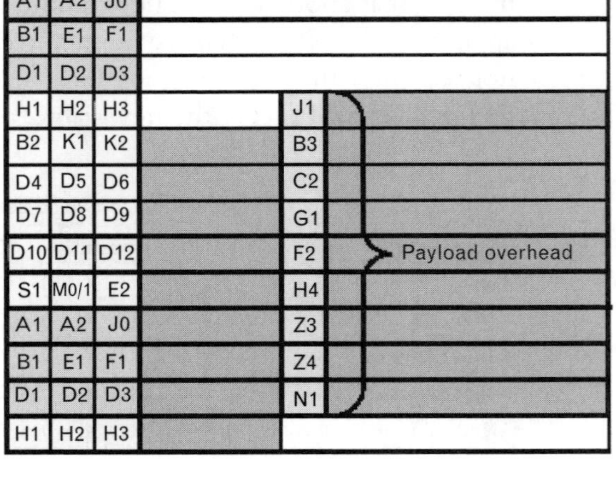

FIGURE 6.17
Payload overhead (POH) is the first column of the synchronous payload envelope (SPE).

A1	A2	J0				
B1	E1	F1				
D1	D2	D3				
H1	H2	H3		J1		
B2	K1	K2		B3		
D4	D5	D6		C2		
D7	D8	D9		G1		
D10	D11	D12		F2		Payload overhead
S1	M0/1	E2		H4		
A1	A2	J0		Z3		
B1	E1	F1		Z4		
D1	D2	D3		N1		
H1	H2	H3				

6.3.2   Payload Overhead

Up to this point, we have talked about the overall structure of the SONET/SDH frame, and how we point to, and track, the first octet of the SPE, which is also the first octet of the POH, but we have not talked about the POH itself and what the POH overhead octets mean. This is the subject of this section.

The POH is the first column of the SPE. (see Figure 6.17). It consists of 9 octets, the functions of which are explained next:

Path trace (J1)—*This octet allows the two ends to verify that the connection is still alive and still connected to the right terminations. It is just like J0 but on a path basis.* This octet is used to transmit repetitively an STS path access point identifier so that a path-receiving terminal can verify its continued connection to the intended transmitter. A 64-byte frame is used for the transmission of path access point identifiers. SDH uses this octet for the same purpose but uses a different message format, defined in G.707/G.831.

Path BIP-8 (B3)—*The B3 octet is used by the receiver to estimate the bit error rate.* This octet is used to provide parity over the payload. The path BIP-8 is calculated over all bits of the previous STS SPE before scrambling. Conceptually, this octet is similar to the B1 and B2 octets. SDH uses this octet for the same purpose but excludes the fixed stuff octets (which will be described when we talk about payload mappings) when calculating the parity. Because of this, the SONET equipment should use stuff octets that will produce the same parity as would be obtained by calculating the parity without the stuff octets [8].

STS path signal label (C2)—*This octet indicates the type of traffic carried in the payload. For our discussion, this octet may have a value of 0x02 to indicate floating VT mode, 0x04 for asynchronous mapping of a DS-3, 0x13 to indicate mapping of ATM, 0x16 for packet over SONET (POS), and 0x1b for generic framing procedure.* There are many other values—see T1.105 for the complete list. This octet is used to identify the construction and content of the STS-level SPE, and for STS path payload defect indication (PDI-P). PDI-P is an application-specific code that indicates to downstream equipment that there is a defect in one or more directly mapped embedded payloads in the STS SPE. SDH does not define codes for PDI-P.

Path status (G1)—*This octet is mainly used to send status back to the transmitter about the path status.* This octet is used to convey back to an originating STS *path terminating equipment* (PTE) the path terminating status and performance. This feature permits the status and performance of the complete duplex path to be monitored at either end or at any point along that path. As illustrated in Figure 6.18, bits 1 through 4 convey the count of interleaved bit blocks that have been detected in error by the path BIP-8 code (B3). This count has nine legal values, namely, 0 to 8 errors. The remaining

FIGURE 6.18
The path status octet,
showing the two fields,
the REI and the
RDI-P.
(Source: Draft
Standard T1.105,
October 2000.)

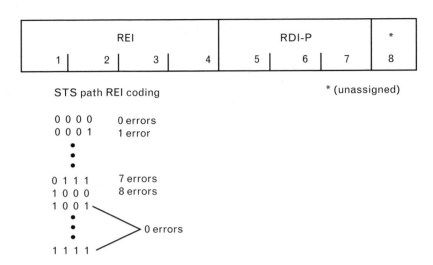

seven possible values represented by these 4 bits can only result from some condition unrelated to the forward path and shall be interpreted as zero errors. Bits 5, 6, and 7 provide codes to indicate both an old version (compliant with earlier versions of this standard) and an enhanced version of the STS path *remote defect indication* (RDI-P). The enhanced version of RDI-P allows differentiation between payload, connectivity, and server defects. Bit 7 is set to the inverse of bit 6 to distinguish the enhanced version of RDI-P from the old version. Overhead bit codes for RDI-P defects are described in T1.105. Bit 8 is unassigned at this time. SDH uses this byte for the same purpose. STS path REI coding is not suppressed when RDI-P is active. If RDI-P is triggered by a server defect, the REI will be undefined. If RDI-P is triggered by a connectivity defect or payload defect, the REI coding will reflect the incoming bit errors. The REI signal is undefined in SDH when RDI is active.

Path user channel (F2)—*This octet is not important.* This octet is allocated for user communications, similar to the F1 octet in the transport overhead. SDH uses this octet for the same purpose.

Multiframe indicator (H4)—*The uses of this octet will be described in later sections.* This octet provides a generalized multiframe indicator for payloads. This indicator is used for two purposes. The first is for VT-structured payloads, which are described in a later section. The second use is for the support of virtual concatenation of STS-1 SPEs. SDH uses this octet for the same purpose.

Growth octets (Z3, Z4)—*Ignore these two octets.* These two octets are reserved for future standardization. SDH defines Z3 as growth. SDH refers to Z4 as K3. Bits 1–4 of K3 are defined as a STS path APS overhead channel. Bits 5–8 of K3 are reserved for growth.

Tandem connection maintenance/tandem connection data link (N1)—
Ignore this octet. This octet is allocated to support *tandem connection mainte-nance* (TCM) and the *tandem connection data link* (TCDL). Bits 1–4 of N1 are used to provide the tandem connection *incoming error count* (IEC). In option 1, bits 5–8 of N1 are used to provide the TCDL. In option 2, bits 5–8 of N1 are used to provide maintenance information including remote error indication, outgoing error indication, remote defect information, outgoing defect information, and tandem connection access point identifier. For more information regarding TCM, refer to T1.105.05. SDH uses this byte for a similar purpose.

The TCDL of option 1 is an optional 32-Kbps data channel available to applications or services (e.g., tandem connection, security, and so on) that span more than one LTE–LTE (LTE = *line terminating equipment*) con-nection, but may be shorter than a PTE–PTE connection. As a result, the B3 byte must be recalculated at each point where this channel is altered. Any errors received at this point must be included in the resultant newly calculated B3 byte. Note that the TCDL can exist independently of TCM. The TCDL will use the LAPD protocol for passing information. Due to their timing requirements, TCM messages get precedence of use for the TCDL and may, therefore, preempt other messages on that channel. Because tandem connection terminating equipment is not required to per-form store-and-forward or layer 2 termination functions on non-TCM messages, some or all of the preempted messages may be lost and require retransmission. Note that the TCDL is not supported in SDH, which uses only option 2.

6.4   Concatenated Payloads

Up to this point, we have been talking about STS-1 (OC-1) streams (at 51.84 Mbps) and a little about SDH STM-1 streams (at 155.52 Mbps). For simplicity, let's continue to concentrate on the SONET situation.

We also talked about interleaving multiple STS-1 streams into a higher stream on the line, for example, three STS-1 streams into an STS-3. But with interleaved streams, each STS-1 stream maintains its own identity and is limited to 51.84 Mbps. Suppose, however, that you need a data stream faster than 51.84 Mbps—how can you get a faster rate from SONET?

The answer is with *concatenated payloads*. With concatenated payloads, multiple STS-1 payloads are joined and treated as a single payload. For example, a common concatenated payload is to join three STS-1 payloads. When this is done, the resulting stream is known as an STS-3c, with the lowercase "c" indicating concatenated. An STS-3c would look like Figure 6.4—it would consist of 270 columns with 9 rows and 27 columns of transport overhead. An important difference between an interleaved

STS-3 and a concatenated STS-3c is that there are three POH columns in the interleaved structure, one for each STS-1, while there is only one POH in the concatenated structure (see Figure 6.6).

Thus, an STS-3c provides a payload that is a little more than the payload of three STS-1s (because only one POH is required, the other two can be used to carry data). Although not precisely accurate, an STS-3c is essentially the same as an SDH STM-1.

A reasonable question at this point is how does the system know that the data stream is concatenated? Looking back at Figure 6.4, row 4, one will see three sets of H1, H2, H3 octets. The pointer is formed by taking the first H1 and the first H2 octets and interpreting the bits as described earlier, but multiplying the value in the last 10 bits by three (because it is an STS-3). The remaining H1 and H2 octets are used to indicate that the payload is concatenated.

The second H1 octet is paired with the second H2 octet to produce a 16-bit string similar to that produced by the first H1, H2 octets. Let's call this 16-bit word the *second pointer*. The third H1 is paired in the same manner with the third H2 octet, and we will call this 16-bit word the *third pointer*. To indicate concatenation, the second and third pointers are set up as follows: The first bits are set to 1001 (the new data flag value), the next 2 bits can be anything, and the last 10 bits are set to all 1's. Because 10 ones form a value greater than 782, which would normally be invalid, it (together with the NDF in the first 4 bits) provides the indicator mechanism that the SPE, which would normally be pointed to by this pointer, is joined to the previous SPE (see Figure 6.7). This "joining" is a chain that is terminated only by a valid pointer value so any level of concatenation can be done, from 2 to 768 SPEs.

SDH indicates concatenation in the same fashion. However, if the payload of an STM-1 consists of three VC-3's, all three sets of the H1, H2, H3 pointers will contain valid pointers, each to a different VC-3. If the SDH STM-1 payload consist of a single payload, the pointers will be as described above for the STS-3c.

6.5 Mapping of SONET/SDH Payloads

One way to look at the payload of SONET/SDH is simply as a bit stream. We could look at an STS-1 stream as providing 86 columns of octets, by 9 rows, 8,000 times per second, for a payload rate of 49.536 Mbps. An STS-3c could be thought of as 260 columns of octets (270 total columns minus 9 columns of transport overhead, minus 1 column of payload overhead), by 9 rows, 8,000 times per second for a payload rate of 149.76 Mbps. Within that payload, we could put any traffic we wanted.

Actually, we would not be that far off if we looked at it that way. However, the payload area has some additional complexities. Remember the statement at the beginning of this chapter about the fact that the designers of SONET/SDH were concerned about carrying their plesiochronous traffic. So we need to look at what facilities were built into the payload area to handle the different DS-N and E rates and how the differing clocks of the plesiochronous traffic are accommodated. Handling the clocks is especially important—SONET/SDH is specified as a synchronous system, which means we should only have to accommodate clock jitter. But plesiochronous networks do not use the same clock—there are definitely differences between clocks on different plesiochronous circuits and also between the plesiochronous traffic and the SONET/SDH clocks.

So the first thing we are going to look at is how plesiochronous traffic is mapped into SONET. Later, we will examine how nonplesiochronous traffic, such as ATM, *packet-over-SONET* (POS), and GFP, is mapped.

6.5.1 Virtual Tributaries

The specific traffic that the designers were interested in carrying is shown in Table 6.2. Although SONET was designed by ANSI, note the specification of a European rate, E1. This was done to reduce problems of cross-border traffic. Let's say that a company in Europe wanted to provide a high-speed connection to an office in the United States They might purchase an E1 leased line (2.048 Mbps). But when the circuit got to the United States, it would either have to be reduced to DS-1 (1.544 Mbps) or two DS-1s would have to be used and the traffic divided between them. Supporting E1 rates in SONET allowed E1 rates to be delivered in the United States. At the time SONET was designed, higher rates were too expensive to use for this type of application.

Unfortunately, this area of mapping plesiochronous traffic into SONET is not simple. Let us look one layer at a time. To begin, the payload area of an STS-1 SONET frame consists of 87 columns by 9 rows, as shown in Figure 6.19.

TABLE 6.2 PLESIOCHRONOUS TRAFFIC MAPPED INTO A SONET STS-1 FRAME

TYPE OF DIGITAL CIRCUIT	BIT RATE (MBPS)
DS-1 (T1)	1.544
E1	2.048
DS-1C	3.152
DS-2	6.312
DS-3 (T3)	44.736

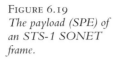

FIGURE 6.19
*The payload (SPE) of
an STS-1 SONET
frame.*

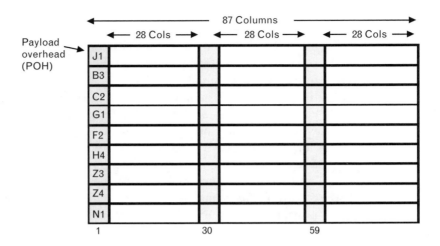

One column is taken by the POH leaving 86 columns. Next, we break the 86 columns into seven groups of 12 columns. Now seven groups of 12 columns is only 84 columns, leaving 2 extra columns. These 2 columns are columns 30 and 59 [9], where the POH is counted as column 1. All mappings of payloads into an STS-1 frame have these 2 columns "blocked out," meaning that the real payload of an STS-1 SPE is really only 84 columns by 9 rows, by 8 bits, 8,000 times per second or 48.384 Mbps.

Each of these seven groups is called a *virtual tributary group* (VTG). The seven VTGs are interleaved into the 84 columns in the same manner as was discussed earlier for interleaving STS-1s into higher levels of SONET, for example, into an STS-3. That is, the first column of the first VTG goes into the column after the POH. Then the first column of the second VTG, then the first column of the third VTG, and so on. After you finish with first column of all seven VTGs, you do the same thing with the second columns of all of the VTGs, and so on. This interleaves the VTGs (see Figure 6.20).

Now let's look closer at one of the VTGs. It consists of 12 columns, equaling a gross bit rate of 6.912 Mbps. Remember that we want to carry a number of different plesiochronous rates, from 1.544 to 6.312 Mbps. The gross bit rate is good for the DS-2 (at 6.312 Mbps), but it would be wasteful for the lower rates.

We can accommodate all of the rates specified in Table 6.2, however, simply by subdividing the 12 columns. For example, three columns, which together are called a VT-1.5, give a gross bit rate of 1.728 Mbps. Four columns, which are called a VT-2, give a gross bit rate of 2.304 Mbps. Six columns, called a VT-3, give a gross bit rate of 3.456 Mbps. And 12 columns, called a VT-6, give a gross bit rate of 6.912 Mbps. So a DS-1 (T1) fits into three columns (a VT-1.5), an E1 fits into four columns (a VT-2), a DS-1C fits into six columns (a VT-3), and a DS-2 fits into 12 columns (a VT-6). One restriction is that a VTG can only contain one type of mapping, that is, four VT-1.5s for four DS-1s, three VT-2s for three E-1s, two VT-3s for

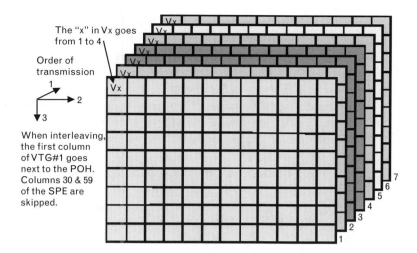

The "x" in Vx goes from 1 to 4

Order of transmission

When interleaving, the first column of VTG#1 goes next to the POH. Columns 30 & 59 of the SPE are skipped.

FIGURE 6.20 *The 84 usable columns of the STS-1 SPE are divided into seven groups of 12 columns. Each group of 12 columns is known as a VTG. The VTGs are interleaved into the SPE in the same fashion as STS-1 are interleaved into an STS-N.*

two DS-1Cs, or one VT-6 for one DS-2. An SPE payload can contain different VTGs with different types of VTs, but within a VTG one can only have one type of traffic (See Figure 6.21).

Next, we are going to look at the overhead octets that are required to make all of this work. We describe the mapping of DS-1 (T1) circuits; the mappings of the other rates is similar—if one understands the DS-1 case, one will be able to read the specification to learn the specifics of how the other rates are handled.

To begin, the C2 octet in the POH signals that the payload contents are VTs by containing the hex code 0x02 (0000 0010). Once this is known, the SONET equipment knows the location of each of the VTGs because they are in fixed positions in the payload.

FIGURE 6.21
A VTG with four VT-1.5s interleaved within it. Each level of shading represents one VT-1.5.

One VT-1.5

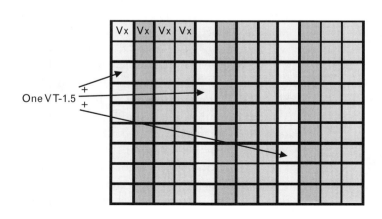

But we immediately run into a problem. The three columns used to transport a DS-1 have only 27 octets per frame (which occurs 8,000 times per second). A DS-1 has 24 octets per frame (which occurs 8,000 times per second) plus an extra framing bit. If we have to use more than three of our VT octets for overhead, we cannot carry the DS-1. This is a problem because we have to implement a pointer system, similar to the H1, H2, H3 octets used in the SONET transport overhead, to handle the differences in clocks between the plesiochronous traffic and the SONET network.

To get around this problem, the designers created the concept of a *superframe*. A superframe is simply four VT frames. If we take one octet from each of the four frames, we will have enough octets to implement a pointer like the H1, H2, H3 pointer system. And, it turns out, that is exactly what we do.

But how are we going to indicate the superframe? We use the last 2 bits in the H4 octet in the POH. These 2 bits count from 00 to 11 and then roll over to 00 again (00, 01,10, 11, 00). There is one special thing about the count in the H4 octet. The value in the H4 octet indicates the superframe number for the *next* payload, not for the payload associated with this H4. So an H4 with the value 00 in the last 2 bits indicates that the first frame of a superframe will occur in the next SONET payload.

So now that we can identify the frames of the superframe, we take the first octet in the VTG (which is also the first octet of the first VT in the VTG) and use it for overhead. Because we have a superframe of four frames, we have four overhead octets, known as V1, V2, V3, and V4. The SONET equipment takes the V1 and V2 octets and creates a 16-bit word. This word is interpreted in a very similar fashion to the way the word is created by combining the H1, H2 octets in the transport overhead (see Figure 6.22).

Note that the layout of the 16-bit word is the same as the H1, H2 octets, except that bits 5 and 6 are now used to indicate the type of traffic within the VTG. Each VT has the first octet taken as overhead (the Vx octets). Because the first VT will occupy the first column of the VTG, and since the SONET equipment knows how to find the VTGs, it can easily find the Vx octet of the first VT. Once it has processed one superframe it will have the V1, V2 octets for that VT. It can look at bits 5 and 6 of the V1, V2 word and find out what kind of VTs are in the VTG. (In our case, it will contain VT-1.5.) Because all the VTs in a VTG must be the same, the SONET equipment will know how many columns are contained in the VT and from that, it can find the other VTs in the VTG [10]. Finding the start of the VTs is easy, of course, because of the interleaving. They are together at the beginning of the VTG. But until the SONET equipment knows what kind of VT, it does not know how many VTs are in the VTG.

The V3 octet is used in the same fashion as the H3 octet. It normally does not contain data but will when a negative justification is done. The

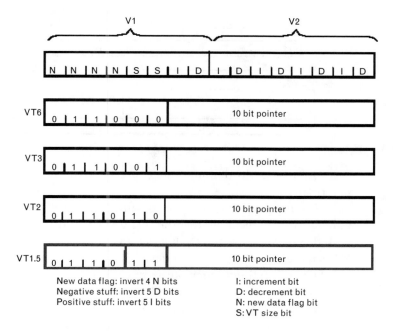

FIGURE 6.22   *The bit definition of the V1, V2 octets in a virtual tributary. The VT-1.5 is highlighted for discussion purposes. (Source: Draft Standard T1.105, October 2000.)*

octet after the V3 octet is used for a positive justification, just like the octet after the H3 octet. The V4 octet has no meaning and is reserved for future standardization.

It was mentioned that the V1, V2 octets form a pointer, but what does this pointer point to? Well, just like the H1, H2 pointer points to the SONET payload (SPE), the V1, V2 pointer points to the VT payload. And just what is this VT payload and where is it? Well, we started with 27 octets in each VT-1.5 and we used one octet for the V1, V2, V3, and V4 overheads. We now have 26 octets remaining in each payload frame of the superframe. So what we have left is four frames of 26 octets each, or a total of 104 octets. Because there are 104 payload octets, the pointer value for a VT-1.5 goes from 0 to 103 (decimal), where 0 points to the first location and 103 points to the last location, exactly like the H1, H2 pointer. The operation of the V1, V2 octets is exactly the same as the H1, H2 octets. To increment (decrement) the pointer, the sender can invert the I (D) bits to signal the receiver. Likewise, the NDF can be used in exactly the same fashion as was described for the H1, H2 octets. Now, let's see what those values mean and where they point (see Figure 6.23).

The V1, V2 pointer points to the first octet of the 104-octet VT superframe. If this octet is located immediately after the V2 octet, the pointer value is 0. Because there are only 104 octets in the VT superframe, the highest value of the pointer is 103, which means that the first

FIGURE 6.23
The value of the pointer for different locations of the VT payload.

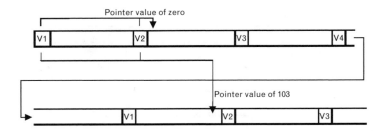

octet of this payload immediately precedes the V2 octet of the next superframe.

This 104-octet payload "floats" just like the SONET payload. The only sort of complexity is that the V1, V2, V3, and V4 octets are mixed in with these 104 octets (the total of the VT payload and the V1, V2, V3, and V4 octets gives 108 octets), but let's ignore this right now.

The SONET equipment keeps track of where the V1, V2, V3, and V4 octets are so we can think of the VT payload as existing in isolation without the V1, V2, V3, and V4 octets. Because we have a four-frame multiframe, the VT payload is divided into four sections, or frames, of 26 octets each (see Figure 6.24).

Each payload frame of the VT superframe contains 1 overhead octet at its beginning, leaving 25 octets in each frame. These overhead octets are the V5, J2, Z6, and Z7 octets and are defined next. Much of the text for the definitions that follow of each octet is taken from the T1.105 draft.

VT Header (V5)—*This octet serves a number of purposes but for our discussion that follows, we are only concerned with certain values in bits 5, 6, and 7, specifically in the value of 010, which indicates "asynchronous mapping" of a DS-1, and in a value of 100, which indicates "byte-synchronous mapping" of a DS-1. Remember that a value of 11 for bits 5 and 6 of the V1, V2 word indicates that we are mapping a DS-1.* This octet is the first octet of the VT payload, is pointed to by the VT pointers (V1, V2), and provides for VT paths the same functions that B3, C2, and G1 provide for STS paths, namely, error checking, signal label, and path status. The bit assignments of the V5 byte are specified in

FIGURE 6.24
The four VT payload frames of a superframe.

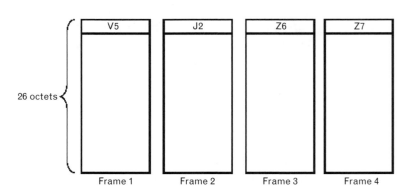

the following paragraphs and are illustrated in Figure 6.25. SDH uses this byte for the same purpose.

Bits 1 and 2 of V5 are used for error performance monitoring. A BIP scheme is specified. Bit 1 is even-parity calculated over all odd-numbered bits (1, 3, 5, and 7) in all bytes in the previous VT SPE. Similarly, bit 2 is even-parity calculated over the even-numbered bits. Note that the calculation of the BIP-2 includes the VT path overhead bytes but excludes the VT pointers.

Bit 3 of V5 is a VT path REI that is sent back toward an originating VT PTE if one or more errors were detected by the BIP-2. REI coding is not suppressed when RDI-V is active. If RDI-V is triggered by a server defect, the REI will be undefined. If RDI-V is triggered by a connectivity defect or payload defect, the REI coding will reflect the incoming BIP errors.

Bit 4 of V5 is reserved for mapping-specific functions. Currently, it is only defined for byte-synchronous DS-1 mapping, in which case it is used for a VT *remote failure indication* (RFI-V). RFI-V is generated by setting bit 4 of V5 to one. RFI-V is cleared by setting bit 4 of V5 to zero.

Bits 5 through 7 of V5 provide a VT signal label. Eight binary values are possible in these 3 bits. See the T1.105 standard for a complete list of meanings for bits 5 through 7 of V5.

Bit 8 of V5, in combination with bits 5, 6 and 7 of Z7, provide codes to indicate both an old version and an enhanced version of the VT Path remote defect indications (RDI-V). The enhanced version of RDI-V allows differentiation between payload, connectivity, and server defects. Bit 8 of V5 is set equal to bit 5 of Z7. This allows old equipment to receive and interpret an RDI-V indication. Bit 7 of Z7 is set to the inverse of bit 6 of Z7, to distinguish the enhanced version of RDI-V from the old version.

VT Path Trace (J2)—*This octet allows the two ends to verify that the connection is still alive and still connected to the right terminations.* This octet is used to transmit repetitively a VT path access point identifier so that a path-receiving terminal can verify its continued connection to the intended transmitter. A 16-byte frame is used for the transmission of path access point identifiers. SDH uses this byte for the same purpose, but uses the format specified in G.707/G.831.

Growth (Z6, Z7)—*Ignore these two octets.* These octets actually have some complex meanings that will not be described here. See T1.105 for additional details.

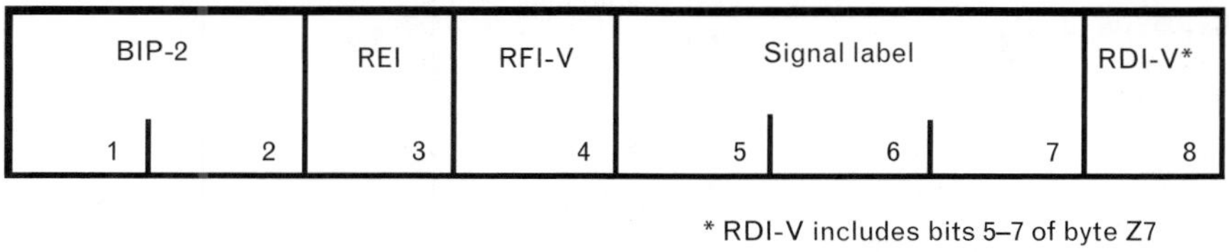

FIGURE 6.25
The bit assignments of the V5 octet. (Source: Draft Standard T1.105, October 2000.)

BIP-2		REI	RFI-V	Signal label			RDI-V*
1	2	3	4	5	6	7	8

* RDI-V includes bits 5–7 of byte Z7

Note that we have 25 octets left for each frame (which occurs each 125 μs). A DS-1 has 24 octets in a 125 μs, but there are a couple of additional problems. A DS-1 can be used for carrying 24 voice calls, or it can be used to carry data, such as Internet traffic. Also, a DS-1 is not just a stream of octets. Every 24 octets (or voice samples when the DS-1 is used for voice), an extra bit is used for framing. Additionally, the voice samples may have associated signaling to indicate the status of the voice call. It turns out that we can communicate all of this by using one octet of each VT frame, leaving 24 octets to carry the DS-1 payload. Let's look at Figure 6.26 to see how this is done.

The P bits are used to indicate the phase of the DS-1 signal. For example, a DS-1 using *superframe* (SF) framing has a superframe of 12 193-bit frames. A DS-1 using *extended superframe* (ESF) framing has a superframe of twenty-four 193-bit frames. The two P bits allow the framing to be divided into groups of six. For example, a value of 00 is used in the P bits to indicate the first six frames of either SF or ESF framing.

The S bits are used to communicate the A, B, C, and D channel signaling bits [11]. The status of four channels can be sent each frame. Because each P period is six frames, the status of a single indicator (A, B, C, or D) can be sent each P period. Thus, for ESF, all channel status bits (A, B, C, and D) can be sent for all 24 channels through one rotation of the P bits (from 00, to 01, to 10, to 11). Table 6.3 gives the meaning of the framing and channel signaling bits for various values of the P bits.

This type of mapping is known as *byte-synchronous mapping*. Byte-synchronous mapping preserves the format of the DS-1 signal, allowing any DS-0 to be extracted anywhere along the communication chain.

Asynchronous mapping, described next, does not maintain the DS-1 payload octet identity—it just takes a group of bits, usually 193 bits, and puts them in a frame of the VT superframe. Asynchronous mapping is very common. A great many DS-1 circuits do not carry channelized voice (i.e., telephone calls)—they carry data, usually IP traffic. Because of this, there is no need to maintain the identity of the DS-1 payload octets; simply carrying the

FIGURE 6.26
Byte synchronous mapping of a DS-1 signal. The octet after the overhead octets is used to transport the SF or ESF bits, as well as the A, B, C, D signaling bits for each DS-0.

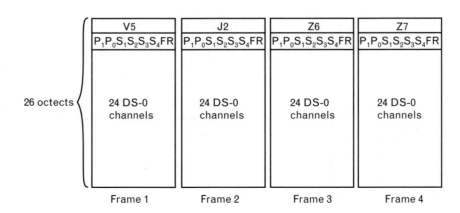

TABLE 6.3 INTERPRETATION OF THE Sn AND SF AND ESF FRAMING BITS FOR DIFFERENT VALUES OF THE P BITS

Signaling			Format		
2 State	4 State	16 State	SF	ESF	
$S_1\ S_2\ S_3\ S_4$	$S_1\ S_2\ S_3\ S_4$	$S_1\ S_2\ S_3\ S_4$	F	F	P_1P_0
$A_1\ A_2\ A_3\ A_4$	$A_1\ A_2\ A_3\ A_4$	$A_1\ A_2\ A_3\ A_4$	F_1	M_1	00
$A_5\ A_6\ A_7\ A_8$	$A_5\ A_6\ A_7\ A_8$	$A_5\ A_6\ A_7\ A_8$	S_1	C_1	00
$A_9\ A_{10}\ A_{11}\ A_{12}$	$A_9\ A_{10}\ A_{11}\ A_{12}$	$A_9\ A_{10}\ A_{11}\ A_{12}$	F_2	M_2	00
$A_{13}\ A_{14}\ A_{15}\ A_{16}$	$A_{13}\ A_{14}\ A_{15}\ A_{16}$	$A_{13}\ A_{14}\ A_{15}\ A_{16}$	S_2	F_1	00
$A_{17}\ A_{18}\ A_{19}\ A_{20}$	$A_{17}\ A_{18}\ A_{19}\ A_{20}$	$A_{17}\ A_{18}\ A_{19}\ A_{20}$	F_3	M_3	00
$A_{21}\ A_{22}\ A_{23}\ A_{24}$	$A_{21}\ A_{22}\ A_{23}\ A_{24}$	$A_{21}\ A_{22}\ A_{23}\ A_{24}$	S_3	C_2	00
$A_1\ A_2\ A_3\ A_4$	$B_1\ B_2\ B_3\ B_4$	$B_1\ B_2\ B_3\ B_4$	F_4	M_4	01
$A_5\ A_6\ A_7\ A_8$	$B_5\ B_6\ B_7\ B_8$	$B_5\ B_6\ B_7\ B_8$	S_4	F_2	01
$A_9\ A_{10}\ A_{11}\ A_{12}$	$B_9\ B_{10}\ B_{11}\ B_{12}$	$B_9\ B_{10}\ B_{11}\ B_{12}$	F_5	M_5	01
$A_{13}\ A_{14}\ A_{15}\ A_{16}$	$B_{13}\ B_{14}\ B_{15}\ B_{16}$	$B_{13}\ B_{14}\ B_{15}\ B_{16}$	S_5	C_3	01
$A_{17}\ A_{18}\ A_{19}\ A_{20}$	$B_{17}\ B_{18}\ B_{19}\ B_{20}$	$B_{17}\ B_{18}\ B_{19}\ B_{20}$	F_6	M_6	01
$A_{21}\ A_{22}\ A_{23}\ A_{24}$	$B_{21}\ B_{22}\ B_{23}\ B_{24}$	$B_{21}\ B_{22}\ B_{23}\ B_{24}$	S_6	F_3	01
$A_1\ A_2\ A_3\ A_4$	$A_1\ A_2\ A_3\ A_4$	$C_1\ C_2\ C_3\ C_4$	F_1	M_7	10
$A_5\ A_6\ A_7\ A_8$	$A_5\ A_6\ A_7\ A_8$	$C_5\ C_6\ C_7\ C_8$	S_1	C_4	10
$A_9\ A_{10}\ A_{11}\ A_{12}$	$A_9\ A_{10}\ A_{11}\ A_{12}$	$C_9\ C_{10}\ C_{11}\ C_{12}$	F_2	M_8	10
$A_{13}\ A_{14}\ A_{15}\ A_{16}$	$A_{13}\ A_{14}\ A_{15}\ A_{16}$	$C_{13}\ C_{14}\ C_{15}\ C_{16}$	S_2	F_4	10
$A_{17}\ A_{18}\ A_{19}\ A_{20}$	$A_{17}\ A_{18}\ A_{19}\ A_{20}$	$C_{17}\ C_{18}\ C_{19}\ C_{20}$	F_3	M_9	10
$A_{21}\ A_{22}\ A_{23}\ A_{24}$	$A_{21}\ A_{22}\ A_{23}\ A_{24}$	$C_{21}\ C_{22}\ C_{23}\ C_{24}$	S_3	C_5	10
$A_1\ A_2\ A_3\ A_4$	$B_1\ B_2\ B_3\ B_4$	$D_1\ D_2\ D_3\ D_4$	F_4	M_{10}	11
$A_5\ A_6\ A_7\ A_8$	$B_5\ B_6\ B_7\ B_8$	$D_5\ D_6\ D_7\ D_8$	S_4	F_5	11
$A_9\ A_{10}\ A_{11}\ A_{12}$	$B_9\ B_{10}\ B_{11}\ B_{12}$	$D_9\ D_{10}\ D_{11}\ D_{12}$	F_5	M_{11}	11
$A_{13}\ A_{14}\ A_{15}\ A_{16}$	$B_{13}\ B_{14}\ B_{15}\ B_{16}$	$D_{13}\ D_{14}\ D_{15}\ D_{16}$	S_5	C_6	11
$A_{17}\ A_{18}\ A_{19}\ A_{20}$	$B_{17}\ B_{18}\ B_{19}\ B_{20}$	$D_{17}\ D_{18}\ D_{19}\ D_{20}$	F_6	M_{12}	11
$A_{21}\ A_{22}\ A_{23}\ A_{24}$	$B_{21}\ B_{22}\ B_{23}\ B_{24}$	$D_{21}\ D_{22}\ D_{23}\ D_{24}$	S_6	F_6	11

Notes:

SF Format: $F_1 - F_6$ = Frame alignment bits
$S_1 - S_6$ = Signaling alignment bits

ESF Format: $F_1 - F_6$ = Frame alignment bits
$C_1 - C_6$ = Cyclic redundancy check-6 bits
$M_1 - M_6$ = Data link bits
$A_1 - A_{24}$ = Signaling bits
$B_1 - B_{24}$ = Signaling bits
$C_1 - C_{24}$ = Signaling bits
$D_1 - D_{24}$ = Signaling bits

Source: ANSI Standard T1.105.02-1995.

stream of bits is sufficient. The terminating equipment at the ends of the circuit will be able to obtain frame synchronization and extract the payload.

Asynchronous mapping is simpler than byte synchronous mapping. Because of this, the bits in the second octet of each VT frame have a different meaning from byte synchronous mapping (see Figure 6.27).

FIGURE 6.27
*Asynchronous
mapping of a DS-1
signal. The octet after
the overhead octets is
used to transport the
framing bits and to
accommodate jitter.*

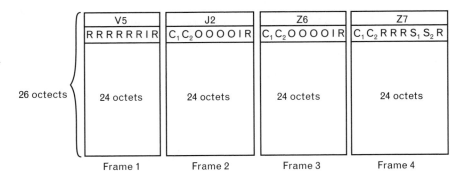

The R bits are unused (fixed stuff). The O bits are reserved for future standardization and should be considered the same as the R bits. The I bits are simply information bits, needed because we normally carry 193 bits in a frame.

Note that the last frame of the superframe has different bit assignments in the second octet. The purpose of these signaling bits is to allow the last frame to carry either 192, 193, or 194 bits to adjust for clock differences. Let's see how this is done.

The three C1 bits are used to control the function of the S1 bit, while the three C2 bits are used to control the function of the S2 bit. If all of the C1 bits are zero, it indicates that the S1 bit should be treated as data and its content included in the output data stream. If all of the C1 bits are one, it indicates that the S1 bit should be treated as a stuff bit and its content *should not* be included in the output data stream. The C2 bits are used in the same way to control the use of the S2 bit. Majority voting applies (two out of three in case of a bit error).

So why does one do this? Normally, the last frame of the VT super-frame should contain 193 bits. By convention, bit S2 is used to carry an information bit under this normal situation. If the DS-1 clock is running fast, however, we use the S1 bit to carry an extra information bit for this one frame of the superframe. The last frame of the VT superframe will carry 194 bits in this case, or 773 bits in the VT superframe.

If the DS-1 clock is running slow, we can signal that the S2 bit is a stuff bit and only carry 192 bits in the last frame of the VT superframe, or 771 bits in the VT superframe. This allows us to make one clock adjustment every VT superframe.

This completes our discussion of how DS-1 signals are mapped into SONET. The mapping of E1 (2.048 Mbps) is very similar to the mapping of DS-1 signals—E1 signals can be transported with byte-synchronous mapping or asynchronous mapping. Higher levels of signals, for example, DS-1C and DS-2 signals, can only be transported via asynchronous mapping.

Now, we move to examine the mapping of some signals that do not require virtual tributaries—these signals fill the STS-1 SPE. The specific signals we will examine are DS-3 (T3), ATM, POS, and GFP.

6.5.2   Support of DS-3 Signals

Carriage of a DS–3 signal is indicated by a value of hex 04 (0x04) in the C2 octet of the POH. When a DS–3 is mapped into an STS-1 SONET payload, columns 30 and 59 of the SPE are stuffed with a fixed value and cannot be used (the same situation as described above when the virtual tributary groups are used). This leaves 84 columns by 9 rows for usable payload. The SPE is broken into 9 frames, one per row (see Figure 6.28).

Note that there are 77 full information payload octets in each row. Additionally, the C1 octet carries 5 information bits, providing a total of 621 information bits per row. Because there are 9 of these frames per SPE and the SPE repeats 8,000 times per second, these 621 bits would provide a bit rate of 44.712 Mbps.

A DS–3 operates at 44.736 Mbps. How are we going to carry that rate when we only have a capacity of 44.712 Mbps? The answer lies in the stuff bit in C3 and the "c" bits in C1, C2, and C3. When the "c" bits in C1, C2, and C3 are zero, the "s" bit is interpreted to be a data bit and its contents are inserted into the output data stream. When the "c" bits in C1, C2, and C3 are one, the "s" bit is interpreted to be a stuff bit and its value is ignored.

If every frame had the stuff bit set as a data bit, each frame would carry 622 bits, which would give a data rate of 44.784 Mbps. So the use of this

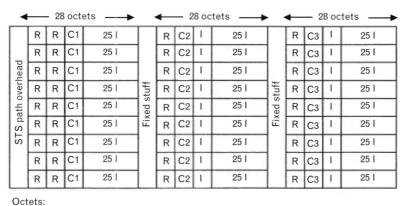

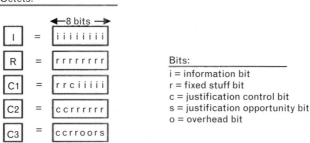

Bits:
i = information bit
r = fixed stuff bit
c = justification control bit
s = justification opportunity bit
o = overhead bit

FIGURE 6.28   *Mapping of a DS-3 signal into an STS-1 SPE (payload). Note that columns 30 and 59 of the SPE cannot be used, just as when VT transport is done. Each row is identical, providing 9 frames every 125 μs. (Source: ANSI Standard T1.105.02-1995.)*

stuff bit allows us to achieve the proper bit rate for a DS-3 and also allows us to accommodate clock differences between the DS-3 signal and the SONET signal.

6.5.3 Support of ATM, POS, and GFP

Compared to carrying plesiochronous traffic, carrying ATM [12], POS [13], or GFP [14] traffic is straightforward. For an STS-1, the payload consists of 9 rows and 87 columns, 1 of which is the POH, and 2 are fixed stuff (columns 30 and 50 numbered from the POH). This leaves 84 columns by 9 rows for payload.

All three types of traffic require that the octet boundary of the traffic be available. That is, the traffic is placed in the SPE with the traffic octet aligned with the SONET payload octets. This is one reason why POS defines a shielding character instead of using zero bit insertion, as is done in ordinary HDLC. If zero bit insertion were done, octet alignment would be lost very quickly [15].

Beyond that one requirement, the payload of the SONET frame is simply viewed as an octet transport mechanism. The traffic is not examined in any way, nor is there any requirement for any kind of alignment on SPE boundaries. As an example, ATM cells are taken one octet at a time with each octet placed in the next available octet in the SPE without regard for any boundaries in the cell or the SPE, other than maintaining octet alignment. POS and GFP are handled in exactly the same way. This topic is discussed in more detail in Chapter 7.

As an aside, note that the SPE of an STS-1 always has columns 30 and 59 of the SPE stuffed and unavailable for payload traffic. If a customer had the option of putting traffic into three STS-1s or one STS-3c, it would be better to choose the STS-3c. Let's see why. The SPE of an STS-3c consists of 261 columns (270 columns minus 9 columns for transport overhead). The POH will take one column of the SPE leaving 260 columns for user traffic.

If the customer used three STS-1s, they would receive three 84 columns of payload, or only 252 columns compared to 260 columns for the STS-3c. Eight columns of payload is equal to a little more than 4.6 Mbps, or the equivalent of about three DS-1s. It is one of the oddities of SONET/SDH that part of this extra bandwidth is only available at STS-3c and not at higher levels of SONET/SDH. For higher levels of SONET there are $(N/3) - 1$ columns of fixed stuff after the POH. This is true for SDH, also. For all levels of SDH greater than STM-1, there are $N - 1$ columns of fixed stuff after the POH (where N indicates the STM level greater than 1).

6.6 Automatic Protection Switching

APS is the function in SONET/SDH that provides the ability to restore service in the case of a failure of an optical fiber line or a network node.

Like many things in SONET/SDH, this area is very complex. To simplify things, we describe how APS works in two situations, both of which involve a ring topology.

SONET/SDH networks can be configured as linear networks, where the SONET/SDH nodes (known as add/drop multiplexers, or ADMs) are just hooked together in a line, as shown in Figure 6.29. There may be only two fiber connections between the ADMs, as shown in the figure, or four fiber connections, with one set serving as a "protection," or backup, pair.

Although linear networks have some applicability, by far the most common topology in the network is the ring, as shown in Figures 6.30 and 6.31. Rings are used because they provide an alternate path to communicate between any two nodes. For example, in the linear network, even if two sets of fiber were used between the nodes, it is possible for all of the fibers to be cut at the same time, unless great pains are taken with routing the fiber. And in most cases, the limitations on rights-of-way do not permit this kind of route diversification.

A two-fiber ring can be operated in either of two ways: (1) as a unidirectional ring or (2) as a bidirectional ring. Let's examine the unidirectional case first. With a unidirectional ring, traffic could be limited to one fiber and always flows the same way around the ring. The second fiber is simply the protection fiber and is used in a special way to provide backup, as explained later.

FIGURE 6.29
A linear SONET/ SDH network.

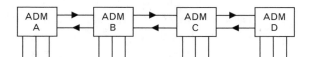

FIGURE 6.30
A two-fiber SONET/SDH ring. The diagram shown here is for a unidirectional fiber ring. In this type of ring, all traffic could be carried on one fiber. A two-fiber ring can also operate as a bidirectional ring.

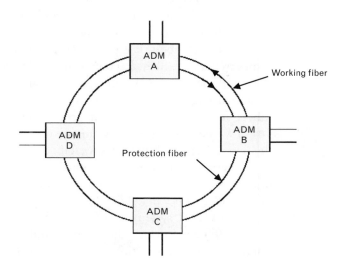

FIGURE 6.31
*A four-fiber (two-pair)
SONET/SDH ring.
This type of ring is
always bidirectional.*

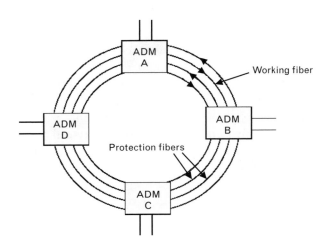

With unidirectional traffic there can be a difference in the transmitting and receiving propagation delay between two nodes. For example, in Figure 6.30, if node B sends traffic to node A, the propagation delay is one link. However, when node A sends traffic to node B, the traffic has to go through nodes D and C to reach node B, leading to a longer propagation delay.

With bidirectional traffic, data are sent on both fibers so the concept of *working fiber* and *protection fiber* does not have any meaning—both fibers are working fibers. When data are sent between nodes A and B, it simply flows over the two fibers connecting the two nodes. Bidirectional rings do not buy us any additional capacity, however. To be able to provide backup, each fiber in a bidirectional ring can only be used to half its capacity—the second half of the capacity is reserved for backup.

Four-fiber rings are always operated as bidirectional rings. In this case, we get the full rate that can be put on the working fibers, but the protection fibers are not used under normal conditions. No matter how we slice it, rings always require twice as much bandwidth as the amount of traffic carried in the ring, if we want to provide full backup [16].

With four fibers, it is possible to do a link recovery if one, and sometimes two, of the fibers fail between two nodes. The more common case is when all the fibers between two nodes are cut. In this situation, the bidirectional ring provides restoration by routing the traffic over the protection fibers, in the opposite direction around the ring.

Two types of backup systems are used on fiber: path and line. We start by describing a path switched system. Path switching can be implemented on either a unidirectional or a bidirectional ring but, today, path switching is always implemented on a unidirectional ring, and is known as a *unidirectional path-switched ring* (UPSR).

In a UPSR system, all of the traffic is transmitted in both directions around the ring, in one direction on the working fiber, and in the other direction on the protection fiber. The *add/drop multiplexers* (ADMs) are the places where traffic enters or leaves the ring. This traffic can be at various levels in the SONET/SDH hierarchy. For example, a DS-1 could be feeding into an ADM.

The ADM would put the DS-1 traffic into a virtual tributary, which would then be put into a VTG, which would then be put into an STS-1 SPE (payload) and this STS-1 might be multiplexed into a higher level signal, such as an STS-48. The DS-1 would be transported to another ADM, where the traffic would be removed. In this example, the VT carrying the DS-1 traffic is a "path."

It is also possible that an external piece of equipment has placed certain traffic into a concatenated STS-3c and is presenting this STS-3c traffic to the ADM, which then multiplexes it into an STS-48. In this case, the STS-3c is considered the "path" for purposes of the SONET/SDH ring.

The only restriction in path switching is that both the entry and exit nodes for a path must operate at the same level.

In UPSR, the SONET/SDH equipment monitors the path traffic on both fibers and selects the traffic that is the "best" (see Figure 6.32). This monitoring is based on a number of things. The BIP octets, available at every level of the multiplexing hierarchy, provide insight into the number of bit errors on the path. More serious errors can occur, such as the failure of one fiber or detection of an AIS on a path.

Because the receiver of each channel is monitoring both paths on both fibers, switching between fibers is immediate, with no loss of data. Additionally, restoration is accomplished by the receiver, without any coordination with the transmitter—no APS communication channel is needed in UPSR.

The description given above is valid for a fiber cut, which is one of the most common failures in SONET/SDH rings. If an ADM fails, or if multiple fiber cuts occur that cut an ADM out of the ring, additional actions are required. The loss of an ADM will be detected by the two ADMs that were connected to it. These ADMs must put an AIS on each path that originated

FIGURE 6.32
Path switching on a unidirectional ring. On the Tx side of the traffic, the path is sent over both counter-rotating rings. On the Rx side, both fibers are monitored and the "best" traffic selected.

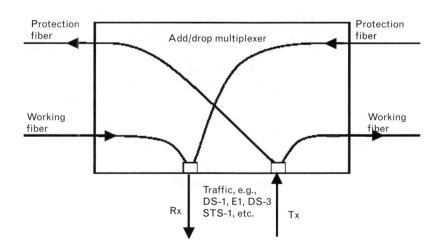

and terminated on the lost ADM. This is an immediate indication to the device on the other end of the path that the data in that path is bad. Another reason for doing this is that paths can be reused in a unidirectional ring. Certain types of failures could occur that could cause data received in a path to be coming from the wrong source. The receiving device may not recognize the misconnection and deliver incorrect data. If AIS is put on the path, there is no question the data are invalid.

Path switching on a unidirectional ring has a number of advantages. It is simple for fiber cuts. No coordination is needed between the receiver and transmitter—it is fully implemented at the receiver (for loss of ADMs, additional processing is required). It provides hitless restoration. No data are lost on a restoration unless an ADM fails or is separated from the ring by multiple fiber failures.

On the negative side, it is expensive because two sets of path equipment are needed wherever a circuit is dropped at an ADM. Also, unidirectional rings have the disadvantage of asymmetric delay—the time it takes to go one way around the ring is usually different than the time it takes to go the other way around the ring. Because of this, buffering must be done at the path-terminating site. And as rings get larger and faster, more buffering must be done. This is the primary reason why unidirectional rings are used for the most part in metropolitan networks and with lower line rates. In the back-bone, especially in large rings, bidirectional rings are much more common.

The defining characteristic of bidirectional rings is that the traffic between two nodes flows in two directions, rather than in only a single direction, as occurs in unidirectional rings. For the simplest case of two nodes, for example, A and B, adjacent to each other, traffic from node A to node B will flow in one direction around the ring, perhaps clockwise, while traffic from node B to node A will flow in the opposite direction.

Bidirectional rings can be either two-fiber or four-fiber rings. In a two-fiber bidirectional ring, each fiber can only carry half its capacity, reserving the other half for backup (also known as protection). In a four-fiber bidirectional ring, two of the fibers are reserved for protection.

When a fiber failure occurs in a two-fiber bidirectional ring, the only recovery possible is a ring switch (or as the standards call it, a line switch), sending data in the opposite direction over the two fibers. This is why the recovery mechanism is called a *bidirectional line switched ring* (BLSR).

In a four-fiber bidirectional ring, a single fiber failure can usually be recovered from by doing a span switch, simply switching to the protection fiber over that one link. Failure of multiple fibers usually requires a ring switch.

Let's discuss what happens in a bidirectional ring in order to accomplish a ring switch. We start with a four-fiber bidirectional ring as shown in Figure 6.33. Assume that a complete break occurs between ADM A and ADM B. Loss of a single fiber is, of course, possible but the recovery is a bit more complex. We will use this example to keep things simple.

FIGURE 6.33
A four-fiber bidirectional ring with a complete fiber failure between nodes A and B. The detailed diagrams show how the nodes reroute the traffic.

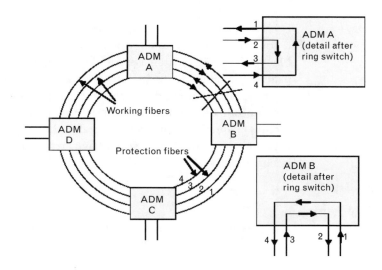

When the break occurs, both ADM A and ADM B will detect the loss because both will see loss of signal on the fibers, including the protection fibers. The two ADMs will then send a signal in the K1, K2 octets "backward" over the ring, with the other ADM's address in the K1 octet. ADMs C and D will simply pass the K1, K2 octets on since the octets are not addressed to them. When ADM A and ADM B receive the failure messages from each other, they bridge the signals as shown in the detailed diagrams of Figure 6.33.

Note what happens once the bridge occurs. Look at ADM B first. The signals arriving on fiber 1 would ordinarily be transmitted to ADM A on fiber 1. Now, however, the signal is put on the protection fiber number 4, which carries it in the reverse direction around the ring to ADM A. ADM A then takes the signal on fiber 4 and bridges it to fiber 1. So the signal on fiber 1 on ADM B still gets to fiber 1 on ADM A, it just does so the long way around. Signals on fiber 2 on ADM A are transported to fiber 2 on ADM B in the same fashion by fiber 3.

This type of bridging means that none of the other ADMs needs to be concerned about the fiber cut between ADM A and ADM B. Those two ADMs handle the fault and the rest of the ring keeps doing the same thing. The specifications require that the ring switch and restoration of service occur within 50 ms.

When all fibers are cut, both ADMs see the failure at the same time and take the same actions. If only one fiber is lost, only one ADM will see the failure. In that case, the ADM detecting the failure notifies the other ADM through the K1, K2 octets, and the other ADM then responds. In the case of a single fiber loss, the two ADMs will attempt a link switch first by attempting to switch to the protection fiber(s) across the link between them. If that is not possible, they will do a ring switch as described earlier.

Recovery on a two-fiber bidirectional ring is essentially the same as on a four-fiber bidirectional ring except that a line switch is not possible—only a ring switch is possible. Signaling with the K1, K2 octets is exactly the same. Traffic is routed back around the ring, not on a separate fiber, but in the unused capacity of the other fiber (which carries traffic in the opposite direction). Just as in the four-fiber bidirectional ring, the protection channels simply bring the traffic the long way around the ring until it gets to the node that it would have arrived at if the fiber failure had not occurred.

ENDNOTES

[1] The framing hardware searches for the boundary between A1 and A2 for octet synchronization. It must find the A1/A2 pattern for a certain number of consecutive frames before it leaves the seek state and enters the synchronized state.

[2] Not all of the payload area of the STS-1 signal can be used to carry voice traffic. And if the voice traffic is mapped as DS-1s or a DS-3, it can only carry 28 DS-1s or one DS-3 for a total of 672 voice channels.

[3] SONET/SDH utilizes octet multiplexing in order to reduce delay. Octet multiplexing is used instead of bit multiplexing because, in SONET/SDH, everything is done in octets instead of bits. Octet multiplexing, therefore, causes the minimum delay (compare it to row multiplexing, for example).

[4] *Stuff* means "extra" or "unused." Stuff octets are, by definition, filler and not customer data.

[5] Some of the text in the descriptions here is taken verbatim from the latest draft T1.105 standard.

[6] The reason for this limitation on the number of A1, A2 octets in STS-768 has to do with DC balance. The A1, A2 octets are DC balanced when taken together but are not DC balanced by themselves. If STS-768 followed the pattern of lower levels of SONET, there would be 768 A1s in a row, followed by 768 A2s (remember that the A1, A2 octets are not scrambled).

[7] They were set to 10 to indicate the type of AU (AU-4, Au-4-Xc, AU-3, or TU-3). In the October 2000 revision of G.783 and G.806, these bits are no longer used to determine the type of AU.

[8] Because there are an even number of stuff octets, one way to achieve this is to put the same value in every stuff octet.

[9] This breaks the SPE into three equal-size groups of 28 columns of payload.

[10] The VTs are interleaved in a VTG but the SONET equipment can easily account for this.

[11] These bits are defined as part of a DS-1 and are used to send call status, such as "on-hook," "off-hook," "answer," and a number of other signaling conditions. See the T.403.02 recommendation for more information. For our discussion, just accept that they need to be transported as part of the DS-1.

[12] Indicated by a value of 0x13 in octet C2 of the POH.

[13] Indicated by a value of 0x16 in octet C2 of the POH

[14] Indicated by a value of 0x1b in octet C2 of the POH. This value was specified in the latest version of G.707.

[15] The other reason is that handling bits at the highest SONET rates would require very expensive circuitry. It is less expensive to handle the traffic on octet boundaries. And most equipment actually handles traffic in 16-bit words.

[16] In some configurations one fiber, or one pair of fibers, backs up more than one working fiber, or pair of working fibers. This is known as 1:*n* backup. These types of systems cannot fully back up a ring, however. If all of the fibers are cut, some traffic must be discarded.

BIBLIOGRAPHY

G.707, *Network Node Interface for the Synchronous Digital Hierarchy (SDH), ITU-T, Geneva, Switzerland.*

G.707(d) Study Group 15, *Report of Working Party 3/15, Multiplexing and Switching (Geneva Meeting, 3-14 April 2000), Part II B.5—Draft Determined Revised Recommendation.*

G.707/Y.1322, Document COM15-R71, available at http://www.itu.int/itu-doc/itu-t/com15/reports/.

G.707(e), *Network Node Interface for the Synchronous Digital Hierarchy (SDH)—Editor's Draft,* ITU Working Party 3 working document g707-2000_ww9.doc, available at http://ties.itu.int/u/tsg15/sg15/wp3/.

G.783, *Characteristics of Synchronous Digital Hierarchy (SDH) Equipment Functional Blocks,* available at http://www.itu.int.

Goralski, W. J., *SONET,* 2nd ed., New York: McGraw- Hill, 2000.

GR-253, *Synchronous Optical Network (SONET) Transport Systems: Common Generic Criteria,* Telcordia Technologies GR-253-CORE, Issue 2, December 1995; revision 2, January 1999.

GR-1400, *SONET Unidirectional Path Switched Ring (UPSR) Equipment Generic Criteria,* Telcordia Technologies GR-1400-CORE, Issue 2, January 1999.

GR-1230, *SONET Bidirectional Line-Switched Ring Equipment Generic Criteria,* Telcordia Technologies GR-1230-CORE, Issue 4, December 1998.

T.105base, *Optical Interface Rates and Formats Specifications (SONET),*" ANSI T1.105-1995, available at http://www.atis.org.

T.105base(d), *Revised Draft T105 SONET Base Standard,* T1X1.5/2000-193R1, available at http://www.t1.org/t1x1/_x1-grid.htm.

T.105.01, *Synchronous Optical Network (SONET)—Automatic Protection Switching,* ANSI.

T1.105.01-1994, available at http://www.atis.org.

T.105.01(d), *Revised Draft T105.01 SONET Automatic Protection Standard,* T1X1.5/ 1999-065R1, available at http://www.t1.org/t1x1/_x1-grid.htm.

T.105.02, *Synchronous Optical Network (SONET)—Payload Mappings,* ANSI T1.105.02-1995, available through http://www.atis.org.

T.105.02(d), *Revised Draft T105.02 Payload Mapping Standard,* T1X1.5/2000-192, available at http://www.t1.org/t1x1/_x1-grid.htm.

T.403.02, *Network and Customer Installation Interfaces—DS-1 Robbed-Bit Signaling State Definitions,* ANSI T1.403.02-1999, available at http://www.atis.org.

T1Rpt36, *A Comparison of SONET (Synchronous Optical NETwork) and SDH (Synchronous Digital Hierarchy),* ANSI T1X2 Working Group, Technical Report 36, available at http://www.atis.org.

Appendix 6A: SDH

When approached for the first time, most people find SDH terminology difficult and convoluted. Perhaps it is just the way the G.707 standard is written. Once one understands the naming conventions, however, one can read the standard and make perfect sense of it [1].

One cause of this difficulty with the terminology is that the base rate of SDH is three times the base rate of SONET. SDH has to do the same things as SONET but it must fit into a frame that is three times the size of the lowest level SONET frame. And the standards people, being good engineers, appear to have provided the maximum flexibility (which leads to maximum confusion).

The basic STM-1 SDH frame is 270 columns by 9 rows, three times the size of the SONET STS-1 frame. This frame has 9 columns of overhead, called *section overhead* (SOH). This is almost the same as the transport overhead in SONET, except that row 4 is excluded from the SOH. So the section overhead is the first 9 columns, less row 4 (see Figure 6A.1). The first 3 rows of the SOH are known as the *regenerator section overhead* (RSOH), whereas the last 5 rows of the SOH are known as the *multiplex section overhead* (MSOH).

The remaining 261 columns, plus row 4 of the first 9 columns, is known as the *administrative unit group* (AUG). An AUG can consist of three administrative units, known AU-3, or one AU-4. As you might suspect, an AU-3 consists of 87 columns, plus 3 octets of pointers (see Figure 6A.2). An AU-4 consists of 261 columns, plus 12 octets of pointers. It causes as extreme amount of confusion to talk about both of these at the same time so we begin by focusing on the AU-3 and how traffic is mapped into it. Later, we discuss the AU-4 and how traffic is mapped into it.

FIGURE 6A.1
An SDH STM-1 frame showing the section overhead and the administrative unit group. The figure is not to scale—the overhead is shown much larger.

|←————————————————— 270 columns —————————————————→|
							←————— 261 columns —————→
A1	A1	A1	A2	A2	A2	J0	Z0	Z0	
B1	X	X	E1	X	X	F1	X	X	
D1	X	X	D2	X	X	D3	X	X	
H1	H1	H1	H2	H2	H2	H3	H3	H3	
B2	B2	B2	K1	X	X	K2	X	X	
D4	X	X	D5	X	X	D6	X	X	
D7	X	X	D8	X	X	D9	X	X	
D10	X	X	D11	X	X	D12	X	X	
S1	Z1	Z1	M0/1	Z2	M2	E2	X	X	

▨ Section overhead

▯ Administrative unit group (AUG)

FIGURE 6A.2
An SDH STM-1 frame with three AU-3s carried in the AUG (not to scale).

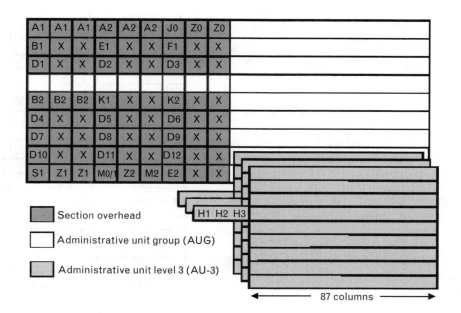

An AU–3 is a fixed logical structure of 87 columns plus 3 octets of pointers—the AU–3 does not "float." To provide the ability for the payload to float, as has been described for SONET, SDH introduces the concept of a virtual container. A VC that fits into an AU–3 is called a VC–3. And just like a SONET STS–1 payload, the VC–3 has one column of POH, and its mapping into an AU–3 includes two fixed stuff columns at columns 30 and 59. However, the VC–3 does not include the fixed stuff columns at columns 30 and 59, so the VC–3 is 85 columns with 1 column of POH at the beginning. When the VC–3 is mapped into an AU–3, columns 30 and 59 are skipped (see Figure 6A.3).

The net effect is that a VC–3, plus its mapping into an AU–3, is the same as the payload of a SONET STS–1. Note that the AU–3s are interleaved into the SDH STM–1 structure in exactly the same fashion as three STS–1s are interleaved into an STS–3. In fact, throughout this section, each structure is interleaved in exactly the same fashion as in SONET.

In the SONET payload we had seven VTGs. In SDH we have same thing except that they are called *tributary unit groups* (TUGs). In this case, they are known as TUG–2, perhaps because multiple (seven) TUG–2s map into a VC–3. Handling TUG–2s is exactly the same as handling VTGs in SONET. Now, in SONET the VTG could contain VT–1.5s, VT–2s, VT–3s, or a VT–6. SDH provides a similar breakdown but calls them by different names. The equivalent of the VT–1.5 is called the *tributary unit 11* (TU–11). The equivalent of the VT–2 is the TU–12. There is no equivalent to the VT–3 in SDH—they must have decided not to carry DS–1C signals for some reason. The equivalent of the VT–6 called the TU–2 (because it fully fills the TUG–2) and is specified to carry a DS–2 signal.

An SDH AU-3 and a VC-3 that floats within the AU-3. Note that the VC-3 only has 85 columns, whereas the AU-3 has 87. Two columns (30 and 59) are fixed stuff (same as in SONET) but are not included in the VC-3. When the VC-3 is mapped into the AU-3, these columns are skipped.

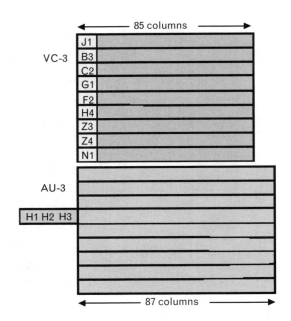

This is a bit of a simplification because there is an additional structure that we skipped. Inside each Tux is a *virtual container level x* (VC-x) that actually carries the information. But the VC-x is simply the TU-x without the pointer.

A DS-1 signal is mapped into the TU-11 in exactly the same fashion as in SONET (for the VT-1.5). Likewise, the E1 signal is mapped into the TU-12 in the same fashion as an E1 is mapped to a VT-2 in SONET. And likewise for the TU-2 and the DS-2 signal.

At this point, you can start to discern the naming conventions used in SDH. The level 1 is the lowest level structure, carrying the PDH information. The information is carried in a "container," which becomes a "virtual container" (VC) with the addition of the POH. The addition of the pointer creates a "tributary unit" (TU). The lowest level of this is the VC-11/TU-11 (level 1 structure, first type) and the VC-12/TU-12 (level 1 structure, second type). The physically smallest gets the lowest number. So a VC-x and a TU-x (when x is the same) are the same structures except that the TU contains the pointer and the VC does not.

When a structure essentially fills the next level structure, it gets the same number, for example, a TU-2 fully fills a TUG-2, and a VC-3 fills an AU-3. TUs then fit into a "tributary unit group" (TUG), which will then fit into a higher level VC or "administrative unit" (AU). AUs then fit into an "administrative unit group" (AUG).

This concludes the discussion of the AU-3 and its mappings. These mappings are very similar to the SONET mappings so they are fairly easy to understand. Now, let's look at how all of this is mapped when we start with an AU-4 (261 columns plus 9 octets of pointers).

We start with the same AUG as we did for the AU–3 case, but this time we only have one AU to map into it. This AU is known as an AU–4 (to differentiate it from the case of three AU–3s) and completely fills the AUG (see Figure 6A.4). The pointers are shown as H1, H1, H1, H2, H2, H2, H3, H3, H3. Note that only the first H1 and H2 will carry valid pointers. The second and third H1s will contain 1001 xx11, the concatenation indicator. The second and third H2 octets will contain all ones, 1111 1111, also the concatenation indicator. The H3 octets will be utilized because pointer adjustments will be made three octets at a time. See the SONET section of this paper for more details on concatenation indicators in pointers.

We want to carry the same plesiochronous traffic as described earlier for the AU–3 so we have to find some way to eventually map the TU–11, TU–12, and TU–2 into the AU–4. In fact, if we can find some way to map VC–3s into an AU–4, we can use the same process to get to the TU–11, and so on, as we used earlier. This is done by defining a structure known as a *tributary unit group level 3* (TUG–3), which turns out to function a lot like an AU–3. A VC–3 consists of 85 columns. The TUG–3 consists of 86 columns, with the extra column carrying pointer octets H1, H2, H3 (see Figure 6A.5). The H1, H2, H3 pointers in the extra column of the TUG–3 allows the VC–3 to "float" exactly as it does in an AU–3 (the AU–3 provides the pointers to the VC–3). Another note on terminology—an AU–3, for example, consists of the payload area plus the pointer. SDH uses an equivalent structure here, known as the *tributary unit level 3* (TU–3). This consists of the VC–3 plus the three pointer octets, H1, H2, H3, which are

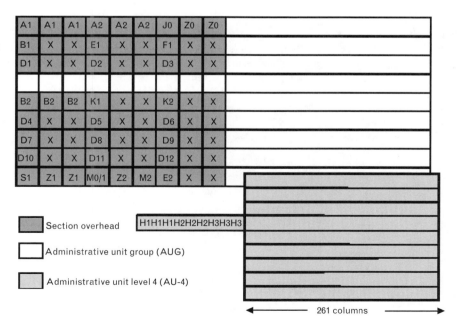

FIGURE 6A.4
An SDH STM-1 frame with one AU-4 carried in the AUG (not to scale).

H1H1H1H2H2H2H3H3H3

Section overhead

Administrative unit group (AUG)

Administrative unit level 4 (AU-4)

261 columns

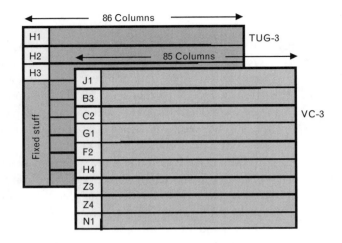

FIGURE 6A.5
Three VC-3s are carried in an AU-4 by introducing a structure known as a TUG-3. Note the extra column in the TUG-3, which carries the pointers that allow the VC-3 to float within the TUG-3.

part of the TUG-3. So if you see TU-3, it is just a VC-3 plus the H1, H2, H3 pointers that are in the TUG-3.

Remember that the AU-4 is a 261-column structure, whereas the TUG-3 is an 86-column structure. Three TUG-3s will take 258 columns. Of the remaining 3 columns of the AU-4, one is dedicated to the POH. The 2 columns following the POH are fixed stuff to make the number of columns come out right. The TUG-3s are interleaved in the remaining 258 columns. And just as we saw earlier, seven TUG-2s are interleaved within the VC-3, and TU-11s, TU-12s, or a TU-2 is carried within the TUG-2, in exactly the same fashion as described earlier. So SDH does the same things as SONET but just calls it by different names.

ENDNOTE

[1] Once you understand the naming conventions, you see that the naming is very regular. However, the G.707 recommendation does not give an explanation of the naming conventions, forcing the reader to discover the conventions by reading the document.

CHAPTER 7

Next Generation SONET, GFP, and Ethernet over SONET

This chapter looks at developments that were taking place at press time to extend the capabilities of SONET in the area of data support. Basically, next generation systems are data–aware SONET NEs that enjoy one or more of the following characteristics: They have direct 100-Mbps, 1–Gbps, or 10-Gbps Ethernet handoffs in addition to traditional handoffs; data interfaces such as ESCON, FICON, and Fibre Channel are supported; they are cheaper than traditional SONET NEs because of advances in technology and because they do not support voice-originated capabilities such as a complex DS-1/DS-2 level VT mechanism (although they may support other kinds of data–focused VT mechanisms); they support full OAM&P while enjoying SNMP access to the management information; they support new trunk-side architectures (for example, they support a packet-based mechanism such as, but not restricted to, RPR) and have OC-x handoffs; they support IP routing functions, including perhaps MPLS; and they obviate the need for ATM, but employ framing based on POS or GFP.

As a technology, next generation SONET systems are not revolutionary, but rather are evolutionary. At the network topology level, these systems are generally similar or identical to traditional SONET/SDH systems. Next generation SONET is not (as of yet) supported by a generalized standardized definition from ANSI or ITU. Rather, it is a collection of vendor-advanced approaches to improve the cost effectiveness of traditional SONET in a broadband data context, in conjunction with some standards such as ITU G.7041 GFP and ITU G.7042 LCAS, which were briefly introduced in Chapter 4. Table 7.1 identifies the scope of our definition for next generation SONET. This chapter explores these new technologies, with a focus on GFP and LCAS.

Current networks' multiple layers can be difficult and costly to provision and manage. Therefore, a much simpler SONET-based approach is advantageous. Solutions include optical Ethernet, namely, the various VLAN/campus-extended LAN technology approach; RPR-based systems; streamlined POS; GFP-based POS; and LCAS. Multiservice solutions afforded by next generation SONET (or by any platform for that

TABLE 7.1   DEFINITION OF NEXT GENERATION SONET

EDGE (USER HANDOFF)	CORE (NETWORK HANDOFF)	TYPE
SONET	SONET	Traditional SONET
Ethernet	SONET	Traditional SONET, and next generation SONET
SONET Ethernet	Packet (RPR)	Next generation SONET
SONET Ethernet	Hybrid: packet (RPR) on some slots and TDM on other slots	Next generation SONET
SONET Ethernet	Ethernet	Ethernet based
Ethernet	Ethernet	Ethernet based

matter) must incorporate QoS, fault tolerance, and the restoration low latency, as well as low latency of SONET. Additionally, they must be able to handle the bursty nature of (some) data networks while offering the scalability and capacity to transport video applications. It is also desirable for these capabilities to be added incrementally because each service provider's timing and business model may be different.

A distinction needs to be made between an architecture that uses Ethernet at the edge (which has been deployed by carriers for more than 10 years, typically using a point-to-point private line, traditional SONET, and/or ATM with distinct edge devices) and an architecture that uses Ethernet at the core (a new technology). In our definition, next generation SONET fits for the most part the former case, but it offers better streamlined data-ready/data-inclusive equipment with integrated *media access control* (MAC) bridging and tunneling and/or IP routing, as well as improved operational efficiency compared to traditional SONET systems. Our definition, however, also includes the reverse case, where the NE is packet based at the core, but it has both Ethernet as well as traditional ITU SONET handoffs at the edge. In the first case, a packet path is emulated over a circuit path; in the latter case, a circuit path is emulated over a packet path.

7.1   GFP

This section [1] looks at the Generic Framing Procedure, which is a new standard that has been developed to overcome inefficiencies and deficiencies in transporting data over existing ATM and packet over SONET/SDH protocols. *Transparent GFP* (GFP-T) is an extension to GFP that has been developed to provide efficient, low-latency support for high-speed WAN applications including *storage area networks* (SANs). Rather than handling data on a frame-by-frame (or packet-by-packet) basis, transparent GFP-T handles the block-coded character streams (e.g., 8B/10B) directly.

7.1.1 Introduction

Several important high-speed LAN protocols use a layer 1 block code in order to communicate both data and control information. The most common block code is the 8B/10B line code used for GbE, ESCON, Single Byte Command Code Sets Connection (SBCON), Fibre Channel, FICON, and Infiniband. These have become increasingly important with the growing popularity of SANs. The 8B/10B line code maps the 28 (256) possible data values into the 210 (1,024) values of the 10-bit code space. The code assignment is done such that the running number of 1's and 0's transmitted on the line (the running disparity) remains balanced. Twelve of the 10-bit codes are reserved for use as control codes that may be used by the data source to signal control information to the data sink. To efficiently transport protocols using the 8B/10B line code through a public transport network such as SONET/SDH or the OTN, it is necessary to carry both the data and the 8B/10B control code information. However, using the 8B/10B coding expands the data bandwidth by 25%, which is undesirable in the transport network. Among the alternative protocols for carrying these LAN signals through the SONET/SDH and OTN networks, the newly developed Generic Framing Procedure [2] provides some advantages over ATM and POS. ATM requires a more complex adaptation process than GFP. POS requires the client signal's layer 2 protocol to be terminated and the signal remapped into the *Point-to-Point Protocol* (PPP) over HDLC. This mapping further requires "escaping" the HDLC control characters found in the data stream, resulting in a nondeterministic bandwidth expansion.

GFP standardization began in the ANSI accredited T1X1 subcommittee, which chose to work with the ITU-T for the final version of the standard that has been published by the ITU-T [2]. The transparent version of GFP has been optimized for transparently carrying block-coded client signals with a minimum of latency. This section describes the transparent GFP protocol along with some of the motivations and target applications that led to its development.

7.1.2 Transparent GFP Description

7.1.2.1 General GFP Overview

The basic GFP frame structure is shown in Figure 7.1. The GFP core header consists of a two-octet length field, which specifies the length of the GFP frame's payload area in octets, and a CRC-16 error check code over this length field. To synchronize alignment to this frame structure, the framer looks for a 32-bit pattern that has the proper zero CRC remainder.

FIGURE 7.1
GFP frame format.

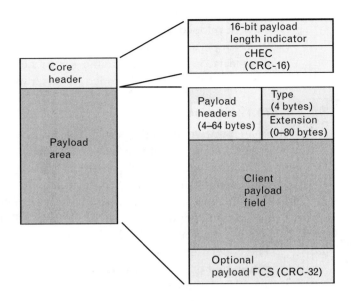

It then confirms that this is the correct frame alignment by verifying that another valid 32-bit sequence exists immediately following where the length field specifies that the current frame ends.

In contrast, protocols such as HDLC rely on specific data patterns for frame delineation or to provide control information. These specific data patterns are not allowed in the payload because they would mimic the reserved characters and interfere with the proper operation of the protocol. The HDLC protocol requires additional escape bits or characters adjacent to the payload strings or bytes that mimic the reserved characters, which increases the amount of bandwidth needed to convey the payload. A major drawback with this scheme is that the bandwidth expansion is nondeterministic. If the client payload data consist entirely of data emulating these reserved characters, then byte-stuffed HDLC protocols like POS will require nearly twice the bandwidth to transmit the packet than if the payload did not contain such characters. GFP avoids this problem by using only the information in its core header for frame delineation. Because no special characters are used for framing, there are no forbidden payload values that require escape characters. In addition, the CRC-16 also provides a level of robustness by allowing a single bit error to be corrected in the core header once frame alignment has been acquired.

In *frame-mapped GFP* (GFP–F), a single client data frame (e.g., an IP packet or Ethernet MAC frame) is mapped into a single GFP frame. The payload length is, therefore, variable for frame-mapped GFP. Furthermore, the client frame must be buffered in its entirety in order to determine its length. For transparent GFP, however, a fixed number of client characters are mapped into a GFP frame of predetermined length. This makes the payload length static for transparent GFP. As its name implies, transparent GFP supports the transparent transport of 8B/10B control characters as well as data characters. This is a primary advantage of transparent GFP over frame-mapped GFP. In addition, frame-mapped GFP incurs the latency associated with buffering an entire client data frame, whereas transparent GFP requires only a few bytes of mapper/demapper latency. This lower latency is critical for SAN protocols, which are very sensitive to transmission delay.

7.1.2.2 Transparent GFP 64B/65B Block Coding

The client 8B/10B codes are decoded into control codes and 8-bit data values. Eight of these decoded characters are then mapped into the eight payload bytes of a 64B/65B code. The leading (flag) bit of the 64B/65B code indicates whether there are any control codes present in that 64B/65B code. The presence of a control code is indicated by flag = 1. The 64B/65B block structure for the various combinations of control codes and payload bytes is illustrated in Table 7.2. A control code byte consists of three fields. The first field consists of a single bit to indicate whether

TABLE 7.2   64B/65B BLOCK CODE STRUCTURE

INPUT CLIENT CHARACTERS	FLAG BIT	64-BIT(8-OCTET FIELD)							
		Octet 1	Octet 2	Octet 3	Octet 4	Octet 5	Octet 6	Octet 7	Octet 8
All data	0	D1	D2	D3	D4	D5	D6	D7	D8
7 data, 1 control	1	0 aaa C1	D1	D2	D3	D4	D5	D6	D7
6 data, 2 control	1	1 aaa C1	0 bbb C2	D1	D2	D3	D4	D5	D6
5 data, 3 control	1	1 aaa C1	1 bbb C2	0 ccc C3	D1	D2	D3	D4	D5
4 data, 4 control	1	1 aaa C1	1 bbb C2	1 ccc C3	0 ddd C4	D1	D2	D3	D4
3 data, 5 control	1	1 aaa C1	1 bbb C2	1 ccc C3	1 ddd C4	0 eee C5	D1	D2	D3
2 data, 6 control		1 aaa C1	1 bbb C2	1 ccc C3	1 ddd C4	1 eee C5	0 fff C6	D1	D2
1 data, 7 control	1	1 aaa C1	1 bbb C2	1 ccc C3	1 ddd C4	1 eee C5	1 fff C6	0 ggg C7	D1
8 control	1	1 aaa C1	1 bbb C2	1 ccc C3	1 ddd C4	1 eee C5	1 fff C6	1 ggg C7	0 hhh C8

Leading bit in a control octet (LCC) is logic 1 if there are more control octets and is logic 0 if this payload octet contains the last control octet in that block.

aaa = 3-bit representation of the 1st control code's original position (1st control code locator).

bbb = 3-bit representation of the 2nd control code's original position (2nd control code locator).

...

hhh = 3-bit representation of the 8th control code's original position (8th control code locator).

Ci = 4-bit representation of the ith control code's original position (control code locator).

Di = 8-bit representation of the ith control code's original position (control code locator).

this byte contains the last control code in this 64B/65B block; it is set to logic 0 if it is the last. The next field is a 3-bit address (aaa – hhh) indicating the original location of that control code in the client data stream relative to the other characters mapped into that 64B/65B block. The last field is a 4-bit code (Cn) representing the control code. Because there are only 12 defined 8B/10B control codes, 4 bits are adequate to represent them. The control codes are placed in the leading bytes of the 64B/65B block, followed by the data bytes. Figure 7.2 illustrates an example of mapping of 2 control and 6 data octets in the 64B/65B block.

Aligning of the 64B/65B payload bytes with the SONET/SDH/ OTN payload bytes simplifies parallel data path implementations and increases the data observability. This is accomplished by combining a group of eight 64B/65B codes into a superblock. The superblock structure, as shown in Figure 7.3, concatenates the payload bytes in order, then takes the leading flag bits of the eight constituent 64B/65B codes and groups them into a trailing byte. This is followed by a CRC-16, calculated over the bits of that superblock. The CRC-16 is discussed further in Section 7.1.2.4.

7.1.2.3   Transport Bandwidth Considerations

The transparent GFP channel sizes are chosen to accommodate the client data stream under worst case clock tolerance conditions—that is, when the

FIGURE 7.2
Example of mapping client byte stream into 64B/65B block.

Octet No	000	001	010	011	100	101	110	111
Client byte stream	D1	K1	D2	D3	D4	K2	D5	D6

Octet No	L	000	001	010	011	100	101	110	111
65B byte stream	1	1 001 C1	0 101 C2	D1	D2	D3	D4	D5	D6

FIGURE 7.3
Superblock structure for mapping 64B/ 65B code components into the GFP frame.

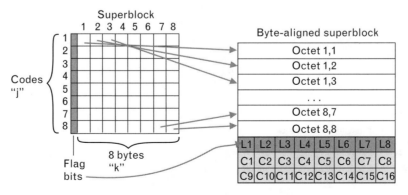

where: Octet j, k is the kth octet and the jth 64B/65B code in the superblock
Lj is the leading (Flag) bit jth 64B/65B code in the superblock
Ci is the ith error control bit

transport clock is running at the slowest rate of its tolerance range, and the client clock is running at the fastest rate of its tolerance range. In the case of a SONET/SDH transport channel, transparent GFP is carried over virtually concatenated signals. Multiple SONET SPEs/SDH VCs are grouped to form a higher bandwidth pipe between the endpoints of the virtually concatenated path. The constituent SPEs/VCs in the virtually concatenated path do not need to be time-slot contiguous. This greatly simplifies the provisioning and increases the flexibility of virtual concatenation. Furthermore, virtual concatenation is transparent to the intermediate nodes; only the endpoints of the virtually concatenated path need to be aware of its existence.

Virtually concatenated signals are indicated with the nomenclature <SPE/VC type>-Xv, where X indicates the number of SPEs/VCs that are being concatenated. For example, STS-3c-7v is the virtual concatenation of seven STS-3c SPEs, which is equivalent to VC-4-7v for SDH. Virtual concatenation is specified in ITU-T [3], ANSI [4] and ETSI [5]. Table 7.3 shows the minimum virtually concatenated channel size that can be used for various transparent GFP clients.

In practice, the SONET/SDH channel must be slightly larger in bandwidth than what is needed to carry the GFP signal. As a result the GFP mapper's client signal ingress buffer will underflow. There are two ways to handle this situation. One approach is to buffer an entire transparent GFP frame's worth of client data characters before starting to transmit that GFP frame. However, this approach increases the mapper's latency and buffer

TABLE 7.3   VIRTUALLY CONCATENATED CHANNEL SIZES FOR VARIOUS TRANSPARENT GFP CLIENTS

Client Signal	Native (Unencoded) Client Signal Bandwith	Minimum Virtually Concatenated Transport Channel Size	Nominal Transport Channel Bandwidth	Minimum Number of Superblocks per GFP Frame	Worst/Best Case Residual Overhead Bandwith[1]	Best Case Client Management Payload Bandwidth[2]
ESCON	160 Mbps	STS-1-4v/ VC-3-4v	193.536 Mbps	1	5.11 Mbps/ 24.8 Mbps	6.76 Mbps
Fiber channel	850 Mbps	STS-3c-6v/ VC-4-6v	898.56 Mbps	13	412 Kbps/ 85.82 Mbps	2.415 Mbps
Gbit Ethernet	1.0 Gbps	STS-3c-7v/ VC-4-4v	1.04832 Gbps	95	281 Kbps/ 1.138 Mbps	376.5 Mbps

Infiniband[3]

Notes:

1. The worst case residual bandwidth occurs when the minimum number of superblocks is used per GFP frame. The best case occurs for the value of N that allows exactly one client management frame per GFP data frame. A 160-bit client management frame was assumed for the best case (with a CRC-32). For both cases, it was assumed that no extension headers were used.

2. The best case client management payload bandwidth assumed 8 "payload" bytes per client management frame and the best case residual overhead bandwidth conditions.

3. The requirements for the transport of Infiniband over transparent GFP have not yet been established.

size. A second approach, which was adopted for the standard, is to use a dummy 64B/65B control code as a 65B_PAD character. Whenever there is no client character available in the ingress buffer, the mapper treats the situation the same as if a client control character were present and inserts the 4-bit 65B_PAD character. This is illustrated in Figure 7.4. The demapper at the other end of the GFP link recognizes this character as a dummy pad and removes it from the reconstituted data stream. By using the 65B_PAD character, the mapper ingress buffer size is reduced to effectively 8 bytes (the amount of data required to form a 64B/65B block) plus the number of bytes that can accumulate during the SONET/SDH overhead and the GFP frame overhead bytes. An 8-byte latency always exists since the mapper cannot complete the 64B/65B block coding until it knows whether there are any control codes present in the 8 characters that will comprise that block.

As we discuss in Section 7.1.2.5, client management frames have been proposed for GFP that would use this "spare" bandwidth for client management applications. These client management frames would be up to 20 bytes in length (including the GFP encapsulation bytes) and, by having a lower priority than the client data, would only be sent when the ingress buffer is nearly empty. To support these client management frames, an additional 20 bytes must be added to the ingress buffer requirements.

7.1.2.4   Error Control Considerations

Error Detection

8B/10B codes have a built-in error detection capability such that a single bit error will always result in an illegal code. But increasing bandwidth efficiency by remapping the data from 8B/10B codes into 64B/65B codes sacrifices much of this error detection capability. There are four situations in which bit errors can cause significant problems with the 64B/65B codes. The first and most serious problem results when the code's leading flag bit is received in error. Because the value of this flag bit indicates whether the block contains control codes and data, or just data, an error here causes the bytes to be misinterpreted. For example, if the original block contained any control codes, these codes will be interpreted as data. If the original block contained only data, some of these data bytes may be interpreted as control codes. The number of data bytes that are erroneously interpreted as control

FIGURE 7.4
*Example of an
insertion of the
65B_PAD character.*

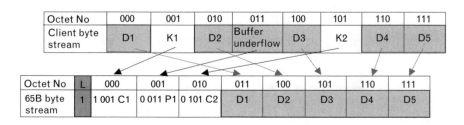

codes depends on the value of the first bit [6] of the bytes and whether the values of the location address bit positions contain increasing values (which is always the case for a legal block). Data that are erroneously converted into control codes could truncate a client data frame, which in turn causes error detection problems for the client data since there is a possibility that the truncated client frame appears to have a correct CRC value. A second, similar problem occurs when a block correctly contains control characters and the last control code indicator bit is affected by an error. The third problem results when errors occur in the control code location address. This will cause it to be placed in the wrong sequence by the demapper. The fourth problem results when errors occur in the 4-bit control code value. This will cause the demapper to generate an incorrect control code. Any error that results in a spurious or incorrect control code can have potentially serious consequences.

To improve the error detection to address these potential problems a CRC-16 is added to each superblock. Once an error is detected, the most reliable mechanism for error control is for the demapper to discard all of the data in the errored superblock. For those clients that have defined such a code, the data are "discarded" by having the demapper output 10B_ERROR 8B/10B codes for all the characters in the superblock. For those clients that do not have error codes defined, the demapper can output another, illegal 8B/10B character for all of the characters in the super-block. Optionally, the CRC-16 can also allow the possibility of single error correction.

Implications of Payload Scrambling
The payload area of the GFP frame is scrambled with a self-synchronous scrambler to address both the physical properties of the transport medium and the desire for robustness in public networks. In SONET/SDH and OTN the data are passed through a SONET/SDH/OTN frame-synchronous scrambler and then transmitted using a NRZ line code. This NRZ line code controls the laser, turning it on for the bit period representing a "1" and turning it off to represent a "0." The NRZ line code is used because it is simple and bandwidth efficient. However, the disadvantage of NRZ is that the receiver clock and data recovery circuits can lose synchronization after a long string of either 0's or 1's. The frame-synchronized scrambler, which is reset at a regular interval based on the SONET/ SDH/OTN frame, randomizes the user data to limit the length of these consecutive strings, and it is adequate to defend against normally occurring user data patterns. Unfortunately, because the user can send any data pattern, it is possible for a malicious user to choose a packet payload that is the same as the frame-synchronized scrambler sequence. If this sequence lines up in the correct position in the frame, it "undoes" the scrambling and results in an adequately long string of 0's or 1's that causes the receiver to lose synchronization. The network's reaction to the resulting loss of synchronization is to take down

the link while the receiver attempts to recover, thus denying the link to other users for the duration of the recovery. This problem was originally discovered in ATM networks and is exacerbated by the longer frames used in POS or GFP. To address this danger, a self-synchronous scrambler was chosen for ATM, POS, and GFP with a scrambler polynomial of $x^{43} + 1$. The scrambler takes each bit of the payload area and exclusively ORs it with the scrambler output bit that precedes it by 43 bit positions, as shown in Figure 7.5(a). The scrambler state is retained between successive GFP frames, making it much more difficult for a user to purposely choose a malicious payload pattern. The decoder's descrambler shown in Figure 7.5(b) reverses this process, restoring the user payload data.

The same payload scrambling technique is used for both frame-mapped and transparent GFP, where all of the GFP payload bits including the superblock CRC bits are scrambled. As a result, the superblock CRC is calculated over the superblock payload bits *prior* to scrambling and is checked at the decoder *after* descrambling. Unfortunately, the drawback of self-synchronous scramblers is that each transmission error produces a pair of errors in the descrambled data. In the case of the $x^{43} + 1$ scrambler polynomial, the errors are 43 bits apart. The superblock CRC must be able to cope with this error multiplication if it is to be of any use. Investigations have shown [7, 8] that a CRC will preserve its error detection capability in such a situation as long as the scrambler polynomial and the CRC generator polynomial have no common factors. However, all of the standard CRC-16 polynomials contain $x + 1$ as a factor, which is also a factor in the $x^{43} + 1$ (or any $x^n + 1$) scrambler polynomial. Therefore, a new CRC generator polynomial is required that preserves the triple error detection capability (which is the maximum achievable over this block size) and does not have any common factors with the scrambler. Furthermore, to perform single error correction, the syndromes for single errors and for double errors spaced 43 bits apart must all be unique [8]. The CRC-16 polynomial $x^{16} + x^{15} + x^{12} + x^{10} + x^4 + x^3 + x^2 + x + 1$ selected for the GFP-T superblock has both of these desired properties, and hence retains its triple error detection and optional single error correction capabilities in the presence of the scrambler [8–10].

FIGURE 7.5
Payload self-synchronous (a) $x^{43} + 1$ scrambler and (b) $x^{43} + 1$ descrambler.

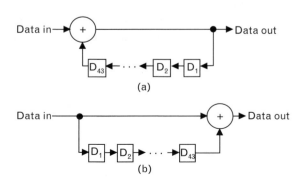

7.1.2.5   Transparent GFP Client Management Frames

As previously mentioned, there is some residual "spare" bandwidth in the SONET/SDH channel for each of the client signal mappings. Table 7.3 shows the amount of this "spare" bandwidth, which depends on the efficiency of the mapping, which in turn is partially a function of the number of superblocks used in each GFP frame. The residual bandwidth can be used as a client management overhead channel for client management functions, as described in [11]. *Client management frames* (CMFs) can also be used for downstream indication of client signal fail, as described in the next subsection.

CMFs have the same structure as GFP client data frames but are identified by the payload type code PTI = 100. Like GFP client data frames, CMFs have a core header and a payload type header (both with 2-byte HEC), and an optional 32-bit FCS. The total CMF payload size in transparent mapped mode is recommended to be no greater than 8 bytes. In applications where the CMF payload is 8 bytes, no FCS is used, and no extension header is used (which would add a total of 20 bytes), the payload efficiency is 50%. Using the FCS and especially using the extension headers will greatly reduce the efficiency of the CMFs.

Client Signal Fail Indication

GFP uses CMGs to indicate *client signal fail* (CSF) to the far-end GFP equipment. When a failure defect is detected in the ingress client signal, a GFP CMF is transmitted immediately after the current frame. This CMF has the payload type header set to PTI = 100, PFI = 0 (no FCS), an appropriate EXI field, and the UPI = 0000 0001 (for loss of client signal) or UPI = 0000 0010 (for loss of client character synchronization). Because a CSF can occur in the middle of a GFP client data frame and the payload length indication has already been transmitted at the beginning of the client data frame, the remainder of the current frame will be filled with 10B_ERR codes to give it the required length. Then the CSF CMF can be sent. Client management frames indicating CSF are sent on a regular interval, every 100 ms$<T<$1000 ms in order to prevent the following:

- Overwhelming the receiver with frequent CSF indications;
- Excessively hogging the bandwidth with CSF indications for just one channel in the case of a frame multiplexed scenario.

When the GFP client sink receives the CSF indication, the receiver declares a sink client signal failure and outputs either 10B_ERROR or another illegal 8B/10B code on the client egress signal. If the CSF condition is a loss of signal and lasts for an extended time, the client egress output signal transmitter (e.g., the laser) can be turned off for protection. The CSF condition at the receiver is cleared when a valid client data frame is received or when less than N CSF indications in $N \times 1,000$ ms is received. A value of 3 is suggested for N.

CMF Potential Uses Far-End Performance Reporting
The most universal CMF application is reporting client-specific performance information from the far end of the GFP link. The GFP receiver can report performance statistics such as the BER or the ratio of good to bad client frames either on a periodic basis or when queried. Far-end client-specific performance reporting allows both ends of the GFP link to see the status of both directions of the GFP link, which is valuable when one of the ends is in an unmanned office or when the link crosses carrier domains.

Remote Management   If both ends of the GFP link are owned by the same carrier and the intervening SONET/SDH/OTN network is owned by another operator, then the opportunity exists to send provisioning commands using CMFs. It is common for IECs to provide the *customer-premises terminal equipment* (CPE) and rely on a *local exchange carrier* (LEC) to provide the connection between this CPE and the interexchange network (see Figure 7.6). Ideally, the IEC would like to manage the CPE as part of its own network, freeing the customer from having to manage the equipment and potentially receiving revenue by providing this management service. Normally, management information is communicated through a SONET/SDH *section data communication channel* (SDCC). However, carriers do not allow SDCC data to cross the network interfaces into their networks, preventing unwanted control access. This also prevents the IEC from exchanging management communications with the CPE through the intervening LEC network. The transparent GFP CMFs provide a mechanism to tunnel the SDCC information through the intervening network.

The last column of Table 7.3 shows the maximum amount of "payload" capacity that can be derived from the client management frames for SDCC tunneling or for other OAM applications. With the assumptions stated in the table notes and assuming a 20-byte client management frame with an 8-byte payload field, adequate bandwidth is available to carry a 192-Kbps SDCC channel for all client signal types.

7.1.3 Potential Extensions to Transparent GFP

There are three main areas where extensions are possible to GFP-T:

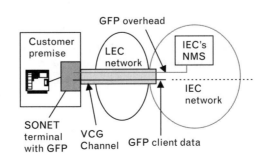

FIGURE 7.6
*SDCC tunneling
application example
with transparent
GFP.*

1. *To support other client signal types.* ETSI standard digital video broadcast has recently been proposed as a GFP–T mapping. Infiniband is another potential LAN signal that could be mapped into GFP–T. If an application arises for it, a recent proposal has shown that it would also be possible to map the 4B/5B 100Base Ethernet signal into GFP–T in a manner similar to the 8B/10B coded mappings [12].

2. *To support other transport media.* Another potential extension is the direct mapping of GFP onto a wavelength in an OTN network (i.e., with no underlying SONET/SDH or OTN transport signal). This extension would require the definition of a GFP physical layer.

3. *To support other client management applications.* Two applications have already been described in the preceding section that make use of the versatile CMFs. Other extensions could include using the CMFs as a trace function to guarantee correct connectivity of the GFP–T stream as it passes through a network element that routes GFP signals.

7.1.4   Conclusions

Transparent GFP provides an efficient mechanism for mapping constant bit rate block coded data signals across a SONET/SDH or OTN network. By performing the mapping on a client character basis rather than a client frame basis, the transport latency is significantly reduced. Reducing latency is a critical issue for SAN protocols including GbE. Translating the client block codes into the more efficient 64B/65B mapping provides a significant bandwidth efficiency increase, while the superblock structure itself provides robust error performance. Transparent GFP also improves the performance monitoring capability for the transport layer, while the ability to tunnel SDCC management information through an intervening network provides a powerful extension to network providers' capabilities.

7.2   Ethernet over SONET

This section [13] describes the use of the new ITU G.7041 GFP and ITU G.7042 LCAS standards to provide Ethernet leased line service over SONET virtual concatenation.

7.2.1   Introduction

The communications landscape is dominated by two differing technologies: (1) Ethernet in the LAN for internal business communications and (2) SONET/SDH in the telco/PTT WAN. When businesses need to communicate with each other or when a business wants to connect its head office and branch offices to the same LAN, interconnection problems arise. Historically, interfacing the LAN to the WAN provided by the telco/PTT required an interworking protocol, because Ethernet is not directly supported over the SONET/SDH network.

Ironically, in today's networks, in order to transport the customer's traffic over the telco network in a "standard" manner, different technologies have been employed (see Figure 7.7). These technologies (frame relay, ATM, POS, ML-PPP to name a few) each requires interworking of the native Ethernet traffic to the transport protocol prior to transmission. Often, the interworking function must terminate the Ethernet and map the underlying IP traffic into a new layer 2 (L2). Alternatively, the Ethernet is encapsulated within another L2 technology. Both of these techniques introduce additional complexity and cost into the WAN interface.

The various technologies also influence the demarcation point between the customer's network and the carrier's network. In some service models, the customer is required to interwork their network traffic prior to handing it off to the public network. In alternative models, the carrier takes complete responsibility for the interworking function. In both cases, network management issues exist as well as the need for additional equipment for interworking. Furthermore, problems can arise when businesses are required to hand off non-Ethernet traffic to the WAN provider. Although the local enterprise IT staff may be experts with Ethernet, they would not have the necessary expertise in the various WAN protocols to deal with issues that may occur.

Given the complexities and expense of the current interworking solutions, why hasn't Ethernet been carried directly over SONET/SDH? Why

FIGURE 7.7
*Traditional
private-line network.*

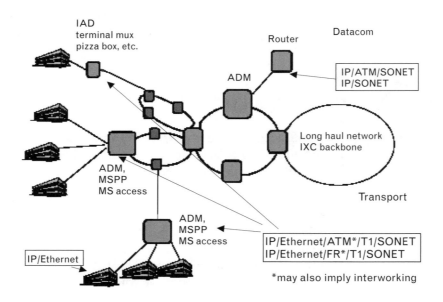

has all the protocol interworking evolved in the first place? In a nutshell, Ethernet rates do not match SONET/SDH rates and encapsulation methods have not been efficient.

As illustrated in Table 7.4, Ethernet rates are typically 10 Mbps, 100 Mbps, or 1 Gbps, always increasing in factors of 10. On the other hand, SONET rates are optimized for telecommunications (or voice traffic) and do not match the optimal rates for transporting the typical Ethernet data stream. These rate mismatches make carrying a single Ethernet connection over a SONET pipe bandwidth inefficient.

To address the inefficiencies of carrying a single Ethernet stream over a SONET link, a multitude of WAN technologies evolved to share the large transport pipes among multiple users' data streams while still providing QoS and bandwidth guarantees. Furthermore, WAN technologies evolved that needed to address the reality of the user bandwidth being distributed across subrates as a result of limitations in the access bandwidth. Often, WAN bandwidth was severely constrained to fractional T1 rates due to cost. As the costs have dropped and demand for bandwidth has increased, the WAN access solutions (ATM or FR) provided via discrete T1 lines are no longer sufficient. One method of increasing bandwidth is by jumping to a T3 access solution (a larger pipe) and incurring the higher costs associated with it. A more cost-effective method currently being employed for acquiring additional bandwidth from a telco-based WAN provider is through multilink services, which combine the T1s into a virtual higher-speed channel either through FR multilink or IMA for ATM. These multilink technologies add inefficiencies due to the encapsulation that counteracts the benefits of bandwidth efficiencies through aggregation.

In spite of these issues, network equipment vendors have developed solutions that transport Ethernet over SONET using a variety of proprietary approaches. Unfortunately, the obvious problem with these proprietary approaches is that interoperability between various vendors' equipment is difficult, if not impossible.

To help optimize the transport of Ethernet over SONET/SDH links, two new technologies have been standardized. The first, virtual concatenation, allows for nonstandard SONET/SDH multiplexing in order to

TABLE 7.4   TYPICAL ETHERNET RATES VERSUS SONET RATES

| | | SONET | |
DATA BIT RATE	SONET RATE	EFFECTIVE PAYLOAD RATE	BANDWIDTH EFFICIENCY
10-Mbps Ethernet	STS-1	~48.4 Mbps	21%
100-Mbps Fast Ethernet	STS-3c	~150 Mbps	67%
1-Gbps Ethernet	STS-48c	~2.4 Gbps	42%

address the bandwidth mismatch problem. The second, GFP, provides deterministic encapsulation efficiency and eliminates interworking, as already covered.

7.2.2   Virtual Concatenation

Virtual concatenation is a technique that allows SONET channels to be multiplexed together in arbitrary arrangements. This permits custom-sized SONET links to be created that are any multiple of the basic rates. Virtual concatenation is valid for STS-1 rates as well as for VT rates. All of the intelligence to handle virtual concatenation is located at the endpoints of the connections, so each SONET channel may be routed independently through the network without it requiring any knowledge of the virtual concatenation. In this manner, virtually concatenated channels may be deployed on the existing SONET/SDH network with a simple endpoint upgrade. All equipment currently in the center of the network need not be aware of the virtual concatenation.

In contrast, arbitrary contiguous concatenation also allows for custom sized channelss to be created but requires that the concatenated pipe be treated as a single entity through the network. This requirement makes deployment of this service virtually impossible over the legacy SONET networks.

Using virtual concatenation, the SONET/SDH transport links may be "right sized" for Ethernet transport. In effect, the SONET link size may be any multiple of 50 Mbps for high-order virtual concatenation (STS-1), or 1.6 Mbps (VT1.5)/2.176 Mbps (VT2) for low-order virtual concatenation (see Table 7.5). Virtual concatenation rates are designated by STS-m-nv for high-order concatenation, where the nv indicates a multiple n of the STS-m base rate. Similarly, low-order virtual concatenation is designated by VT-m-nv.

TABLE 7.5   TYPICAL ETHERNET RATES VERSUS SONET/SDH
RATES USING VIRTUAL CONCATENATION

| | SONET | | |
DATA BIT RATE	SONET RATE	EFFECTIVE PAYLOAD RATE	BANDWIDTH EFFICIENCY %
10-Mbps Ethernet	VT-1.5-7v	~11.2 Mbps	89
10-Mbps Ethernet	VT-2.0-7v	~10.88 Mbps	92
100-Mbps Fast Ethernet	STS-1-2v	~96.77 Mbps	103
1-Gbps Ethernet	STS-1-21v	~1.02 Gbps	98
1-Gbps Ethernet	STS-3c-7v	~1.05 Gbps	95

Virtual concatenation provides flexibility in choosing the transport size to better match the desired bandwidth requirements. In addition to sizing the transport paths to handle the peak bandwidth expected, virtual concatenation may be used to create an arbitrarily sized transport channel. The channel may be sized for the average bandwidth consumed for a single connection, or it may be sized to provide a statistically multiplexed transport pipe.

In virtual concatenation, data are striped over the multiple channels in the virtual concatenation group. This is illustrated in Figure 7.8. Control packets, which contain the necessary information for reassembling the original data stream, are inserted in some of the currently unused SONET overhead bytes. This information contains the sequence order of the channels and a frame number, which is used as a time stamp.

The receiving endpoint is then responsible for reassembling the original byte stream. This includes compensating for differential delay that may have occurred by different routings or paths that the channels took through the network (see Figure 7.9).

7.2.2.1   Dynamic Bandwidth Allocation

Along with virtual concatenation, the capability to dynamically change the amount a bandwidth used for a VC channel is being developed. This capability is commonly referred to as the link capacity adjustment scheme. Signaling messages are exchanged within the SONET overhead in order to change the number of tributaries being used by a *virtually concatenated group* (VCG). The number of tributaries may be either reduced or increased, and the resulting bandwidth change may be applied without loss of data in the absence of network errors.

FIGURE 7.8
Virtual concatenation.

Virtual concatenation stripes data over multiple STS-1 or STS-3c channels
A VC channel constructed of STS-1s is an STS-1-nV
A VC channel constructed of STS-3s is an STS-3-nV

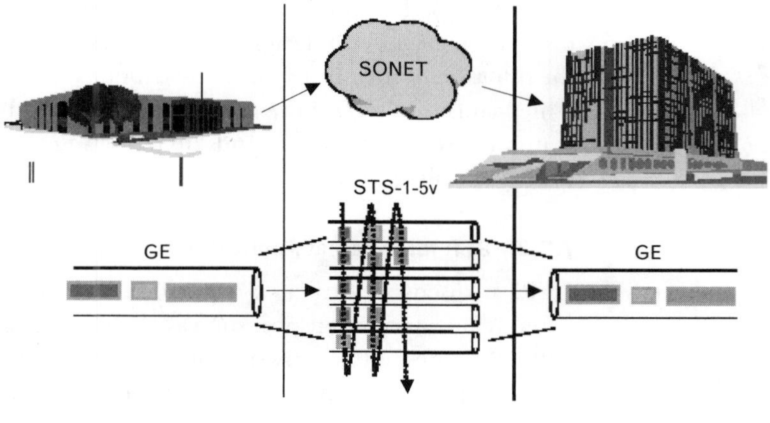

FIGURE 7.9
Differential delay.

Individual STS-1's or STS-3c's subchannels can take different paths through the SONET network. This can introduce *differential delay*. Buffering at the far end is required to align the subchannels and extract the original frames.

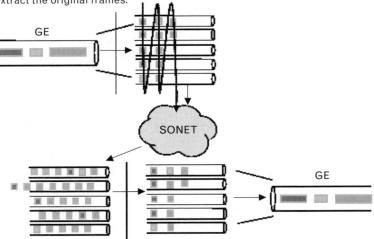

The ability to change the amount of bandwidth allows for further engineering of the data network and provision of new services. Bandwidth can be adjusted based on time-of-day demands and seasonal fluctuations. For example, businesses can subscribe to higher bandwidth connections (say, for backup) when the demand for bandwidth is low and hence the cost is lower.

LCAS can further provide "tuning" of the allocated bandwidth. If the initial bandwidth allocation is only for the average amount of traffic rather than the full peak bandwidth, and the average bandwidth usage changes over time, the allocation can be modified to reflect this change. This tunability can then be used to provide (and charge for) only as much bandwidth as the customer requires (see Figure 7.10).

LCAS is also useful for fault tolerance and protection since the protocol has the ability to remove failed links from the VCG. As the data stream is octet-striped across the tributaries in the VCG, without such a mechanism if one of the tributaries has errors, the entire data stream has errors for the duration of the error within the tributary. The LCAS protocol provides a mechanism to detect the tributary in error and automatically remove it from the group. The VCG ends up operating at a reduced bandwidth, but the VCG still continues to carry data that are error free.

7.2.3 GFP for Ethernet Applications

GFP, as discussed earlier, is a protocol for mapping packet data into an octet-synchronous transport such as SONET. Unlike HDLC-based protocols, GFP does not use any special characters for frame delineation. Instead, it has adapted the cell delineation protocol used by ATM to encapsulate

FIGURE 7.10
*Dynamic bandwidth
adjustment.*

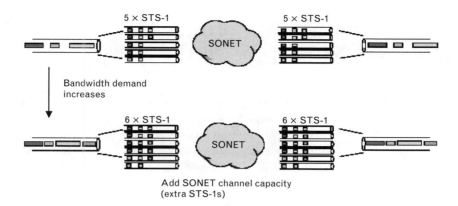

variable-length packets. A fixed amount of overhead is required by the
GFP encapsulation that is independent of the contents of the packets.
In contrast to HDLC, whose overhead is data dependent, the fixed
amount of overhead per packet allows deterministic matching of bandwidth between the Ethernet stream and the virtually concatenated
SONET stream (see Figure 7.11). The GFP overhead can consist of up to
three headers:

- A *core header* containing the packet length and a CRC, which is used
  for packet delineation;

- A *type header* identifying the payload type;

- An *extension header,* which is optional.

Frame delineation is performed on the core header. The core header
contains the 2-byte packet length and a CRC. The receiver would hunt
for a correct CRC and then use the received packet length to predict the
location of the start of the next packet. Within GFP, two different mapping
modes are defined: frame-based mapping and transparent mapping. Each
mode is optimized for providing different services.

FIGURE 7.11
*GFP encapsulation
format.*

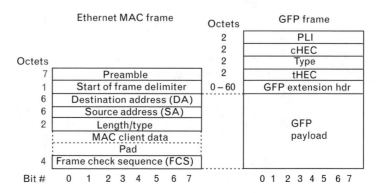

7.2.3.1 Frame-Based GFP

Frame-based GFP is used for connections where efficiency and flexibility are key. To support the frame delineation mode utilized within GFP, the frame length must be known and prepended to the head of the packet. In many protocols, this forces a store-and-forward encapsulation architecture in order to buffer the entire frame and determine its length. This buffering may add undesirable latency. Frame-based GFP is good for subrate services and statistically multiplexed services because the entire overhead associated with the line coding and *interpacket gap* (IPG) is discarded and not transported.

7.2.3.2 Transparent GFP

Transparent GFP is useful for applications that are sensitive to latency or for unknown physical layers, as discussed earlier in this chapter. In this encapsulation, all code words from the physical interface are transmitted. Currently, only physical layers that use 8B/10B encoding are supported. To increase efficiency, the 8B/10B line codes are transcoded into a 64B/65B block code and then the block codes are encapsulated into fixed-size GFP packets. This coding method is primarily targeted at SANs where latency is very important and the delays associated with frame-based GFP cannot be tolerated.

7.2.4 What New Services Are Enabled?

The advent of virtual concatenation, LCAS, and GFP encapsulation will act as enabling technologies for deployment of some new Ethernet-based services from the carriers. These types of services are available today but require that the traffic be interworked to a different layer 2 technology prior to transport and switching. This interworking increases the complexity and cost of these services.

7.2.4.1 Private Leased Line

Private leased line services are used to interconnect various business locations. They are widespread today and are typically provided via ATM, FR, or multilink FR. An Ethernet-based leased line service could be carried through the currently deployed SONET network using the GFP encapsulation and virtual concatenation technologies. Ethernet private lines may be provisioned at various service rates from 50 Mbps to 1 Gbps utilizing STS-1 concatenation and from 1.6 to 100 Mbps utilizing VT1.5 concatenation.

Ethernet private lines deployed over SONET offer the reliability and broad service area coverage associated with the carrier infrastructure. Because it is a private line, data rate guarantees and security are key offerings, as well as upgradable bandwidth utilizing the LCAS protocol to adjust the bandwidth supplied.

7.2.4.2   Virtual Leased Line

Virtual leased lines or VPNs would be deployed in a manner similar to that of private leased lines. The main difference is that a virtual leased line is a shared service where many customers share the same transport bandwidth. This leads to more efficient use of the transport bandwidth via statistical multiplexing and, thus, lower costs. Because the transport bandwidth is shared, this service is generally a more economical service than a private leased line but generally does not have the QoS provided by the private leased line. Instead, service parameters are controlled with *service level agreements* (SLAs).

7.2.5   What Products Use These Techniques?

The discussion of mapping Ethernet into SONET/SDH has thus far focused on the technology and the advantages of the techniques. The next question that arises is "What products would use these techniques?" Before answering this question it is necessary to look at a simplified network, such as that shown in Figure 7.12, and discuss the typical products in the network.

The telco/PTT networks are based on SONET/SDH rings. Multiple rings are connected to provide complete connectivity around a

FIGURE 7.12
*Simplified network
definition.*

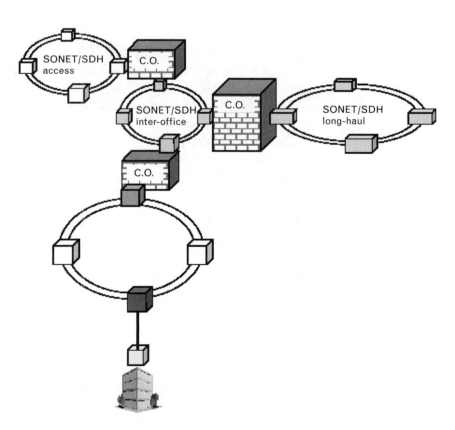

metropolitan area (city) and from city to city. Four basic building blocks, or types of equipment, are used to provide this connectivity, as defined in Figure 7.13.

7.2.5.1   Add/Drop Multiplexer and Multiservice Provisioning Platform

Some forms of Multiservice Provisioning Platform (MSPP) are ADMs (see Figure 7.14) with data interfaces, some are routers with SONET interfaces and switching. For the purposes of this discussion, MSPPs are assumed to be ADMs with Ethernet interfaces.

ADMs have traditionally been used to provide PDH (T1/E1/T3, and so on) and SONET/SDH drops to connect to specific customers. These platforms provide a good place for Ethernet over SONET/SDH customer interfaces.

ADMs may be placed in large office buildings or in a more geographically central location at a CO. When providing Ethernet over SONET/SDH from the CO the interface is most likely to be a fiber connection (i.e., Ethernet/fiber) due to the potentially long paths between the CO and the customer. When placed in the telephone closet of a multiple-tenant office building, the ADM could provide CAT5 cable interfaces to the customer.

7.2.5.2   Terminal Multiplexer

TMs (see Figure 7.15) are similar to an ADM except that a TM terminates the SONET/SDH path (there is no through path). TMs are traditionally used to provide PDH (T1/E1/T3, and so on) and SONET/SDH drops to specific customers. These platforms also provide a good place for Ethernet over SONET/SDH customer interfaces.

TMs are often placed in multiple-tenant office buildings. These products are likely to be found in the telephone closet of a multiple-tenant office building with CAT5 cable interfaces to the customer.

FIGURE 7.13
Definition of connectivity equipment.

▨ DCS (digital cross-connect switch) provides grooming and connectivity between rings.
Has no drops; only serves to provide connection between rings.

▨ ADM (add/drop multiplexer) provides grooming and connectivity for access equipment.
Has interfaces to PDH and/or optical services. Typically layer 1 only, and closer to network edge than DCS.

▢ TM (terminal multiplexer) provides aggregation of optical and PDH services for transport on a typically unprotected optical uplink to a higher order ADM.

▨ MSPP (multiservice provisioning platform) adds layer 2 and/or layer 3 functionality to the ADM.
Usually resides at the metro edge (might take the form of a POS card, and so on.)

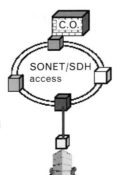

FIGURE 7.14
ADM functional block diagram.

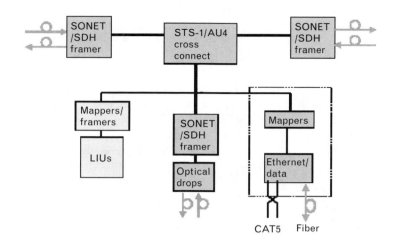

FIGURE 7.15
Terminating multiplexer functional block diagram.

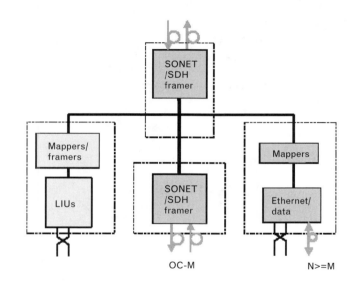

ENDNOTES

[1]   This section is based on a white paper by Steve Gorshe, "Transparent Generic Framing Procedure (GFP) Technology White Paper," May 2002, PMC-Sierra, Inc., http:/www. pmc-sierra.com. Mr. Gorshe is a senior member of the IEEE and coeditor for the IEEE Communications magazine's *Broadband Access Series*. He is the chief editor for the ANSI T1X1 Subcommittee and is responsible for SONET and optical network interface standards. He has also been a technical editor for T1.105, T1.105.01, T1.105.02, and T1.105.07 within the SONET standard series as well as the ITU-T SG15 G.7041 (GFP) recommendation. PMC-Sierra develops high-speed broadband communications semiconductors and MIPS-based processors for Access, Metro Transport, and Optical Transport network equipment.

[2]   ITU-T Recommendation G.7041/Y.1303, *Generic Framing Procedure*.

[3]   ITU-T Recommendation G.707/Y.1322, *Network Node Interface for the Synchronous Digital Hierarchy (SDH)*.

[4]   ANSI Standard T1.105, *Synchronous Optical Network (SONET)—Basic Description Including Multiplex Structure, Rates and Formats.*

[5]   ETSI Standard EN 300 417-9-1, *Synchronous Digital Hierarchy (SDH) Concatenated Path Layer Functions.*

[6]   Recall that this bit position indicates the last control code.

[7]   T1X1.5/2001-094, "Impact of $x^{43} + 1$ Scrambler on the Error Detection Capabilities of Ethernet CRC," Standards Contribution from N. Figueira, Nortel Networks, March 2001.

[8]   Gorshe, S., "CRC-16 Polynomials Optimized for Applications Using Self-Synchronous Scramblers," paper accepted for publication in *Proc. ICC2002.*

[9]   T1X1.5/2001-125, "Recommended CRC-16 Polynomial for the Transparent GFP Superblock and Associated New Text," Standards Contribution from S. Gorshe, PMC-Sierra, June 2001.

[10]   T1X1.5/2001-174, "Optimum CRC-16 Polynomial for the Transparent GFP Superblock," Standards Contribution from S. Gorshe, PMC-Sierra, September 2001.

[11]   T1X1.5/2001-148, "Proposed Draft Text for GFP OAM Frames," Standards Contribution from T. Wilson, Nortel Networks; M. Scholten, AMCC; and S. Gorshe PMC-Sierra;, June 2001.

[12]   T1X1.5/2001-177, "Proposed ITU-T Contribution for Adding a 4B/5B Ethernet Mapping to Transport GFP," Standards Contribution from P. Thaler, Agilent, and S. Gorshe, PMC-Sierra; September 2001.

[13]   This section is based on a white paper by Mimi Dannhardt, "Ethernet Over SONET Technology White Paper," PMC-Sierra, Inc., Issue 1, February 2002, http://www.pmc-sierra.com.

CHAPTER 8

Wireless Technologies: WPAN, WLAN, and WWAN

The trend toward mobility is ever increasing. Already more than 750 million people worldwide have wireless telephones at the time of this writing. It is forecast that by 2007, about 2 billion people will subscribe to wireless services and carry wireless telephone sets [1]. The demand for wireless services has experienced major growth in the past 20 years: Since the introduction of cellular service in 1981, it has experienced a CAGR of around 40%. (By comparison, regular telephone service has only experienced a 5% annual growth.) From 2001 to 2005, the CAGR is forecast by most observers to be around 20% [1, 2]. Although the initial application of wireless was to support voice, major interest now revolves around delivering high-speed data and Internet applications to subscribers. The first edition of this book recognized the importance of data transmission over radio links and two substantive chapters were dedicated to the topic. We continue to view wireless as one of the three key pillars of the telecom industry for the rest of the decade in terms of importance and relevance, the other two pillars being broadband (optics and so on) and new applications. Mobility is viewed by proponents as the "killer application."

This chapter focuses on three areas: (1) wireless wide area IP networks (which we also call *nomadic networks*), with data service available continuously everywhere in a large region, nation, or continent; (2) location-specific hotspot networks that support data service in defined environments (e.g., hotels, libraries, and airports); hotspot services are often available in various locations that may or may not be contiguous; and (3) new IEEE 802.16 fixed broadband wireless access systems. The emphasis of the chapter, however, is on mobile networks. Figure 8.1 depicts the taxonomy of the various technologies discussed in this chapter.

The term *mobile* refers to an entity that is in motion during a (data) transmission or session; movement could be at a low speed (for example, a pedestrian) or high speed (for example, a car or train). The term *portable* refers to the ability to access information while at a remote location; typically, there is no motion during the session. The term *nomadic* (wanderer) has the connotation of not being part of a fixed community, such as a specific company (enterprise); however, the user will eventually have to pass

FIGURE 8.1
*Taxonomy of wireless
technologies.*

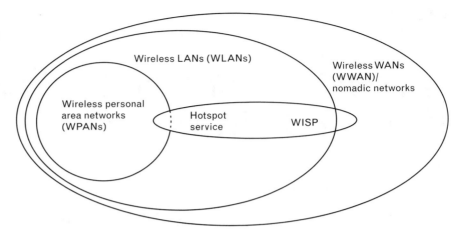

some sort of authentication test, implying that the user is ultimately a member of the community of registered subscribers. The term *fixed wireless* refers to point-to-point or point-to-a-few-points transmission over a (directional) radio link. The term *wireless* can apply to all of these scenarios, but it most often implies mobility. Wireless communication can take place within a building (the range is called a *picocell*), within a city (the range is called a *microcell*), in greater metropolitan areas including suburban locations (the range is called a *macrocell*), or globally (the range is called a *worldcell*).

8.1   Overview

The trend toward wireless data services has already resulted in major deployments of WLANs in corporate settings: Approximately 5% of all business PCs (about 10 million) were already on WLANs as of press time. (Note that WLANs started to appear in the early 1990s.) WPANs and, what one could consider to be the early generation of wireless wide area networks, WWANs are also being deployed, but the penetration of these is currently smaller than is the case for WLANs. Work is also being conducted in the area of *wireless metropolitan area networks* (WMANs). Topologically, WMANs are a subset of WWANs, although the technologies used are generally different. Some, including this author, have also defined *neighborhood area networks* (NANs); these can also be served by wireless technologies [3, 4]. *Wireless NANs* (WNANs) are also subset of WWANs. WNANs are generally slightly larger than a *public access location* (PAL)/ hotspot network (these are known as hotspot networks). PALs that support public hotspot services [also called *location-based services* (LBS)] in specific localized environments are now emerging in many developed countries.

Until now, the emphasis in the industry for wireless data has been on private WLANs. According to observers, many events have contributed to the success of WLANs: The advent of the IEEE 802.11b standard that achieved nearly Ethernet-equivalent speeds, the creation of the *Wireless Ethernet Compatibility Alliance* (WECA) as an industry forum that focused on wireless-fidelity (Wi-Fi) interoperability among equipment vendors, the regulatory environment that permits the use of unlicensed spectrum, and the decision by major notebook PC makers to integrate WLANs into mobile PCs for the mass market all played pivotal roles. Such efforts and trends will likely continue for 802.11a and for the future of WLANs [5].

The desire by users to have continuous access (i.e., systems always on and always connected) will drive the deployment of location-specific hotspot services in the near term. Information is only valuable to the potential consumer when it is readily accessible. Information can have location-based and time-based values. Hotspot services support both kinds of values, but focus on the former. Gartner/Dataquest estimated that there were more than 4,000 public wireless hotspots in the United States by the end of 2001 and, to foster this deployment even more, that in the following 3 years 30% of professional notebook PCs would have WLAN cards (Figure 8.2 shows a forecast for PCs, laptops, and PDAs), while by 2007 approximately 70% of these kinds of devices would have wireless NICs. At the time of this writing, it was reported that Korea Telecom planned to roll out about 10,000 IEEE 802.11-based hotspots in Korea in phase 1 and more locations later. Providers of hotspot services are known as Wi-Fi operators, Wi-Fi being the marketing name for WLAN (IEEE 802.11b) technology that is employed to support these services. Wi-Fi, along with other emerging license-exempt communications standards, will gain significant market share against licensed frequency systems, with the possibility of radically reforming several industries.

WWANs that can provide service anywhere in a metro area, state, country, or continent are expected to emerge. As noted, the majority of traffic on WWAN networks today is voice oriented, but the demand for data and Internet services is becoming more pronounced. Figure 8.3 depicts the expected growth in wireless subscribers in the United States for major WWAN and other wireless technologies, based on 2002 information. Approximately 105 million people are expected to use mobile data services in 2005 [with 80 million on second generation (2G)/midway between second generation and third generation (2.5G) systems and 25 million on 3G systems]. As can be seen in this figure, major opportunities exist for all types of data-enabled wireless services, spanning the WPAN, WLAN, WWAN, and hotspot arena. Third generation (3G) services are expected to be widely

FIGURE 8.2
Laptops, PC, and PDA forecast. (Source: Instat/MDR.)

PC market forcast						
	2001	2002	2003	2004	2005	2006
Notebook PC shipments (millions)	25.5	2.7.6	34.1	40	46	51.6
Desktop PC shipments (millions)	101.1	105.2	121.9	133.3	143.5	151.6
PDA shipments (millions)	8.1	9.5	12.4	16.1	19.5	23.4

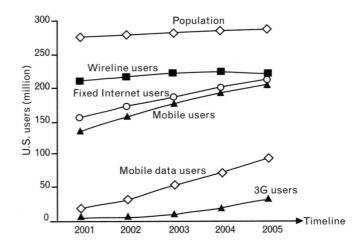

available in the United States from 2005 onward. It is anticipated that there will be 25 million subscribers in 2005 in the United States out of a base of 201 million mobile users who will be on 3G systems in 2005, and the number is expected to grow to 171 million out of a base of 229 million by 2010. Table 8.1 provides worldwide information on WWAN and other wireless technologies. Environments with highly developed wired networks have less demand for wireless services, just as the availability of cable TV in major urban locations lowers the demand for DirecTV for city dwellers (however, in more rural environments, direct broadcast satellite services are more in demand). This is why demand for Web access via cell phones has been more pronounced, so far, in Asia and Europe. Wireless has taken off in countries where telecom charges are high because it is more available and less expensive than the wireline alternatives.

Whereas WLANs and hotspot services are relatively inexpensive to deploy, large-area 2.5G and 3G systems will cost hundreds of billions of dollars to deploy on a broad scale (2.5G systems were being deployed in the early 2000s and 3G systems were expected to be deployed in the mid- to late 2000s.) According to a press-time *Wall Street Journal* article [6], "European mobile-phone companies are retreating from their investments in third generation (3G) wireless technology… after spending more than $150 billion on 3G licenses and infrastructure, European operators have grown much less optimistic about potential returns of offering 3G services…. Some of those … are likely to pull out or drastically reduce rollout costs to minimize their losses." It is worth noting that there have been major "battles" during the late 1990s with regard to which would be the wireless technology of choice for 3G: TDMA or CDMA. At least in terms of the technology selection and standards, that choice was made in the early part of the decade in favor of CDMA. In our opinion, *generally available* 3G is relatively far in the future (e.g., 2006 and beyond), because of the tremendous expenses involved; hotspot services on the other hand are much cheaper and faster to deploy.

TABLE 8.1 WORLDWIDE FORECAST OF WIRELESS SERVICES BY
TECHNOLOGY AND GEOGRAPHY: TWO DIFFERENT VIEWS

	2001	2002	2003	2004	2005
Total subscriptions at January 1 (millions)	703	940	1,180	1,406	1,616
Total analog	80	61	37	20	13
Total digital	623	879	1,143	1,386	1,603
GSM	408	557	732	878	994
CDMA	86	128	167	202	230
PDC	48	53	60	63	66
D-AMPS	81	130	182	228	274
Third generation	0	11	2	15	39

Source: Mobile@ovum.

REGION	YE 2001	YE 2002	YE 2003	YE 2004	YE 2005
Africa	29.0	48.2	67.4	84.5	101.6
Americas	98.7	140.6	181.3	214.7	240.5
Asia-Pacific	330.9	444.5	564.2	678.2	780.9
Europe: Eastern	44.5	60.3	75.8	89.9	102.3
Europe: Western	367.2	494.1	607.5	694.4	754.5
Middle East	14.9	20.9	29.0	38.6	48.5
USA/Canada	139.9	165.2	191.6	216.7	239.7
World	1,025.3	1,373.8	1,716.8	2,016.9	2,268.0

Note: Units are in millions.
Source: EMC World Cellular Database; June 2001 forecast based on actual figures
through the end of March 2001.

WPANs consist of devices that operate within a short range (e.g.,
33 ft), and may support traditional computing devices as well as new IP
appliances, including "wearable" computers. Functionally, the WPAN
space is a subset of the WLAN space. Nomadic services rely on the contin-
uum of WLAN-MAN-WAN connectivity, whereas PALs tend to rely
more on WLAN connectivity. Figures 8.4 and 8.5 depict examples of
wireless-ready laptops and PDAs; Figure 8.6 shows two possible intranet
arrangements, one using WPAN and WLAN technology and the other
using WLAN technology exclusively. The latter is the more likely long-
term scenario. Currently, most of the penetration of wireless data is in the
WLAN arena.

In closing this introductory section, note that there is an entire vocabu-
lary of systems, specifications, and technologies, particularly in the WWAN

FIGURE 8.4
*Example of Wi-Fi-
based laptop.*

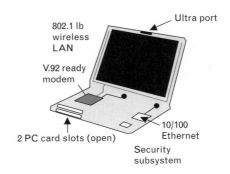

FIGURE 8.5
*Example of Wi-Fi-
based PDA.*

FIGURE 8.6
*Example of
WLAN/WPAN
connectivity.*

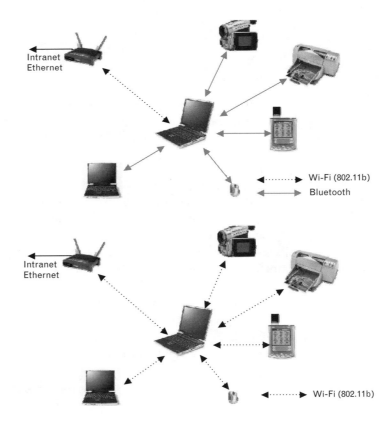

(2G, 2.5G, and 3G) space. This chapter does not aim to provide an exhaustive treatment of this topic, for which one needs perhaps several books, but only to highlight some key trends and representative technologies.

8.2 Standards

8.2.1 Overview

As implied in the introduction, wireless technologies are currently being deployed for personal, home, local, and wide area networks. Standardization is important in order to support interoperability and reduce costs. In this section we look at key WLAN, WPAN, and WWAN standards. Table 8.2 summarizes some of the pertinent specifications, while Figure 8.7 depicts graphically some of the applications of the standards. Table 8.3 provides a snapshot of key aspects of these IEEE standards, and Table 8.4 compares these standards to other proposals. The IEEE 802.11 family is important in the WLAN and hotspot arena. In addition to the key standards identified in Table 8.2, the IEEE is involved in a number of related activities, as seen in Figure 8.8.

8.2.2 IEEE 802.11 Family of Wireless Standards

IEEE 802.11 specifications are focused on the *physical layer* (PHY) and *medium access control* (MAC) sublayer of WLANs. The MAC is consistent with the IEEE 802.3 Ethernet standard. The IEEE standard developed by Working Group 802.11 was accepted by the IEEE board in 1997 and became IEEE Standard 802.11–1997 (see Table 8.5). The standard defines three different WLAN physical implementations (signaling techniques and modulations), a MAC function, and a management function. All of the implementations support data rates of 1 Mbps and, optionally, 2 Mbps (IEEE 802.11b extends the speed to 11 Mbps). Security, roaming, and QoS are also considered, although major improvements to the security apparatus have been shown to be necessary (and have in fact evolved as practical add-ons). The three physical implementations are as follows:

- Direct sequence spread spectrum radio (DSSS) in the 2.4–GHz band (the most commonly deployed technology);
- Frequency hopping spread spectrum radio (FHSS) in the 2.4–GHz band;
- Infrared light (IR).

The DSSS and FHSS PHY options were designed specifically to conform to FCC regulations for operation in the 2.4–GHz *Industrial, Scientific, and Medical* (ISM) band, which has worldwide allocation for unlicensed operation. Both FHSS and DSSS PHYs support 1 and 2 Mbps; all 11–Mbps

TABLE 8.2   KEY WLAN, WPAN, AND WWAN STANDARDS

WLAN-RELATED STANDARDS

IEEE 802.1x	Security framework for all IEEE 802 networks, this is one of the key components of future multivendor interoperable wireless security systems, but implementation will not be simple.
IEEE 802.11	Basic standard for WLANs, which was developed in the late 1990s, supporting speeds of up to 2 Mbps.
IEEE 802.11b	Basic standard for WLANs. An extention of the IEEE 802.11 specifications, supporting speeds of 1, 2, 5.5, and 11 Mbps. Operates in the 2.4-GHz band.
IEEE 802.11a	High-speed WLAN (6 Mbps through 54 Mbps ranges), operating at 5 GHz.
IEEE 802.11e	Revision of 802.11 Media Access Control (MAC) standards, this provides QoS capabilities needed for real-time applications like IP telephonoy and video.
IEEE 802.11g	A new standard for 2.4-GHz WLANs, this provides an increase in the data rate to 54 Mbps, but backward-compatible products were not expected to arrive soon.
IEEE 802.11i	Mired in technical debate and politics at the time of this writing, this is critical to WLAN market expansion, but delays and indecisiveness may make it meaningless if de facto standards emerge.

WPAN-RELATED STANDARDS

Bluetooth/IEEE 802.15	Derivative of Bluetooth 1.x spec and more meaningful standards developments relate to Bluetooth application profiles.

WWAN-RELATED STANDARDS (SUBSET)

Code Division Multiple Access (CDMA) 2000 1x	2.5G standard for wireless WANs, this provides more effecient voice and packet-switched data services than TDMA with peak data rates of 153 Kbps.
CDMA 2000 1xEV	Qualcomm was advocating 1xEV as an evolution of 1x technology. 1xEV uses a 1.25-MHz CDMA radio channel dedicated to and optimized for packet data, and has throughputs of more than 2 Mbps.
CDMA 2000 3x	3G standard for WWANs, this uses the same architecture as 1x. It offers 384-Kbps outdoors and 3-Mbps indoors, but operators will likely nned to wait for new spectrum.
Enhanced Data Rates for Global Evolution (EDGE)	Advocates the General Packet Radio Service (GPRS) data rate to 384 Kbps, but upgrades may be costly for carriers.
General Packet Radio Service (GPRS)	The 2.5G standard for WWANs based on Global System for Moblie Communications (GSM) systems deployed throughout Europe and in other parts of the world. GPRS is an IP-based, packet-data system providing theoretical peak data rates of up to 160 Kbps.
Wideband-CDMA (W-CDMA)	3G standard similar to CDMA2000 but uses wider 5-MHz radio channels. It provides data rates up to 2 Mbps, but more spectrum needs to be allocated in some areas.
IEEE 802.16	Defines physical and MAC standards for fixed point-to-multipoint broadband wireless access (BWA) systems.
Wireless Internet Service Provider Roaming (WISPR)	Driven by the Wireless Ethernet Compatibility Association (WECA) this represents the industry's first effort to provide transparent roaming and billing across public WLANs.

FIGURE 8.7
Positioning of technologies and standards.

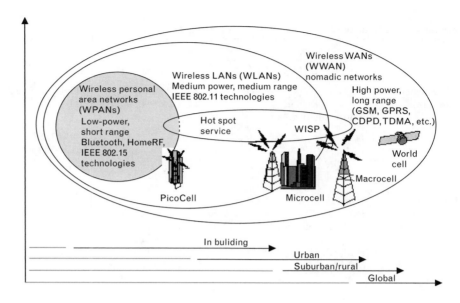

TABLE 8.3   SNAPSHOT OF KEY IEEE STANDARDS

	802.11	802.11a	802.11b
Standard approved	July 1997	September 1999	September 1999
Available bandwidth	83.5 MHz	300 MHz	83.5 MHz
Unlicensed frequencies of operation	2.4–2.4835 GHz	5.15–5.35 GHz, 5.725–5.82 GHz	2.4–2.4835 GHz
Number of nonoverlapping channels	3 (Indoor/outdoor)	4 (Indoor) 4 (Indoor/outdoor) 4 (Indoor/outdoor)	3 (Indoor/ outdoor)
Data rate per channel	1, 2 Mbps	6, 9, 12, 18, 24, 36, 48, 54 Mbps	1, 2, 5.5, 11 Mbps

radios are DSSS. DSSS is the implementation of choice in WLANs. The following is the global spectrum allocation around 2.4 GHz:

United States	2.4–2.4835 GHz
Europe	2.4–2.4835 GHz
Japan	2.471–2.497 GHz
France	2.4465–2.4835 GHz
Spain	2.445–2.475 GHz

There is significant penetration of IEEE 802.11b technologies at this time. As noted, the marketing and promotion of the IEEE WLAN

TABLE 8.4   COMPARING IEEE 802.11 WITH OTHER TECHNOLOGIES

STANDARD	IEEE 802.11b, WI-FI WLAN	IEEE 802.11a, WLAN	BLUETOOTH	HIPERLAN/2 (EUROPE)	HOMERF
Frequency spectrum	2.4 GHz (2.400–2.4835 in North America)	5 GHz	2.4 GHz (2.400–2.4835★)	5 GHz (5.15–5.3 GHz)	2.4 GHz
Data bandwidth	11, 5.5, 2, 1 Mbps	54, 48, 36, 24, 12, 6 Mbps	v1.1: 721 Kbps v1.2: 10 Mbps	6, 9, 12, 18, 27, 36, 54 Mbps	10, 5, 1.6, 0.8 Mbps (future plans: 20 Mbps)
Security apparatus	*Wired Equivalent Privacy* (WEP), WEP-2, dynamic WEP keys, IETF EAP, IEEE 802.11x	WEP, WEP-2, dynamic WEP keys, IETF EAP, IEEE 802.11x	Public address that is unique for each user, two secret keys, and a random number that is different for each new transaction	Encryption– decryption scheme for optional use in the HiperLAN/2	128-bit encryption, frequency hopping, 48-bit network ID
Nominal operating range	150 ft indoors, 300 ft outdoors	150 ft indoors, 300 ft outdoors	10m (30 ft)	150m maximum	Domicile
Ideal application(s)	Laptops, desktops where running cable is difficult, PDAs	Laptops, desktops where running cable is difficult, PDAs	Phone hands-free headset, laptops, PDA devices	Packetized voice, video, and Internet communications	Laptops and cable modems with wireless gateways built in
Example of devices using the standard	Cisco Aironet, Apple Airport, Dell TrueMobile, Linksys, D-Link	Intel, Proxim, Cisco	Ericsson HDH-10 hands-free headset, Widcomm Hand- spring Springboard module		Cayman Systems, Compaq, Intel, Motorola, Proxim

★ Each country assigns radio-frequencies differently. For example, in France Bluetooth operates on 2.4465–2.4835 GHz. This implies that Bluetooth products sold in one country will not interoperate with products distributed in another country.
Source: Table partially based on Marks, L. V., "Surveying the Wireless Landscape," IBM White Paper, May 2001, retrieved from http://www.ibm.com.

technology is addressed by WECA's Wi-Fi efforts. The differentiation of IEEE 802.11b with Ericsson's Bluetooth WPAN technology is that the latter is a low-power, short-range radio link, while IEEE technology has higher throughput and range. IEEE 802.11b has been a successful technology with a large base, good user experience, and tested interoperability. IEEE 802.11a offers even higher throughput (on a different frequency band), but there are issues related to range, battery life, cost, and spectrum. In any event, backward compatibility is a requirement.

After the initial promulgation of the wireless LAN standard, the 802.11 Working Group considered additions to the standard to provide higher data rates (5.5 and 11 Mbps) in the 2.4-GHz band, and to allow WLANs to operate in a 5-GHz band at 54 Mbps. Specifically, IEEE 802.11a uses the 5-GHz band called the *Unlicensed National Information Infrastructure* (UNII)

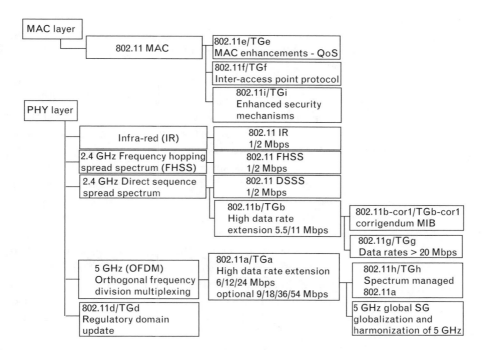

FIGURE 8.8 *IEEE Standards Groups.*

in the United States; it supports 54 Mbps thanks to the higher frequency and greater bandwidth allocation. IEEE 802.11a uses *orthogonal frequency-division multiplexing* (OFDM) modulation. For comparison:

- 802.11a supports 6, 12, and 24 Mbps (mandatory), 9, 18, 36, 48, and 54 Mbps (optional) using 5-GHz OFDM;

TABLE 8.5 IEEE STANDARDS

Standard	Description
IEEE 802.11, 1999 Edition (ISO/IEC 88.2-11: 1999)	IEEE Standard for Information Technology—Telecommunications and Information Exchange Between Systems—Local and Metropolitan Area Network—Specific Requirements—Part 11: Wireless LAN Medium Access Control (MAC) and Physical (PHY) Layer Specifications
IEEE 802.11a-1999 (8802-11:1999/Amd 1:2000(E))	IEEE Standard for Information Technology—Telecommunications and Information Exchange Between Systems—Local and Metropolitan Area Network—Specific Requirements—Part 11: Wireless LAN Medium Access Control (MAC) and Physical (PHY) Layer Specifications—Amendment 1: High-speed Physical Layer in the 5-GHz Band
IEEE 802.11b-1999	Supplement to 802.11—1999, Wireless LAN MAC and PHY Specifications: Higher-Speed Physical Layer(PHY) Extention in the 2.4-GHz Band

• 802.11b supports 1 and 2 Mbps with phase shift keying techniques and 5.5 and 11 Mbps using 2.4–GHz *complimentary code keying* (CCK) [7].

The IEEE 802.11a specification progressed relatively rapidly and chip-makers quickly brought out chipsets. According to IEEE 802.11a proponents, the interference and performance issues at 2.4–GHz have the WLAN industry headed for the 5–GHz frequency band, where the opportunity exists for a cleaner wireless networking environment. Similar to the 2.4–GHz band, the 5–GHz spectrum does not require a license for use throughout much of the world. In addition to being free of interference from microwave ovens and other sources, the 5–GHz region has more than twice the bandwidth available at 2.4 GHz, thereby allowing for higher data throughput. Certain inherent advantages of 802.11a are evident in terms of more frequency spectrum, higher data rates, and more advanced modulation techniques. For example, studies indicate that 802.11a systems have ranges similar to those of 802.11b systems in a typical office environment but with a two to five times higher data rate and throughput performance (e.g., see Figure 8.9) [8]. However, the higher density of hubs and the high price on early equipment still make 802.11b and, by 2003 or 2004, 820.11g the more affordable choice for the majority of enterprise and hotspot applications at this time.

Power is a key consideration in the design of network devices, given the fact that most of the wireless applications are produced for devices that are powered by batteries. Power is essentially proportional to throughput at a given range, so achieving 50 Mbps takes approximately five times the power of 10 Mbps. The result is that 54–Mbps, 5–GHz designs must be more power efficient to achieve similar range or power usage as 11–Mbps, 2.4–GHz designs [9]; this, in general, presents certain engineering challenges. Developers are now looking into *radio-on-a-chip* (RoC) designs that are highly integrated, all–CMOS–based, end-to-end solutions that improve performance and power efficiency, while also reducing the cost of high-speed wireless connectivity.

IEEE 802.11g extends 802.11b to speeds of 54 Mbps for WLANs. IEEE 802.11g was under development at press time, with completion mid–2003. When complete, this specification will extend the IEEE 802.11 family of standards, with data rates up to 54 Mbps in the 2.4–GHz band (not the 5.0–GHz band).

IEEE 802.15 WPAN effort aims at developing consensus standards for WPANs or short-distance wireless networks. (See Figure 8.10 for a view of the protocol model.) These WPANs address the wireless networking of portable and mobile computing devices such as PCs, PDAs, peripherals, cell phones, pagers, and consumer electronics, enabling these devices to communicate and interoperate with one another. The goal of the working group has been to publish standards, recommended practices, or guides that have broad market applicability and deal effectively with the issues of

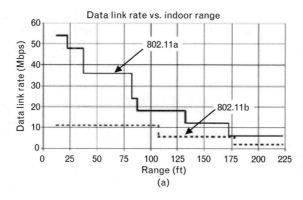

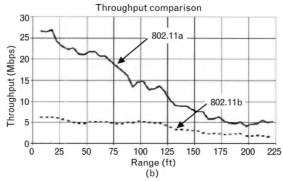

FIGURE 8.9   *Performance of 802.11a. (a) Measured median range performance data for 1,500-byte data packets i ndicates that the range of 802.11a is similar to that of 802.11b up to 225 ft in a typical office environment. At 225 ft, 802.11a systems were measured at 6 Mbps, whereas 802.11b systems were measured at 2 Mbps. (b) Averaged throughput performance data for 1,500-byte data packets. The results indicate that 802.11a throughputs are always at least a factor of 2 times and up to 4.5 times larger than 802.11b systems up to 225 ft. (Courtesy James C. Chen and Jeffrey M. Gilbert, Ph.D., http://www.atheros.com.)*

coexistence and interoperability with other wired and wireless networking solutions. IEEE 802.15 (building on Ericsson's Bluetooth) is a 10-m-radius, low-power technology. IEEE 802.15 Task Group 1 (TG1) is deriving a WPAN standard based on the Bluetooth v1.x foundation specifications. The scope and purpose are as follows:

1. To define PHY and MAC specifications for wireless connectivity with fixed, portable, and moving devices within or entering a *personal operating space* (POS). A goal of the WPAN group is to achieve a level of interoperability that could allow the transfer of data between a WPAN device and an 802.11 device. A POS is the space around a person or object that typically extends up to 10m in all directions and envelops the person whether stationary or in motion. The proposed WPAN standard will be developed to ensure coexistence with all 802.11 networks.

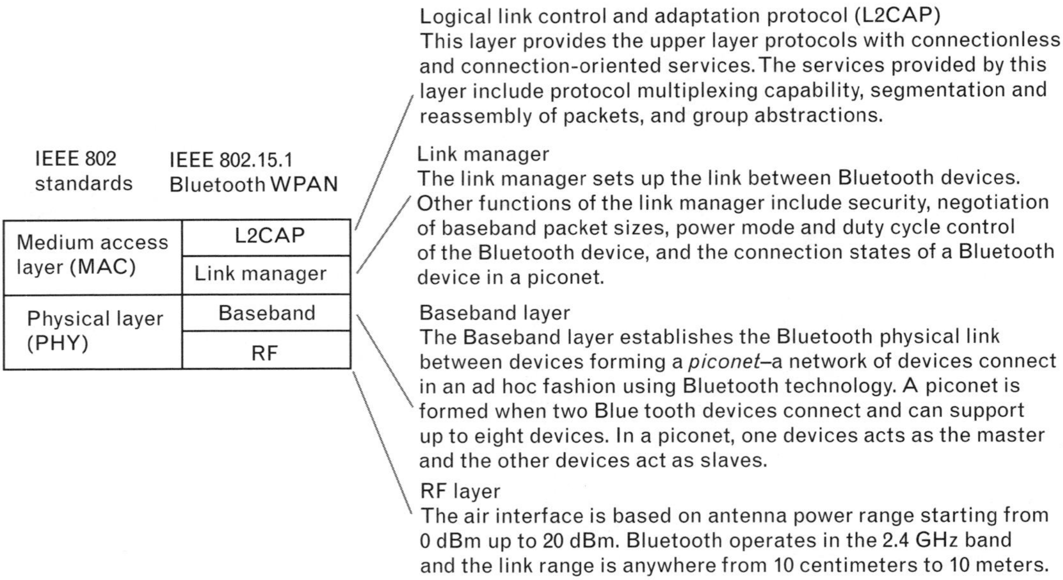

IEEE 802 standards	IEEE 802.15.1 Bluetooth WPAN
Medium access layer (MAC)	L2CAP
	Link manager
Physical layer (PHY)	Baseband
	RF

Logical link control and adaptation protocol (L2CAP)
This layer provides the upper layer protocols with connectionless and connection-oriented services. The services provided by this layer include protocol multiplexing capability, segmentation and reassembly of packets, and group abstractions.

Link manager
The link manager sets up the link between Bluetooth devices. Other functions of the link manager include security, negotiation of baseband packet sizes, power mode and duty cycle control of the Bluetooth device, and the connection states of a Bluetooth device in a piconet.

Baseband layer
The Baseband layer establishes the Bluetooth physical link between devices forming a *piconet*–a network of devices connect in an ad hoc fashion using Bluetooth technology. A piconet is formed when two Blue tooth devices connect and can support up to eight devices. In a piconet, one devices acts as the master and the other devices act as slaves.

RF layer
The air interface is based on antenna power range starting from 0 dBm up to 20 dBm. Bluetooth operates in the 2.4 GHz band and the link range is anywhere from 10 centimeters to 10 meters.

FIGURE 8.10   *IEEE 802.15 WPAN protocol stack.*

2. To provide a standard for low-complexity, low-power-consumption wireless connectivity to support interoperability among devices within or entering the POS. This includes devices that are carried, worn, or located near the body. The project addresses QoS to support a variety of traffic classes. Examples of devices that can be networked include computers, PDAs, *handheld personal computers* (HPCs), printers, microphones, speakers, headsets, bar code readers, sensors, displays, pagers, and cellular and PCS phones.

The IEEE 802.16 family of standards is targeted to fixed wireless applications and may support the interconnection of hotspot cells (such as picocells and microcells). They also have more general applicability. The 802.16-2001 standard specifies the PHY and MAC layer of the air interface of interoperable, fixed, point-to-multipoint *broadband wireless access* (BWA) systems [10]. The specification enables the transport of data, video, and voice services. It applies to systems operating in the vicinity of 30 GHz but is broadly applicable to systems operating between 10 and 66 GHz. The project purpose is to enable the rapid worldwide deployment of cost-effective, interoperable multivendor BWA products. The goal is to facilitate competition in broadband access by providing alternatives to wireline broadband access. Another goal is to facilitate coexistence studies, encourage consistent worldwide allocation, and accelerate the commercialization of BWA spectrum.

The Standards Board of the *IEEE Standards Association* (IEEE-SA) formally approved IEEE Standard 802.16 in December 2001. The approval sets the stage for the widespread deployment of 10- to 66-GHz wireless metropolitan area networks as an economical method of high-speed "last-mile" connection to public networks. The global IEEE 802.16 WirelessMAN air interface standard is the first broadband wireless access standard from an accredited standards body. The WirelessMAN standard is a development that could changes the landscape for providers and customers of high-speed networks, according to the chair of the 802.16 Working Group on BWA. The standard makes efficient use of bandwidth and supports voice, video, and data applications with the quality that customers demand. The 802.16 standard creates a platform on which to build a broadband wireless industry using high-rate systems that install rapidly without extensive metropolitan cable infrastructures. It was created in a 2-year, open-consensus process that involved hundreds of engineers from the world's leading operators and vendors. The standard enables interoperability among devices from multiple manufacturers. It includes a MAC that supports multiple physical layer specifications. The physical layer is optimized for bands from 10 to 66 GHz. Extensions to the 2- to 11-GHz bands were expected to be completed by press time in the working group's 802.16a amendment. The companion IEEE Standard 802.16.2, *IEEE Recommended Practice for Local and Metropolitan Area Networks—Coexistence of Fixed Broadband Wireless Access Systems*, was published by IEEE in September 2001 [11]. This document provides guidelines for system deployment and is expected to be a valuable source of planning information for operators wishing to deploy IEEE 802.16 systems. The IEEE 802.16 Working Group on BWA has 178 members and 52 official observers; the working group is a unit of the IEEE 802 LAN/MAN Standards Committee [12].

8.2.3 WWAN Standardization Activities

WWANs have received a lot of attention in recent years. WWANs have been around for a number of years, but data support was rather limited and expensive until the late 1990s. Figure 8.11 depicts the progression of the wireless data technologies during the past 25 years. In the late 1990s the growth in traffic and in the number of subscribers in the mobile networks was greater than anticipated (in fact, in many places mobile penetration has grown to the point that mobile subscribers represent more than 50% of the telephone users); in the 2000s, however, the growth has been less than first anticipated. WWANs/mobility services give the user the ability to move around town or around the country while still remaining connected to the Internet or the company intranet at high speed (0.1–2 Mbps). As such, they require an extensive infrastructure, precisely like (or based on) cellular telephony with its numerous tall towers all over town, switches, central offices, gateways, and other infrastructure. These networks and systems

FIGURE 8.11
Evolution of WWAN
technologies.

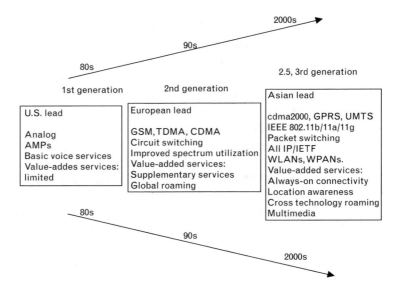

Technology			Features
1G	AMPS	Advanced mobile phone service	Analog voice service
2G	CDMA	Code division multiple access	Digital voice service CDMA, TDMA 9.6 Kbit/sec to 14.4 Kbit/sec.
	TDMA	Time-division multiple access	
	GSM	Global system for mobile communications	
	PDC	Personal digital cellular	
2.5G	GPRS	General packet radio service	115–307 kbps
	CDMA1x	Code division multiple access	
3G	W-CDMA	Wideband code division multiple access	Broadband data services, video and multimedia Up to 2M bit/sec. Always-on data
	CDMA2000	Based in the Interim Standard-95 CDMA standard	
	TD-SCDMA	Time-division synchronous code-division multiple-access	

tend to be the purview of major carriers such as AT&T Wireless, Sprint, Verizon, and so on. Mobile telephones are also becoming portable Internet terminals. PDAs, laptops, and data–ready phones benefit from mobile Internet access, although the speeds started out at the low end (compared to LAN connections). Major advances were seen in the area of standard efforts with regards to 2.5G and 3G WWAN services during the mid-1990s and early 2000s.

The range of a WAN is typically measured in dozen or hundreds of miles. As noted, communication over such distances requires relatively high-power transmissions and, because of that, a license for a specific frequency band. In most instances, carriers pay a fee for a license to transmit at certain power levels in a particular frequency spectrum. High-power transmission also leads to trade-offs between power consumption and data rates in WWANs. Typical data rates for today's cellular networks are relatively slow due largely to the transmission power needed to reach the cellular tower from a handset. Higher data rates at these same levels of power transmission are impractical using early 2000s battery technologies. 3G cellular systems will support higher data rates; however, to maintain power consumption at reasonable levels, 3G systems will require cellular towers to be much closer together.

As of yet, however, no single killer application has emerged beyond the basic convenience of mobility because individual interests are varied. For example, e-mail has limits (not everyone wants to read e-mails while shopping at the mall, and so on); stock quote/trading via wireless has been available since the mid-1980s; in addition, people's investments should not, for the overwhelming majority, require hour-by-hour or minute-by-minute reference to one's account status, since such investment modality is not sustainable. Enterprise applications (secure Web-enabled intranet access to ERP systems via wireless) may be of increasing importance in the near future.

Today's typical WWAN technologies are based on infrastructure in common use for cellular communications, such as GSM, TDMA, and CDMA. WWANs piggyback on wireless (cellular) telephone systems; hence, any discussion of WWANs will entail references to these systems. In this context, Table 8.6 summarizes the key technologies [13], while Table 8.7 provides a more inclusive glossary. Many second generation mobile technologies exist today, each having an influence in specific parts of the world. GSM, TDMA (IS-54, but more particularly, IS-136), and CDMA (IS-95) are the main technologies in the second generation mobile market. European providers have deployed systems based on the GSM standard. In the United States carriers have utilized a variety of standards including CDMA, TDMA, and GSM. TDMA is the most widely used wireless technology in the Americas. To most U.S. consumers, TDMA equates to a particular mobile system, the one known as the *Digital-American Mobile Phone System* (D-AMPS) or IS-136. GSM by far has been the most successful standard in terms of its coverage. All of these systems have different features and capabilities. Although both GSM- and TDMA-based networks use TDM on the air interfaces, their channel sizes, structures, and core networks are different. CDMA has a different air interface [14]. Each technology has its own codec (coder/decoder), and, as is the case with all compression, there is a trade-off between signal quality and signal coding efficiency. GSM's codec has a data

TABLE 8.6   MAJOR WWAN SYSTEMS

GLOBAL SYSTEM FOR MOBILE COMMUNICATION (GSM)	GSM service was introduced in 1991. As of 1997 it was already available in more than 100 countries and has become the de facto standard in Europe and Asia. GSM is used on the 900- and 1,800-MHz frequencies in Europe, Asia, and Australia, and the- 1,900-MHz frequency in North America and Latin America. GSM allows eight simultaneous calls on the same radio frequency and uses narrowband TDMA.
TIME DIVISION MULTIPLE ACCESS (TDMA)	TDMA technology is used for digital transmissions such as moving a signal between a mobile phone and a base station. With TDMA, a frequency band is subdivided into several channels or time slots that are then stacked into shorter time units, facilitating the sharing of a single channel by several calls. GSM actually uses narrowband TDMA, enabling eight simultaneous calls on the same radio-frequency. In November 2001, the number of TDMA users world-wide was estimated at 82 million—a more than 50% increase in one year. Standard: TDMA (IS-136).
CODE DIVISION MULTIPLE ACCESS (CDMA)	A (wireless) transmission method in which signals are encoded using a random sequence, or code, to define a channel. CDMA offers improved spectral efficiency over analog transmission in that it allows for greater frequency reuse. Characteristics of CDMA systems include fewer dropped calls, increased battery life, and better security. CDMA was originally a military technology first used in early form during World War II. Because Qualcomm Inc. created communications chips for CDMA technology, it was privy to the classified information that later became public. Qualcomm has since claimed patents on the technology and was the first to commercialize it. From Qualcomm... "CDMA works by converting speech into digital information, which is then transmitted as a radio signal over a wireless network. Using a unique code to distinguish each different call, CDMA enables many more people to share the airwaves at the same time—without static, cross-talk or interference." Standard: CDMA (IS-95).
GENERAL PACKET RADIO SERVICE (GPRS)	A packet-related technology that enables high-speed wireless Internet and other communica- tions over a GSM network, GPRS is well suited for sending and receiving small bursts of data. GPRS enables information to be sent or received immediately and users are considered to be "always connected" because no dial-up modem connection is required. Benefits include faster data speeds and "always on" mobility; an almost instantaneous connection setup; connection to an abundance of data sources around the world; thorough support for multiple protocols, including IP; and a step toward full 3G services. GPRS provides users with fast file downloads and effective Internet searching. GPRS subscribers are charged only for data sent and received, not for time on-line.

Source: http://www.wirelessdevnet.com/newswire-less/feb012002.html.

rate of 13 Kbps, and as might be expected, gives better results than a lower rate coder. The GSM codec produces a 260-bit packet every 20 ms. The D-AMPS codec requires only 8 Kbps, which is lower quality but still intelligible. (Vendors have recently developed "half-rate"[1] codecs for both systems, with the aim of doubling the voice capacity of a network, though so far these are rarely used. The CDMA deployed in the U.S. also has a 13-Kbps codec with an 8-Kbps variant that can be used for traffic capacity relief.

1. "Half-rate is a term that entered the mainstream discourse on or after 1993—prior to that time the term "low rate vocoding" was in use.

TABLE 8.7   GLOSSARY OF KEY CONCEPTS IN CELLULAR AND WWAN SERVICES

ANALOG CELLULAR TECHNOLOGIES

AMPS	Advanced Mobile Phone System. Developed by Bell Labs in the 1970s and first used commercially in the United States in 1983. It operates in the 800-MHz band and is currently the world's largest cellular standard.
C-450	Installed in South Africa during the 1980s. Uses 450-MHz band. Much like C-Netz. Now known as Motorphone and run by Vodacom SA.
C-Netz	Older cellular technology found mainly in Germany and Austria. Uses 450 MHz.
Comvik	Launched in Sweden in August 1981 by the Comvik network.
N-AMPS	Narrowband Advanced Mobile Phone System. Developed by Motorola as an interim technology between analog and digital. It has some three times greater capacity than AMPS and operates in the 800-MHz range.
NMT450	Nordic Mobile Telephones/450. Developed specially by Ericsson and Nokia to service the rugged terrain that characterizes the Nordic countries. Range 25 km. Operates at 450 MHz. Uses FDD FDMA.
NMT900	Nordic Mobile Telephones/900. The 900-MHz upgrade to NMT 450 developed by the Nordic countries to accommodate higher capacities and handheld portables. Range 25 km. Uses FDD FDMA technology.
NMT-F	French version of NMT900.
NTT	Nippon Telegraph and Telephone. The old Japanese analog standard. A high-capacity version is called HICAP.
RC2000	Radiocom 2000, a French system launched in November 1985.
TACS	Total Access Communications System. Developed by Motorola and is similar to AMPS. It was first used in the United Kingdom in 1985, although in Japan it is called JTAC. It operates in the 900-MHz frequency range.

DIGITAL CELLULAR TECHNOLOGIES

A1-Net	Austrian name for GSM 900 networks.
B-CDMA	Broadband CDMA. Now known as W-CDMA (see below). To be used in UMTS.
Composite CDMA/TDMA	Wireless technology that uses both CDMA and TDMA. For large-cell licensed band and small-cell unlicensed band applications. Uses CDMA between cells and TDMA within cells. Based on Omnipoint technology.
CDMA	Code Division Multiple Access. There are now a number of variations of CDMA, in addition to the original Qualcomm-invented N-CDMA (originally just "CDMA," also known in the US as IS-95; see N-CDMA below). Latest variations are B-CDMA, W-CDMA, and composite CDMA/TDMA. Developed originally by Qualcomm, CDMA is characterized by high capacity and small cell radius, employing spread-spectrum technology and a special coding scheme. It was adopted by the Telecommunications Industry Association (TIA) in 1993. The first CDMA-based networks are now operational. B-CDMA is the basis for 3G UMTS (see below).
cdmaOne	First generation narrowband CDMA (IS-95). See above.
CDMA2000	The new second generation CDMA MoU spec for inclusion in UMTS.
CT-2	A second generation digital cordless telephone standard. CT2 has 40 carriers 1 duplex bearer per carrier supports 40 voice channels.
CT-3	A third generation digital cordless telephone, which is very similar and a precursor to DECT.
CTS	GSM Cordless Telephone System. In the home environment, GSM-CTS phones communicate with a CTS *home base station* (HBS), which offers perfect indoor radio coverage. The CTS-HBS hooks up to the fixed network and offers the best of the fixed and mobile worlds: low cost and high quality of the *Public Switched Telephone Network* (PSTN), services and mobility of the GSM.

TABLE 8.7   GLOSSARY OF KEY CONCEPTS IN CELLULAR AND WWAN SERVICES (CONTINUED)

D-AMPS (IS-54)	Digital AMPS, a variation of AMPs. Uses a three-time-slot variation of TDMA; also known as IS-54. An upgrade to the analog AMPS. Designed to address the problem of using existing channels more efficiently, DAMPS (IS-54) employs the same 30-kHz channel spacing and frequency bands (824–849 and 869–894 MHz) as AMPS. By using TDMA instead of FDMA, IS-54 increases the number of users from 1 to 3 per channel (up to 10 with enhanced TDMA). An AMPS/D-AMPS infrastructure can support use of either analog AMPS phone or digital D-AMPS phones. This is because the FCC mandated only that digital cellular in the United States must act in a dual-mode capacity with analog. Both operate in the 800-MHz band.
DCS 1800	Digital Cordless Standard. Now known as GSM 1800. GSM operated in the 1,800-MHz range. It is a different frequency version of GSM, and (900-MHz) GSM phones cannot be used on DCS 1800 networks unless they are dual band.
DECT	Digital European Cordless Telephone. Uses 12-time-slot TDMA. This started off as Ericsson's CT-3, but developed into ETSI's Digital European Cordless Standard. It is intended to be a more flexible standard than the CT2 standard, in that it has more RF channels (10 RF carriers × 12 duplex bearers per carrier = 120 duplex voice channels). It also has a better multimedia performance because 32 Kbps bearers can be concatenated. Ericsson has developed a dual GSM/DECT handset.
EDGE	UWC-136, the next generation of data heading toward third generation and personal multimedia environments builds on GPRS and is known as Enhanced Data rate for GSM Evolution (EDGE). It allows GSM operators to use existing GSM radio bands to offer wireless multimedia IP-based services and applications at theoretical maximum speeds of 384 Kbps with a bit-rate of 48 Kbps per time slot and up to 69.2 Kbps per time slot in good radio conditions.
E-Netz	The German name for GSM 1800 networks.
FDMA	Frequency Division Multiple Access.
GMSS	Geostationary Mobile Satellite Standard, a satellite air interface standard developed from GSM and formed by Ericsson, Lockheed Martin, U.K. Matra Marconi Space, and satellite operators Asia Cellular Satellite and Euro-African Satellite Telecommunications.
GSM	Global System for Mobile Communications. The first European digital standard, developed to establish cellular compatibility throughout Europe. Its success has spread to all parts of the world and more than 80 GSM networks are now operational. It operates at 900 MHz.
IDEN	Integrated Digital Enhanced Network (IDEN). Launched by Motorola in 1994, this is a private mobile radio system from Motorola's Land Mobile Products Sector (LMPS). iDEN technology, currently available in the 800-MHz, 900-MHz, and 1.5-GHz bands. It utilizes a variety of advanced technologies, including state-of-the-art vocoders, M16QAM modulation, and TDMA. It allows *Commercial Mobile Radio Service* (CMRS) operators to maximize the dispatch capacity and provides the flexibility to add optional services such as full-duplex telephone interconnect, alphanumeric paging, and data/fax communication services (e.g., Nextel uses iDEN).
IMT DS	Wideband CDMA, or W-CDMA.
IMT MC	Widely known as CDMA2000 and consisting of the 1X and 3X components.
IMT TC	Called UTRA TDD or TD-SCDMA.
IMT SC	Called UWC-136 and widely known as EDGE.
IMTFT	Well known as DECT.
Inmarsat	Acronym derived from International Maritime Satellite System. It uses a number of GEO satellites. Available as Inmarsat A, B, C, and M.
Iridium	Mobile Satellite phone/pager network launched November 1998. Uses TDMA for intersatellite links. Uses 2-GHz band.
IS-54	TDMA-based technology used by the D-AMPS system at 800 MHz.
IS-95	CDMA-based technology used at 800 MHz.

TABLE 8.7   GLOSSARY OF KEY CONCEPTS IN CELLULAR AND WWAN SERVICES (CONTINUED)

IS-136	TDMA-based technology. TDMA IS-136 is an evolved form of TDMA IS-54.
JS-008	CDMA based standard for 1,900 MHz.
N-CDMA	Narrowband Code Division Multiple Access, or plain old original "CDMA." Also known in the United States as IS-95. Developed by Qualcomm and characterized by high capacity and small cell radius. Has a 1.25-MHz spread-spectrum air interface. It uses the same frequency bands as AMPS and supports AMPS operation, employing spread-spectrum technology and a special coding scheme. It was adopted by the TIA in 1993. The first CDMA-based networks are now operational.
PACS-TDMA	An eight-time-slot TDMA-based standard, primarily for pedestrian use. Derived from Bellcore/Telcordia's wireless access spec for licensed band applications. Motorola supported.
PCS	Personal Communications Service. The PCS frequency band is 1,850 to 1,990 MHz, which encompasses a wide range of new digital cellular standards like N-CDMA and GSM 1900. Single-band GSM 900 phones cannot be used on PCS networks. PCS networks operate throughout the North America.
PDC	Personal Digital Cellular is a TDMA-based Japanese standard operating in the 800- and 1,500-MHz bands.
PHS	Personal Handy System. A TDD TDMA Japanese-centric system that offers high-speed data services and superb voice clarity. A wireless local loop (WLL) system with only 300-m to 3-km coverage.
SDMA	Space Division Multiple Access, thought of as a component of third generation digital cellular/UMTS.
TDMA	Time Division Multiple Access. The first U.S. digital standard to be developed. It was adopted by the TIA in 1992. The first TDMA commercial system began in 1993. A number of variations exist.
Telecentre-H	A proprietary WLL system by Krone. Range 30 km, in the 350- to 500-MHz and 800- to 1,000-MHz range. Uses FDD FDM/FDMA and TDM/TDMA technologies.
TETRA	TErrestrial Trunked RAdio (TETRA) is a new open digital trunked radio standard that is defined by ETSI to meet the needs of the most demanding professional mobile radio users.
TETRA-POL	Proprietary TETRA network from Matra and AEG. Does not conform to TETRA MoU specifications.
UltraPhone 110	A proprietary WLL system by IDC. Range 30 km, in the 350- to 500-MHz range. Uses FDD FDM/TDMA technologies. The UltraPhone system allows four conversations to operate simultaneously on every 25 khz-spaced channel. A typical UP 24-channel WLL system can support 95 full-duplex voice circuits in 1.2 kHz of spectrum.
UMTS	Universal Mobile Telephone Standard, the next generation of global cellular, which should be in place by 2004. Proposed data rates of <2 Mbps, using combination TDMA and W-CDMA. Operates at around 2 GHz.
W-CDMA	One of the latest components of UMTS, along with TDMA and cdma2000. It has a 5-MHz air interface and is the basis of higher bandwidth data rates.
WLL	Wireless local loop limited-number systems are usually found in remote areas where fixed-line usage is impossible. Most modern WLL systems use CDMA technology.

Source: Retrieved from http://www.astalavista.com/mobile/wct.shtml.

8.2.3.1   Review of 2G Cellular Systems in Place

This section surveys existing 2G cellular systems mostly from a standard telephony perspective. Of the many quality reference materials available, we base our discussion in this subsection on a white paper by Puneet Gupta [14].

Global System for Mobile Communication

In 1984 a project was endorsed by the European Commission that paved the way for a digital wireless solution for the 900-MHz frequency band. In 1987, operators from 13 countries signed a *memorandum of understanding* (MoU). The new standard was supposed to employ TDMA, a technology supported by Nokia, Ericsson, and Siemens. After validation tests, the MoU operators signed an invitation–to–tender in 1988. In 1989, ETSI was formed and accorded equal status to administrators, operators, and manu-facturers, which resulted in the publication of the GSM 900 specifications in 1990. The United Kingdom's *personal communications network* (PCN) decided to adopt the GSM specification for their Digital Cellular System 1900 (DCS 1900) development, which was later renamed GSM 1800. In 1992, Australia's Telstra signed on with the MoU and just 4 years later they had 1 million subscribers, which was 5.6% of the population. One year later, GSM had also expanded to India, Africa, Asia, and the Arab world. In 1994, the FCC made the 1,900-MHz band available in the United States. In 1995, the MoU had 156 members serving 12 million customers in 86 countries. At the same time, phase 2 of GSM and demonstrations of fax, video, and data communications via GSM appeared [15].

The *GSM-ANSI-136 Interoperability Team* (GAIT) is working on developing mobile phones that work on both GSM and TDMA networks while providing overseas roaming. This is of interest to companies such as Cingular and AT&T Wireless, operators that still have TDMA networks and are currently deploying GSM networks. GAIT phase 1 is the integra-tion of GSM and TDMA voice and data technology into one handset with the option to select which system to use. The supporting organization is the *Universal Wireless Communications Consortium* (UWCC). The UWCC was founded in 1996 and works closely with other global organizations such as ITU, 3GPP, and the *Universal Mobile Telecommunications System* (UMTS) Forum.

GSM's air interface is based on narrowband TDMA technology, where the available frequency bands are divided into time slots. Each user has access to one time slot at regular intervals. Narrowband TDMA allows eight simultaneous communications on a single 200-kHz carrier and is designed to support 16 half-rate channels. The fundamental unit of time in this TDMA scheme is called a *burst period* and it lasts 15/26 ms (or approx. 0.577 ms). Eight burst periods are grouped into a *TDMA frame* (120/26 ms, or approx. 4.615 ms), which forms the basic unit for the definition of logi-cal channels. One physical channel is one burst period per TDMA frame. A GSM mobile can seamlessly roam nationally and internationally. Roaming requires that registration, authentication, call routing, and location updat-ing functions exist and be standardized in GSM networks.

A TDMA cell phone transmits and receives in only one slot, remaining silent until its time slot comes around again. The number of slots, the cycle length, and the frequency width all depend on the particular technology. For example, D-AMPS (IS-54/IS-136) station is only active for one-third of the time, and a GSM for one-eighth. GSM uses wider frequencies than other systems, with much shorter time slots. GSM's wide frequencies support scalability, but the short time slots can cause phone synchronization problems. Radio signals have a round-trip delay of around 0.4 ms for a telephone set 40 miles from the base station. The GSM time slot only lasts 0.577 ms (156 bits, which in theory amounts to a capacity of 33.8 Kbps for each of the eight channels); this propagation delay is enough to create a risk for the phone to miss its slot entirely. In practice, GSM phones cannot be used more than 35 km (22 miles) from a BTS, no matter how strong the signal.

The European version of GSM operates at the 900-MHz frequency (and now at the newer 1,800-MHz frequency). Since the North American version of GSM operates at the 1,900-MHz frequency, the telephones sets are not interoperable, but the *subscriber identification modules* (SIMs) are. (SIM cards are small smart cards that fit inside phones based on the GSM technology; SIMs contain personalized information about their users including the network activation and even phone book entries; a user can put his or her SIM card in another GSM phone and use it as if it was his or her own phone.) Dual-band 900–1,800 and 900–1,900 phones are already released and in production. Tri-band 900–1,800–1,900 GSM phones are expected to be manufactured in the next few years, which will allow interoperability between Europe and North America.

As shown in Figure 8.12, the GSM network consists of mobile stations communicating with the base transceiver station, on the Um interface.

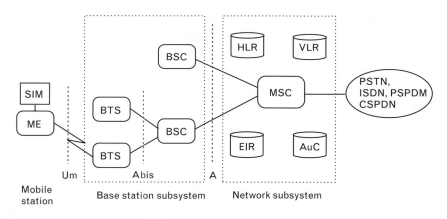

SIM: Subscriber identity module BSC: Base station controller MSC: Mobile services switching center
ME: Mobile equipment HLR: Home location register EIR: Equipment identity register
BTS: base station transceiver VLR: Visitor location register AuG: Authentication center

FIGURE 8.12 *GSM architecture.*

Many *base transceiver stations* (BTSs) are connected to a *base station controller* (BSC) via the Abis interface and the BSC connect to the mobile service switching center (MSC; the core switching network) via the A interface. The signaling protocol in GSM is structured into three general layers depending on the interface, as shown in Figure 8.13. Layer 1 is the physical layer, which uses the channel structures discussed above over the air interface. Layer 2 is the data link layer. Across the Um interface, the data link layer is a modified version of the LAPD protocol used in ISDN, called LAPDm. Across the A interface, the message transfer part layer 2 of Signaling System Number 7 (SS7) is used. Layer 3 of the GSM signaling protocol is itself divided into three sublayers [16]:

- *Radio resources management.* Controls the setup, maintenance, and termination of radio and fixed channels, including handovers.

- *Mobility management.* Manages the location updating and registration procedures, as well as security and authentication.

- *Connection management.* Handles general call control, similar to ITU Recommendation Q.931, and manages supplementary services and the short message service.

Signaling between the different entities in the fixed part of the network, such as between the home location register (HLR) and visitor location register (VLR), is accomplished via the mobile application part (MAP). MAP is built on top of the transaction capabilities application part (TCAP), which is the top layer of SS7.

HLR and VLR provide customized subscriber services and allow seamless movement from one cell to another. The authentication register and the equipment register provide security and authentication. An *operations and maintenance center* (OMC) and a cell broadcast center allow configuration of the network and provide the cell broadcast service in the GSM network (not shown in the diagram). The voice transmitted on the air interface can be encrypted. The speech is coded at 13 Kbps over the air interface. Using *enhanced full-rate coding* (EFR), the voice quality

FIGURE 8.13
GSM interface.

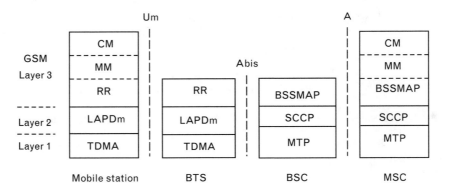

approaches the quality of a land line. Developments like *adaptive multirate coding* (AMR) allow speech coding and channel coding to be dynamically adjusted, giving acceptable performance even in case of bad radio conditions. The GSM network supports automatic handovers. Because the mobiles are not transmitting or receiving at all times, battery consumption can be conserved. Furthermore, by using DTX and DRX (discontinuous transmission and reception, in which a mobile terminal transmits or receives only when it detects voice activity), battery power can be conserved even more. Also since the mobile is not transmitting or receiving at all times, the mobile can listen to control channels and provide useful information about other channels back to the cell.

In GSM, a *traffic channel* (TCH) is used to carry speech and data traffic (Figure 8.14). Traffic channels are defined using a 26-frame multiframe; a multiframe is group of 26 TDMA frames. The temporal length of a 26-frame multiframe is 120 ms, which is how the length of a burst period is defined (120 ms divided by 26 frames divided by 8 burst periods per frame); 24 are used for traffic, 1 is used for the *slow associated control channel* (SACCH), and 1 is unused at this time. TCHs for the uplink and downlink are separated in time by three burst periods; this implies that the mobile station does not have to transmit and receive simultaneously, consequently simplifying the electronics. In addition to these *full-rate* TCHs, *half-rate* TCHs are also defined, although they have not yet been implemented [16]. These capabilities effectively double the capacity of a system once half-rate speech coders are specified (i.e., speech coding at around 7 Kbps, in place of 13 Kbps).

GSM offers a variety of data services: users can send and receive data, at rates up to 9,600 bps, to users on *plain old telephone service* (POTS), ISDN, packet switched public data networks, and circuit switched public data networks using a variety of access methods and protocols, such as X.25 or

FIGURE 8.14
GSM channel.

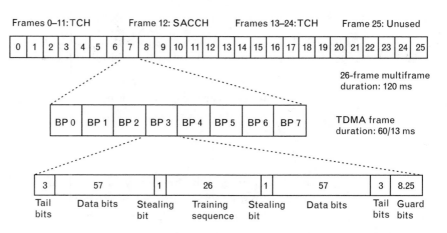

X.32. Other data services include Group 3 facsimile, as described in ITU-T recommendation T.30 (fax is supported by use of an appropriate fax adapter.) A unique feature of GSM, not found in older analog systems, is the short message service. SMS is a bidirectional service for short alphanumeric (up to 160 bytes) messages. Messages are transported in a store-and-forward fashion. For point-to-point SMS, a message can be sent to another subscriber to the service, and an acknowledgment of receipt is provided to the sender. SMS can also be used in a cell-broadcast mode, for sending messages such as traffic updates or news updates. Messages can also be stored in the SIM card for later retrieval. Recent developments and initiatives include the following:

- GSM association together with the UWCC, which represents the interests of the TDMA community, are working toward interstandard roaming between GSM and TDMA (ANSI-136) networks.

- The majority of European GSM operators were planning to implement GPRS technology as their network evolution path to third generation.

- MExE will allow operators to provide customized, user-friendly interfaces to a host of services from GSM, through GPRS and eventually UMTS. The first implementations of MExE are expected to support the WAP and Java applications. MExE can extend the capabilities that currently exist within WAP by enabling a more flexible user interface, more powerful features, and better security.

- The GSM cordless telephony system can provide a small home base station to work with a standard GSM mobile phone in a mode similar to that of a cordless phone. The base station would be connected to the PSTN.

- Number portability will allow customers to retain their mobile numbers when they change operators or service providers.

- Location services are needed to standardize the methods for determining a GSM subscriber's physical location.

- Tandem free operation where the compressed speech is passed unchanged over the 64-Kbps links between the transcoders, hence improving the voice quality.

TDMA IS-136 Technology

In TDMA systems the frequency bands available to the network are partitioned into time slots, with each user having access to one time slot at regular intervals. Three users share a 30-kHz bandwidth (IS-136) by splitting a 30-kHz carrier into three time slots. TDMA makes more efficient use of available bandwidth than the previous generation analog technology.

TDMA was first specified as a standard in EIA/TIA Interim Standard 54 (IS-54). IS-136, an evolved version of IS-54, is the U.S standard for TDMA for both the cellular (850-MHz) and PCS (1.9-GHz) spectrums. Unlike IS-54, IS-136 utilizes TDM for both voice and control channel transmissions. Digital control channels allow residential and in-building coverage, increased battery standby time, several messaging applications, over the air activation, and expanded data applications (GSM also has the same characteristics). The digital control channel allows for the creation of microcell applications, making it suitable for wireless PBX and paging applications. TDMA networks transmit at a higher data rate on a relatively low bandwidth channel resulting in the possibility of cochannel interference occurring. As described earlier for GSM, the time-slot structure allows the mobiles to conserve battery power and to collect information about other channels. IS-136 specifies a "sleep mode" that instructs the compatible cellular phones to conserve power. IS-136 handsets are not compatible with IS-54.

TDMA IS-136 exists in North America at both the 800- and 1,900-MHz bands. IS-136 TDMA normally coexists with analog channels on the same network. One advantage of this dual-mode technology is that users can benefit from the broad coverage of established early analog networks while IS-136 TDMA coverage grows within, and at the same time take advantage of the more advanced technology of IS-136 TDMA where it exists. TDMA networks have increased the capacity of the analog networks (using the same bandwidth) by three times.

As noted earlier, UWCC is a group of more than 100 telecom carriers and vendors of mobile products and services that focuses on efforts to develop services based on IS-136 TDMA and IS-41 WIN. The *Global TDMA Forum* (GTF) of UWCC focuses on both technical and market-led developments. IS-136 revision A has introduced several new features like *adaptive channel allocation* (ACA) depending on the instantaneous channel quality determined by the level of interference; the *private system identification* (PSID), which allows development of large-scale corporate private systems either as multilocation or in-building closed user groups; and two-way SMS (256 characters). IS-136 revision B includes all IS-136+ proposals from the UWC-136 Radio Transmission Technology (RTT) proposal for voice and circuit switched features. Notable features are packet data service, mobile assisted handoff, improved SMS, and intelligent roaming. Major U.S. carriers using TDMA are AT&T Wireless Services, Bell South, and Southwestern Bell.

CDMA Technology (IS-95) (cdmaOne)

The original CDMA standard is known as cdmaOne. cdmaOne is still common in cellular telephones in the United States. CDMA is a coding scheme, used as a modulation technique, in which multiple channels are independently coded for transmission over a single wideband channel.

Spread spectrum is a means of transmission in which the signal occupies a bandwidth in excess of the minimum necessary to send the information; the band spread is accomplished by means of a code that is independent of the data; and a synchronized reception with the code at the receiver is used for despreading and subsequent data recovery [17]. In some communication systems, CDMA is used as an access method that permits carriers from different stations to use the same transmission equipment by using a wider bandwidth than the individual carriers. On reception, each carrier can be distinguished from the others by means of a specific modulation code, thereby allowing for the reception of signals that were originally overlapping in frequency and time. Thus, several transmissions can occur simultaneously within the same bandwidth, with the mutual interference reduced by the degree of orthogonality of the unique codes used in each transmission. CDMA permits a uniform distribution of energy in the emitted bandwidth [18]. Spread–spectrum communications is distinguished by three key elements [19]:

1. The signal occupies a bandwidth much greater than minimally necessary to send the information. This results in many benefits, such as immunity to interference and jamming and multiuser access, which we will discuss later.

2. The bandwidth is spread by means of a code that is independent of the data. The independence of the code distinguishes this from standard modulation schemes in which the data modulation will always spread the spectrum somewhat.

3. The receiver synchronizes to the code to recover the data. The use of an independent code and synchronous reception allows multiple users to access the same frequency band at the same time.

To protect the signal, the code used is pseudorandom. It appears random, but is actually deterministic, so that the receiver can reconstruct the code for synchronous detection. This pseudorandom code is also called *pseudonoise* (PN). There are three ways to spread the bandwidth of the signal [19]:

• *Frequency hopping*. The signal is rapidly switched between different frequencies within the hopping bandwidth pseudorandomly, and the receiver knows beforehand where to find the signal at any given time.

• *Time hopping*. The signal is transmitted in short bursts pseudorandomly, and the receiver knows beforehand when to expect the burst.

• *Direct sequence*. The digital data are directly coded at a much higher frequency. The code is generated pseudorandomly and the receiver knows how to generate the same code and correlates the received signal with that code to extract the data.

CDMA is a DSSS system. The CDMA technology used in North America is based on the IS-95 protocol standard first developed by Qualcomm. CDMA differs from the other two technologies in its use of spread-spectrum techniques for transmitting voice or data over the air. Rather than partitioning the RF spectrum into separate user channels by frequency slices or time slots, spread-spectrum technology separates users by assigning them digital codes within the same broad spectrum. Advantages of CDMA technology include high user capacity and immunity from interference by other signals. Like TDMA IS-136, CDMA operates in the 1,900-MHz band as well as the 800-MHz band.

Work on developing the CDMA standard is conducted mainly by the *CDMA Development Group* (CDG), a consortium of the main CDMA manufacturers and operators formed to standardize and promote CDMA technology. Although work to develop CDMA as a third generation technology has attracted a great deal of attention, the CDG has also been working to improve the current performance of CDMA as a second generation technology. CDG has formally adopted the cdmaOne name and logo as a technology designator for all IS-95-based CDMA systems. The term represents the end-to-end wireless system and the necessary specifications that govern its operation. cdmaOne incorporates the IS-95 CDMA air interface, the ANSI-41 network standard for switch interconnection, and many other standards that make up a complete wireless system (see Figure 8.15).

cdmaOne supports data transmission speeds of only up to 14.4 Kbps in its single-channel form and up to 115 Kbps in an eight-channel form. CDMA2000 and W-CDMA support data transmission at higher speeds. W-CDMA is a CDMA channel that is four times wider than the current channels that are typically used in 2G networks in North America.

The CDMA technology, used in the Interim Standard IS-95, maximizes spectrum efficiency and enables more calls to be carried over a single 1.25-MHz channel. In a CDMA system, each digitized voice is assigned a binary sequence that directs the proper response signal to the corresponding user. The receiver demodulates the signal using the appropriate code.

FIGURE 8.15
CDMA designs.

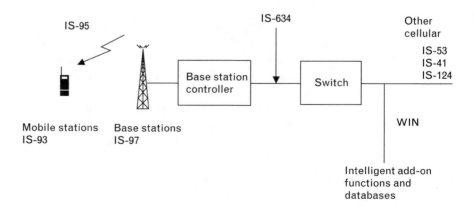

The resulting audio signal contains only the intended conversation, eliminating any background noise. This allows more calls to occupy the same space in the communication channel, thereby increasing capacity. As a simple example, let us assume a user is talking over a CDMA network. The transmitted portion of a voice signal has frequency components from approximately 300 to 3,400 Hz. This analog signal is digitally encoded, using quadrature phase shift keying (QPSK) at 9,600 bps. The signal is then spread to approximately 1.23 Mbps using special codes that add redundancy. Some of these codes include a device ID that is unique to the phone (like a serial number). Next the signal is broadcast over the channel. When broadcast, the signal is added to the signals of the other users in the channel. On the receiving end, the same code is used to decode the incoming signal. The 9,600-bps signal is obtained and the original analog signal is reconstructed. When the same code is used on another user's signal, the redundancy is not removed and the signal remains at 1.23 Mbps.

To address voice quality issues, leading wireless carriers are using 13-Kbps vocoders; this improves quality but at the expense of capacity. The technology has been widely adopted by major cellular and PCS carriers in the United States and also internationally. CDMA networks provide operators with reliable digital systems that offer higher capacity, large coverage area and improved voice quality and above all a good 3G upgrade path, CDMA2000. It also offers simplified system planning—through the use of the same frequency in every sector of every cell. Factors contributing to CDMA's capacity gains are as follows:

- Frequency reuse;
- Soft handoffs;
- Power control;
- Variable rate vocoders.

cdmaOne technology improves QoS through the use of soft handoffs, which reduce the number of dropped calls and ensure a smooth transition between cells. In soft handoff, a connection is made to the new cell while maintaining the connection with the original cell. This transition between cells is one that is almost undetectable to the subscriber. cdmaOne technology also takes advantage of multipath to enhance communications and voice quality. Using a rake receiver and other improved signal processing techniques, each mobile station selects the three strongest multipath signals and coherently combines them to produce an enhanced signal.

The cdmaOne data capabilities are based on IS-95A, which can provide data speeds of 14.4 Kbps. IS-95B and IS-95C are designed to enhance CDMA's data capability. IS-95B can provide data speeds of up to 64 Kbps by aggregating existing channels. IS-95B can provide these enhanced data rates through software upgrades only. IS-95C aims to offer a minimum of 24.4 Kbps per channel and aggregated data speeds of more than 115 Kbps.

It is expected that IS-95C will define CDMA's capability as a third generation system. CDMA already supports asynchronous data and faxing (IS-99) and has standardized packet data (IS-657). Some of the benefits of using cdmaOne are as follows:

- Capacity gains of 8 to 10 times that of AMPS analog systems;
- Improved call quality, with better and more consistent sound as compared to AMPS systems;
- Simplified system planning through the use of the same frequency in every sector of every cell;
- Enhanced privacy through the spreading of voice signals;
- Improved coverage characteristics, allowing for fewer cell sites;
- Increased talk time for portables due to closed-loop power control (which also contributes to increased system capacity by reducing the noise floor in adjoining cells reusing the same spectrum).

The major development initiatives being taken by the CDG for 2G CDMA systems enhancements include enhanced roaming, which enables transparent roaming across cellular and PCS networks, with selection of networks and location services. Enhanced roaming will provide roaming between CDMA systems similar to that on GSM: Registration, authentication, and credit-checking are automatically carried out between the networks without users having to do anything more than switch on their mobiles. Roaming agreements will still be needed between operators.

At the end of 2001, there were about 500 million GSM subscribers worldwide, which is 65% of the world's wireless market. According to the coverage and the subscriber numbers worldwide, GSM comes out as a clear winner in digital technologies. Today, 89% of all cellular subscribers are using digital technology and around 65% of these are GSM subscribers, with CDMA and D-AMPS having about 14% of digital subscribers each. As the number of digital subscribers grows (by almost 100% by 2003) the subscriber ratio is expected to remain the same. 3G should begin showing its presence but with a very small number of subscribers until 2003 and a subscriber base of under 2.5% till 2005.

Second generation technologies, thanks to the continuous improvements and 2.5G overlays, should be viable in the medium term and continue to win market share for at least the next 5 years. Also, the 3G technologies should not have any major impact until 2004 and then coexist with 2G technologies for another 2 to 3 years before gaining prominence. 3G related work was still going on at the time of this writing, and Japan was expected to be among the very first countries with commercial 3G roll out. Figure 8.16 shows the distribution of cellular subscribers across various regions of the world and forecasts for the future.

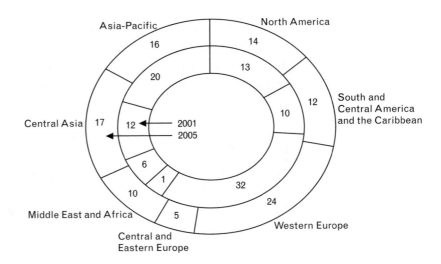

FIGURE 8.16
Distribution of cellular subscribers across various regions of the world. (Source: Mobile@Ovum.)

What is clear is that after coexisting with the digital technologies for 3 to 5 years, we will see a disappearance of the analog technologies. According to the latest data, the number of digital cellular subscribers is expected to double from around 750 million in early 2001 to more than 1.2 billion at the beginning of 2003 and may move up to 1.6 to 2 billion at the beginning of 2005. At the same time, we will see the number of analog cellular subscriptions falling from 80 million at the beginning of 2001 to 37 million by the beginning of 2003 and 14 million by the beginning of 2005.

8.2.3.2 Existing Baseline Data Services (2G and 2.5G)

This section surveys existing 2G and 2.5G systems from a more focused data transmission/Internet perspective. Figure 8.17, building on Table 8.2, depicts some of the existing WWAN 2G/2.5G/3G standards, along with a taxonomy and a possible transition path. Table 8.8 provides a summary of the key technologies, and Table 8.9 identifies key technical aspects of these technologies. Mobile networks are difficult to design because they combine the problems of the Internet—delays and routing problems—with the problems of the mobile—loss of signal, handovers, and congestion [20].

U.S. Technologies Data

As noted, existing 2G systems now offer low data rates, such as 14.4-Kbps circuit-switched services. We indicated that each GSM time slot carries 156 bits, which equates to a capacity of 33.8 Kbps for each of the eight GSM channels. Unfortunately, most of this is not available for data applications because the GSM protocol has a nontrivial overhead, requiring a portion of the bandwidth for such functions as signaling and encipherment. (Every transmission is encrypted using a secret algorithm known as A5, which is thought to use keys of up to 56 bits.) The bandwidth of a D-AMPS channel is only 16 Kbps; however, less bandwidth is used for

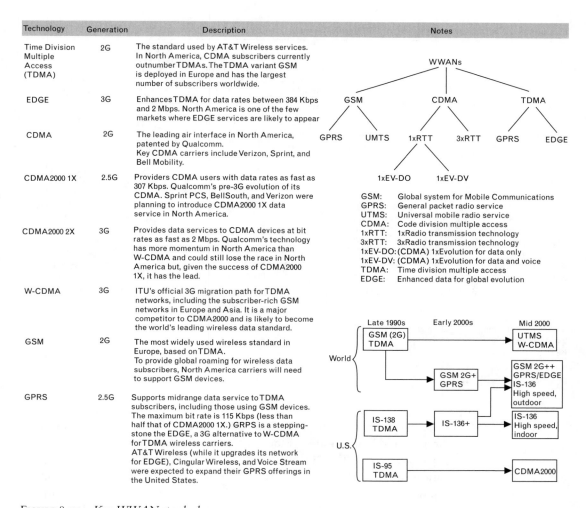

Technology	Generation	Description	Notes
Time Division Multiple Access (TDMA)	2G	The standard used by AT&T Wireless services. In North America, CDMA subscribers currently outnumber TDMAs. The TDMA variant GSM is deployed in Europe and has the largest number of subscribers worldwide.	
EDGE	3G	Enhances TDMA for data rates between 384 Kbps and 2 Mbps. North America is one of the few markets where EDGE services are likely to appear	
CDMA	2G	The leading air interface in North America, patented by Qualcomm. Key CDMA carriers include Verizon, Sprint, and Bell Mobility.	
CDMA2000 1X	2.5G	Providers CDMA users with data rates as fast as 307 Kbps. Qualcomm's pre-3G evolution of its CDMA. Sprint PCS, BellSouth, and Verizon were planning to introduce CDMA2000 1X data service in North America.	
CDMA2000 2X	3G	Provides data services to CDMA devices at bit rates as fast as 2 Mbps. Qualcomm's technology has more momentum in North America than W-CDMA and could still lose the race in North America but, given the success of CDMA2000 1X, it has the lead.	
W-CDMA	3G	ITU's official 3G migration path for TDMA networks, including the subscriber-rich GSM networks in Europe and Asia. It is a major competitor to CDMA2000 and is likely to become the world's leading wireless data standard.	
GSM	2G	The most widely used wireless standard in Europe, based on TDMA. To provide global roaming for wireless data subscribers, North America carriers will need to support GSM devices.	
GPRS	2.5G	Supports midrange data service to TDMA subscribers, including those using GSM devices. The maximum bit rate is 115 Kbps (less than half that of CDMA2000 1X.) GRPS is a stepping-stone the EDGE, a 3G alternative to W-CDMA for TDMA wireless carriers. AT&T Wireless (while it upgrades its network for EDGE), Cingular Wireless, and Voice Stream were expected to expand their GPRS offerings in the United States.	

GSM:	Global system for Mobile Communications
GPRS:	General packet radio service
UTMS:	Universal mobile radio service
CDMA:	Code division multiple access
1xRTT:	1xRadio transmission technology
3xRTT:	3xRadio transmission technology
1xEV-DO:	(CDMA) 1xEvolution for data only
1xEV-DV:	(CDMA) 1xEvolution for data and voice
TDMA:	Time division multiple access
EDGE:	Enhanced data for global evolution

FIGURE 8.17 *Key WWAN standards.*

protocol overhead. (In addition, call setup and location monitoring are handled by a separate analog channel, for compatibility with the older AMPS networks.)

Besides throughput, other 2G limitations include the fact that the link is not always on and that applications do not have access to a packetized

TABLE 8.8   SUMMARY OF WWLAN TECHNOLOGIES

GMS, TDMA, and cdmaOne IS-95A are 2G technologies.

GPRS and cdmaOne IS-95B are 2.5G technologies.

EDGE, W-CDMA, CDMA2000 1x, CDMA2000 1xEV-DO, and CDMA2000 1xEV-DV are 3G technologies.

TABLE 8.9   CAPABILITIES OF VARIOUS WWAN TECHNOLOGIES

	GSM	CDMA IS-95	GPRS	CDMA2000 (1xRTT)	UMTS (Wideband CDMA [W-CDMA])	Universal Mobile Telecommunications Service (UMTS) (W-CDMA)	CDMA2000 (HDR,...)
Generation	2G	2G	2.5G	2.5G	3G	3G	3G
Core network architecture	TDM circuit switched (CS)	TDM CD	TDM CS + packet switched PS overlay	TDM CS + PS/overlay	Third Generation Partnership Project (3GPP) R'99 (CS + PS)	3GPP R'00 All-IP	3GPP2 All-IP
Voice equipment	Mobile switching center (MSC)	MSC	MSC	MSC	MSC	SGSN/GGSN MGW	WAG (PDSN) MGW
Data equipment	Interworking function (IWF)	IWF	Serving GPRS service node/ Gateway GPRS support node (SGSN/GGSN)	PDSN	SGSN/GGSN	SGSN/GGSN	WAG (PDSN)
Signaling equipment	MSC	MSC	MSC + SGSN/GGSN	MSC + PDSN	MSC + SGSN/GGSN	CSCF SGSN/GGSN	Session manager
Signaling protocols	SS7 based	SS7 based	SS7 + IP based	SS7 + IP based	SS7 + IP based	Session Initialization Protocol (SIP) and others	SIP and others
Mobility management for data	Through MSC	Through MSC	GPRS mobility	Mobile IP	GPRS mobility	GPRS mobility	Mobile IP

Source: WaterCove Networks.

stream, which is intrinsically more desirable for data applications than a circuit-mode link. Some of the early packet-mode services such as *cellular digital packet data* (CDPD) have proven to be viable mechanisms for application support; these systems will be subsumed by more widely adopted standards. Improved solutions (called 2.5G) are emerging. As carriers enhance their networks to prepare for 3G, they can offer early adopters upgraded data services. Labeled 2.5G because they fall between current 2G and 3G technology, these intermediary technologies (namely, GPRS and CDMA2000 1X) deliver data at rates between 115 and 307 Kbps. (2.5G WWAN systems supporting 50–150 Kbps was expected to be available by 2003 and systems supporting 300–1,000 Kbps were expected to be available between 2003 and 2005.) When 3G is finally deployed, 2.5G systems can be replaced, but both carriers and end users must budget to replace the 2.5G devices and infrastructure.

For example, *1x Evolution for Data Only* (1xEV-DO) is a CDMA data protocol for faster speeds over dedicated channels. The maximum theoretical data rate is 2.4 Mbps. Although it is unlikely one can achieve the maximum rate, 1xEV-DO should offer faster data rates than its precursors. Currently, *1xRadio Transmission Technology* (1xRTT) (voice plus data) provides speeds of about 40 to 70 Kbps, depending on the implementation and the operator that is implementing it. Korean subscribers are able to receive the higher data rate. In the United States, Verizon Wireless and Sprint PCS have indicated that they were planning to offer 40 to 60 Kbps at the outset [21]. Sprint and/or Verizon may be offering 1xEV-DO service by late 2004; these carriers are still, however, in the early stages of 1xRTT. This will be the main priority for some time. (Sprint PCS was reportedly planning to offer 1xRTT service in 2003—there are various challenges with cellular data, but the industry is progressing.)

CDMA2000 is the name used by the TIA to refer to 3G CDMA. CDMA2000 is now an ITU standard and is included in the evolution part from cdmaOne. CDMA2000 1x will double the voice capacity and allow packet-based data transfer speeds of up to 307 Kbps. CDMA2000 1xEV, the next step, will allow data transfer rates up to 2.4 Mbps (for 1xEV-DO, data only) and 4.8 Mbps in phase 2 (for 1xEV-DV, data and voice). The evolution process will be complex. The CDMA path to 3G is through CDMA2000 1X and then to 1xEV-DO and 1xEV-DV. The first phase of CDMA2000, also known as 1X, enables operators with existing IS–95 systems to double the overall system capacity, yielding data rates up to 153.6 Kbps. We discuss evolution further in the pages that follow.

With the new GSM, GPRS, and CDMA 1xRTT packet data networks starting to be deployed in the United States in 2003 and going forward, a number of products should become available to end users—*some* end users that is because *the penetration and coverage footprint remains a question*. For example, when this author's relatives in *Johnstown, PA*, can use a *plethora* of 3G services from a *choice of providers* is, clearly, an open question at this time. Because all of this deployment is very capital intensive, and because questions remain about customers' interest in the offered services, in the pricing, and in the rate, deployment will have to be done (probably less by POTS ratepayers and more) by investment risk takers and entrepreneurs. However, no venture capitalist needs to "buy" the story from any company, no matter how polished the presentation and the management team are, that "these investments will return 80% gross margins, be EBITDA positive in 3 years, and IPO at $22 in 3.75 years." Almost without consuming cornea cells to read a business plan, on can be "almost sure" that these investments will return no more than 40% gross margins, be EBITDA positive in no less than 6 years, and IPO at no more than $11 in no less than 7.5 years, when done by a competent (if not as well-polished) team whose team members know what they are doing.

GPRS Technologies Data

GPRS is a packet-switched wireless data network operations in the GSM environment that enables data to be sent and received using GPRS devices in a more cost efficient and quicker way than was possible over the GSM cellular system. Users can secure data download rates up to 53.6 Kbps over GPRS compared to 14.4 Kbps via circuit-switched data over GSM. GPRS is a 2.5G wireless technology standard that was expected to improve the data services that can be added to GSM. ETSI defined GPRS in 1997 with the goal of providing packet-mode data services in GSM. GPRS is an over-the-air system for transmitting data on GSM networks that converts data into standard IP packets, enabling interoperability between the Internet and GSM network. In GPRS, a single time slot may be shared by multiple users to transfer packet data. GSM service providers are enhancing their service capabilities through the support of GPRS for packet data, as we discuss later in the transition section. GPRS wireless technology employs authentication and encryption via standard GSM algorithms.

With rapid session setup (0.5–1 sec) and a data rate between 9 and 171.2 Kbps, GPRS is often said to be a very important step toward 3G. The speed of a GPRS-capable device depends on how many time slots (one to eight, shared by active users) are used within a TDMA frame. Uplink and downlink are treated separately and various radio channel coding schemes allow bit rates from 9 to more than 171.2 Kbps per user (if a device were to use all eight channels without error correction). To make the most efficient use of network resources, GPRS provides dynamic sharing between speech and data services as well as for several profiles to make QoS decisions on how to treat packages (for example, video packages might be more time critical than packages resulting from normal WAP browsing) [22]. GPRS also supports the option for bandwidth-based billing. GPRS utilizes the same method for authentication and encryption used by GSM, but its techniques are optimized for packet data transmission. It also enables SMS transfer over GPRS radio channels. The technical specifications for GPRS are generated by the 3GPP. Applications of GPRS include laptops or handhelds connected through a GPRS-capable cell phone or dedicated GPRS mode, mobile phones with a WAP browser, and dedicated equipment such as mobile credit-card swipers.

ETSI has also defined a 2.5G standard called *Enhanced Data Rates for GSM Evolution* (EDGE) to support higher speed data services in GSM. It is meant to be the basis for a 3G wireless standard that can be utilized by TDMA and GSM operators. EDGE is an evolution of GPRS that allows up to three times higher throughput compared to GSM using the same bandwidth. EDGE, in combination with GPRS, will deliver data services up to 384 Kbps in the near future in specific geographic areas. It works with a new modulation technique that enables better usage of existing frequencies. EDGE can be deployed on existing GSM networks and is part of the UWC-136 standard that TDMA carriers have proposed as their 3G standard of choice. EDGE enables data transfer speeds of up to 384 Kbps. This results in 48 Kbps per time slot enabling three times the data speeds of GPRS.

Other systems are available or evolving. For example, *high-speed circuit-switched data* (HSCSD) is another enhancement to GSM networks that enables users to send and receive data at higher speeds. This improvement is achieved by increasing the data rate of one channel and providing the option of utilizing several channels at the same time. These enhancements result in speeds of about 60 Kbps. The improved performance makes HSCSD useful to people who wish to access the Internet at higher speeds than the current GSM phones offer. The higher data speed per channel is achieved by utilizing a new coding scheme that has lower bandwidth requirements for error correction. A GSM channel provides a bit rate of 22.8 Kbps, but about 13.3 Kbps of this bandwidth is used for error correction. With HSCSD, only 8.6 Kbps is used for error correction. This results in higher speeds in cases where users are close to GSM cells where there is high-quality reception (the need for error correction increases with distance/noise). Several manufacturers already have HSCSD-enabled mobile phones on the market. The hope is that ultimately all of these will fold into 3G.

8.2.3.3   3G Efforts

As alluded to earlier, 3G mobile communications is a concept outlined in a set of proposals called *International Mobile Telecommunications-2000* (IMT-2000) to define an anywhere–anytime standard for the future of universal personal communications. The ITU has given support to two 3G technologies: W-CDMA and CDMA2000. W-CDMA appears to be poised to take the lead in high-speed wireless services. North American providers have leaned in favor of CDMA2000 and EDGE. The ITU has recommended that 3G wireless devices work on W-CDMA and CDMA2000 networks to make worldwide data roaming possible. Carriers were planning commercial rollouts in 2003 or 2004. However, as we noted, the global deployment of wireless has been hampered recently by a downturn in the telecom industry. Market fragmentation is also an issue impacting ubiquitous deployment, particularly in North America.

Generally speaking, 3G seeks to provide up to 2 Mbps of data (for example, on the Internet) to cell phones and mobile devices. By implementing 3G wireless services, carriers plan to upgrade their infrastructures to high-speed data. 3G supports bit rates as fast as 2 Mbps. Although this is less than the 11 Mbps possible with an IEEE 802.11b WLAN or the 54 Mbps possible with the IEEE 802.11a WLAN, the reach (distance) is much greater with 3G. As noted earlier, 1G mobile telecommunication systems that were introduced in the 1980s were analog. These systems, which are still in use, do not intrinsically have data transport capabilities. To provide data services in these analog systems, a capability such as CDPD has to be added to the analog. Naturally, this arrangement supports only slow-speed data. 2G systems (IS-136, cdmaOne, and GSM) are digital and have intrinsic data transport capabilities. However, the data support is still limited: GSM supports short messaging services and data at rates only up to 9.6 Kbps, and IS-95B provides data rates in the neighborhood of 64 to 115 Kbps in increments of 8 Kbps over a 1.25-MHz channel. 2.5G GPRS was developed in the late 1990s as a further enhancement: In GPRS, each slot can handle up to 20 Kbps and since each user may be allocated up to 8 slots, data rates up to about 160 Kbps per user are therefore achievable. Yet, none of these operate higher than 256 Kbps. The goal of 3G is to do just that. Table 8.10 provides for comparison nominal data rates for some 2.5G/3G WWAN technologies.

Multiple groups are involved in standards development (see Figure 8.18), particularly the ITU and other related bodies such as ETSI, IETF, ANSI T1, *Third Generation Partnership Project* (3GPP), *Third Generation Partnership Project 2* (3GPP2), and so on. 3GPP is a GSM-originated group. Initially, 3GPP was to produce globally applicable technical specifications and technical reports for 3G based on evolved GSM core networks. This charter was amended to include the maintenance and development of GSM and evolved into radio access technologies (GPRS and EDGE). 3GPP2 is a CDMA-oriented 3G standards group under the auspices of ANSI. 3GPP supports efforts for Europe and Asia, whereas 3GPP2 supports efforts for North America. The major blueprint for 3G was pulled together by the ITU.

ITU has advanced the concept of IMT-2000 as its generalized view of future wireless/nomadic networks. The scope of IMT-2000 outlined in Figure 8.19. IMT-2000 is the ITU globally coordinated definition of 3G

FIGURE 8.18
Some of the standards groups working on 3G.

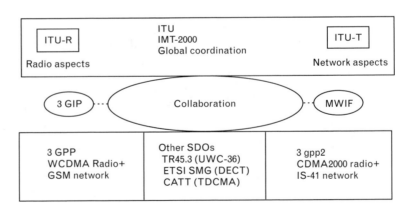

TABLE 8.10   DATA RATES FOR 2.5G/3G TECHNOLOGIES

ASPECT	SYSTEM				
	UTMS	EDGE	CDMA 3G-1X RTT	CDMA 3G-1X EVDO Evolution (Data Only)	CDMA 3G-3X
Technology	Code division	Time division	Code division	Code division	Code division
Nominal data rate (Kbps)	2,000/384/144	167/200/ 349/470	144–300	2,400 (downstream)	2,000

FIGURE 8.19
ITU IMT-2000.

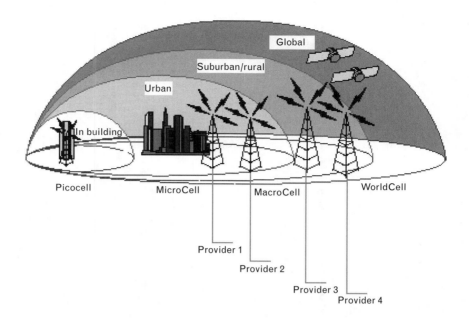

covering key issues such as frequency spectrum use and technical standards. 3G is not a technology per se but a term encompassing all aspects of future wireless networks. 3G adjoins high-speed radio access and IP-based services to enable subscribers to be "always on," that is, "always be on-line." Some of the basic desiderata for IMT-2000 are as follows:

- Speeds of 2 Mbps (indoors) and 144 Kbps (outdoors) or better. (Increased speeds may not be achieved in one shot in 3G: The data rate supported may initially be only 144 Kbps, it may be 384 Kbps in the second phase, and it may reach 2.048 Mbps in the final phase; phases are designed to be backward compatible.).
- Circuit- and packet-switched (IP) services.
- Good voice quality comparable with wire-line quality (with a *mean opinion score* [MOS] of 4.0 or thereabouts).
- Increased capacity and improved spectrum efficiency.
- Global roaming between different operational environments.
- Support different types of user traffic: constant-bit-rate traffic, such as high-quality audio speech, video telephony, and video, that need QoS since they are sensitive to delays and delay variation; real-time, variable-bit-rate traffic, such as variable-bit-rate audio, *Moving Pictures Expert Group* (MPEG)/ISO video, and so on. This type of traffic also requires QoS, being sensitive to delays and delay variation; nonreal-time, variable-bit-rate traffic that can tolerate delays or delay variations.

Eventually, after major industry involvement, four systems for 3G mobile communications materialized (see Figure 8.20):

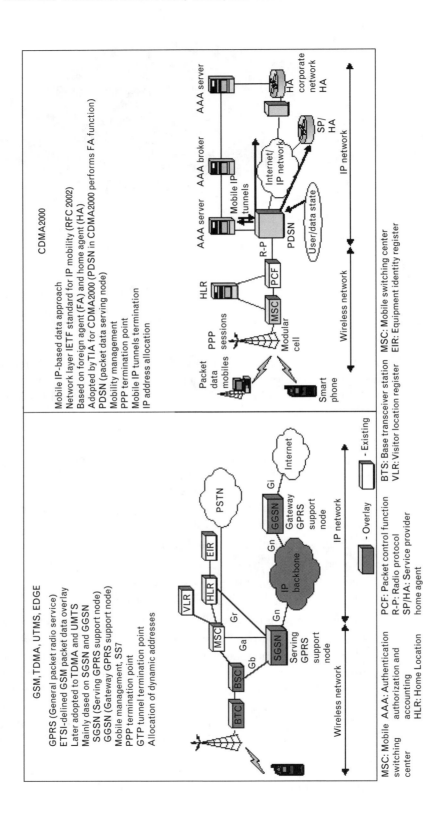

FIGURE 8.20   *Simplified target networks. (Source: Adapted from Steve Akers, Lucent Technologies.)*

1. W–CMDA UMTS FDD (frequency-division duplexing);

2. CDMA2000;

3. W–CDMA UMTS TDD (time-division duplexing);

4. UWC–136 (ITM–SC single-carrier EDGE).

Recommendations on these systems were published in 1999 by ITU–R as a harmonized standard with the four modes just listed. As we have noted, CDMA is a high-capacity cellular technology that employs spread-spectrum technology and a unique digital coding system rather than separate channels to differentiate subscribers. CDMA was developed by Qualcomm and introduced commercially in the mid-1990s. In 1999, CDMA was adopted as the basis for 3G wireless systems by the ITU: CDMA is used as the basis of UMTS and W–CDMA. CDMA2000 must comply with EIA/ TIA IS–41, and W–CDMA UMTS must comply with GSM *Manufacturing Automation Protocol* (MAP) intersystem networking standards. UWC–136 is based on TDMA, while the other three use *direct-sequence code-division multiple access* (DS–CDMA). CDMA2000 is a multi-carrier, direct-sequence CDMA FDD system with a single carrier that has a bandwidth of 1.25 MHz (later on it may have as many as three carriers.) CDMA2000 is an evolution of the existing North American CDMA system cdmaOne (IS–95 standards). CDMA2000 supports packet-mode data services. UMTS W–CDMA FDD is a DS–CDMA system with a nominal bandwidth of 5 MHz. UMTS W–CDMA TDD also uses CDMA with a bandwidth of 5 MHz; however, the frequency band is time shared in both directions. (Half the time it is used for transmission in the forward direction, and the other half it is used in the reverse direction.) The TDMA version of the 3G system for use in North America is known as UWC–136. UWC–136 is a TDMA scheme in which each physical channel is partitioned into a number of fixed time slots; each user is assigned one or more slots. The UWC–136 system is planned to be introduced in three stages:

1. IS–136+ with a bandwidth of 30 kHz (provides voice and up to 64 Kbps of data);

2. IS–136 HS (vehicular/outdoor) with a bandwidth of 200 kHz (provides data rates up to 384 Kbps for outdoor/vehicular operations);

3. IS–136 HS (indoor) with a bandwidth of 1.6 MHz (users can get a data rate of up to 2 Mbps).

IS–136+ is an enhancement of the existing IS–136 and uses improved modulation techniques. IS–136 HS (vehicular/outdoor) supports data rates up to 384 Kbps and has parameters similar to those of EDGE. IS–136 HS (indoor) provides data rates up to about 2 Mbps [23].

Enhancing the data capability of CDMA is the 1xRTT CDMA standard, which forms the basis for one of the 3G wireless standards. 1xRTT is designed to support always-on data transmission speeds up to an order of magnitude faster than typically available in the early 2000s up to a maximum of 153.6 Kbps [24].

All of the various technologies build on a so-called *core network*. The technologies in the GSM evolution are based on a so-called GSM MAP core network and the others on the IS-41 core network. With the new IMT-2000 standards, all radio options (such as CDMA) should work on all core networks (such as GSM MAP—currently CDMA does not work over GSM MAP). This might make later changes in the network radio interfaces easier to implement [22, 25].

Several prototypes showing possible 3G terminal designs have appeared, but the first commercial introduction of 3G terminals had yet to take place by press time. One important aspect is that terminals need to be dual mode (3G and 2G) to enable efficient network usage.

At the time of this writing, 27 countries were awarded 3G licenses. Since Finland awarded four 3G licenses in March 1999, an additional 100 3G licenses have been awarded in another 26 countries, earning governments more than $112 billion in revenues. Unsurprisingly, given recent developments in the telecommunications industry, only eight 3G licenses (five in Taiwan and three in Luxembourg) were issued in 2002. Recently, funds funneled into license fees have drastically reduced. As of June 2002, the highest total fee, almost $9.4 billion, was that paid by Vodafone in the United Kingdom in April 2000, equivalent to $158/POP(ulation), although the average license fee of $7.7 billion paid in Germany in August 2000 exceeded the average price of $7.1 billion paid in the United Kingdom. Since January 2001, however, the average license fee in Europe has dropped to $672 million with an associated decline in price per POP. This figure includes the original license fees of EUR 4.95 billion that were paid by Cegetel/SFR and France Telecom in the French 3G "beauty contest" that have since been restructured to EUR 619 million each. Apart from Europe and three licenses in Israel, the only other region to award 3G licenses has been Asia Pacific. Of the 28 licenses issued in this region, only four were awarded before 2001, making it the focus of recent licensing activity. In Eastern Europe, Croatia, Estonia, Latvia, and the Slovak Republic have all announced their intentions to hold auctions. Figure 8.21 shows licensing information [26].

8.2.3.4   Transition to 3G

Before discussing transitions to 3G, we discuss a related topic. For a number of years the question has been "Which is the best cellular technology?" As one might well imagine, there are no absolute answers: One size does not fit all. One must take into account the location, the service plans of the mobile service providers for the present and for the future, the evolution options available for the operator at the technical level, the embedded base and the investments made so far, and other considerations. For

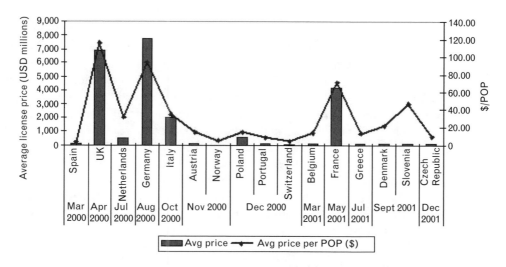

FIGURE 8.21   *3G license information. (Source: EMC world cellular licenses.)*

any technology to be successful at the commercial level there are at least three essential requirements: (1) technology availability from multiple providers, (2) optimized resource management/system capacity (the radio spectrum is the only place where bandwidth optimally is a true requirement), and (3) ease and simplicity in planning, engineering, deploying, and maintaining the system (that is, OAM&P).

The debate basically revolves around TDMA and CDMA. TDMA systems such as GSM have the advantages of (1) enjoying large market penetration, (2) being in operation for almost 15 years, (3) being easy and inexpensive to upgrade to packet-based data services, (4) supporting roaming over large areas of the globe, (5) taking advantage of recent developments regarding roaming across TDMA and GSM networks, and (6) exploring the possibility of a common air interface and subsequent evolution toward 3G after implementing 2.5G techniques (such as EDGE). Claims are made by proponents that TDMA networks are more rugged compared to CDMA; for example, CDMA suffers from problems such as deteriorating speech quality when the traffic load increases.

CDMA makes the claim for improvements in capacity, security, and speech quality. Specifically, more users can be supported over the same bandwidth in CDMA as compared to TDMA. A CDMA system uses a combination of frequency division and code division to provide multiplexing. Although the capacity of a CDMA system is not unlimited, its limitations are considerably higher than those of a TDMA system. Hence, it is able to support an order of magnitude more users than traditional FDMA/CDMA. CDMA (e.g., IS-95) has a simple and well-defined evolutionary path to 3G; this is directly due to the fact that 3G networks use CDMA in the air interface. According to observers, although the evolution path to 2.5G may be cheaper for GSM/TDMA, the total costs involved in moving to 3G will be substantially higher for GSM/TDMA.

At the practical commercial level, however, service providers appear to continue to be going toward earlier (TDMA) technology. A major concern about CDMA is the fact that it has relatively little field experience. "Time-to-market" considerations are important and many providers choose to invest in TDMA systems that have already been developed and proven. The idea is to minimize risks at the technology, vendor, and investment (rate-of-return, dividends) levels. This has resulted in the practical outcome, so far, of going with a time-proven technology such as TDMA. Also, developments and extensions in the 2G networks, through the use of 2.5G capabilities, have added value to the carrier's investments and have so far been able to keep up with the demands for higher capacities and data applications. The expectation is that CDMA will acquire larger importance toward the later part of the decade. However, equipment vendors have to be constantly reminded that providers do not swap out a technology just because the new one reduces their equipment costs by 10%. This author always gets the vendors' attention when he states to them over business conversations that the equipment cost reduction has to be in the 40% to 60% range for the proposition to be compelling (in many cases), particularly when keeping in mind that the total cost of the operator is dominated by OAM&P costs, not equipment costs. And, as has been discovered, many carriers do not spend, say, an allocated $8 billion in capital, all in equipment: Expenses may (inappropriately) be charged against capital budgets, as are "legitimate" capitalized labor expenses. Perhaps as little as 50% of the stated capital budget actually goes to buy equipment.

A number of subtending transition trends are expected to occur over time that are technology independent to some degree; these trends include a move toward distributed networks (as illustrated in Figure 8.22) and a move to IP-based infrastructures (including voice-over-IP). This evolution is expected to be as follows (also see Figure 8.23):

- cdmaOne IS-95A (to cdmaOne IS-95B) to CDMA2000 1x to CDMA2000 1x EV-DO or EV-DV;

- TDMA to CDMA2000 1x;

- TDMA to GSM;

- GSM to GPRS to EDGE to W-CDMA.

As we have seen, several evolutionary paths to 3G exist. GSM operators can enhance their networks with GPRS and later deploy EDGE, which is already defined as a 3G technology by the IMT-2000. These networks will then evolve to future 3G networks based on W-CDMA, the standard technology for the UTMS band. TDMA operators can either switch to GSM and continue with that approach or go on to CDMA2000.

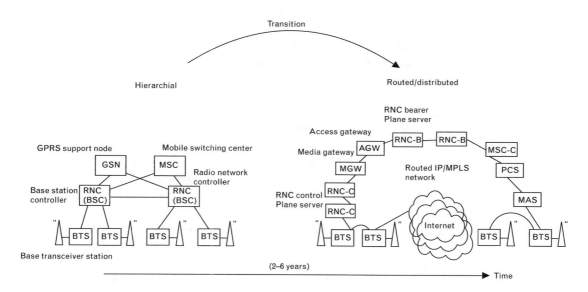

FIGURE 8.22   *Transition of WWANs over time.*

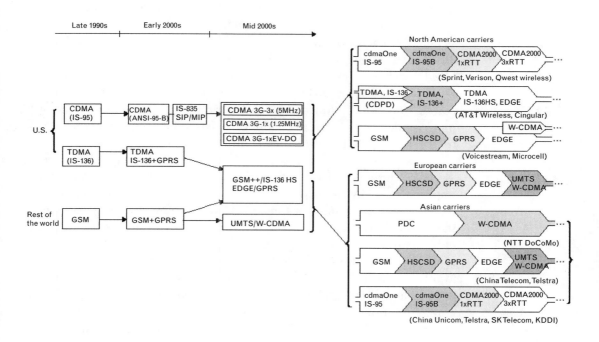

FIGURE 8.23   *Some of the possible evolutions to 3G.*

The Japanese PDC standard will evolve to W-CDMA [22]. Forecasters make the claim that with the upgrade to W-CDMA, GSM will likely hold an 80% market share in 2005. Work is also under way to interwork 2.5G/3G WWANs with WLANs.

As we move to 3G, there is a strong desire to move to an all-IP environment. The transition from the current embedded base, however, will not be quick, inexpensive, or easy. For example, with millions and even billions of mobile users, IP address space could become an issue. IP version 6 (IPv6) provides more space, but the adoption of IPv6 by service providers has been slow.

The commercially reported migration paths to 3G in North America are as follows [27]. Cingular, AT&T Wireless, and VoiceStream (T-Mobile) have made a commitment to GSM/UMTS; this represents 48% of the U.S. wireless market at the time of this writing. Verizon and Sprint PCS have committed to CDMA2000; this represents 45% of the market. Nextel, which uses iDEN, has not yet announced a 3G (it has, however, conducted tests with CDMA2000). Deployment of 3G was being planned in the United States at the time of this writing, but it is likely that the initial technology and services will fall short of expectations.

Although UMTS has been championed by companies such as NTT DoCoMo, AT&T Wireless, VoiceStream (T-Mobile), and European operators (bound to this technology by the terms of their government licenses), CDMA2000 (backed by KDDI, Qualcomm, Sprint, and Verizon Wireless) has achieved an early lead. As of June 2002, 10 million subscribers were using CDMA2000 worldwide, compared with 112,000 for UMTS. There were 100 handset models for CDMA2000, but only 7 handset models for UMTS; the former cost $65 to $120, while the latter cost $240 to $400. Nonetheless, the world's largest operators and equipment vendors have invested billions of dollars in UMTS, so that the industry still favors UMTS as the long-term leader.

About 56% of the total growth in digital cellular subscribers in the Americas during 2001 came from GSM and TDMA carriers. While that implies that CDMA-based technologies accounted for nearly half of the region's growth during that time, GSM will continue to grow "at a substantial rate" as more operators in North, Central, and South America either remain committed or switch to the platform. Overall, GSM subscribers rose by 50.4% in 2001 to nearly 17 million in the Americas. Worldwide as of 2002, there were more than 650+ million GSM customers—including 100 million in 27 countries with access to high-speed data communications. In North America, TDMA operators AT&T Wireless, Cingular Wireless, and Rogers Wireless have implemented their GSM evolutionary migration, while in Latin America, Telcel Radiomovil in Mexico, Telecom Personal in Argentina, Entel Movil in Bolivia, and CTE in El Salvador have committed to a similar GSM migration strategy [28].

The success of GPRS has been seen as a precursor to the uptake of data services and beyond to 3G. The industry as a whole is watching the progress made by European and Asian operators launching GPRS services.

Since the first commercial launch of GPRS in 2000, a total of 106 systems in 46 countries had been launched by 2002. Figures from EMC indicate that uptake for GPRS-based applications has been slow (see Figure 8.24). A top-end estimate, based on primary research with operators, suggests that a little over 2 million subscribers are regular users of GPRS—this is less than 1% of the 650+ million GSM subscribers worldwide [29]. Not surprisingly, Western Europe dominates the GPRS landscape, accounting for 51% of networks launched by 2002. Asia Pacific and Eastern Europe have also seen a significant number of launches. EMC expects a further 40 operators to launch services in 2002, including the remainder of Western Europe's large operators (T-Mobile UK, SFR, and Bouygues Telecom France) as well as the world's largest operator, China Mobile. Early tariff plans are structured around buckets of data, rather than being application based. As a result, most usage is in a business environment. The prepaid market accounts for 62% of Western Europe's subscribers, but with prepaid GPRS being offered only by a few operators, the potential market of GPRS is being heavily restricted.

8.3 Technology Basics

This section provides some additional details on WLANs, WPANs, WMANs, and WWANs.

8.3.1 WLAN Technologies [30]

The 1997 completion of the IEEE 802.11 standard for WLANs was an important step in the evolutionary development of wireless networking technologies. IEEE 802.11 is focused in scope to the PHY layer and MAC sublayer, with MAC based on the IEEE 802.3 Ethernet standard. The

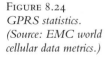

FIGURE 8.24
*GPRS statistics.
(Source: EMC world
cellular data metrics.)*

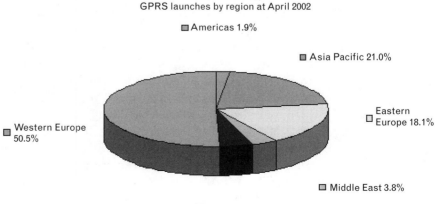

IEEE 820.11 standard was developed to maximize interoperability between differing brands of wireless LANs as well as to introduce a variety of performance improvements and benefits. In 1999, the IEEE 802.11 committee ratified a revision of the 802.11 standard, called 802.11 *High Rate* (HR), that provides much higher data rates, while maintaining the 802.11 protocol. In addition to providing high performance and robust systems, 802.11 protocols also promise multivendor interoperability among product with the same PHY layers. This means that customers are freer to mix and match vendors to meet their requirements for each given application. Furthermore, standardization also delivers lower cost components, which will translate into lower prices for users. Because of this, almost all WLAN vendors have now moved to IEEE compliance. With the newer high-speed IEEE 802.11b specification, most of these vendors have announced high-speed 802.11–compliant products as well. The standard specifies a choice of different PHYs. Vendors' implementations have either used DSSS or FHSS, both RF based. In 802.11, the DSSS PHY specifies a 2-Mbps peak data rate with optional fallback to 1 Mbps in very noisy environments. HR supports rates at 11 Mbps.

The IEEE 802.11 standard defines the FHSS PHY as operating at 1 Mbps and allows for optional 2-Mbps operation in "clean" environments. Most vendors chose to implement DSSS since 802.11b HR (11 Mbps) standard is based on DSSS as well. This makes migration from a 2-Mbps 802.11 DSSS system to an 11-Mbps 802.11 HR system easy, given that the underlying modulation scheme is very similar. The 2-Mbps 802.11 DSSS systems will be able to coexist with 11-Mbps 802.11 HR systems, enabling a smooth transition to the higher data rate technology. This is similar to migrating from 10-Mbps Ethernet to 100-Mbps Ethernet, enabling a large performance improvement while maintaining the same protocol.

The 802.11 MAC protocol is robust and feature rich. It includes sequence control and retry fields supporting a feature called *MAC-layer acknowledge* that minimizes interference and maximizes usage of the bandwidth available on the wireless channel. Type/subtype and duration fields ensure reliable communications in the presence of hidden stations. WEP fields result in data security that was intended, in the original design, to be equal to that achievable with standard Ethernet. When WEP is enabled, all data transmitted over the wireless network in encrypted. (However, after its introduction, WEP was been found to be problematic, as discussed in the next subsection.) Sequence control and "more frag" fields support a concept called *fragmentation* that can allow a WLAN to operate in the presence of interference or signal fading. The 802.11 MAC can work seamlessly with standard Ethernet, via a bridge or *access point* (AP), to ensure that wireless and wired nodes on an enterprise LAN can interoperate with each other. The WLAN standard uses a *carrier sense multiple access with collision avoidance* (CSMA-CA) MAC scheme, whereas standard Ethernet uses a *carrier sense multiple access with collision detection* (CSMA-CD) scheme.

The 802.11 protocol allows a client to roam among multiple APs that can be operating on the same or separate channels. For example, every 100 ms, an AP might transmit a beacon signal that includes a time stamp for client synchronization, a traffic indication map, an indication of supported data rates, and other parameters. Roaming clients use the beacon to gauge the strength of their existing connection to an AP. If the connection is judged weak, the roaming station can attempt to associate itself with a new AP.

The 802.11 standard adds features to the MAC that can maximize battery life in portable clients via power management schemes. Power management causes problems with WLAN systems because typical power management schemes place a system in sleep mode (low or no power) when no activity occurs for some specific or user-definable time period. Unfortunately, a sleeping system can miss critical data transmissions. To support clients that periodically enter sleep mode, the 802.11 specified that APs include buffers to queue messages. Sleeping clients are required to awaken periodically and retrieve any messages. The APs are permitted to dump unread messages after a specified time passes and the messages go unretrieved.

One of the key gains from the 802.11 standard is the ability for products from different vendors to interoperate with each other. This was not the case with WLAN products available throughout the 1990s. This means that as a user, one can purchase a wireless LAN card from one vendor and a wireless LAN card from another vendor and they can communicate with each other, independent of the brand of access point utilized. This gives the user the choice to choose the system that best meets the needs for each application. As a supplement to the 11-Mbps interoperability testing that will be preformed through WECA, a number of vendors have successfully tested interoperability together at the University of New Hampshire Interoperability Lab (http://www.iol.unh.edu/consortiums/wireless).

As mentioned earlier, the 802.11 committee has been moving rapidly to provide continuous improvements, as has been seen within the Ethernet 802.3 environment. The next step in the evolution was the ratification of the 802.11b or HR, providing data rates of 11 Mbps. The standard's 11-Mbps PHY layer uses CCK technology. This standard is based on DSSS technology and provides speeds up to 11 Mbps with fallback rates of 5.5, 2, and 1 Mbps. It uses the same bandwidth as the 2-Mbps DSSS standard and thus interoperates with legacy IEEE DSSS systems. As in the wired world, higher speeds are continuously desired for applications such as streaming video, telephony, and multimedia. Moreover, faster peak rates will allow more nodes to effectively connect to a WLAN via a single channel. Also, progress has been made on the 54-Mbps IEEE 802.11a product space. In 2002, WECA begun certification testing for IEEE 802.11a wireless LAN products in its certification laboratory in San Jose, California [31].

8.3.1.1   Technical Details [32]

The basic topology of an 802.11 network is shown in Figure 8.25. A *basic service set* (BSS) consists of two or more wireless nodes, or stations (STAs), that have recognized each other and have established communications. In the most basic form, stations communicate directly with each other on a peer-to-peer level sharing a given cell coverage area. This type of network is often formed on a temporary basis and is commonly referred to as an *ad hoc network*. In most practical instances, the BSS contains an AP. The function of an AP is to form a bridge between wireless and wired LANs (the AP is analogous to a base station used in cellular phone networks.) A BSS in this configuration is said to be operating in the *infrastructure mode*. When an AP is present, all communications between stations or between a station and a wired network client go through the AP. In most instances the APs are not mobile and form part of the wired network infrastructure. The *extended service set* (ESS) shown in Figure 8.25 consists of a series of overlapping BSSs (each containing an AP) connected by means of a *distribution system* (DS). Although the DS could be any type of network, it is often an Ethernet LAN. Mobile nodes can roam between APs and seamless campuswide coverage is possible. All station clocks within a BSS are synchronized by periodic transmission of time-stamped beacons. In the infrastructure mode, the AP serves as the timing master and generates all timing beacons. Synchronization is maintained to within 4 μs plus propagation delay. Timing beacons also play an important role in power management. Two power saving modes are defined: *awake* and *doze*. In the *awake* mode, stations are fully powered and can receive packets at any time. Nodes must inform the AP before entering doze. In this mode, nodes

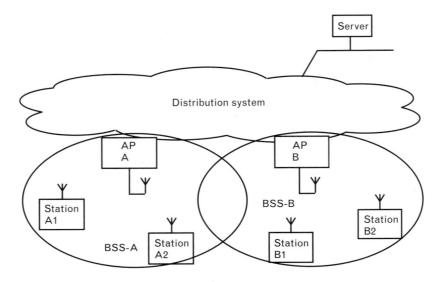

FIGURE 8.25
*ESS provides
campuswide coverage.*

must "wake up" periodically to listen for beacons that indicate the AP has queued messages.

As noted, IEEE 802.11 provides for two variations of the PHY. The DSSS and FHSS PHY options were designed specifically to conform to FCC regulations for operation in the 2.4-GHz ISM band, which has worldwide allocation for unlicensed operation. Both FHSS and DSSS PHYs currently support 1 and 2 Mbps. However, all 11-Mbps radios are DSSS. In DSSS each information bit is combined via an XOR function with a longer PN sequence. The result is a high-speed digital stream that is then modulated onto a carrier frequency using *differential phase shift keying* (DPSK). Note that differential encoding implies the need for differential decoding; in general, the decoding can be of two types: coherently detected DPSK and differentially detected DPSK, with the latter being more common.

When receiving the DSSS signal, a matched filter correlator is used. The correlator removes the PN sequence and recovers the original data stream. At the higher data rates of 5.5 and 11 Mbps, DSSS receivers employ different PN codes and a bank of correlators to recover the transmitted data stream. The high rate modulation method is called complimentary code keying. The PN sequence spreads the transmitted bandwidth of the resulting signal (thus the term *spread spectrum*) and reduces *peak* power. Note, however, that *total* power is unchanged. Upon reception, the signal is correlated with the same PN sequence to reject narrowband interference and recover the original binary data. Regardless of whether the data rate is 1, 2, 5.5, or 11 Mbps, the channel bandwidth is about 20 MHz for DSSS systems. Therefore, the ISM band will accommodate up to three nonoverlapping channels, as seen in Figure 8.26.

The basic access method for 802.11 is the DCF that uses CSMA-CA. This requires each station to listen for other users. If the channel is idle, the station may transmit. However, if it is busy, each station waits until transmission stops and then enters into a random back-off procedure. This prevents multiple stations from seizing the medium immediately after completion of the preceding transmission. Packet reception in DCF requires acknowledgment as shown in Figure 8.27. The period between completion of packet transmission and start of the ACK frame is one *short interframe space* (SIFS). ACK frames have a higher priority than other traffic. Fast acknowledgment is one of the salient features of the 802.11 standard, because it requires ACKs to be handled at the MAC sublayer. Transmissions other than ACKs must wait at least one *DCF interframe space* (DIFS) before transmitting data. If a transmitter senses a busy medium, it determines a random back-off period by setting an internal timer to an integer number of slot times. Upon expiration of a DIFS, the timer begins to decrement. If the timer reaches zero, the station may begin transmission. However, if the channel is seized by another station before the timer

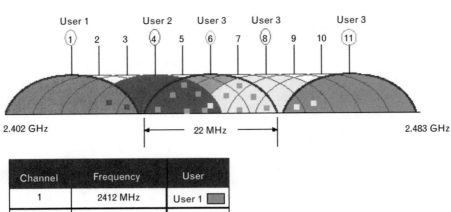

Channel	Frequency	User
1	2412 MHz	User 1
2	2417 MHz	
3	2422 MHz	
4	2427 MHz	User 2
5	2432 MHz	
6	2437 MHz	User 3
7	2442 MHz	
8	2447 MHz	User 3
9	2452 MHz	
10	2457 MHz	
11	2462 MHz	User 3

Note: Ch. 1, 6, and 11 are nonoverlapping

Example of three collocated users (in hotspot setting) sharing the ISM band. In this example (a system actually set up by the author), User 1 requires one frequency, User 2 requires one frequency, and User 3 requires three frequencies because they set up a three-sector antenna system. Some interference will result. If there were only three antennas (systems) in this location, then channels 1, 6, and 11 would be nonoverlapping.

FIGURE 8.26   *DSSS channels in the ISM band multiple access.*

FIGURE 8.27
Back-off algorithm.

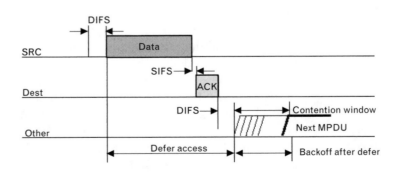

reaches zero, the timer setting is retained at the decremented value for sub-sequent transmission. The method described above relies on the *physical carrier sense.*

The underlying assumption is that every station can "hear" all other stations. This is not always the case. To address this problem, a second carrier sense mechanism is available. *Virtual carrier sense* enables a station to reserve the medium for a specified period of time through the use of request to send/clear to send (RTS/CTS) frames. Consider the case where STA-A sends an RTS frame to the AP and where the RTS will not be heard by STA-B. The RTS frame contains a duration/ID field that specifies the period of time for which the medium is reserved for a subsequent transmission. The reservation information is stored in the *network allocation vector* (NAV) of all stations detecting the RTS frame. Upon receipt of the RTS, the AP responds with a CTS frame, which also contains a duration/ID field specifying the period of time for which the medium is reserved. Although STA-B did not detect the RTS, it will detect the CTS and update its NAV accordingly. Thus, collision is avoided even though some nodes are hidden from other stations. The RTS/CTS procedure is invoked according to a user-specified parameter. It can be used always, never, or for packets that exceed an arbitrarily defined length.

As mentioned earlier, DCF is the basic media access control method for 802.11 and it is mandatory for all stations. The *point coordination function* (PCF) is an optional extension to DCF. PCF provides a time-division duplexing capability to accommodate time-bounded, connection-oriented services such as cordless telephony.

8.3.2 Security Considerations for WLANs

IEEE 802.11 provides for security via two mechanisms: *authentication* and *encryption*. Authentication is the process by which one station is verified to have authorization to communicate with other stations or APs in a given coverage area. In the infrastructure mode, authentication is established between an AP and each station. Authentication can be either *open system* or *shared key*. In an open system, any STA may request authentication. The STA receiving the request may grant authentication to any request or only to those from stations on a user-defined list. In a shared key system, only stations that possess a secret encrypted key can be authenticated. Shared key authentication is available only to systems having the optional encryption capability. Encryption is intended to provide a level of security comparable to that of a wired LAN. The specific method utilized in 802.11 is WEP. WEP uses the RC4 PRNG algorithm from RSA Data Security, Inc. [32].

Unfortunately, WEP has been found to be problematic. WEP and shared key protocols have been shown to have significant cryptographic flaws that allow cryptographic attack on both the confidentiality and access control functions (for example, see the Wagner/Goldberg paper at http://www.isaac.cs.berkeley.edu/isaac/wep-faq.html, and the Arbaugh paper at http://www.cs.umd.edu/~waa/attack/v3dcmnt.htm). Although

WEP and shared key are flawed, they should still be turned on because they do provide some baseline protection. What follows next is a position taken by WECA on the issue, as given by suggestions 1 through 6 below [33].

The recently published approach to compromising WEP is the most effective reported to date. However, Wi-Fi wireless LANs with appropriate security measures are a proven technology that brings significant benefits to large and small businesses, schools, home consumers, and many other settings. It is important that they be deployed in a manner that is consistent with the same sound security practices used to secure wired LANs and dial-up access connections. The following information should be considered when deploying this or any other network technology.

1. WEP was part of the original IEEE 802.11 standard. From the outset, the goal for WEP has been to provide an equivalent level of privacy as is ordinarily present with a wired LAN. Traditional wired LANs such as IEEE 802.3 (Ethernet) are ordinarily protected by the physical security mechanisms within a facility (such as controlled entrances to a building) and, therefore, IEEE wired LAN standards did not incorporate encryption. WLANs may not be protected by a physical boundary since their transmissions penetrate walls. As a result, WEP encryption was added to the IEEE 802.11 standard to provide an equivalent level of privacy similar to a physical boundary (like a wall).

2. By far, the biggest threat to the security of a wireless LAN is the failure to use any form of security. This is the most significant risk today. WEP should be used as the "first line of defense" to deter the casual intruder. If the value of the data justifies it, other more advanced security techniques should be deployed. For users in smaller organizations, at home, or where the value of the data does not justify extensive additional measures, WECA recommends one or more of the following:

   · Turn WEP on and manage the WEP key by changing the default key and, subsequently, changing the WEP key, daily to weekly.

   · Password protect drives and folders.

   · Change the default SSID (wireless network name).

   · Use session keys if available in your product.

   · Use MAC address filtering if available in your product.

   · Use a VPN system. Though it would require a VPN server, the VPN client is already included in many operating systems such as Windows 98 Second Edition, Windows 2000, and Windows XP.

3. For larger organizations, or those where the value of the data justifies strong protection, users should set up additional security methods. Some examples of these methods are RADIUS- or Kerberos-based access control; end-to-end encryption; password protection; user authentication; *Extensible Authentication Protocol* (EAP), possibly a vendor-specific implementation in the short term; dynamic WEP-key rotation, possibly a vendor-specific implementation in the short term; VPNs; *secure socket layer* (SSL); and firewalls. Wi-Fi technology integrates seamlessly with these and other security approaches.

4. IEEE 802.11 Task Group I (IEEE 802.11i) has been working on extensions to WEP for incorporation within a future version of the standard. The enhancements currently proposed include an entirely different privacy algorithm and provisions for enhanced authentication. The work of Task Group I benefits from recent work on RC4 and WEP and will incorporate new mechanisms for dealing with the threats described.

5. Wi-Fi certification will include requirements for implementing IEEE 802.11i after it is an approved standard. WECA expects to include the new security enhancements from IEEE 802.11i in the next update to Wi-Fi certification testing in 2002.

6. WECA has formed a security subcommittee that is working closely with the participants of IEEE Task Group I. The goal of the subcommittee is to develop an interim solution that will be secure against all the known attacks, run on existing Wi-Fi-certified hardware, and be available before the IEEE 802.11i standard is complete.

It appears, however, that the new Wi-Fi security, based on IEEE's 802.1x (IEEE standard 802.1X-2001 7.9, *Use of EAPOL in Shared Media LANs*) can also be problematic. Strengthened by 802.1x, Wi-Fi networks promised secure and swift enterprise data on mobile devices, allowing for message exchange, as well as other services. But then came the news that, despite all precautions, Wi-Fi's next generation security protocol had been compromised in government-funded (NIST) research at the University of Maryland [34, 35]. 802.1x appears to have three security vulnerabilities:

1. *Session hijacking.* Wi-Fi is especially vulnerable when using a "public hotspot" and without even using the essential WEP protocol. If WEP is not involved, it effectively exploits the so-called race conditions between the 802.1x and 802.11 state machines, in which a thief who wants to steal the session is racing the owner to the access point "door." If he wins, he owns everything.

2. *Man-in-the-middle vulnerability*. This was supposedly fixed, according to WECA, but discussions continue on whether or not that is a fact. To exploit this vulnerability seems even easier than session hijacking, as the attacker stays in the middle acting as an AP to the real user, and as a user to the actual AP.

3. *DOS (denial of service)*. All wireless communication is vulnerable to DOS because someone can deliberately jam the frequency.

Providing encryption over the client and AP communication helps lower the vulnerability, but the problem of an intruder breaking the encryption and gaining access to everything persists. And the question of how to make a more effective network protocol design remains. WECA itself, aware of the 802.1x security issues, is committed to improving it by introducing the next generation of security called TKIP, which is backward-compatible with current Wi-Fi products and upgradable with software. TKIP is a rapid rekeying protocol that changes the encryption key about every 10,000 frames; vendor-specific implementations were already available at press time. WECA is also preparing to introduce AES, the latest security protocol, but which so far requires additional hardware acceleration for encryption using a coprocessor—something mobile devices have so far lacked. But AES has its critics, among whom is William Arbaugh, the professor who discovered 802.1x's vulnerabilities. Arbaugh, of the University of Maryland, has said in industry reports that relying on a confidentiality mechanism (such as AES) is bad design.

Pragmatic suggestions of experts have been as follows: "If even LAN communication has its own security flaws, do we stop 'networking'? Of course not. We need to be aware of the benefits of mobile clients communicating with one other and accessing enterprise data. We also must be aware of how data and communication could be compromised—then we will know how to lower the risks. Just as the benefits of protocols like 802.1x make us want to use them, any weaknesses in those protocols should only provide an invitation to companies, start-ups, and developers to come up with better, more secure solutions. Clients, companies, the government, and individuals will certainly make a market for such solutions, especially with the exponential growth of mobile devices" [35].

In conclusion, vendors are responding to the flawed protocols with fixes in several stages. In the short term, vendors are adding new authentication/key management protocols that provide secure authentication, and that provide new WEP keys for each card, for each session. In addition, in the near term, vendors are working on a tweak to WEP to make attacks more difficult, and they are also working on a long-term complete fix [36]. Essentially, security has to be implemented end to end, as with transport protocols.

8.3.3 IEEE 802.11a Details

802.11a is viewed as a "new(er)" specification that represents the next generation of enterprise-class wireless LANs. Among the advantages it has over current technologies are greater scalability, better interference immunity, and significantly higher speed, up to 54 Mbps and beyond, which simultaneously allows for higher bandwidth applications and more users. This section provides some highlights of the technology [37].

8.3.3.1 Physical Layer

5-GHz Frequency Band

802.11a utilizes 300 MHz of bandwidth in the 5-GHz UNII band. Though the lower 200 MHz is physically contiguous, the FCC has divided the total 300 MHz into three distinct 100-MHz domains, each with a different legal maximum power output. The "low" band operates from 5.15 to 5.25 GHz, and has a maximum of 50 mW. The "middle" band is located from 5.25 to 5.35 GHz, with a maximum of 250 mW. The "high" band utilizes 5.725 to 5.825 GHz, with a maximum of 1 W (see Figure 8.28). Because of the high power output, devices transmitting in the high band will tend to be building-to-building products. The low and medium bands are more suited to in-building wireless products. One requirement specific to the low band is that all devices must use integrated antennas.

Different regions of the world have allocated different amounts of spectrum, so geographic location will determine how much of the 5-GHz band is available. In the United States, the FCC has allocated all three bands for unlicensed transmissions. In Europe, however, only the low and middle bands are free. Though 802.11a is not yet certifiable in Europe, efforts are currently under way between IEEE and ETSI to rectify this. In Japan, only the low band may be used. This will result in more contention for signal, but will still allow for very high performance.

The frequency range used currently for most enterprise-class unlicensed transmissions, including 802.11b, is the 2.4-GHz ISM band. This highly populated band offers only 83 MHz of spectrum for all wireless traffic, including cordless phones, building-to-building transmissions, and microwave ovens. In comparison, the 300 MHz offered in the UNII band represents a nearly fourfold increase in spectrum—all the more impressive when considering there is limited wireless traffic in the band today.

FIGURE 8.28
Power ranges for
IEEE 802.11a.

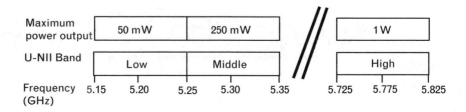

OFDM Modulation Scheme

802.11a uses OFDM, a new(er) encoding scheme that offers benefits over spread spectrum in channel availability and data rate. Channel availability is significant because the more independent channels that are available, the more scalable the wireless network becomes. The high data rate is accomplished by combining many lower speed subcarriers to create one high-speed channel. 802.11a uses OFDM to define a total of eight nonoverlapping 20-MHz channels across the two lower bands; each of these channels is divided into 52 subcarriers, each approximately 300 kHz wide. By comparison, 802.11b uses three nonoverlapping channels (as noted earlier in the chapter) (see Figure 8.29).

A large (wide) channel can transport more information per transmission than a small (narrow) one. As described earlier, 802.11a utilizes channels that are 2 MHz wide, with 52 subcarriers contained within. The subcarriers are transmitted in "parallel," meaning that they are sent and received simultaneously. The receiving device processes these individual signals, each one representing a fraction of the total data that, together, make up the actual signal. With this many subcarriers comprising each channel, a tremendous amount of information can be sent at once.

With so much information per transmission, it becomes important to guard against data loss. FEC was added to the 802.11a specification for this purpose (FEC does not exist in 802.11b). At its simplest, FEC consists of sending a secondary copy (or additional bits) along with the primary information. If part of the primary information is lost, insurance then exists to help the receiving device recover (through sophisticated algorithms) the lost data. This way, even if part of the signal is lost, the information can be recovered so data are received as intended, eliminating the need to retransmit. Because of its high speed, 802.11a can accommodate this overhead with negligible impact on performance.

Another threat to the integrity of the transmission is *multipath reflection,* also called *delay spread.* When a radio signal leaves the "sending" antenna, it radiates outward, spreading as it travels. If the signal reflects off a flat surface, the original signal and the reflected signal may reach the "receiving" antenna simultaneously. Depending on how the signals overlap, they can either augment or cancel each other (see Figure 8.30). A baseband processor, or equalizer, unravels the divergent signals. However, if

FIGURE 8.29
OFDM subcarriers.

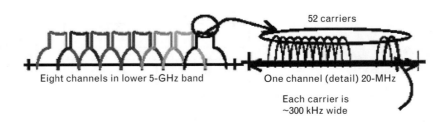

Eight channels in lower 5-GHz band

52 carriers

One channel (detail) 20-MHz

Each carrier is
~300 kHz wide

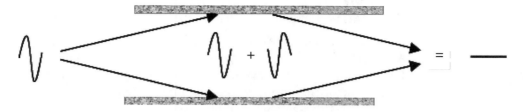

FIGURE 8.30 *Multipath interference occurs when reflected signals cancel each other. 802.11a uses a slower symbol rate to minimize multipath interference.*

the delay is long enough, the delayed signal spreads into the next transmission. OFDM specifies a slower symbol rate to reduce the chance that a signal will encroach on the following signal, minimizing multipath interference.

Data Rates and Range
Devices utilizing 802.11a are required to support speeds of 6, 12, and 24 Mbps. Optional speeds go up to 54 Mbps, but will also typically include 48, 36, 18, and 9 Mbps. These differences are the result of implementing different modulation techniques and FEC levels. To achieve 54 Mbps, a mechanism called *64-level quadrature amplitude modulation* (64-QAM) is used to pack the maximum amount of information possible (allowable by the standard) on each subcarrier. Just as with 802.11b, as an 802.11a client device travels farther from its AP, the connection will remain intact but speed decreases (falls back). As Figure 8.31 illustrates, at any range, 802.11a has a significantly higher signaling rate than 802.11b.

FIGURE 8.31
Example of range/data rate differences between 802.11a and 802.11b.

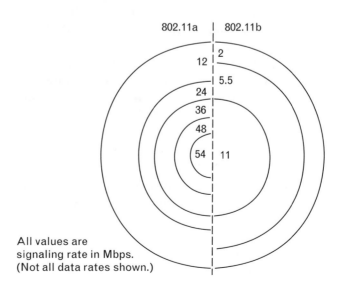

All values are signaling rate in Mbps. (Not all data rates shown.)

8.3.3.2 MAC Layer

802.11a uses the same MAC layer technology as 802.11b, namely, CSMA/CA. CSMA/CA is a basic protocol used to avoid signals colliding and canceling each other. It works by requesting authorization to transmit for a specific amount of time prior to sending information. The sending device broadcasts an RTS frame with information on the length of its signal. If the receiving device permits it at that moment, it broadcasts a CTS frame. Once the CTS goes out, the sending machine transmits its information. Any other sending devices in the area that "hear" the CTS realize another device will be transmitting and allow that signal to go out uncontested.

8.3.3.3 Relation to HiperLAN/2

HiperLAN/2 is a wireless specification developed by ETSI, and it has some similarities to 802.11a at the physical layer. It also uses OFDM technology and operates in the 5-GHz frequency band. The MAC layers are different, however. Whereas 802.11a uses CSMA/CA, HiperLAN/2 utilizes TDMA.

Because the 5-GHz UNII equivalent bands have been reserved for HiperLAN/2 systems in Europe, 802.11a is not yet certifiable in Europe by ETSI. In an effort to rectify this, two additions to the IEEE 802.11a specification have been proposed to allow both 802.11a and HiperLAN/2 to coexist. *Dynamic channel selection* (DCS) and *transmit power control* (TPC) allow clients to detect the most available channels and use only the minimum output power necessary if interference is evident. The implementation of these additions will significantly increase the likelihood of European 802.11a certification.

8.3.3.4 Compatibility with 802.11b

Although 802.11a and 802.11b share the same MAC layer technology, there are significant differences at the physical layer. 802.11b, using the ISM band, transmits in the 2.4-GHz range, whereas 802.11a, using the UNII band, transmits in the 5-GHz range. Because their signals travel in different frequency bands, one significant benefit is that they will not interfere with each other. A related consequence, therefore, is that the two technologies are not compatible. There are various strategies for migrating from 802.11b to 802.11a, or even using both on the same network concurrently.

8.3.3.5 Summary

802.11a represents the next generation of enterprise-class wireless LAN technology, with many advantages over current options. At speeds of 54 Mbps and greater, it is faster than any other unlicensed solution. 802.11a and 802.11b both have a similar range, but 802.11a provides higher speed throughout the entire coverage area. The 5-GHz band in which it operates is not highly populated, so there is less congestion to cause interference or signal contention. In addition, the eight nonoverlapping channels allow for a highly scalable and flexible installation. 802.11a is the most reliable and efficient medium by which to accommodate high-bandwidth applications for numerous users.

Chip manufacturers such as Broadcom (http://www.broadcom.com) have introduced simultaneous dual-band 802.11a/b WLAN chipsets. These all–CMOS, dual-band solutions typically consist of a base-band/medium access control IC, which supports the 802.11b and 802.11a protocols as well as IEEE 802.1x security, a 2.4-GHz radio IC and a 5-GHz radio IC. The chipset dynamically selects the best performance available at either 2.4 or 5 GHz. These chipsets also offer the ability to operate simultaneously in both frequency bands. The solution incorporates hardware support for WEP and AES and system support for the leading security protocols, TKIP and 802.1x, and its software can be upgraded to the forthcoming 802.11i security standard.

8.3.4 WPAN Technologies

WPANs saw strong advocacy on the part of Ericsson during the late 1990s and early 2000s. Ericsson developed the Bluetooth technology (so named after a Viking King who lived more than 1,000 years ago). At press time, Bluetooth was an industry specification for short-range RF–based connectivity for portable personal devices. The Bluetooth Special Interest Group, established in 1998, made their specifications publicly available in the middle of 1999. After this release, the IEEE 802.15 group took the Bluetooth work and developed a vendor-independent standard. Some see Bluetooth/IEEE 802.15 and IEEE 802.11b as competitors; others see them as complementary, particularly give the different technical optimization points (weight, power, range, applications, and so on) for these two technologies. The technologies share some characteristics and overlap in some usage models, but they address basically different purposes. The following are characteristics of a WPAN (also see Figure 8.32):

- Short range;
- Low power;
- Low cost;

FIGURE 8.32
Comparison of topologies for IEEE 802.11b and IEEE 802.15 environments.

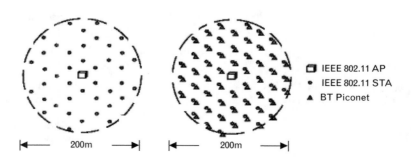

□ IEEE 802.11 AP
• IEEE 802.11 STA
▲ BT Piconet

|← 200m →| |← 200m →|

• Small networks;

• Communication of devices within a POS.

As we have done earlier in the chapter, one way to segment wireless communication technologies is by the geographic area they cover. The range of a WAN is typically measured in miles, as seen in Table 8.11. Communication over such distances requires relatively high-power transmissions and, almost invariably, a government license for a specific frequency band. Service providers are required to pay a fee for the license. There is a trade-off between power consumption and data rates. Typical data rates for 2G/2.5G cellular networks are relatively low (20–200 Kbps), mostly related to the transmission power needed to reach the service tower from a handset; higher data rates at these same levels of power transmission are impractical using current battery technologies. 3G cellular systems will have significantly higher data rates (200–2,000 Kbps); however, to maintain power consumption at workable levels, 3G systems will require cellular towers to be much closer. WLANs generally span distances from 30 to a few hundred feet (often less than 750 feet). This smaller footprint permits lower power transmissions and the use of unlicensed frequency bands. IEEE 802.11b, for example, has a nominal range of 330 ft and data rates up to 11 Mbps (higher ranges are possible at the lower 802.11 speeds). This combination of footprint and data rate leads to moderate-to-high power consumption; therefore, the types of devices normally used with WLANs are ones that have a robust computing platform and power supply, such as a PC or laptop computers. WPANs cover distances on the order of 30 ft and are intended to be used to connect personal portable devices without using cables. This peer-to-peer device communication generally requires data

TABLE 8.11   TYPICAL PERFORMANCE VALUES FOR WIRELESS COMMUNICATION

CHANNEL MEDIUM	CARRIER FREQUENCY (MHz)	MAXIMUM BIT RATE (KBPS)	RANGE (M)	ADVANTAGES	PROBLEMS	APPLICATION
RF UHF	300–1,900	2,000	60,000	Efficient for distance	FCC regulations, licenses	Cellular telephony, 2.5G/3G
RF microwave; ISM band	2,400	1,000	10–100	No license needed	Interference	WLANs, WPANs
RF microwave; UNII band	5,150–5,825	54,000	10–100	No license needed	Distance limitations	WLANs
Optical infrared over the air	Infrared light band	1,544–622,000	5,000	No license needed, inexpensive	Directional, subject to atmospheric conditions (such as fog, but newer 10-μm systems are more impervious to weather)	Temporary links

rates on the order of 1 Mbps. The small footprint and relatively low data rates leads to low power consumption, making WPAN technologies suitable for use with small mobile devices such as PDAs, mobile phones, and so on. In addition, low-power transmission allows for the use of unlicensed ISM frequency bands (the 2.4-GHz spectrum [38]). In summary, Bluetooth technology defines specifications for small-form-factor, low-cost wireless radio communications among notebook computers, personal digital assistants, cellular phones and other portable, handheld devices, and connectivity to the Internet.

Bluetooth technology is optimized by design for WPANs, whereas IEEE 802.11 is optimized by design for WLANs. Bluetooth supports the physical link between devices forming a *piconet*, which is a network of devices connected in an ad hoc fashion (a piconet can support up to eight devices). Bluetooth technology was developed to replace cables between small personal devices (PDAs, cell phones, and so on). As such, Bluetooth technology is optimized for short-range, low-power voice and data communication. The Bluetooth SIG (http://www.bluetooth.com) is driving development of the technology and bringing it to market. The SIG is comprised of telecommunications, computing, network, and consumer electronics industry leaders and includes promoter group companies 3Com Corporation, Ericsson Technology Licensing AB, IBM Corporation, Intel Corporation, Agere Systems, Microsoft Corporation, Motorola, Nokia Corporation, Toshiba Corporation, as well as hundreds of associate and adopter member companies. Table 8.12 provides a forecast of the penetration of WPAN/Bluetooth technology.

While some Bluetooth profiles describe methods to connect personal devices to networks, Bluetooth technology is not a full-fledged networking technology: Devices are limited to the use of PPP for dial-up

TABLE 8.12   FORECAST OF THE PENETRATION OF WPAN/BLUETOOTH TECHNOLOGY

	2001	2002	2003	2004	2005	2006
NOTEBOOK PC SHIPMENTS (MILLIONS)	25.5	27.6	34.1	40	46	51.6
% Bluetooth	1%	7%	19%	36%	57%	69%
Total Bluetooth	0	2	6	15	26	36
DESKTOP PC SHIPMENTS (MILLIONS)	101.1	105.2	121.9	133.3	143.5	151.6
% Bluetooth	0%	1%	4%	11%	22%	38%
Total Bluetooth	0.0	1.1	4.8	14.8	31.9	57.9
PDA SHIPMENTS (MILLIONS)	8.1	9.5	12.4	16.1	19.5	23.4
% Bluetooth	70%	10%	17%	22%	23%	26%
Total Bluetooth	5.7	0.9	2.0	3.5	4.5	6.0

Source: 4/02 Instat/MDR.

networking or LAN access. Each has strengths that make it well suited to its primary domain and that are, in turn, weaknesses in the other's domain. For example, there would be little need in a WPAN for transmission power capable of reaching a device 100m away, and the associated power consumption to do so would be a drawback to using IEEE 802.11 for a WPAN. Note the following [39]:

1. Both technologies operate in the 2.4-GHz band and, therefore, interfere with each other when used in the same place at the same time. Groups within the IEEE and the Bluetooth SIG are performing studies and developing recommendations to understand and address this and other issues related to RF interference.

2. There is some potential overlap in a few usage models between the two technologies, with each having the capability to do some of the things the other does, albeit nonoptimally.

3. They are developed by different industry groups, so they must be at odds with each other.

In 1999 the IEEE established a committee to develop a standard called 802.15 *Wireless Personal Area Networks,* which will use the Bluetooth specification as its base. The IEEE 802.15 WPAN work aims at standardizing the MAC and PHY layers of Bluetooth. The goal of the 802.15 standard is to facilitate wider adoption of WPANs, and to deal with issues such as coexistence and interoperability within the networks. From the bottom up, the sublayers of the IEEE 802.15 work are as follows:

- RF layer (PHY);
- Baseband layer (PHY);
- Link manager (MAC);
- *Logical Link Control and Adaptation Protocol* (L2CAP) (MAC).

The functionality is as follows [40]:

RF layer: The air interface is based on antenna power range starting from 0 dBm up to 20 dBm, 2.4-GHz band, and the link range from 0.1 to 10m.

Baseband layer: The baseband layer establishes the Bluetooth *piconet.* The piconet is formed when two Bluetooth devices connect. In a piconet one device acts as the master and the other devices act as slaves.

Link manager: The link manager establishes the link between Bluetooth devices. Additional functions include security, negotiation of baseband packet sizes, power mode and duty cycle control of the Bluetooth device, and the connection states of a Bluetooth device in a piconet.

L2CAP: This sublayer provides the upper layer protocols with connectionless and connection-oriented services. The services provided by this layer include protocol multiplexing capability, segmentation and reassembly of packets, and group abstractions.

The 802.15 WPAN committee was originally comprised of four subcommittees [40, 41], as discussed next.

8.3.4.1   802.15 WPAN Task Group 1

The scope and focus of TG1 are to define PHY and MAC specifications for wireless connectivity between devices that are either fixed or portable within the personal operating space. The goal is to allow low-complexity, low-power consumption wireless connectivity to support data transfer to and from a WPAN device and an 802.11 device. The standard aims at taking into account coexistence with all 802.11 devices. The 802.15 WPAN TG1 utilized the Bluetooth v1.0 specifications to derive the WPAN standard. The IEEE Project 802.15.1 has derived a WPAN standard based on the Bluetooth v1.1 foundation specifications. The IEEE Std 802.15.1-2002 was published in June 2002. The IEEE Std 802.15.12002 standard is an additional resource for those who implement Bluetooth devices.

IEEE Standard 802.15.1, *Wireless MAC and PHY Specifications for Wireless Personal Area Networks*, is adapted from portions of the Bluetooth wireless specification. The lower transport layers (L2CAP, LMP, baseband, and radio) of the Bluetooth wireless technology are defined. The IEEE has reviewed and provided a standard adaptation of the Bluetooth Specification v1.1 Foundation MAC (L2CAP, LMP, and baseband) and PHY (radio). Also specified is a clause on SAPs that includes a LLC/MAC interface for the ISO/IEC 8802-2 LLC. Also specified is a normative annex that provides a *protocol implementation conformance statement* (PICS) pro forma. Also specified is an informative high-level behavioral ITU-T Z.100 *specification and description language* (SDL) model for an integrated Bluetooth MAC sublayer. The IEEE licensed wireless technology from Bluetooth SIG, Inc., to adapt and copy a portion of the Bluetooth specification as base material for IEEE Standard 802.15.1-2002. The approved IEEE 802.15.1 standard is compatible with the Bluetooth v1.1 specification. The new standard gives the Bluetooth specification greater validity and support in the market and is an additional resource for those who implement Bluetooth devices.

The IEEE standard also added a major clause on SAPs, which includes an LLC/MAC interface for the ISO/IEC 8802-2 LLC and, as mentioned, a normative annex that provides a PICS pro forma, and an informative, high-level behavioral ITU-T Z.100 SDL model for an integrated Bluetooth MAC sublayer. This SDL model offers an extensive overview (more than 500 pages long) of a significant portion of the Bluetooth protocols for example, baseband, LMP, L2CAP, and the link manager [using the *host controller interface* (HCI)].

The IEEE-SA also plans to further develop the 802.15.1 SDL model source to support the standard. The SDL code, which will be available on CD-ROM, will include a computer model for use with any SDL tool that supports the SDL–88, SDL–92, or SDL–2000 update of ITU–T Recommendation Z.100. The IEEE 802.15.1 Working Task Group used the SDL to translate the natural language of the Bluetooth specification into a formal specification that defines how the Bluetooth protocols react to events in the environment that are communicated to a system by signals.

8.3.4.2 802.15 WPAN Task Group 2

The IEEE 802.15 Coexistence Task Group 2 (TG2) for WPANs is developing recommended practices to facilitate coexistence of WPANs (802.15) and WLANs (802.11). TG2 is developing a coexistence model to quantify the mutual interference of a WLAN and a WPAN. The task group is also developing a set of coexistence mechanisms to facilitate coexistence of WLAN and WPAN devices.

8.3.4.3 802.15 WPAN Task Group 3

The IEEE P802.15.3 High Rate Task Group (TG3) for WPANs is chartered to draft and publish a new standard for high-rate (20 Mbps or greater) WPANs. Besides a high data rate, the new standard will provide for low-power, low-cost solutions addressing the needs of portable consumer digital imaging and multimedia applications. TG3 MAC and PHY are designed to meet the demanding requirements of portable consumer imaging and multimedia applications. TG3 MAC and PHY features are as follows:

- *Data rates:* 11, 22, 33, 44, and 55 Mbps;
- QoS isochronous protocol;
- Ad hoc peer-to-peer networking;
- Security;
- Low power consumption;
- Low cost.

The current status is as follows:

- *TG3 HR MAC and PHY proposals selected.* TG3 selected final MAC and PHY proposals for 802.15.3 at the November 2000, IEEE 802 Plenary Meeting.
- *TG3 HR MAC and PHY proposals approved.* Merging key components from multiple proposals, a hybrid MAC was developed and adopted for 802.15.3. The selected MAC enjoyed a 100% confidence vote from the WPAN HR Task Group and was affirmed by the

802.15 WPAN Working Group with only one dissenting vote. Driven by consensus, TG3 adopted a modified PHY proposal based on a 2.4-GHz OQPSK radio design. The selected PHY proposal was confirmed by a consensus vote of the WPAN HR Task Group and affirmed by the 802.15 WPAN Working Group with only one dissenting vote.

8.3.4.4   802.15 WPAN Task Group 4

Task Group 4's (TG4) scope and focus is to determine a solution with a low data rate and long battery life, potentially months to years, with very low complexity. The solution determined would need to operate within an unlicensed and global frequency band. Potential applications are sensors, interactive toys, smart badges, remote controls, and home automation. The start of sponsor ballot activity was September 2002. IEEE 802.15 TG4 features include the following:

- Data rates of 250 and 40 Kbps;
- Up to 254 using short addressing or more using IEEE addresses;
- Support for critical latency devices, such as joysticks;
- CSMA/CA channel access;
- Automatic network establishment by the coordinator;
- Dynamic device addressing;
- Fully handshaked protocol for transfer reliability;
- Power management to ensure low-power consumption;
- Sixteen channels in the 2.4-GHz ISM band, 10 channels in the 915-MHz band, and one channel in the 868-MHz band.

8.3.5   WMAN Technologies

IEEE Standard 802.16-2001 [10], completed in October 2001 and published on April 8, 2002, defines the WirelessMAN air interface specification for WMANs. The completion of this standard heralds the entry of broadband wireless access as a major new tool in the effort to link homes and businesses to core telecommunications networks worldwide [42].

As currently defined through IEEE Standard 802.16, a WMAN provides network access to buildings through exterior antennas communicating with central radio BSs. The wireless MAN offers an alternative to cabled access networks, such as fiber optic links, coaxial systems using cable modems, and DSL links. Because wireless systems have the capacity to address broad geographic areas without the costly infrastructure development required in deploying cable links to individual sites, the technology may prove less expensive to deploy and may lead to more ubiquitous broadband access. Such systems have been in use for several years, but the development of the new standard marks the maturation of the industry and forms the basis of new industry success using second generation equipment.

In this scenario, with WirelessMAN technology bringing the network to a building, users inside the building will connect to it with conventional in-building networks such as, for data, Ethernet (IEEE Standard 802.3) or wireless LANs (IEEE Standard 802.11). However, the fundamental design of the standard may eventually allow for the efficient extension of the WirelessMAN networking protocols directly to the individual user. For instance, a central BS may someday exchange MAC protocol data with an individual laptop computer in a home. The links from the BS to the home receiver and from the home receiver to the laptop would likely use quite different physical layers, but design of the WirelessMAN MAC could accommodate such a connection with full QoS. With the technology expanding in this direction, it is likely that the standard will evolve to support nomadic and increasingly mobile users. For example, it could be suitable for a stationary or slow-moving vehicle.

IEEE Standard 802.16 was designed to evolve as a set of air interfaces based on a common MAC protocol but with physical layer specifications dependent on the spectrum of use and the associated regulations. The standard, as approved in 2001, addresses frequencies from 10 to 66 GHz, where extensive spectrum is currently available worldwide but at which the short wavelengths introduce significant deployment challenges. The IEEE was expecting to complete an amendment denoted IEEE 802.16a [43] in 2003. This document extends the air interface support to lower frequencies in the 2- to 11-GHz band, including both licensed and license-exempt spectra. Compared to the higher frequencies, such spectra offer the opportunity to reach many more customers less expensively, although at generally lower data rates. This suggests that such services will be oriented toward individual homes or small to medium-sized enterprises.

8.3.5.1   The 802.16 Working Group

Development of IEEE Standard 802.16 and the included WirelessMAN air interface, along with associated standards and amendments, is the responsibility of IEEE Working Group 802.16 on Broadband Wireless Access Standards (http://WirelessMAN.org). The working group's initial interest was the 10- to 66-GHz range. The 2- to 11-GHz amendment project that led to IEEE 802.16a was approved in March 2000. The 802.16a project primarily involves the development of new physical layer specifications, with supporting enhancements to the basic MAC. In addition, the working group has completed IEEE Standard 802.16.2 [11] (*Recommended Practice for Coexistence of Fixed Broadband Wireless Access Systems*) to address 10- to 66-GHz coexistence and, through the amendment project

802.16.2a, is expanding its recommendations to include licensed bands from 2 to 11 GHz.

Historically, the 802.16 activities were initiated at an August 1998 meeting called by the *National Wireless Electronics Systems Testbed* (N-WEST) of the U.S. National Institute of Standards and Technology. The effort was welcomed in IEEE 802, which opened a study group. The 802.16 working group has held weeklong meetings at least bimonthly since July 1999. More than 700 individuals have attended a session. Membership, which is granted to individuals based on their attendance and participation, currently stands at 130. The work has been closely followed; for example, the IEEE 802.16 Web site received over 2.8 million file requests in 2000.

8.3.5.2 Technology Design Issues

MAC

The IEEE 802.16 MAC protocol was designed for point-to-multipoint broadband wireless access applications. It addresses the need for very high bit rates, both uplink (to the BS) and downlink (from the BS). Access and bandwidth allocation algorithms must accommodate hundreds of terminals per channel, with terminals that may be shared by multiple end users. The services required by these end users are varied in their nature and include legacy TDM voice and data, IP connectivity, and packetized *voice over IP* (VoIP). To support this variety of services, the 802.16 MAC must accommodate both continuous and bursty traffic. Additionally, these services expect to be assigned QoS in keeping with the traffic types.

The 802.16 MAC provides a wide range of service types analogous to the classic ATM service categories as well as newer categories such as *guaranteed frame rate* (GFR). The 802.16 MAC protocol must also support a variety of backhaul requirements, including both ATM and packet-based protocols. Convergence sublayers are used to map the transport-layer-specific traffic to a MAC that is flexible enough to efficiently carry any traffic type. Through such features as payload header suppression, packing, and fragmentation, the convergence sublayers and MAC work together to carry traffic in a form that is often more efficient than the original transport mechanism.

Issues of transport efficiency are also addressed at the interface between the MAC and the PHY layer. For example, the modulation and coding schemes are specified in a burst profile that may be adjusted adaptively for each burst to each subscriber station. The MAC can make use of bandwidth-efficient burst profiles under favorable link conditions but shift to more reliable, although less efficient, alternatives as required to support the planned 99.999% link availability.

The request-grant mechanism is designed to be scalable, efficient, and self-correcting. The 802.16 access system does not lose efficiency when presented with multiple connections per terminal, multiple QoS levels per terminal, and a large number of statistically multiplexed users. It takes advantage of a wide variety of request mechanisms, balancing the stability of contentionless access with the efficiency of contention-oriented access.

While extensive bandwidth allocation and QoS mechanisms are provided, the details of scheduling and reservation management are left unstandardized and provide an important mechanism for vendors to differentiate their equipment.

Along with the fundamental task of allocating bandwidth and transporting data, the MAC includes a privacy sublayer that provides authentication of network access and connection establishment to avoid theft of service, and it provides key exchange and encryption for data privacy. To accommodate the more demanding physical environment and different service requirements of the frequencies between 2 and 11 GHz, the 802.16a project is upgrading the MAC to provide *automatic repeat request* (ARQ) and support for mesh, rather than only point-to-multipoint, network architectures.

The Physical Layer

10 to 66 GHz   In the design of the PHY specification for 10 to 66 GHz, line-of-sight propagation was deemed a practical necessity. With this condition assumed, single-carrier modulation was easily selected; the air interface is designated *WirelessMAN-SC*. Many fundamental design challenges remained, however. Because of the point-to-multipoint architecture, the BS basically transmits a TDM signal, with individual subscriber stations allocated time slots serially. Access in the uplink direction is by TDMA. Following extensive discussions regarding duplexing, a burst design was selected that allows both TDD, in which the uplink and downlink share a channel but do not transmit simultaneously, and FDD, in which the uplink and downlink operate on separate channels, sometimes simultaneously. This burst design allows both TDD and FDD to be handled in a similar fashion. Support for half-duplex FDD subscriber stations, which may be less expensive since they do not simultaneously transmit and receive, was added at the expense of some slight complexity. Both TDD and FDD alternatives support adaptive burst profiles in which modulation and coding options may be dynamically assigned on a burst-by-burst basis. Figure 8.33 shows the downlink subframe structure.

2 to 11 GHz   The 2- to 11-GHz bands, both licensed and license exempt, are addressed in IEEE Project 802.16a. The standard was in ballot at the time of this writing. The draft specifies that compliant systems implement one of three air interface specifications, each of which provides for interoperability. Design of the 2- to 11-GHz physical layer is driven by the need for *nonline-of-sight* (NLOS) operation. Because residential applications are expected, rooftops may be too low for a clear sight line to a BS antenna,

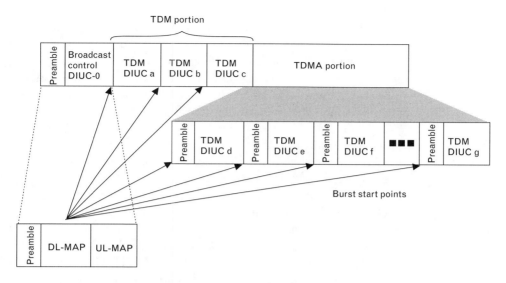

FIGURE 8.33 *The downlink subframe structure.*

possibly due to obstruction by trees. Therefore, significant multipath propagation must be expected. Furthermore, outdoor-mounted antennas are expensive due to both hardware and installation costs.

The three 2- to 11-GHz air interface specifications in 802.16a Draft 3 are as follows:

1. *WirelessMAN-SC2:* This uses a single-carrier modulation format.

2. *WirelessMAN-OFDM:* This uses OFDM with a 256-point transform. Access is by TDMA. This air interface is mandatory for license exempt bands.

3. *WirelessMAN-OFDMA:* This uses OFDM access with a 2,048-point transform. In this system, multiple access is provided by addressing a subset of the multiple carriers to individual receivers.

Because of the propagation requirements, the use of advanced antenna systems is supported.

It is premature to speculate on further specifics of the 802.16a amendment prior to its completion. While the draft seems to have reached a level of maturity, the contents could change in balloting. Modes could even be deleted or added.

Physical Layer Details
The PHY specification defined for the 10- to 66-GHz band uses burst single-carrier modulation with adaptive burst profiling in which transmission parameters, including the modulation and coding schemes, may be

adjusted individually to each *subscriber station* (SS) on a frame-by-frame basis. Both TDD and burst FDD variants are defined. Channel bandwidths of 20 or 25 MHz (typical U.S. allocation) or 28 MHz (typical European allocation) are specified, along with Nyquist square-root raised-cosine pulse shaping with a rolloff factor of 0.25. Randomization is performed for spectral shaping and to ensure bit transitions for clock recovery.

The FEC used is Reed–Solomon GF(256), with variable block size and error correction capabilities. This is paired with an inner block convolutional code to robustly transmit critical data, such as frame control and initial accesses. The FEC options are paired with QPSK, 16-QAM, and 64-QAM to form burst profiles of varying robustness and efficiency. If the last FEC block is not filled, that block may be shortened. Shortening in both the uplink and downlink is controlled by the BS and is implicitly communicated in the uplink map (UL-MAP) and downlink map (DL-MAP).

The system uses a frame of 0.5, 1, or 2 ms. This frame is divided into physical slots for the purpose of bandwidth allocation and identification of PHY transitions. A physical slot is defined to be 4 QAM symbols. In the TDD variant of the PHY, the uplink subframe follows the downlink subframe on the same carrier frequency. In the FDD variant, the uplink and downlink subframes are coincident in time but are carried on separate frequencies. The downlink subframe is shown in Figure 8.33.

The downlink subframe starts with a frame control section that contains the DL-MAP for the current downlink frame as well as the UL-MAP for a specified time in the future. The downlink map specifies when physical layer transitions (modulation and FEC changes) occur within the downlink subframe. The downlink subframe typically contains a TDM portion immediately following the frame control section. Downlink data are transmitted to each SS using a negotiated burst profile. The data are transmitted in order of decreasing robustness to allow SSs to receive their data before being presented with a burst profile that could cause them to lose synchronization with the downlink.

In FDD systems, the TDM portion may be followed by a TDMA segment that includes an extra preamble at the start of each new burst profile. This feature allows better support of half-duplex SSs. In an efficiently scheduled FDD system with many half-duplex SSs, some may need to transmit earlier in the frame than they receive. Due to their half-duplex nature, these SSs lose synchronization with the downlink. The TDMA preamble allows them to regain synchronization.

Due to the dynamics of bandwidth demand for the variety of services that may be active, the mixture and duration of burst profiles and the presence or absence of a TDMA portion vary dynamically from frame to frame. Since the recipient SS is implicitly indicated in the MAC headers rather than in the DL-MAP, SSs listen to all portions of the downlink subframe they are capable of receiving. For full-duplex SSs, this means receiving all burst profiles of equal or greater robustness than they have negotiated with the BS.

A typical uplink subframe for the 10- to 66-GHz PHY is shown in Figure 8.34. Unlike the downlink, the UL-MAP grants bandwidth to specific SSs. The SSs transmit in their assigned allocation using the burst profile specified by the *uplink interval usage code* (UIUC) in the UL-MAP entry granting them bandwidth. The uplink subframe may also contain contention-based allocations for initial system access and broadcast or multicast bandwidth requests. The access opportunities for initial system access are sized to allow extra guard time for SSs that have not resolved the transmit time advance necessary to offset the round-trip delay to the BS.

Between the PHY and MAC is a transmission convergence (TC) sublayer (see Figure 8.35). This layer performs the transformation of variable-length MAC *protocol data units* (PDUs) into the fixed-length FEC blocks (plus possibly a shortened block at the end) of each burst. The TC layer has a PDU sized to fit in the FEC block currently being filled. It starts with a pointer indicating where the next MAC PDU header starts within the FEC block. This is shown in Figure 8.36.

The TC PDU format allows resynchronization to the next MAC PDU in the event that the previous FEC block had irrecoverable errors. Without the TC layer, a receiving SS or BS would potentially lose the entire remainder of a burst when an irrecoverable bit error occurred.

MAC Details

The MAC includes service-specific convergence sublayers that interface to higher layers, above the core MAC common part sublayer that carries out

FIGURE 8.34
The uplink subframe structure.

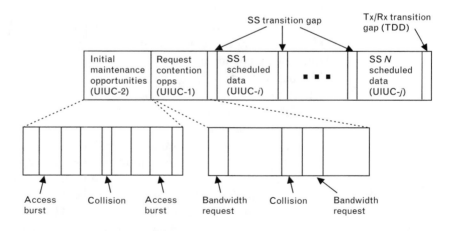

FIGURE 8.35
TC PDU format.

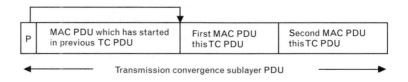

FIGURE 8.36
*Format of generic
header for MAC
PDU.*

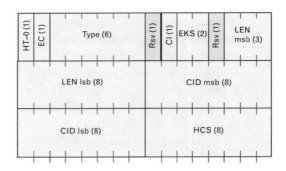

the key MAC functions. Below the common part sublayer is the privacy sublayer.

Service-Specific Convergence Sublayers

IEEE Standard 802.16 defines two general service-specific convergence sublayers for mapping services to and from 802.16 MAC connections. The ATM convergence sublayer is defined for ATM services, and the packet convergence sublayer is defined for mapping packet services such as IPv4, IPv6, Ethernet, and VLAN. The primary task of the sublayer is to classify *service data units* (SDUs) to the proper MAC connection, preserve or enable QoS, and enable bandwidth allocation. The mapping takes various forms depending on the type of service. In addition to these basic functions, the convergence sublayers can also perform more sophisticated functions such as payload header suppression and reconstruction to enhance airlink efficiency.

Common Part Sublayer

Introduction and General Architecture   In general, the 802.16 MAC is designed to support a point-to-multipoint architecture with a central BS handling multiple independent sectors simultaneously. On the downlink, data to SSs are multiplexed in TDM fashion. The uplink is shared between SSs in TDMA fashion.

The 802.16 MAC is connection oriented. All services, including inherently connectionless services, are mapped to a connection. This provides a mechanism for requesting bandwidth, associating QoS and traffic parameters, transporting and routing data to the appropriate convergence sublayer, and all other actions associated with the contractual terms of the service. Connections are referenced with 16-bit connection identifiers (CIDs) and may require continuously granted bandwidth or bandwidth on demand. As will be described, both are accommodated.

Each SS has a standard 48-bit MAC address, but this serves mainly as an equipment identifier, since the primary addresses used during operation are the CIDs. Upon entering the network, the SS is assigned three management connections in each direction. These three connections reflect the three different QoS requirements used by different management levels. The first of these is the basic connection, which is used for the transfer of short, time-critical MAC and *radio link control* (RLC) messages. The primary management connection is used to transfer longer, more delay-tolerant messages such as those used for authentication and connection setup. The secondary management connection is used for the transfer of standards-based management messages such as *Dynamic Host Configuration Protocol* (DHCP), *Trivial File Transfer Protocol* (TFTP), and SNMP. In addition to these management connections, SSs are allocated transport connections for the contracted services. Transport connections are unidirectional to facilitate different uplink and downlink QoS and traffic parameters; they are typically assigned to services in pairs.

The MAC reserves additional connections for other purposes. One connection is reserved for contention-based initial access. Another is reserved for broadcast transmissions in the downlink as well as for signaling broadcast contention-based polling of SS bandwidth needs. Additional connections are reserved for multicast, rather than broadcast, contention-based polling. SSs may be instructed to join multicast polling groups associated with these multicast polling connections.

MAC PDU Formats   The MAC PDU is the data unit exchanged between the MAC layers of the BS and its SSs. A MAC PDU consists of a fixed-length MAC header, a variable-length payload, and an optional *cyclic redundancy check* (CRC). Two header formats, distinguished by the HT field, are defined: the generic header and the bandwidth request header.

Except for bandwidth request MAC PDUs, which contain no payload, MAC PDUs contain either MAC management messages or convergence sublayer data.

Three types of MAC subheader may be present. The *grant management subheader* is used by an SS to convey bandwidth management needs to its BS. The *fragmentation subheader* contains information that indicates the presence and orientation in the payload of any fragments of SDUs. The *packing subheader* is used to indicate the packing of multiple SDUs into a single PDU. The grant management and fragmentation subheaders may be inserted in MAC PDUs immediately following the generic header if so indicated by the type field. The packing subheader may be inserted before each MAC SDU if so indicated by the type field. More details are provided in the following subsections.

Transmission of MAC PDUs   The IEEE 802.16 MAC supports various higher layer protocols such as ATM or IP. Incoming MAC SDUs from corresponding convergence sublayers are formatted according to the MAC PDU format, possibly with fragmentation and/or packing, before being conveyed over one or more connections in accordance with the MAC protocol. After traversing the airlink, MAC PDUs are reconstructed into the original MAC SDUs so that the format modifications performed by the MAC layer protocol are transparent to the receiving entity.

IEEE 802.16 takes advantage of incorporating the packing and fragmentation processes with the bandwidth allocation process to maximize the flexibility, efficiency, and effectiveness of both. Fragmentation is the process by which a MAC SDU is divided into one or more MAC SDU fragments. Packing is the process in which multiple MAC SDUs are packed into a single MAC PDU payload. Both processes may be initiated by either a BS for a downlink connection or an SS for an uplink connection.

IEEE 802.16 allows simultaneous fragmentation and packing for efficient use of the bandwidth.

PHY Support and Frame Structure   The IEEE 802.16 MAC supports both TDD and FDD. In FDD, both continuous and burst downlinks are supported. Continuous downlinks allow for certain robustness enhancement techniques, such as interleaving. Burst downlinks (either FDD or TDD) allow the use of more advanced robustness and capacity enhancement techniques, such as subscriber-level adaptive burst profiling and advanced antenna systems.

The MAC builds the downlink subframe starting with a frame control section containing the DL-MAP and UL-MAP messages. These indicate PHY transitions on the downlink as well as bandwidth allocations and burst profiles on the uplink.

The DL-MAP is always applicable to the current frame and is always at least two FEC blocks long. The first PHY transition is expressed in the first FEC block, to allow adequate processing time. In both TDD and FDD systems, the UL-MAP provides allocations starting no later than the next downlink frame. The UL-MAP can, however, allocate starting in the current frame as long as processing times and round-trip delays are observed. The minimum time between receipt and applicability of the UL-MAP for an FDD system is shown in Figure 8.37.

RLC   The advanced technology of the 802.16 PHY layer requires equally advanced RLC, particularly the capability of the PHY to transition from

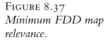

FIGURE 8.37
Minimum FDD map relevance.

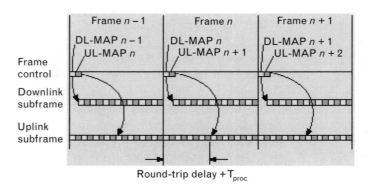

one burst profile to another. The RLC must control this capability as well as the traditional RLC functions of power control and ranging.

RLC begins with periodic BS broadcast of the burst profiles that have been chosen for the uplink and downlink. The particular burst profiles used on a channel are chosen based on a number of factors, such as rain region and equipment capabilities. Burst profiles for the downlink are each tagged with a DIUC. Those for the uplink are each tagged with an UIUC.

During initial access, the SS performs initial power leveling and ranging using *ranging request* (RNG-REQ) messages transmitted in initial maintenance windows. The adjustments to the SS's transmit time advance, as well as power adjustments, are returned to the SS in ranging response (RNG-RSP) messages. For ongoing ranging and power adjustments, the BS may transmit unsolicited RNG-RSP messages commanding the SS to adjust its power or timing.

During initial ranging, the SS also requests to be served in the downlink via a particular burst profile by transmitting its choice of DIUC to the BS. The choice is based on received downlink signal quality measurements performed by the SS before and during initial ranging. The BS may confirm or reject the choice in the ranging response. Similarly, the BS monitors the quality of the uplink signal it receives from the SS. The BS commands the SS to use a particular uplink burst profile simply by including the appropriate burst profile UIUC with the SS's grants in ULMAP messages.

After initial determination of uplink and downlink burst profiles between the BS and a particular SS, RLC continues to monitor and control the burst profiles. Harsher environmental conditions, such as rain fades, can force the SS to request a more robust burst profile. Alternatively, exceptionally good weather may allow an SS to temporarily operate with a more efficient burst profile. The RLC continues to adapt the SS's current UL and DL burst profiles, ever striving to achieve a balance between robustness and efficiency. Because the BS is in control and directly monitors the uplink signal quality, the protocol for changing the uplink burst profile for an SS is simple: The BS merely specifies the profile's associated UIUC whenever granting the SS bandwidth in a frame. This eliminates the need for an acknowledgment, since the SS will always receive either both the UIUC and the grant or neither. Hence, no chance of uplink burst profile mismatch between the BS and SS exists.

In the downlink, the SS is the entity that monitors the quality of the receive signal and therefore knows when its downlink burst profile should change. The BS, however, is the entity in control of the change. Two methods are available to the SS to request a change in downlink burst profile, depending on whether the SS operates in the *grant per connection* (GPC) or *grant per SS* (GPSS) mode (see the "Bandwidth Requests and Grants" section later in this chapter). The first method would typically apply (based on the discretion of the BS scheduling algorithm) only to GPC SSs. In this case, the BS may periodically allocate a station maintenance interval to the

SS. The SS can use the RNG-REQ message to request a change in down-link burst profile. The preferred method is for the SS to transmit a *downlink burst profile change request* (DBPC-REQ). In this case, which is always an option for GPSS SSs and can be an option for GPC SSs, the BS responds with a *downlink burst profile change response* (DBPC-RSP) message confirming or denying the change.

Because messages may be lost due to irrecoverable bit errors, the protocols for changing an SS's downlink burst profile must be carefully structured. The order of the burst profile change actions is different when transitioning to a more robust burst profile than when transitioning to a less robust one. The standard takes advantage of the fact that an SS is always required to listen to more robust portions of the downlink as well as the profile that was negotiated. Figure 8.38 shows a transition to a more robust burst profile. Figure 8.39 shows a transition to a less robust burst profile.

Uplink Scheduling Services   Each connection in the uplink direction is mapped to a *scheduling service*. Each scheduling service is associated with a set of rules imposed on the BS scheduler responsible for allocating the uplink capacity and the request-grant protocol between the SS and the BS. The detailed specification of the rules and the scheduling service used for a particular uplink connection are negotiated at connection setup time.

The scheduling services in IEEE 802.16 are based on those defined for cable modems in the DOCSIS standard [44].

Unsolicited grant service (UGS) is tailored for carrying services that generate fixed units of data periodically. Here the BS schedules regularly, in a preemptive manner, grants of the size negotiated at connection setup, without an explicit request from the SS. This eliminates the overhead and latency of bandwidth requests in order to meet the delay and delay jitter requirements

FIGURE 8.38
Transition to a more robust burst profile.

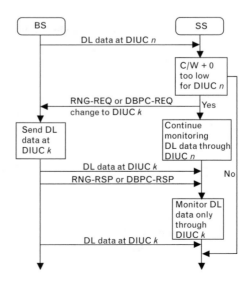

FIGURE 8.39
Transition to a less robust burst profile.

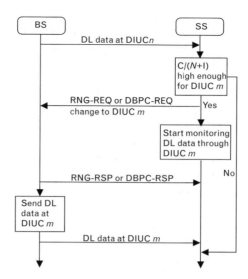

of the underlying service. A practical limit on the delay jitter is set by the frame duration. If more stringent jitter requirements are to be met, output buffering is needed. Services that typically would be carried on a connection with UGS service include ATM CBR and E1/T1 over ATM.

When used with UGS, the grant management subheader includes the poll-me bit (see the "Bandwidth Requests and Grants" section later in this chapter) as well as the slip indicator flag, which allows the SS to report that the transmission queue is backlogged due to factors such as lost grants or clock skew between the IEEE 802.16 system and the outside network. The BS, upon detecting the slip indicator flag, can allocate some additional capacity to the SS, allowing it to recover the normal queue state. Connections configured with UGS are not allowed to utilize random access opportunities for requests.

The real-time polling service is designed to meet the needs of services that are dynamic in nature, but offers periodic dedicated request opportunities to meet real-time requirements. Because the SS issues explicit requests, the protocol overhead and latency are increased, but this capacity is granted only according to the real need of the connection. The real-time polling service is well suited for connections carrying services such as VoIP or streaming video or audio.

The nonreal-time polling service is almost identical to the real-time polling service except that connections may utilize random access transmit opportunities for sending bandwidth requests. Typically, services carried on these connections tolerate longer delays and are rather insensitive to delay jitter. The nonreal-time polling service is suitable for Internet access with a minimum guaranteed rate and for ATM GFR connections.

A best effort service has also been defined. Neither throughput nor delay guarantees are provided. The SS sends requests for bandwidth in either random access slots or dedicated transmission opportunities. The occurrence of dedicated opportunities is subject to network load, and the SS cannot rely on their presence.

Bandwidth Requests and Grants   The IEEE 802.16 MAC accommodates two classes of SS, differentiated by their ability to accept bandwidth grants simply for a connection or for the SS as a whole. Both classes of SS request bandwidth per connection to allow the BS uplink scheduling algorithm to properly consider QoS when allocating bandwidth. With the GPC class of SS, bandwidth is granted explicitly to a connection, and the SS uses the grant only for that connection. RLC and other management protocols use bandwidth explicitly allocated to the management connections.

With the GPSS class, SSs are granted bandwidth aggregated into a single grant to the SS itself. The GPSS SS needs to be more intelligent in its handling of QoS. It will typically use the bandwidth for the connection that requested it, but need not. For instance, if the QoS situation at the SS has changed since the last request, the SS has the option of sending the higher QoS data along with a request to replace this bandwidth stolen from a lower QoS connection. The SS could also use some of the bandwidth to react more quickly to changing environmental conditions by sending, for instance, a DBPC-REQ message.

The two classes of SS allow a trade-off between simplicity and efficiency. The need to explicitly grant extra bandwidth for RLC and requests, coupled with the likelihood of more than one entry per SS, makes GPC less efficient and scalable than GPSS. Additionally, the ability of the GPSS SS to react more quickly to the needs of the PHY and those of connections enhances system performance. GPSS is the only class of SS allowed with the 10- to 66-GHz PHY layer.

With both classes of grants, the IEEE 802.16 MAC uses a self-correcting protocol rather than an acknowledged protocol. This method uses less bandwidth. Furthermore, acknowledged protocols can take additional time, potentially adding delay. There are a number of reasons the bandwidth requested by an SS for a connection may not be available:

- The BS did not see the request due to irrecoverable PHY errors or collision of a contention-based reservation.

- The SS did not see the grant due to irrecoverable PHY errors.

- The BS did not have sufficient bandwidth available.

- The GPSS SS used the bandwidth for another purpose.

In the self-correcting protocol, all of these anomalies are treated the same. After a time-out appropriate for the QoS of the connection (or immediately, if the bandwidth was stolen by the SS for another purpose), the SS simply requests again. For efficiency, most bandwidth requests are incremental; that is, the SS asks for more bandwidth for a connection. However, for the self-correcting bandwidth request–grant mechanism to work correctly, the bandwidth requests must occasionally be aggregate; that is, the SS informs the BS of its total current bandwidth needs for a connection. This allows the BS to reset its perception of the SS's needs without a complicated protocol acknowledging the use of granted bandwidth.

The SS has a plethora of ways to request bandwidth, combining the determinism of unicast polling with the responsiveness of contention-based requests and the efficiency of unsolicited bandwidth. For continuous bandwidth demand, such as with CBR T1/E1 data, the SS need not request bandwidth; the BS grants it unsolicited.

To short circuit the normal polling cycle, any SS with a connection running UGS can use the poll–me bit in the grant management subheader to let the BS know it needs to be polled for bandwidth needs on another connection. The BS may choose to save bandwidth by polling SSs that have unsolicited grant services only when they have set the poll–me bit.

A more conventional way to request bandwidth is to send a bandwidth request MAC PDU that consists of simply the bandwidth request header and no payload. GPSS SSs can send this in any bandwidth allocation they receive. GPC terminals can send it in either a request interval or a data grant interval allocated to their basic connection. A closely related method of requesting data is to use a grant management subheader to piggyback a request for additional bandwidth for the same connection within a MAC PDU.

In addition to polling individual SSs, the BS may issue a broadcast poll by allocating a request interval to the broadcast CID. Similarly, the standard provides a protocol for forming multicast groups to give finer control to contention-based polling. Due to the nondeterministic delay that can be caused by collisions and retries, contention-based requests are allowed only for certain lower QoS classes of service.

Channel Acquisition   The MAC protocol includes an initialization procedure designed to eliminate the need for manual configuration. Upon installation, an SS begins scanning its frequency list to find an operating channel. It may be programmed to register with a specified BS, referring to a programmable BS ID broadcast by each. This feature is useful in dense deployments where the SS might hear a secondary BS due to selective fading or when the SS picks up a sidelobe of a nearby BS antenna.

After deciding on which channel or channel pair to attempt communication, the SS tries to synchronize to the downlink transmission by detecting the periodic frame preambles. Once the physical layer is synchronized, the SS will look for the periodically broadcast DCD and UCD messages that enable the SS to learn the modulation and FEC schemes used on the carrier.

Initial Ranging and Negotiation of SS Capabilities Upon learning what parameters to use for its initial ranging transmissions, the SS will look for initial ranging opportunities by scanning the UL–MAP messages present in every frame. The SS uses a truncated exponential backoff algorithm to determine which initial ranging slot it will use to send a ranging request message. The SS will send the burst using the minimum power setting and will try again with increasingly higher transmission power if it does not receive a ranging response.

Based on the arrival time of the initial ranging request and the measured power of the signal, the BS commands a timing advance and a power adjustment to the SS in the ranging response. The response also provides the SS with the basic and primary management CIDs. Once the timing advance of the SS transmissions has been correctly determined, the ranging procedure for fine-tuning the power can be performed using invited transmissions.

All transmissions up to this point are made using the most robust, and thus least efficient, burst profile. To avoid wasting capacity, the SS next reports its PHY capabilities, including the modulation and coding schemes it supports, and whether, in an FDD system, it is half duplex or full duplex. The BS, in its response, can deny the use of any capability reported by the SS.

SS Authentication and Registration Each SS contains both a manufacturer-issued factory-installed X.509 digital certificate and the certificate of the manufacturer. These certificates, which establish a link between the 48-bit MAC address of the SS and its public RSA key, are sent to the BS by the SS in the authorization request and authentication information messages. The network is able to verify the identity of the SS by checking the certificates and can subsequently check the level of authorization of the SS. If the SS is authorized to join the network, the BS will respond to its request with an authorization reply containing an *authorization key* (AK) encrypted with the SS's public key and used to secure further transactions.

Upon successful authorization, the SS will register with the network. This will establish the secondary management connection of the SS and determine capabilities related to connection setup and MAC operation. The version of IP used on the secondary management connection is also determined during registration.

IP Connectivity After registration, the SS attains an IP address via DHCP and establishes the time of day via the Internet Time Protocol. The DHCP server also provides the address of the TFTP server from which the SS can request a configuration file. This file provides a standard interface for providing vendor-specific configuration information.

Connection Setup IEEE 802.16 uses the concept of service flows to define unidirectional transport of packets on either downlink or uplink. Service flows are characterized by a set of QoS parameters such as latency and jitter. To most efficiently utilize network resources such as bandwidth and memory, 802.16 adopts a two-phase activation model in which resources assigned to a particular admitted service flow may not be actually committed until the service flow is activated. Each admitted or active service flow is mapped to a MAC connection with a unique CID.

In general, service flows in IEEE 802.16 are preprovisioned, and setup of the service flows is initiated by the BS during SS initialization. However, service flows can also be dynamically established by either the BS or the SS. The SS typically initiates service flows only if there is a dynamically signaled connection, such as a *switched virtual connection* (SVC) from an ATM network. The establishment of service flows is performed via a three-way handshaking protocol in which the request for service flow establishment is responded to and the response acknowledged.

In addition to dynamic service establishment, IEEE 802.16 also supports dynamic service changes in which service flow parameters are renegotiated. Like dynamic service flow establishment, service flow changes also follow a similar three-way handshaking protocol.

Privacy Sublayer IEEE 802.16's privacy protocol is based on the *Privacy Key Management* (PKM) protocol of the DOCSIS BPI+ specification [45], but has been enhanced to fit seamlessly into the IEEE 802.16 MAC protocol and to better accommodate stronger cryptographic methods, such as the recently approved advanced encryption standard.

Security Associations PKM is built around the concept of *security associations* (SAs). The SA is a set of cryptographic methods and the associated keying material; that is, it contains the information about which algorithms to apply, which key to use, and so on. Every SS establishes at least one SA during initialization. Each connection, with the exception of the basic and primary management connections, is mapped to an SA either at connection setup time or dynamically during operation.

Cryptographic Methods Currently, the PKM protocol uses X.509 digital certificates with RSA public key encryption for SS authentication and authorization key exchange. For traffic encryption, the *Data Encryption Standard* (DES) running in the *cipher block chaining* (CBC) mode with 56-bit keys is currently mandated. The CBC initialization vector is dependent on the frame counter and differs from frame to frame. To reduce the number of computationally intensive public key operations during normal operation, the transmission encryption keys are exchanged using 3DES with a key exchange key derived from the authorization key.

The PKM protocol messages themselves are authenticated using the *Hashed Message Authentication Code* (HMAC) protocol [46] with SHA-1 [47]. In addition, message authentication in vital MAC functions, such as the connection setup, is provided by the PKM protocol.

8.3.5.3   Summary and Conclusion

The WirelessMAN air interface specified in IEEE Standard 802.16 provides a platform for the development and deployment of standards-based metropolitan area networks providing broadband wireless access in many regulatory environments. The standard is intended to allow for multiple vendors to produce interoperable equipment. However, it also allows for extensive vendor differentiation. For instance, the standard provides the base station with a set of tools to implement efficient scheduling. However, the scheduling algorithms that determine the overall efficiency will differ from vendor to vendor and may be optimized for specific traffic patterns. Likewise, the adaptive burst profile feature allows great control to optimize the efficiency of the PHY transport. Innovative vendors will introduce clever schemes to maximize this opportunity while maintaining interoperability with compliant subscriber stations.

The publication of IEEE Standard 802.16 is a defining moment in which broadband wireless access moves to its second generation and begins its establishment as a mainstream alternative to broadband access. Through the dedicated service of many volunteers, the IEEE 802.16 working group succeeded in quickly designing and forging a standard based on forward-looking technology. IEEE Standard 802.16 is the foundation of the wireless metropolitan area networks of the next few decades.

8.3.6   WWAN Technologies: 2.5G and 3G

In this section we look at a particular aspect of 2.5G/3G support: the Wireless Application Protocol.

8.3.6.1   WAP

WAP is a standard developed by the WAP Forum, a group founded by Nokia, Ericsson, Motorola, and others; the WAP Forum now includes Microsoft, Oracle, IBM, and Intel along with several hundred other companies. This section briefly introduces the concept [48]. The goals of WAP are to be:

- Independent of wireless network standard;
- Open to all;
- Proposed to the appropriate standards bodies;
- Scalable across transport options;

- Scalable across device types;
- Extensible over time to new networks and transports.

WAP is designed to be accessible to (but not limited to) the following:

- GSM-900, GSM-1800, GSM-1900;
- CDMA IS-95;
- TDMA IS-136;
- 3G systems: IMT-2000, UMTS, W-CDMA, wideband IS-95.

WAP defines a communications protocol as well as an application environment. In essence, it is a standardized technology for cross-platform distributed computing. WAP is very similar to the combination of HTML and HTTP except that it adds one very important feature: optimization for low-bandwidth, low-memory, and low-display capability environments. These types of environments include PDAs, wireless phones, pagers, and virtually any other communications device.

8.3.6.2   WAP and the Web

From a certain viewpoint, the WAP approach to content distribution and the Web approach are virtually identical in concept. Both concentrate on distributing content to remote devices using inexpensive, standardized client software. Both rely on back-end servers to handle user authentication, database queries, and intensive processing. Both use markup languages derived from SGML for delivering content to the client. In fact, as WAP continues to grow in support and popularity, it is likely that WAP application developers will make use of their existing Web infrastructure (in the form of application servers) for data storage and retrieval.

WAP (and its parent technology, XML) will serve to highlight the Web's status as the premier *n*-tier application in existence today. WAP allows a further extension of this concept because the existing "server" layers can be reused and extended to reach out to the vast array of wireless devices in business and personal use today. Note that XML, as opposed to HTML, contains no screen formatting instructions; instead, it concentrates on returning structured data that the client can use as it sees fits.

8.3.6.3   How Does It Work?

WAP client applications make requests very similar in concept to the URL concept in use on the Web. As a general example, consider the following explanation (exact details may vary on a vendor-to-vendor basis): A WAP request is routed through a WAP gateway that acts as an intermediary between the "bearer" used by the client (GSM, CDMA, TDMA, and so on) and the computing network that the WAP gateway resides on (TCP/IP in most cases). The gateway then processes the request, retrieves contents or calls CGI scripts, Java servlets, or some other dynamic mechanism, then formats data for return to the client. This data is formatted as *Wireless Markup Language* (WML), a markup language based directly on XML. Once the WML has been prepared (known as a deck), the gateway then sends the completed request back (in binary form due to bandwidth restrictions) to the client for display and/or processing. The client retrieves the first card off of the deck and displays it on the monitor.

The *deck of cards* metaphor is designed specifically to take advantage of small display areas on handheld devices. Instead of continually requesting and retrieving cards (the WAP equivalent of HTML pages), each client request results in the retrieval of a deck of one or more cards. The client device can employ logic via embedded WMLScript (the WAP equivalent of client-side JavaScript) for intelligently processing these cards and the resultant user inputs.

To sum up, the client makes a request. This request is received by a WAP gateway that then processes the request and formulates a reply using WML. When ready, the WML is sent back to the client for display. As mentioned earlier, this is very similar in concept to the standard stateless HTTP transaction involving client Web browsers.

8.3.6.4 Communications Between Client and Server

The WAP Protocol Stack is implemented via a layered approach (similar to the OSI network model). This approach consists of these layers (from top to bottom):

- *Wireless Application Environment* (WAE);
- *Wireless Session Protocol* (WSP);
- *Wireless Transaction Protocol* (WTP);
- *Wireless Transport Layer Security* (WTLS);
- *Wireless Datagram Protocol* (WDP);
- Bearers (GSM, IS-136, CDMA, GPRS, CDPD, and so on).

WSP offers means to:

- Provide HTTP/1.1 functionality: extensible request-reply methods, composite objects, content-type negotiation;
- Exchange client and server session headers;
- Interrupt transactions in process;
- Push content from server to client in an unsynchronized manner;
- Negotiate support for multiple, simultaneous asynchronous transactions.

WTP provides the protocol that allows for interactive browsing (request/response) applications. It supports three transaction classes: unreliable with no result message, reliable with no result message, and reliable with one reliable result message. Essentially, WTP defines the transaction environment in which clients and servers will interact and exchange data.

The WDP layer operates above the bearer layer used by your communications provider. Therefore, this additional layer allows applications to operate transparently over varying bearer services. While WDP uses IP as the routing protocol, unlike the Web, it does not use TCP. Instead, it uses the *User Datagram Protocol* (UDP), which does not require messages to be split into multiple packets and sent out only to be reassembled on the client. Due to the nature of wireless communications, the mobile application must be talking directly to a WAP gateway (as opposed to being routed through myriad WAP access points across the wireless Web), which greatly reduces the overhead required by TCP.

For secure communications, WTLS is available to provide security. It is based on SSL and TLS.

8.3.6.5   WML

WML is a markup language that is based on XML. The official WML specification is developed and maintained by the WAP Forum. This specification defines the syntax, variables, and elements used in a valid WML file. The actual WML 1.1 document type definition is available at: http://www.wapforum.org/DTD/wml_1.1.xml. A valid WML document must correspond to this DTD or it cannot be processed. Included in the WML specification are elements that fall into the following categories: decks/cards, events, tasks, variables, user input, anchors/images/timers, and text formatting. See the WML tutorial for specific examples on using these elements to build applications.

XML is a markup language that has garnered enormous support due its ability to describe data (HTML, meanwhile, is used to describe the *display* of data). Whereas HTML predefines a "canned" set of tags guaranteed to be understood and displayed in a uniform fashion by a Web browser, XML allows the document creator to define any set of tags he or she wishes to. This set of tags is then grouped into a set of grammar "rules" known as the *document type definition* (DTD).

If a phone or other communications device is said to be *WAP capable,* this means that it has a piece of software loaded onto it (known as a *microbrowser*) that fully understands how to handle all entities in the WML 1.1 DTD.

The first statement within an XML document is known as a *prolog.* While the prolog is optional, it consists of two lines of code: the XML declaration (used to define the XML version) and the document type declaration (a pointer to a file that contains this document's DTD). A sample prolog is as follows:

```
<xml version='1.0'>
 <!DOCTYPE wml PUBLIC "-//WAPFORUM//DTD WML 1.1//EN"
    "http://www.wapforum.org/DTD/wml_1.1.xml">
```

Following the prolog, every XML document contains a single element that contains all other subelements and entities. Like HTML all elements are bracketed by the

```
<>
```

and

```
</>
```

characters. As an example,

```
<code><element>datadatadata</element></code>
```

There can only be one document element per document. With WML, the document element is

```
<code><wml></code>
```

All other elements are contained within it.

The two most common ways to store data within an XML document are elements and attributes. Elements are structured items within the document that are denoted by opening and closing element tags. Elements can also contain subelements as well. Attributes, meanwhile, are generally used to describe an element. As an example, consider the following code snippet:

```
<!— This is the Login Card —>
<card id="LoginCard" title="Login">
Please select your user name.
</card>
```

In the code above, the card element contains the id and title attributes. (On a side note, a comment in WML must appear between the tags.) We will make use of the WML-defined elements and their attributes later as we build our examples.

8.3.6.6   Valid WML Elements

WML predefines a set of elements that can be combined together to create a WML document. These elements include can be broken down into two groups: the Deck/Card elements and the Event elements.

Deck/Card Elements
```
wml card template head access meta
```

Event Elements
```
do ontimer onenterforward onenterbackward onpick onevent
postfield
```

Tasks
```
go prev refresh noop
```

Variables
```
setvar
```

User input
```
input select option optgroup fieldset
```

Anchors, Images, and Timers
```
a anchor img timer
```

Text Formatting
```
br p table tr td
```

Each of these elements is entered into the document using the following syntax:

```
<element> element value </element>
```

If an element has no data between it (as is often the case with formatting elements such as <br>), you can save space by entering one tag appended with a \ character (for instance, <br/>).

8.3.6.7   Additional Intelligence via WMLScript

The purpose of WMLScript is to provide client-side procedural logic. It is based on ECMAScript (which is based on Netscape's JavaScript language), however, it has been modified in places to support low-bandwidth communications and thin clients. The inclusion of a scripting language into the base standard was an absolute must. While many Web developers regularly choose not to use client-side JavaScript due to browser incompatibilities (or clients running older browsers), this logic must still be replaced by additional server-side scripts. This involves extra round-trips between clients and servers, which is something all wireless developers want to avoid. WMLScript allows code to be built into files transferred to mobile client so that many of these round-trips can be eliminated. According to the WMLScript specification, some capabilities supported by WMLScript that are not supported by WML are as follows:

• Check the validity of user input.

- Access to facilities of the device. For example, on a phone, allow the programmer to make phone calls, send messages, add phone numbers to the address book, access the SIM card, and so on.

- Generate messages and dialogs locally, thus reducing the need for expensive round-trip to show alerts, error messages, confirmations, and so on.

- Allow extensions to the device software and configuring a device after it has been deployed.

WMLScript is a case-sensitive language that supports standard variable declarations, functions, and other common constructs such as IF–THEN statements, and FOR–WHILE loops. Among the standard's more interesting features are the ability to use external compilation units (via the use URL pragma), access control (via the access pragma), and a set of standard libraries defined by the specification (including the Lang, Float, String, URL, WMLBrowser, and Dialogs libraries). The WMLScript standard also defines a bytecode interpreter since WMLScript code is actually compiled into binary form (by the WAP gateway) before being sent to the client.

8.3.6.8   The Business Case

WAP's biggest business advantage is the prominent communications vendors who have lined up to support it. The ability to build a single application that can be used across a wide range of clients and bearers makes WAP pretty much the only option for mobile handset developers at the current time. Whether this advantage will carry into the future depends on how well vendors continue to cooperate and also on how well standards are followed. Already vendor toolkits are offering proprietary tags that will only work with their microbrowser. Given the history of the computing industry and competition, in general, this was to be expected. However, further differentiation between vendor products and implementations may lead to a fragmented wireless Web.

WAP also could be found lacking if compared to more powerful GUI platforms such as Java, for instance. For now, processor speeds, power requirements, and vendor support are all limiting factors to Java deployment but it's not hard to imagine a day in the near future where Java and WAP exist side by side just as Java and HTML do today. In that circumstance, Java would hold a clear advantage over WAP due to the fact that a single technology could be used to build applications for the complete range of operating devices. Of course, on the flip side, the world is not all Java and there will always be a place for markup languages in lieu of full-blown object-oriented platforms.

8.3.6.9 Conclusion

Some critics have pondered the need for a technology such as WAP in the marketplace. With the widespread proliferation of HTML, is yet another markup language really required? WAP's use of the deck of cards "pattern" and use of binary file distribution meshes well with the display size and bandwidth constraints of typical wireless devices. Scripting support gives us support for client-side user validation and interaction with the portable device again helping to eliminate round-trips to remote servers. WAP is a young technology that is certain to mature as the wireless data industry as a whole matures; however, even as it exists today, it can be used as a powerful tool in every software developer's toolbox.

ENDNOTES

[1] EMC placed the worldwide cellular subscriptions at the 1 billion mark at the end of 2002, up from 722 million at year-end 2000. While acknowledging that cellular subscription growth in Western Europe has been slowing down as indicated by the latest operator reports, the remainder of the world continues to grow with the largest growth being seen in Latin America and the Asia-Pacific regions.

[2] A CAGR of 20% would increase the population from 750 million in 2001 to 2 billion in 2007.

[3] Minoli, D., P. Johnson, and E. Minoli, *Ethernet-Based Metro Area Networks*, New York: McGraw-Hill, Jan. 2002.

[4] Pozar, T., and S. Oliver, *Neighborhood Area Networks*, retrieved from http://www.lns.com/papers/nans101.

[5] Chen, J. C., and J. M. Gilbert, "Measured Performance of 5-GHz 802.11a Wireless LAN Systems," Atheros Communications White Paper, retrieved from http://www.atheros.com, August 2001.

[6] Latour, A., et al., "Wireless Concerns in Europe Suspend '3G' Investments," *Wall Street Journal*, July 26, 2002, p. A3.

[7] For 1-Mbps transmission, BPSK is used (one phase shift for each bit). To accomplish 2-Mbps transmission, QPSK is used. QPSK uses four phase changes (0, 90, 180, and 270 degrees) to encode 2 bits of information in the same bandwidth as BPSK encodes 1. In 1998, Lucent Technologies and Harris Semiconductor jointly proposed to the IEEE a standard called CCK. To achieve 11 Mbps, the vendors had to change the way they went about encoding the data and use a series of codes called *complementary sequences*. Because 64 unique code words can be used to encode the signal, up to 6 bits can be represented by any one particular code word (instead of the 1 bit represented by a Barker symbol). The CCK code word is then modulated with the QPSK technology used in 2-Mbps wireless DSSS radios. This allows for an additional 2 bits of information to be encoded in each symbol. Eight chips are sent for each 6 bits, but each symbol encodes 8 bits because of the QPSK modulation. The spectrum math for 1-Mbps transmission works out as 11 megachips per second times 2 MHz (the null-to-null bandwidth of a BPSK signal) equals 22 MHz of spectrum. Likewise, at 2 Mbps, one is modulating 2 bits per symbol with QPSK, 11 megachips per second, and thus have 22 MHz of spectrum. To send 11 Mbps, one would send 11 million bits per second times 8 chips/8 bits, which equals 11 megachips per second times 2 MHz for QPSK encoding, yielding 22 MHz of frequency spectrum.

From Conover, J., "Anatomy of IEEE 802.11b Wireless," *Network Computing,* Aug. 7, 2000.

[8]   Material from [5].

[9]   Marks, L. M., "Surveying the Wireless Landscape," IBM White Paper, retrieved from http://www.ibm.com, May 2001.

[10]   IEEE 802.16-2001, *IEEE Standard for Local and Metropolitan Area Networks—Part 16: Air Interface for Fixed Broadband Wireless Access Systems,* April 8, 2002.

[11]   IEEE 802.16.2-2001, *IEEE Recommended Practice for Local and Metropolitan Area Networks—Coexistence of Fixed Broadband Wireless Access Systems,* Sept. 10, 2001.

[12]   IEEE 802/16 press release, Dec. 7, 2001.

[13]   The Wireless Developer Network (http://www.wirelessdevnet.com/newswireless/feb012002.html) is an on-line community for information technology professionals interested in mobile computing and communications. WDN's mission is to assist developers, strategists, and managers in bridging the gap between today's desktop and enterprise applications and tomorrow's mobile users communicating via wireless networks; they are interested in supporting the deployment of these evolving technologies through high-quality technical information, news, industry coverage, and commentary.

[14]   Gupta, P., "Mobile Wireless Communications Today," retrieved from http://www.wirelessdevnet.com/channels/wireless/training/mobilewirelesstoday.html.

[15]   Retrieved from www.gsmworld.com/about/history_page1.html.

[16]   Scourias, J., "Overview of the Global System for Mobile Communications," retrieved from http://ccnga.uwaterloo.ca/~jscouria/GSM/gsmreport.html, 1997.

[17]   Pickholdz, R. L., D. L. Schilling, and L. B. Milstein, "Theory of Spread Spectrum Communications—A Tutorial," *IEEE Trans. on Communications,* Vol. COM-30, No. 5, May 1982, pp. 855–860.

[18]   Definition from http://www.atis.org/tg2k/_code-division_multiple_access.html.

[19]   Retrieved from http://www.bee.net/mhendry/vrml/library/cdma/cdma.htm; Viterbi, A., *CDMA: Principles of Spread Spectrum Communication* Reading, MA: Addison-Wesley, 1995; Pickholtz, R. L., D. L. Schilling, and L. B. Milstein, "Revisions to 'Theory of Spread-Spectrum Communications—A Tutorial.'" *IEEE Trans. on Communications,* Vol. COM-32, No. 2, Feb. 1984, pp. 211–212; and also [18].

[20]   Hayes, N., "When Is 3G not 3G?" retrieved from http://www.wirelessdevnet.com, July 25, 2002.

[21]   Retrieved from http://reiter.weblogger.com.

[22]   Thylmann, O., retrieved from www.infosync.no/show.php?id_1246, September 2, 2001.

[23]   Karim, M. R., and M. Sarraf, *W-CDMA and cdma2000 for 3G Mobile Networks,* New York: McGraw-Hill, 2002.

[24]   Retrieved from http://www.novatelwireless.com/pressreleases/2001/story103.htm.

[25]   Retrieved from http://www.imt-2000.org.

[26]   Figures and information contained in this report can be found in EMC's World Cellular License Database at http://www.emc-database.com/website.nsf/index/pr020606#this-page. EMC is a leading provider of specialist market intelligence for the wireless industry.

[27]   Based on information from Siemens, Merrill Lynch, Dresdner Kleinwort, JP Morgan, and Legg Mason.

[28]   "GSM, TDMA Numbers Show America's Growth," *Wireless Week,* Aug. 4, 2002.

[29]   Figures and information contained in this report can be found in [26].

[30]   Material supplied by WECA, specifically their release by Champness, A., *IEEE 802.11 DSSS: The Path to High Speed Wireless Data Networking.*

[31]   Retrieved from http://www.wi-fi.org, July 8, 2002.

[32]   Some of this material is based on Zyren, J., and A. Petrick, "IEEE 802.11 Tutorial," retrieved from WECA Web site, http://www.weca.net.

[33]   Wireless Ethernet Compatibility Alliance, "WEP Security Statement," Sept. 7, 2001.

[34]   "A University of Maryland professor and his graduate student have apparently uncovered serious weaknesses in the next-generation Wi-Fi (Wireless Fidelity) security protocol known as 802.1x. In a paper, 'An Initial Security Analysis of the IEEE 802.1X Standard,' funded by the National Institute of Standards, Professor William Arbaugh and his graduate assistant Arunesh Mishra outline two separate scenarios that nullify the benefits of the new standard and leave Wi-Fi networks wide open to attacks. The use of public access "hotspots" are particularly vulnerable to session hijacking because these locations do not even deploy the rudimentary WEP protocol. 'This problem exists whether you use WEP or not, but it is trivial to exploit if not using WEP,' said Arbaugh." Schwartz, E., "Researchers Claim to Crack Wireless Security," retrieved from http://www.cnn.com/2002/TECH/industry/02/18/wifi .security.idg.

[35]   Vichr, R., "Security and the Prevention of Breaches Therein," Etensity Corporation, April 2002.

[36]   IBM sources, in support of their Wireless Security Auditor (WSA) system.

[37]   Section based on a "802.11a: A Very-High-Speed, Highly Scalable Wireless LAN Standard," Proxim White Paper, retrieved from http://www.proxim.com/learn/library/whitepapers/wp2001-09-highspeed.html. Proxim introduced 802.11a products into its Harmony product family in 2002.

[38]   Keep in mind that other wireless technologies operate in this band, including HomeRF, some cordless phones, microwave ovens, and some specialized lighting fixtures.

[39]   Miller, B., "IEEE 802.11 and Bluetooth Wireless Technology," IBM White Paper, retrieved from http://www.ibm.com, Oct. 2001.

[40]   Malhotra, V., "An Update on the 802.15 WPAN Committee's Work," Etensity, Inc., White Paper, retrieved from http://www. etensity.com, Sept. 2001.

[41]   IEEE 802.15 materials from the group's Web site.

[42]   Material based on published IEEE material, used with permission. Eklund, C., et al., "IEEE Standard 802.16: A Technical Overview of the WirelessMAN™ Air Interface for Broadband Wireless Access," *IEEE Communications Magazine,* June 2002, pp. 102 ff. ©2002, IEEE. All rights reserved.

[43]   IEEE P802.16a/D3-2001, *Draft Amendment to IEEE Standard for Local and Metropolitan Area Networks—Part 16: Air Interface for Fixed Wireless Access Systems—Medium Access Control Modifications and Additional Physical Layers Specifications for 2–11 GHz,* March 25, 2002.

[44]   SCTE DSS 00-05, Data-Over-Cable Service Interface Specification (DOCSIS) SP-RFIv1.1-I05-000714, *Radio Frequency Interface 1.1 Specification,* July 2000.

[45]   SCTE DSS 00-09, DOCSIS SP-BPI+-I06-001215, *Baseline Privacy Plus Interface Specification,* Dec. 2000.

[46]   Krawczyk, H., M. Bellare, and R. Canetti, *HMAC: Keyed-Hashing for Message Authentication,* IETF RFC 2104, Feb. 1997.

[47]   Federal Information Processing Standards Publication 180–1, *Secure Hash Standard,* April 1995.

[48]   Section based on a white paper by the Wireless Developer Network, retrieved from http://www.wirelessdevnet.com/channels/wap/training/wapoverview.html.

WAN Data Services and Technologies: FR, ATM, and MPLS

This chapter looks at basic switched data services available to enterprise users, such as FR service, cell relay service (i.e., ATM), and MPLS connectivity service. Quite a lot has been written on these topics over the years; we focus more on FR, because it is an ubiquitous technology and a widely deployed enterprise networking service. The chapter looks at these topics from the perspective of representing services available to end users. The concept is further addressed in Chapter 10 from a transition perspective to higher speeds. The first edition of this book recognized the importance of switched data services and a major chapter was dedicated to the topic, particularly FR, but also broadband ISDN (ATM) [1]. We continue to view advanced user services as one of the three key pillars of the telecom industry for the rest of the decade in terms of importance and relevance, the other two pillars being broadband (optics and so on) and wireless; this is the motivation for covering these topics in the book.

9.1   Recent Advances in FR

9.1.1   Introduction and Scope

In 2001, FR services had 1.78 million port-equivalents [2] deployed worldwide and $12.7 billion in yearly revenue [3]. More than 30,000 enterprises throughout the world rely on FR to support their day-to-day operations. With market revenue of $12.7 billion in 2001 and $1.7 billion in 1991 when FR services were first launched, these revenue figures represent a CAGR of 22.2% over the 10-year period. FR technology is now ubiquitous and services are offered by hundreds of IXCs, ILECs, CLECs, PTTs, global carriers, and other vendors throughout the world.

The worldwide market for FR services continues to grow steadily with revenue projected to reach $21 billion in 2004, a 19% CAGR from the 2000 revenue levels (see Figure 9.1). Although the United States will continue to represent the most sizable market for FR services, rest of world (ROW) markets are expected to grow more rapidly during the same time

FIGURE 9.1
*Worldwide FR
services.*

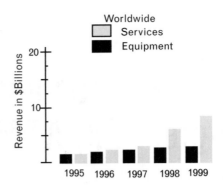

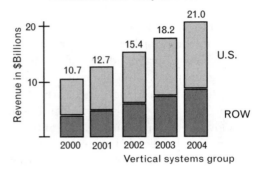

period [4]. U.S. FR revenue is projected to grow at a 16% CAGR to $11.9 billion in 2004, up from $6.6 billion in 2000. Revenue growth in the U.S. market is being driven by the expansion of existing customer networks and by the demand for higher speed ports, new customer installations, and global connectivity. Revenue for the non-U.S. segment will increase based on a CAGR of 22% from $4.1 billion in 2000 to $9.2 billion in 2004. Europe represents the largest regional market in this segment, but countries in the Asia-Pacific and ROW regions are spurring revenue growth. Service providers continue to deploy FR at more carrier nodes worldwide and enterprises are taking advantage of this expansion, contributing to non-U.S. market growth. Frame relay services, however, are not recession-proof: Year-to-year revenue growth did slow worldwide from 2000 to 2001 due to softening economic conditions that delayed planned port additions, particularly in the United States, Canada, and Europe. Frame relay surpassed X.25 in 1997 as the world's second most deployed data networking service behind *private-line* (PL) services. Recently ISPs have used FR as a customer access technology, for port aggregation, and for local traffic switching [5]. ILEC frame relay services are generally tariffed in a distance-insensitive manner. These requirements made switches sold to ISPs the fastest growing segment of the FR equipment market in the late 1990s.

During the next few years, FR faces market challenges from emerging technologies such as DSL, dedicated IP VPNs, GbE-based WAN services [6], and other access options, which are services optimized to handle the increasing amount of IP traffic on today's networks. However, observers see limited erosion of the FR customer base to the other technologies through 2004. The fact remains that, fundamentally, FR is a low-speed service, so other low-speed technologies such as DSL become competitive technologies. Hundreds (if not thousands) of articles appeared in the early 2000s about alleged bandwidth glut; however, such a glut has not translated at all into increased affordable bandwidth to the end user. The upshot of this is that users continue to have to rely, in large measure, on sub–T1 services, such as FR.

FR was developed by the ITU-T and by carriers as an "improved" packet technology that could be offered in conjunction with ISDN, as a replacement for the more protocol-burdensome X.25 technology. Soon after its development in the late 1980s, ISDN and FR went "their separate ways." In the 1990–1992 time frame an advocacy industry consortium comprised of key equipment manufacturers and service providers was launched; this consortium is the *FR Forum* (FRF) (now known as the MPLS/Frame Relay Alliance). What made FR so successful in the United States was the fact that the *regional Bell Operating Companies* (RBOCs) in the late 1980s were very concerned about the possibility of being "bypassed," thereby losing the revenue that they enjoyed from the significant deployment of PL services that had occurred up to that point. To counter that possibility, carriers priced FR services aggressively, thereby giving these services the impetus they required for large-scale commercial introduction. Driven by cost savings relative to PLs or dedicated X.25 services, FR steadily gained market acceptance as an enterprise data networking solution. Today, FR is a mature global transport service with an outlook for continued growth, particularly in conjunction with some enhancements described in Chapter 10. FR technology continues to provide secure networking for point-to-point and point-to-multipoint applications and for private and public enterprises. Considering that market and economic conditions are exerting negative pressure on buying decisions and new technology deployment, some would say that FR is even more viable today. Customers need cost–effective, simple, and manageable communications solutions; FR is the de facto solution that satisfies this need.

According to the FR Forum [7], FR will continue to gain acceptance. In recent years both enterprise customers and service providers have gravitated toward retaining their installed base of legacy circuits and transport infrastructures. A majority of service providers have shown themselves to be reluctant to take chances on the deployment of new technologies that require significant investments for themselves and for their enterprise customers. Planners at end–user organizations are also wary of new entrants and new carriers, considering the carnage of the early 2000s. While adhering to the existing offerings of service providers, enterprises are attempting to balance their cost of operations in meeting consumer demands. Frame relay satisfies the service providers' and the enterprises' current requirements, and at the same time, it provides a migration path to new technologies. In addition to the intranet applications of the past, FR provides a service platform and backbone for provisioning access to the Internet. FR has five key applications of focus:

1. LAN-to-LAN interconnection;
2. SNA connectivity;
3. Internet access;
4. Voice/fax/video support;
5. Branch connectivity to central sites.

The FRF is working with other forums to establish possible migration strategies. The FRF's work with the MPLS Forum, the ATM Forum, the ITU, and the IETF will augment the adoption and continuing emphasis of FR. The refinements and enhancements that the FRF is undertaking with the ATM Forum are aimed at increasing the enterprise's comfort level with the integration of FR and ATM networks. The FRF is also working with the MPLS Forum to create an interoperability architecture; this effort is similar to the FRF/ATM Forum work. These efforts provide the FR-centric enterprise with the ability to move in its own time frame to advanced technologies that are being implemented in core networks. FRF/MPLS work enables the enterprise to take advantage of the QoS mechanisms that will be in place in the core network. Enterprise users will have greater latitude to support their telecommunication needs, for example, video and voice, over a FR network.

In the discussion that follows, we focus on FRF advances regarding this technology. However, the reader should keep in mind that the fundamental standardization work is undertaken by the ITU-T.

9.1.2   Basic FR Concepts

Frame relay technology is based on the concept of VCs [8]. VCs are two-way, software-defined data paths between two ports that act as PL replacements in the network. Customers have a physical link to the closest provider's node, typically a T1 line [9]; on the customer premises this link terminates on a router [or possibly on a *FR packet assembler/disassembler* (FRAD)]. This physical connectivity to the "network cloud" happens at both ends of the link. The "network cloud" is deployed and maintained by the carrier, and it consists of FR/ATM switches interconnected with shared high-speed (ATM) trunks. The customer only has a virtual connection, not a physical connection through the "network cloud"; the only physical resources assigned to the user are at the access level. Figure 9.2 depicts baseline FR reference models.

Although two types of VC FR connections are currently defined, *switched virtual circuits* (SVCs) and *permanent virtual circuits* (PVCs), PVCs were the original service offering [10]. As a result, PVCs have been more commonly used because they are superior to SVCs for intranet/enterprise applications. SVC products and services, however, are gaining some interest for more "diffused" applications. PVCs are set up by a network operator—whether a private network or a service provider—via a network management system. PVCs are initially defined as a connection between two sites or endpoints. New PVCs may be added when there is a demand for new sites, additional bandwidth, alternate routing, or when new applications require existing ports to talk to one another. PVCs are fixed paths; they are not available on demand or on a call-by-call basis. Although the actual path taken through the network may vary from time to time, such as when automatic rerouting takes place, the beginning and end of the circuit will not change. In this sense, the PVC is like a dedicated point-to-point PL circuit. PVCs are popular because they provide a cost-effective alternative to leased lines. Provisioning of PVCs requires thorough planning, a knowledge of traffic patterns, and bandwidth utilization. Lead times for the turnup of a PVC are fixed, which limits the flexibility of adding service when required for short usage periods.

SVCs are available on a call-by-call (session-by-session) basis. Establishing a call by using the SVC signaling protocol (Q.933) is comparable to setting up an ISDN call, in which users specify a destination address similar to a phone number. This process, however, can be resource intensive. First, the network must dynamically establish connections based on requests by many users (as opposed to PVCs in which a central network operator configures the network) and allocate bandwidth based on the

FIGURE 9.2
Rereference models.

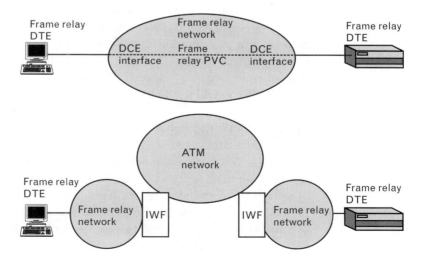

users' requests. Secondly, the network must track the calls and bill according to the amount of service provided. While PVCs offer the statistical bandwidth gain of FR, SVCs deliver the any-to-any connectivity that can be utilized in some applications. Implementing SVCs in the network is more complex than using PVCs, although this implementation is more transparent to end users on the intranet. As noted, although SVCs were defined in the initial FR specifications, they were not implemented by the first carriers or vendors of FR. More recently, SVC-based service have seen some deployment.

The issue related to SVC service is that a router serving a connectionless (IP-based) session is, arguably, ill served by a connection-oriented service that requires a connection to be established across the network every time a user packet, a short DNS query, a UDP packet, or a routing update packet is transacted, and then dropped, only to be, in all probability, reestablished a few milliseconds later. This behavior would be extremely resource intensive on the switch, causing it to be fairly expensive because of the additional processing capacity, memory, and so on. Because of the industry bandwidth glut (or simply a reduced per-bit transmission cost), there are no imperatives to be extremely parsimonious in the use of allocated bandwidth-based resources; users may as well "nail-up" a PVC across the network. Fundamentally, anyone familiar with data applications as well as with signaling [11] understands that in the data world (in enterprise networking application) the sources and sinks of information (that is, the boundary routers) are located at fixed and well-known geographic locations, at least for all but the outermost tier of users. For example, a company may have five fixed operational centers: a headquarters in New York City; a back-office in Hoboken, New Jersey; a disaster recovery site in Roseland, New Jersey; a major branch in Chicago; and a West Coast location in San Francisco. The data always flow between these five locations. The routes are well established. Data environments are very much different from voice environments (where in fact the signaling protocols for FR originated): In a voice environment, the user (say, of a cell phone) could be making what could appear to be random calls to all sorts of locations. Maybe a bill came in the mail, and having a question about it, the user calls the establishment that issued it. Maybe a "missed call" from an unknown party shows up on the cell phone, and the user calls back to see who it was. Maybe the user saw an advertisement for a "buy 2 days in a hotel in Hawaii and get 2 days free," so he calls the islands to check this out, and so on. This is *not* the way an enterprise data network works. The concept of SVC was developed in the mid-1970s, before the development of router technology. SVCs do not easily support routers or router-based backbones, which are the staple of today's Internet and intranets. If anything, SVC could find use in edge-focused applications such as point of sale and mobile users. Finally, the idea of a router existing in an SVC environment by nailing up "long-duration SVCs" is counterproductive to the very concept of an SVC. Furthermore, companies do not want to be in a situation where they have instructions such as "the first person to arrive in the morning should fire up SVC1 from X to Y and SVC2 from Z to W, and SVC3 from P to Q, and so on."

To support the relay service both the *data terminating equipment* (DTE) and *data circuit-terminating equipment* (DCE) must support a protocol stack across the UNI that covers layers 1 and 2 in the *user plane* (U-plane) and layers 1 through 7 in the *control plane* (C-plane—this for SVC handling). FRF1.2 IA, *PVC User-to-Network Interface (UNI) Implementation Agreement—July 2000*, describes the flow of information from the network to the user from the perspective of the user device. Refer to Figure 9.3 for an illustration of the UNI/NNI reference model.

Physical Layer Interfaces Guidelines

Physical layer interface guidelines are provided in the Physical Interface Implementation Agreement (FRF.14).

Data Transfer

Implementations for the FR UNI U-plane are based on ITU-T Q.922 Annex A. A maximum FR information field size of 1,600 octets is expected to be supported by the network and the user. In addition, maximum information field sizes of less than or greater than 1,600 octets may be agreed to between networks and users at subscription time. The 2-octet address format is supported with *data link connection identifier* (DLCI) values. The 4-octet address format may optionally be supported, except that the range 1–15 is reserved, and virtual circuit identification begins at DLCI 16. Other address structure variables, such as the *command/response* (C/R), *discard eligibility indicator* (DE), *forward explicit congestion notification* (FECN), and *backward explicit congestion notification* (BECN) bits, and their usage are as specified in Q.922 Annex A.

Congestion Control

The congestion control strategy for FR is defined in ITU I.370. The following implementation agreements apply to user equipment and network equipment respectively:

- I.370 Section 1.5.2 Network Response to Congestion & Explicit Congestion Signals (Q.922 §A.6.2.1): Mandatory procedures of ITU

FIGURE 9.3
UNI/NNI reference model.

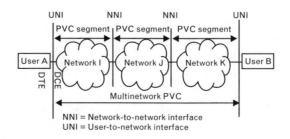
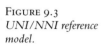

NNI = Network-to-network interface
UNI = User-to-network interface

I.370 are implemented. When implemented, rate enforcement using the DE indicator and/or setting of the FECN and BECN indicators should be implemented according to ITU I.370.

- I.370 Section 1.5.3 User Response to Congestion & Q.922 §§ A.6.1, A.6.2: User equipment reaction is dependent on the protocols operating over the DLC sublayer. The procedures of ITU Q.922 Appendix I should be implemented where appropriate.

PVC Procedures

User devices (and the network) must implement the mandatory procedures of the Q.933 Annex A (1995). By bilateral agreement, optional procedures of Annex A of the revised ITU Q.933 may be implemented. Note that the number of PVCs that can be supported by Annex A is limited by the maximum frame size that can be supported by the user device and the network on the bearer channel (e.g., when the maximum FR information field size is 1,600 octets, then a maximum of 317 PVC status information elements may be encoded in the STATUS message).

SVC Procedures

These are covered in FRF.4.1, FR SVC UNI Implementation Agreement. The material that follows amplifies some of these concepts.

9.1.2.1   The FR Header and DLCI

Figure 9.4 illustrates the format of some of the most common protocol PDUs, while Figure 9.5 shows the FR frame structure and its header. The FR header contains a 10-bit field, called the DLCI. The DLCI is the FR VC number (with local significance) that corresponds to a particular destination. (In the case of LAN–WAN internetworking, the DLCI denotes the port to which the destination LAN is attached.) As shown in Figure 9.6, the routing tables at each intervening FR switch in the private or carrier FR network route the frames to the proper destination. In figures illustrating FR networks, the user devices are often shown as LAN routers, since this is a common FR application. They could also be LAN bridges, hosts, mainframe front-end processors, FRADs, or any other device with a FR interface.
The DLCI allows data arriving at a FR switch to be sent across the network using a simple, three-step process:

1. Check the integrity of the frame using the *frame check sequence* (FCS); if it indicates an error, discard the frame.

2. Look up the DLCI in a table; if the DLCI is not defined for this link, discard the frame.

3. Relay the frame toward its destination by sending it out the port or trunk specified in the table.

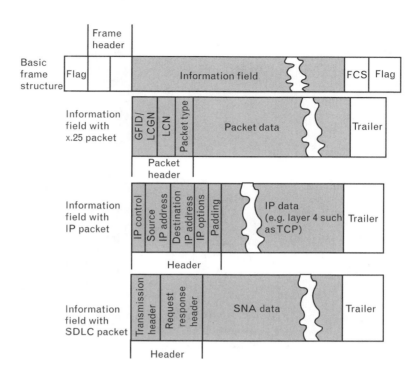

FIGURE 9.4
*Basic frame structure
for several popular
synchronous
communications
protocols.*

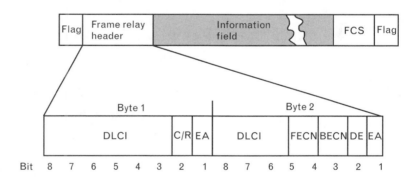

FIGURE 9.5
*Frame structure and
header for FR.*

DLCI = Data link connection identifier
C/R   = Command/response field bit (application specific-not modified by network)
FECN = Forward explicit congestion notification
BECN = Backward explicit congestion notification
DE   = Discard eligibility indicator
EA   = Extension bit (allows indication of 3 or 4 byte header)

When FR was first proposed, it was based on a simple rule: Keep the network protocol simple and let the higher layer protocols of the end devices worry about the other problems. To simplify FR as much as possible, one simple rule exists: If there is any problem with a frame, simply discard it. There are two principal reasons why FR data might be discarded:

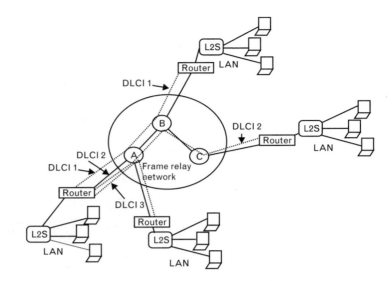

detection of errors in the data and congestion (the network is overloaded). The question then becomes "But how can the network discard frames without impacting the integrity of the communications?" The answer lies in the existence of intelligence in the endpoint devices, such as PCs, workstations, and hosts. These endpoint devices operate with multilevel protocols that detect and recover from loss of data in the network, specifically TCP.

9.1.2.2   The Signaling Mechanisms

Signaling mechanisms address three important issues:

1.   Allowing the network to signal that congestion exists;
2.   Telling the status of connections (PVCs);
3.   Setting up sessions (SVCs).

Although these mechanisms add complexity to FR, the standards have an important provision that allows basic FR to remain simple: The use of signaling mechanisms is optional. Without the signaling mechanisms, the resulting FR interface would still be compliant with the standard; with the signaling mechanisms, however, the throughput of the network, the response time to users, and the efficiency of line and host usage are improved.

9.1.2.3   Congestion Notification Mechanisms

Packet-based networks, which rely on statistical multiplexing and over-booking, typically utilize congestion management. In frame relay, congestion management mechanisms, like the other signaling mechanisms, are optional for compliance, but they will affect performance. The traffic entering the network is called the *offered load*. As the offered load increases, the actual network throughput typically increases linearly. The beginning of congestion occurs when the network cannot keep up with the entering traffic and begins flow control. If the entering traffic continues to increase, it reaches a state of severe congestion, in which the actual effective throughput of the network starts to decrease due to the number of retransmissions. This causes a frame to be sent into the network multiple times before successfully making it through. In severe congestion, the overall network throughput can diminish, and the only way to recover is for the user devices to reduce their traffic. For that reason, several mechanisms have been developed to notify the user devices, such as routers, that congestion is occurring and that they should reduce their offered load. The network should be able to detect when it is approaching congestion, rather than waiting until the point of saturation is reached before notifying the end devices to reduce traffic. Early notification can avoid severe congestion altogether. Two types of mechanisms are used to minimize, detect, and recover from congestion situations, in effect providing flow control: (1) explicit congestion notification and (2) discard eligibility.

Another mechanism that may be employed by end-user devices is explicit congestion notification. These mechanisms use specific bits contained within the header of each frame. The locations of these specific bits (FECN, BECN, and DE) are shown in Figure 9.4.

Explicit Congestion Notification

The first mechanism uses two *explicit congestion notification* (ECN) bits in the FR header, the FECN and the BECN bits, as introduced earlier. Figure 9.7 depicts the use of these bits. Let us suppose node B is approaching a congestion condition. This could be caused by a temporary peak in traffic coming into the node from various sources or by a peak in the amount of traffic on the link between B and C. Here is how forward congestion notification would occur:

FIGURE 9.7
The use of FECN and BECN in explicit congestion notification.

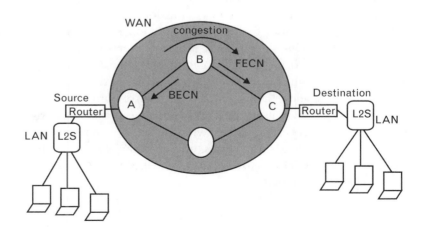

- Node B would detect the onset of congestion based on internal measures such as memory buffer usage or queue length.

- Node B would signal node C (the downstream node, toward the destination) of the congestion by changing the FECN contained within the frames destined for node C from 0 to 1.

- All interim downstream nodes, as well as the attached user device, would thus learn that congestion is occurring on the DLCI(s) affected.

Depending on the protocols used and the capabilities of the CPE device and the network switches, it is sometimes more useful to notify the source of the traffic that there is congestion, so the source can slow down until congestion subsides. (This assumes that the source is capable of responding to receipt of the congestion notification signals.) This is called *backward congestion notification*. This is how backward congestion notification occurs:

- Node B watches for frames coming in the other direction on the connection.

- Node B sets the backward ECN bit within those frames to signal the upstream node(s) and the attached user device.

The FECN and BECN process can take place simultaneously on multiple DLCIs in response to congestion on a given line or node, thus notifying multiple sources and destinations. The ECN bits represent an important tool for minimizing serious congestion conditions.

Implicit Congestion Notification
Some upper layer protocols, such as TCP, operating in the end devices have an implicit form of congestion detection. These protocols can infer that congestion is occurring by an increase in round-trip delay or by detection of the loss of a frame, for example. Reliance on network traffic characteristics to indicate congestion is known as *implicit congestion notification*. These upper layer protocols were developed to run effectively over networks whose capacity was undetermined. Such protocols limit the rate at which they send traffic onto the network by means of a "window," which allows only a limited number of frames to be sent before an acknowledgment is received. When it appears that congestion is occurring, the protocol can reduce its window size, which reduces the load on the network. As congestion abates, the window size is gradually increased. The same window-size adjustment is also the normal way for the end-user devices to respond to explicit congestion notification—FECN and BECN. Implicit and explicit congestion notification are complementary and can be used together for best results.

Discard Eligibility

Frame relay standards state that the user device should reduce its traffic in response to congestion notification. Implementation of the recommended actions by the user device will result in a decrease in the traffic into the network, thereby reducing congestion. However, if the user device is incapable of responding to the signaling mechanisms, it might simply ignore the congestion signal and continue to transmit data at the same rate as before. This would lead to continued or increased congestion. In this case, how does the network protect itself? The answer is found in the basic rule of FR: If there is a problem, discard the data. Therefore, if congestion causes an overload, more frames will be discarded. This will lengthen response times and reduce overall network throughput, but the network will not fail.

When congestion does occur, the nodes must decide which frames to discard. The simplest approach is to select frames at random. The drawback of this approach is that it maximizes the number of endpoints that must initiate error recovery due to missing frames. A better method is to predetermine which frames can be discarded. This approach is accomplished through the use of the *committed information rate* (CIR). CIR is the average information capacity of the VC. When one subscribes to a FR service from a carrier, one specifies a CIR value that is chosen depending on how much information capacity one determines the network will need.

CIR is the main traffic engineering tool users have to affect the performance of the network. CIR choice can impact delay because it is used to handle congestion, which can impact throughput and delay [12]. CIR is the committed rate (in bits per second) at which the ingress access interface and egress access interfaces transfer information to the destination FR end system under normal conditions. The rate is averaged over a minimum time interval Tc. When users send data over a FR access interface, the amount of data sent during this time interval is counted. If the amount of data sent exceeds the CIR, the traffic exceeding the CIR is marked to be discard eligible, indicating the network may potentially discard this traffic in the event of congestion. CIR is a tool used by FR networks to:

- Regulate the flow of traffic;
- Allow users some choice in user throughput rate;
- Potentially determine the pricing structure for the FR service (e.g., many FR services charge less for lower CIRs than for higher CIRs).

In each frame header, there is a bit called the *discard eligibility bit* (see Figure 9.8). A DE bit is set to 1 by the CPE device or the network switch when the frame is above the CIR. When the DE bit is set to l, it makes the frame eligible for discard in response to situations of congestion. A frame with a DE bit of 1 is discarded in advance of nondiscard-eligible data (those frames with a DE bit set to 0). When the discard of DE-eligible data, by itself, is not sufficient to relieve severe congestion, additional incoming

FIGURE 9.8
*PVC status signaling
per LMI specification.*

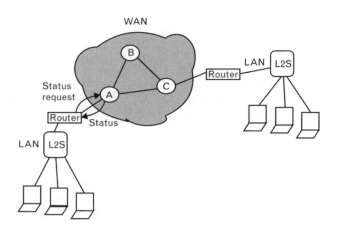

frames are discarded without regard to the setting of the DE bit. The DE bit can be set by the user or by the network.

Frame relay networks are engineered to deliver CIR-conformant traffic with a higher level of assurance than discard-eligible traffic. Therefore, users need to select CIR values high enough to ensure their mission-critical applications get through and only burst above CIR for traffic that is potentially eligible for discard and/or for a very short period of time.

Note that the CIR can be different for traffic in each direction of the VC. In fact, the VC can be thought of as two simplex VCs that are provisioned in each direction to create a full-duplex path. Also note that other implementation-specific mechanisms may be present that can mark traffic that exceeds the bandwidth contract (specifically, CIR, burst) in addition to the inband DE bit; these implementation-specific mechanisms can be used to make decisions about how traffic might be treated and dropped in the presence of congestion in the network.

9.1.2.4   Status of Connections (PVCs and SVCs)

The next type of optional signaling mechanism defines how the two sides of a FR interface (e.g., the network and the router) communicate with each other regarding the status of the interface and the various PVCs on that interface. Again, these are optional parameters. It is possible to implement a FR interface and pass data without implementing these parameters. This signaling mechanism simply enables the user to retrieve more information about the status of the network connection.

This status information is accomplished through the use of special management frames with a unique DLCI address that may be passed between the network and the access device. These frames monitor the status of the connection and provide the following information:

- Whether the interface is still active—this is called a "keep alive" or "heartbeat" signal;

• The valid DLCIs defined for that interface;

• The status of each virtual circuit; for example, if it is congested or not.

The connection status mechanism is termed the *local management interface* (LMI) specification. There are currently three versions of the LMI specification, as shown in Table 9.1.

Although LMI was used colloquially for the FRF.1 IA, it may also be used as a generic term to refer to any and all of the protocols. The revised FR Forum IA FRF.1.1 calls for the mandatory implementation of Annex A of ITU Q.933. Each version includes a slightly different use of the management protocol. Virtually all equipment vendors support LMI and most support Annex D, whereas Annex A is supported by fewer vendors. To ensure interoperability when an organization's network consists of equipment from different vendors, the same version of the management protocol must be used at each end of the FR link. FRF.1 was superceded by FRF.1.1, ANSI T1.617, and ITU Q.933 referenced in FRF.1.1.

The first definition for PVC status signaling was in the LMI specification. The protocol defined for the LMI provides for a "status inquiry" message that the user device (e.g., router) can send, either simply as a "keep alive" message to inform the network that the connection to the router is still up, or as a request for a report on the status of the PVCs on that port. The network then responds with a "status" message, either in the form of a "keep alive" response or in the form of a full report on the PVCs (see Figure 9.8). An additional optional message, "status update," is also defined that enables the network to provide an unsolicited report of a change in PVC status. Notice that the LMI status query only provides for one-way querying and one-way response, meaning that only the user device (e.g., router) can send a "status inquiry" message, and only the network can respond with a "status" message. Although this approach was simple to implement, it resulted in some limitations in functionality. Using status inquiries in this manner, both sides of the interface are unable to provide the same commands and responses. Most notably, it addressed only the UNI and would not work in a NNI due to the one-way communications of the interface. UNI provides the end device interface to the network. NNI provides the ability for networks to query and respond to one

TABLE 9.1   LMI Specifications

Protocol	Specification
LMI	Frame Relay Forum Implementation Agreement (IA)FRF.1 superceded by FRF.1.1
Annex D	ANSI T1.617
Annex A	ITU Q.933 referenced in FRF.1.1

another. When only UNIs are available, this could lead to problems within hybrid private/public networks, in which a private network node would have a FR NNI interface to a public FR service.

Therefore, just before final approval of the standard for FR signaling, ANSI extended the standard to provide a bidirectional mechanism for PVC status signaling that is symmetric. To ensure interoperability in a multivendor network environment, the same version of management protocol must be at each end of the FR link. The bidirectional mechanism provides the ability for both sides of the interface to issue the same queries and responses. This mechanism is contained in Annex D of T1.617, known simply as Annex D. Annex D works in both the UNI and NNI interfaces. In contrast to the LMI (which uses DLCI 1023), Annex D reserves DLCI 0 for PVC status signaling. The current requirement in FRF.1.1, Annex A signaling, is similar to Annex D and also uses DLCI 0.

9.1.2.5 SVC Implementation Agreement

The SVC implementation agreement of the Frame Relay Forum is based on existing SVC standards in ANSI and ITU-T. The applicable SVC standards are ANSI T1.617 and ITU-T Q.933. (These two documents are the basis for Q.2931, the standard for access signaling for ATM as well as for the PVC management procedures for FR.) The SVC implementation agreement can enable expanded service in FR networks. Use in internal networks involves implementing SVCs that are internal to a public or private network. The SVCs would remain transparent to the users who maintain their user-to-network interface PVCs, for example, in the case of disaster recovery. In wide area networks SVCs may be used over large geographic areas such as transatlantic applications, where cost continues to remain an issue.

9.1.2.6 ISDN and Switched Access for PVCs and SVCs

Access on demand for PVCs and SVCs, whether via ISDN or switched access, is another method to reach the FR network. Access on demand holds a great deal of promise for remote locations accessing high-speed FR implementations. In switched access, a circuit-switched connection to the FR switch can be established using the existing voice network. An indication is then sent to the switch that a FR call is being established; the switch makes the connection and bills the call appropriately. The customer pays only for the use of the local loop when needed, without requiring PVCs at the user-to-network interface. The same benefits are true for ISDN access and E.164 addressing and lead to true, any-to-any connectivity through ISDN or switched access.

9.1.2.7 Interworking

Frame relay users often have a requirement to interwork other technologies with their FR network, for example, to address increasing bandwidth needs and the requirements of delay-sensitive traffic. Because very few networks today are based on a single technology, the implementation agreements and standards that provide for interoperability among technologies have become more important than ever. Some of the most common scenarios for corporate and carrier networks include interworking FR and ATM as well as Internet access. The vast majority of carrier FR networks today are based on frame-to-ATM network interworking. In this case, the end user uses FR, and the network transports the FR traffic over an ATM network. The FR Forum has a number of implementation agreements that specify the components for interworking between FR and ATM (these documents were written in cooperation with the ATM Forum and the ITU to ensure compatibility and adherence to related specifications):

- FR/ATM Network Interworking Implementation Agreement (FRF.5) refers to FR service user traffic transported over an ATM network. The FR user does not (and need not) know that the underlying transport technology being used is ATM.

- FR/ATM Service Interworking (FRASI) Implementation Agreement for PVC (FRF.8.1) refers to the situation where the FR service user interworks with an ATM service user.

9.1.3   Key FRF Implementation Agreements

The industry consensus about the need for FR to supplement existing switching technologies resulted in rapid development of industry standards. As implied earlier, two major standards organizations are active in this area: ANSI and ITU-T, which was formerly called the Consultative Committee for International Telephone and Telegraph (CCITT). The following is a list of key ITU-T standards:

- CCITT Recommendation I.122, *Framework for Providing Additional Packet Mode Bearer Services,* ITU, Geneva, 1988.

- CCITT Recommendation I.233.1, *FR Bearer Services,* 1991.

- Recommendation Q.922, *ISDN Data Link Layer Specification for Frame Mode Bearer Services,* ITU, Geneva, 1993.

- ITU-T Recommendation Q.921, *ISDN User-Network Interface-Data Link Layer Specification,* ITU, Geneva, 1997.

- ITU Recommendation Q.933, *ISDN Signaling Specifications for Frame Mode Switched and Permanent Virtual Connections Control and Status Monitoring,* ITU, Geneva, 1995.

- ITU-T Recommendation Q.931, *ISDN User-Network Interface Layer 3 Specification for Basic Call Control,* ITU, Geneva, 1993.

- ITU Recommendation I.370, *Congestion Management for the ISDN FRing Bearer Service,* 1991.
- ITU-T Recommendation I.372, *Frame Relaying Bearer Service Network-to-Network Interface Requirements,* ITU, Geneva, 1993.
- ITU-T I.555, Frame Relaying Bearer Service Interworking, ITU, Geneva, 1997.
- *ITU-T Implementor's Guide for Recommendation Q.922,* ITU, Geneva, December 2000.
- ITU-T Recommendation E.164 / I.331, *The International Public Telecommunication Numbering Plan,* ITU, Geneva, 1997.
- ITU-T Recommendation X.121, *International Numbering Plan for Public Data Networks,* ITU, Geneva, 1996.

The ITU-T (then called CCITT) approved Recommendation I.122, *Framework for Additional Packet Mode Bearer Services,* in 1988. I.122 was part of a series of ISDN-related specifications. ISDN developers had been using LAPD to carry the signaling information on the D-channel of ISDN. (LAPD is defined in ITU Recommendation Q.921.) Developers recognized that LAPD had characteristics that could be very useful in other applications. One of these characteristics is that it has provisions for multiplexing virtual circuits at level 2, the frame level, instead of level 3, the packet level, as in X.25. Therefore, I.122 was written to provide a general framework outlining how such a protocol might be used in applications other than ISDN signaling. At that point, rapid progress began, led by an ANSI committee known as T1S1. This work resulted in a set of standards defining FR. The essential FR standards are shown in Table 9.2.

T1.606 was approved early in 1990. Thanks to the work of the ANSI committee, coupled with a clear mandate from the market, the remaining ANSI standards sped through the stages of the standards process to receive complete approval in 1991. The fast pace of FR standards work at ANSI was matched by an outstanding degree of cooperation and consensus in the international arena. As a result, the ITU-T recommendations for FR are in

TABLE 9.2   FR STANDARDS

DESCRIPTION	ANSI STANDARD	STATUS	ITU STANDARD	STATUS
Service Description	T1.606	Standard	1.233	Approved
Core Aspects	T1.618 (previously known as T1.6ca)	Standard	Q.922 Annex A	Approved
Access Signaling	T1.617 (previously known as T1.6fr)	Standard	Q.933	Approved

alignment with the ANSI standards and have also moved rapidly through the approval process. However, protocol detail clarifications always seem to be needed; this issue is addressed via implementer's agreements.

When the FR Forum was established in 1991, its goals were identified in its charter [13]: To promote the acceptance of FR, to support the development of national and international standards (through liaison with accredited standards–developing entities), and to promote the interoperability of service between carrier and vendor equipment and services. The fulfillment of those goals translated into work by the FR Forum's technical committee to develop *implementation agreements* (IAs) specifying the communications procedures for basic FR networking. At that time, the consensus was that once IAs had been created to address the basic procedures such as the UNI and the NNI, the majority of the work of the technical committee would be complete. Regional committees operate in Europe and Asia, and interest has been expressed in creating another regional committee in Latin America.

Table 9.3 shows the key FRF IAs that have been developed over the years. As FR has evolved into a true multiservice technology with worldwide application, the FR Forum has also evolved. In recent years the technical committee work has focused on advanced applications with more

TABLE 9.3   FRF IMPLEMENTATION AGREEMENTS, CIRCA 2002

FRF.1.2, *PVC User-to-Network Interface (UNI) Implementation Agreement,* July 2000

FRF.2.2, *Relay Network-to-Network Interface (NNI) Implementation Agreement,* March 2002

FRF.3.2, *FR Multiprotocol Encapsulation Implementation Agreement,* April 2000

FRF.4.1, *SVC User-to-Network Interface (UNI) Implementation Agreement,* January 2000

FRF.5, *FR/ATM PVC Network Interworking Implementation,* December 1994

FRF.6, *FR Service Customer Network Management Implementation Agreement (MIB),* March 1994

FRF.7, *FR PVC Multicast Service and Protocol Description Implementation Agreement,* October 1994

FRF.8.1, *FR/ATM PVC Service Interworking Implementation Agreement,* February 2000

FRF.9, *Data Compression Over FR Implementation Agreement,* January 1996

FRF.10.1, *FR Network-to-Network SVC Implementation Agreement,* September 1996

FRF.11.1, *Voice over FR Implementation Agreement,* May 1997; *Annex J* added March 1999

FRF.12, *FR Fragmentation Implementation Agreement,* December 1997

FRF.13, *Service Level Definitions Implementation Agreement,* August 1998

FRF.14, *Physical Layer Interface Implementation Agreement,* December 1998

FRF.15, *End-to-End Multilink FR Implementation Agreement,* August 1999

FRF.16.1, *Multilink FR UNI/NNI Implementation Agreement,* May 2002

FRF.17, *FR Privacy Implementation Agreement,* January 2000

FRF.18, *Network-to-Network FR/ATM SVC Service Interworking Implementation Agreement,* April 2000

FRF.19, *FR Operations, Administration and Maintenance Implementation Agreement,* March 2001

FRF.20, *FR IP Header Compression Implementation Agreement,* June 2001 sophisticated underlying network requirements. For example, IAs ratified over the years include FRF.11, *Voice over FR*, and FRF.12, *Frame Relay Fragmentation*. In addition, updates and modifications have been made to existing IAs that add support for higher speed physical interfaces, such as DS-3 support for UNI and NNI connections. Technical committee work in recent years continues this trend toward increased functionality for FR, addressing issues such as service level definitions, FR/ATM SVC interworking, FR OA&M, and SONET/SDL support for physical layer interfaces. Providers are also looking to MPLS-based technologies to provide next generation core networks. There is, therefore, a desire to interwork MPLS and FR. As noted, the FRF and the MPLS Forum have been working on an FR/MPLS interoperability IA to address the overall architecture and data transfer aspects of interoperability (this topic will be revisited after we introduce MPLS).

FRF.20 describes how compressed IP datagrams are encapsulated within a FR frame, and how IP header compression algorithms and their parameters are negotiated over FR VCs. The specification applies to both IPv4 and IPv6. A *FR IP Header Compression Control Protocol* (FRIHCP) is defined that negotiates the use of IP header compression in each direction of the VC. FRIHCP is based on the Link Control Protocol (RFC 1661). FRIHCP uses a simple handshake (using configure-request and configure-ack packets only) to enable the header compression algorithms and negotiate their parameters. The algorithms are described in RFC 2507 (*IP Header Compression*) and RFC 2508 (*Compressing IP/UDP/RTP Headers for Low-Speed Serial Links*). Unlike data compression and encryption, there is no separate PPP control protocol for IP header compression. In that case, IP header compression is negotiated over a PPP link as part of the *PPP IP Control Protocol* (RFC 1332) and *IPv6 over PPP* (RFC 2472). The FRIHCP, however, is a single control protocol to negotiate header compression for both IPv4 and IPv6. FRIHCP relies on phases similar to PPP.

FRF.19 is an OA&M protocol and procedures IA for a FR service. This IA provides a means to test, diagnose, and measure the quality of FR services. This OA&M protocol and procedures document is designed to either supplement, or replace, I.620. Figure 9.9 illustrates a reference FR network that is interworking with an ATM network. A number of example monitoring points are indicated. In this reference network, the following examples of circuit connections may be indicated:

- A VC that spans a single FR access network section;
- A VC that spans two FR access network sections connected by an NNI;
- A VC that spans two FR access network sections connected by a FR transit network;
- A VC that spans two FR access network sections connected by an ATM transit network section using network interworking;

FIGURE 9.9
*Network reference
model.*

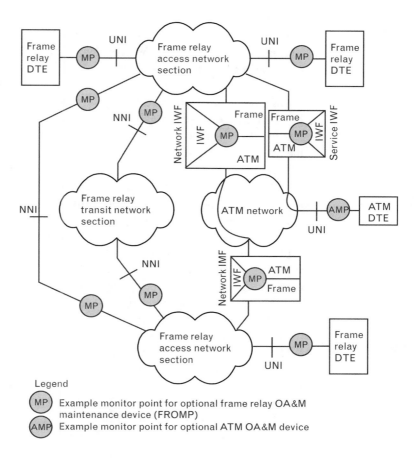

Legend

(MP) Example monitor point for optional frame relay OA&M
     maintenance device (FROMP)
(AMP) Example monitor point for optional ATM OA&M device

• An FR circuit that spans a single FR access network section but is terminated as an ATM circuit using service interworking.

FRF.18, *Network-to-Network FR/ATM SVC Service Interworking Implementation Agreement,* defines SVC and *soft permanent virtual connections*
(SPVC) service interworking between FR and ATM technologies at the
NNI [14]. Enterprise customers can be confident in knowing that the
information they send via a FR virtual connection will be secure, and that
the data will be read only by the intended recipient.

FRF.17, the privacy IA, provides authentication and encryption facilities that may be used over a FR virtual connection, both permanent and
switched, to eliminate unauthorized data eavesdropping.

FRF.15 and FRF.16 pave the way for higher bandwidth FR networks.
These implementation agreements provide protocols and procedures to
allow support for splitting a high-speed stream of FR traffic over multiple
FR interfaces [15]. Completion of these IAs is significant because they
enable service providers and equipment vendors to offer a new range
of standards–based fractional T3/E3 FR solutions. Market demand for

high-speed FR services is growing, but until now users have been limited to using (costly) DS-3s, using proprietary FT3 implementations, or implementing ATM for network sites that have high-bandwidth requirements. As a result of the new IAs, enterprise customers will benefit from a new "class" of cost-effective service offerings that will meet the increasing demand for broadband applications, and with the corresponding network bandwidth. Given the popularity of FR, end users, carriers, and vendors have been striving to support additional bandwidth options for accessing FR networks. The market for high-bandwidth services such as NxT1 and NxE1 is growing, and with the new implementation agreements, users no longer need to be concerned about limitations associated with proprietary vendor implementations to support these higher speeds.

During the past several years, carriers and vendors have introduced support for FR access above DS-1 rates with NxT1, but until now there were no standards to provide interoperability guidelines among different devices for aggregating this type of traffic. The two new FRF implementation agreements each support a specific mode of multilink aggregation; FRF.15 supports end-to-end multilink (DTE-to-DTE) and FRF.16 supports UNI and NNI multilink. These new technical capabilities support FR's access platform for VPN services, an emerging application that has significantly contributed the continued growth of the FR market. Multilink technologies address fundamental problems with the scarcity of available last-mile access. The demand for broadband FR has been there for a number of years, but one could not provision it due to the unavailability of the transport (and the standards for the CPE and for the switch). FRF.15 and FRF.16 provide a means to address this using the common vehicle of the T1 span. They are specific to FR and accompany ML-PPP that already existed for Internet access using PPP.

FRF.15 and FRF.16 provide physical interface or virtual circuit emulation for FR devices, consisting of one or more physical links or virtual circuits aggregated to form a single interface of bandwidth. This service provides a frame-based inverse multiplexing function, sometimes referred to as an *IMUX*. Multilink FR solutions overcome the lack of required bandwidth availability due to facility constraints (e.g., no E3/T3 service in a geographical region) or service offering restrictions (e.g., no fractional E1/T1 service). The implementation agreements also resolve the issue of the physical interface or virtual circuit being an inflexible pool of bandwidth and a single point of failure. FRF.15 and FRF.16 give service providers the ability to design flexible network services to meet individual customer requirements. By combining multiple physical interfaces or virtual circuits, a network operator can design a FR interface that can support more bandwidth than is available from any single physical interface or virtual circuit. Furthermore, these agreements provide techniques that support the dynamic addition and removal of these interfaces to change the total bandwidth available on the interface. Resilience is provided when multiple physical interfaces or virtual circuits are provisioned so that when some of the physical interfaces or virtual circuits fail, the emulated interface continues to support the FR service. This topic is revisited in Chapter 10.

FRF.14 is the physical layer interface IA. The significance of FRF.14 is that FR can now operate over optical media. This includes the North American standard SONET and the closely related SDH covered in earlier chapters. FRF.14 adds support for the direct mapping of FR frames into SONET/SDH SPEs at speeds of OC-3c/STM-1 (155 Mbps), OC-12c/STM-4 (622 Mbps), and OC-48c/STM-16 (2.5 Gbps). (The FRF.14 IA also consolidates the descriptions of all FRF-defined interfaces formerly found in the different interface implementation agreements.) The physical interface descriptions included in the interface IAs and now in the new FRF.14 IA are nonbinding. The FRF has never used the physical interface specifications for conformance tests or as a mandatory element of a service. In fact, no one physical interface is mandatory and no undefined interfaces are prohibited. Instead, these descriptions document desired physical interface characteristics that, if supported on both ends of the physical interface, result in the greatest degree of interoperability. This is particularly valuable for cases where a vendor has choices in implementing options of a physical interface specification [16].

FRF.14 provides specifications for a wide range of interface types and speeds. (Table 9.4 provides a complete list of the interfaces included in FRF.14.) Interfaces are provided for different regions of the world, such as E1/E3 for Europe and T1/T3 for North America. Popular new interfaces may be added in the future. The SONET/SDH transport provides a

TABLE 9.4   INTERFACES INCLUDED IN FRF.14

INTERFACE	MAXIMUM SPEED
V.35	2 Mbps
V.36/V.37	2 Mbps
EIA-530	2 Mbps
X.21	2 Mbps
HSSI (TIA-613)	53 Mbps
56/64 Kbps (ANSI T1.410)	64 Kbps
ISDN (I.430)	2 Mbps
T1 (ANSI T1.403)	1.544 Mbps
T3 (ANSI T1.404)	45 Mbps
E1 (G.703)	2 Mbps
E3 (G.703)	34 Mbps
SONET/SDH (ANSI T1.105/ITU G.707)	155 Mbps to 2.5 Gbps (higher with SDH STM)

full-duplex octet-oriented synchronous link. The technique adopted by the FR Forum technical committee matches the PPP over SONET/SDH and frame-based ATM proposals being considered standardized by the IETF and ATM Forum, respectively. By standardizing on a single consistent HDLC-based SONET/SDH mapping technique, customers can potentially select the appropriate data transport technology for an application without swapping out physical interfaces. Each frame is inserted into the SPE on an octet boundary. HDLC flags (01111110) are transmitted when no FR frames are present. With support for SONET/SDH interfaces, FR provides high-speed data connections without the overhead of the ATM cell "tax."

FRF.13 was adopted in 1998 to help FR users make informed decisions regarding the selection and use of FR services. FRF.13 helps the end user understand the services they procure, not just the underlying technology. FRF.13 allows users to select the offering that best meets their business requirements from various services. As end users increasingly depend on FR networks for business-critical applications, it has become apparent in recent years that one needs a way to define and measure network performance for end-user service level agreements. This type of interoperability agreement sets FR apart from other networking technologies by defining FR service performance in terms that end users can understand.

The *service level definitions* (SLDs) of FRF.13 define user information transfer parameters that describe FR service performance. End users of FR services can utilize these parameters to compare services from different FR service providers, to measure the QoS of specific FR service offerings, and to enforce contractual service level agreement commitments. Defining network performance includes not only a single network, but NNIs and multiple service provider networks. The SLD IA provides a common set of terminology related to FR service performance that can be utilized by service providers and equipment vendors in support of that requirement, even in a completely heterogeneous networking environment. The focus of the SLD IA is the definition of a set of *transfer parameters* for frame transfer delay, frame delivery ratio, data delivery ratio, and service availability. These transfer parameters are used to describe FR service performance during the information transfer phase for FR circuits. Also defined are the *connection components* or the building blocks that represent the structure of any end-to-end connection, and reference points along the connection path that the information frames transit between source and destination. Measurement and reporting of data collected at these reference points provides the basis for common performance and availability comparisons. The SLD IA provides a basis for making an analytical comparison of FR service levels offered by different network service providers. Compliance to the SLD IA is based on the use of the defined parameters and all applicable requirements as defined within the agreement. Work was under way within the FR Forum's technical committee to extend the agreement's practical use by defining a common mechanism to be used in performing the measurements. The SLD IA is an important step in providing the ability to provide true QoS for FR, but it is only the first step.

The FR Forum has also made amendments in recent years to four other preexisting IAs to further the scope and versatility of FR: FRF.1.2, FRF.3.2, FRF.4.1, and FRF.8.1.

FRF.1.2, the PVC UNI implementation agreement, has been ratified by the FR Forum in order to enhance interoperability by describing an agreed-on set of features for FR PVCs that will be implemented industrywide. It defines the protocol peers in the DTE and DCE to support PVC services. The 1.2-version enhancements make it possible to handle the large number of PVCs that will be common on high-speed interfaces, for example, when FR is used over SONET/SDH. Another important enhancement to FRF.1.2 is related to PVC configuration. Until now, provisioning a new PVC or changing an existing PVC was a time-consuming and error-prone process; a new feature of FRF.1.2 streamlines the process, making provisioning faster, easier, and more accurate. This feature enables routers and other FR user's devices to be automatically configured from the network with the appropriate traffic and QoS parameters.

FRF.3.2, the multiprotocol encapsulation IA, has been passed to expand the number of other protocols that may be carried within FR, which further extends FR's utility as an all-purpose carrier of other data networking protocols, such as IP, SNA, *voice over FR* (VoFR), and X.25. This agreement also simplifies the process of supporting more protocols in the future and adds support for multiprotocol encapsulation to FR SVCs in addition to FR PVCs, which were already supported by FRF.3.1.

With the latest version of FRF.4.1, the FR SVC UNI implementation agreement, the FR Forum added new features to support FR QoS and improve interworking with ATM SVC.

The FR Forum has also made some enhancements to FRF.8.1, the FR/ATM PVC service interworking IA, to improve the flexibility of the translation between FR and ATM data transfer protocols.

FRF.2.2, the NNI implementation agreement, describes FR PVC NNI agreements. The NNI is concerned with the transfer of C-plane and U-plane information between two network nodes belonging to two different FR networks.

9.1.4   CIR Considerations

As we noted earlier, CIR is a key engineering tool for the designer of the enterprise network. When FR is used primarily for LAN interconnect traffic, CIR engineering is relatively straightforward: For a small office LAN interconnect traffic is bursty and reasonably tolerant of delay. (Note, however, that it is not bursty when an organization has 100, 500, or 1,000 users accessing the remote location via the aggregation router in question.)

A typical user with a low-volume FR site can choose perhaps a 56-Kbps FR port and access circuit and a 16- or 32-Kbps CIR for the router-to-router PVC. The site would utilize excess capacity in the FR network by bursting over the CIR as needed. For a user site with a high number of LAN users, one would choose a 1,024-Kbps FR port and access circuit and a 768-Kbps CIR for the PVC or 768-Kbps FR port and access circuit and a 384-Kbps CIR. Designers have to keep in mind that well-designed networks are engineered for the *busy hour*, namely, for the hour of the day when, say, 1,000 corporate users are all on the enterprise network and are trying to get through and conduct business. At such time, the traffic is *not bursty* (see Figure 9.10 as a pictorial example); such LAN-to-LAN traffic could indeed be bursty in the early hours of the morning, or at lunch time, or toward evening, but, frankly, who cares? Those are not times with which the design is concerned. The design must be made for the busy hour, at which time the traffic is *generally not bursty*. More recently, in addition to LAN interconnect, users are implementing many different types of applications on their FR networks, including Internet/intranet, *systems network architecture* (SNA) networking, and voice, video, and fax applications. Table 9.5 highlights some of the characteristics of different applications.

Delay-intolerant or session-oriented applications, such as voice or SNA, may require a higher CIR than more delay-tolerant applications (e.g., LAN-to-LAN interconnections) [17]. In addition, applications that have high-bandwidth requirements and are delay intolerant, such as video, may require a higher relative CIR. One approach may be to use separate PVCs for the various applications and choose different relative CIRs for each PVC. In this situation, the recommendation is to implement all session-oriented or delay-intolerant applications on PVCs with higher

FIGURE 9.10
Proper network design criterion.

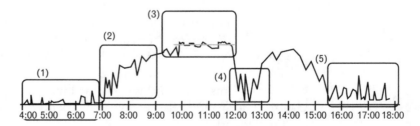

(1) Low night traffic some burstiness
(2) Increasing number of users: some burstiness due to the fact that few people are on the system
(3) Busy hours(s): relatively low burstiness, since there are many users on the system. The Law of Large Numbers guarantees that the traffic is smoothed out to a high average arrival rate
(4) Burstiness due to the fact that few people are on the system
(5) Burstiness due to the fact that few people are on the system

Proper network design rule: Design for the Busy Hour; at such Busy Hour the traffic will **NOT** be bursty for networks of nontrivial size

TABLE 9.5   TRAFFIC CHARACTERISTICS FOR SOME TRADITIONAL
APPLICATIONS

	BANDWIDTH REQUIRED	BURSTINESS (WHEN IN ACTUAL SESSION)	DELAY TOLERANCE
LAN-to-LAN	High (64 Kbps to DS3)	High in small offices / Low in medium-to-large offices	High
SNA	Generally love (56 Kbps)	Low	Medium to low
Internet/intranet	High	High in small offices / Low in medium-to-large offices	High
Voice	Generally low	Low	Low
Video	Medium to High	Low	Low

relative CIRs, and to utilize PVCs with lower relative CIRs for bursty applications, such as Internet access. This would give the delay-intolerant applications a higher probability of delivery due to more traffic being sent within CIR. However, many users do not choose to implement different applications on different PVCs. In these cases, when delay-intolerant applications are combined with other applications, it is advisable to choose the CIR for the most sensitive application.

Frame relay standards have focused on the access to the FR network, with network implementation largely left to the carrier. Public FR service providers allow traffic to burst into their networks at speeds higher than the subscribed CIR. Whether the bursty traffic gets through depends on overall network congestion and how it is handled by the network. Two key network design factors impact the choice of CIR: the congestion management algorithm and the carrier's overall network capacity.

In turn, two types of congestion control algorithms are used in FR networks: closed loop and open loop. Closed-loop algorithms prevent frames from entering the network unless there is an extremely high probability that the frames will be accepted by the network without discards. Open-loop systems send data into the network regardless of potential congestion and assume the end stations can retransmit if the network drops frames. Because closed-loop algorithms implement "flow control" at the entry point of the network, a higher CIR may be required to ensure that data are accepted into the network. An open-loop implementation potentially enables a lower CIR, assuming there is adequate capacity in the network.

Beyond the basic algorithm, the actual handling of CIR and DE traffic in a congested network will vary based on the FR switch; however, there are some common processes. Figure 9.11 illustrates a typical FR switch

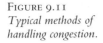

FIGURE 9.11
Typical methods of
handling congestion.

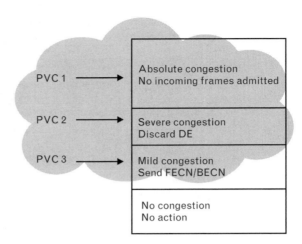

handling congestion. Depending on the level of congestion, the switch will utilize, as noted earlier, congestion notification to notify an interface device of potential congestion and then manage buffer capacity based on CIR and DE. When the congestion levels become high enough, DE traffic will be discarded. If congestion becomes severe enough, all frames could be discarded. Remember too, that to the switch, all CIR and all DE traffic is considered the same, without regard to which PVC it came from.

Another key factor in the choice of CIR is the carrier's overall network capacity. Although congestion management algorithms determine the way congestion is handled, the overall capacity or sizing of the network—traffic, ports, and CIR—will determine whether or not congestion will occur. Typically, carriers size their networks to a target utilization level, but metrics may vary. Some carriers base their network capacity on the total CIR subscribed by customers. Other carriers look at overall port bandwidth or traffic levels to determine capacity.

Customers utilizing a carrier network design based on total CIR will find that the CIR they select is more important in the determination of whether their traffic gets through, since the possibility does exist, though remotely, that all customers will burst above CIR simultaneously. Carriers deploying more bandwidth between FR nodes (either relative to CIR or overall) may create an environment where the difference between CIR and DE is less significant. This is why some carriers are able to "assure" [18] delivery of DE traffic. Users will need to weigh their overall costs relative to their carrier network design and their performance requirements to determine the right CIR for them.

Overall subscription relative to port speed is another element to consider in the choice of CIR. Oversubscription is defined by the aggregate of CIR values on all PVCs connected to the port divided by the port speed (see Figure 9.12). By definition, oversubscription implies that the CIR values alone will exceed port speed. However, most carriers permit some level of

FIGURE 9.12
*Comparison of
networks with and
without
oversubscription.*

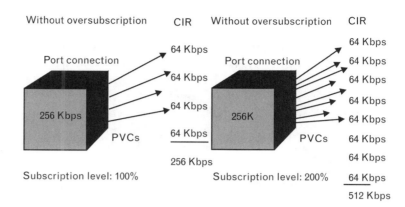

oversubscription because the network usually has excess capacity and, typically, not all applications are active at the same time. The user needs to look at the traffic flows and timing of applications since it is only possible to transmit up to port speed. Carriers also oversubscribe trunks between FR nodes.

Many carriers "assure" the delivery of CIR traffic at levels greater than 99%. Customers who choose higher CIRs can be assured of very high levels of delivery within the public FR network. Some carriers will even "assure" the delivery of DE traffic, albeit at lower levels, enabling users to feel confident about buying lower CIRs and bursting to some degree.

For users with requirements greater than CIR and DE can support, more advanced solutions are available, although they are proprietary. Many customer-premises equipment vendors have prioritization or other traffic shaping capabilities that enable users to optimize the different traffic flows beyond setting higher or lower CIRs. These types of capabilities are also available in FR switches and can be offered by public carriers. In this implementation, traffic can be prioritized on a PVC-by-PVC basis and handled differently beyond the CIR or DE designation. In some implementations, the buffer queues within a FR consider PVCs high, medium, or low priority and further delineate between the priorities of both CIR and DE on those PVCs.

The service provider will require some appropriate mechanism to take the CIRs associated with the various incoming (half-)PVCs and allocate or reserve capacity on the network to service the contracts. These mechanisms are typically specific to the switch and/or service provider. For example, if the network is built out of Cascade/Ascend/Lucent 9000 FR switches, there is a provisioning mechanism because the switches actually implement a constrained-SPF-based routing protocol to place the VCs on trunking resources. This technique was later applied to MPLS-TE using RSVP-TE and a link-state-based IGP to carry the path attribute information. (The Cascade switches did this earlier with OSPF as an internal protocol on their switches.) Other FR platforms, like those from Wellfleet/Bay/Nortel, simply encapsulated the FR frame inside IP and used a simple minimum-hop, destination-only-based IP forwarding mechanism. Here "performance" was "assured" by overprovisioning the carrier's FR network. This approach may be acceptable, except when taking into consideration trunk or node failure. The C–SPF approach does a better job of rerouting the PVCs and attempting to support the bandwidth contracts in the failure mode scenario because the rerouting algorithms are aware of the constraints. The gross overprovisioning was probably why services with "0–CIR" worked as well as they did.

9.1.5 Service Level Metrics

As noted, FR Forum's IA FRF.13 provides a common language for the development of an SLA; it spells out the terminology to be used to define the quality provided. It also provides explicit reference points for measurements to be made, as well as the quality parameters to be measured. Within FRF.13 one finds the definitions of four transfer parameters such as delay, frame delivery rate, and availability. These metrics constitute the main elements of a SLA and are the benchmarks by which we measure network performance and availability, whether the FR network is private or provided by a carrier [19]. (Unfortunately, there are no recommendations that state the maximum network latency FR networks must meet; the delay parameters are network specific.) One can then trust the service provider to deliver the required SLA, or, if the application is mission critical, one can use DSU/CSU equipment that implements FRF.13 parameter measurements. With these types of DSU/CSUs, one will be able to confirm that the SLA requirements are or are not being meet. Figure 9.13 illustrates an overall, enterprisewide service level management model [20] within which fit the FR service level metrics.

9.1.5.1 Delay

Delay metrics describe the time required to transport data from one end of the network to the other. Measuring delay involves three interdependent elements: access line speed, frame size, and WAN delay. Measuring and

FIGURE 9.13
*Enterprise application
verification.*

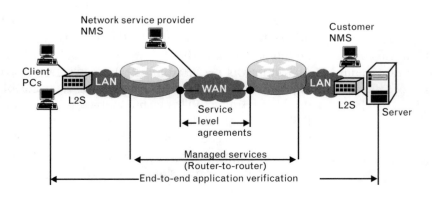

reporting delay should be in the context of these elements to be useful. Access line speed refers to the delay that arises from the data rate of the line from the user site to the FR network at both the local and remote ends of the network. Access line delay can contribute significantly to the overall delay of the network. (Delay is also introduced by the router or FRAD, but for our purposes we will concentrate on network delay.) In the network with a 64-Kbps access link, a 4,000-byte packet takes approximately 500 ms to cross a 64-Kbps line. If the local and remote access lines are 64 Kbps, the access lines alone add nearly a second of delay. Delay caused by the access line can be managed by increasing line speed or segmenting the data into smaller frames, which is usually handled by the router or FRAD. In the previous example, if a voice frame, which is typically 50 bytes or so in length, has to wait for a large frame (e.g., 4,000 bytes) to be transmitted, the effect is added delay. Changing the frame size to 128 bytes reduces the access line delay to approximately 16 ms. Alternatively, increasing the line speed to T1 (1,544 Kbps), reduces the 4,000-byte frame delay to a more manageable 20 ms. The third element of delay, network delay, is difficult to manage in a public network environment. However, measurement of delay across the WAN, separate from the delay imposed by the access line, can help the user pinpoint performance problems. Eliminating WAN delay will help the user focus on other causes of inadequate performance, for example, configuration or application difficulties, which account for as much as 70% of performance problems.

9.1.5.2 Frame Delivery Rate

The *frame delivery rate* (FDR) must also be viewed in relative terms. Frame relay networks typically categorize frames in two ways (there are other distinctions, but these two are the most common): below the CIR and above CIR. To provide a valid FDR, it must be determined if the measurement is for frames within CIR, in excess of CIR, or for the total number of frames presented to the network for delivery. FDR statistics are particularly helpful for future network planning.

9.1.5.3 Connection Availability

Having the performance and throughput required to successfully run applications across the network is negated if the network is not available when the user needs it. The metric of connection availability measures the percentage of time the network connection is accessible to support the communications needs of the network. There are several elements to connection availability: overall availability, mean time to repair in the event a connection is lost, and mean time between service outages.

Overall availability refers to the total time the network connection is available, compared to the total measured time. If a network did not experience any services outages in a 30-day period, then its availability would be expressed as 100%. If it were down for 6 hours, the availability would be 99.17%.

Availability can be measured networkwide (all sites together), or individual measurements can be taken for each site and brought together to reflect a total network calculation. Connection *mean time to repair* (MTTR), or mean time to restore, has a direct impact on availability because the longer it takes to repair a connection, the longer the service is unavailable. Most SLAs have specific measurements for MTTR. One method of reducing the impact of a service outage, usually caused by a failure in the local loop, is the use of ISDN as a backup service.

The degree of impact caused by service outages is more apparent if the time between outages is measured. *Mean time between service outages* (MTBSO) measures the availability time between outages. Having four 6-hour outages in one day has a greater impact than one 6-hour outage a week for a month. MTBSO gives the network manager the information needed to evaluate the other availability metrics. SLAs provide agreed-on metrics for measuring and reporting network performance. After the FR Forum technical committee's work to provide industry standard definitions of key SLA metrics was completed, it started to work on the development of the measurement and reporting mechanisms.

OA&M mechanisms incorporated in FRF.19 provide two fundamental capabilities: a vendor-independent method of monitoring an FRF.13-compliant SLA and new diagnostic tools for frame networks [21]. Frame relay circuits often consist of multiple providers, and each provider may want to monitor the section it administers. In addition, the customer may want to independently monitor the circuit. OA&M supports the overlapping segmentation of the network, allowing independent measurements to be made. An example of a VC with overlapping domains is shown in Figure 9.14. In this example, measurements made in FR provider domain A are independent of those in domain B. The end-user domain is not allowed to address OA&M components in domain A or B, but may pass OA&M messages through the FR provider equipment to reach the far end. Security of OA&M is provided by the introduction of a selective administrative boundary located at the edge of a domain. This boundary will protect the OA&M of one administration from being interfered with by

FIGURE 9.14
Multidomain service.

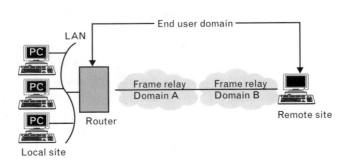

another, yet allow outer domains, such as the end–user domain shown in Figure 9.14, to pass through.

Of the four measurement parameters defined in FRF.13, an OA&M protocol is needed to measure all except availability. This is because VC status is already propagated via end–to–end via link management procedures.

To measure *frame transfer delay* (FTD), occasional test frames will travel round-trip between two measurement points. The *turnaround time* at the far end is removed in the measurement, allowing the OA&M frames to use low-priority queuing at the measurement endpoints. By using a round-trip measurement, and dividing it by two, the need to synchronize the clocks used by the two OA&M devices is eliminated.

To measure the *frame delivery ratio* (FDR) and *data delivery ratio* (DDR) between two points, marker OA&M frames are sent containing the current transmitted frame and octet counters for the circuit. The receiving OA&M device will compare differences between these counters with the state of its own receive counters to determine the ratio of transmitted-to-received frames in the interval. This measurement is made for both frames within the committed information rate and for excess burst frames. The measurement is made independently for each direction.

To improve the ability to test networks, two diagnostic tools are included. The tool first is a virtual channel loopback, as shown in Figure 9.15. This allows an individual VC to be temporarily taken out of service for testing, without affecting other VCs sharing the same link. The second tool is a *virtual bit error rate test* (virtual BERT). Together, these tools provide the means to test both the layer 1 and 2 subsystems between two points. They can also be used to quantify how the network will respond to a specific traffic load pattern.

OA&M introduces a test status (in addition to active and inactive) and will propagate fault location and cause information when available. Many frame networks currently have ATM backbone sections capable of providing this information to the FR gateway.

9.1.6   Other Services, Capabilities, and Features

9.1.6.1   VoFR

Voice over FR has been working successfully for many years [22]. Two major accomplishments in 1997 were FRF.11, the VoFR IA, and FRF.12,

FIGURE 9.15
Loopback.

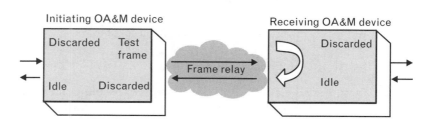

the FR fragmentation IA, both of which relate to delay-sensitive voice traffic. However, there has been significant interest in introducing VoIP into corporate and enterprise networks as a convergent solution at the IP layer. Some providers have introduced a service access that has been optimized to ensure low delay.

As alluded to earlier [23], when a user transports voice packets over an IP network, each voice packet is comprised of the voice content (payload) and a header that contains additional protocol information. IP voice calls use three layers of protocols: RTP, UDP, and IP, with the header consuming a substantial portion of the total voice packet. While the voice payload itself can be as little as 20 bytes in length, the VoIP header can be as large as 40 bytes. FRF.20 provides a FR IP header compression IA that defines how IP header compressed packets should be handled over a FR interface, including how the compression is negotiated between compressor and decompressor and how packets are encapsulated. IP header compression is used per VC between end-user systems (i.e., DTE to DTE). The header compression procedure is transparent to the FR network. IP header compression offers benefits for the installed base of FR users: Remote and branch offices around the world will realize bandwidth efficiencies and cost savings for their low-speed FR VoIP calls with the adoption of this protocol. IP header compression examines the 40 bytes of header and finds redundancy and other opportunities to send only a reference to the context, rather than the entire header. IP header compression reduces the VoIP header from 40 bytes to a minimum of 2 to 4 bytes, offering bandwidth efficiencies for enterprises, service providers, and providers of managed services networks. An ordinary 64-Kbps link can transport only two uncompressed G.729 VoIP calls. Using IP header compression, the same link can carry five concurrent calls. For the enterprise that has 56-, 64-, or 128-Kbps FR links connecting a large number of remote sites, such as a banking or retail environment, IP header compression can provide significant improvement in bandwidth efficiencies and cost savings. IP header compression uses a combination of techniques to achieve the compression of headers, including the following:

- Once the complete header has been sent once, it is only necessary to pass a reference to the context in subsequent packets.
- Only the changes in the fields are sent, reducing the space required.
- FRF.20 exploits field changes that are constant from packet to packet; as a result, the entire field can be reconstructed by the decompressor with knowledge of the original value, and the change packet to packet.

9.1.6.2   Fax

FRF.11, the VoFR IA, provides recommendations for transporting voice, fax, and data. Included in the agreement is Annex D, *Fax Transfer Relay*

Syntax, describing the format for transporting fax data in FR networks. The continued rise in fax usage and the characteristics of fax transmission make it a candidate for transport over FR. According to observers, fax transmissions will continue to grow despite the introduction of e-mail. Some argue that fax will be overtaken by newer image technologies; this may be true, but it will not happen overnight since there is such a large embedded base of fax terminals around the world generating millions of pages daily. A turn-of-the-decade study by the International Data Corporation expected the number of transmitted fax pages to grow at a compound annual growth rate of 22.6%, to nearly 11 billion pages in 2003. Fax over the Internet is also growing, and although this does not directly impact the growth of fax over frame, it does contribute to the demand of FR to handle the increased IP traffic. Six basic steps are used for the transmission of a fax over the PSTN [24]:

1. An image is scanned using a *charge coupled device* (CCD).
2. The CCD sends a signal to the A/D converter. The strength of the signal's amplitude represents the brightness of each element recorded on the CCD. The digital stream, with each element represented by 1 bit, is then sent to be coded.
3. The digital stream is converted to a series of code words using *modified Huffman-modified Read* (MH/MR), which reduces the amount of digital information by up to 1/20.
4. The compressed data are then converted by a modem for delivery over the PSTN.
5. The receiving fax machine expands the compressed MH/MR back into a digital stream.
6. The digital stream is then used to reproduce the original image—a facsimile.

In transmitting fax over frame, the FR network replaces the PSTN (step 4). The FRAD detects the fax signal from the modem and demodulates (recreates) the digital stream the modem received from the MH/MR compression stage (step 3). The sending FRAD transports the fax data in a series of frames to the receiving FRAD, which extracts the fax data and remodulates the data for the modem in the receiving fax terminal. The characteristics of fax transmission make it a good candidate for FR transport, which can result in cost savings over the traditional PSTN. Packet network issues that often impact voice quality—delay, lost packets, and jitter—are more easily tolerated by fax, a fact not lost on users. Group 3 fax recommendations contained in T.30, *Procedures for Document Facsimile Transmission in the General Switched Telephone Network,* provide application level procedures for the transmitting and, in the case of problems, the retransmission of fax information. An error will mean that the offending packet is simply discarded.

9.1.6.3   VPN

Providers have also targeted customers looking to create the flexibility of VPN networks while leveraging the inherent security and dependable performance of FR technology.

A VPN is a means of transmitting information in a partitioned, secure, reliable manner over an overlay shared network, typically some public network such as the Internet. In practical terms, VPNs can be achieved as overlays to layer 2 networks such as FR and ATM, or as overlays to layer 3 networks, such as X.25 (PLP) or IP. Layer 2 VPNs have been available since the mid- to late 1980s, whereas layer 3 VPNs are more of a mid- to late 1990s construct. PVCs available in ATM and FR inherently provide a "tunnel" in that the VPI/VCIs or DLCIs associated with a different access device are established by the network operator (within the company or service provider) and eliminate any possibility of exchanging information with unauthorized locations; if one couples this with an encryption mechanism (by either encrypting the layer 2 payload or even just the layer 3 payload), the communication will be fairly secure. Although VPNs can include remote access, this section focuses on intranets and LAN-to-LAN connectivity.

Proponents of either technology advance the advantages of their choice of technology. The potential savings of any VPN technology of course should be examined on the basis of total cost of network ownership. The bottom line is that there are advantages and disadvantages with each approach, and to a large extent the company's decision should be made based on the answer to this question: "To which technology does the company have the most visible IT boundary interface: IP or frame/ATM?" In other words, if the IT interface is principally to a layer 3 capability (for example, in the case of an ISP), then the organization should probably consider IP VPNs. If the IT interface is principally to a layer 2 capability (e.g., with SNA being carried over a FR or even a PL network), then the organization should more probably consider layer 2 VPNs. The answer to the question of whether FR and IP VPN technologies are competitive or complementary is that they are "both."

It appears that VPNs strive mostly to replicate the current dominant design or popular networking models, but more cheaply. History seems to repeat itself: A few years ago, cost savings fueled the growth of FR, a cost-effective alternative (depending on network topologies and geographic coverage) to leased lines [25].

IP VPNs (see Figure 9.16 for an example) are fairly inexpensive; however, performance can often be an issue. Service providers do not (and cannot) offer any guarantee on service levels on a per tunnel basis. The IP VPN promise of any-to-any connectivity is attractive in extranet and branch/mobile access environments. However, in IP VPNs the network administrators must configure the rules by which individual users communicate with each other, requiring a quadratic number of configurations.

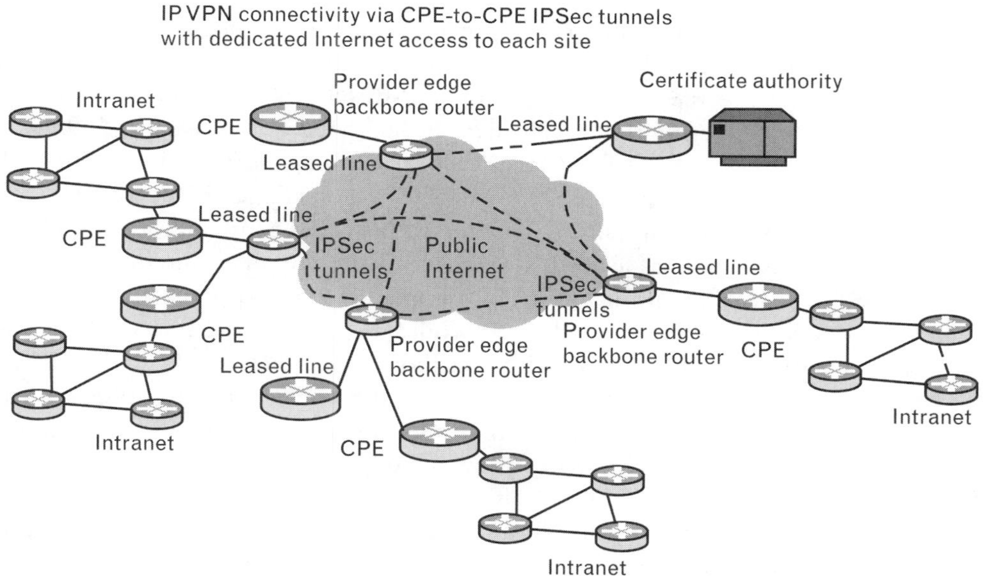

IP VPN connectivity via CPE-to-CPE IPSec tunnels
with dedicated Internet access to each site

FIGURE 9.16   *Internet VPN CPE-to-CPE encryption.*

This can become administratively costly. There still are IPX and SNA/ SDLC users out there. This less than homogeneous environment is a problem because IPSec (IPSecure) can support only IP. [Multiprotocol support for IPSec requires a proxy server to handle the necessary protocol translation to IP, but this represents additional burden of overhead; another approach is to combine IPSec with L2TP tunneling, or utilize (Data Link Switching (DLSw).]

A layer 2 VPN service provisioned using PVCs on a FR access platform eliminates the need to acquire special tunneling and performance enhancing options to accompany it [26]. Current FR networks support thousands of concurrent active PVCs with high levels of availability, scaling and forwarding performance, and minimal, predictable delay and jitter. SVCs can theoretically provide a way to extend the reach of FR services to remote users and extranet sites. However, as noted, FR SVC services have not been deployed or priced attractively by all carriers yet. In actual usage, point-to-point and star topologies are still the primary implementation of PVCs in the recent past, as indicated by the results of a 1999 Distributed Network Associate Survey, in which about 80% of the respondents stated they were still using the two types of network topologies.

CPE supporting FR SVCs is available today, but SVC service adoption is solely dependent on the way service providers price and position SVCs compared to PVCs. Examples exist in Europe where major service providers offer parity pricing for SVC and PVC services (for equivalent aggregate

CIRs). For end users, the trend and the preference is *against* usage-based pricing in order to establish billing stability. Carriers should remember the adverse reaction to ISDN minute-based pricing. Again, the trend is for fixed-fee Internet access (e.g., $19.95 MRC on dial-up or $39.95 MRC on DSL); fixed-fee cellular calling plans (e.g., 1,000 minutes for a $39.95 MRC); and fixed fee IXC voice packages (e.g., an MRC of $9.95 for 400 minutes). Naturally, a dynamic VPN would be useful for a mobile workforce that has a need to connect to the corporate intranet to upload or download information. To dial up a FR connection from a coffee shop, a hotel, an airport, and so on would be problematic. IP VPN better serves these applications. IP VPNs are more secure than simple FR VPNs because they utilize encryption techniques. (Of course, an enterprise user can still utilize its own end-to-end IPSec by running the IP—specifically the IPSec—over the carrier-provided unprotected FR service.)

In favor of FR VPNs, one notes that performance management and monitoring are generally a strong requirement for companies to address different internal needs. As noted, current FR performance management tools provide the means to quantify performance on a given link down to the virtual circuit and determine if performance is equal to levels specified in a contract with a service provider, or in agreement between users and IT departments. FRF.13 capabilities enable service level verification and management, capacity planning and trending, bandwidth optimization, preemptive warnings of oversubscription, automatic troubleshooting, and more.

The issue of ensuring certain levels of service is tied closely to that of traffic engineering and capacity planning, and equipment interoperability and performance as well. Service providers face the nontrivial challenge of anticipating the growth and demand in bandwidth, which compounds the difficulty of providing predictable and reliable services. Designing and tuning their networks for optimum performance and operation while minimizing complexity and operating costs is a significant task. Initially the IP VPN market was characterized by the lack of standards and proprietary approaches and by poor quality. The connectionless nature of IP makes it more difficult for service providers to determine traffic patterns; this makes capacity planning and traffic-engineering a nontrivial challenge. For the foreseeable future, enterprises will use either one of these two technologies, in a complementary manner.

9.1.6.4 DSL Access

Implementation of DSL for FR access represents an opportunity for FR network service providers alike. This is true primarily because DSL access is much more affordable than traditional TDM access. The marriage of DSL access to FR services is an attractive combination due to the high cost of FR access, which, at press time, represents approximately 36% of the total cost of FR service. In the United States, the cost for TDM access averages anywhere between $200 and $485 for T1 access. DSL access, with variable speeds that range between 128 Kbps up to and beyond 1.536 Mbps, are generally priced significantly lower than TDM access pricing; they can, in fact, be 50% less than average TDM rates or even lower. But with the affordability comes service performance and delivery challenges. With DSL access, service providers are faced with implementation of multiple technologies in multiple networks to deliver end-to-end services. This can impact network performance and availability, service delivery, and service management. Because DSL access technologies generally leverage ATM capability at the backplane of the *DSL access multiplexer* (DSLAM), statistical multiplexing is introduced to the access portion of FR network [27].

9.1.6.5 FR Backup Approaches

Mission-critical applications often require a service-continuity backup mechanism. Enterprises with mission-critical operations require the flexibility of directing data traffic to one or more backup data centers to guard against large regional geographic failures. Subscribing to the services of a *disaster recovery vendor* (DRV) enables redirection of traffic to any number of geographically diverse sites; use of a DRV can be economical because it enables use of the vendor's shared secondary site port and local access channel circuit. Circuit redirection to the secondary host site provides a cost-effective and flexible approach; it is certainly cheaper than subscribing to duplicate, static circuits to the secondary site. Several options for PVC redirection are available today that offer cost savings as well as speed and flexibility [28]:

1. The most popular and economic method of primary site protection via PVC redirection is having predetermined but inactive secondary PVCs, called *backup PVCs* (BPVCs). This method is ideal for users who employ the services of a DRV to provide a backup data center via a shared access port. Under normal conditions, one PVC path is active between the remote locations and the primary site. At the time of failure, the PVCs that are directed to the failed primary site are deactivated, and the PVCs directed to the backup site are activated using predefined recovery scenarios. Redirection can usually be performed within minutes per PVC. With BPVCs the network administrator can avoid doing remote router maintenance at the time of redirection if one is prudent with the scenario development and network addressing schema. Another advantage is that this feature reuses the IXC's bandwidth while also getting the economic benefit of using a shared port and access channel at the third-party vendor's secondary location. A potential disadvantage is that the customer must coordinate with the DRV for use of the DRV's shared port. In general, DRVs adequately size their shared pool of ports and access channels to satisfy multiple simultaneous users. Some DRVs even have the capability to backhaul

circuits on their network to another secondary site if there is contention at the secondary host site.

2. Another approach is to establish dual PVCs provisioned at a lower CIR than their associated primary circuits; this is clearly more cost effective than having full duplicate active circuits. Under normal (nonfailure) conditions, two simultaneous PVC paths are active: the higher CIR path to the primary location site and the lower CIR path to the secondary (backup) location site. At the time of failure, the customer requests the IXC technician to deactivate the PVCs directed to the failed primary site and increase the CIR of the already-active secondary PVCs. Depending on which IXC is used, this is done within minutes per PVC if the user has a predefined plan. This configuration is commonly referred to in the industry as *growable PVCs* (GPVCs) and is appropriate for users who have multiple dedicated data processing sites. GPVCs provide the comfort of having an active secondary circuit that is priced at a much lower rate and is used for load sharing to the secondary host location. The principal disadvantages are the additional cost for a dedicated secondary recovery site with dedicated ports.

3. Only a few IXCs have the technical capability for ATM and FR/ATM interworking of PVC redirection. FR/ATM interworking of PVCs is similar to the other alternatives except that all of the redirection takes place on the ATM side of the gateway interface to the interworking networks (see Figure 9.17). Today all circuit redirections from the primary to secondary site are done on a "like-for-like" port basis. Configuration parameters of the primary and secondary ports must match—or redirection will be difficult. Carriers were planning to allow the indiscriminate redirection across platform types.

4. In a total VPN redirection approach, the dedicated secondary port is configured with a dedicated *customer edge router* (CER) and with full-time connection to a *provider edge router* (PER) in the IXC network (see Figure 9.18). During normal operations, the network sees the simultaneous advertisements from both primary and secondary locations. It will forward any matching IP packets only to the primary site based on the fact that is has a lower cost metric since *as-path prepend* (AS) is utilized to increase the number of AS hops to the secondary location. The result is that the IXC network sees fewer hops to the primary location than the secondary location. Hence, all packets destined to the IP address are forwarded to the primary location. The secondary location can also advertise additional addresses if other applications are active at this site. In case of an activation (test or actual disaster), IP address advertisements from the primary site will cease and the IXC's network will

FIGURE 9.17
*Primary site recovery
via FR/ATM
interworking.*

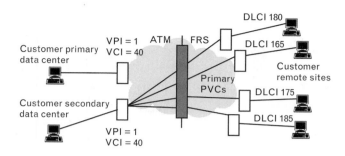

FIGURE 9.18
*VPN redirection to a
dedicated secondary
site port.*

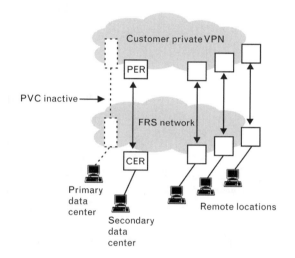

see only the advertisements from the secondary site. The network will begin forwarding all matching packets to the secondary site since the host machine equipment at the secondary site is replicated and has the same addresses at the primary site.

5.   Yet another approach is for an enterprise to contract with a DRV for an alternate data center on a shared basis and redirect their VPN (see Figure 9.19). Redirection of the remote sites to the shared secondary site requires that both protocol layer 3 and layer 2 functional capabilities be used in the recovery. Because sequencing is important, layer 3 activities should be complete before requesting layer 2 redirection by the IXC technician. Before any IP routing changes can affect the way the IXC network forwards IP packets, a PVC connection at layer 2 must be built between the PER for the primary site and the secondary site CER. In essence, the IXC will redirect the preprovisioned PVC originally between the primary site's PER and CER while maintaining all operational parameters of the original PVC. BGP routing takes effect as soon

FIGURE 9.19
VPN redirection to a
shared secondary site
port.

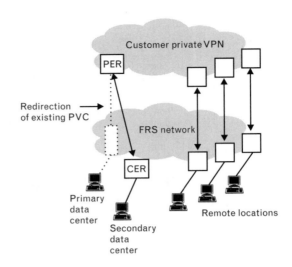

as the PVC is established. In this case, the configuration at the secondary site is usually copied from the primary site.

9.1.7   European Outlook

In Europe there is widespread use of legacy X.25 public data services [29]. It follows that the drivers for FR deployment in Europe are different from those that encouraged the North American FR growth of the past few years. Bandwidth in the United States is considerably cheaper than in Europe (on average, a T1 is only 30% to 40% of the price of an E1 over similar distance), and will remain so until deregulation and a fully competitive market take full effect in Europe. Such tariffing means that historically many European businesses have been much more cost constrained in WAN deployment than their North American counterparts. As an alternative, they chose usage-based, connection-oriented services—typically X.25. The X.25 technology is still generally appropriate for many applications given its SVC mechanisms, per-VC parameter negotiation, and full global reach through intercarrier gateways. During the 1980s and early 1990s, European data service investment was focused on meeting robust demand for X.25, while North America deployed private-line networks. By 1992, FR service began to expand to meet the demand for high-speed, high-performance networks required by increasing numbers of LAN deployments. It follows that, while European FR services are now well established in many countries as the prime choice for medium-speed LAN interconnect, users are often faced with transitioning applications from X.25 rather than private lines. Often, this is more complex and expensive than simply switching a router from PPP over leased lines to public FR service [30]. Europe has a significant installed base of equipment, applications, business processes, and associated investment, all of which is tied to X.25 services. Typical applications include financial services and banking, retail and point of sale, and travel and airline services. Many of these mission–critical applications are not likely to disappear or migrate onto new technology in the short term. The growth of FR deployment in Europe, as in the United States, is driven primarily by the need for LAN interconnects.

The approach and level of sophistication to integrating X.25 and FR varies from simple HDLC encapsulation of the X.25 link level using proprietary techniques or FRF.3/RFC 1490, through fully integrated X.25 level 3 switching and feature support. To understand the differences, we can compare the FR structure with that of X.25. The simplest techniques essentially provide tunneling by encapsulating the X.25 packets at the link level (level 2) and placing the packet within a DLCI. This is similar to the basic approach taken in routers where X.25 is encapsulated within IP, which is then carried in PPP or FR. The IP tunneling approach adds extra load to the routers and significant layers of overhead. The drawbacks of these HDLC encapsulation approaches is that all *logical channel numbers* (LCNs are analogous to DLCIs) must pass through the FR network between the point of encapsulation and the point of decapsulation. This does not integrate the X.25 networks' real switching functionality, but is useful for limited transport of X.25 point-to-point links. Perhaps the biggest drawback is that the problem of managing and maintaining outdated X.25 switches is not solved because they are still required for X.25 switching [30].

At the other end of the interworking spectrum are implementations that combine full X.25 and FR functionality on a single card. This, combined with the simple HDLC transport techniques already described, allows complete replacement of the switches and full integration of a virtual X.25 network within the FR service infrastructure. Of necessity, this approach must maintain standards compatibility with existing access equipment and applications in order to be cut over as a transparent replacement to the current X.25 infrastructure. Such an approach also means that the service port can be converted to FR without additional investment by the service provider. With the more sophisticated approach, additional considerations relate to the extended features of standard X.25. The dominance of SVCs, built-in security functions such as *network user ID* (NUI), and usage-based billing mean that service providers must link operational support systems to routing and billing engines [30].

9.2   Recent Advances in ATM

9.2.1   Overview and Scope

In this section we provide a summary of key ATM capabilities. Quite a bit has been written on this topic [31] (including writings by the present author, who coauthored the first book on IP carriage on ATM [32] and several other books on ATM); therefore, the coverage provided herewith is relatively brief. The material that follows is based liberally on ATM Forum materials. The ATM Forum is a nonprofit international organization of more than 325 organizations representing all sectors of the world's computer and communication industries, as well as government agencies, research organizations, and end users. Established in 1991, the Forum's mission is twofold: (1) to support the acceptance and implementation of ATM and other broadband technologies, applications, and services, based on national and international standards; and (2) to support the interoperability, advancement, and convergence of technologies that enhance ATM and other broadband markets. The ATM Forum has been accelerating the availability of ATM specifications in cooperation with other forums/standards bodies such as the ITU, ETSI, IETF, DSL Forum, and OIF.

ATM is a means of digital communications that is capable of transmission and switching speeds that reach into the gigabit–per–second range per connection. It is the "statistical multiplexing technology" of choice, for which agreement has been reached on what standards to use to support multiplexing. ATM services are used for the transport of voice, video, data, and images. The basic service obtainable on the ATM platform is cell relay, which ensures the delivery of cells in a reliable and consistent manner at the destination. This standards–based transport medium is widely used within the core, at the access, and on the edge of telecommunications systems to transmit data, video, and voice at high speeds. The basic ITU specifications are as follows (among many others):

ITU Q.2100   *B-ISDN Signaling ATM Adaptation Layer Overview;*

ITU Q.2110   *B-ISDN Adaptation Layer—Service Specific Connection Oriented Protocol;*

ITU Q.2130   *B-ISDN Adaptation Layer—Service Specific Connection Oriented Function for Support of Signaling at the UNI;*

ITU Q.2931   The signaling standard for ATM to support-switched virtual connections. This is based on the signaling standard for ISDN.

ATM is best known for its ability to integrate with other technologies and for its sophisticated management features that allow carriers to "assure" QoS [33]. Initially, ATM was supposed to be the overarching technology that "carried" all other services; now it presents itself more as a technology that can *integrate* with other services. QoS and traffic management features are built into the different layers of ATM, giving the protocol an inherently robust set of controls. ATM uses short, fixed-length 53-octet packets called *cells* for transport and switching. Information is divided among these cells, transmitted, and then reassembled at their final destination, as depicted in Figures 9.20 and 9.21.

FIGURE 9.20
ATM scope.

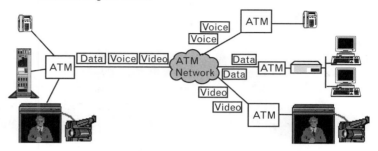

ATM Vision

The ultimate integrated services network

- ATM network moves cells (fixed length packets) with low delay and low delay variation at high speeds
- Devices at ends translate(e.g., segment and reassemble) between cells and original traffic

FIGURE 9.21
Typical IP application.

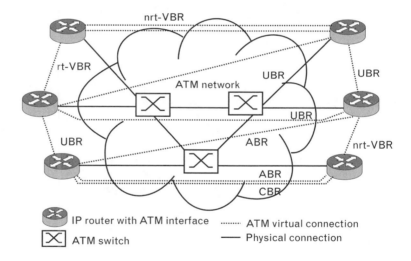

Small PDUs are utilized to maximize transmission bandwidth efficiency. Short, fixed-length cells lend themselves to inexpensive, hardware-based switching. Also, short, fixed-length cells allow tight control on delay at a muxing point. For example, the waiting time on a OC–1 line for a high-priority packet is as follows: If the packet in service is 64 kbytes the delay is 10 ms; if the packet in service is 53 bytes the delay is 9 μs. However, the fixed-length ATM cell has two impacts on overhead as follows:

1. Five of every 53 bytes is cell overhead (this equates to 9.4%).

2. An "application" message is padded to a 48-byte multiple. For example, a 1,500-byte IP packet fits in 32 cells (with 8 AAL-5 trailer

bytes and 28 pad bytes). This equates to an overhead of 2.3%. However, many IP-based packets are smaller, for example, DNS resolves, pings, and so on. For example, a 64-byte IP packet would have to fit in the two cells C1 = 5 + 8 + 40 and C2 = 5 + 24 + 24, where the last 24 bytes are for padding. This is an overhead of 24 on 64, or 37.5%.

Overall, studies show ATM uses 15% to 20% of bandwidth for overhead; this is the so called *cell tax*.

Table 9.6 depicts some of the main distinctions between IP and ATM, and Table 9.7 identifies some of the initial services available under ATM (a new service category called "Guarantee" Frame Rate has been added of late [34]). ATM has seen some end-user deployments in North America during the past 10 years. According to press-time research, 33% of ATM ports installed by customers are implemented with frame to ATM service interworking. Here, the end user has both frame relay and ATM services at different sites in their network and the network performs the translation between frames and cells.

TABLE 9.6   COMPARING IP AND ATM

IP	ATM
Layer 3	Layer 2
Connectionless	Connection orientated
Minimal QoS	Strong QoS
Universal connectivity	ATM islands
Variable-length packet (40–64k bytes)	53-Byte cell

TABLE 9.7   ATM SERVICES AND APPLICATIONS

REAL-TIME SERVICE CATAGORIES

CBR	Voice, circuit emulation
rt-VBR	Compressed voice and video

NON-REAL-TIME SERVICE CATEGORIES

nrt-VBR	VPN, Frame Relay interworking
ABR	Network feedback, low loss
UBR	No QoS, best effort, tradional Internet traffic

Like FR, ATM is connection oriented. Table 9.8 provides a basic lexicon of the types of channels and connections that can be supported in ATM. Like FR, ATM has a layered-protocol architecture that has two (OSI) layers in the U-plane and seven (OSI) layers in the C-plane. Figures 9.22 and 9.23 depict the layered architecture, and Figure 9.24 depicts the interfaces of interest and the network environment. The first layer—known as the *adaptation layer*—aims at providing a transparent substructure to higher layer protocols (IP) or applications (e.g., digital video). The ATM layer handles multiplexing and the formation (assemblage) of the cell (stream). This section directs the transmission. Lastly, the physical layer attaches the electrical elements and network interfaces.

9.2.2 Applications

A telecommunications network is often designed in a series of tiers. A typical configuration may utilize a mix of TDM, FR, ATM, and/or IP tiers. Each of these technologies is better suited for a different section of the network, for example, access, core, international, and so on. Within a network, carriers often extend the capabilities of ATM by blending it with other technologies, such as ATM over SONET/SDH, or DSL over ATM. By doing so, they extend the management features of ATM to other platforms in a cost-effective manner. The vast majority (roughly 80%) of the world's carriers use ATM in the core of their networks. ATM has been widely adopted because of its flexibility in supporting an array of technologies, as we just noted, including DSL, IP Ethernet, FR, SONET/SDH, Cable TV and wireless platforms. It also acts a bridge between legacy equipment and the new generation of operating systems and platforms: ATM communicates with both, allowing carriers to maximize their infrastructure investment.

ATM in the WAN

A blend of ATM, IP, and Ethernet options abounds in the WAN. Carriers turn to ATM when they need midspeed layer 2 transport in the core, coupled with the security of an assured level of QoS. FR backbones are one example of such use of ATM at the core. Distance can be a problem for some high-speed platforms; this is not the case with ATM. The integrity of the transport signal is maintained even when different kinds of traffic are traversing the same network. And because of its ability to scale up to OC-48, different services can be offered at varying speeds and at a range of performance levels.

ATM in the MAN

The MAN is important because it is in the proximity of the user. The typical MAN configuration is a point of convergence for many different types of traffic that are generated by many different sources. The advantage of

TABLE 9.8   BASIC CONNECTIVITY LEXICON

VC	A communications channel that provides for the sequential unidirectional transport of ATM cells.
VCC	*Virtual channel connection.* A concatenation of VCLs that extends between the points where the ATM service users access the ATM layer. The points at which the ATM cell payload is passed to, or received from, the users of the ATM layer (i.e., a higher layer or ATM entity) for processing signify the endpoints of a VCC. VCCs are unidirectional.
VCI	*Virtual channel identifier.* A unique numerical tag as defined by a 16-bit field in the ATM cell header that identifies a virtual channel over which the cell is to travel.
VCL	*Virtual channel link.* A means of unidirectional transport of ATM cells between the point where a VCI value is assigned and the point where that value is translated or removed.
VP	*Virtual path.* A unidirectional logical association or bundle of VCs.
VPC	*Virtual path connection.* A concatenation of VPLs between virtual path terminators. VPCs are unidirectional.
VPI	*Virtual path identifier.* An 8-bit field in the ATM cell header that indicates the virtual path over which the cell should be routed. VPI value is assigned at the point where that value is translated or removed.
VPL	*Virtual path link.* A means of unidirectional transport of ATM cells between the point where a VPI value is assigned and the point where that value is translated or removed.
VPT	*Virtual path terminator.* A system that unbundles the VCs of a VP for independent processing of each VC.
PVC	*Permanent virtual circuit.* This is a link that has the static route defined in advance, usually by manual setup.
PVCC	*Permanent virtual channel connection.* A VCC is an ATM connection in which switching is performed on the VPI/VCI fields of each cell. A PVCC is one that is provisioned through some network management function and left up indefinitely.
PVPC	*Permanent virtual path connection.* A VPC is an ATM connection in which switching is performed on the VPI field only of each cell. A PVPC is one that is provisioned through some network management function and left up indefinitely.
SVC	*Switched virtual circuit.* A connection established via signaling. The user defines the endpoints when the call is initiated.
SVCC	*Switched virtual channel connection.* A VCC is an ATM connection where switching is performed on the VPI/VCI fields of each cell. A SVCC is one that is established and taken down dynamically through control signaling.
SVPC	*Switched virtual path connection.* A VPC is an ATM connection in which switching is performed on the VPI field only of each cell. A SVPC is one that is established and taken down dynamically through control signaling.
Switched connection	A connection established via signaling.
Multipoint-to-multipoint connection	A multipoint-to-multipoint connection is a collection of associated ATM VC or VP links, and their associated nodes, with the following properties: (1) All modes in the connection, called endpoints, serve as root nodes in a point-to-multipoint connection to all of the $(N-1)$ remaining endpoints. (2) Each of the endpoints on the connection can send information directly to any other endpoint, but the receiving endpoint cannot distinguish which of the endpoints is sending information without additional (e.g., higher layer) information.
Multipoint-to-point connection	A point-to-multipoint connection may have zero bandwidth from the root node to the leaf nodes, and nonzero return bandwidth from the leaf nodes to the root node. Such a connection is also known as a multipoint-to-point connection. Note that UNI 4.0 does not support this connection type.

TABLE 9.8   BASIC CONNECTIVITY LEXICON (CONTINUED)

Point-to-multipoint connection	A point-to-multipoint connection is a collection of associated ATM VC or VP links, with associated endpoint nodes, with the following properties: (1) One ATM link, called the root link, serves as the root in a simple tree topology. When the root node sends information, all of the remaining nodes on the connection, called leaf nodes, receive copies of the information. (2) Each of the leaf nodes on the connection can send information directly to the root node. The root node cannot distinguish which leaf is sending information without additional (higher layer) information. (UNI 4.0 does not support traffic sent from a leaf to the root; the leaf nodes cannot communicate directly to each other with this connection type.)
Point-to-point connection	A connection with only two endpoints.

FIGURE 9.22
*ATM protocol model
(U-plane).*

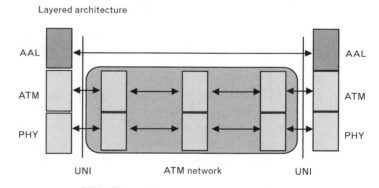

PHY = Physical layer
UNI = User network interface
AAL = ATM adaption layer
  Different AAL protocols for different traffic types
  (e.g., data, voice, video)

FIGURE 9.23
*End-to-end protocol
model, pure ATM
link.*

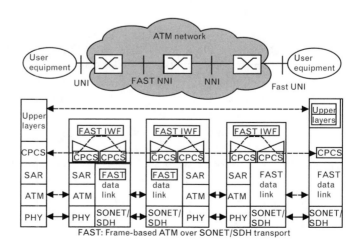

FIGURE 9.24
ATM environments.

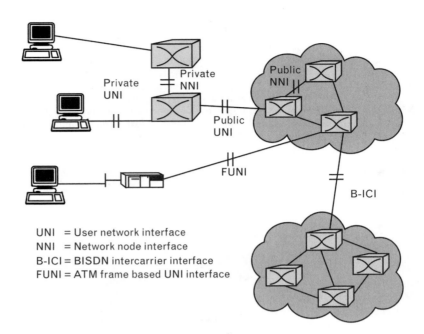

UNI   = User network interface
NNI   = Network node interface
B-ICI = BISDN intercarrier interface
FUNI = ATM frame based UNI interface

ATM in the MAN is that it easily accommodates these divergent transmissions, oftentimes supporting the coexistence of legacy equipment with high-speed networks. Today, ATM scales from T1 to OC-48. (Beyond this, systems operating at 10 Gbps are in relatively limited use and systems operating up to 40 Gbps have been utilized in trials—in particular it is difficult to find low-cost high-bandwidth SAR implementations on router ports operating, say, above OC-48.) However, this area has also seen competition such as gigabit metro Ethernet; furthermore, the advantages of multiplexing are more pronounced in the long-haul and international networks.

9.2.3   ATM Overview

ATM standards establish UNI specifications at layer 1 and 2 for the U-plane, and layers 1 through 7 for the control plane (C-plane) when SVCs are involved. Figure 9.25 depicts the PHY layers supported, and Figure 9.26 depicts the well-known layer 2 cell layout.

Figures 9.27, 9.28, and 9.29 show additional details on the lower layers; in particular, Figure 9.29 depicts the concept of *inverse multiplexing with ATM* (IMA), which allows users to put in place a number of joined T1s to achieve higher combined access speeds without the need (wait) for a DS-3 or fiber facility at the customer's site.

Figures 9.30, 9.31, and 9.32 depict connectivity at layer 2, illustrating some of the concepts that were introduced in Table 9.8.

Figures 9.33 and 9.34 illustrate SVC features, also illustrating some of the concepts that were introduced in Table 9.8. Figure 9.35 defines the addressing mechanisms that are used in conjunction with SVC services.

FIGURE 9.25
ATM PHY layers.

ATM forum physical layer UNI interfaces

	Frame format	Bit rate/line rate	Transmission media
Private UNI	Cell stream	25.6 Mbps / 32 Mbaud	UTP-3
	STS-1	51.84 Mbps	UTP-3
	FDDI	100 Mbps / 125 Mbaud	MMF
	STS-3c, STM-1	155.52 Mbps	UTP-5, STP
	STS-3c, STM-1	155.52 Mbps	SMF, MMF, coax pair
	Cell stream	155.52 Mbps / 194.4 Mbaud	MMF/STP
	STS-3c, STM-1	155.52 Mbps	UTP-3
	STS 12, STM/	622.08 Mbps	SMF, MMF
	STS-48, STV-16	2,488.32 Mbps	SMF
Public UNI	DS1	1.544 Mbps	Twisted pair
	DS3	44.736 Mbps	Coax pair
	STS-3c, STM-1	155.520 Mbps	SMF
	E1	2,048 Mbps	Twisted pair, coax pair
	E3	34.368 Mbps	Coax pair
	J2	6.312 Mbps	Coax pair
	N×T1	N×1.544 Mbps	Twisted pair
	N×E1	N×2.048 Mbps	Twisted pair

FIGURE 9.26
ATM cell.

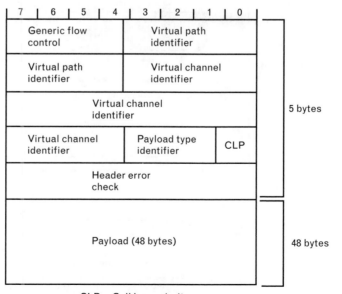

ATM UNI cell

CLP = Cell loss priority

FIGURE 9.27
Typical cell mappings.

155 Mbps, SONET STS-3c/SDH STM-1

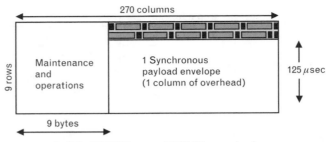

- 9 x 260 x 8/125 125 μsec= 149.76 Mbps payload

1.5 Mbps, DS1

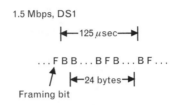

- (24 bytes x 8 bits/byte)/125 μsec = 1.536 Mbps of payload
- Cell delineation by HEC detection as with SONET
- Cell payload = 1.536 Mbps x (48/53) = 1.391 Mbps

FIGURE 9.28
Additional mappings.

44 Mbps, DS3

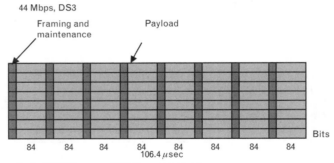

- 7 x 8 x 84/106.4 μsec = 44.21 Mbps of payload

- TCS must delimit 53 byte cells within payload bits

DS3 cell delineation (PLCP mapped)

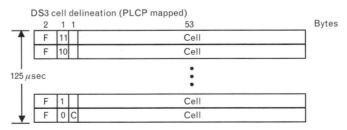

- F is framing pattern
- 12 cells/125 μsec means 96,000 cells per second or
  36.864 Mbps of cell payload
- Stuff is either 13 or 14 nibbles and is indicated by C
- HEC delineation, instead of PLCP, is also specified

FIGURE 9.29
IMA.

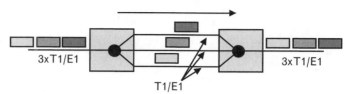

Inverse multiplexing with ATM (IMA)

3xT1/E1   3xT1/E1

T1/E1

• Uses n parallel T1E1 links
• Preserves cell order
• Provides finer granularity of bandwidth use over the WAN

FIGURE 9.30
Example of ATM VCs.

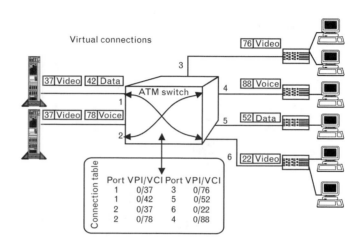

FIGURE 9.31
VC and VP switching.

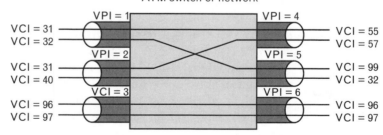

ATM QoS is defined on an end-to-end basis in terms of the following attributes of the end-to-end ATM connection:

• Cell loss ratio;

• Cell transfer delay;

• Cell delay variation.

FIGURE 9.32
PVC setup.

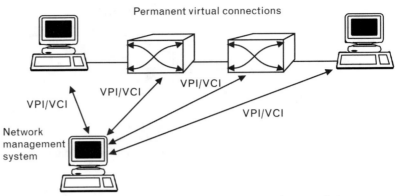

Permanent virtual connections

VPI/VCI

VPI/VCI

VPI/VCI

VPI/VCI

VPI/VCI

Network management system

Long setup time (especially with human intervention) means that connections are left active for long periods of time, e.g, days, weeks

VPI/VCI tables are setup in terminals and switches

FIGURE 9.33
Signaling.

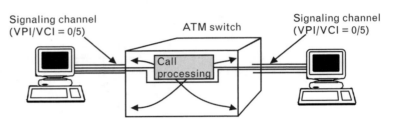

Signaling channel (VPI/VCI = 0/5)

ATM switch

Signaling channel (VPI/VCI = 0/5)

Call processing

Switch and terminal exchange signaling messages using the predefined signaling channel, VPI/VCI = 0/5

Table 9.9 describes the connectivity QoS categories that that can be achieved over the PVCs and/or SVCs. Figure 9.36 provides a pictorial summary, and Figure 9.37 lists service features.

Table 9.10 summarizes the functionality of the various *ATM adaptation layers* (AALs) required to support the service categories. QoS is secured via *traffic management* (TM). Figures 9.38 and 9.39 depict basic mechanisms required for TM. TM is the aspect of the traffic control and congestion control procedures for ATM. *ATM layer traffic control* refers to the set of actions taken by the network to avoid congestion conditions. *ATM layer congestion control* refers to the set of actions taken by the network to minimize the intensity, spread, and duration of congestion. The following functions form a framework for managing and controlling traffic and congestion in ATM networks and may be used in appropriate combinations:

- Connection admission control;
- Feedback control;
- Usage parameter control;

FIGURE 9.34
Setting up a call.

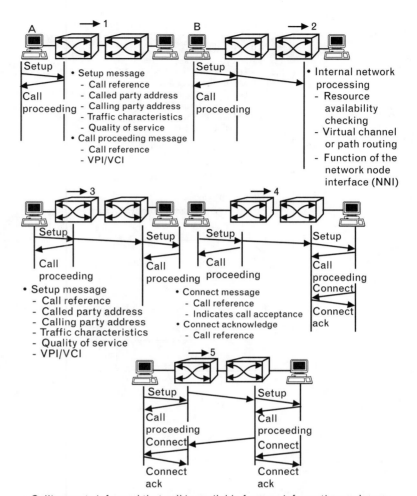

Calling party informed that call is available for user information exchange

- Priority control;

- Traffic shaping;

- Network resource management;

- Frame discard;

- ABR flow control.

Figure 9.40 illustrates AAL-1; Figure 9.41 shows AAL-2; and Figure 9.42 depicts AAL-5. Figure 9.43 gives a pictorial view of congestion management to support traffic types such as ABR.

Figures 9.44, 9.45, and 9.46 depict aspects of the new GFR. Finally, Figure 9.47 shows the ATM management model.

FIGURE 9.35
Addressing in ATM.

Address formats

• Private networks
  - 20 byte address
  - Format modeled after OSI NSAP (network service access point)
  - Mechanisms for administration exist
  - Hierarchical structure will facilitate virtual connection
    routing in large ATM networks
• Public networks
  - E.164 numbers (telephone numbers)
  - Up to 15 digits
• Private networks
  - LAN Mac address will be encapsulated within NSAP

Private address formats

Data country code

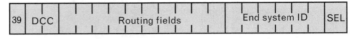

International code designator

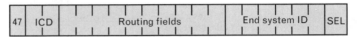

E.164 private address

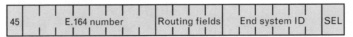

TABLE 9.9   ATM QoS CATEGORIES

ABR   *Available bit rate.* ABR is an ATM layer service category for which the limiting ATM layer transfer characteristics provided by the network may change subsequent to connection establishment. A flow control mechanism is specified that supports several types of feedback to control the source rate in response to changing ATM layer transfer characteristics. It is expected that an end system that adapts its traffic in accordance with the feedback will experience a low cell loss ratio and obtain a fair share of the available bandwidth according to a network-specific allocation policy. Cell delay variation is not controlled in this service, although admitted cells are not delayed unnecessarily.

CBR   *Constant bit rate.* An ATM service category that supports a constant or guaranteed rate to transport services such as video or voice as well as circuit emulation that requires rigorous timing control and performance parameters.

GFR   *Guaranteed frame rate.* GFR optimizes the handling of packet-based LAN traffic that otherwise relies on UBR service across ATM backbones. ATM Forum TM 5.0 GFR delivers entire frames through network, and not partial frames.

UBR   *Unspecified bit rate.* UBR is an ATM service category that does not specify traffic-related service guarantees. Specifically, UBR does not include the notion of a per-connection negotiated bandwidth. No numerical commitments are made with respect to the cell loss ratio experienced by a UBR connection, or as to the cell transfer delay experienced by cells on the connection.

VBR   *Variable bit rate.* An ATM Forum-defined service category that supports variable-bit-rate data traffic with average and peak traffic parameters.

FIGURE 9.36
Service classes.

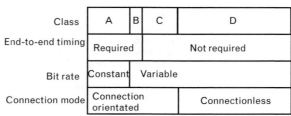

- Class A:
  - 64-Kbps digital voice
- Class B:
  - Variable bit rate encoded video?
- Class C:
  - Frame relay over ATM
- Class D:
  - CCITT I.364 (SMDS) over ATM
- Class X:
  - Raw cell service (e.g., proprietary AAL)

FIGURE 9.37
*ATM service
categories.*

ATM service categories

- CBR
  - Constant bit rate
  - Continous flow of data with tight bounds on delay and delay variation
- rt-VBR
  - Real-time variable bit rate
  - Variable bandwidth with tight bounds on delay and delay variation
- nrt-VBR
  - Nonreal-time variable bit rate
  - Variable bandwidth with tight bound on cell loss
- UBR
  - Unspecified bit rate
  - No guarantees (i.e., best effort delivery)
- ABR
  - Available bit rate
  - Flow control on source with tight bound on cell loss

TABLE 9.10   AAL-RELATED DEFINITIONS

AAL	*ATM adaptation layer.* The standards layer that allows multiple applications to have data converted to and from the ATM cell. A protocol is used that translates higher layer services into the size and format of an ATM cell.
AAL connection	Association established by the AAL between two or more next higher layer entities.
AAL-1	*ATM adaptation layer type 1.* AAL functions in support of constant-bit-rate, time-dependent traffic such as voice and video.
AAL-2	*ATM adaptation layer type 2.* This AAL is still undefined by the international standards bodies. It is a placeholder for variable-bit-rate video transmission.
AAL-3/4	*ATM adaptation layer type 3/4.* AAL functions in support of variable-bit-rate, delay-tolerant data traffic requiring some sequencing and/or error detection support. Originally was two AAL types (i.e., connection-oriented and connectionless), but they have been combined.
AAL-5	*ATM adaptation layer type 5.* AAL functions in support of variable-bit-rate, delay-tolerant connection-oriented data traffic requiring minimal sequencing or error detection support.

FIGURE 9.38
*Traffic management
mechanism.*

Cell loss priority

- Cells with bit set should be discarded before those with bit not set
- Can be set by the terminal
- Can be set by ATM switches for internal network control
    - Virtual channels/paths with low quality of service
    - Cells that violate traffic management contract
- Key to ATM traffic management

FIGURE 9.39
*Traffic management
methods.*

Traffic management

- Problem: providing quality of service
    - How should ATM network resources be allocated to ensure good performance including preventing congestion, e.g., how many virtual channels should be assigned to a particular transmission link?
- Solution: traffic management
    - Specify the traffic "contract" on each virtual channel/path
    - Route (including rejecting setup request) each virtual channel/path along a path with adequate resources (admission control)
    - Mark (via cell loss priority bit) for loss all cells that violate the contract (traffic policing)

Generic cell rate algorithm

- For a sequence of cell arrival times,$\{t_k\}$, determines which cells conform to the traffic contract

    - A counter scheme based on two parameters denoted GCRA(I,L)
        - Increment parameter: I
            · affects cell rate
        - Limit parameter: L
            · affects cell bursts
    - "Leaky bucket"
        - A cell that would cause the bucket to overflow is nonconforming

I for each cell arrival

L + 1

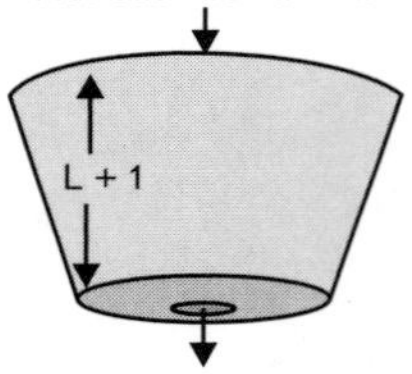

One unit leak per
unit of time

9.2.4   ATM Standardization

The standardization of ATM started in ITU-T about 15 years ago during the study period from 1984 to 1988, and resulted in the I.121 recommendation on *Broadband Aspects of ISDN*. Since then, ATM standards have rapidly evolved and are defined in numerous recommendations that describe the major characteristics of the technology. The ATM Forum produces specifications that other groups often use and convert them into standards. Table 9.11 depicts a list of ATM Forum works in progress. Table 9.12 provides a listing of all specifications completed and approved by the ATM Forum since its inception in 1991.

9.2.5   Interworking

As ATM evolved, other emerging technologies have come into play in the telecommunications industry. Many of these technologies are complementary to ATM. Several organizations have found improved solutions in

FIGURE 9.40
AAL-1.

AAL1 for class A

1 byte	47 bytes
AAL1 header	Payload

Header functions include:
- Lost cell detection
  · Used by adaptive clock method
- Byte alignment
  · Allows channelized circuit emulation, e.g., channelized DS1, E1, etc.
- Time stamp
  · Used for end-to-end clock synchronization, e.g.,
    synchronous residual time stamp method

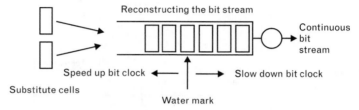

Received cells

Reconstructing the bit stream

Continuous bit stream

Speed up bit clock ◄——— ———► Slow down bit clock

Substitute cells

Water mark

Bit stream rate is independent of ATM network
and (theoretically) can be any value

Cell delay variation is critical to buffer sizing and bit clock jitter

FIGURE 9.41
AAL-2.

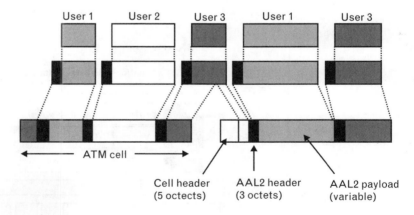

AAL2 for class B

Small payload to reduce packetization delay

User 1   User 2   User 3   User 1   User 3

◄——— ATM cell ———►

Cell header   AAL2 header   AAL2 payload
(5 octets)   (3 octets)   (variable)

combining ATM with other technologies. This is an overlay model, where IP rides on a service such as ATM. However, some purists want to operate IP over SONET (e.g., POS) or IP over optics (as we have discussed in earlier chapters of this book). There is a role for both approaches: Medium-

FIGURE 9.42
AAL-5.

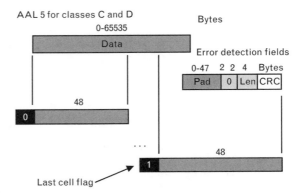

AAL 5 for classes C and D

0-65535

Data

Bytes

Error detection fields

0-47 2 2 4 Bytes

Pad | 0 | Len | CRC

48

0

. . .

48

1

Last cell flag

- 48 bytes of data per cell
- Uses a PTI bit to indicate last cell
- Only one packet at a time on a virtual connection

FIGURE 9.43
ABR feedback.

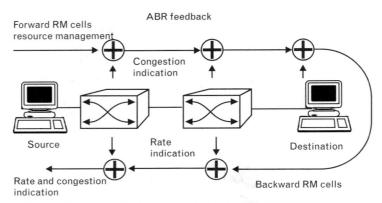

ABR feedback

Forward RM cells
resource management

Congestion
indication

Source

Rate
indication

Destination

Rate and congestion
indication

Backward RM cells

- Source sets actual cell rate based on rate and congestion feedback

FIGURE 9.44
ATM frames.

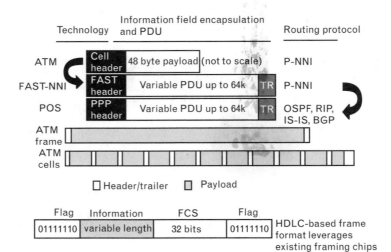

Technology

Information field encapsulation
and PDU

Routing protocol

ATM | Cell header | 48 byte payload (not to scale) | | P-NNI

FAST-NNI | FAST header | Variable PDU up to 64k | TR | P-NNI

POS | PPP header | Variable PDU up to 64k | TR | OSPF, RIP, IS-IS, BGP

ATM frame

ATM cells

☐ Header/trailer   ▨ Payload

Flag | Information | FCS | Flag

01111110 | variable length | 32 bits | 01111110

HDLC-based frame
format leverages
existing framing chips

FIGURE 9.45
GFR.

Router-interconnect

Service category	Strengths	Weaknesses
UBR	simple source, queuing, CAC	no QoS
nrt-VBR	strict CLR control	rigid source model, market perception problem no frame awareness
ABR	MCR, good utilization	complex

• Needed: minimum guaranteed throughput, high utilization, frame awareness, simple operation

• Solution: guaranteed frame rate (GFR)
  • "Frame" = IP packet
  • Segmented into many cells then reassembled
  • Deliver entire frame or discard entire frame
  • Need to define "QoS-eligible" traffic (within MCR) uniformly for all cells of a frame
  • New tool - Frame-based leaky bucket (F-GCRA):
    - check credit when first cell of frame arrives,
    - whole frame passes or whole frame is best effort

GFR is like...(depending on point of view):

  • UBR with a minimum rate
  • VBR with MCR acting like SCR
  • ABR with no feedback

GFR adds frame awareness to the conformance definition

FIGURE 9.46
GFR example.

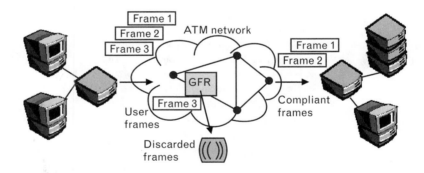

FIGURE 9.47
ATM management.

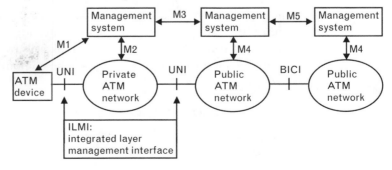

TABLE 9.11   WORK IN PROGRESS ITEMS AS OF MARCH 2002

TECHNICAL WG	MAJOR WORK EFFORTS	STATUS	SPEC FORECAST
Control signaling	Loop Detection	Final ballot	April 2002
	PNNI 1.1	Final ballot	April 2002
	Signaling Congestion Control	Final ballot	April 2002
	UNI 4.1 Signaling	Final ballot	April 2002
	Call Processing	Final ballot	April 2002
	Signaling and Routing Support of ATM-MPLS Network Interworking	Work in progress	TBD
	Routing Resynchronization	Work in progress	TBD
	Proxy Enhancements	Work in progress	TBD
	Fast SVC Restoration	Work in progress	TBD
	Routing Rate Control	Work in progress	TBD
	Policy Routing	Work in progress	TBD
	SPVC-ILMI Interworking	Work in progress	TBD
	PNNI for Transport Networks	Work in progress	TBD
Network management	SNMP MIB for ATM Layer APS	Final ballot	April 2002
	Data Transfer Format	Straw ballot	July 2002
	M4 Network View CORBA Model	Straw ballot	July 2002
Security	Security Renegotiation with OAM Cells	Final ballot	April 2002
	Methods for Security Managing ATM Network Elements—Implementation Agreement	Final ballot	April 2002
	ATM Connection Filtering MIB and Audit Log	Straw ballot	July 2002
	Addendum to Security Specification v1.1, Secure CBR Traffic in a Policed Network	Straw ballot	July 2002
	Addendum to Security Specification v1.1 In-Band Security for Simplex Connections	Straw ballot	July 2002
Testing	Conformance Abstract Test Suite for Signaling (UNI 3.1) for the Network Side v2.0 Errata	Straw ballot	June 2002
	Performance Testing Specification v1.1	Baseline/living list	TBD
	RMON MIB Upgrade	Work in progress	
	UNI SIG Performance Test Suite Enhancements	Living list	
	Conf ATS UNI 4.0 SIG for the Network Side	Baseline	TBD
Voice and multimedia over ATM	Loop Emulation Service PICS	Straw ballot	July 2002
	LES Using AAL2—H.248 Signaling Addendum	Straw ballot	July 2002

range applications can utilize ATM, particularly if this is priced aggressively compared to a PL solution. High-end applications may want to take the IP over optics route. In this section, we look at how ATM interworks with various technologies, the advantage of the combination, and the progress of standards work.

9.2.5.1   ATM and IP Interworking

ATM and IP can be seen as complementary platforms. The discussion often becomes which of the two are more beneficial, but the best case

TABLE 9.12   APPROVED ATM FORUM SPECIFICATIONS

TECHNICAL WORKING GROUP	APPROVED SPECIFICATIONS	APPROVED DATE
AIC/ATM-IP Collaboration (formerly LANE)	ATM-MPLS Network Interworking 1.0	August 2001
	LAN Emulation over ATM 1.0	January 1995
	LAN Emulation Client Management Specification	September 1995
	LANE 1.0 Addendum	December 1995
	LANE Servers Management Specification v1.0	March 1996
	LANE v2.0 LUNI Interface	July 1997
	LAN Emulation Client Management Specific Specification v2.0	October 1998
	MultiProtocol over ATM Specification v1.0	July 1997
	MultiProtocol over ATM v1.0 MIB	July 1998
	Multiprotocol over ATM Specification, v1.1	May 1999
	MPOA v1.1 Addendum on VPN Support	October 1999
B-ICI	B-ICI 1.0	September 1993
	B-ICI 1.1	
	B-ICI 2.0 (delta spec to B-ICI 1.1)	December 1995
	B-ICI 2.0 (integrated specification)	December 1995
	B-ICI 2.0 Addendum or 2.1	November 1996
Control signaling	PNNI Addendum on PNNI/B-QSIG Interworking and Generic Functional Protocol for the Support of Supplementary Services	October 1998
	Addressing Addendum for UNI Signaling 4.0	February 1999
	PNNI Transported Address Stack, v1.0	May 1999
	PNNI v1.0 Security Signaling Addendum	May 1999
	UNI Signaling 4.0 Security Addendum	May 1999
	ATM Inter-Network Interface (AINI) Specification	July 1999
	PNNI Addendum for Generic Application Transport v1.0	July 1999
	PNNI SPVC Addendum v1.0	July 1999
	PHY/MAC Identifier Addendum to UNI Signaling 4.0	November 1999
	Network Call Correlation Identifier v1.0	March 2000
	PNNI Addendum for Path and Connection Trace, v1.0	March 2000
	Operation of the Bearer Independent Call Control (BICC) Protocol with SIG 4.0/PNNI 1.0-AINI	July 2000
	UBR with MDCR Addendum to UNI 4.0/PNNI 1.0 AINI	July 2000
	Behavior Class Selector Signaling v1.0	October 2000
	Modification of Traffic Parameters for an Active Connection Signaling Specification (PNNI, AINI, and UNI) v2.0	May 2001
	Guaranteed Frame Rate (GFR) Signaling (PNNI, AINI, and UNI) v1.0	August 2001
	Domain-Based Rerouting for Active Point-to-Point Calls, v1.0	August 2001
	Loop Detection, v1.0	April 2002
	Signaling Congestion Control, v1.0	April 2002
	Call Processing Priority, v1.0	April 2002
	Private Network–Network Interface Specification v1.1	April 2002
	ATM User Network Interface (UNI) Signaling Specification v4.1	April 2002
Data exchange interface	Data Exchange Interface v1.0	August 1993

TABLE 9.12   APPROVED ATM FORUM SPECIFICATIONS (CONTINUED)

TECHNICAL WORKING GROUP	APPROVED SPECIFICATIONS	APPROVED DATE
Directory and naming services	ATM Named System v2.0	July 2000
Frame-based ATM	Frame-based ATM Transport over Ethernet (FATE)	March 2000
	Frame-Based ATM over SONET/SDH	July 2000
ILMI	ILMI 4.0	September 1996
Network Management	Customer Network Management (CNM) for ATM Public Network Service	October 1994
	M4 Interface Requirements and Logical MIB	October 1994
	M4 Interface Requirements and Logical MIB: ATM Network Element View	October 1998
	CMIP Specification for the M4 Interface	September 1995
	CMIP Specification for the M4 Interface: ATM Network Element View, v2	July 1999
	M4 Public Network View	March 1996
	M4 Interface Requirements and Logical MIB: ATM Network View, v2	May 1999
	M4 "NE View"	January 1997
	Circuit Emulation Service Interworking Requirements, Logical and CMIP MIB	January 1997
	M4 Network View CMIP MIB Spec v1.0	January 1997
	M4 Network View Requirements & Logical MIB Addendum	January 1997
	ATM Remote Monitoring SNMP MIB	July 1997
	SNMP M4 Network Element View MIB	July 1998
	Network Management M4 Security Requirements and Logical MIB	January 1999
	Autoconfiguration of PVCs	May 1999
	Requirements and Logical MIB for Management of Path and Connection Trace	April 2001
	ATM Usage Measurement Requirements	November 2000
Physical layer	Issued as part of UNI 3.1: 44.736 DS3 Mbps Physical Layer 100 Mbps Multimode Fiber Interface Physical Layer 155.52 Mbps SONET STS-3c Physical Layer 155.52 Mbps Physical Layer	
	ATM Physical Medium Dependent Interface Specification for 155 Mbps over Twisted Pair Cable	September 1994
	DS1 Physical Layer Specification	September 1994
	UTOPIA	March 1994
	Mid-Range Physical Layer Specification for Category 3 UTP	September 1994
	6,312 Kbps UNI Specification	June 1995
	E3 UNI	August 1995
	UTOPIA Level 2	June 1995
	Physical Interface Specification for 25.6 Mbps over Twisted Pair	November 1995
	Cell-Based Transmission Convergence Sublayer for Clear Channel Interfaces	January 1996
	622.08 Mbps Physical Layer	January 1996
	155.52 Mbps Physical Layer Specification for Category 3 UTP (see also UNI 3.1, af-uni-0010.002)	November 1995
	120 Ohm Addendum to ATM PMD Interface Specification for 155 Mbps over TP	January 1996

TABLE 9.12   APPROVED ATM FORUM SPECIFICATIONS (CONTINUED)

TECHNICAL WORKING GROUP	APPROVED SPECIFICATIONS	APPROVED DATE
	DS3 Physical Layer Interface Spec	March 1996
	155 Mbps over MMF Short Wave Length Lasers, Addendum to UNI 3.1	July 1996
	WIRE (PMD to TC layers)	July 1996
	E-1 Physical Layer Interface Specification	September 1996
	155 Mbps over Plastic Optical Fiber (POF) v1.0	May 1997
	155 Mbps Plastic Optical Fiber and Hard Polymer Clad Fiber PMD Specification v1.1	January 1999
	Inverse ATM Mux v1.0	July 1997
	Inverse Multiplexing for ATM (IMA) Specification v1.1	March 1999
	Physical Layer High Density Glass Optical Fiber Annex	February 1999
	622 and 2488 Mbps Cell-Based Physical Layer	July 1999
	ATM on Fractional E1/T1	October 1999
	2.4 Gbps Physical Layer Specification	October 1999
	Physical Layer Control	October 1999
	Utopia 3 Physical Layer Interface	November 1999
	Specification of the Device Control Protocol (DCP) v1.0	March 2000
	Multiplexed Status Mode (MSM3)	March 2000
	Frame-Based ATM Interface (Level 3)	March 2000
	UTOPIA Level 4	March 2000
	Cell-Based 1,000 Mbps (CB1G)Physical Layer Specification over Single-Mode or Multimode Fiber and Category 6 Twisted Pair Copper Cabling	April, 2001
P-NNI	Interim Inter-Switch Signaling Protocol	December 1994
	P-NNI V1.0	March 1996
	PNNI 1.0 Addendum (soft PVC MIB)	September 1996
	PNNI ABR Addendum	January 1997
	PNNI v1.0 Errata and PICs	July 1997
	Private Network-Network Interface Specification v1.1	April 2002
Routing and addressing	PNNI Augmented Routing (PAR) v1.0	January 1999
	ATM Forum Addressing: User Guide v1.0	January 1999
	ATM Forum Addressing: Reference Guide	February 1999
	PNNI Addendum for Mobility Extensions v1.0	May 1999
	ATM Bi-Level Addressing Document, v1.0	April 2001
	Addendum to PNNI, v1.0-Secure Routing	November 2001
Residential broadband	Residential Broadband Architectural Framework	July 1998
	RBB Physical Interfaces Specification	January 1999
Service aspects and applications	Frame UNI	September 1995
	Circuit Emulation	September 1995
	Native ATM Services: Semantic Description	February 1996
	Audio/Visual Multimedia Services: Video on Demand v1.0	January 1996
	Audio/Visual Multimedia Services: Video on Demand v1.1	March 1997
	ATM Names Service	November 1996
	FUNI 2.0	July, 1997

TABLE 9.12   APPROVED ATM FORUM SPECIFICATIONS (CONTINUED)

TECHNICAL WORKING GROUP	APPROVED SPECIFICATIONS	APPROVED DATE
	Native ATM Services DLPI Addendum v1.0	February 1998
	API Semantics for Native ATM Services	February 1999
	FUNI Extensions for Multimedia	February 1999
	H.323 Media Transport over ATM	July 1999
Security	ATM Security Framework v1.0	February 1998
	ATM Security Specification v1.0	February 1999
	ATM Security Specification v1.1	March 2001
	Security Specification v1.1 Protocol Implementation Conformance Statement (PICS) Proforma Specification	March 2001
	Control Plane Security	November 2001
	Methods of Securely Managing ATM Network Elements-Implementation Agreements, v1.1	April 2002
	Security Services Renegotiation Addendum to Security, v1.1	March 2002
	Addendum to Security Specification v1.1, In-Band Security for Simplex Connections	July 2002
	ATM Connection Filtering MIB and Audit Log	July 2002
	Addendum to Sec 1.1 Secure CBR Traffic in a Policed Network	July 2002
Signaling	UNI Signaling 4.0	July 1996
	Signaling ABR Addendum	January 1997
	ATM User Network Interface (UNI) Signaling Specification v4.1	April 2002
Testing	Introduction to ATM Forum Test Specifications	December 1994
	PICS Proforma for the DS3 Physical Layer Interface	September 1994
	PICS Proforma for the SONET STS-3c Physical Layer Interface	September 1994
	PICS Proforma for the 100 Mbps Multimode Fibre Physical Layer Interface	September 1994
	PICS Proforma for the ATM Layer (UNI 3.0)	April 1995
	Conformance Abstract Test Suite for the ATM Layer for Intermediate Systems (UNI 3.0)	September 1995
	Interoperability Test Suite for the ATM Layer (UNI 3.0)	April 1995
	Interoperability Test Suites for Physical Layer: DS-3, STS-3c, 100 Mbps MMF (TAXI)	April 1995
	PICS Proforma for the DS-1 Physical Layer	April 1995
	Conformance Abstract Test Suite for the ATM Layer (End Systems) UNI 3.0	January 1996
	PICS for AAL5 (ITU spec)	January 1996
	PICS Proforma for the 51.84 Mbps Mid-Range PHY Layer Interface	January 1996
	Conformance Abstract Test Suite for the ATM Layer of Intermediate Systems (UNI 3.1)	January 1996
	PICS for the 25.6 Mbps over Twisted Pair Cable (UTP-3) Physical Layer	March 1996
	Conformance Abstract Test Suite for the ATM Adaptation Layer (AAL) Type 5 Common Part (Part 1)	March 1996
	PICS for ATM Layer (UNI 3.1)	July 1996
	Conformance Abstract Test Suite for the UNI 3.1 ATM Layer of End Systems	June 1996
	Conformance Abstract Test Suite for the SSCOP Sub-Layer (UNI 3.1)	September 1996

TABLE 9.12   APPROVED ATM FORUM SPECIFICATIONS (CONTINUED)

TECHNICAL WORKING GROUP	APPROVED SPECIFICATIONS	APPROVED DATE
	SSCOP Conformance Abstract Test Suite, v1.1	May 1999
	PICS for the 155 Mbps over Twisted Pair Cable (UTP-5/STP-5) Physical Layer	November 1996
	PICS for Direct Mapped DS-3	July 1997
	Conformance Abstract Test Suite for Signaling (UNI 3.1) for the Network Side	September 1997
	Abstract Test Suite for UNI 3.1 ATM Signaling for the Network Side v 2.0	March 2000
	ATM Test Access Function (ATAF) Specification v1.0	February 1998
	PICS for Signaling (UNI v3.1)—User Side	April 1998
	Interoperability Test for PNNI v1.0	February 1999
	PICS Proforma for UNI 3.1 Signaling (Network Side)	May 1999
	ATM Forum Performance Testing Specification	October 1999
	Implementation Conformance Statement (ICS) Proforma Style Guide	March 2000
	Conformance ATS for PNNI Routing	October 2000
	Conformance ATS for PNNI Signaling	October 2000
	Conformance ATS for ABR Source and Destination Behaviors	January 2001
	UNI Signaling Performance Test Suite	October 2000
	Introduction to ATM Forum Test Specifications, v2.0	October 2001
Traffic Management	Traffic Management 4.0	April 1996
	Traffic Management ABR Addendum	January 1997
	Traffic Management 4.1	March 1999
	Addendum to TM 4.1: Differentiated UBR	July 2000
	Addendum to Traffic Management v4.1: Optional Minimum Desired Cell Rate Indication for UBR	July 2000
Voice and Telephony over ATM	Circuit Emulation Service 2.0	January 1997
	Voice and Telephony over ATM to the Desktop	May 1997
	Voice and Telephony over ATM to the Desktop	February 1999
	(DBCES) Dynamic Bandwidth Utilization in 64 KBPS Time Slot Trunking Over ATM—Using CES	July 1997
	ATM Trunking Using AAL1 for Narrow Band Services v1.0	July 1997
	ATM Trunking Using AAL2 for Narrowband Services	February 1999
	Low Speed Circuit Emulation Service	May 1999
	ICS for ATM Trunking Using AAL2 for Narrowband Services	May 1999
	Low Speed Circuit Emulation Service (LSCES) Implementation Conformance Statement Performance	October 1999
	Loop Emulation Service Using AAL2	July 2000
	Loop Emulation Service Using AAL2 File Transfer Addendum	October 2001
	Loop Emulation Service Using AAL2 CP-IWF MIB Addendum	October 2001
UNI	ATM User-Network Interface Specification V2.0	June 1992
	ATM User-Network Interface Specification V3.0	September 1993
	ATM User-Network Interface Specification V3.1	1994

scenario is actually built on their unified strengths. To support reliable and cost-effective communication, a certain well-defined set of functions needs to be undertaken. Some may want to place these functions at layer 2, whereas others want to place these functions at layer 3. That would be analogous to someone placing a certain job function under the CTO of a company, while in another company, the same function could be placed under the CIO. What is important is to not needlessly replicate the function. Naturally, the technical issues are more complex, but the point is that design flexibility exists in terms of where some functions can be located in the protocol stack.

Through mass adoption of the Internet and the associated protocols, IP has become the de facto interface for desktop applications. Most IP/data carriers maintain (and have maintained for years) an ATM layer over which IP traffic rides, because of the traffic engineering and QoS features that are inherent to ATM and lacking in traditional connectionless IP. Also, ATM has strong OAM&P support. Using ATM, carriers can grow their IP networks to offer:

- Scalability;
- Traffic engineering;
- Service differentiation;
- High availability;
- Value-added applications such as VPNs.

ATM's shortfall in data environments of adding a lot of overhead in packet-oriented networks by introducing the so called *cell tax* has been overcome through the ATM Forum's adoption of the *frame-based ATM over SONET/SDH transport* (FAST) specification as an industry standard for heavily data-oriented ATM networks. It is in this way that ATM can add value to IP networks and enable them to scale while simultaneously enabling other non-IP applications and services to reside on the same core infrastructure. As noted earlier, the availability of high-speed/low-cost SAR technology for routers remains an issue; this predicament tends to drive packet-based (e.g., MPLS) alternatives for OC-192 and higher applications.

This discussion does not dogmatically imply that IP over ATM is the only way to go. It greatly depends of the application at hand. This author could cite an (actual) anecdotal example from the early 1980s, when the prevailing attitude was "I've got COBOL, what can I do with it?" instead of "I've got a (complex, demanding) application that needs to be developed with requirement 1, requirement 2, requirement 3, and so on. Which programming language of the various ones that are available is best suited to support this application?" The answer is not always COBOL; neither is it always ATM.

The ATM Forum recognized the merits of IP in private as well as public networks early (see Table 9.13) and has since developed a number of specifications within the ATM Forum to interwork the two services. More specifications were developed in conjunction with other standards bodies to ensure that ATM continues to add value to IP-based services and applications. Of the many applications and specifications developed in order to support data and specifically IP over ATM, the best known are these:

- *MultiProtocol over ATM* (MPOA) as an addition to *LAN emulation* (LANE) for LAN/MAN environments;
- Additions to ATM's routing protocol PNNI;
- The adoption of FAST for service provider environments.

Driving and steering all of the efforts surrounding ATM/IP interworking within the ATM Forum and in cooperation with other standards bodies is the ATM/IP Collaboration (AIC) Working Group, established to ensure that ATM continues to add value to IP-based services and applications.

Another industry development approaching mainstream implementation is MPLS. The protocol work for MPLS is conducted in the IETF with which the ATM Forum has an ongoing working relationship regarding MPLS/ATM integration issues. Additional efforts surrounding MPLS and ATM/MPLS originate from the MPLS Forum and ITU-T with which the ATM Forum has established informal or working relationships. All of these efforts translate into ATM switches being one of the generic devices that have evolved to become MPLS *label switch routers* (LSRs) or, more precisely, hybrid devices that act both as ATM switch as well as MPLS ATM LSRs.

ATM's position in service provider networks is increasing in the broadband access, aggregation layer, and mobile networks. The core is evolving to OTN/ASTN/ASON. All of this industry's efforts center

TABLE 9.13   IP ENVIRONMENT

IP is critically important at Layer 3

ATM has Qos, routing, etc., that provide good layer 2 support to IP

Improvements to ATM allow it to "play well":

Frame-based ATM (FAST) to avoid cell tax

Guaranteed frame rate (GFR): QoS and best effort on frames, simple to implement

Differentiated UBR: relative service based on "behavior classes" of UBR connections around a new, unified control plane, right from the access layer into the core based on MPLS and named generalized MPLS (GMPLS) or other mechanism, as discussed earlier in the book.

In the light of these developments, ATM clearly establishes itself as the technology of choice in the access layer, the aggregation layer, and, depending on the size of the network, even in the core.

9.2.5.2   ATM and FR Interworking

As noted earlier in the chapter, simplicity is the hallmark of FR. This connection-oriented packet-switching technology is designed to efficiently transmit data traffic at data rates of up to approximately E3/T3 without the connection and traffic management overhead of its predecessor, X.25. Multiservice integration and higher level management functions are left to ATM in the network core. What FR lacks, ATM delivers. Interfacing with FR was one of the ATM Forum's first missions, working closely with the FR Forum to standardize interworking between FR and ATM. It began with FR networks being able to use an ATM core for FR–FR connectivity and has since evolved to full interworking capabilities between ATM and FR nodes and end users, allowing ATM and FR nodes to communicate directly with each other.

ATM gave FR an evolution path, extending to higher data rates of up to 10 Gbps today from E3/T3 rates available with FR and adding critical core features and applications that banned FR from evolving to a true core network technology. However, FR's simplicity served (and continues to do so) operators well in the lower speed portion of packetized data-oriented networks. In the core network ATM complements FR by adding many of the applications and features required in a multiservice switched core network such as QoS as well as class of service, traffic engineering, different service classes, and so on. As such, ATM and FR continue to be complementary technologies and continue to be deployed as such.

FR is one of the largest consumers of ATM technologies. The blending of the two protocols continues to bridge legacy data devices and next generation networks with a cost-effective architecture that employs end-to-end management techniques. The FR/ATM combo can expect continued success in filling the technology gap as the world strives toward universal communication protocols.

9.2.5.3   ATM and GbE Interworking

GbE enjoys success both in private as well as in some public networks. Widely accepted Ethernet has been in operation for decades. Only recently, though, has it been extended to gigabit rates beginning at speeds of 1.25 Gbps and now 10 Gbps. Although the ever-popular Ethernet is an open, easy-to-use technology, the high speeds of GbE can be enhanced when coupled with ATM. At 10 Gbps (10GbE), however, the Ethernet standard allows a SONET-like transport link in the WAN that bypasses ATM.

Ethernet's key benefits are its simplicity, wide deployment, cost effectiveness, and acceptance as an enterprise and in some instances metro solution. However, Ethernet lacks the robustness of carrier-grade technologies required to deliver the 99.999% service availability required in carrier networks. Ethernet also lacks, among other features, traffic engineering capabilities, resilience, hard QoS, and the ability to carry the whole portfolio of traffic types that exist in today's operator networks. But the combination of the two technologies unites the best of both worlds—the simplicity and wide deployment of Ethernet (making it the ideal service interface and/or cost-effective solution for noncritical, cost-effective, data-only metro networks) with the breadth of multiservice, carrier-grade features and robustness of ATM (for the broadband access, aggregation, and core network).

ATM has embedded features such as automatic and controlled reroute on failure, traffic engineering, QoS/CoS, carrier class management features, maturity, and the available associated *operational support systems* (OSSs) and *business support systems* (BSSs). These features add to the value of a combined ATM/GbE solution, allowing ATM to complement Ethernet's simplicity and cost effectiveness with carrier-grade functionality and features. This also applies to the critical and multiservice portions of the network.

Even with the advance of MPLS on GbE and 10GbE, ATM and ATM/MPLS hybrid solutions continue to enhance LAN and MAN solutions based on Ethernet while adding true multiservice capabilities and carrier-grade core infrastructure that Ethernet continues to lack for the foreseeable future.

9.2.5.4 ATM and DSL Interworking

ATM and DSL have been used in tandem in recent years. ATM offers a breadth of services and applications with which DSL easily interfaces. It allows different types of traffic to be transported over the same network. DSL employs ATM's chameleon-like personality to reliably deliver a variety of services to the user via one of the DSL variants such as ADSL, G-Lite, HDSL, or the latest high-speed version, VDSL.

ATM is particularly well suited for DSL because it allows bandwidth to be dynamically allocated between an array of voice and data services reliably and with minimal overhead. It also gives carriers the management tools required to meet specified levels of performance common in SLAs as well as the required BSS and OSS required to enable carrier-grade, scalable, cost-effective and reliable large-scale deployment.

The wide adoption of ATM by most DSL providers is extending the benefits of ATM from the last mile into the core network, while also allowing for interoperability with any other technology present (e.g., TDM, GbE, POS/IP, and FR) at any part in the network. This flexibility gives the operator the investment protection, flexibility, and freedom of choice required in this challenging, fast changing, and competitive segment of the market and through this flexibility and adaptability significantly reduces cost of ownership, capital expenditure, and operational cost.

The wide adoption of ATM by most DSL providers is extending the benefits of ATM that are dominant in core networks into the access arena. The evolving integration of *Voice over DSL* (VoDSL) in the local loop could spur further ATM investment in the access and core network. The evolution of ATM switches to MPLS ATM LSRs further enables reliable, cost-effective, and carrier-grade evolution to MPLS for core IP transport. The already available breadth of existing ATM applications and services provides additional sources of revenue for DSL operators from their existing infrastructure and with minimal additional investment. Global ATM deployment, enabling broadband (DSL), will continue to rise for the foreseeable future, allowing new, additional applications to evolve in the local loop.

9.2.5.5   ATM and Wireless Interworking

As we covered in Chapter 8, wireless applications have matured well beyond the voice-only age of early cellular. Today, the first wireless Internet applications are entering the market, giving subscribers access to a variety of on-line services and applications as well as delivering on the promised of true, unlimited mobility. Enabling 3G mobile services, also known as UMTS, are a variety of transport and switching mediums. In UMTS infrastructures ATM serves in the *radio access network* (RAN) as well as the core network, carrying both voice and data traffic efficiently, reliably, and with the required QoS. In addition to enabling UMTS, ATM also serves as an integration platform for 2G (GSM), 2.5G (GPRS), and 3G (UMTS) on a common, multiservice access and core network, giving the operators increased flexibility and investment protection while lowering capital expenditures and operational costs by eliminating the need for separate infrastructures and enabling even further applications and services beyond mobile if required.

Always-on wireless IP data is the new component in 3G. ATM approaches this burgeoning industry with well-entrenched footing in both worlds. By using ATM as the switching layer, UMTS/wireless carriers employ AAL-2 to carry both voice and data traffic in the RAN. In the core network AAL-2 is used for voice and AAL-5 and/or an ATM/MPLS hybrid for IP. The latter is possibly due to the fact that per definition ATM switches can run multiple control planes and, as defined in the IETF MPLS specifications, can evolve to become a hybrid ATM switch/ATM MPLS LSR. The maturity and flexibility of ATM and the widely deployed and tested ATM switch infrastructures and OSS/BSS further ease and speed deployment of these new networks. Lastly, the deployment of ATM takes a lot of risk out of 3G deployments, which together with the increased speed and lowered cost of deployment address operators' key concerns in this highly competitive marketplace. Jointly UMTS/ATM deliver an unprecedented bandwidth of up to 2 Mbps always-on IP to the mobile users.

Wireless applications are expected to firmly take root during this decade. The first 3G networks are launching in Japan followed by Europe. As 3G emerges beyond the first set of UMTS specifications, ATM will extend its reach farther into the wireless world and continue to play a key role.

9.2.5.6   ATM and SONET/SDH Interworking

ATM and SONET/SDH have been the two cornerstones of reliable WAN networks. In the 1990s ATM over SONET/SDH developed a proven track record for being the most reliable, flexible, efficient, and scalable bandwidth management approach. The benefits are traced to the strong attributes of each technology. The standards-based SONET/SDH approach delivers automatic protection switching for instantaneously self-healing in the event of a cable cut. That reinforcement strengthens protection for the ATM switch transmissions. ATM uses statistical multiplexing to compress signals from a variety of sources, thereby maximizing the use of bandwidth generated by SONET/SDH. Different types of applications are more easily muxed over the SONET/SDH equipment, reducing capital expenditures incurred with add-on equipment.

Much of the latest work in blending ATM and SONET/SDH seeks to achieve greater network efficiencies by, for instance, allowing ATM to emulate SONET/SDH in certain parts of the network. This eliminates the need for SONET/SDH and therefore reduces the number of network layers, further reducing operational complexity and cost as well as capital expenditure. This is being done while also paying attention to the need for more data-oriented network characteristics in addition to traditional TDM fixed-bandwidth circuit-switched characteristics. With photonics or DWDM establishing itself in the core networks today where they deliver large amounts of capacity, ATM is being moved out to the aggregation and access layers of the network where its core strengths lie.

Further in the future the industry discussions currently circle around the establishment of a common control plane protocol for access, aggregation, and core layer technologies based on MPLS and GMPLS that can also be applied to SONET/SDH, DWDM, and other technologies. At the time of this writing no trial implementations of GMPLS—let alone large-scale productive deployments—have been undertaken, but once this occurs ATM or currently available ATM/MPLS hybrid implementations will continue to have a key role to play in the unified, multiservice new public network.

9.3   MPLS Technology

This section provides an overview of MPLS-based technology, services, and standards. As we saw earlier in the chapter, during the past 25 years corporations have sought improved packet technologies to support intranets, extranets, and public switched data networks such as the Internet. The progression went from X.25 packet-switched technology to FR technology and also, on a parallel track, cell relay/ATM technology. In the meantime, throughout the 1980s and 1990s, IP-based connectionless packet services (a layer 3 service) continued to make major inroads. IP, however, has limited QoS capabilities by itself. Therefore, the late 1990s and early 2000s saw the development of MPLS. MPLS is a hybrid layer 2/layer 3 service that attempts to bring together the best of both worlds: layer 2, layer 3, ATM, and IP. A question might be [35] "Where does MPLS fit in the OSI reference model?" Some might argue that MPLS does not fit in the OSI reference model. The fact that MPLS is a framework that contains enhancements to the current layer 3 and layer 2 technologies makes it hard to fit MPLS within one layer of the OSI model. MPLS alone cannot be considered a layer in the OSI sense, since it does not have a unified format for the transport of data from the layer above: It uses a shim header over SONET or Ethernet, it uses the existing VPI/VCI of ATM, and so on. However, an individual MPLS function could be categorized as either an OSI layer 3 or layer 2 function.

9.3.1 Overview

MPLS was a late 1990s set of specifications that provides a link-layer-independent transport mechanism for IP. The work was carried out by the IETF. MPLS protocols allow high-performance label switching of IP packets: Network traffic is forwarded using a simple label apparatus as described in RFC 3031 [36]. By combining the attributes of layer 2 switching and layer 3 routing into a single entity, MPLS provides [37] (1) enhanced scalability by way of switching technology, (2) support of CoS- and QoS-based services (differentiated services/*diffserv*, as well as integrated services/*intserv*), (3) elimination of the need for an IP-over-ATM overlay model and its associated management overhead, and (4) enhanced traffic shaping and engineering capabilities. In addition, MPLS provides a gamut of features in support of VPNs. Table 9.14 provides a snapshot of some key technical accomplishments in this arena as of press time [38].

The basic idea of MPLS involves assigning short fixed-length labels to packets at the ingress to an MPLS cloud (based on the concept of forwarding equivalence classes). Throughout the interior of the MPLS domain, the labels attached to packets are used to make forwarding decisions (usually without recourse to the original packet headers). A set of powerful constructs to address many critical issues in the (eventually) emerging *diffserv* Internet can be devised from this relatively simple paradigm. One of the most significant initial applications of MPLS is in traffic engineering. (Note that even though the focus is on Internet backbones, the capabilities described in MPLS TE are equally applicable to traffic engineering in enterprise networks [39]).

TABLE 9.14   RECENT DEVELOPMENTS IN MPLS

BGP/MPLS VPN

Multivendor interoperable implementations

Deployment of advanced features (interprovider, carrier's carrier; QoS support)

Integration with various access tunnel technologies (e.g., IPSec, L2TP) for CE–PE link

More attention to the network management aspect

LAYER 2 FRAMES OVER MPLS

Multivendor interoperable implementations

Deployed by multiple service providers

FAST REROUTES

Converging to a common set of mechanisms to support a range of options (one-to-one backup, facility based)

Handles both link and node failures

Paves the way for multivendor interoperable implementations

DIFFSERV TRAFFIC ENGINEERING

Came from realization that certain types of QoS require not just queuing, but routing as well

Converging to single proposal

Supports up to eight class types

Small extensions to ISIS/OSPF TE

Very little scalability impact compared to the existing MPLS TE

Backward compatibility with MPLS TE

Meets all the requirements produced by the IETF TE working group

Paves the way for multivendor interoperable implementations

GRACEFUL RESTART WITH MPLS

Allows restart of the MPLS control plane without disrupting the MPLS data plane (supports planned or unplanned restart)

Simplifies LDP graceful restart (assumes availability of ISIS/OSPF graceful restart)

RSVP-TE graceful restart (specified as a part of GMPLS signaling, but could be used with MPLS as well)

BGP graceful restart with MPLS (simple extensions to BGP graceful restart)

VIRTUAL PRIVATE LAN SERVICES

Alternative to point-to-point layer 2 VPNs from the customer point of view (simpler configuration, familiar technology)

Several proposals (noninteroperable): DNS for autodiscovery, LDP for signaling, BGP for autodiscovery, LDP for signaling, BGP for autodiscovery, BGP for signaling

TROUBLESHOOTING PACKET-BASED MPLS FORWARDING PLANE

Option 1: Use a combination of MPLS ICMP traceroute, LSP-ping, GTTP, and SNMP MIBs (IP centric tools can be used to troubleshoot LSPs that do not carry IP).

Option 2: Develop a single unifying approach.

Sources: Rekhter, Y., Juniper Networks, and http://www.mplsworld.com.

The key MPLS RFCs are named next. RFC 2702, *Requirements for Traffic Engineering over MPLS*, identifies the functional capabilities required to implement policies that facilitate efficient and reliable network operations in an MPLS domain; these capabilities can be used to optimize the utilization of network resources and to enhance traffic-oriented performance characteristics. RFC 3031, *Multiprotocol Label Switching Architecture*, specifies the architecture of MPLS. RFC 3032, *MPLS Label Stack Encoding*, specifies the encoding to be used by a LSR in order to transmit labeled packets on PPP data links, on LAN data links, and possibly on other data links. Also, RFC3032 specifies rules and procedures for processing the various fields of the label stack encoding. An array of supplementary Internet drafts support the various aspects of MPLS.

MPLS runs over ATM, FR, Ethernet and point-to-point packet-mode links. MPLS-based networks use existing IP mechanisms for addressing of elements and for routing of traffic. MPLS adds connection-oriented capabilities to the connectionless IP architecture. It is the industry-accepted manifestation of the "network layer/layer 3/tag/IP switching" technology that was developed by various constituencies in the mid- to late 1990s [40]. MPLS integrates the label swapping forwarding paradigm with network layer routing. In an MPLS environment, when a stream of data traverses a common path, a *label-switched path* (LSP) can be established using MPLS signaling protocols. At the ingress LSR, each packet is assigned a label and is transmitted downstream. At each LSR along the LSP, the label is used to forward the packet to the next hop. To deliver reliable service, MPLS requires a set of procedures to provide protection of the traffic carried on different paths. This requires that the LSRs support fault detection, fault notification, and fault recovery mechanisms, and that MPLS signaling support the configuration of recovery [41]. QoS support is where MPLS can find its sweet technical spot in supporting multimedia and voice applications. The improved traffic management, the QoS capabilities (that is, hooks to QoS mechanisms), and the expedited packet forwarding via the label mechanism can be a significant technical advantage to delay-sensitive applications.

The LSRs know what label values to use because this information will have been propagated by a label distribution protocol. (The MPLS architecture allows several different methods and protocols for label distribution, which we discuss this later.) In essence, the LSP is established once all of the LSRs have valid label entries in their forwarding tables. In Figure 9.48, LSR1 and LSR4 represent the ingress/egress to the MPLS network [42].

1. An IP packet arrives at LSR1.

2. LSR1 examines the IP header and the IP destination address.

3. The IP packet is then labeled; the label value given to the packet is associated with a LSP across the network to the egress point (LSR4).

FIGURE 9.48
MPLS network.

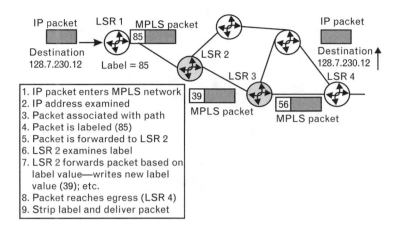

MPLS network

IP packet   LSR 1   MPLS packet

Destination
128.7.230.12   Label = 85

LSR 2

LSR 3

IP packet

Destination
128.7.230.12

LSR 4

1. IP packet enters MPLS network
2. IP address examined
3. Packet associated with path
4. Packet is labeled (85)
5. Packet is forwarded to LSR 2
6. LSR 2 examines label
7. LSR 2 forwards packet based on
   label value—writes new label
   value (39); etc.
8. Packet reaches egress (LSR 4)
9. Strip label and deliver packet

39

MPLS packet

56

MPLS packet

4. Now that the packet has been labeled, the label associates the packet with a particular path. There is no longer any need to look up the IP address inside the network because the other LSRs can forward the labeled packet based on the label value.

5. A point to note is that the labels are rewritten at each switch. This is shown in the figure, where the label changes from 85 to 39 to 56.

6. When the packet reaches the last or egress LSR (LSR 4), the label is stripped (popped), thus exposing the IP packet.

7. LSR 4 can then deliver the IP packet to the destination using normal IP forwarding.

Some of the features of MPLS that make it a useful networking technology are as follows:

- *Aggregation of PDU streams.* In MPLS the label stacking mechanism can be used to perform the aggregation within layer 2 itself. Typically, when multiple streams have to be aggregated for forwarding into a switched path, processing is required at both layer 2 and layer 3. The top label of the MPLS label stack is used to switch PDUs along the label switched path, while the rest of the label stack is "application specific."

- *Explicit/improved routes.* MPLS supports explicit routes, which are routes that have not been set up by normal IP hop-by-hop routing; instead, an ingress/egress node has specified all or some of the downstream nodes of that route).

- *Improved performance.* MPLS enables higher data transmission performance due to simplified packet forwarding and switching mechanisms.

- *Link layer independence.* MPLS works with any type of link layer medium such as ATM, FR, packet-over-SONET, and Ethernet.

- *QoS support.* Explicit routes provide a mechanism for QoS/constraint routing and so on.

- *Scalability of network layer routing.* A key MPLS desideratum was to achieve an improved and more efficient transfer of PDUs in the current IP networks. Combining the routing knowledge at layer 3 with the ATM switching capability in ATM devices results in a better solution. In the MPLS scenario, it is sufficient to have adjacencies with the immediate peers. The edge LSRs interact with the adjacent LSRs and this is sufficient for the creation of LSPs for the transfer of data.

- *Traffic engineering.* MPLS supports traffic engineering, a process of selecting the paths chosen by data traffic in order to balance the traffic load on the various links, routers, and switches in the network. Key performance objectives of TE are (1) traffic oriented, including those aspects that enhance the QoS of traffic streams; and (2) resource oriented, including those aspects that pertain to the optimization of resource utilization.

- *VPN support.* VPN is an application that uses the label stacking mechanisms. At the VPN ingress node, the VPN label is mapped onto the MPLS label stack and packets are label switched along the LSP within the VPN until they emerge at the egress. At the egress node, the label stack is used to determine further forwarding of the PDUs.

In particular, TE is concerned with performance optimization of operational networks. In general, it encompasses the application of technology and scientific principles to the measurement, modeling, characterization, and control of Internet traffic, and the application of such knowledge and techniques to achieve specific performance objectives. The aspects of TE that are of interest concerning MPLS are measurement and control. A major goal of Internet TE is to facilitate efficient and reliable network operations while simultaneously optimizing network resource utilization and traffic performance. Traffic engineering has become an indispensable function in many large autonomous systems because of the high cost of network assets and the commercial and competitive nature of the Internet. These factors emphasize the need for maximal operational efficiency [43]. Note that MPLS is not a routing protocol. In fact, MPLS needs the reachability information provided by the current routing protocols in order to calculate the paths that it uses. MPLS augments the functionality of the routing protocols, but does not replace them.

Originally, the main benefit of label switching was facilitating high-speed switching in layer 3 devices. However, this is no longer perceived as the main benefit of MPLS, because nowadays ASIC-based routers can perform line speed routing on most interfaces. Now, the major benefits of MPLS are perceived to be [35]:

- *Simplifying packet forwarding.* Since the routing decision is made only once at the edge of the network, the core could keep only minimal routing information, thus reducing the overall complexity of the network (e.g., BGP could be run at the edge only, but there would be no need for it in the core).

- *Traffic engineering.* MPLS offers the tools to control the paths taken by different flows. Using these tools, traffic could be rerouted to avoid congestion points in a network.

- *Delivering QoS and differentiated services.* Using MPLS's inherent mechanisms for traffic prioritization and traffic path control, a service provider could create a network that delivers QoS, facilitates offering differentiated services to customers, and fulfills the offered service level agreements.

- *Network scalability.* Using MPLS's label stacking capability, MPLS domains could be arranged in a hierarchy, offering multiple levels of abstraction and, therefore, scalability.

- *Supporting VPNs.* Because MPLS provides tunneling of packets from an ingress point to an egress point, VPN applications that leverage this capability can be created easily.

9.3.2 Key Elements

MPLS requires a set of procedures for augmenting network layer packets with *label stacks,* thereby turning them into *labeled packets.* Routers that support MPLS are known as *label-switching routers.* To transmit a labeled packet on a particular data link, an LSR must support an encoding technique which, given a label stack and a network layer packet, produces a labeled packet [44]. MPLS can be logically and functionally partitioned into two elements to provide the label-switching functionality: (1) MPLS forwarding/label-switching mechanism and (2) MPLS label distribution mechanism, both of which are discussed in the following subsections.

9.3.2.1 MPLS Forwarding/Label-Switching Mechanism

The key mechanism of MPLS is the forwarding/label-switching function. This is an advanced form of packet forwarding that replaces the conventional longest-address-match-forwarding with a more efficient label-swapping forwarding algorithm. The IP header analysis is performed once at the ingress of the LSP for the classification of PDUs. PDUs that are forwarded via the same next hop are grouped into a *forwarding equivalence class* (FEC) based on one or more of the following parameters: address prefix, host address, or host address and QoS.

The FEC to which the PDU belongs is encoded at the edge LSRs as a short fixed-length value known as a *label* (see Figure 9.49). When the PDU is forwarded to its next hop, the label is sent along with it. At downstream hops, there is no further analysis of the PDU's network layer header. Instead, the label is used as an index into a table; the entry in the table specifies the next hop and a new label. The incoming label is replaced with this outgoing label, and the PDU is forwarded to its next hop. Labels usually have local significance and are used to identify FECs based on the type of underlying network. For example, in ATM networks, the VPI and VCI are used in generating the MPLS label; in FR networks, the DLCI is used. In ATM environments, the labels assigned to the FECs (PDUs) are the VPI/VCI of the virtual connections established as a part of the LSP. In FR environments, the labels assigned to the FECs (PDUs) are the DCLIs.

So, an FEC is a class of packets that should be forwarded in the same manner (i.e., over the same path) [35]. A FEC is not a packet, nor is it a label. A FEC is a logical entity created by the router to represent a class (category) of packets. When a packet arrives at the ingress router of an MPLS domain, the router parses the packet's headers, and checks to see if the packet matches a known FEC (class). Once the matching FEC is determined, the path and outgoing label assigned to that FEC are used to forward the packet. FECs are typically created based on the IP destinations known to the router, so for each different destination a router might create a different FEC, or if a router is doing aggregation, it might represent multiple destinations with a single FEC (for example, if those destinations are reachable through the same immediate next hop anyway). The MPLS framework, however, allows for the creation of FECs using advanced criteria such as source and destination address pairs, destination address and TOS, and so on.

Label switching has been designed to leverage the layer 2 switching function done in the current data link layers such as ATM and FR. It follows that the MPLS forwarding mechanism should be able to update the switching fabric(s) in ATM and FR hardware in the LSR for the relevant sets of LSPs, which can be switched at the hardware level [45]. In the Ethernet-based networks, the labels are short headers placed between the data link headers and the data link layer PDUs.

FIGURE 9.49
MPLS label.

```
 0 1 2 3 4 5 6 7 8 9 0 1 2 3 4 5 6 7 8 9 0 1 2 3 4 5 6 7 8 9 0 1
+-+-+-+-+-+-+-+-+-+-+-+-+-+-+-+-+-+-+-+-+-+-+  Label
|            Label               |Exp|S|    TTL        |
+-+-+-+-+-+-+-+-+-+-+-+-+-+-+-+-+-+-+-+-+-+-+  Entry
```

Label: Label value, 20 bits
Exp: Experimental use, 3 bits
S: Bottom of stack, 1 bit
TTL: Time to live, 8 bits

9.3.2.2 MPLS Label Distribution Mechanism

In an MPLS environment the distribution of labels in MPLS is accomplished in two ways (see Figure 9.50):

1. By utilizing the RSVP signaling mechanism to distribute labels mapped to the RSVP flows;

2. By utilizing the LDP.

Label Distribution Using RSVP

RSVP [46] defines a "session" to be a data flow with a particular destination and transport-layer protocol. From the early 1990s to the late 1990s, RSVP was being considered for QoS support in IP networks. When RSVP and MPLS are combined, a flow or session can be defined with greater generality. The ingress node of an LSP can use a variety of means to determine which PDUs are assigned a particular label. Once a label is assigned to a set of PDUs, the label effectively defines the "flow" through the LSP. Such an LSP is referred to as an *LSP tunnel* because the traffic flowing though it is "opaque" to intermediate nodes along the label-switched path. The label request information for the labels associated with RSVP flows will be carried as part of the RSVP *Path* messages and the label mapping information for the labels associated with RSVP flows will be carried as part of the RSVP *Resv* messages [45]. The initial implementers of MPLS chose to extend RSVP into a signaling protocol to support the creation of LSPs that could be automatically routed away from network failures and congestion. An Internet draft defines the extension to RSVP for establishing LSPs in MPLS networks [47].

FIGURE 9.50
MPLS protocols.

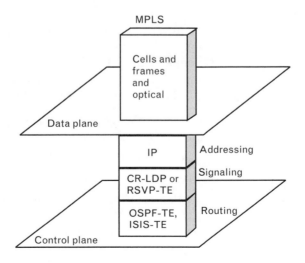

ISIS: Intermediate system-to-intermediate system
CR-LDP: Constraint-based routed label distribution protocol
RSVP-TE: Resource reservation protocol-traffic engineering

The use of RSVP as a signaling protocol for traffic engineering is different from that envisioned by its original developers in the mid-1990s [48, 49]:

- A number of extensions were added to the base RSVP specification (RFC 2205 and RFC 2209) to support the establishment and maintenance of explicitly routed LSPs.

- RSVP signaling takes place between pairs of routers (rather than pairs of hosts) that act as the ingress and egress points of a traffic trunk. Extended RSVP installs state that applies to a collection of flows that share a common path and a common pool of shared network resources, rather than a single host-to-host flow. By aggregating numerous host-to-host flows into each LSP tunnel, extended RSVP significantly reduces the amount of RSVP state that needs to be maintained in the core of a service provider's network.

- RSVP signaling installs distributed state related to packet forwarding, including the distribution of MPLS labels.

- The scalability, latency, and traffic overhead concerns regarding RSVP's soft state model are addressed by a set of extensions that reduce the number of refresh messages and the associated message processing requirements.

- The path established by RSVP signaling is not constrained by conventional destination-based routing, so it is a good tool to establish traffic engineering trunks.

The initial implementers of MPLS had a number of reasons to choose to extend RSVP rather than design an entirely new signaling protocol to support traffic engineering requirements [48]:

- By implementing the proposed extensions, RSVP provides a unified signaling system that delivers everything that network operators needed to dynamically establish LSPs.

- Extended RSVP creates an LSP along an explicit route to support the traffic engineering requirements of large service providers.

- Extended RSVP establishes LSP state by distributing label-binding information to the LSRs in the LSP.

- Extended RSVP can reserve network resources in the LSRs along the LSP (the traditional role of RSVP). Extended RSVP also permits an LSP to carry best effort traffic without making a specific resource reservation.

Hence, RSVP can serve a dual role in MPLS: for label distribution and for QoS support.

LDP

LDP is a set of procedures and messages by which LSRs establish LSPs through a network by mapping network layer routing information directly to data link layer switched paths. These LSPs may have an endpoint at a directly attached neighbor (this being comparable to IP hop-by-hop forwarding), or they may have an endpoint at a network egress node, enabling switching via all intermediary nodes. LDP associates an FEC with each LSP it creates. The FEC associated with an LSP specifies which PDUs are "mapped" to that LSP. LSPs are extended through a network as each LSR "splices" incoming labels for an FEC to the outgoing label assigned to the next hop for the given FEC. The messages exchanged between the LSRs are classified into four categories:

1. *Discovery messages,* which are used to announce and maintain the presence of an LSR in a network;

2. *Session messages,* which are used to establish, maintain, and terminate sessions between LSP peers;

3. *Advertisement messages*, which are used to create, change, and delete label mappings for FECs;

4. *Notification messages*, which are used to provide advisory information and to signal error information.

The LDP uses the TCP for session, advertisement, and notification messages. TCP is utilized to provide reliable and sequenced messages. Discovery messages are transmitted using the UDP. These messages are sent to the LSP port at the "all routers on this subnet" group multicast address.

Discovery messages provide a mechanism for the LSRs to indicate their presence in a network. LSRs send the *Hello* message periodically. When an LSR chooses to establish a session with another LSR discovered via the *Hello* message, it uses the LDP initialization procedure (this is done using TCP). On successful completion of the initialization procedure, the two LSRs are LSP peers, and may exchange *advertisement* messages. The LSR requests a label mapping from a neighboring LSR when it needs one, and advertises a label mapping to a neighboring LSR when it wishes the neighbor to use a label.

9.3.3 MPLS Architecture Overview

This section discusses the architecture for MPLS in detail, as laid out in RFC 3031 [50]. As a packet of a traditional connectionless network layer protocol travels from one router to the next, each router makes an independent forwarding decision for that packet. That is, each router analyzes the packet's header, and each router runs a network layer routing algorithm. Each router independently chooses a next hop for the packet, based on its analysis of the packet's header and the results of running the routing algorithm.

Packet headers contain considerably more information than is needed simply to choose the next hop. Choosing the next hop can, therefore, be thought of as the composition of two functions. The first function partitions the entire set of possible packets into a set of FECs. The second maps each FEC to a next hop. Insofar as the forwarding decision is concerned, different packets that get mapped into the same FEC are indistinguishable. All packets that belong to a particular FEC and that travel from a particular node will follow the same path (or if certain kinds of multipath routing are in use, they will all follow one of a set of paths associated with the FEC).

In conventional IP forwarding, a particular router will typically consider two packets to be in the same FEC if there is some address prefix X in that router's routing tables such that X is the "longest match" for each packet's destination address. As the packet traverses the network, each hop in turn reexamines the packet and assigns it to an FEC.

In MPLS, the assignment of a particular packet to a particular FEC is done just once, as the packet enters the network. The FEC to which the packet is assigned is encoded as a short fixed-length value known as a *label*. When a packet is forwarded to its next hop, the label is sent along with it; that is, the packets are "labeled" before they are forwarded. At subsequent hops, there is no further analysis of the packet's network layer header. Rather, the label is used as an index into a table that specifies the next hop and a new label. The old label is replaced with the new label, and the packet is forwarded to its next hop.

In the MPLS forwarding paradigm, once a packet is assigned to a FEC, no further header analysis is done by subsequent routers; all forwarding is driven by the labels. This has a number of advantages over conventional network layer forwarding.

- MPLS forwarding can be done by switches that are capable of doing label lookup and replacement, but are either not capable of analyzing the network layer headers or are not capable of analyzing the network layer headers at adequate speed.

- Because a packet is assigned to a FEC when it enters the network, the ingress router may use, in determining the assignment, any information it has about the packet, even if that information cannot be gleaned from the network layer header. For example, packets arriving on different ports may be assigned to different FECs. Conventional forwarding, on the other hand, can only consider information that travels with the packet in the packet header.

- A packet that enters the network at a particular router can be labeled differently than the same packet entering the network at a different router, and as a result forwarding decisions that depend on the ingress router can be easily made. This cannot be done with conventional forwarding, since the identity of a packet's ingress router does not travel with the packet.

- The considerations that determine how a packet is assigned to a FEC can become ever more and more complicated, without any impact at all on the routers that merely forward labeled packets.

- Sometimes it is desirable to force a packet to follow a particular route that is explicitly chosen at or before the time the packet enters the network, rather than being chosen by the normal dynamic routing algorithm as the packet travels through the network [51]. This may be done as a matter of policy, or to support traffic engineering. In conventional forwarding, this requires the packet to carry an encoding of its route along with it (*source routing*). In MPLS, a label can be used to represent the route, so that the identity of the explicit route need not be carried with the packet.

Some routers analyze a packet's network layer header not merely to choose the packet's next hop, but also to determine a packet's "precedence" or "class of service." They may then apply different discard thresholds or scheduling disciplines to different packets. MPLS allows (but does not require) the precedence or class of service to be fully or partially inferred from the label. In this case, one may say that the label represents the combination of a FEC and a precedence or CoS.

MPLS's techniques are applicable to any network layer protocol. In this discussion, however, we focus on the use of IP as the network layer protocol. Table 9.15 identifies key terms used in MPLS.

9.3.4 MPLS Basics

In this section, we introduce some of the basic concepts of MPLS and describe the general approach to be used.

9.3.4.1 Labels

As noted, a label is a short, fixed-length, locally significant identifier that is used to identify a FEC. The label given to a particular packet represents the FEC to which that packet is assigned. Most commonly, a packet is assigned to a FEC based (completely or partially) on its network layer destination address. However, the label is never an encoding of that address.

If Ru and Rd are LSRs, they may agree that when Ru transmits a packet to Rd, Ru will label the packet with label value L if and only if the packet is a member of a particular FEC F. That is, they can agree to a "binding" between label L and FEC F for packets moving from Ru to Rd. As a result of such an agreement, L becomes Ru's "outgoing label" representing FEC F, and L becomes Rd's "incoming label" representing FEC F. Note that L does not necessarily represent FEC F for any packets other than

TABLE 9.15   KEY TERMS USED IN MPLS

DLCI	A label used in FR networks to identify frame relay circuits.
Forwarding equivalence class	A group of IP packets that is forwarded in the same manner (e.g., over the same path, with the same forwarding treatment).
Frame merge	Label merging, when it is applied to operation over frame-based media, so that the potential problem of cell interleave is not an issue.
Label	A short fixed-length physically contiguous identifier that is used to identify a FEC, usually of local significance.
Label merging	The replacement of multiple incoming labels for a particular FEC with a single outgoing label.
Label stack	An ordered set of labels.
Label swap	The basic forwarding operation consisting of looking up an incoming label to determine the outgoing label, encapsulation, port, and other data handling information.
Label swapping	A forwarding paradigm that allows streamlined forwarding of data by using labels to identify classes of data packets that are treated indistinguishably when forwarding.
Label-switched hop	The hop between two MPLS nodes, on which forwarding is done using labels.
Label-switched path	The path through one or more LSRs at one level of the hierarchy followed by a packets in a particular FEC.
Label-switching router	An MPLS node that is capable of forwarding native L3 packets.
Layer 2	The protocol layer under layer 3 (which therefore offers the services used by layer 3). Forwarding, when done by the swapping of short fixed-length labels, occurs at layer 2 regardless of whether the label being examined is an ATM VPI/VCI, a frame relay DLCI, or an MPLS label.
Layer 3	The protocol layer at which IP and its associated routing protocols operate link layer synonymous with layer 2.
Loop detection	A method of dealing with loops in which loops are allowed to be set up, and data may be transmitted over the loop, but the loop is later detected.
Loop prevention	A method of dealing with loops in which data are never transmitted over a loop.
Merge	Point of a node at which label merging is done.
MPLS domain	A contiguous set of nodes that operates MPLS routing and forwarding and which are also in one routing or administrative domain.
MPLS edge node	An MPLS node that connects an MPLS domain with a node that is outside of the domain, either because it does not run MPLS or because it is in a different domain. Note that if an LSR has a neighboring host that is not running MPLS, then that LSR is an MPLS edge node.
MPLS egress node	An MPLS edge node in its role in handling traffic as it leaves an MPLS domain.
MPLS ingress node	An MPLS edge node in its role in handling traffic as it enters an MPLS domain.
MPLS label	A label that is carried in a packet header and represents the packet's FEC.
MPLS node	A node that is running MPLS. An MPLS node will be aware of MPLS control protocols, will operate one or more L3 routing protocols, and will be capable of forwarding packets based on labels. An MPLS node may optionally be also capable of forwarding native L3 packets.
Network layer	Synonymous with layer 3.
Stack	Synonymous with label stack.
Switched path	Synonymous with label-switched path.
VC merge	Label merging where the MPLS label is carried in the ATM VCI field (or combined VPI/VCI field), so as to allow multiple VCs to merge into one single VC.
Virtual circuit	A circuit used by a connection-oriented layer 2 technology such as ATM or FR, requiring the maintenance of state information in layer 2 switches.

TABLE 9.15 KEY TERMS USED IN MPLS (CONTINUED)

VP merge	Label merging where the MPLS label is carried in the ATM VPI field, so as to allow multiple VPs to be merged into one single VP. In this case two cells would have the same VCI value only if they originated from the same node. This allows cells from different sources to be distinguished via the VCI.
VPI/VCI	A label used in ATM networks to identify circuits.

those that are being sent from Ru to Rd. L is an arbitrary value whose binding to F is local to Ru and Rd.

When one speaks above of packets "being sent" from Ru to Rd, one does not imply either that the packet originated at Ru or that its destination is Rd. Rather, one means to include packets that are "transit packets" at one or both of the LSRs. For an arriving packet carrying label L, it may sometimes be difficult or even impossible for Rd to tell that the label L was placed in the packet by Ru, rather than by some other LSR. (This will typically be the case when Ru and Rd are not direct neighbors.) In such cases, Rd must make sure that the binding from label to FEC is one to one. That is, Rd must not agree with Ru1 to bind L to FEC F1, while also agreeing with some other LSR Ru2 to bind L to a different FEC F2, *unless* Rd can always tell, when it receives a packet with incoming label L, whether the label was put on the packet by Ru1 or whether it was put on by Ru2. It is the responsibility of each LSR to ensure that it can uniquely interpret its incoming labels.

9.3.4.2 Upstream and Downstream LSRs

Suppose Ru and Rd have agreed to bind label L to FEC F, for packets sent from Ru to Rd. Then with respect to this binding, Ru is the "upstream LSR," and Rd is the "downstream LSR." To say that one node is upstream and one is downstream with respect to a given binding means only that a particular label represents a particular FEC in packets traveling from the upstream node to the downstream node. This is not meant to imply that packets in that FEC would actually be routed from the upstream node to the downstream node.

9.3.4.3 Labeled Packet

A *labeled packet* is a packet into which a label has been encoded. In some cases, the label resides in an encapsulation header that exists specifically for this purpose. In other cases, the label may reside in an existing data link or network layer header, as long as there is a field that is available for that purpose. The particular encoding technique to be used must be agreed to by both the entity that encodes the label and the entity that decodes the label.

9.3.4.4   Label Assignment and Distribution

In the MPLS architecture, the decision to bind a particular label L to a particular FEC F is made by the LSR that is downstream with respect to that binding. The downstream LSR then informs the upstream LSR of the binding. Thus, labels are "downstream-assigned," and label bindings are distributed in the "downstream to upstream" direction. If an LSR has been designed so that it can only look up labels that fall into a certain numeric range, then it merely needs to ensure that it only binds labels that are in that range.

9.3.4.5   Attributes of a Label Binding

A particular binding of label L to FEC F, distributed by Rd to Ru, may have associated "attributes." If Ru, acting as a downstream LSR, also distributes a binding of a label to FEC F, then under certain conditions, it may be required to also distribute the corresponding attribute that it received from Rd.

9.3.4.6   Label Distribution Protocols

A label distribution protocol is a set of procedures by which one LSR informs another of the label/FEC bindings it has made. Two LSRs that use a label distribution protocol to exchange label/FEC binding information are known as *label distribution peers* with respect to the binding information they exchange. If two LSRs are label distribution peers, we will speak of there being a *label distribution adjacency* between them. (Note that two LSRs may be label distribution peers with respect to some set of bindings, but not with respect to some other set of bindings.)

The label distribution protocol also encompasses any negotiations in which two label distribution peers need to engage in order to learn of each other's MPLS capabilities.

The architecture does not assume that there is only a single label distribution protocol. In fact, a number of different label distribution protocols are being standardized. Existing protocols have been extended so that label distribution can be piggybacked on them (see, e.g., [52, 53]). New protocols have also been defined for the explicit purpose of distributing labels (see, e.g., [54, 55]).

In this section, the acronym *LDP* refers specifically to the protocol defined in [54]; when speaking of label distribution protocols in general, we try to avoid the acronym.

9.3.4.7   Unsolicited Downstream Versus Downstream-on-Demand Label Distribution

The MPLS architecture allows an LSR to explicitly request, from its next hop for a particular FEC, a label binding for that FEC. This is known as *downstream-on-demand* label distribution. The MPLS architecture also allows an LSR to distribute bindings to LSRs that have not explicitly requested them. This is known as *unsolicited downstream* label distribution.

It is expected that some MPLS implementations will provide only downstream-on-demand label distribution, some will provide only unsolicited downstream label distribution, and some will provide both. Which is provided may depend on the characteristics of the interfaces that are supported by a particular implementation. However, both of these label distribution techniques can be used in the same network at the same time. On any given label distribution adjacency, the upstream LSR and the downstream LSR must agree on which technique is to be used.

9.3.4.8 Label Retention Mode

An LSR Ru may receive (or have received) a label binding for a particular FEC from an LSR Rd, even though Rd is not Ru's next hop (or is no longer Ru's next hop) for that FEC.

Ru then has the choice of whether to keep track of such bindings or to discard them. If Ru keeps track of such bindings, then it may immediately begin using the binding again if Rd eventually becomes its next hop for the FEC in question. If Ru discards such bindings, then if Rd later becomes the next hop, the binding will have to be reacquired.

If an LSR supports the *liberal label retention mode,* it maintains the bindings between a label and a FEC that are received from LSRs that are not its next hop for that FEC. If an LSR supports the *conservative label retention mode,* it discards such bindings.

The liberal label retention mode allows for quicker adaptation to routing changes, but the conservative label retention mode requires an LSR to maintain many fewer labels.

9.3.4.9 The Label Stack

So far, we have spoken as if a labeled packet carries only a single label. As we shall see, it is useful to have a more general model in which a labeled packet carries a number of labels, organized as a last-in, first-out stack. We refer to this as a *label stack.*

Although, as we shall see, MPLS supports a hierarchy, the processing of a labeled packet is completely independent of the level of hierarchy. The processing is always based on the top label, without regard for the possibility that some number of other labels may have been "above it" in the past or that some number of other labels may be below it at present.

An unlabeled packet can be thought of as a packet whose label stack is empty (i.e., whose label stack has depth 0). If a packet's label stack is of depth $m,$ we refer to the label at the bottom of the stack as the level 1 label, to the label above it (if such exists) as the level 2 label, and to the label at the top of the stack as the level m label. The utility of the label stack will become clear when we introduce the notion of the LSP tunnel and the MPLS hierarchy in Section 9.3.4.27.

9.3.4.10   The Next Hop Label Forwarding Entry

The *next hop label forwarding entry* (NHLFE) is used when forwarding a labeled packet. It contains (1) the packet's next hop and (2) the operation to perform on the packet's label stack, which will be one of the following: (a) replace the label at the top of the label stack with a specified new label, (b) pop the label stack, or (c) replace the label at the top of the label stack with a specified new label, and then push one or more specified new labels onto the label stack. It may also contain (d) the data link encapsulation to use when transmitting the packet, (e) the way to encode the label stack when transmitting the packet, or (f) any other information needed in order to properly dispose of the packet.

Note that at a given LSR, the packet's "next hop" might be that LSR itself. In this case, the LSR would need to pop the top level label, and then "forward" the resulting packet to itself. It would then make another forwarding decision, based on what remains after the label stacked is popped. This may still be a labeled packet, or it may be the native IP packet. This implies that in some cases the LSR may need to operate on the IP header in order to forward the packet. If the packet's "next hop" is the current LSR, then the label stack operation must be to "pop the stack."

9.3.4.11   Incoming Label Map

The *incoming label map* (ILM) maps each incoming label to a set of NHLFEs. It is used when forwarding packets that arrive as labeled packets. If the ILM maps a particular label to a set of NHLFEs that contains more than one element, exactly one element of the set must be chosen before the packet is forwarded. The procedures for choosing an element from the set are beyond the scope of this document. Having the ILM map a label to a set containing more than one NHLFE may be useful if, for example, the user wants to do load balancing over multiple equal-cost paths.

9.3.4.12   FEC-to-NHLFE Map

The *FEC-to-NHLFE* (FTN) maps each FEC to a set of NHLFEs. It is used when forwarding packets that arrive unlabeled, but which are to be labeled before being forwarded.

If the FTN maps a particular label to a set of NHLFEs that contains more than one element, exactly one element of the set must be chosen before the packet is forwarded. The procedures for choosing an element from the set are beyond the scope of this document. Having the FTN map a label to a set containing more than one NHLFE may be useful if, for example., it is desired to do load balancing over multiple equal-cost paths.

9.3.4.13   Label Swapping

Label swapping is the use of the following procedures to forward a packet. To forward a labeled packet, a LSR examines the label at the top of the label stack. It uses the ILM to map this label to an NHLFE. Using the information in the NHLFE, it determines where to forward the packet, and performs an operation on the packet's label stack. It then encodes the new label stack into the packet and forwards the result.

To forward an unlabeled packet, a LSR analyzes the network layer header, to determine the packet's FEC. It then uses the FTN to map this to an NHLFE. Using the information in the NHLFE, it determines where to forward the packet and performs an operation on the packet's label stack. (Popping the label stack would, of course, be illegal in this case.) It then encodes the new label stack into the packet, and forwards the result.

It is important to note that when label swapping is in use, the next hop is always taken from the NHLFE; this may in some cases be different from what the next hop would be if MPLS were not in use.

9.3.4.14 Scope and Uniqueness of Labels

A given LSR Rd may bind label L1 to FEC F, and distribute that binding to label distribution peer Ru1. Rd may also bind label L2 to FEC F, and distribute that binding to label distribution peer Ru2. Whether or not L1 = L2 is not determined by the architecture; rather, this is a local matter.

A given LSR Rd may bind label L to FEC F1 and distribute that binding to label distribution peer Ru1. Rd may also bind label L to FEC F2, and distribute that binding to label distribution peer Ru2 if (and only if) Rd can tell, when it receives a packet whose top label is l, whether the label was put there by Ru1 or by Ru2; in such case, the architecture does not require that F1 = F2. In such cases, we may say that Rd is using a different "label space" for the labels it distributes to Ru1 than for the labels it distributes to Ru2.

In general, Rd can only tell whether it was Ru1 or Ru2 that put the particular label value L at the top of the label stack if the following conditions hold:

- Ru1 and Ru2 are the only label distribution peers to which Rd distributed a binding of label value L;
- Ru1 and Ru2 are each directly connected to Rd via a point-to-point interface.

When these conditions hold, an LSR may use labels that have "per interface" scope, that is, that are only unique per interface. We may say that the LSR is using a "per-interface label space." When these conditions do not hold, the labels must be unique over the LSR that has assigned them, and we may say that the LSR is using a "per-platform label space."

If a particular LSR Rd is attached to a particular LSR Ru over two point-to-point interfaces, then Rd may distribute to Ru a binding of label L to FEC F1, as well as a binding of label L to FEC F2, F1 $\neq$ F2, if and only if each binding is valid only for packets that Ru sends to Rd over a particular one of the interfaces. In all other cases, Rd must not distribute to Ru bindings of the same label value to two different FECs.

This prohibition holds even if the bindings are regarded as being at different levels of the hierarchy. In MPLS, there is no notion of having a different label space for different levels of the hierarchy; when interpreting a label, the level of the label is irrelevant.

The question arises as to whether it is possible for an LSR to use multiple per-platform label spaces or to use multiple per-interface label spaces for the same interface. This is not prohibited by the architecture. However, in such cases the LSR must have some means, not specified by the architecture, of determining, for a particular incoming label, to which label space that label belongs. For example, [44] specifies that a different label space is used for unicast packets than for multicast packets, and uses a data link layer code point to distinguish the two label spaces.

9.3.4.15   Label-Switched Path, LSP Ingress, LSP Egress

A LSP of level m for a particular packet P is a sequence of routers, <R1, R2, ..., Rn> with the following properties:

1. R1, the LSP ingress, is an LSR that pushes a label onto P's label stack, resulting in a label stack of depth m.

2. For all i, $1 < i < n$, P has a label stack of depth m when received by LSR Ri.

3. At no time during P's transit from R1 to R$[n-1]$ does its label stack ever have a depth of less than m.

4. For all i, $1 < i < n$, Ri transmits P to R$[i+1]$ by means of MPLS, that is, by using the label at the top of the label stack (the level m label) as an index into an ILM.

5. For all i, $1 < i < n$, if a system S receives and forwards P after P is transmitted by Ri but before P is received by R$[i+1]$ (e.g., Ri and R$[i+1]$ might be connected via a switched data link subnetwork, and S might be one of the data link switches), then S's forwarding decision is not based on the level m label or on the network layer header. This may be because (a) the decision is not based on the label stack or the network layer header at all or (b) the decision is based on a label stack on which additional labels have been pushed (i.e., on a level $m + k$ label, where $k > 0$).

In other words, we can speak of the level m LSP for packet P as the sequence of routers:

1. Which begins with an LSR (an LSP ingress) that pushes on a level m label;

2.   All of whose intermediate LSRs make their forwarding decision by label switching on a level m label;

3.   Which ends (at an LSP egress) when a forwarding decision is made by label switching on a level $m-k$ label, where $k > 0$, or when a forwarding decision is made by "ordinary" non-MPLS forwarding procedures.

A consequence (or perhaps a presupposition) of this is that whenever an LSR pushes a label onto an already labeled packet, it needs to make sure that the new label corresponds to a FEC whose LSP egress is the LSR that assigned the label, which is now second in the stack.

We will call a sequence of LSRs the "LSP for a particular FEC F" if it is an LSP of level m for a particular packet P when P's level m label is a label corresponding to FEC F.

Consider the set of nodes that may be LSP ingress nodes for FEC F. Then there is an LSP for FEC F that begins with each of those nodes. If a number of those LSPs have the same LSP egress, then one can consider the set of such LSPs to be a tree, whose root is the LSP egress. (Because data travel along this tree toward the root, this may be called a *multipoint-to-point tree*.) We can thus speak of the *LSP tree* for a particular FEC F.

9.3.4.16   Penultimate Hop Popping

Note that according to the definitions of Section 9.3.4.15, if <R1, R2, ..., Rn> is a level m LSP for packet P, P may be transmitted from R$[n-1]$ to Rn with a label stack of depth $m-1$. That is, the label stack may be popped at the penultimate LSR of the LSP, rather than at the LSP Egress.

From an architectural perspective, this is perfectly appropriate. The purpose of the level m label is to get the packet to Rn. Once R$[n-1]$ has decided to send the packet to Rn, the label no longer has any function and no longer needs to be carried.

There is also a practical advantage to doing penultimate hop popping. If one does not do this, then when the LSP egress receives a packet, it first looks up the top label, and determines as a result of that lookup that it is indeed the LSP egress. Then it must pop the stack and examine what remains of the packet. If there is another label on the stack, the egress will look this up and forward the packet based on this lookup. (In this case, the egress for the packet's level m LSP is also an intermediate node for its level $m-1$ LSP.) If there is no other label on the stack, then the packet is forwarded according to its network layer destination address. Note that this would require the egress to do two lookups, either two label lookups or a label lookup followed by an address lookup.

If, on the other hand, penultimate hop popping is used, then when the penultimate hop looks up the label, it determines (1) that it is the penultimate hop and (2) who the next hop is.

The penultimate node then pops the stack, and forwards the packet based on the information gained by looking up the label that was previously at the top of the stack. When the LSP egress receives the packet, the label that is now at the top of the stack will be the label that it needs to look up in order to make its own forwarding decision. Or, if the packet was only carrying a single label, the LSP egress will simply see the network layer packet, which is just what it needs to see in order to make its forwarding decision. This technique allows the egress to do a single lookup and also requires only a single lookup by the penultimate node.

The creation of the forwarding "fastpath" in a label-switching product may be greatly aided if it is known that only a single lookup is ever required:

- The code may be simplified if it can assume that only a single lookup is ever needed.
- The code can be based on a "time budget" that assumes that only a single lookup is ever needed.

In fact, when penultimate hop popping is done, the LSP egress need not even be an LSR. However, some hardware switching engines may not be able to pop the label stack, so this cannot be universally required. There may also be some situations in which penultimate hop popping is not desirable. Therefore, the penultimate node pops the label stack only if this is specifically requested by the egress node, or if the next node in the LSP does not support MPLS. (If the next node in the LSP does support MPLS, but does not make such a request, the penultimate node has no way of knowing that it in fact is the penultimate node.)

An LSR that is capable of popping the label stack at all must do penultimate hop popping when so requested by its downstream label distribution peer. Initial label distribution protocol negotiations must allow each LSR to determine whether its neighboring LSRS are capable of popping the label stack. A LSR must not request a label distribution peer to pop the label stack unless it is capable of doing so.

The question might arise of whether the egress node can always interpret the top label of a received packet properly if penultimate hop popping is used. As long as the uniqueness and scoping rules of Section 9.3.4.14 are obeyed, it is always possible to interpret the top label of a received packet unambiguously.

9.3.4.17   LSP Next Hop

The LSP next hop for a particular labeled packet in a particular LSR is the LSR that is the next hop, as selected by the NHLFE entry used for forwarding that packet. The LSP next hop for a particular FEC is the next hop as selected by the NHLFE entry indexed by a label that corresponds to that FEC. Note that the LSP next hop may differ from the next hop that would be chosen by the network layer routing algorithm. We will use the term *L3 next hop* when we refer to the latter.

9.3.4.18 Invalid Incoming Labels

What should an LSR do if it receives a labeled packet with a particular incoming label, but has no binding for that label? It is tempting to think that the labels can just be removed, and the packet forwarded as an unlabeled IP packet. However, in some cases, doing so could cause a loop. If the upstream LSR thinks the label is bound to an explicit route, and the downstream LSR does not think the label is bound to anything, and if the hop by hop routing of the unlabeled IP packet brings the packet back to the upstream LSR, then a loop is formed.

It is also possible that the label was intended to represent a route that cannot be inferred from the IP header. Therefore, when a labeled packet is received with an invalid incoming label, it must be discarded, unless it is determined by some means (not within the scope of this chapter) that forwarding it unlabeled cannot cause any harm.

9.3.4.19 LSP Control: Independent Versus Ordered

Some FECs correspond to address prefixes that are distributed via a dynamic routing algorithm. The setup of the LSPs for these FECs can be done in one of two ways: independent LSP control or ordered LSP control.

In independent LSP control, each LSR, on noting that it recognizes a particular FEC, makes an independent decision to bind a label to that FEC and to distribute that binding to its label distribution peers. This corresponds to the way in which conventional IP datagram routing works; each node makes an independent decision as to how to treat each packet and relies on the routing algorithm to converge rapidly so as to ensure that each datagram is correctly delivered.

In ordered LSP control, an LSR only binds a label to a particular FEC if it is the egress LSR for that FEC, or if it has already received a label binding for that FEC from its next hop for that FEC.

If one wants to ensure that traffic in a particular FEC follows a path with some specified set of properties (e.g., that the traffic does not traverse any node twice, that a specified amount of resources is available to the traffic, that the traffic follows an explicitly specified path, and so on), ordered control must be used. With independent control, some LSRs may begin label switching of traffic in the FEC before the LSP is completely set up, and thus some traffic in the FEC may follow a path that does not have the specified set of properties. Ordered control also needs to be used if the recognition of the FEC is a consequence of the setting up of the corresponding LSP. Ordered LSP setup may be initiated either by the ingress or the egress.

Ordered control and independent control are fully interoperable. However, unless all LSRs in an LSP are using ordered control, the overall effect on network behavior is largely that of independent control, since one cannot be sure that an LSP is not used until it is fully set up.

This architecture allows the choice between independent control and ordered control to be a local matter. Because the two methods interwork, a given LSR need support only one or the other. Generally speaking, the choice of independent versus ordered control does not appear to have any effect on the label distribution mechanisms that need to be defined.

9.3.4.20 Aggregation

One way of partitioning traffic into FECs is to create a separate FEC for each address prefix that appears in the routing table. However, within a particular MPLS domain, this may result in a set of FECs such that all traffic in all those FECs follows the same route. For example, a set of distinct address prefixes might all have the same egress node, and label swapping might be used only to get the traffic to the egress node. In this case, within the MPLS domain, the union of those FECs is itself a FEC. This creates a choice: Should a distinct label be bound to each component FEC, or should a single label be bound to the union, and that label applied to all traffic in the union?

The procedure of binding a single label to a union of FECs that is itself a FEC (within some domain), and of applying that label to all traffic in the union, is known as *aggregation*. The MPLS architecture allows aggregation. Aggregation may reduce the number of labels needed to handle a particular set of packets and may also reduce the amount of label distribution control traffic needed.

Given a set of FECs that can be aggregated into a single FEC, it is possible to (1) aggregate them into a single FEC, (2) aggregate them into a set of FECs, or (3) not aggregate them at all. Thus we can speak of the *granularity* of aggregation, with (1) being the "coarsest granularity" and (3) being the "finest granularity."

When order control is used, each LSR should adopt, for a given set of FECs, the granularity used by its next hop for those FECs. When independent control is used, it is possible that two adjacent LSRs, Ru and Rd, for example, will aggregate some set of FECs differently.

If Ru has finer granularity than Rd, this does not cause a problem. Ru distributes more labels for that set of FECs than Rd does. This means that when Ru needs to forward labeled packets in those FECs to Rd, it may need to map n labels into m labels, where $n > m$. As an option, Ru may withdraw the set of n labels that it has distributed, and then distribute a set of m labels, corresponding to Rd's level of granularity. This is not necessary to ensure correct operation, but it does result in a reduction of the number of labels distributed by Ru, and Ru is not gaining any particular advantage by distributing the larger number of labels. The decision whether to do this or not is a local matter.

If Ru has coarser granularity than Rd (i.e., Rd has distributed n labels for the set of FECs, while Ru has distributed m, where $n > m$), it has two choices:

1. It may adopt Rd's finer level of granularity. This would require it to withdraw the m labels it has distributed and instead distribute n labels. This is the preferred option.

2. It may simply map its m labels into a subset of Rd's n labels, if it can determine that this will produce the same routing. For example, suppose that Ru applies a single label to all traffic that needs to pass through a certain egress LSR, whereas Rd binds a number of different labels to such traffic, depending on the individual destination addresses of the packets. If Ru knows the address of the egress router, and if Rd has bound a label to the FEC that is identified by that address, then Ru can simply apply that label.

In any event, every LSR needs to know (by configuration) what granularity to use for labels that it assigns. Where ordered control is used, this requires each node to know the granularity only for FECs that leave the MPLS network at that node. For independent control, best results may be obtained by ensuring that all LSRs are consistently configured to know the granularity for each FEC. However, in many cases this may be done by using a single level of granularity that applies to all FECs (such as "one label per IP prefix in the forwarding table" or "one label per egress node").

9.3.4.21 Route Selection

Route selection refers to the method used for selecting the LSP for a particular FEC. The MPLS protocol architecture supports two options for route selection: (1) hop by hop routing or (2) explicit routing. Hop-by-hop routing allows each node to independently choose the next hop for each FEC. This is the usual mode today in existing IP networks. A hop-by-hop routed LSP is an LSP whose route is selected using hop-by-hop routing. In an explicitly routed LSP, each LSR does not independently choose the next hop; rather, a single LSR, generally the LSP ingress or the LSP egress, specifies several (or all) of the LSRs in the LSP. If a single LSR specifies the entire LSP, the LSP is "strictly" explicitly routed. If a single LSR specifies only some of the LSP, the LSP is "loosely" explicitly routed.

The sequence of LSRs followed by an explicitly routed LSP may be chosen by configuration or may be selected dynamically by a single node (for example, the egress node may make use of the topological information learned from a link state database in order to compute the entire path for the tree ending at that egress node).

Explicit routing may be useful for a number of purposes, such as policy routing or traffic engineering. In MPLS, the explicit route needs to be specified at the time that labels are assigned, but the explicit route does not have to be specified with each IP packet. This makes MPLS explicit routing much more efficient than the alternative of IP source routing.

The procedures for making use of explicit routes, either strict or loose, are beyond the scope of this chapter.

9.3.4.22 Lack of Outgoing Label

When a labeled packet is traveling along an LSP, it may occasionally happen that it reaches an LSR at which the ILM does not map the packet's incoming label into an NHLFE, even though the incoming label is itself valid. This can happen due to transient conditions, or due to an error at the LSR, which should be the packet's next hop. It is tempting in such cases to strip off the label stack and attempt to forward the packet further via conventional forwarding, based on its network layer header. However, in general this is not a safe procedure:

- If the packet has been following an explicitly routed LSP, this could result in a loop.

- The packet's network header may not contain enough information to enable this particular LSR to forward it correctly.

Unless it can be determined (through some means outside the scope of this chapter) that neither of these situations occurs, the only safe procedure is to discard the packet.

9.3.4.23 Time-to-Live

In conventional IP forwarding, each packet carries a *time-to-live* (TTL) value in its header. Whenever a packet passes through a router, its TTL gets decremented by 1; if the TTL reaches 0 before the packet has reached its destination, the packet gets discarded.

This provides some level of protection against forwarding loops that may exist due to misconfigurations or due to failure or slow convergence of the routing algorithm. TTL is sometimes used for other functions as well, such as multicast scoping, and supporting the "traceroute" command. This implies that MPLS needs to deal with two TTL- related issues: (1) TTL as a way to suppress loops and (2) TTL as a way to accomplish other functions, such as limiting the scope of a packet.

When a packet travels along an LSP, it should emerge with the same TTL value that it would have had if it had traversed the same sequence of routers without having been label switched. If the packet travels along a hierarchy of LSPs, the total number of LSR hops traversed should be reflected in its TTL value when it emerges from the hierarchy of LSPs.

The way in which TTL is handled may vary depending on whether the MPLS label values are carried in an MPLS-specific "shim" header [46], or if the MPLS labels are carried in an L2 header, such as an ATM header [56] or a frame relay header [57]. If the label values are encoded in a "shim" that sits between the data link and network layer headers, then this shim *must have* a TTL field that *should* be initially loaded from the network layer header TTL field, should be decremented at each LSR hop, and should be copied into the network layer header TTL field when the packet emerges from its LSP.

If the label values are encoded in a data link layer header (e.g., the VPI/VCI field in ATM's AAL-5 header), and the labeled packets are forwarded by an L2 switch (e.g., an ATM switch), and the data link layer (like ATM) does not itself have a TTL field, then it will not be possible to decrement a packet's TTL at each LSR hop. An LSP segment that consists of a sequence of LSRs that cannot decrement a packet's TTL will be called a *non-TTL LSP segment.*

When a packet emerges from a non-TTL LSP segment, it *should,* however, be given a TTL that reflects the number of LSR hops it traversed. In the unicast case, this can be achieved by propagating a meaningful LSP length to ingress nodes, enabling the ingress to decrement the TTL value before forwarding packets into a non–TTL LSP segment.

Sometimes it can be determined, upon ingress to a non–TTL LSP segment, that a particular packet's TTL will expire before the packet reaches the egress of that non–TTL LSP segment. In this case, the LSR at the ingress to the non–TTL LSP segment must not label switch the packet. This means that special procedures must be developed to support traceroute functionality, for example, traceroute packets may be forwarded using conventional hop-by-hop forwarding.

9.3.4.24 Loop Control

On a non–TTL LSP segment, by definition, TTL cannot be used to protect against forwarding loops. The importance of loop control may depend on the particular hardware being used to provide the LSR functions along the non–TTL LSP segment.

Suppose, for instance, that ATM switching hardware is being used to provide MPLS switching functions, with the label being carried in the VPI/VCI field. Because ATM switching hardware cannot decrement TTL, there is no protection against loops. If the ATM hardware is capable of providing fair access to the buffer pool for incoming cells carrying different VPI/VCI values, this looping may not have any deleterious effect on other traffic. If the ATM hardware cannot provide fair buffer access of this sort, however, then even transient loops may cause severe degradation of the LSR's total performance.

Even if fair buffer access can be provided, it is still worthwhile to have some means of detecting loops that last "longer than possible." In addition, even where TTL and/or per-VC fair queuing provides a means for surviving loops, it still may be desirable where practical to avoid setting up LSPs that loop. All LSRs that may attach to non-TTL LSP segments will therefore be required to support a common technique for loop detection; however, use of the loop detection technique is optional. The loop detection technique is specified in [54, 56].

9.3.4.25   Label Encodings

To transmit a label stack along with the packet whose label stack it is, we need to define a concrete encoding of the label stack. The architecture supports several different encoding techniques; the choice of encoding technique depends on the particular kind of device being used to forward labeled packets.

MPLS-Specific Hardware and Software

If one is using MPLS-specific hardware and/or software to forward labeled packets, the most obvious way to encode the label stack is to define a new protocol to be used as a "shim" between the data link layer and network layer headers. This shim would really be just an encapsulation of the network layer packet; it would be "protocol independent" such that it could be used to encapsulate any network layer. Hence, we will refer to it as the *generic MPLS encapsulation*. The generic MPLS encapsulation would in turn be encapsulated in a data link layer protocol. The MPLS generic encapsulation is specified in [44].

ATM Switches as LSRs

Note that MPLS forwarding procedures are similar to those of legacy "label swapping" switches such as ATM switches. ATM switches use the input port and the incoming VPI/VCI value as the index into a cross-connect table, from which they obtain an output port and an outgoing VPI/VCI value. Therefore, if one or more labels can be encoded directly into the fields that are accessed by these legacy switches, then the legacy switches can, with suitable software upgrades, be used as LSRs. We refer to such devices as *ATM- LSRs*. There are three obvious ways to encode labels in the ATM cell header (presuming the use of AAL-5):

1. *SVC encoding*. Use the VPI/VCI field to encode the label that is at the top of the label stack. This technique can be used in any network. With this encoding technique, each LSP is realized as an ATM SVC, and the label distribution protocol becomes the ATM "signaling" protocol. With this encoding technique, the ATM-LSRs cannot perform "push" or "pop" operations on the label stack.

2. *SVP encoding*. Use the VPI field to encode the label that is at the top of the label stack, and the VCI field to encode the second label

on the stack, if one is present. This technique has some advantages over the previous one in that it permits the use of ATM "VP switching." That is, the LSPs are realized as ATM SVPs, with the label distribution protocol serving as the ATM signaling protocol. However, this technique cannot always be used. If the network includes an ATM virtual path through a non-MPLS ATM network, then the VPI field is not necessarily available for use by MPLS. When this encoding technique is used, the ATM-LSR at the egress of the VP effectively does a "pop" operation.

3. *SVP multipoint encoding.* Use the VPI field to encode the label that is at the top of the label stack, use part of the VCI field to encode the second label on the stack, if one is present, and use the remainder of the VCI field to identify the LSP ingress. If this technique is used, conventional ATM VP-switching capabilities can be used to provide multipoint-to-point VPs. Cells from different packets will then carry different VCI values. As we shall see in a later section, this enables us to do label merging, without running into any cell interleaving problems, on ATM switches that can provide multipoint-to-point VPs, but that do not have the VC merge capability. This technique depends on the existence of a capability for assigning 16-bit VCI values to each ATM switch such that no single VCI value is assigned to two different switches. (If an adequate number of such values could be assigned to each switch, it would be possible to also treat the VCI value as the second label in the stack.) If there are more labels on the stack than can be encoded in the ATM header, the ATM encodings must be combined with the generic encapsulation.

Interoperability Among Encoding Techniques

If <R1, R2, R3> is a segment of a LSP, it is possible that R1 will use one encoding of the label stack when transmitting packet P to R2, but R2 will use a different encoding when transmitting a packet P to R3. In general, the MPLS architecture supports LSPs with different label stack encodings used on different hops. Therefore, when we discuss the procedures for processing a labeled packet, we speak in abstract terms of operating on the packet's label stack. When a labeled packet is received, the LSR must decode it to determine the current value of the label stack, then must operate on the label stack to determine the new value of the stack, and then encode the new value appropriately before transmitting the labeled packet to its next hop.

Unfortunately, ATM switches have no capability for translating from one encoding technique to another. The MPLS architecture therefore requires that, whenever it is possible for two ATM switches to be successive LSRs along a level *m* LSP for some packet, those two ATM switches must use the same encoding technique.

Naturally some MPLS networks will contain a combination of ATM switches operating as LSRs, and other LSRs that operate using an MPLS shim header. In such networks some of the LSRs may have ATM interfaces as well as MPLS shim interfaces. This is one example of an LSR with different label stack encodings on different hops. Such an LSR may swap off an ATM encoded label stack on an incoming interface and replace it with an MPLS shim header encoded label stack on the outgoing interface.

9.3.4.26   Label Merging

Suppose that an LSR has bound multiple incoming labels to a particular FEC. When forwarding packets in that FEC, one would like to have a single outgoing label that is applied to all such packets. The fact that two different packets in the FEC arrived with different incoming labels is irrelevant; one would like to forward them with the same outgoing label. The capability to do so is known as *label merging*.

Let us say that an LSR is capable of label merging if it can receive two packets from different incoming interfaces, and/or with different labels, and send both packets out the same outgoing interface with the same label. Once the packets are transmitted, the information that they arrived from different interfaces and/or with different incoming labels is lost.

Let us say that an LSR is not capable of label merging if, for any two packets that arrive from different interfaces or with different labels, the packets must either be transmitted out different interfaces or must have different labels. ATM-LSRs using the SVC or SVP encodings cannot perform label merging. This is discussed in more detail in the next section.

If a particular LSR cannot perform label merging, then if two packets in the same FEC arrive with different incoming labels, they must be forwarded with different outgoing labels. With label merging, the number of outgoing labels per FEC need only be 1; without label merging, the number of outgoing labels per FEC could be as large as the number of nodes in the network.

With label merging, the number of incoming labels per FEC that a particular LSR needs can never be larger than the number of label distribution adjacencies. Without label merging, the number of incoming labels per FEC that a particular LSR needs is as large as the number of upstream nodes that forward traffic in the FEC to the LSR in question. In fact, it is difficult for an LSR to even determine how many such incoming labels it must support for a particular FEC.

The MPLS architecture accommodates both merging and nonmerging LSRs, but allows for the fact that some LSRs may not support label merging. This leads to the issue of ensuring correct interoperation between merging LSRs and nonmerging LSRs. The issue is somewhat different in the case of datagram media versus the case of ATM. The different media types will therefore be discussed separately.

Nonmerging LSRs

The MPLS forwarding procedures is very similar to the forwarding procedures used by such technologies as ATM and FR. That is, a unit of data arrives, a label (VPI/VCI or DLCI) is looked up in a "cross-connect table," on the basis of that lookup an output port is chosen, and the label value is rewritten. In fact, it is possible to use such technologies for MPLS forwarding; a label distribution protocol can be used as the "signaling protocol" for setting up the cross-connect tables.

Unfortunately, these technologies do not necessarily support the label merging capability. In ATM, if one attempts to perform label merging, the result may be the interleaving of cells from various packets. If cells from different packets get interleaved, it is impossible to reassemble the packets. Some FR switches use cell switching on their backplanes. These switches may also be incapable of supporting label merging, for the same reason—cells of different packets may get interleaved, and there is then no way to reassemble the packets.

One can support two solutions to this problem. First, MPLS will contain procedures that allow the use of nonmerging LSRs. Second, MPLS will support procedures which allow certain ATM switches to function as merging LSRs.

Since MPLS supports both merging and nonmerging LSRs, MPLS also contains procedures to ensure correct interoperation between them.

Labels for Merging and Nonmerging LSRs

An upstream LSR that supports label merging needs to be sent only one label per FEC. An upstream neighbor that does not support label merging needs to be sent multiple labels per FEC. However, there is no way of knowing a priori how many labels it needs. This will depend on how many LSRs are upstream of it with respect to the FEC in question.

In the MPLS architecture, if a particular upstream neighbor does not support label merging, it is not sent any labels for a particular FEC unless it explicitly asks for a label for that FEC. The upstream neighbor may make multiple such requests, and is given a new label each time. When a downstream neighbor receives such a request from upstream, and the downstream neighbor does not itself support label merging, then it must in turn ask its downstream neighbor for another label for the FEC in question.

Some nodes that support label merging may only be able to merge a limited number of incoming labels into a single outgoing label. Suppose, for example, that due to some hardware limitation a node is capable of merging four incoming labels into a single outgoing label. Suppose, however, that this particular node has six incoming labels arriving at it for a particular FEC. In this case, this node may merge these into two outgoing labels.

Whether label merging is applicable to explicitly routed LSPs needs further study.

Merge over ATM

Methods of Eliminating Cell Interleave   Several methods can be used to eliminate the cell interleaving problem in ATM, thereby allowing ATM switches to support stream merge:

1. *VP merge using SVP multipoint encoding.* When VP merge is used, multiple virtual paths are merged into a virtual path, but packets from different sources are distinguished by using different VCIs within the VP.

2. *VC merge.* When VC merge is used, switches are required to buffer cells from one packet until the entire packet is received (this may be determined by looking for the AAL-5 end of frame indicator).

VP merge has the advantage that it is compatible with a higher percentage of existing ATM switch implementations. This makes it more likely that VP merge can be used in existing networks. Unlike VC merge, VP merge does not incur any delays at the merge points and also does not impose any buffer requirements. However, it has the disadvantage that it requires coordination of the VCI space within each VP. This can be accomplished in a number of ways. Selection of one or more methods requires further study.

This trade-off between compatibility with existing equipment versus protocol complexity and scalability implies that it is desirable for the MPLS protocol to support both VP merge and VC merge. To do so, each ATM switch participating in MPLS needs to know whether its immediate ATM neighbors perform VP merge, VC merge, or no merge.

Interoperation: VC Merge, VP Merge, and Nonmerge   The interoperation of the various forms of merging over ATM is most easily described by first describing the interoperation of VC merge with nonmerge. In the case where VC merge and nonmerge nodes are interconnected, the forwarding of cells is based in all cases on a VC (i.e., the concatenation of the VPI and VCI). For each node, if an upstream neighbor is doing VC merge, then that upstream neighbor requires only a single VPI/VCI for a particular stream. (This is analogous to the requirement for a single label in the case of operation over frame media.) If the upstream neighbor is not doing merge, then the neighbor will require a single VPI/VCI per stream for itself, plus enough VPI/VCIs to pass to its upstream neighbors. The number required will be determined by allowing the upstream nodes to request additional VPI/VCIs from their downstream neighbors. (This is again analogous to the method used with frame merge.)

A similar method is possible to support nodes that perform VP merge. In this case the VP merge node, rather than requesting a single VPI/VCI or a number of VPI/VCIs from its downstream neighbor, instead may request a single VP (identified by a VPI) but several VCIs within the VP.

Furthermore, suppose that a nonmerge node is downstream from two different VP merge nodes. This node may need to request one VPI/VCI (for traffic originating from itself) plus two VPs (one for each upstream node), each associated with a specified set of VCIs (as requested from the upstream node).

To support all of the VP merge, VC merge, and nonmerge processes, it is therefore necessary to allow upstream nodes to request a combination of zero or more VC identifiers (consisting of a VPI/VCI), plus zero or more VPs (identified by VPIs), each containing a specified number of VCs (identified by a set of VCIs that are significant within a VP). VP merge nodes would therefore request one VP, with a contained VCI for traffic that it originates (if appropriate) plus a VCI for each VC requested from above (regardless of whether or not the VC is part of a containing VP). The VC merge node would request only a single VPI/VCI (since they can merge all upstream traffic into a single VC). Nonmerge nodes would pass on any requests that they get from above, plus request a VPI/VCI for traffic that they originate (if appropriate).

9.3.4.27   Tunnels and Hierarchy

Sometimes a router Ru takes explicit action to cause a particular packet to be delivered to another router Rd, even though Ru and Rd are not consecutive routers on the hop-by-hop path for that packet, and Rd is not the packet's ultimate destination. For example, this may be done by encapsulating the packet inside a network layer packet whose destination address is the address of Rd itself. This creates a "tunnel" from Ru to Rd. We refer to any packet so handled as a *tunneled packet*.

Hop-by-Hop Routed Tunnel

If a tunneled packet follows the hop-by-hop path from Ru to Rd, we say that it is in a *hop-by-hop routed tunnel* whose transmit endpoint is Ru and whose receive endpoint is Rd.

Explicitly Routed Tunnel

If a tunneled packet travels from Ru to Rd over a path other than the hop-by-hop path, we say that it is in an *explicitly routed tunnel* whose transmit endpoint is Ru and whose receive endpoint is Rd. For example, we might send a packet through an explicitly routed tunnel by encapsulating it in a packet that is source routed.

LSP Tunnels

It is possible to implement a tunnel as a LSP and use label switching rather than network layer encapsulation to cause the packet to travel through the tunnel. The tunnel would be a LSP <R1, R2, ..., Rn>, where R1 is the transmit endpoint of the tunnel, and Rn is the receive endpoint of the tunnel. This is called a *LSP tunnel*.

The set of packets to be sent through the LSP tunnel constitutes a FEC, and each LSR in the tunnel must assign a label to that FEC (i.e., must assign a label to the tunnel). The criteria for assigning a particular packet to an LSP tunnel is a local matter at the tunnel's transmit endpoint. To put a packet into an LSP tunnel, the transmit endpoint pushes a label for the tunnel onto the label stack and sends the labeled packet to the next hop in the tunnel.

If it is not necessary for the tunnel's receive endpoint to be able to determine which packets it receives through the tunnel, as discussed earlier, the label stack may be popped at the penultimate LSR in the tunnel.

A *hop-by-hop routed LSP tunnel* is a tunnel that is implemented as a hop-by-hop routed LSP between the transmit endpoint and the receive endpoint. An *explicitly routed LSP tunnel* is a LSP tunnel that is also an explicitly routed LSP.

Hierarchy: LSP Tunnels Within LSPs

Consider a LSP <R1, R2, R3, R4>. Let us suppose that R1 receives unlabeled packet P, and pushes on its label stack the label to cause it to follow this path, and that this is in fact the hop-by-hop path. However, let us further suppose that R2 and R3 are not directly connected, but are "neighbors" by virtue of being the endpoints of an LSP tunnel. So the actual sequence of LSRs traversed by P is <R1, R2, R21, R22, R23, R3, R4>.

When P travels from R1 to R2, it will have a label stack of depth 1. R2, switching on the label, determines that P must enter the tunnel. R2 first replaces the incoming label with a label that is meaningful to R3. Then it pushes on a new label. This level 2 label has a value that is meaningful to R21. Switching is done on the level 2 label by R21, R22, R23. R23, which is the penultimate hop in the R2–R3 tunnel, pops the label stack before forwarding the packet to R3. When R3 sees packet P, P has only a level 1 label, having now exited the tunnel. Because R3 is the penultimate hop in P's level 1 LSP, it pops the label stack, and R4 receives P unlabeled. The label stack mechanism allows LSP tunneling to nest to any depth.

Label Distribution Peering and Hierarchy

Suppose that packet P travels along a level 1 LSP <R1, R2, R3, R4>, and when going from R2 to R3 travels along a level 2 LSP <R2, R21, R22, R3>. From the perspective of the level 2 LSP, R2's label distribution peer is R21. From the perspective of the level 1 LSP, R2's label distribution peers are R1 and R3. One can have label distribution peers at each layer of hierarchy. We will see in Sections 9.3.5.6 and 9.3.5.7 some ways to make use of this hierarchy. Note that in this example, R2 and R21 must be IGP neighbors, but R2 and R3 need not be.

When two LSRs are IGP neighbors, we will refer to them as *local label distribution peers*. When two LSRs may be label distribution peers, but are not IGP neighbors, we will refer to them as *remote label distribution peers*. In the preceding example, R2 and R21 are local label distribution peers, but R2 and R3 are remote label distribution peers.

The MPLS architecture supports two ways to distribute labels at different layers of the hierarchy: explicit peering and implicit peering.

Label distribution is performed with one's local label distribution peer by sending label distribution protocol messages that are addressed to the peer. Label distribution with remote label distribution peers is accomplished in one of two ways:

1. *Explicit peering.* In explicit peering, one distributes labels to a peer by sending label distribution protocol messages that are addressed to the peer, exactly as one would do for local label distribution peers. This technique is most useful when the number of remote label distribution peers is small, or the number of higher level label bindings is large, or the remote label distribution peers are in distinct routing areas or domains. Of course, one needs to know which labels to distribute to which peers; this is addressed in a later section, as are examples of the use of explicit peering.

2. *Implicit peering.* In implicit peering, one does not send label distribution protocol messages that are addressed to one's peer. Rather, to distribute higher level labels to one's remote label distribution peers, one encodes a higher level label as an attribute of a lower level label, and then distributes the lower level label, along with this attribute, to one's local label distribution peers. The local label distribution peers then propagate the information to their local label distribution peers. This process continues until the information reaches the remote peer. This technique is most useful when the number of remote label distribution peers is large. Implicit peering does not require an n-square peering mesh to distribute labels to the remote label distribution peers because the information is piggybacked through the local label distribution peering. However, implicit peering requires the intermediate nodes to store information in which they might not be directly interested. An example of the use of implicit peering is given in Section 9.3.5.3.

9.3.4.28 Label Distribution Protocol Transport

A label distribution protocol is used between nodes in an MPLS network to establish and maintain the label bindings. In order for MPLS to operate correctly, label distribution information needs to be transmitted reliably, and the label distribution protocol messages pertaining to a particular FEC need to be transmitted in sequence. Flow control is also desirable, as is the capability to carry multiple label messages in a single datagram. One way to meet these goals is to use TCP as the underlying transport, as is done in [52, 54].

9.3.4.29   Why More Than One Label Distribution Protocol?

This architecture does not establish hard and fast rules for choosing which label distribution protocol to use in which circumstances. However, it is possible to point out some of the considerations.

BGP and LDP

In many scenarios, it is desirable to bind labels to FECs that can be identified with routes to address prefixes (see Section 9.3.5.1). If there is a standard, widely deployed routing algorithm that distributes those routes, it can be argued that label distribution is best achieved by piggybacking the label distribution on the distribution of the routes themselves.

For example, BGP distributes such routes, and if a BGP speaker needs to also distribute labels to its BGP peers, using BGP to do the label distribution [52] has a number of advantages. In particular, it permits BGP route reflectors to distribute labels, thus providing a significant scalability advantage over using LDP to distribute labels between BGP peers.

Labels for RSVP Flowspecs

When RSVP is used to set up resource reservations for particular flows, it can be desirable to label the packets in those flows, so that the RSVP filter-spec does not need to be applied at each hop. It can be argued that having RSVP distribute the labels as part of its path/reservation setup process is the most efficient method of distributing labels for this purpose.

Labels for Explicitly Routed LSPs

In some applications of MPLS, particularly those related to traffic engineering, it is desirable to set up an explicitly routed path, from ingress to egress. It is also desirable to apply resource reservations along that path. One can imagine two approaches to this:

- Start with an existing protocol that is used for setting up resource reservations and extend it to support explicit routing and label distribution.

- Start with an existing protocol that is used for label distribution and extend it to support explicit routing and resource reservations.

The first approach has given rise to the protocol specified in [53], the second to the approach specified in [55].

9.3.5   Some Applications of MPLS

9.3.5.1   MPLS and Hop-by-Hop Routed Traffic

A number of uses of MPLS require that packets with a certain label be forwarded along the same hop-by-hop routed path that would be used for forwarding a packet with a specified address in its network layer destination address field.

Labels for Address Prefixes

In general, router R determines the next hop for packet P by finding the address prefix X in its routing table that is the longest match for P's destination address. That is, the packets in a given FEC are just those packets that match a given address prefix in R's routing table. In this case, a FEC can be identified with an address prefix. Note that a packet P may be assigned to FEC F, and FEC F may be identified with address prefix X, even if P's destination address does not match X.

Distributing Labels for Address Prefixes

Label Distribution Peers for an Address Prefix   LSRs R1 and R2 are considered to be label distribution peers for address prefix X if and only if one of the following conditions holds:

1. R1's route to X is a route that it learned about via a particular instance of a particular IGP, and R2 is a neighbor of R1 in that instance of that IGP.

2. R1's route to X is a route that it learned about by some instance of routing algorithm A1, and that route is redistributed into an instance of routing algorithm A2, and R2 is a neighbor of R1 in that instance of A2.

3. R1 is the receive endpoint of an LSP tunnel that is within another LSP, and R2 is a transmit endpoint of that tunnel, and R1 and R2 are participants in a common instance of an IGP and are in the same IGP area (if the IGP in question has areas), and R1's route to X was learned via that IGP instance or is redistributed by R1 into that IGP instance.

4. R1's route to X is a route that it learned about via BGP, and R2 is a BGP peer of R1.

In general, these rules ensure that if the route to a particular address prefix is distributed via an IGP, the label distribution peers for that address prefix are the IGP neighbors. If the route to a particular address prefix is distributed via BGP, the label distribution peers for that address prefix are the BGP peers. In other cases of LSP tunneling, the tunnel endpoints are label distribution peers.

Distributing Labels   To use MPLS for the forwarding of packets according to the hop-by-hop route corresponding to any address prefix, each LSR must do the following:

1. Bind one or more labels to each address prefix that appears in its routing table.

2. For each such address prefix X, use a label distribution protocol to distribute the binding of a label for X to each of its label distribution peers for X.

    There is also one circumstance in which an LSR must distribute a label binding for an address prefix, even if it is not the LSR that bound that label to that address prefix:

3. If R1 uses BGP to distribute a route to X, naming some other LSR R2 as the BGP next hop to X, and if R1 knows that R2 has assigned label L to X, then R1 must distribute the binding between L and X to any BGP peer to which it distributes that route.

These rules ensure that labels corresponding to address prefixes that correspond to BGP routes are distributed to IGP neighbors if and only if the BGP routes are distributed into the IGP. Otherwise, the labels bound to BGP routes are distributed only to the other BGP speakers.

These rules are intended only to indicate which label bindings must be distributed by a given LSR to which other LSRs.

Using the Hop-by-Hop Path as the LSP

If the hop-by-hop path that packet P needs to follow is <R1, ..., Rn>, then <R1, ..., Rn> can be an LSP as long as:

1. There is a single address prefix X, such that, for all i, $1 <= i < n$, X is the longest match in Ri's routing table for P's destination address;

2. For all i, $1 < i < n$, Ri has assigned a label to X and distributed that label to R$[i - 1]$.

Note that a packet's LSP can extend only until it encounters a router whose forwarding tables have a longer best match address prefix for the packet's destination address. At that point, the LSP must end and the best match algorithm must be performed again.

Suppose, for example, that packet P, with destination address 10.2.153.178 needs to go from R1 to R2 to R3. Suppose also that R2 advertises address prefix 10.2/16 to R1, but R3 advertises 10.2.153/23, 10.2.154/23, and 10.2/16 to R2. That is, R2 is advertising an "aggregated route" to R1. In this situation, packet P can be label-switched until it reaches R2, but since R2 has performed route aggregation, it must execute the best match algorithm to find P's FEC.

LSP Egress and LSP Proxy Egress

An LSR R is considered to be an "LSP egress" LSR for address prefix X if and only if one of the following conditions holds:

1. R has an address Y, such that X is the address prefix in R's routing table which is the longest match for Y;

2. R contains in its routing tables one or more address prefixes Y such that X is a proper initial substring of Y, but R's "LSP previous hops" for X do not contain any such address prefixes Y; that is, R is a "deaggregation point" for address prefix X.

An LSR R1 is considered to be an *LSP proxy egress* LSR for address prefix X if and only if:

1. R1's next hop for X is R2, and R1 and R2 are not label distribution peers with respect to X (perhaps because R2 does not support MPLS);

2. R1 has been configured to act as an LSP proxy egress for X.

The definition of LSP allows for the LSP egress to be a node that does not support MPLS; in this case the penultimate node in the LSP is the proxy egress.

The Implicit NULL Label

The implicit NULL label is a label with special semantics that an LSR can bind to an address prefix. If LSR Ru, by consulting its ILM, sees that labeled packet P must be forwarded next to Rd, but that Rd has distributed a binding of implicit NULL to the corresponding address prefix, then instead of replacing the value of the label on top of the label stack, Ru pops the label stack and forwards the resulting packet to Rd.

LSR Rd distributes a binding between implicit NULL and an address prefix X to LSR Ru if and only if:

1. The rules of the preceding section titled "Distributing Labels for Address Prefixes" indicate that Rd distributes to Ru a label binding for X;

2. Rd knows that Ru can support the implicit NULL label (i.e., that it can pop the label stack);

3. Rd is an LSP egress (not proxy egress) for X.

This causes the penultimate LSR on a LSP to pop the label stack. This is quite appropriate; if the LSP egress is an MPLS egress for X, then if the penultimate LSR does not pop the label stack, the LSP egress will need to look up the label, pop the label stack, and then look up the next label (or look up the L3 address, if no more labels are present). By having the penultimate LSR pop the label stack, the LSP egress is saved the work of having to look up two labels in order to make its forwarding decision.

However, if the penultimate LSR is an ATM switch, it may not have the capability to pop the label stack. Hence a binding of implicit NULL may be distributed only to LSRs that can support that function.

If the penultimate LSR in an LSP for address prefix X is an LSP proxy egress, it acts just as if the LSP egress had distributed a binding of implicit NULL for X.

Option: Egress–Targeted Label Assignment

In some situations, an LSP ingress, Ri, will know that packets of several different FECs must all follow the same LSP, terminating at, say, LSP egress Re. In such a case, proper routing can be achieved by using a single label for all such FECs; it is not necessary to have a distinct label for each FEC. If (and only if) the following conditions hold:

1. The address of LSR Re is itself in the routing table as a "host route;"
2. There is some way for Ri to determine that Re is the LSP egress for all packets in a particular set of FECs.

Then Ri may bind a single label to all FECS in the set. This is known as *egress-targeted label assignment.*

How can LSR Ri determine that an LSR Re is the LSP egress for all packets in a particular FEC? There are a number of possible ways:

- If the network is running a link state routing algorithm, and all nodes in the area support MPLS, then the routing algorithm provides Ri with enough information to determine the routers through which packets in that FEC must leave the routing domain or area.
- If the network is running BGP, Ri may be able to determine that the packets in a particular FEC must leave the network via some particular router that is the *BGP next hop* for that FEC.
- It is possible to use the label distribution protocol to pass information about which address prefixes are "attached" to which egress LSRs. This method has the advantage of not depending on the presence of link state routing.

If egress-targeted label assignment is used, the number of labels that need to be supported throughout the network may be greatly reduced. This may be significant if one is using legacy-switching hardware to do MPLS, and the switching hardware can support only a limited number of labels.

One possible approach would be to configure the network to use egress-targeted label assignment by default, but to configure particular LSRs to *not* use egress-targeted label assignment for one or more of the address prefixes for which it is an LSP egress. We impose the following rule:

If a particular LSR is *not* an LSP egress for some set of address prefixes, then it should assign labels to the address prefixes in the same way as is done by its LSP next hop for those address prefixes. That is, suppose Rd is Ru's LSP next hop for address prefixes X1 and X2. If Rd assigns the same label to X1 and X2, Ru should as well. If Rd assigns different labels to X1 and X2, then Ru should as well.

For example, suppose one wants to make egress-targeted label assignment the default, but to assign distinct labels to those address prefixes for which there are multiple possible LSP egresses (i.e., for those address prefixes that are multihomed.) One can configure all LSRs to use egress-targeted label assignment, and then configure a handful of LSRs to assign distinct labels to those address prefixes that are multihomed. For a particular multihomed address prefix X, one would only need to configure this in LSRs that are either LSP egresses or LSP proxy egresses for X.

It is important to note that if Ru and Rd are adjacent LSRs in an LSP for X1 and X2, forwarding will still be done correctly if Ru assigns distinct labels to X1 and X2 while Rd assigns just one label to the both of them. This just means that R1 will map different incoming labels to the same outgoing label, an ordinary occurrence.

Similarly, if Rd assigns distinct labels to X1 and X2, but Ru assigns to them both the label corresponding to the address of their LSP egress or proxy egress, forwarding will still be done correctly. Ru will just map the incoming label to the label that Rd has assigned to the address of that LSP egress.

9.3.5.2 MPLS and Explicitly Routed LSPs

There are a number of reasons why it may be desirable to use explicit routing instead of hop-by-hop routing. For example, this allows routes to be based on administrative policies and also allows the routes that LSPs take to be carefully designed to allow traffic engineering [39].

Explicitly Routed LSP Tunnels
In some situations, the network administrators may desire to forward certain classes of traffic along certain prespecified paths, where these paths differ from the hop-by-hop path that the traffic would ordinarily follow. This can be done in support of policy routing or in support of traffic engineering. The explicit route may be a configured one or it may be determined dynamically by some means, for example, by constraint-based routing. MPLS allows this to be easily done by means of explicitly routed LSP tunnels. All that is needed is:

1. A means of selecting the packets that are to be sent into the explicitly routed LSP tunnel;

2. A means of setting up the explicitly routed LSP tunnel;

3. A means of ensuring that packets sent into the tunnel will not loop from the receive endpoint back to the transmit endpoint.

If the transmit endpoint of the tunnel wishes to put a labeled packet into the tunnel, it must first replace the label value at the top of the stack with a label value that was distributed to it by the tunnel's receive endpoint. Then it must push on the label that corresponds to the tunnel itself, as distributed to it by the next hop along the tunnel. To allow this, the tunnel endpoints should be explicit label distribution peers. The label bindings they need to exchange are of no interest to the LSRs along the tunnel.

9.3.5.3 Label Stacks and Implicit Peering

Suppose a particular LSR Re is an LSP proxy egress for 10 address prefixes, and it reaches each address prefix through a distinct interface. One could assign a single label to all 10 address prefixes. Then Re is an LSP egress for all 10 address prefixes. This ensures that packets for all 10 address prefixes get delivered to Re. However, Re would then have to look up the network layer address of each such packet in order to choose the proper interface on which to send the packet. Alternatively, one could assign a distinct label to each interface. Then Re is an LSP proxy egress for the 10 address prefixes. This eliminates the need for Re to look up the network layer addresses in order to forward the packets. However, it can result in the use of a large number of labels.

Another alternative would be to bind all 10 address prefixes to the same level 1 label (which is also bound to the address of the LSR itself), and then to bind each address prefix to a distinct level 2 label. The level 2 label would be treated as an attribute of the level 1 label binding, which we call the *stack attribute*. We impose the following rules:

When LSR Ru initially labels a hitherto unlabeled packet, if the longest match for the packet's destination address is X, and Ru's LSP next hop for X is Rd, and Rd has distributed to Ru a binding of label L1 to X, along with a stack attribute of L2, then:

1. Ru must push L2 and then L1 onto the packet's label stack, and then forward the packet to Rd.
2. When Ru distributes label bindings for X to its label distribution peers, it must include L2 as the stack attribute.
3. Whenever the stack attribute changes (possibly as a result of a change in Ru's LSP next hop for X), Ru must distribute the new stack attribute.

Note that although the label value bound to X may be different at each hop along the LSP, the stack attribute value is passed unchanged, and it is set by the LSP proxy egress. Thus the LSP proxy egress for X becomes an *implicit peer* with each other LSR in the routing area or domain. In this case, explicit peering would be too unwieldy, because the number of peers would become too large.

9.3.5.4 MPLS and Multipath Routing

If an LSR supports multiple routes for a particular stream, then it may assign multiple labels to the stream, one for each route. Thus the reception of a second label binding from a particular neighbor for a particular address prefix should be taken as meaning that either label can be used to represent that address prefix.

If multiple label bindings for a particular address prefix are specified, they may have distinct attributes.

9.3.5.5 LSP Trees as Multipoint-to-Point Entities

Consider the case of packets P1 and P2, each of which has a destination address whose longest match, throughout a particular routing domain, is address prefix X. Suppose that the hop-by-hop path for P1 is <R1, R2, R3>, and the hop-by-hop path for P2 is <R4, R2, R3>. Let's suppose that R3 binds label L3 to X, and distributes this binding to R2. R2 binds label L2 to X, and distributes this binding to both R1 and R4. When R2 receives packet P1, its incoming label will be L2. R2 will overwrite L2 with L3, and send P1 to R3. When R2 receives packet P2, its incoming label will also be L2. R2 again overwrites L2 with L3, and sends P2 on to R3.

Note then that when P1 and P2 are traveling from R2 to R3, they carry the same label, and as far as MPLS is concerned, they cannot be distinguished. Thus instead of talking about two distinct LSPs, <R1, R2, R3> and <R4, R2, R3>, we might talk of a single *multipoint-to-point LSP tree,* which we might denote as <{R1, R4}, R2, R3>.

This creates a difficulty when we attempt to use conventional ATM switches such as LSRs. Because conventional ATM switches do not support multipoint-to-point connections, procedures must be in place to ensure that each LSP is realized as a point-to-point VC. However, if ATM switches that do support multipoint-to-point VCs are in use, then the LSPs can be most efficiently realized as multipoint-to-point VCs. Alternatively, if the SVP multipoint encoding (see earlier section) can be used, the LSPs can be realized as multipoint-to-point SVPs.

9.3.5.6 LSP Tunneling Between BGP Border Routers

Consider the case of an autonomous system A that carries transit traffic between other autonomous systems. Autonomous system A will have a number of BGP border routers, and a mesh of BGP connections among them, over which BGP routes are distributed. In many such cases, it is desirable to avoid distributing the BGP routes to routers that are not BGP border routers. If this can be avoided, the "route distribution load" on those routers is significantly reduced. However, we need some means of ensuring that the transit traffic will be delivered from border router to border router by the interior routers.

This can easily be done by means of LSP tunnels. Suppose that BGP routes are distributed only to BGP border routers, and not to the interior routers that lie along the hop-by-hop path from border router to border router. LSP tunnels can then be used as follows:

1. Each BGP border router distributes, to every other BGP border router in the same autonomous system, a label for each address prefix that it distributes to that router via BGP.

2. The IGP for the autonomous system maintains a host route for each BGP border router. Each interior router distributes its labels for these host routes to each of its IGP neighbors.

3. Suppose that (a) BGP border router B1 receives an unlabeled packet P; (b) address prefix X in B1's routing table is the longest match for the destination address of P; (c) the route to X is a BGP route; (d) the BGP next hop for X is B2; (e) B2 has bound label L1 to X, and has distributed this binding to B1; (f) the IGP next hop for the address of B2 is I1; (g) the address of B2 is in B1's and I1's IGP routing tables as a host route; and (h) I1 has bound label L2 to the address of B2, and distributed this binding to B1.

   Then before sending packet P to I1, B1 must create a label stack for P, then push on label L1, and then push on label L2.

4. Suppose that BGP border router B1 receives a labeled packet P, where the label on the top of the label stack corresponds to an address prefix X, to which the route is a BGP route, and that conditions 3b, 3c, 3d, and 3e all hold. Then before sending packet P to I1, B1 must replace the label at the top of the label stack with L1, and then push on label L2.

With these procedures, a given packet P follows a level 1 LSP, all of whose members are BGP border routers, and between each pair of BGP border routers in the level 1 LSP, it follows a level 2 LSP. These procedures effectively create a hop-by-hop routed LSP tunnel between the BGP border routers.

Because the BGP border routers are exchanging label bindings for address prefixes that are not even known to the IGP routing, the BGP routers should become explicit label distribution peers with each other.

It is sometimes possible to create hop-by-hop routed LSP tunnels between two BGP border routers, even if they are not in the same autonomous system. Suppose, for example, that B1 and B2 are in AS 1. Suppose that B3 is an EBGP neighbor of B2, and is in AS2. Finally, suppose that B2 and B3 are on some network that is common to both autonomous systems (a "demilitarized zone" so to speak). In this case, an LSP tunnel can be set up directly between B1 and B3 as follows:

- B3 distributes routes to B2 (using EBGP), optionally assigning labels to address prefixes.
- B2 redistributes those routes to B1 (using IBGP), indicating that the BGP next hop for each such route is B3. If B3 has assigned labels to address prefixes, B2 passes these labels along, unchanged, to B1.
- The IGP of AS1 has a host route for B3.

9.3.5.7   Other Uses of Hop-by-Hop Routed LSP Tunnels

The use of hop-by-hop routed LSP tunnels is not restricted to tunnels between BGP next hops. Any situation in which one might otherwise have used an encapsulation tunnel is one in which it is appropriate to use a hop-by-hop routed LSP tunnel. Instead of encapsulating the packet with a new header whose destination address is the address of the tunnel's receive endpoint, the label corresponding to the address prefix that is the longest match for the address of the tunnel's receive endpoint is pushed on the packet's label stack. The packet that is sent into the tunnel may or may not already be labeled.

If the transmit endpoint of the tunnel wishes to put a labeled packet into the tunnel, it must first replace the label value at the top of the stack with a label value that was distributed to it by the tunnel's receive endpoint. Then it must push on the label that corresponds to the tunnel itself, as distributed to it by the next hop along the tunnel. To allow this, the tunnel endpoints should be explicit label distribution peers. The label bindings they need to exchange are of no interest to the LSRs along the tunnel.

9.3.5.8   MPLS and Multicast

Multicast routing proceeds by constructing multicast trees. The tree along which a particular multicast packet must get forwarded depends in general on the packet's source address and its destination address. Whenever a particular LSR is a node in a particular multicast tree, it binds a label to that tree. It then distributes that binding to its parent on the multicast tree. (If the node in question is on a LAN, and has siblings on that LAN, it must also distribute the binding to its siblings. This allows the parent to use a single label value when multicasting to all children on the LAN.)

When a multicast labeled packet arrives, the NHLFE corresponding to the label indicates the set of output interfaces for that packet, as well as the outgoing label. If the same label encoding technique is used on all the outgoing interfaces, the very same packet can be sent to all the children.

9.3.6   Label Hop-by-Hop Distribution Procedures

In this section, we consider only label bindings that are used for traffic to be label switched along its hop–by–hop routed path. In these cases, the label in question will correspond to an address prefix in the routing table.

9.3.6.1   The Procedures for Advertising and Using Labels

A number of different procedures can be used to distribute label bindings. Some are executed by the downstream LSR and some by the upstream LSR. The downstream LSR must perform (1) the distribution procedure and (2) the withdrawal procedure. The upstream LSR must perform (1) the request procedure, (2) the not available procedure, (3) the release procedure, and (4) the label use procedure.

The MPLS architecture supports several variants of each procedure. However, the MPLS architecture does not support all possible combinations of all possible variants. The set of supported combinations is described in Section 9.3.6.2, where the interoperability between different combinations is also discussed.

Downstream LSR Distribution Procedure
The distribution procedure is used by a downstream LSR to determine when it should distribute a label binding for a particular address prefix to its label distribution peers. The architecture supports four different distribution procedures.

Irrespective of the particular procedure that is used, if a label binding for a particular address prefix has been distributed by a downstream LSR Rd to an upstream LSR Ru, and if at any time the attributes (as defined above) of that binding change, then Rd must inform Ru of the new attributes.

If an LSR is maintaining multiple routes to a particular address prefix, it is a local matter as to whether that LSR binds multiple labels to the address prefix (one per route), and hence distributes multiple bindings.

PushUnconditional   Let Rd be an LSR. Assume the following:

1.   X is an address prefix in Rd's routing table.

2.   Ru is a label distribution peer of Rd with respect to X.

Whenever these conditions hold, Rd must bind a label to X and distribute that binding to Ru. It is the responsibility of Rd to keep track of the bindings that it has distributed to Ru and to make sure that Ru always has these bindings. This procedure would be used by LSRs that are performing unsolicited downstream label assignment in the independent LSP control mode.

PushConditional   Let Rd be an LSR. Assume the following:

1. X is an address prefix in Rd's routing table.

2. Ru is a label distribution peer of Rd with respect to X.

3. Rd is either an LSP egress or an LSP proxy egress for X, or Rd's L3 next hop for X is R*n*, where R*n* is distinct from Ru, and R*n* has bound a label to X and distributed that binding to Rd.

Then as soon as these conditions all hold, Rd should bind a label to X and distribute that binding to Ru.

Whereas PushUnconditional causes the distribution of label bindings for all address prefixes in the routing table, PushConditional causes the distribution of label bindings only for those address prefixes for which one has received label bindings from one's LSP next hop, or for which one does not have an MPLS-capable L3 next hop. This procedure would be used by LSRs that are performing unsolicited downstream label assignment in the ordered LSP control mode.

PulledUnconditional   Let Rd be an LSR. Assume the following:

1. X is an address prefix in Rd's routing table.

2. Ru is a label distribution peer of Rd with respect to X.

3. Ru has explicitly requested that Rd bind a label to X and distribute the binding to Ru.

Then Rd should bind a label to X and distribute that binding to Ru. Note that if X is not in Rd's routing table, or if Rd is not a label distribution peer of Ru with respect to X, then Rd must inform Ru that it cannot provide a binding at this time.

If Rd has already distributed a binding for address prefix X to Ru, and it receives a new request from Ru for a binding for address prefix X, it will bind a second label and distribute the new binding to Ru. The first label binding remains in effect. This procedure would be used by LSRs performing downstream-on-demand label distribution using the independent LSP control mode.

PulledConditional   Let Rd be an LSR. Assume the following:

1. X is an address prefix in Rd's routing table.

2. Ru is a label distribution peer of Rd with respect to X.

3. Ru has explicitly requested that Rd bind a label to X and distribute the binding to Ru.

4. Rd is either an LSP egress or an LSP proxy egress for X, or Rd's L3 next hop for X is Rn, where Rn is distinct from Ru, and Rn has bound a label to X and distributed that binding to Rd.

Then as soon as these conditions all hold, Rd should bind a label to X and distribute that binding to Ru. Note that if X is not in Rd's routing table and a binding for X is not obtainable via Rd's next hop for X, or if Rd is not a label distribution peer of Ru with respect to X, then Rd must inform Ru that it cannot provide a binding at this time. However, if the only condition that fails to hold is that Rn has not yet provided a label to Rd, then Rd must defer any response to Ru until such time as it has receiving a binding from Rn.

If Rd has distributed a label binding for address prefix X to Ru, and at some later time any attribute of the label binding changes, then Rd must redistribute the label binding to Ru, with the new attribute. It must do this even though Ru does not issue a new request.

This procedure would be used by LSRs that are performing downstream-on-demand label allocation in the ordered LSP control mode.

In Section 9.3.6.2, we discuss how to choose the particular procedure to be used at any given time, and how to ensure interoperability among LSRs that choose different procedures.

Upstream LSR: Request Procedure
The request procedure is used by the upstream LSR for an address prefix to determine when to explicitly request that the downstream LSR bind a label to that prefix and distribute the binding. There are three possible procedures that can be used.

RequestNever   Never make a request. This is useful if the downstream LSR uses the PushConditional procedure or the PushUnconditional procedure, but is not useful if the downstream LSR uses the PulledUnconditional procedure or the PulledConditional procedures.

This procedure would be used by an LSR when unsolicited downstream label distribution and liberal label retention mode are being used.

RequestWhenNeeded   Make a request whenever the L3 next hop to the address prefix changes, or when a new address prefix is learned, and one doesn't already have a label binding from that next hop for the given address prefix. This procedure would be used by an LSR whenever conservative label retention mode is being used.

RequestOnRequest   Issue a request whenever a request is received, in addition to issuing a request when needed (as described in Section 5.1.2.2). If Ru is not capable of being an LSP ingress, it may issue a request only when it receives a request from upstream.

If Rd receives such a request from Ru, for an address prefix for which Rd has already distributed Ru a label, Rd shall assign a new (distinct) label, bind it to X, and distribute that binding. (Whether Rd can distribute this binding to Ru immediately or not depends on the distribution procedure being used.)

This procedure would be used by an LSR that is doing downstream-on-demand label distribution, but is not doing label merging, for example, an ATM-LSR that is not capable of VC merge.

Upstream LSR: NotAvailable Procedure
If Ru and Rd are respectively upstream and downstream label distribution peers for address prefix X, and Rd is Ru's L3 next hop for X, and Ru requests a binding for X from Rd, but Rd replies that it cannot provide a binding at this time because it has no next hop for X, then the NotAvailable procedure determines how Ru responds. There are two possible procedures governing Ru's behavior:

RequestRetry   Ru should issue the request again at a later time. That is, the requester is responsible for trying again later to obtain the needed binding. This procedure would be used when downstream-on-demand label distribution is used.

RequestNoRetry   Ru should never reissue the request, instead assuming that Rd will provide the binding automatically when it is available. This is useful if Rd uses the PushUnconditional procedure or the PushConditional procedure, that is, if unsolicited downstream label distribution is used.

Note that if Rd replies that it cannot provide a binding to Ru, because of some error condition, rather than because Rd has no next hop, the behavior of Ru will be governed by the error recovery conditions of the label distribution protocol, rather than by the NotAvailable procedure.

Upstream LSR: Release Procedure   Suppose that Rd is an LSR that has bound a label to address prefix X, and has distributed that binding to LSR Ru. If Rd does not happen to be Ru's L3 next hop for address prefix X, or has ceased to be Ru's L3 next hop for address prefix X, then Ru will not be using the label. The release procedure determines how Ru acts in this case. There are two possible procedures governing Ru's behavior:

ReleaseOnChange   Ru should release the binding and inform Rd that it has done so. This procedure would be used to implement the conservative label retention mode.

NoReleaseOnChange   Ru should maintain the binding, so that it can use it again immediately if Rd later becomes Ru's L3 next hop for X. This procedure would be used to implement the liberal label retention mode.

Upstream LSR: LabelUse Procedure

Suppose Ru is an LSR that has received label binding L for address prefix X from LSR Rd, and Ru is upstream of Rd with respect to X, and in fact Rd is Ru's L3 next hop for X. Ru will make use of the binding if Rd is Ru's L3 next hop for X. If, at the time the binding is received by Ru, Rd is *not* Ru's L3 next hop for X, Ru does not make any use of the binding at that time. Ru may, however, start using the binding at some later time if Rd becomes Ru's L3 next hop for X.

The labelUse Procedure determines just how Ru makes use of Rd's binding. There are two procedures that Ru may use.

UseImmediate   Ru may put the binding into use immediately. At any time when Ru has a binding for X from Rd, and Rd is Ru's L3 next hop for X, Rd will also be Ru's LSP next hop for X. This procedure is used when loop detection is not in use.

UseIfLoopNotDetected   This procedure is the same as UseImmediate, unless Ru has detected a loop in the LSP. If a loop has been detected, Ru will discontinue the use of label L for forwarding packets to Rd. This procedure is used when loop detection is in use. This will continue until the next hop for X changes or until the loop is no longer detected.

Downstream LSR: Withdraw Procedure

In this case, there is only a single procedure. When LSR Rd decides to break the binding between label L and address prefix X, then this unbinding must be distributed to all LSRs to which the binding was distributed.

The unbinding of L from X is required to be distributed by Rd to a LSR Ru before Rd distributes to Ru any new binding of L to any other address prefix Y, where X != Y. If Ru were to learn of the new binding of L to Y before it learned of the unbinding of L from X, and if packets matching both X and Y were forwarded by Ru to Rd, then for a period of time, Ru would label both packets matching X and packets matching Y with label L.

The distribution and withdrawal of label bindings is done via a label distribution protocol. All label distribution protocols require that a label distribution adjacency be established between two label distribution peers (except implicit peers). If LSR R1 has a label distribution adjacency to LSR R2 and has received label bindings from LSR R2 via that adjacency, then if adjacency is brought down by either peer (whether as a result of failure or as a matter of normal operation), all bindings received over that adjacency must be considered to have been withdrawn.

As long as the relevant label distribution adjacency remains in place, label bindings that are withdrawn must always be withdrawn explicitly. If a second label is bound to an address prefix, the result is not to implicitly withdraw the first label, but to bind both labels; this is needed to support multipath routing. If a second address prefix is bound to a label, the result is not to implicitly withdraw the binding of that label to the first address prefix, but to use that label for both address prefixes.

9.3.6.2 MPLS Schemes: Supported Combinations of Procedures

Consider two LSRs, Ru and Rd, which are label distribution peers with respect to some set of address prefixes, where Ru is the upstream peer and Rd is the downstream peer. The MPLS scheme that governs the interaction of Ru and Rd can be described as a quintuple of procedures: <Distribution Procedure, Request Procedure, NotAvailable Procedure, Release Procedure, labelUse Procedure>. (Because there is only one Withdraw Procedure, it need not be mentioned.) A "★" appearing in one of the positions is a wildcard, meaning that any procedure in that category may be present; an "N/A" appearing in a particular position indicates that no procedure in that category is needed.

Only the MPLS schemes specified next are supported by the MPLS architecture. Other schemes may be added in the future, if a need for them is shown.

Schemes for LSRs that Support Label Merging
If Ru and Rd are label distribution peers, and both support label merging, one of the following schemes must be used:

1. <PushUnconditional, RequestNever, N/A, NoReleaseOnChange, UseImmediate>. This is unsolicited downstream label distribution with independent control, liberal label retention mode, and no loop detection.

2. <PushUnconditional, RequestNever, N/A, NoReleaseOnChange, UseIfLoopNotDetected>. This is unsolicited downstream label distribution with independent control, liberal label retention, and loop detection.

3. <PushConditional, RequestWhenNeeded, RequestNoRetry, ReleaseOnChange, ★>. This is unsolicited downstream label distribution with ordered control (from the egress) and conservative label retention mode. Loop detection is optional.

4. <PushConditional, RequestNever, N/A, NoReleaseOnChange, ★>. This is unsolicited downstream label distribution with ordered control (from the egress) and liberal label retention mode. Loop detection is optional.

5. <PulledConditional, RequestWhenNeeded, RequestRetry, ReleaseOnChange, ★>. This is downstream-on-demand label distribution with ordered control (initiated by the ingress), conservative label retention mode, and optional loop detection.

6. <PulledUnconditional, RequestWhenNeeded, N/A, ReleaseOnChange, UseImmediate>. This is downstream-on-demand label distribution with independent control and conservative label retention mode, without loop detection.

7. <PulledUnconditional, RequestWhenNeeded, N/A, ReleaseOnChange, UseIfLoopNotDetected>. This is downstream-on-demand label distribution with independent control and conservative label retention mode, with loop detection.

Schemes for LSRs That Do Not Support Label Merging
Suppose that R1, R2, R3, and R4 are ATM switches that do not support label merging, but are being used as LSRs. Suppose further that the L3 hop-by-hop path for address prefix X is <R1, R2, R3, R4>, and that packets destined for X can enter the network at any of these LSRs. Because there is no multipoint-to-point capability, the LSPs must be realized as point-to-point VCs, which means that there needs to be three such VCs for address prefix X: <R1, R2, R3, R4>, <R2, R3, R4>, and <R3, R4>.

Therefore, if R1 and R2 are MPLS peers, and either is an LSR that is implemented using conventional ATM switching hardware (i.e., no cell interleave suppression) or is otherwise incapable of performing label merging, the MPLS scheme in use between R1 and R2 must be one of the following:

1. <PulledConditional, RequestOnRequest, RequestRetry, ReleaseOnChange, *>. This is downstream-on-demand label distribution with ordered control (initiated by the ingress), conservative label retention mode, and optional loop detection. The use of the RequestOnRequest procedure will cause R4 to distribute three labels for X to R3; R3 will distribute two labels for X to R2, and R2 will distribute one label for X to R1.

2. <PulledUnconditional, RequestOnRequest, N/A, ReleaseOnChange, UseImmediate>. This is downstream-on-demand label distribution with independent control and conservative label retention mode, without loop detection.

3. <PulledUnconditional, RequestOnRequest, N/A, ReleaseOnChange, UseIfLoopNotDetected>. This is downstream-on-demand label distribution with independent control and conservative label retention mode, with loop detection.

Interoperability Considerations
It is easy to see that certain quintuples do *not* yield viable MPLS schemes. For example:

• <PulledUnconditional, RequestNever, *, *, *>, <PulledConditional, RequestNever, *, *, *>. In these MPLS schemes, the downstream LSR Rd distributes label bindings to upstream LSR Ru only upon request from Ru, but Ru never makes any such requests. Obviously, these schemes are not viable, since they will not result in the proper distribution of label bindings.

• <*, RequestNever, *, *, ReleaseOnChange>. In these MPLS schemes, Rd releases bindings when it isn't using them, but it never asks for them again, even if it later has a need for them. These schemes thus do not ensure that label bindings get properly distributed.

In this section, we specify rules to prevent a pair of label distribution peers from adopting procedures that lead to infeasible MPLS schemes. These rules require either the exchange of information between label distribution peers during the initialization of the label distribution adjacency or a priori knowledge of the information (obtained through a means outside the scope of this chapter).

1. Each must state whether it supports label merging.

2. If Rd does not support label merging, Rd must choose either the PulledUnconditional procedure or the PulledConditional procedure. If Rd chooses PulledConditional, Ru is forced to use the RequestRetry procedure. That is, if the downstream LSR does not support label merging, its preferences take priority when the MPLS scheme is chosen.

3. If Ru does not support label merging, but Rd does, Ru must choose either the RequestRetry or RequestNoRetry procedure. This forces Rd to use the PulledConditional or PulledUnConditional procedure respectively. That is, if only one of the LSRs does not support label merging, its preferences take priority when the MPLS scheme is chosen.

4. If both Ru and Rd both support label merging, then the choice between liberal and conservative label retention mode belongs to Ru. That is, Ru gets to choose either to use RequestWhen-Needed/ReleaseOnChange (conservative), or to use RequestNever/NoReleaseOnChange (liberal). However, the choice of "push" versus "pull" and "conditional" versus "unconditional" belongs to Rd. If Ru chooses liberal label retention mode, Rd can choose either PushUnconditional or PushConditional. If Ru chooses conservative label retention mode, Rd can choose PushConditional, PulledConditional, or PulledUnconditional.

These choices together determine the MPLS scheme in use.

9.3.7   Security Considerations

Some routers may implement security procedures that depend on the network layer header being in a fixed place relative to the data link layer header. The MPLS generic encapsulation inserts a shim between the data link layer header and the network layer header. This may cause any such security procedures to fail.

An MPLS label has its meaning by virtue of an agreement between the LSR that puts the label in the label stack (the *label writer*), and the LSR that interprets that label (the *label reader*). If labeled packets are accepted from untrusted sources, or if a particular incoming label is accepted from an LSR to which that label has not been distributed, then packets may be routed in an illegitimate manner [36].

9.3.8   Encoding the Label Stack Introduction

RFC 3032 [44] specifies the encoding to be used by an LSR in order to transmit labeled packets on PPP data links and on LAN data links. The specified encoding may also be useful for other data links as well. The RFC also specifies rules and procedures for processing the various fields of the label stack encoding. Because MPLS is independent of any particular network layer protocol, the majority of such procedures are also protocol independent. A few, however, do differ for different protocols. RFC 3032 specifies the protocol-independent procedures, and we specify the protocol-dependent procedures for IPv4 and IPv6. LSRs that are implemented on certain switching devices (such as ATM switches) may use different encoding techniques for encoding the top one or two entries of the label stack. When the label stack has additional entries, however, the encoding technique described in this document *must* be used for the additional label stack entries.

9.3.9   The Label Stack

9.3.9.1   Encoding the Label Stack [58]

The label stack is represented as a sequence of label stack entries. Each label stack entry is represented by 4 octets. The label stack entries appear *after* the data link layer headers, but *before* any network layer headers. The top of the label stack appears earliest in the packet, and the bottom appears latest. The network layer packet immediately follows the label stack entry, which has the S bit set.

Each label stack entry is broken down into the following fields:

1. *Bottom of stack (S).* This bit is set to one for the last entry in the label stack (i.e., for the bottom of the stack), and zero for all other label stack entries.

2. *TTL.* This 8-bit field is used to encode a TTL value. The processing of this field is described in Section 9.3.9.4.

3. *Experimental use.* This 3-bit field is reserved for experimental use.

4. *Label value.* This 20-bit field carries the actual value of the label.

When a labeled packet is received, the label value at the top of the stack is looked up. As a result of a successful lookup one learns (1) the next hop to which the packet is to be forwarded and (2) the operation to be performed on the label stack before forwarding; this operation may be to replace the top label stack entry with another, or to pop an entry off the label stack, or to replace the top label stack entry and then to push one or more additional entries on the label stack.

In addition to learning the next hop and the label stack operation, one may also learn the outgoing data link encapsulation and possibly other information that is needed in order to properly forward the packet. There are several reserved label values:

- A value of 0 represents the *IPv4 explicit NULL label.* This label value is only legal at the bottom of the label stack. It indicates that the label stack must be popped, and the forwarding of the packet must then be based on the IPv4 header.

- A value of 1 represents the *router alert label.* This label value is legal anywhere in the label stack except at the bottom. When a received packet contains this label value at the top of the label stack, it is delivered to a local software module for processing. The actual forwarding of the packet is determined by the label beneath it in the stack. However, if the packet is forwarded further, the *router alert label* should be pushed back onto the label stack before forwarding. The use of this label is analogous to the use of the *router alert option* in IP packets [59]. Because this label cannot occur at the bottom of the stack, it is not associated with a particular network layer protocol.

- A value of 2 represents the *IPv6 explicit NULL label.* This label value is only legal at the bottom of the label stack. It indicates that the label stack must be popped, and the forwarding of the packet must then be based on the IPv6 header.

- A value of 3 represents the *implicit NULL label.* This is a label that an LSR may assign and distribute, but which never actually appears in the encapsulation. When an LSR would otherwise replace the label at the top of the stack with a new label, but the new label is *implicit NULL,* the LSR will pop the stack instead of doing the replacement. Although this value may never appear in the encapsulation, it needs to be specified in the label distribution protocol, so a value is reserved.

- Values 4–15 are reserved.

9.3.9.2   Determining the Network Layer Protocol

When the last label is popped from a packet's label stack (resulting in the stack being emptied), further processing of the packet is based on the packet's network layer header. The LSR that pops the last label off the stack must therefore be able to identify the packet's network layer protocol. However, the label stack does not contain any field that explicitly identifies the network layer protocol. This means that the identity of the network layer protocol must be inferred from the value of the label that is popped from the bottom of the stack, possibly along with the contents of the network layer header itself.

Therefore, when the first label is pushed onto a network layer packet, either the label must be one that is used *only* for packets of a particular network layer, or the label must be one that is used *only* for a specified set of network layer protocols, where packets of the specified network layers can be distinguished by inspection of the network layer header. Furthermore, whenever that label is replaced by another label value during a packet's transit, the new value must also be one that meets the same criteria. If these conditions are not met, the LSR that pops the last label off a packet will not be able to identify the packet's network layer protocol.

Adherence to these conditions does not necessarily enable intermediate nodes to identify a packet's network layer protocol. Under ordinary conditions, this is not necessary, but under some error conditions it is desirable. For instance, if an intermediate LSR determines that a labeled packet is undeliverable, it may be desirable for that LSR to generate error messages that are specific to the packet's network layer. The only means the intermediate LSR has for identifying the network layer is inspection of the top label and the network layer header. So if intermediate nodes are to be able to generate protocol-specific error messages for labeled packets, all labels in the stack must meet the criteria specified above for labels that appear at the bottom of the stack.

If a packet cannot be forwarded for some reason (e.g., it exceeds the data link MTU), and either its network layer protocol cannot be identified, or there are no specified protocol-dependent rules for handling the error condition, then the packet *must* be silently discarded.

9.3.9.3   Generating ICMP Messages for Labeled IP Packets

Sections 9.3.9.4 and 9.3.10 discuss situations in which it is desirable to generate ICMP messages for labeled IP packets. In order for a particular LSR to be able to generate an ICMP packet and have that packet sent to the source of the IP packet, two conditions must hold:

1. It must be possible for that LSR to determine that a particular labeled packet is an IP packet.

2. It must be possible for that LSR to route to the packet's IP source address.

Condition 1 is discussed in Section 9.3.9.2. The following two subsections discuss condition 2. However, in some cases condition 2 does not hold at all, and in these cases it will not be possible to generate the ICMP message.

Tunneling Through a Transit Routing Domain
Suppose one is using MPLS to "tunnel" through a transit routing domain, where the external routes are not leaked into the domain's interior routers. For example, the interior routers may be running OSPF, and may only know how to reach destinations within that OSPF domain. The domain might contain *several autonomous system border routers* (ASBRs), which talk BGP to each other. However, in this example the routes from BGP are not distributed into OSPF, and the LSRs that are not ASBRs do not run BGP.

In this example, only an ASBR will know how to route to the source of some arbitrary packet. If an interior router needs to send an ICMP message to the source of an IP packet, it will not know how to route the ICMP message.

One solution is to have one or more of the ASBRs inject "default" into the IGP. (Note that this does *not* require that there be a "default" carried by BGP.) This would then ensure that any unlabeled packet that must leave the domain (such as an ICMP packet) gets sent to a router that has full routing information. The routers with full routing information will label the packets before sending them back through the transit domain, so the use of default routing within the transit domain does not cause any loops.

This solution only works for packets that have globally unique addresses and for networks in which all the ASBRs have complete routing information. The next subsection describes a solution that works when these conditions do not hold.

Tunneling Private Addresses Through a Public Backbone
In some cases where MPLS is used to tunnel through a routing domain, it may not be possible to route to the source address of a fragmented packet at all. This would be the case, for example, if the IP addresses carried in the packet were private (i.e., not globally unique) addresses, and MPLS were being used to tunnel those packets through a public backbone. Default routing to an ASBR will not work in this environment.

In this environment, in order to send an ICMP message to the source of a packet, one can copy the label stack from the original packet to the ICMP message and then label switch the ICMP message. This will cause the message to proceed in the direction of the original packet's destination, rather than its source. Unless the message is label switched all the way to the destination host, it will end up, unlabeled, in a router that does know how to route to the source of the original packet, at which point the message will be sent in the proper direction.

This technique can be very useful if the ICMP message is a *time exceeded* message or a *destination unreachable because fragmentation needed and DF set* message.

When copying the label stack from the original packet to the ICMP message, the label values must be copied exactly, but the TTL values in the label stack should be set to the TTL value that is placed in the IP header of the ICMP message. This TTL value should be long enough to allow the circuitous route that the ICMP message will need to follow.

Note that if a packet's TTL expiration is due to the presence of a routing loop, then if this technique is used, the ICMP message may loop as well. Because an ICMP message is never sent as a result of receiving an ICMP message, and because many implementations throttle the rate at which ICMP messages can be generated, this is not expected to pose a problem.

9.3.9.4   Processing the TTL Field

Definitions

The *incoming TTL* of a labeled packet is defined to be the value of the TTL field of the top label stack entry when the packet is received. The *outgoing TTL* of a labeled packet is defined to be the larger of (1) one less than the incoming TTL or (2) zero.

Protocol-Independent Rules

If the outgoing TTL of a labeled packet is 0, then the labeled packet *must not* be further forwarded, nor may the label stack be stripped off and the packet forwarded as an unlabeled packet. The packet's lifetime in the network is considered to have expired.

Depending on the label value in the label stack entry, the packet may be simply discarded, or it may be passed to the appropriate "ordinary" network layer for error processing (e.g., for the generation of an ICMP error message, see Section 9.2.3).

When a labeled packet is forwarded, the TTL field of the label stack entry at the top of the label stack must be set to the outgoing TTL value. Note that the outgoing TTL value is a function solely of the incoming TTL value and is independent of whether any labels are pushed or popped before forwarding. There is no significance to the value of the TTL field in any label stack entry that is not at the top of the stack.

IP-Dependent Rules

We define the IP TTL field to be the value of the IPv4 TTL field, or the value of the IPv6 hop limit field, whichever is applicable. When an IP packet is first labeled, the TTL field of the label stack entry must be set to the value of the IP TTL field. (If the IP TTL field needs to be decremented, as part of the IP processing, it is assumed that this has already been done.)

When a label is popped, and the resulting label stack is empty, then the value of the IP TTL field should be replaced with the outgoing TTL value, as defined earlier. In IPv4 this also requires modification of the IP header checksum.

It is recognized that in some situations a network administration might prefer to decrement the IPv4 TTL by one as it traverses an MPLS domain, instead of decrementing the IPv4 TTL by the number of LSP hops within the domain.

Translating Between Different Encapsulations

Sometimes an LSR may receive a labeled packet over, for example, a *label switching controlled ATM* (LC-ATM) interface [56] and may need to send it out over a PPP or LAN link. Then the incoming packet will not be received using the encapsulation specified in RFC 3032, but the outgoing packet will be sent using the encapsulation specified in the RFC.

In this case, the value of the incoming TTL is determined by the procedures used for carrying labeled packets on, for example, LC-ATM interfaces. TTL processing then proceeds as described earlier.

Sometimes an LSR may receive a labeled packet over a PPP or a LAN link, and may need to send it out, say, an LC-ATM interface. Then the incoming packet will be received using the encapsulation specified in RFC 3032, but the outgoing packet will not be sent using the encapsulation specified in the RFC. In this case, the procedure for carrying the value of the outgoing TTL is determined by the procedures used for carrying labeled packets on, for example, LC-ATM interfaces.

9.3.10 Fragmentation and Path MTU Discovery

Just as it is possible to receive an unlabeled IP datagram that is too large to be transmitted on its output link, it is possible to receive a labeled packet that is too large to be transmitted on its output link.

It is also possible that a received packet (labeled or unlabeled) that was originally small enough to be transmitted on that link becomes too large by virtue of having one or more additional labels pushed onto its label stack. In label switching, a packet may grow in size if additional labels get pushed on. Thus if one receives a labeled packet with a 1,500-byte frame payload, and pushes on an additional label, one needs to forward it as frame with a 1,504-byte payload.

This section specifies the rules for processing labeled packets that are "too large." In particular, it provides rules that ensure that hosts implementing path MTU discovery [60] and hosts using IPv6 [61, 62] will be able to generate IP datagrams that do not need fragmentation, even if those datagrams get labeled as they traverse the network.

In general, IPv4 hosts that do not implement path MTU discovery [60] send IP datagrams that contain no more than 576 bytes. Because the MTUs in use on most data links today are 1,500 bytes or more, the probability that such datagrams will need to get fragmented, even if they get labeled, is very small.

Some hosts that do not implement path MTU discovery [60] will generate IP datagrams containing 1,500 bytes, as long as the IP source and destination addresses are on the same subnet. These datagrams will not pass through routers and, hence, will not get fragmented.

Unfortunately, some hosts will generate IP datagrams containing 1,500 bytes, as long the IP source and destination addresses have the same classful network number. This is the one case in which there is any risk of fragmentation when such datagrams get labeled. (Even so, fragmentation is not likely unless the packet must traverse an Ethernet of some sort between the time it first gets labeled and the time it gets unlabeled.)

RFC 3032 specifies procedures that allow one to configure the network so that large datagrams from hosts that do not implement path MTU discovery get fragmented just once, when they are first labeled. These procedures make it possible (assuming suitable configuration) to avoid any need to fragment packets that have already been labeled.

9.3.10.1 Terminology

With respect to a particular data link, we can use the following terms:

- *Frame payload:* The contents of a data link frame, excluding any data link layer headers or trailers (e.g., MAC headers, LLC headers, 802.1Q headers, PPP header, frame check sequences, and so on). When a frame is carrying an unlabeled IP datagram, the frame payload is just the IP datagram itself. When a frame is carrying a labeled IP datagram, the frame payload consists of the label stack entries and the IP datagram.

- *Conventional maximum frame payload size:* The maximum frame payload size allowed by data link standards. For example, the conventional maximum frame payload size for Ethernet is 1,500 bytes.

- *True maximum frame payload size:* The maximum size frame payload that can be sent and received properly by the interface hardware attached to the data link. On Ethernet and 802.3 networks, it is believed that the true maximum frame payload size is 4 to 8 bytes larger than the conventional maximum frame payload size (as long as neither an 802.1Q header nor an 802.1p header is present, and as long as neither can be added by a switch or bridge while a packet is in transit to its next hop). For example, it is believed that most Ethernet equipment could correctly send and receive packets carrying a payload of 1,504 or perhaps even 1,508 bytes, at least, as long as the Ethernet header does not have an 802.1Q or 802.1p field. On PPP links, the true maximum frame payload size may be virtually unbounded.

- *Effective maximum frame payload size for labeled packets:* This is either the conventional maximum frame payload size or the true maximum frame payload size, depending on the capabilities of the equipment on the data link and the size of the data link header being used.

- *Initially labeled IP datagram:* Suppose that an unlabeled IP datagram is received at a particular LSR, and that the LSR pushes on a label before forwarding the datagram. Such a datagram will be called an initially labeled IP datagram at that LSR.

- *Previously labeled IP datagram:* An IP datagram that had already been labeled before it was received by a particular LSR.

9.3.10.2 Maximum Initially Labeled IP Datagram Size

Every LSR that is capable of (1) receiving an unlabeled IP datagram, (2) adding a label stack to the datagram, and (3) forwarding the resulting labeled packet should support a configuration parameter known as the *maximum initially labeled IP datagram size,* which can be set to a non-negative value. If this configuration parameter is set to zero, it has no effect. If it is set to a positive value, it is used in the following way: If (1) an unlabeled IP datagram is received and (2) that datagram does not have the DF bit set in its IP header, and (3) that datagram needs to be labeled before being forwarded, and (4) the size of the datagram (before labeling) exceeds the value of the parameter, then (a) the datagram must be broken into fragments, each of whose size is no greater than the value of the parameter, and (b) each fragment must be labeled and then forwarded. For example, if this configuration parameter is set to a value of 1,488, then any unlabeled IP datagram containing more than 1,488 bytes will be fragmented before being labeled. Each fragment will be capable of being carried on a 1,500-byte data link, without further fragmentation, even if as many as three labels are pushed onto its label stack.

In other words, setting this parameter to a nonzero value allows one to eliminate all fragmentation of previously labeled IP datagrams, but it may cause some unnecessary fragmentation of initially labeled IP datagrams.

Note that the setting of this parameter does not affect the processing of IP datagrams that have the DF bit set; hence, the result of path MTU discovery is unaffected by the setting of this parameter.

9.3.10.3 When Are Labeled IP Datagrams Too Big?

A labeled IP datagram whose size exceeds the conventional maximum frame payload size of the data link over which it is to be forwarded *may* be considered to be "too big."

A labeled IP datagram whose size exceeds the true maximum frame payload size of the data link over which it is to be forwarded *must* be considered to be "too big." A labeled IP datagram that is not "too big" *must* be transmitted without fragmentation.

9.3.10.4   Processing Labeled IPv4 Datagrams That Are Too Big

If a labeled IPv4 datagram is too big, and the DF bit is not set in its IP header, then the LSR MAY silently discard the datagram. Note that discarding such datagrams is a sensible procedure only if the maximum initially labeled IP datagram size is set to a nonzero value in every LSR in the network that is capable of adding a label stack to an unlabeled IP datagram.

If the LSR chooses not to discard a labeled IPv4 datagram that is too big, or if the DF bit is set in that datagram, then it *must* execute the following algorithm:

1. Strip off the label stack entries to obtain the IP datagram.

2. Let N be the number of bytes in the label stack (i.e., four times the number of label stack entries).

3. If the IP datagram does *not* have the Don't Fragment bit set in its IP header, then (a) convert it into fragments, each of which must be at least N bytes less than the effective maximum frame payload size, (b) prepend each fragment with the same label header that would have been on the original datagram had fragmentation not been necessary, and (c) forward the fragments.

4. If the IP datagram *does have* the Don't Fragment bit set in its IP header, then (a) the datagram *must not* be forwarded; (b) create an ICMP destination unreachable message: (i) set its code field [63] to *fragmentation required and DF set* and (ii) set its next hop MTU field [60] to the difference between the effective maximum frame payload size and the value of N; and (c) if possible, transmit the ICMP destination unreachable message to the source of the discarded datagram.

9.3.10.5   Processing Labeled IPv6 Datagrams That Are Too Big

To process a labeled IPv6 datagram that is too big, an LSR *must* execute the following algorithm:

1. Strip off the label stack entries to obtain the IP datagram.

2. Let N be the number of bytes in the label stack (i.e., four times the number of label stack entries).

3. If the IP datagram contains more than 1,280 bytes (not counting the label stack entries), or if it does not contain a fragment header, then (a) create an ICMP packet too big message, and set its

next-hop MTU field to the difference between the effective maximum frame payload size and the value of N; (b) if possible, transmit the ICMP packet too big message to the source of the datagram; and (c) discard the labeled IPv6 datagram.

4. If the IP datagram is not larger than 1,280 octets, and it contains a fragment header, then (a) convert it into fragments, each of which *must* be at least N bytes less than the effective maximum frame payload size; (b) prepend each fragment with the same label header that would have been on the original datagram had fragmentation not been necessary; and (c) forward the fragments.

Reassembly of the fragments will be done at the destination host.

9.3.10.6 Implications with Respect to Path MTU Discovery

The procedures described above for handling datagrams that have the DF bit set, but that are "too large," have an impact on the path MTU discovery procedures of RFC 1191 [60]. Hosts that implement these procedures will discover an MTU that is small enough to allow n labels to be pushed on the datagrams, without need for fragmentation, where n is the number of labels that actually get pushed on along the path currently in use.

In other words, datagrams from hosts that use path MTU discovery will never need to be fragmented due to the need to put on a label header or to add new labels to an existing label header. (Also, datagrams from hosts that use path MTU discovery generally have the DF bit set, and so will never get fragmented anyway.)

Note that path MTU discovery will only work properly if, at the point where a labeled IP datagram's fragmentation needs to occur, it is possible to cause an ICMP destination unreachable message to be routed to the packet's source address.

If it is not possible to forward an ICMP message from within an MPLS "tunnel" to a packet's source address, but the network configuration makes it possible for the LSR at the transmitting end of the tunnel to receive packets that must go through the tunnel, but are too large to pass through the tunnel unfragmented, then:

• The LSR at the transmitting end of the tunnel *must* be able to determine the MTU of the tunnel as a whole. It *may* do this by sending packets through the tunnel to the tunnel's receiving endpoint and by performing path MTU discovery with those packets.

• Any time the transmitting endpoint of the tunnel needs to send a packet into the tunnel, and that packet has the DF bit set, and it exceeds the tunnel MTU, the transmitting endpoint of the tunnel *must* send the ICMP destination unreachable message to the source, with the code fragmentation required and DF set and the next-hop MTU field set as described earlier.

9.3.11 Transporting Labeled Packets over PPP

The PPP [64] provides a standard method for transporting multiprotocol datagrams over point-to-point links. PPP defines an extensible LCP and proposes a family of network control protocols for establishing and configuring different network layer protocols. This section defines the network control protocol for establishing and configuring label switching over PPP.

9.3.11.1 Introduction

PPP has three main components:

1. A method for encapsulating multiprotocol datagrams;
2. A LCP for establishing, configuring, and testing the data link connection;
3. A family of network control protocols for establishing and configuring different network layer protocols.

To establish communications over a point-to-point link, each end of the PPP link must first send LCP packets to configure and test the data link. After the link has been established and optional facilities have been negotiated as needed by the LCP, PPP must send *MPLS control protocol packets* (MPLSCP) to enable the transmission of labeled packets. Once the MPLSCP has reached the opened state, labeled packets can be sent over the link.

The link will remain configured for communications until explicit LCP or MPLSCP packets close the link down or until some external event occurs (an inactivity timer expires or network administrator intervention).

9.3.11.2 A PPP Network Control Protocol for MPLS

The MPLSCP is responsible for enabling and disabling the use of label switching on a PPP link. It uses the same packet exchange mechanism as the LCP. MPLSCP packets may not be exchanged until PPP has reached the network layer protocol phase. MPLSCP packets received before this phase is reached should be silently discarded. The MPLSCP is exactly the same as the LCP [64] with the following exceptions:

1. *Frame modifications.* The packet may utilize any modifications to the basic frame format that have been negotiated during the link establishment phase.
2. *Data link layer protocol field.* Exactly one MPLSCP packet is encapsulated in the PPP information field, where the PPP protocol field indicates type hex 8281 (MPLS).
3. *Code field.* Only codes 1 through 7 (Configure-Request, Configure-Ack, Configure-Nak, Configure-Reject, Terminate-

Request, Terminate- Ack, and Code-Reject) are used. Other codes should be treated as unrecognized and should result in Code-Rejects.

4. *Time-outs.* MPLSCP packets may not be exchanged until PPP has reached the network layer protocol phase. An implementation should be prepared to wait for authentication and link quality determination to finish before timing out waiting for a Configure-Ack or other response. It is suggested that an implementation give up only after user intervention or a configurable amount of time.

5. *Configuration option types.* None.

9.3.11.3   Sending Labeled Packets

Before any labeled packets may be communicated, PPP must reach the network layer protocol phase, and the MPLSCP must reach the opened state. Exactly one labeled packet is encapsulated in the PPP information field, where the PPP protocol field indicates either type hex 0281 (MPLS unicast) or type hex 0283 (MPLS multicast). The maximum length of a labeled packet transmitted over a PPP link is the same as the maximum length of the information field of a PPP encapsulated packet. The format of the information field itself is as defined earlier.

Note that two code points are defined for labeled packets; one for multicast and one for unicast. Once the MPLSCP has reached the opened state, both label switched multicasts and label switched unicasts can be sent over the PPP link.

9.3.11.4   Label-Switching Control Protocol Configuration Options

There are no configuration options.

9.3.12   Transporting Labeled Packets over LAN Media

Exactly one labeled packet is carried in each frame. The label stack entries immediately precede the network layer header, and follow any data link layer headers, including, for example, any 802.1Q headers that may exist. The ethertype value 8847 hex is used to indicate that a frame is carrying an MPLS unicast packet.

The ethertype value 8848 hex is used to indicate that a frame is carrying an MPLS multicast packet. These ethertype values can be used with either the Ethernet encapsulation or the 802.3 LLC/SNAP encapsulation to carry labeled packets. The procedure for choosing which of these two encapsulations to use is beyond the scope of this chapter.

9.3.13   IANA Considerations

Label values 0 to 15 inclusive have special meaning, as specified in the RFC or as further assigned by IANA. In this RFC, label values 0 to 3 are specified in Section 9.3.9.1. Label values 4 to 15 may be assigned by IANA, based on IETF consensus.

9.3.14   Security Considerations

The MPLS encapsulation that is specified herein does not raise any security issues that are not already present in either the MPLS architecture [38] or in the architecture of the network layer protocol contained within the encapsulation. There are two security considerations inherited from the MPLS architecture that may be pointed out here:

- Some routers may implement security procedures that depend on the network layer header being in a fixed place relative to the data link layer header. These procedures will not work when the MPLS encapsulation is used, because that encapsulation is of a variable size.

- An MPLS label has its meaning by virtue of an agreement between the LSR that puts the label in the label stack (the label writer), and the LSR that interprets that label (the label reader). However, the label stack does not provide any means of determining who the label writer was for any particular label. If labeled packets are accepted from untrusted sources, the result may be that packets are routed in an illegitimate manner [44].

9.3.15   FR and MPLS

Proponents see FR and MPLS interacting and interworking. As we saw earlier, MPLS combines the flexibility of the IP network layer with the benefits of a connection-oriented approach to networking. MPLS, like frame relay, is a label-switched system that can carry multiple network layer protocols. Similar to frame, MPLS sends information in frames or packets. Each frame/packet is labeled; the network uses the label to decide the destination of the frame. In an MPLS network we can define explicit paths or let IP routing decide the path. MPLS networks can use FR, ATM, and PPP as the link layer. These different link layers can be employed because data are switched according to a label, not an IP address. Therefore, as long as a suitable label structure can be defined, any transmission/switching system could be used. Put another way, MPLS separates the task of transmitting packets (forwarding) from network control or routing. This makes MPLS extensible to many environments including SDH and optical networks [65].

Frame relay can be used to access an MPLS backbone. The need for frame relay-to-ATM interworking became clear as service providers migrated to ATM core backbones. Similarly, as core networks embrace MPLS, frame relay is expected to need interworking. In the frame relay/MPLS network interworking scenario, the MPLS core network transports FR frames. That is, the frames from an FR access network are delivered to the ingress of an MPLS network and transported across the MPLS network to another FR network. Simply stated, this is "frames in and frames out."

In Figure 9.51, we see an example of tunneling frame relay across an MPLS network. Assume that the frame relay CPE has a routing table that relates the destination IP address to a DLCI.

1. An IP packet arriving at the CPE will be encapsulated in an FR frame with a DLCI required for the PVC to the IP destination. (In the example above, DLCI = 67 is associated with a destination in the network at the top right.)

2. The frame is transported across to the ingress LSR, which then maps the incoming DLCI value (67) to label switched path (LSP A).

3. Frames are tunneled across the MPLS network to the egress LSR.

4. The egress LSR then maps them to another DLCI value for delivery to the destination.

This scenario could also be used to link FR service providers over an MPLS core network. In this case, the MPLS network maintains a mesh of label switched paths connecting each ingress and egress label switched router. This is a good alternative to using the FRF implementation agreement (FRF.2.1) NNI since it can use MPLS network features to provide resilient connections.

Frame relay to MPLS network interworking is the subject of a new work item in the MPLS Forum's technical committee, and the FRF technical committee has established a liaison to the MPLS Forum. The MPLS Forum work item includes the following:

• The use of an evolving IETF protocol for transporting FR over MPLS;

• Supporting soft PVCs in the MPLS network;

FIGURE 9.51
Example of tunneling frame relay across an MPLS network.

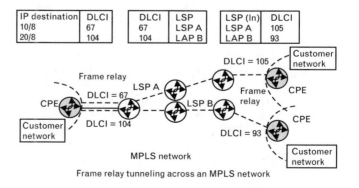

Frame relay tunneling across an MPLS network

· The provision of PVC status management;

· OA&M support;

· Mapping of congestion indications (forward and backward);

· Support of FR priorities.

ENDNOTES

[1]   This author published the *first ever* nonedited, nonmarket-report-based book with a frame relay title: *Enterprise Networking: Frame Relay, ATM, and SONET*, Norwood, MA: Artech House, 1992.

[2]   Based on this author's hands-on experience, some carriers count physical ports; other carriers count DS-0 equivalents. In this latter scenario a T1-based port with a CIR of 1024 Kbps would be counted as 16 "ports."

[3]   Vertical Systems Group statistics. This equates to an MRC of 595 per port equivalent.

[4]   Dunne, E., Frame Relay Forum News, Vertical Systems Group, http://www.verticalsystems.com, Summer 2002.

[5]   Some, for example, UUNET, used frame relay switches in the mid-1990s, operating at DS-3, as the basis of a traffic engineering fabric for an Internet backbone core network; this later evolved (in the late 1990s) into the use of ATM switches operating at OC-12.

[6]   Minoli, D., P. Johnson, and E. Minoli, *Ethernet-Based Metro Area Networks—Planning and Designing the Provider Network,* New York: McGraw-Hill, 2002.

[7]   The FR material in this chapter is based expressly on Frame Relay Forum materials, newsletters, articles, and so on.

[8]   A VC may consist of several sections and components administered by multiple organizations. The portions of a VC that are administered by the same organization(s) create an *administrative domain.*

[9]   While frame relay is widely sold as a "low-speed" service at T1 rates and below, it can in principle operate just fine at DS-3 rates. Some claim that if switch vendors had built systems with OC3/12 interfaces, MPLS might not have happened or been necessary.

[10]   This section is based in part on the reference *The Basic Guide to Frame Relay Networking*, Fremont, CA: The MPLS/Frame Relay Alliance, 1998.

[11]   Minoli, D., and G. Debroowski, *Signaling Principles for Frame Relay and Cell Relay Services,* Norwood, MA: Artech House, 1994.

[12]   Hanssen, M., *How Do I Choose CIR?* MCI WorldCom, MPLS/Frame Relay AllianceNews, 4th Quarter, 1998.

[13]   The MPLS/Frame Relay Alliance News, 3rd Quarter, 1998.

[14]   "Frame Relay Forum Announces Two New Implementation Agreements, Four Existing Agreements Amended to Enhance Scope and Versatility," Fremont, CA: The MPLS/Frame Relay Alliance, September 12, 2000.

[15]   "Frame Relay Forum Announces Multilink Implementation Agreements to Support Broadband Applications, New Agreements Provide Interoperability if High-Bandwidth Frame Relay Networks to Support Broadband Applications," Press Release, Fremont, CA: The MPLS/Frame Relay Alliance, September 13, 1999.

[16]   Rehbehn, K., "FRF.14 Physical Layer IA Adds High Speed Interfaces, Visual Networks," The MPLS/Frame Relay Alliance News, 1st Quarter, 1999.

[17] Some of the material in this section is based on [12].

[18] In the past people used the word "guarantee." Consistent with a general desire for more transparency and factual representations, the term "guarantee" must be eliminated from the packet services lexicon, because no one is guarantying anything: Carriers cannot "guarantee" that data will get to the other end.

[19] This section is based on Nicoll, C., *A Closer Look at Service Level Metrics,* Current Analysis, Inc., MPLS/Frame Relay Alliance News, 2nd Quarter, 1998.

[20] McGuire, T., *Applications Management Driving Service Level Management to New Heights,* Paradyne, MPLS/Frame Relay Alliance News, 1st Quarter, 1999.

[21] This and the next few paragraphs are based on Mangan, T., *OA&M: How a Frame Relay SLA Is Measured,* Sync, MPLS/Frame Relay Alliance News, 3rd Quarter, 1999.

[22] For example, the first book on this topic was Minoli, D., and E. Minoli, *Delivering Voice over Frame Relay and ATM,* New York: Wiley, 1998.

[23] Press Release, Fremont, CA: MPLS/Frame Relay Alliance, July 2, 2001.

[24] Kite, P., *Fax over Frame,* Newbridge Networks Corporation, MPLS/Frame Relay Alliance News, 4th Quarter, 1998.

[25] Sif, M., *IP-VPNs and Frame Relay—Competitive or Complementary?* Nortel Networks, MPLS/Frame Relay Alliance News, 3rd Quarter, 1999.

[26] Some portions of this discussion are based on. Sif, M., *Challenging the IP-VPN Drivers,* Nortel Networks, MPLS/Frame Relay Alliance News, 2nd Quarter, 1999.

[27] Archuleta, J., *DSL Access to FR—Opportunities and Challenges,* Paradyne, Frame Relay Forum News, 4th Quarter, 1999.

[28] The rest of this section is based on Fisher, R. A., *FR Network Recovery Options,* AT&T, MPLS/Frame Relay Alliance News, Fall 2001.

[29] X.25 network services generated $2.7 billion of revenue worldwide in 1999 according to Vertical Systems Group.

[30] FRF News, 1st Quarter, 1998.

[31] An Internet search on a large bookseller's site for the term *ATM asynchronous transfer mode* revealed 158 book titles.

[32] Minoli, D., and J. Amoss, *IP Applications with ATM,* New York: McGraw-Hill, 1994.

[33] Any attorney reviewing a brochure describing ATM services will not allow his or her client to use the word "guarantee" in the literature. So why should the technical discourse use a term that has no business acceptance? Telecom needs to graduate up from the raw technology steeped in hundreds of acronyms to at least a slightly better level of reality. There are no "guarantees" in ATM or frame relay; hence, one should stop abusing the language, and use more common sense, plain English phraseology. Don't promise something to the customer that you can't deliver. "High assurance" may be a more appropriate term than "guarantee."

[34] The quotes have been added by this author, consistent with the comments in [33].

[35] Retrieved from http://www.foundrynet.com/technologies/mpls/faqs.html.

[36] Rosen, E., A. Viswanathan, and R. Callon, *Multiprotocol Label Switching Architecture,* RFC 3031, January 2001. Copyright 2001 The Internet Society. This document and translations of it may be copied and furnished to others, and derivative works that comment on or otherwise explain it or assist in its implementation may be prepared, copied, published and distributed, in whole or in part, without restriction of any kind, provided that the above copyright notice and this paragraph are included on all such copies and derivative works.

[37]   Pulley, R., and P. Christensen, "A Comparison of MPLS Traffic Engineering Initiatives," NetPlane Systems White Paper, retrieved from http://www.netplane.com.

[38]   Retrieved from http://www.mplsworld.com/archi_news/indus/I020206.htm.

[39]   Awduche, D., et al., *Requirements for Traffic Engineering over MPLS*, RFC 2702, September 1999.

[40]   The original "tag switching et al." proposals used a simple forwarding tag as a means of lowering the cost of the forwarding decision from an IP address longest-prefix-match problem, to a simple table index. This was occasioned by the "underpowered hardware" of the time. (Parenthetically, and on the same issue, what is one to make of the recent soft switch hoopla? Isn't that a deja vu situation with the late 1970s situation of X.25 packet switches running of DEC's PDP-11s?) Later router vendors demonstrated that running line rate forwarding of IP address lookups was no longer a problem. The initial driver for what became MPLS, as implemented by Juniper and Cisco, was traffic engineering capabilities. Some state that MPLS is essentially a reinvented frame relay protocol stack with standardized internal network protocols in addition (or the same as) the UNI.

[41]   Retrieved from http://www.ietf/.../draft-ietf-mpls-recovery-frmwrk-03.txt.

[42]   Hopkins, H., *Frame Relay and MPLS, and How They Fit Together*, Accent on Networks, MPLS/Frame Relay Alliance, 4th Quarter, 2001.

[43]   Awduche, D., et al., *MPLS Traffic Engineering*, RFC 2702, September 1999.

[44]   Rosen, E., et al., *MPLS Label Stack Encoding*, RFC 3032, January 2001. Copyright 2001 The Internet Society. This document and translations of it may be copied and furnished to others, and derivative works that comment on or otherwise explain it or assist in its implementation may be prepared, copied, published and distributed, in whole or in part, without restriction of any kind, provided that the above copyright notice and this paragraph are included on all such copies and derivative works.

[45]   "MultiProtocol Label Switching," Future Software Limited White Paper, Chennai, India, retrieved from http://www.futsoft.com, 2001.

[46]   S. Braden, et al., *Resource ReSerVation Protocol (RSVP)—Version 1 Functional Specification*, RFC 2205, September 1997, IETF.

[47]   Awduche, D., et al., *Extensions to RSVP for LSP Tunnels*, work in progress, draft-ietf-mpls-rsvp-lsp-tunnel-08.txt, February 2001, IETF.

[48]   Semeria, C., "RSVP Signaling Extensions for MPLS Traffic Engineering," Juniper Networks White Paper, retrieved from http://www.juniper.net, 2000.

[49]   RSVP code existed and was debugged, although the use of RSVP for QoS in public networks has seen no use to date.

[50]   Sections 9.3.3 through 9.3.7 are based directly on [36], which provides a lucid explanation of the topic.

[51]   For example, there may be policy reasons for doing so; example: law enforcement access for wiretaps at a convenient location.

[52]   Rekhter, Y., and Rosen, E., *Carrying Label Information in BGP-4*, work in progress.

[53]   Awduche, D., et al., *Extensions to RSVP for LSP Tunnels*, work in progress.

[54]   Andersson, L., et al., *LDP Specification*, RFC 3036, January 2001.

[55]   Jamoussi, (Ed.), *Constraint-Based LSP Setup Using LDP*, work in progress.

[56]   Davie, B., et al., *MPLS Using LDP and ATM VC Switching*, RFC 3035, January 2001.

[57]   Conta, A., Doolan, P., and A. Malis, *Use of Label Switching on FR Networks Specification*, RFC 3034, January 2001.

[58] The rest of this section is based on [44].

[59] Katz, D., *IP Router Alert Option,* RFC 2113, February 1997, IETF.

[60] Mogul, J., and S. Deering, *Path MTU Discovery,* RFC 1191, November 1990, IETF.

[61] Conta, A., and S. Deering, *Internet Control Message Protocol (ICMPv6) for the Internet Protocol Version 6 (IPv6) Specification,* RFC 1885, December 1995.

[62] McCann, J., Deering, S. and J. Mogul, *Path MTU Discovery for IP Version 6,* RFC 1981, August 1996.

[63] Postel, J., *Internet Control Message Protocol,* STD 5, RFC 792, September 1981, IETF.

[64] Simpson, W. (Ed.), *The Point-to-Point Protocol (PPP),* STD 51, RFC 1661, July 1994.

[65] This section is based on [42].

CHAPTER 10

Moving Beyond T1: Multilink FR and Other Approaches

This chapter covers what we perceive to be an important pragmatic topic for future intranet development with regard to communication support. This is the topic of *inverse multiplexing* (imux). Inverse multiplexing allows several low-speed channels to be combined transparently into one higher speed channel. Typically, a few T1s are combined into a 3.2- or 6.4-Mbps stream. Although there has been a lot of hype about gigabit (Ethernet) communications in the metro area, users continue to have plain and simple needs that hover around the equivalent of a few T1s. The author has documented in other publications (e.g., [1]) a quantitatively demonstrable end-user need for a typical office environment of 6 Mbps at this time and around 66 Mbps in 5 years, based on traffic scenarios. This implies that while a T1 line's worth of speed is potentially limiting at this time, user needs are not at the gigabit range but at the single-digit megabit range. A large bandwidth, price, and availability gap exists between T1 and the next step up in network access technology, fiber-based DS-3 (45-Mbps) facilities.

This chapter looks at one promising inverse multiplexing technology that can support graceful growth to higher speeds while keeping cost effectiveness in mind. The basic arrangement is depicted in Figure 10.1. Figure 10.2 depicts actual aggregate requirements for the city of Boston, highlighting the point about actual requirements being at the lower end. The fate of several gigabit-Ethernet ELECs (Ethernet LECs), such as Yipes and Telseon, reinforces the veracity of this figure. One way to solve the DS-1–DS-3 gap is to bundle multiple T1 lines into one larger channel. The author installed dozens of multimegabit inverse multiplexed transparent LAN services based on N×T1 technologies in the mid- to late 1980s.

FIGURE 10.1
Basic concept of a generic inverse multiplexer.

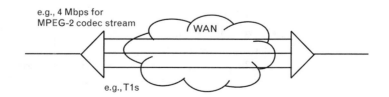

483

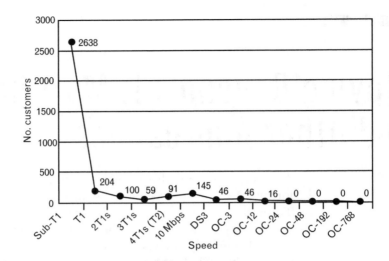

Inverse multiplexing can be supported by a number of technologies, including point-to-point TDM equipment; it is more advantageous, however, when it is used with a switched service such as FR or ATM (see Figure 10.3). *Inverse multiplexed ATM* (IMA) has seen some deployment in the late 1990s. This chapter focuses on FR, because, as covered in the previous chapter, there is a significant embedded base of applications and customers that may benefit from an upgrade. FRF.16.1, *Multilink Frame Relay (MFR) UNI/NNI*, was ratified by the FRF board of directors in 2002 and contains improvements, clarifications, and corrections to FRF.16, which provides economical multilink solutions for increasing bandwidth without using a higher bandwidth transmission facility. CPE devices and CO or POP concentrators have been interoperability based on the FRF.16 MFR specification. Users of MFR can substantially increase their bandwidth without making costly changes to their network and can move forward with confidence that FRF.16-compliant equipment is truly interoperable. CO concentrators enable the carrier to add FR services without changing the switch. Carrier tariffs make MFR an attractive alternative to moving up the digital hierarchy to the next faster access loop; it is less expensive to add a few T1s than to move up to a DS-3 facility. This is especially important for medium-sized hub sites and/or branch locations, and interest seems greatest around 6 to 9 Mbps.

Done correctly, an N×T1 solution will enable network service providers to address the emerging demand for multimegabit service while maximizing the return on investment in the existing T1 infrastructure. After all, T1 lines are affordable and readily available, tariffs already exist, and T1 provides guaranteed symmetrical transport and secure connectivity. Existing bundled T1 solutions, however, were developed for niche markets, are not plug-and-play, and were not intended for the large-scale service deployment required for the Internet. Bit-interleaved inverse

FIGURE 10.3
*Inverse multiplexing
with various WAN
technologies.*

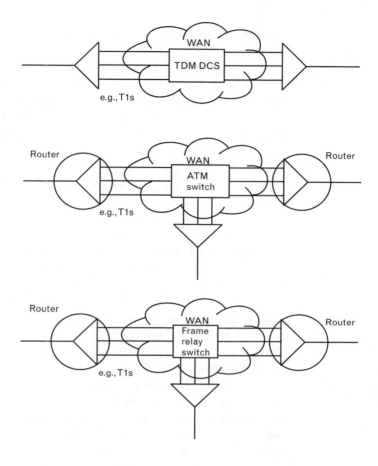

multiplexing, for example, is a technology implemented in proprietary products that typically makes point-to-point connections between two routers at speeds faster than T1. Products that employ this proprietary technology do not support the switching and multipoint connectivity shown in Figure 10.3. Load balancing, used in most routers to intelligently direct packets down parallel paths, can provide higher capacity WAN access for up to two T1s. However, this solution becomes less efficient as more WAN circuits are used. Neither of these solutions is ideal for providing the type of scalable, standards-based solution that the market needs [2].

10.1 Motivations and Scope

The FRF created multilink frame relay as a standard for inverse multiplexing over standard T1 or E1 circuits from the customer's location to the FR service provider POP. The MFR implementation agreements were created by the FRF in response to end-user requests for a standards-based solution to bridge the bandwidth gap that exists between T1/E1 and DS-3/E3 [3]. Figure 10.4 depicts a typical inverse multiplexed FR environment [2].

The continued success of FR technology as a proven, low-latency, access service has driven customer demand for more flexibility in implementations. For a few years, however, network operators have noticed a problem at certain sites: Local loops at those sites are not fast enough to meet the demands of growing traffic volume and new applications [4]. A site that overloads a single 56- or 64-Kbps frame relay access might afford an upgrade to T1, but in some locations additional 56/64-Kbps lines could be more readily available or less expensive than a T1/E1. When the T1/E1 is filled, it a major financial step up to a T3/E3, even if these facilities are available where the user requires them. Changing out the local loop for the next line rate up the traditional digital hierarchy (56 Kbps to 1.5 Mbps to 45 Mbps, or 64 Kbps to 2 Mbps to 34 Mbps) represents a large cost increase—by a factor of almost 30. Many corporate planners cannot justify the jump in the MRC for network access, particularly at smaller branch locations. The 3 times to 8 times increase for the local loop is typical; higher port charges on the switch and other fees can raise the network access portion of the monthly bill by as much as 10 times. Monthly access charges for T1 have dropped below $300 or less in many locations around the United States. DS-3 local access circuits, on the other hand, typically cost $4,000 or more per month, and most business customers do not require the speed of a DS-3. Moreover, local access DS-3 circuits are unavailable to most businesses in the United States and, where available, may take months to provision [2]. Many of today's high-speed access solutions either do not address the T1–DS-3 gap or are not appropriate for business subscribers. ATM imposes a "cell tax" that can reduce circuit payload. ADSL, with its low uplink speeds, is aimed at consumers.

Until recently, a second frame relay access line (Figure 10.5) represented not only just more bandwidth, but also a separate network path. That approach required reengineering of the traffic, assigning part to each

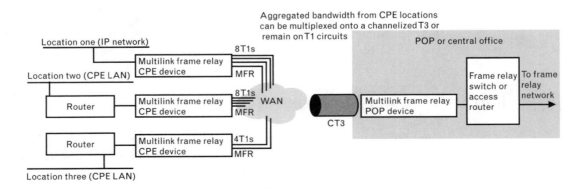

FIGURE 10.4   *A typical inverse multiplexed FR environment.*

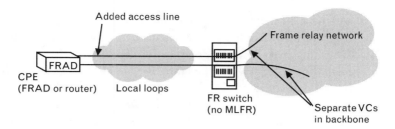

FIGURE 10.5 *Without MFR, adding separate links to increase the bandwidth for access to a frame relay backbone leaves each application restricted to one link; the top burst speed is the speed of that link, not the total speed of all links. (Source: [4]. The Burton Group MPLS/Frame Relay Alliance, on-line newsletter.)*

local loop. An application was confined to its assigned link and could not burst to the top speed represented by the total of the links. If one link were idle, applications assigned to another link could not use the bandwidth. If a router were involved, it would need another IP subnet.

A separate second access line does not provide the maximum reliability that is theoretically possible with two lines. Any plan to shift traffic from a failed link to another active link requires a specific configuration in the CO switch and in the CPE. There may be a charge from the carrier to provide this feature, where available. Fail-over usually depends on detecting the loss of a physical link or virtual circuit, which takes about 45 sec under default settings for the link management polling protocol. If the customer equipment is routers, their routing protocols then have to discover the topology change, which can take another 30 sec up to several minutes. This form of protection switching is not nearly as simple as the basic frame relay service. Hence, adding another line increases bandwidth, but not exactly in the form desired. Addressing that specific part of the problem, carriers offer various proprietary forms of physical layer inverse multiplexing based on multiple loops (56/64 Kbps or T1/E1). These solutions, in practice, are difficult to order and slow to install. The handoff to the user is a HSSI, one of the more expensive router cards that is typically low density (only one port per slot). Some forms of inverse muxing do not load balance; that is, if one link fails, the entire aggregate goes down. Conversely, adding an additional link to increase bandwidth involves taking down and reprovisioning the aggregate on the inverse multiplexer—which disrupts service.

Frame relay connections at access speeds of 1.544 Mbps (T1) or 2.048 Mbps (E1) are widely available from service providers today. Nevertheless, with applications such as audio/video streaming, conferencing, or distributed software development, bandwidth requirements have grown, in many cases, beyond the capacity of that single T1 or E1 link. The obvious solution is to upgrade the customer's frame relay access link, into the "cloud," to the next higher speed in the digital hierarchy: DS-3 (45 Mbps) for North America or E3 (34 Mbps) offered in Europe, South America, and Asia. In many cases, these transmission facilities are too expensive or may not be available in many small- to medium-sized metropolitan areas. Assuming that these lines may be locally tariffed, the required bandwidth, from the enterprise, may be far below DS-3/E3 speeds and the link would, therefore, be underutilized. Unless the telecommunications carrier is offering specific PL services at speeds between T1/E1 and DS-3/E3, which is rare, upgrading the link to the higher speed is often not feasible, economically and logistically. Hence, there is a gap in the continuity of speed solutions for FR access connections greater than T1 or E1 speeds but less than DS-3/E3 [5].

10.2   MFR

The solution to achieve greater than T1/E1 bandwidth is to "bond" or "bundle" up to eight parallel T1/E1 circuits and create a "logical" link with a maximum wire-line speed of 12 Mbps (T1) or 16 Mbps (E1). What is critical, though, is that the technology does this in a vendor-independent manner. Of late, the FRF has developed an implementation agreement for inverse multiplexing over standard T1 or E1 circuits from the customer's location to the FR service provider's POP. This approach will allow CPE, CO concentrator, and FR switch vendors to all interoperate with the same N×T1/E1 inverse muxing protocol and provide customers with an economical and more widely available high bandwidth solution alternative [5].

The FRF standardized two forms of inverse multiplexing, defined in IAs: FRF.15 and FRF.16. Both IAs show how to add increments of bandwidth to network access at a site while preserving the ability of any application to burst to the aggregate speed of all the physical links (it is still one logical pipe).

- *FRF.15* is a point-to-point method between CPE devices; the carrier does not participate. That is, the end devices know that they are doing MFR, but the carrier knows only that they are moving frames across the customer's links.

- *FRF.16* is single-ended solution, between one site and the frame relay network or at the NNI; MFR terminates on the serving switch.

Both of these IAs define how an enterprise user, alone or in cooperation with its frame relay carrier, can increase the effective speed of the access link into a site—exactly what is wanted. To accomplish this goal, MFR aggregates multiple physical links into a single logical path.

FRF.15, the end-to-end MFR implementation agreement, and FRF.16, the MFR UNI/NNI implementation agreement, were published in August 1999 and created the basis for the MFR protocol. MFR specifies the procedures and frame format to be used by CPE to offer an *aggregated virtual circuit* (AVC) service. AVC service allows frame relay CPE to use multiple VCs for transport of a single stream of sequenced frames. The UNI/NNI implementation agreement provides physical interface emulation for frame relay devices, which consists of one or more physical links aggregated into a single bundle of bandwidth. The MFR standard ensures that the multimegabit access solution for frame relay integrates seamlessly with existing network hardware, creating an aggregated path that looks and feels like T1 to end users, but is faster. The bundle will be compatible with the frame relay infrastructure, and will result in additional bandwidth and increased transport resiliency by supporting distribution of data traffic over multiple underlying VCs. These circuits will not require load balancing or future maintenance to get the most efficient use of each T1 circuit. Service providers and business users will be able deploy the MFR solution with minimal changes to their existing network [2].

MFR connections add and drop links gracefully. During a change in the number of constituents in an aggregate, frames flow without interruption. The ability to migrate smoothly between links also creates at least a potential to shop for less expensive or more reliable capacity. The real benefit (besides bandwidth) is resiliency: The aggregate channel can change bandwidth "on the fly" without interrupting traffic flow. Load balancing shifts traffic automatically from a failed link to active links. MFR's inherent load balancing across multiple paths provides resiliency that recovers quickly from a lost link. For optimum protection, access links at both ends will have diverse routing and, for FRF.15 end-to-end MFR, the backbone will have diverse routing as well.

Individual frames may take any of the physical links; therefore, successive frames could take different paths with different delays. The MFR function applies a fragmentation header (per FRF.12) to ensure proper order of the frames on both sides of the aggregate path. Due to the effect of fragmentation, each frame sees reduced latency as well as increased bandwidth. The multilink solution adds some complexity that is similar to adding an additional line without MFR. However, MFR preserves the logical simplicity of a single virtual path, allowing traffic to burst to a speed that is almost the total of all the constituent links. (The fragmentation headers in MFR require a modest amount of additional overhead.)

Adding bandwidth at a site using FRF.16 MFR at the UNI, as shown in Figure 10.6, means placing one or more additional local loops between the serving frame relay switch and the CPE. MFR software, in both the switch and CPE, combines a bundle of physical links into one virtual or logical link. Any PVC the application sees in the CPE appears in the backbone as the same single PVC. The MFR virtual link can be used in any way one would employ a real link. Specifically, one can add a PVC to the network without changing the MFR configuration. The arrangement with the carrier—bundling the physical links into a logical link—is transparent

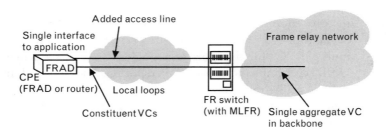

FIGURE 10.6 *Applications see a faster service on the (virtual) aggregate link that MFR at the UNI (FRF.16) creates from the constituent local loops. Because applications cannot see the MFR function, only the aggregate, loss of a physical link does not disrupt an individual user's virtual circuit. (Source: [4]. The Burton Group, MPLS/Frame Relay Alliance, on-line newsleter.)*

to applications, which see normal but faster connections. This UNI access arrangement (FRF.16) depends on carrier support in the serving switch, which has started to roll out but is not yet widespread. There will be a small premium for the MFR termination service in the switch.

There is no need to wait for a carrier to offer MFR or to pay a special MFR fee if the user's CPE supports the end-to-end version of MFR (FRF.15). It works over any FR network—or more than one (Figure 10.7). FRF.15 aggregates multiple links between CPE devices at two user sites, creating a large logical link.

Opting for end-to-end MFR, and not involving the carrier, may have drawbacks for some applications precisely because FRF.15 is a point-to-point connection. This means that CPE at the remote site must also support MFR (FRF.15) to terminate those PVCs as an aggregate. In contrast

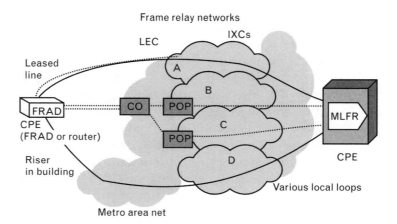

FIGURE 10.7 *End-to-end MFR can preserve a frame relay connection in the face of a failure in a PVC. Diverse routing of the FRF.15 constituent virtual circuits over different carriers protects against the loss of an entire network: The impact is same as the loss of one local loop out of a bundle of links in the FRF.16 form of MFR at the UNI. (Source: [4]. The Burton Group, MPLS/Frame Relay Alliance, on-line newsleter.)*

to FRF.15, FRF.16 at the UNI carries a connection across the backbone on a single virtual circuit. When the carrier terminates MFR, and deals with a single fast PVC, it can more easily:

- Route to any location, including a site whose access link is fast enough not to need MFR;
- Apply services like rerouting to an alternate data center;
- Provision a connection to another frame relay carrier at an NNI.

The logical bundling aspect of MFR can provide sites with more bandwidth, and the resiliency aspect adds reliability. These two factors make MFR a very powerful access option.

To ensure low latency within a virtual multimegabit access circuit, packets are segmented into individual fragments, and each fragment is transported over a separate member of the T1 bundle. The size of the packet fragments can be adjusted to optimize bandwidth efficiency and latency. The fragments are reassembled at the other end of the T1 link using MFR sequence numbers, thus ensuring packet order. If a T1 circuit within a MFR connection fails, the bandwidth is downshifted but service is not interrupted. When the T1 line comes back up, it is automatically added back to the bundle. In addition, the T1 lines in a bundle can be connected to the POP and CPE in any order, eliminating the risk of intrabundle wiring problems throughout the network. Service can be made more robust by using T1 circuits from different carriers. This enables continuous operation of the bundle if a carrier experiences problems with its T1 circuits. Bundling multiple copper lines to create a virtual multimegabit circuit may introduce differential delay between the different T1 links, particularly if different carriers are used within the bundle. The ability to handle these delays and still deliver complete packets, in sequence, is built into MFR protocols [2].

10.3   Comparing Alternatives

Selecting an economical access mechanism to the core network involves some decision making [6]. IMA was one of the first solutions that claimed to deliver cost-effective customer-based multiple services with QoS over a high-speed logical link bundle of multiple T1/E1s. More recently, however, two newer technologies are providing a smoother ride with none of the speed bumps: IP over multilink PPP (MLPPP) and IP over MFR.

10.3.1   IMA

IMA uses ATM technology over multiple point-to-point T1 or E1 links, usually between four and eight T1s or three and six E1s (6 or 12 Mbps). In most cases, the IMA feature is an add-on that requires upgrades to an integrated router located at the customer's site. ATM is used to provide an integrated solution to set priorities and classes of service for simultaneous digital voice and data transmission. Once the data are packed into ATM cells, they are then shipped into an ATM core network that handles the SVC or PVC requirements of each cell it processes. An ATM cell consists of 53 bytes, but only 44 bytes are available for payload. The rest is used for transport and cell *segmentation and reassembly* (SAR)—a heavy tax of 9 bytes or 17% depending on the ATM AAL being used. Giving up bandwidth for overhead is not problematic when it involves fiber optic speeds, which operate at the core of an ATM network. However, added overhead at the edge equates to proportionally wasted bandwidth. In addition, Ethernet data—upward to 1,500 bytes/frame—have to be mapped into those 44-byte cells. Therefore, additional processing effort is required not only for that task but also to evenly distribute the ATM cells over the parallel T1/E1 links. (Figure 10.8 shows a typical IMA CPE to service provider edge solution.)

10.3.2 MLPPP

Traditionally, MLPPP technology was deployed in the ISDN (2B+D and 23B+D/30B+D) market segment. Taking multiple PPP links and "bonding" them into a logical multilink PPP bundle was a way to derive more bandwidth from separate channels on a physical digital connection. An N×T1/E1 MLPPP interface, compliant with RFC 1990, appears as a single high-speed data link to higher layer protocols (e.g., to a single IP subnet). MLPPP takes the incoming Ethernet frames from the LAN and strips them of their MAC addresses. It then adds an MLPPP header along with packet sequence numbers. Using a round-robin line-loading technique, the MLPPP frames are then distributed across all of the facilities that comprise the multilink "bundle." The MLPPP interface, at the CO end of the link, then resequences the incoming stream and forwards the reassembled data to the core side of the CO edge device. MLPPP offers several advantages over IMA:

FIGURE 10.8
*Typical IMA
connectivity.*

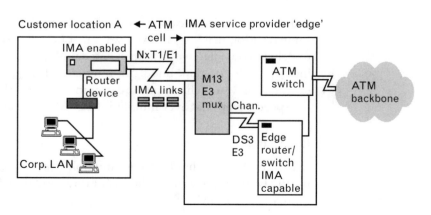

- Maps Ethernet frames into MLPPP frames more efficiently, resulting in less processing overhead.

- Facilitates a QoS algorithm that can prioritize traffic delivery into the cloud by application type or IP source/destination address.

- Supports an all-IP connectionless environment for VPNs.

- Uses an overhead, on average, of only 2% to 3% of the customer's access bandwidth.

Using MLPPP, ATM can then be more efficiently used as an internode transport technology in the high-speed network core, thus eliminating the cell tax at the edge. (Figure 10.9 shows a typical MLPPP CPE-to-service provider edge solution.)

10.3.3 MFR

As noted earlier, MFR (FRF.16) is a standards-based multilink approach, developed by the FRF, to allow fatter logical channels to be brought into a frame relay customer location with minimal management bandwidth overhead. Having completed a successful technical interoperability lab event in late July 2001, multiple vendors such as Lucent, Nortel, Quick Eagle, Larscom, ASC, Adtran, and Tiara Networks have proven the viability of this emerging technology. MFR combines the PVC capabilities of IMA with the low transport overhead found in MLPPP. Using a new generation of multiservice switch platforms, at the POP edge, the MFR approach will allow, for example, a central site frame relay customer node to have a higher aggregate access speed (up to 12 Mbps) into the FR/ATM core. With MFR, the customer can now have CIRs that well exceed an existing T1/E1 line rate. Because FR is based on a PVC model, each LAN host IP address can be mapped to individual FR DLCIs, each with its own SLA based on the frame relay standard FRF.13.

FIGURE 10.9
*Typical N× T1/E1
MLPPP connectivity.*

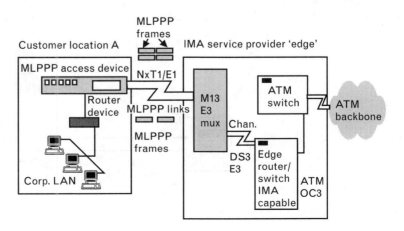

The mapping of Ethernet frames into an FR frame is fairly straightforward since an MFR frame can vary in size from 64 bytes to 8 Kbytes. In addition, varying lengths of FRF.16 frames can be fragmented into fairly equal sizes, as specified in the FRF.12 standard. This approach significantly reduces latency and differential delay, a crucial consideration when engineering a FR network to carry delay-sensitive traffic, such as voice, music streaming, and video. MFR has clear advantages over IMA:

• Support for variable frame sizes and fragmentation;

• Low latency;

• Minimal management bandwidth overhead of 2% to 3%;

• Standards-based SLAs using FRF.13.

In Figure 10.10, FRF.8 is being used at the multiservice switch POPs as the standard FR-to-ATM interworking protocol that, once again, keeps ATM at the high-speed core of the network. In addition, by using intelligent MFR and single-link CPE network access devices, a management solution can be used to validate a customer's SLA. This would give the service provider an opportunity to market an additional service that would allow customers to validate that they are receiving the monthly network

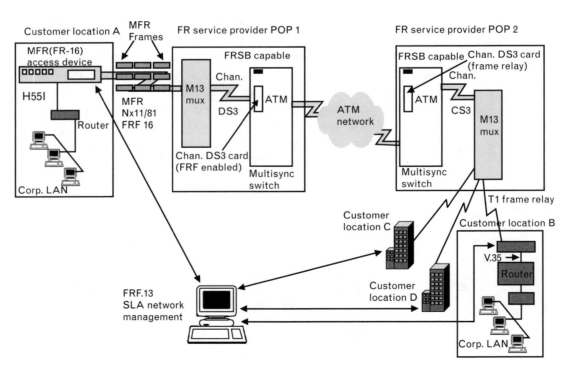

FIGURE 10.10   *Typical MFR connectivity.*

performance for which they are paying. IMA carries nontrivial recurring costs in order to implement the desired bandwidth:

- Typical monthly cost of a T1 line: $500;

- Typical monthly cost of an E1 line: $735;

- Typical multilink bandwidth desired: 6 Mbps (approx.) = 4 T1 or 3 E1;

- IMA potential overhead/bundle is 17% based on the 9 cells out 53 tax;

- Bandwidth sacrificed for 4 T1 or 3 E1 access: 17% 6 Mbps = 1.02 Mbps (that is over 66% of the entire bandwidth of a single T1 line or over 51% of an E1 line);

- Three-year cost of an additional T1 line required to provide desired bandwidth: 36 × $500=$18,000;

- Three-year cost of an additional E1 line required to provide desired bandwidth: 36 × 735=$26,460.

From a recurring cost standpoint, IMA is not well suited to efficiently deliver high-speed services using standard T1/E1 lines from the POP edge to the CPE. Whatever may be saved in an integrated solution at the CPE is spent supporting the wasted bandwidth. In addition to cost, a number of support and reliability factors should be considered when implementing a multilink solution:

- *Installation.* A multilink solution needs to be easily implemented at the customer site. ATM is a very sophisticated protocol that requires a high skill set to provision properly. Many configuration options are available that can cause a lengthy and expensive service provisioning process.

- *Compatibility.* Not all end-to-end IMA solutions are fully interoperable, which would lock the customer and service provider into a single vendor. This could very well cause the customer to have to purchase a new CPE router solution from the ground up. Additionally, the service vendor could not secondary source a CPE access device if it is part of the monthly service contract.

- *Autofallback and recovery/bundle management.* A solid ML solution provides inherent redundancy and data integrity if a link should fail. When a T1/E1 link fails in an IMA network, it will bring down the entire logical bundle. MLPPP and MFR simply remove the "bad apple from the barrel" until the problem is resolved. During the link outage, data are still flowing to/from the cloud, albeit at a reduced data rate. Once the outage has been resolved, the errant T1/E1 can automatically rejoin the bundle.

ENDNOTES

[1]   Minoli, D., P. Johnson, and E. Minoli, *SONET-Based Metro Area Networks,* New York: McGraw-Hill, 2002.

[2]   Hudson, E., *Multilink Frame Relay: Expanding the Limits of T1,* Tiara Networks, MPLS/Frame Relay Alliance News, 4th Quarter, 1999, www.frforum.com.

[3]   FRF.16 *Multilink FR UNI/NNI Implementation Agreement,* Fremont, CA: MPLS/ Frame Relay Alliance, retrieved from http://www.frforum.com/5000/5000index.html.

[4]   Portions of this section and Figures 10.5, 10.6, and 10.7 are based on Flanagan, W. A., *The Case for Multilink Frame Relay Access, Outgrown Your T1, But a T3 Is Too Much? MFR Fills the Gap in Access Speeds, Adds Resiliency to Improve Uptime,* The Burton Group, MPLS/Frame Relay Alliance News, Winter 2000, www.frforum.com.

[5]   Ruby, R. J., *Giving Frame Relay Users Higher Speed On and Off Ramps to "The Cloud,"* MPLS/Frame Relay Alliance News, Autumn 2001, www.frforum.com.

[6]   This section is based directly on Ruby, R. J., *MFR Provides Smooth Ride,* Quick Eagle Networks, MPLS/Frame Relay Alliance News, Winter 2001, www.frforum.com.

CHAPTER 11

Broadband LAN Technologies: GbE and 10GbE

In recent years, major progress has been made in developing high-speed LAN systems to support in-building corporate enterprise connectivity. In the late 1990s, Ethernet operating at gigabit per second speeds was standardized; we refer to these systems here as GbE. In the early 2000s, systems operating at 10 Gbps were standardized; we refer to these here as 10GbE. Work is under way to increase the speed to 40 and possibly 100 Gbps.

This chapter provides some basic information on LANs, GbE, and 10GbE. The information in this chapter is based liberally on IEEE sources. The IEEE is the standardization body (in conjunction with ISO/IEC) that handles LAN standardization. Figure 11.1 depicts the basic IEEE LAN protocol model that is consistent across all LAN technologies.

11.1 LAN Introduction

11.1.1 Standards

Over the years, the IEEE has published a comprehensive set of international standards for LANs employing CSMA/CD as the access method. These standards are embodied under the IEEE 802.3 family. CSMA/CD is a low-complexity multiplexing scheme that allows multiple users to share

FIGURE 11.1
*Basic IEEE LAN
protocol model.*

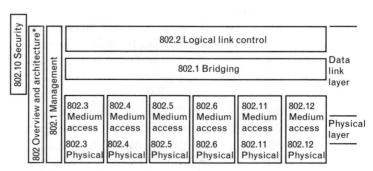

* Formerly IEEE std 802.1

the common medium (e.g., the four-wire bus that comprises the LAN); this is accomplished using channel-contention schemes. More recently, switched configurations that eliminate the need for multiplexing and, hence, contention, have emerged. A majority of installations now use these switched configurations operating at 10- or 100-Mbps speeds. LANs were invented in the 1970s. Digital Equipment Corporation, Intel, and Xerox (DIX) brought out the first "standard" version of a LAN, calling this version *Ethernet*. The formal standardization that followed in the early 1980s by the IEEE was based on DIX-advanced technology.

IEEE 802.3 is intended to encompass several media types and techniques for signal rates from 1 to 10,000 Mbps. The latest edition of the IEEE 802.3 standard provides the necessary specifications for the following families of systems: a 1-Mbps baseband system, 10-Mbps baseband and broadband systems, a 100-Mbps baseband system, a 1,000-Mbps baseband system, and a 10,000-Mbps baseband system. In addition, the standard specifies a method for linearly incrementing a system's data rate by aggregating multiple physical links of the same speed into one logical link.

The IEEE Project 802 develops LAN and MAN standards, mainly for the lowest two layers of the OSI reference model. It coordinates with other national and international standards groups, with some standards now published by ISO as international standards. There is strong international participation in the IEEE standardization activities, and some meetings are held outside the United States [1]. The first meeting of the IEEE Computer Society Local Network Standards Committee (LNSC), Project 802, was held in February 1980. (The project number, 802, was simply the next number available at that time in the sequence being issued by the IEEE for standards projects.) There was going to be one LAN standard, with speeds from 1 to 20 MHz. It was divided into media or PHY, MAC, and higher level interface (HILI). The access method was similar to that for Ethernet, as well as the bus topology. By the end of 1980, a token access method was added, and a year later there were three MACs: CSMA/CD, token bus, and token ring. In the years since, other MAC and PHY groups have been added, as well as one for LAN security. The unifying theme has been a common upper interface to the LLC sublayer, common data framing elements, and some commonality in media interface. The scope of work has grown to include MANs and WANs, and higher data rates have been added. An organizational change gave the team the "LAN/MAN Standards Committee" name and more involvement in the standards sponsorship and approval process. Figure 11.2 depicts the organization of the IEEE 802 committee; also see Table 11.1. Table 11.2 identifies the key standards that have been produced over the years.

Figure 11.3 depicts the various sublayers that comprise the MAC and PHY apparatus. The basic IEEE 802.3 standard provides for two distinct modes of operation: half duplex and full duplex. A given IEEE 802.3 instantiation operates in either half- or full-duplex mode at any one time.

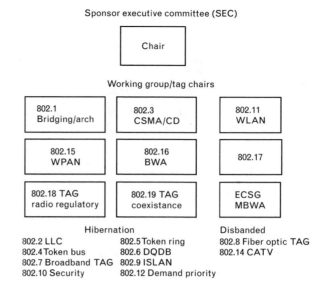

FIGURE 11.2
*IEEE 802
Organization.*

The term *CSMA/CD MAC* is used throughout the standard synonymously with *802.3 MAC,* and may represent an instance of either a half-duplex or full-duplex mode DTE, even though full-duplex mode DTEs do not implement the CSMA/CD algorithms traditionally used to arbitrate access to shared-media LANs.

11.1.1.1 Half-Duplex Operation

In half-duplex mode, the CSMA/CD media access method is the means by which two or more stations share a common transmission medium. To transmit, a station waits (defers) for a quiet period on the medium (i.e., when no other station is transmitting) and then sends the intended message in bit-serial form. If, after initiating a transmission, the message collides with that of another station, then each transmitting station intentionally transmits for an additional predefined period to ensure propagation of the collision throughout the system. The station remains silent for a random amount of time (backoff) before attempting to transmit again. Each aspect of this access method process is specified in detail in subsequent clauses of the IEEE 802.3 standard. Half-duplex operation can be used with all media and configurations.

11.1.1.2 Full-Duplex Operation

Full-duplex operation allows simultaneous communication between a pair of stations using point-to-point media (dedicated channel). Full-duplex operation does not require that transmitters defer, nor do they monitor or react to receive activity, because there is no contention for a shared medium in this mode. Full-duplex mode can only be used when all of the following are true:

TABLE 11.1   IEEE LAN WORKING GROUPS
_____

ACTIVE WORKING GROUPS AND TECHNICAL ADVISORY GROUPS

  802.1   High Level Interface (HILI) Working Group

  802.3   CSMA/CD Working Group

  802.11   Wireless LAN (WLAN) Working Group

  802.15   Wireless Personal Area Network (WPAN) Working Group

  802.16   Broadband Wireless Access (BBWA) Working Group

  802.17   Resilient Packet Ring (RPR) Working Group

  802.18   Radio Regulatory Technical Advisory Group

  802.19   Coexistence Technical Advisory Group

Executive Committee Study Group, Mobile Broadband Wireless Access

HIBERNATING WORKING GROUPS (STANDARDS PUBLISHED, BUT INACTIVE)

  802.2   Logical Link Control (LLC) Working Group

  802.4   Token Bus Working Group

  802.5   Token Ring Working Group

  802.6   Metropolitan Area Network (MAN) Working Group

  802.7   Broadband Technical Adv. Group (BBTAG)

  802.9   Integrated Services LAN (ISLAN) Working Group

  802.10   Standard for Interoperable LAN Security (SILS) Working Group

  802.12   Demand Priority Working Group

DISBANDED WORKING GROUPS (DID NOT PUBLISH A STANDARD)

  802.8   Fiber Optics Technical Adv. Group (FOTAG)

  802.14   Cable-TV Based Broadband Communication Network Working Group

- The physical medium is capable of supporting simultaneous transmission and reception without interference.
- There are exactly two stations connected with a full-duplex point-to-point link. Because there is no contention for use of a shared medium, the multiple access (i.e., CSMA/CD) algorithms are unnecessary.
- Both stations on the LAN are capable of, and have been configured to use, full-duplex operation.

The most common configuration envisioned for full-duplex operation consists of a central bridge (also known as a switch) with a dedicated LAN connecting each bridge port to a single device. Full-duplex operation constitutes a proper subset of the MAC functionality required for half-duplex operation.

TABLE 11.2 KEY IEEE LAN EFFORTS

IEEE 802.1

The IEEE 802.1 Working Group is chartered to concern itself with and develop standards and recommended practices in the following areas: 802 LAN/MAN architecture, internetworking among 802 LANs, MANs and other wide area networks, 802 overall network management, and protocol layers above the MAC and LLC layers.

ACTIVE PROJECTS

802a	Ethertypes
802.1s	Multiple Spanning Trees
802.1y–802.1D	Maintenance
802.1z–802.1Q	Maintenance
802.1aa– 802.1X	Maintenance

PROJECTS UNDER DISCUSSION

802.1AB	Station and Media Access Control Connectivity Discovery
802.1ac	Media Access Control Service revision

ARCHIVED PROJECTS

802	Overview and Architecture
802.1D	MAC Bridges
802.1G	Remote MAC Bridging
802.1Q	Virtual LANs
802.1t–802.1D	Maintenance
802.1u–802.1Q	Maintenance
802.1v	VLAN Classification by Protocol and Port
802.1w	Rapid Reconfiguration of Spanning Tree
802.1x	Port Based Network Access Control

WITHDRAWN PROJECTS

802.1r	GARP Proprietary Attribute Registration Protocol (GPRP)

IEEE 802.3 CSMA/CD (ETHERNET)

The IEEE 802.3 Working Group develops standards for CSMA/CD (Ethernet)-based LANs. They have a number of active projects as listed below:

P802.3ah	Ethernet in the First Mile
P802.3af	DTE Power via MDI
P802.3	Static Discharge in Copper Cables Ad Hoc

11.1.2 Key Concepts

11.1.2.1 Compatibility Interfaces

Five important compatibility interfaces are defined within what is architecturally the physical layer (refer to Figure 11.3):

1. *Medium-dependent interfaces (MDIs).* To communicate in a compatible manner, all stations will adhere rigidly to the exact specification of physical media signals defined in Clause 8 (and beyond)

FIGURE 11.3
*LAN standard
relationship to the
ISO/IEC OSI
reference model.*

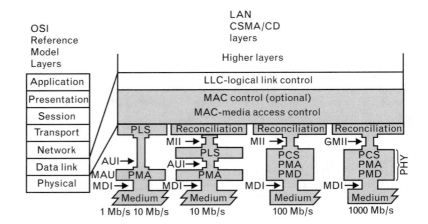

AUI - Attachment unit interface
MDI - Medium dependent interface
MII - Media independent interface
GMII - Gigabit media independent interface
MAU - Medium attachment unit

PLS = Physical layer signaling
PCS = Physical coding sublayer
PMA = Physical medium attachment
PHY = Physical layer device
PMD = Physical medium dependent

in the IEEE 802.3 standard, and to the procedures that define correct behavior of a station. The medium–independent aspects of the LLC sublayer and the MAC sublayer should not be taken as detracting from this point; communication by way of the ISO/IEC 8802-3 [ANSI/IEEE Std 802.3] LAN requires complete compatibility at the physical medium interface (that is, the physical cable interface).

2. *Attachment unit interface (AUI).* It is anticipated that most DTEs will be located some distance from their connection to the physical cable. A small amount of circuitry will exist in the *medium attachment unit* (MAU) directly adjacent to the physical cable, while the majority of the hardware and all of the software will be placed within the DTE. The AUI is defined as a second compatibility interface. Although conformance with this interface is not strictly necessary to ensure communication, it is highly recommended, because it allows maximum flexibility in intermixing MAUs and DTEs. The AUI may be optional or not specified for some implementations of the IEEE 802.3 standard that are expected to be connected directly to the medium and so do not use a separate MAU or its interconnecting AUI cable. The *physical layer signaling* (PLS) and *physical medium attachment* (PMA) are then part of a single unit, and no explicit AUI implementation is required.

3. *Media-independent interface (MII).* It is anticipated that some DTEs will be connected to a remote PHY and/or to different medium-dependent PHYs. The MII is defined as a third compatibility

interface. Although conformance with implementation of this interface is not strictly necessary to ensure communication, it is highly recommended, because it allows maximum flexibility in intermixing PHYs and DTEs. The MII is optional.

4. *Gigabit media-independent interface (GMII).* The GMII is designed to connect a gigabit-capable MAC or repeater unit to a gigabit PHY. Although conformance with implementation of this interface is not strictly necessary to ensure communication, it is highly recommended, because it allows maximum flexibility in intermixing PHYs and DTEs at gigabit speeds. The GMII is intended for use as a chip-to-chip interface. No mechanical connector is specified for use with the GMII. The GMII is optional.

5. *10-bit interface (TBI).* The TBI is provided by the 1000BASE-X PMA sublayer as a physical instantiation of the PMA service interface. The TBI is highly recommended for 1000BASE-X systems, because it provides a convenient partition between the high-frequency circuitry associated with the PMA sublayer and the logic functions associated with the PCS and MAC sublayers. The TBI is intended for use as a chip-to-chip interface. No mechanical connector is specified for use with the TBI. The TBI is optional.

11.1.2.2 Layer Interfaces

In the architectural model used here, the layers interact by way of well-defined interfaces, providing services to each other. In general, the interface requirements are as follows:

- The interface between the MAC sublayer and its client includes facilities for transmitting and receiving frames, and provides per-operation status information for use by higher layer error recovery procedures.

- The interface between the MAC sublayer and the physical layer includes signals for framing (carrier sense, receive data valid, transmit initiation) and contention resolution (collision detect), facilities for passing a pair of serial bit streams (transmit, receive) between the two layers, and a wait function for timing.

The service of a layer or sublayer is the set of capabilities that it offers to a user in the next higher (sub)layer. Abstract services are specified by describing the service primitives and parameters that characterize each service. This definition of service is independent of any particular implementation (see Figure 11.4). Specific implementations may also include provisions for interface interactions that have no direct end-to-end effects. Examples of such local interactions include interface flow control, status requests and

FIGURE 11.4
*Service primitive
notation.*

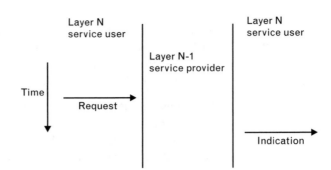

indications, error notifications, and layer management. Specific implementation details are omitted from the IEEE 802.3 service specification both because they will differ from implementation to implementation and because they do not impact the peer-to-peer protocols. Primitives are of two generic types:

1. REQUEST. The request primitive is passed from layer N to layer $N-1$ to request that a service be initiated.

2. INDICATION. The indication primitive is passed from layer $N-1$ to layer N to indicate an internal layer $N-1$ event that is significant to layer N. This event may be logically related to a remote service request, or may be caused by an event internal to layer $N-1$.

The service primitives are an abstraction of the functional specification and the user–layer interaction. The abstract definition does not contain local detail of the user–provider interaction. Each primitive has a set of zero or more parameters, representing data elements that will be passed to qualify the functions invoked by the primitive. Parameters indicate information available in a user–provider interaction; in any particular interface, some parameters may be explicitly stated (even though not explicitly defined in the primitive) or implicitly associated with the service access point. Similarly, in any particular protocol specification, functions corresponding to a service primitive may be explicitly defined or implicitly available.

11.1.2.3   Example for MAC Layer

The services provided by the MAC sublayer allow the local MAC client entity to exchange LLC data units with peer LLC sublayer entities. Optional support may be provided for resetting the MAC sublayer entity to a known state. The optional MAC control sublayer provides an additional service for controlling MAC operation. This may be used to provide flow control between peer MAC client entities across the underlying channel. The primitives (see Figure 11.5) are as follows:

• MA_DATA.request;

FIGURE 11.5
*Service specification
primitive relationships
(optional MAC
control sublayer
implemented).*

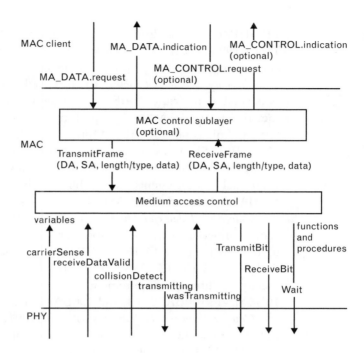

- MA_DATA.indication;

- MA_CONTROL.request (used by optional MAC Control sublayer);

- MA_CONTROL.indication (used by optional MAC Control sublayer).

11.1.2.4 Cabling

Over the years a variety of cabling technologies have been used for LANs, as defined next. Cabling can be shielded or unshielded. Almost universally, nearly all networks today use Category 5 cabling. Market surveys indicate that Category 5 balanced copper cabling is the predominant installed intra-building horizontal networking media today.

Unshielded twisted-pair cable (UTP) is an electrically conducting cable, comprising one or more pairs, none of which is shielded. There may be an overall shield, in which case the cable is referred to as unshielded twisted pair with overall shield.

Shielded twisted-pair (STP) cable is an electrically conducting cable, comprising one or more elements, each of which is individually shielded. There may be an overall shield, in which case the cable is referred to as shielded twisted-pair cable with an overall shield (from ISO/IEC 11801:1995). Specifically for IEEE 802.3 100BASE-TX, it has 150Ω balanced inside the cable with performance characteristics specified to 100 MHz (i.e., performance to Class D link standards per ISO/IEC

11801:1995). In addition to the requirements specified in ISO/IEC 11801:1995, IEEE 802.3 Clauses 23 and 25 provide additional performance requirements for 100BASE-T operation over STP.

Category 3 balanced cabling has balanced 100- and 120-Ω cables and associated connecting hardware whose transmission characteristics are specified up to 16 MHz (i.e., performance meets the requirements of a Class C link as per ISO/IEC 11801:1995). It is commonly used in IEEE 802.3 10BASE-T installations. In addition to the requirements outlined in ISO/IEC 11801:1995, IEEE 802.3 Clauses 14, 23, and 32 specify additional requirements for cabling when used with 10BASE-T, 100BASE-TX, and 1000BASE-T.

Category 4 balanced cabling has balanced 100- and 120-Ω cable and associated connecting hardware whose transmission characteristics are specified up to 20 MHz per ISO/IEC 11801:1995. In addition to the requirements outlined in ISO/IEC 11801:1995, IEEE 802.3 Clauses 14, 23, and 32 specify additional requirements for this cabling when used with 10BASE-T, 100BASE-T4, and 100BASE-T2, respectively.

Category 5 balanced cabling has balanced 100- and 120-Ω cables and associated connecting hardware whose transmission characteristics are specified up to 100 MHz (i.e., cabling components meet the performance specified in ISO/IEC 11801:1995). In addition to the requirements outlined in ISO/IEC 11801:1995, IEEE 802.3 Clauses 14, 23, 25, and 40 specify additional requirements for this cabling when used with 10BASE-T and 100BASE-T.

11.1.2.5   Other Key Definitions

Table 11.3 summarizes some key definitions used in the IEEE 802.3 standards.

11.1.3   MAC Frame Format

Two frame formats are specified in IEEE 802.3: (a) a basic MAC frame format and (2) an extension of the basic MAC frame format for tagged MAC frames, that is, frames that carry Qtag prefixes.

Figure 11.6 shows the nine fields of a frame: the preamble, *start frame delimiter* (SFD), the addresses of the frame's source and destination, a length or type field to indicate the length or protocol type of the following field that contains the MAC client data, a field that contains padding if required, the frame check sequence field containing a cyclic redundancy check value to detect errors in a received frame, and an extension field if required (for 1,000-Mbps half-duplex operation only). Of these nine fields, all are of fixed size except for the data, pad, and extension fields, which may contain an integer number of octets between the minimum and maximum values that are determine by the specific implementation of the CSMA/CD MAC.

TABLE 11.3   KEY DEFINITIONS USED IN IEEE 802.3 STANDARDS

CONCEPT	DEFINITION	CLAUSE
10BASE2	IEEE 802.3 physical layer specification for a 10-Mbps CSMA/CD LAN over RG 58 coaxial cable.	Described in IEEE 802.3 Clause 10.
10BASE5	IEEE 802.3 physical layer specification for a 10-Mbps CSMA/CD LAN over coaxial cable (i.e., thicknet).	Described in IEEE 802.3 Clause 8.
10BASE-F	IEEE 802.3 physical layer specification for a 10-Mbps CSMA/CD LAN over fiber optic cable.	Described in IEEE 802.3 Clause 15.
10BASE-FB port	A port on a repeater that contains an internal 10BASE-FB MAU that can connect to a similar port on another repeater.	Described in IEEE 802.3 Clause 9.
10BASE-FB segment	A fiber optic link segment providing a point-to-point connection between two 10BASE-FB ports on repeaters.	
10BASE-FL segment	A fiber optic link segment providing point-to-point connection between two 10BASE-FL MAUs.	
10BASE-FP segment	A fiber optic mixing segment, including one 10BASE-FP Star and all of the attached fiber pairs.	
10BASE-FP Star	A passive device that is used to couple fiber pairs to form a 10BASE-FP segment. Optical signals received at any input port of the 10BASE-FP Star are distributed to all of its output ports (including the output port of the optical interface from which it was received). A 10BASE-FP Star is typically comprised of a passive-star coupler, fiber optic connectors, and a suitable mechanical housing.	Described in IEEE 802.3 Clause 16.5.
10BASE-T	IEEE 802.3 physical layer specification for a 10-Mbps CSMA/CD LAN over two pairs of twisted-pair telephone wire.	Described in IEEE 802.3 Clause 14.
100BASE-FX	IEEE 802.3 physical layer specification for a 100-Mbps CSMA/CD LAN over two optical fibers.	Described in IEEE 802.3 Clauses 24 and 26.
100BASE-T	IEEE 802.3 physical layer specification for a 100-Mbps CSMA/CD LAN.	Described in IEEE 802.3 Clauses 22 and 28.
100BASE-T2	IEEE 802.3 specification for a 100-Mbps CSMA/CD LAN over two pairs of Category 3 or better balanced cabling.	Described in IEEE 802.3 Clause 32.
100BASE-T4	IEEE 802.3 physical layer specification for a 100-Mbps CSMA/CD LAN over four pairs of Category 3, 4, and 5 UTP wire.	Described in IEEE 802.3 Clause 23.
100BASE-TX	IEEE 802.3 physical layer specification for a 100-Mbps CSMA/CD LAN over two pairs of Category 5 UTP or STP wire.	Described in IEEE 802.3 Clauses 24 and 25.
100BASE-X	IEEE 802.3 physical layer specification for a 100-Mbps CSMA/CD LAN that uses the PMD sublayer and MDI of the ISO/IEC 9314 group of standards developed by ASC X3T12 (FDDI).	Described in IEEE 802.3 Clause 24.
1000BASE-CX	1000BASE-X over specialty shielded balanced copper jumper cable assemblies.	Described in IEEE 802.3 Clause 39.
1000BASE-LX	1000BASE-X using long-wavelength laser devices over multimode and single-mode fiber.	Described in IEEE 802.3 Clause 38.
1000BASE-SX	1000BASE-X using short-wavelength laser devices over multimode fiber.	Described in IEEE 802.3 Clause 38.

TABLE 11.3   KEY DEFINITIONS USED IN IEEE 802.3 STANDARDS (CONTINUED)

CONCEPT	DEFINITION	CLAUSE
1000BASE-T	IEEE 802.3 physical layer specification for a 1000-Mbps CSMA/CD LAN using four pairs of Category 5 balanced copper cabling.	Described in IEEE 802.3 Clause 40.
1000BASE-X	IEEE 802.3 physical layer specification for a 1,000-Mbps CSMA/CD LAN that uses a physical layer derived from ANSI X3.230-1994 (FC-PH).	Described in IEEE 802.3 Clause 36.
10BROAD36	IEEE 802.3 physical layer specification for a 10-Mbps CSMA/CD LAN over a single road and cable.	Described in IEEE 802.3 Clause 11.
1BASE5	IEEE 802.3 physical layer specification for a 1-Mbps CSMA/CD LAN over two pairs of twisted-pair telephone wire.	Described in IEEE 802.3 Clause 12.
4D-PAM5	The symbol encoding method used in 1000BASE-T. The four-dimensional quinary symbols (4D) received from the 8B1Q4 data encoding are transmitted using five voltage levels (PAM5). Four symbols are transmitted in parallel in each symbol period.	Described in IEEE 802.3 Clause 40.
8B/10B transmission code	A dc-balanced octet-oriented data encoding.	
8B1Q4	For IEEE 802.3, the data encoding technique used by 1000BASE-T when converting GMII data (8B-8 bits) to four quinary symbols (Q4) that are transmitted during one clock (1Q4).	Described in IEEE 802.3 Clause 40.
Attachment unit interface (AUI)	In 10-Mbps CSMA/CD, the interface between the MAU and the DTE within a data station. Note that the AUI carries encoded signals and provides for duplex data transmission.	Described in IEEE 802.3 Clauses 7 and 8.
Bit time (BT)	The duration of one bit as transferred to and from the MAC. The bit time is the reciprocal of the bit rate. For example, for 100BASE-T the bit rate is 10^{-8} sec or 10 ns.	
Bridge	A layer 2 interconnection device that does not form part of a CSMA/CD collision domain but conforms to the ISO/IEC 15802-3:1998 [ANSI/IEEE 802.1D, 1998 Edition] international standard. A bridge does not form part of a CSMA/CD collision domain, but instead appears as a MAC to the collision domain.	Described in IEEE Std 100-1996.
Collision	A condition that results from concurrent transmissions from multiple DTE sources within a single collision domain.	
Collision domain	A single, half-duplex mode CSMA/CD network. If two or more MAC sublayers are within the same collision domain and both transmit at the same time, a collision will occur. MAC sublayers separated by a repeater are in the same collision domain. MAC sublayers separated by a bridge are within different collision domains.	
Collision presence	A signal generated within the physical layer by an end station or hub to indicate that multiple stations are contending for access to the transmission medium.	Described in IEEE 802.3 Clauses 8 and 12.
Physical coding sublayer (PCS)	Within IEEE 802.3, a sublayer used in 100BASE-T, 1000BASE-X, and 1000BASE-T to couple the MII or GMII and the PMA. The PCS contains the functions to encode data its into code groups, which can be transmitted over the physical medium. Three PCS structures are defined for 100BASE-T—one for 100BASE-X, one for 100BASE-T4, and one for 100BASE-T2. (Described in IEEE 802.3, Clauses 23, 24, and 32.) One PCS structure is defined for 1000BASE-X and one PCS structure is defined for 1000BASE-T.	Described in IEEE 802.3 Clauses 36 and 40.

TABLE 11.3 KEY DEFINITIONS USED IN IEEE 802.3 STANDARDS (CONTINUED)

CONCEPT	DEFINITION	CLAUSE
Physical layer entity (PHY)	Within IEEE 802.3, the portion of the physical layer between the MDI and the MII, or between the MDI and GMII, consisting of the PCS, the PMA, and, if present, the PMD sublayers. The PHY contains the functions that transmit, receive, and manage the encoded signals that are impressed on and recovered from the physical medium.	Described in IEEE 802.3 Clauses 23–26, 32, 36, and 40.
Physical medium attachment (PMA) sublayer	Within 802.3, that portion of the physical layer that contains the functions for transmission, reception, and (depending on the PHY) collision detection, clock recovery, and skew alignment.	Described in IEEE 802.3 Clauses 7, 12, 14, 16, 17, 18, 23, 24, 32, 36, and 40.
Physical medium dependent (PMD) sublayer	In 100BASE-X, that portion of the physical layer responsible for interfacing to the transmission medium. The PMD is located just above the MDI.	Described in IEEE 802.3 Clause 24.
Repeater	Within IEEE 802.3, a device that is used to extend the length, topology, or interconnectivity of the physical medium beyond that imposed by a single segment, up to the maximum allowable end-to-end transmission line length. Repeaters perform the basic actions of restoring signal amplitude, waveform, and timing applied to the normal data and collision signals. For wired star topologies, repeaters provide a data distribution function. In 100BASE-T, it is a device that allows the interconnection of 100BASE-T PHY network segments using similar or dissimilar PHY implementations (e.g., 100BASE-X to 100BASE-X, 100BASE-X to 100BASE-T4, and so on). Repeaters are only for use in half-duplex mode networks.	Described in IEEE 802.3 Clauses 9 and 27.

FIGURE 11.6
MAC frame format.

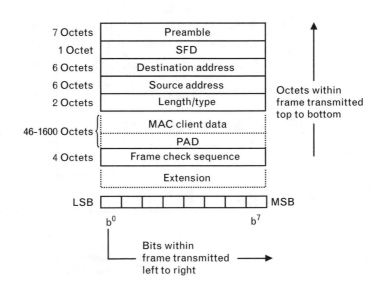

11.1.3.1 Elements of the MAC Frame

Preamble Field

The preamble field is a 7-octet field that is used to allow the PLS circuitry to reach its steady-state synchronization with the received frame's timing.

SFD Field

The SFD field is the sequence 10101011. It immediately follows the preamble pattern and indicates the start of a frame.

Address Fields

Each MAC frame contains two address fields: the destination address field and the source address field, in that order. The destination address field will specify the destination addressee(s) for which the frame is intended. The source address field will identify the station from which the frame was initiated. The representation of each address field is as follows (see Figure 11.7):

- Each address field will be 48 bits in length. While IEEE 802 specifies the use of either 16- or 48-bit addresses, no conformant implementation of IEEE 802.3 uses 16-bit addresses. The use of 16-bit addresses is specifically excluded by the IEEE 802.3 standard.

- The first bit (LSB) will be used in the destination address field as an address type designation bit to identify the destination address either as an individual or as a group address. If this bit is 0, it indicates that the address field contains an individual address. If this bit is 1, it indicates that the address field contains a group address that identifies none, one or more, or all of the stations connected to the LAN. In the source address field, the first bit is reserved and set to 0.

- The second bit will be used to distinguish between locally or globally administered addresses. For globally administered (or U, universal) addresses, the bit is set to 0. If an address is to be assigned locally, this bit will be set to 1. Note that for the broadcast address, this bit is also a 1.

- Each octet of each address field will be transmitted LSB first.

A MAC sublayer address is one of two types:

1. *Individual address.* The address associated with a particular station on the network;

FIGURE 11.7
Address field format.

I/G	U/L	48-bit address

I/G = 0 individual address
I/G = 1 group address
U/L = 0 globally administered address
U/L = 1 locally administered address

2. *Group address.* A multidestination address, associated with one or more stations on a given network.

There are two kinds of multicast address:

1. *Multicast-group address.* An address associated by higher level convention with a group of logically related stations;

2. *Broadcast address.* A distinguished, predefined multicast address that always denotes the set of all stations on a given LAN.

All 1's in the destination address field will be predefined to be the broadcast address. This group will be predefined for each communication medium to consist of all stations actively connected to that medium; it is used to broadcast to all of the active stations on that medium. All stations will be able to recognize the broadcast address. It is not necessary that a station be capable of generating the broadcast address. The address space will also be partitioned into locally administered and globally administered addresses:

1. *Destination address field.* The destination address field specifies the station(s) for which the frame is intended. It may be an individual or multicast (including broadcast) address.

2. *Source address field.* The source address field specifies the station sending the frame. The source address field is not interpreted by the CSMA/CD MAC sublayer.

Length/Type Field
This two-octet field takes one of two meanings, depending on its numeric value. For numerical evaluation, the first octet is the most significant octet of this field.

1. If the value of this field is less than or equal to the value of max-ValidFrame, then the length/type field indicates the number of MAC client data octets contained in the subsequent data field of the frame (length interpretation).

2. If the value of this field is greater than or equal to 1,536 decimal (equal to 0600 hexadecimal), then the length/type field indicates the nature of the MAC client protocol (type interpretation). The length and type interpretations of this field are mutually exclusive.

When used as a type field, it is the responsibility of the MAC client to ensure that the MAC client operates properly when the MAC sublayer pads the supplied data.

Regardless of the interpretation of the length/type field, if the length of the data field is less than the minimum required for proper operation of the protocol, a PAD field (a sequence of octets) will be added at the end of the data field but prior to the FCS field, specified later. The length/type field is transmitted and received with the high-order octet first.

Data and PAD Fields

The data field contains a sequence of n octets. Full data transparency is provided in the sense that any arbitrary sequence of octet values may appear in the data field up to a maximum number specified by the implementation of the standard that is used. A minimum frame size is required for correct CSMA/CD protocol operation and is specified by the particular implementation of the standard. If necessary, the data field is extended by appending extra bits (that is, a pad) in units of octets after the data field but prior to calculating and appending the FCS. The size of the pad, if any, is determined by the size of the data field supplied by the MAC client and the minimum frame size and address size parameters of the particular implementation. The maximum size of the data field is determined by the maximum frame size and address size parameters of the particular implementation.

The length of PAD field required for MAC client data that is n octets long is max [0, minFrameSize − (8 × n + 2 × addressSize + 48)] bits. The maximum possible size of the data field is maxUntaggedFrameSize − (2 × addressSize + 48)/8 octets.

FCS Field

A cyclic redundancy check is used by the transmit and receive algorithms to generate a CRC value for the FCS field. The FCS field contains a 4-octet (32-bit) CRC value. This value is computed as a function of the contents of the source address, destination address, length, LLC data, and pad (that is, all fields except the preamble, SFD, FCS, and extension). The encoding is defined by the following generating polynomial.

$$G(x) = x^{32} + x^{26} + x^{23} + x^{22} + x^{16} + x^{12} + x^{11}$$
$$+ x^{10} + x^{8} + x^{7} + x^{5} + x^{4} + x^{2} + x + 1$$

Extension Field

The extension field follows the FCS field and is made up of a sequence of extension bits that are readily distinguished from data bits. The length of the field is in the range of zero to (slotTime − minFrameSize) bits, inclusive. The contents of the extension field are not included in the FCS computation. Implementations may ignore this field altogether if the number of bit times in the slotTime parameter is equal to the number of bits in the minFrameSize parameter.

11.1.3.2 Invalid MAC Frame

An invalid MAC frame will be defined as one that meets at least one of the following conditions:

- The frame length is inconsistent with a length value specified in the length/type field. If the length/type field contains a type value, then the frame length is assumed to be consistent with this field and should not be considered an invalid frame on this basis.

- It is not an integral number of octets in length.

- The bits of the incoming frame (exclusive of the FCS field itself) do not generate a CRC value identical to the one received.

The contents of invalid MAC frames will not be passed to the LLC or MAC control sublayers, but the occurrence of invalid MAC frames may be communicated to network management.

11.1.3.3 Elements of the Tagged MAC Frame

Figure 11.8 shows the format of a tagged MAC frame. This format is an extension of the basic MAC frame. The extensions for tagging are as follows:

- A 4-octet QTag prefix is inserted between the end of the source address and the MAC client length/type field of the MAC frame. The QTag prefix comprises two fields: (1) a 2-octet constant length/type

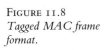

FIGURE 11.8
*Tagged MAC frame
format.*

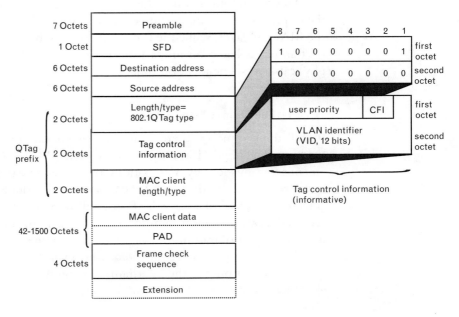

field value consistent with the type interpretation and equal to the value of the 802.1Q tag protocol type (802.1QTagType), and (2) a 2-octet field containing tag control information.

- Following the QTag prefix is the MAC client length/type field, MAC client data, pad (if necessary), FCS, and extension (if necessary) fields of the basic MAC frame.

- The length of the frame is extended by 4 octets by the QTag prefix.

11.1.4 CSMA/CD MAC Method

The procedural model used here is based on seven cooperating concurrent processes. The frame transmitter process and the frame receiver process are provided by the clients of the MAC sublayer (which may include the LLC sublayer) and make use of the interface operations provided by the MAC sublayer. The other five processes are defined to reside in the MAC sublayer. The seven processes are as follows:

1. Frame transmitter process;
2. Frame receiver process;
3. Bit transmitter process;
4. Bit receiver process;
5. Deference process;
6. BurstTimer process;
7. SetExtending process.

This organization of the model is illustrated in Figure 11.9 and reflects the fact that the communication of entire frames is initiated by the client of the MAC sublayer, while the timing of collision backoff and of individual bit transfers is based on interactions between the MAC sublayer and the physical-layer-dependent bit time.

11.1.4.1 Full-Duplex Transmission

In full-duplex mode, there is never contention for a shared physical medium. The physical layer may indicate to the MAC that there are simultaneous transmissions by both stations, but since these transmissions do not interfere with each other, a MAC operating in full-duplex mode must not react to such physical layer indications. Full-duplex stations do not defer to received traffic, nor abort transmission, jam, backoff, and reschedule transmissions as part of transmit media access management.

Transmissions may be initiated whenever the station has a frame queued, subject only to the interframe spacing required to allow recovery for other sublayers and for the physical medium.

FIGURE 11.9
MAC procedure.

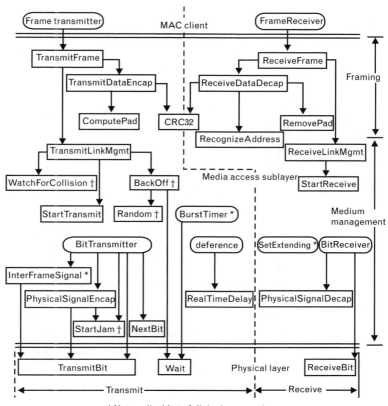

† Not applicable to full duplex operation
* Applicable only to half duplex operation at > 100 Mbps

11.1.4.2 Frame Bursting (Half-Duplex Mode Only)

At operating speeds above 100 Mbps, an implementation may optionally transmit a series of frames without relinquishing control of the transmission medium. This mode of operation is referred to as *burst mode*. Once a frame has been successfully transmitted, the transmitting station can begin transmission of another frame without contending for the medium because all of the other stations on the network will continue to defer to its transmission, provided that it does not allow the medium to assume an idle condition between frames.

The transmitting station fills the interframe spacing interval with extension bits, which are readily distinguished from data bits at the receiving stations and which maintain the detection of carrier in the receiving stations. The transmitting station is allowed to initiate frame transmission until a specified limit, referred to as burstLimit, is reached. Figure 11.10 shows an example of transmission with frame bursting.

The first frame of a burst will be extended, if necessary. Subsequent frames within a burst do not require extension. In a properly configured network, and in the absence of errors, collisions cannot occur during a

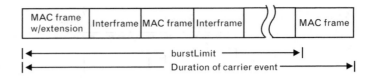

FIGURE 11.10
Frame bursting.

burst at any time after the first frame of a burst (including any extension) has been transmitted. Therefore, the MAC will treat any collision that occurs after the first frame of a burst, or that occurs after the slotTime has been reached in the first frame of a burst, as a late collision.

11.1.4.3   Carrier Extension (Half-Duplex Mode Only)

At operating speeds above 100 Mbps, the slotTime employed at slower speeds is inadequate to accommodate network topologies of the desired physical extent. Carrier extension provides a means by which the slotTime can be increased to a sufficient value for the desired topologies, without increasing the minFrameSize parameter, because this would have deleterious effects. Nondata bits, referred to as extension bits, are appended to frames that are less than slotTime bits in length so that the resulting transmission is at least one slotTime in duration. Carrier extension can be performed only if the underlying physical layer is capable of sending and receiving symbols that are readily distinguished from data symbols, as is the case in most physical layers that use a lock encoding/decoding scheme. The maximum length of the extension is equal to the quantity (slotTime − minFrameSize). Figure 11.11 depicts a frame with carrier extension. The MAC continues to monitor the medium for collisions while it is transmitting extension bits, and it will treat any collision that occurs after the threshold (slotTime) as a late collision.

11.1.5   Allowable Implementations

Table 11.4 identifies the parameter values that are used in the 10-Mbps implementation of a CSMA/CD MAC procedure. Table 11.5 identifies the parameter values that are used in the 100-Mbps implementation of a CSMA/CD MAC procedure. Table 11.6 identifies the parameter values that are used in the 1,000-Mbps implementation of a CSMA/CD MAC procedure.

FIGURE 11.11
Frame carrier extension.

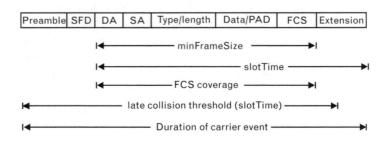

TABLE 11.4   10-MBPS ETHERNET PARAMETERS

PARAMETERS	VALUES
slotTime	512 bit times
interFrameGap	9.6 μs
atemptLimit	16
backoffLimit	10
jamSize	32 bits
maxUntaggedFrameSize	1,518 octets
minFrameSize	512 bits (64 octets)
burstLimit	Not applicable

TABLE 11.5   100-MBPS ETHERNET PARAMETERS

PARAMETERS	VALUES
slotTime	512 bit times
interFrameGap	0.96 μs
atemptLimit	16
backoffLimit	10
jamSize	32 bits
maxUntaggedFrameSize	1,518 octets
minFrameSize	512 bits (64 octets)
burstLimit	Not applicable

TABLE 11.6   1,000-MBPS ETHERNET PARAMETERS

PARAMETERS	VALUES
slotTime	4,096 bit times
interFrameGap	0.096 μs
atemptLimit	16
backoffLimit	10
jamSize	32 bits
maxUntaggedFrameSize	1,518 octets
minFrameSize	512 bits (64 octets)
burstLimit	65,536 bits

Note: The spacing between two noncolliding packets, from the last bit of the FCS field of the first packet to the first bit of the preamble of the second packet, can have a minimum value of 64 BT (bit times), as measured at the GMII receive signals at the DTE. This interFrameGap shrinkage may be caused by variable delays, added preamble bits, and clock tolerances.

11.2   Gigabit Ethernet/IEEE 802.3z

11.2.1   Motivations and Goals

The standard applicable to GbE is the IEEE 802.3z-1998, which has become Clauses 34 thru 42 of the IEEE 802.3 specification. Most of the standardization efforts focused on PHY layers, since the goal of GbE is to retain Ethernet principles and existing upper layer protocols down to the data link layer. Both a full-duplex version of the technology was developed as well as a shared (half-duplex) version that uses CSMA/CD. Table 11.7 identifies the objectives of the standardization work. The GbE specification was approved by the IEEE at the July 1998 IEEE Standards Board meeting, after five drafts (draft D5 became the standard). The work of the IEEE P802.3ab 1000BASE-T Task Force for a UTP version was completed with the approval of IEEE 802.3ab-1999 at the June 1999 IEEE Standards Board meeting. This section is based on the IEEE standard and IEEE materials, but interested developers must refer to the complete IEEE specification.

As the volume of in-building network traffic increases, bandwidth offered by traditional 10-Mbps Ethernet LANs becomes inadequate for a growing number of server/desktop computing environments. These

TABLE 11.7   802.3Z OBJECTIVES

Reach speed of 1,000 Mbps at the MAC/PLS service interface.

Use 802.3/Ethernet frame format.

Meet 802 *functional requirements* (FR), with the possible exception of Hamming distance.

Provide simple forwarding between 1,000-, 100-, 10-Mbps systems.

Preserve min and maxFrameSize of current 802.3 standard.

Provide full- and half-duplex operation.

Support star-wired topologies.

Use CSMA/CD access method with support for at least 1 repeater/collision domain.

Support fiber media and if possible copper media.

Use ANSI Fibre Channel FC-1 and FC-0 as basis for work.

Provide a family of physical layer specifications that support a link distance of (1) at least 25m on copper (100m preferred), (2) at least 500m on multimode fiber, and (3) at least 3 km on single-mode fiber.

Support maximum collision domain diameter of 200m.

Support media selected from ISO/IEC 11801.

Adopt flow control based on 802.3x.

Include a specification for an optional MII.

bottlenecks have been alleviated by the introduction of switched Ethernet; however, this can place a burden on the vertical/campus backbone. Advances in end-user applications drive overall requirements for increased bandwidth at the campuswide and WAN levels. Faster CPUs, graphic applications, and general increases in network traffic are forcing the development of new LAN technologies with higher bandwidth. Most Ethernet networks today incorporate devices and cabling capable of 100-Mbps operation. State-of-the-art network servers today can generate network loads of more than 400 Mbps. A late 1990s survey of 100 companies by Currid & Company found that more than 40% of the respondents indicated that they would need 100- to 1,000-Mbps backbone solutions by 2000. In fact, many network backbones already require bandwidth in excess of 1,000 Mbps; some applications do as well.

At the campus (and building) level, evolving needs will be addressed by GbE systems in developing a new LAN capability. Users have expressed a desire to retain a familiar technology base as faster systems are developed; specifically there has been a desire to retain Ethernet principles. In answer to these needs, GbE allows a graceful transition to higher speeds. Ethernet developers have sought in earnest an extension of Ethernet technology, specifically providing switched segments to the desktops, thereby creating a need for an even higher speed network technology at the *local backbone* and *server* levels. Broadband technology should provide a smooth upgrade path, be cost effective, support QOS, and not require staff retraining and new tools. Gigabit Ethernet provides high-speed connectivity, but does not by itself provide a full set of services such as QoS, automatic redundant failover, or higher level routing services; these are added via other standards.

Gigabit Ethernet supports new full-duplex operating modes for switch-to-switch and switch-to-end station connections, and half-duplex operating modes for shared connections using repeaters and the CSMA/CD access method. Gigabit Ethernet is able to utilize Category 5 UTP cabling. Figure 11.12 depicts the PHYs that are available in GbE. The market acceptance of 100BASE-TX is a clear indication that copper cabling will continue to be the medium of choice for in-building systems wherever it can be applied. This is for simple reasons of economics: Not only is UTP cheaper that a fiber-based horizontal-wiring system, but also replacing the nearly 200 million UTP/LAN runs now deployed in U.S. corporations at an average cost of $150 would cost the industry $30 billion. 1000BASE-T is the natural extension of this evolution and can expect broad market acceptance as the demand for network speed increases. 1000BASE-T continues the Ethernet tradition of providing balanced-cost solutions. Gigabit Ethernet offers an upgrade path for current Ethernet installations, leveraging existing end stations, protocol stacks, management tools, and staff training. Many applications and environments benefit from GbE capabilities:

FIGURE 11.12
GbE PHYs.

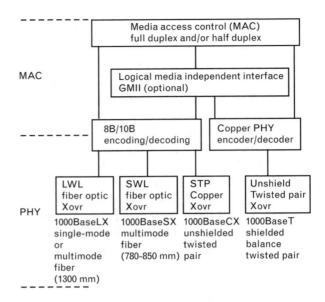

• Building-level backbone, server, and gateway connectivity;

• Multimedia, distributed processing, imaging, medical, and CAD/ CAM applications;

• Aggregation of 100–Mbps switches (e.g., see Figure 11.13);

• Upgrade for large installed base of 10/100 Ethernet;

• Upgrade for large installed base of Category 5 cabling and Class D links.

Vendor support of GbE in general and of 1000BASE–T in particular has been has been strong during the past few years, and a plethora of

FIGURE 11.13
*Migration path
supported by GbE.*

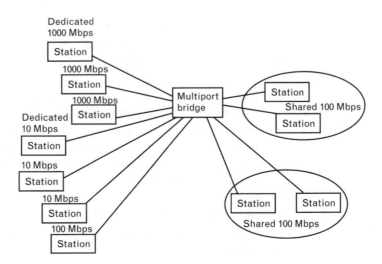

products are now available. Proponents also endeavored to extend GbE to the metro access/metro core space and a couple of dozen vendors have products in this space; this topic is treated in Chapters 12 and 13.

11.2.2   Compatibility with IEEE Standard 802.3

The development goals have been to support (1) conformance with CSMA/CD MAC, PLS, (2) conformance with 802.2, and (3) conformance with 802 frame. The standard conforms to the CSMA/CD MAC, appropriately adapted for 1,000-Mbps use (e.g., packet bursting). In a fashion similar to the 100BASE-T standard, the current physical layers were extended with new PHYs, as appropriate for 1,000-Mbps operation over the defined link—a link that meets the link requirements of 4-pair Category 5 balanced copper component specifications as specified in ANSI/TIA/EIA-568-A or ISO/IEC 11801:1995 and the channel specifications of TIA/EIA568A Annex E.

1000BASE-T offers the maximum compatibility with the current installed base of tens of millions of CSMA/CD devices, most of which utilize copper cabling systems. Support of 802.3 autonegotiation ensures that 802.3 UTP solutions will continue to be autoconfiguring.

The *management information base* (MIB) for 1000BASE-T maintains consistency with the 802.3 MIB for 10/100-Mbps operation, allowing a consistent management model across all operating speeds.

Conformance with IEEE 802.2 is provided by the overlying 802.3z MAC sublayer. The standard conforms to the 802 FR document. The 1000BASE-T PHY conforms to the GMII specified in 802.3z. Two-kilometer network spans, while not supported specifically by 1000BASE-T, will be supported by combination with other members of the 802.3 family of 1,000-Mbps standards.

The standard is a 1,000-Mbps upgrade for 802.3 users based on the 802.3 CSMA/CD MAC. It is the only balanced copper solution for 1,000 Mbps that is capable of providing service over the defined link. As such, it offers a (nominally) easy upgrade for the millions of users who have installed Category 5 cable plants. It is substantially different from other 802.3 copper solutions in that it supports 1-Gbps operation over the defined link. The standard is a supplement to the existing IEEE 802.3.

In spite of the increased data rate, the signaling bandwidth required for 1000BASE-T is no greater than that required for 100BASE-TX. Experience with equipment using ADSL, HDSL, and QAM based signaling techniques in the field has demonstrated the reliability of signaling techniques such as those used in 1000BASE-T. Similar modulation systems are widely used in cable modems, hard disk drives, and so on.

The cost/performance ratio for 1000BASE-T vis-à-vis 100BASE-TX is about the same as that offered by 100BASE-TX vis- à -vis 10BASE-T at the time of initial introduction (1994). The cost of proposed implementations is expected to be two to four times that of 100BASE-TX. The cost model for horizontal copper cabling is well established. A variety of surveys conducted during the past 5 years have demonstrated that Category 5 cabling is the dominant cabling in place today. Table 11.8 depicts key features of the two different capabilities of GbE.

Carrier extension is a way of maintaining 802.3 minimum/maximum frame size with targeted cabling distances. Nondata symbols are included in a "collision window." The entire extended frame is considered for collision decisions and dropped as appropriate. Carrier extension is a straightforward solution, but it is bandwidth inefficient, since up to 488 padding octets may be sent for small packets. For a large number of small packets, throughput improvement is small when compared to 100-Mbps Ethernet.

For contention-mode applications, changes that can impact small-packet performance have been offset by incorporating a new feature, called *packet bursting,* into the CSMA/CD algorithm. Packet bursting, which we have already discussed, allows servers, switches, and other devices to send bursts of small packets in order to fully utilize available bandwidth. Devices that operate in full-duplex mode (switches and *buffered distributors*) are not subject to carrier extension, slot time extension, or packet bursting changes: Full-duplex devices will continue to use regular Ethernet 96-bit IPG and 64-byte minimum packet size. Packet bursting is an elaboration of carrier extension. When a station has a number of frames (packets) to send, the first packet is padded to slot time, if needed, using carrier extension. Packets that follow are sent back to back, with a minimum IPG until a timer (called a *burst timer*) of 1,500 octets expires. Gigabit Ethernet will perform at gigabit wire speeds: As packet size increases, GbE will exceed performance of fast Ethernet by an order of magnitude; as packet bursting is implemented, GbE will become more efficient at handling small packets.

TABLE 11.8   GbE Configurations and Objective

Item	Objective for Switched Configuratons	Objective for Shared Configuratons
Media	Multimode fiber Single-mode fiber Copper	Multimode fiber Copper
Transmission modes	Full duplex Half duplex	Half duplex
Applications	Campus backbone Buliding backbone Wiring closet uplink Servers	Desktop Servers
Performance	High throughput Long distance	Low cost Short distance

Repeaters can be used to connect segments of a network medium together into a single collision domain; different physical signaling systems (e.g., 1000BASE-CX, 1000BASE-SX, 1000BASE-LX,1000BASE-T) can be joined into a common collision domain using an appropriate repeater. Bridges can also be used to connect different portions of the network or different signaling systems; however, if a bridge is so used, each LAN connected to the bridge will comprise a separate collision domain (which is often desirable).

11.2.3 GbE Standard Details

The 802.3z standard is a large 300-page document, with the following sections:

- IEEE 802.3 Section 34: Introduction to 1000-Mbps Baseband network;

- IEEE 802.3 Section 35: Reconciliation Sublayer (RS) and Gigabit Media Independent Interface (GMII);

- IEEE 802.3 Section 36: Physical Coding Sublayer (PCS) and Physical Medium Attachment (PMA) Sublayer, Type 1000BASE-X;

- IEEE 802.3 Section 37: Auto-Negotiation Function, Type 1000BASE-X;

- IEEE 802.3 Section 38: Physical Medium Dependent (PMD) Sublayer and Baseband Medium, Type 1000BASE-LX (Long Wavelength Laser) and 1000BASE- X (Short Wavelength Laser);

- IEEE 802.3 Section 39: Physical Medium Dependent (PMD) Sublayer and Baseband Medium, Type 1000BASE-CX (Short-Haul Copper);

- IEEE 802.3 Section 40: Physical Coding Sublayer (PCS), Physical Medium Attachment (PMA) Sublayer and Baseband Medium, Type 1000BASE-T;

- IEEE 802.3 Section 41: Repeater for 1,000-Mbps Baseband Networks;

- IEEE 802.3 Section 42: System Considerations for Multisegment 1,000-Mbps Networks.

Gigabit Ethernet couples an extended version of the ISO/IEC 8802-3 (CSMA/CD MAC) to a family of 1,000-Mbps physical layers. The relationships among GbE, the extended ISO/IEC 8802-3 (CSMA/CD MAC), and the ISO/IEC OSI reference model are shown in Figure 11.14. Gigabit Ethernet uses the extended ISO/IEC 8802-3 MAC layer interface, connected through a GMII layer to physical layer entities (PHY sublayers) such as 1000BASE-LX, 1000BASE-SX, 1000BASE-CX, and 1000BASE-T.

FIGURE 11.14
GbE protocol model.

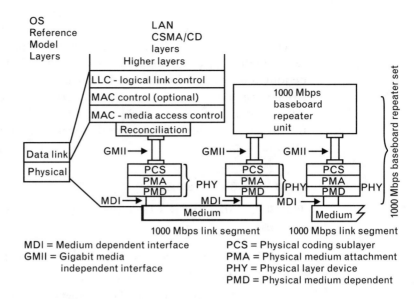

MDI = Medium dependent interface
GMII = Gigabit media
            independent interface

PCS = Physical coding sublayer
PMA = Physical medium attachment
PHY = Physical layer device
PMD = Physical medium dependent

In GbE the bit rate is faster, and the bit times are shorter—both in proportion to the change in bandwidth. In full-duplex mode, the minimum packet transmission time has been reduced by a factor of 10. Achievable topologies for 1,000–Mbps full-duplex operation are comparable to those found in 100BASE-T full-duplex mode. In half-duplex mode, the minimum packet transmission time has been reduced, but not by a factor of 10. Cable delay budgets are similar to those in 100BASE-T. The resulting achievable topologies for the half-duplex 1,000–Mbps CSMA/CD MAC are similar to those found in half-duplex 100BASE-T.

GMII provides an interconnection between the MAC sublayer and PHY entities and between PHY and STA entities. This GMII supports 1,000–Mbps operation through its 8-bit-wide (octet-wide) transmit and receive paths. The *reconciliation sublayer* (RS) provides a mapping between the signals provided at the GMII and the MAC/PLS service definition.

The specification defines the logical and electrical characteristics for the RS and the GMII between CSMA/CD media access controllers and various PHYs. While the AUI was defined to exist between the PLS and PMA sublayers for 10-Mbps DTEs, the GMII maximizes media independence by cleanly separating the data link and physical layers of the ISO/IEC seven-layer reference model. This allocation also recognizes that implementations can benefit from a close coupling between the PLS or PCS sublayer and the PMA sublayer. This interface has the following characteristics:

• It is capable of supporting 1,000–Mbps operation.

• Data and delimiters are synchronous to clock references.

• It provides independent 8-bit-wide transmit and receive data paths.

• It provides a simple management interface.

- It uses signal levels that are compatible with common CMOS digital ASIC processes and some bipolar processes.

- It provides for full–duplex operation.

Each direction of data transfer is serviced by data (an 8-bit bundle), delimiter, error, and clock signals. Two media status signals are provided: One indicates the presence of a carrier, and the other indicates the occurrence of a collision. The RS maps the signal set provided at the GMII to the PLS service primitives provided to the MAC.

This GbE standard specifies a family of physical layer implementations. The generic term 1,000-Mbps MAC refers to any use of the 1,000-Mbps ISO/IEC 8802-3 CSMA/CD MAC (the Gigabit Ethernet MAC) coupled with any physical layer implementation. The term *1000BASE-X* refers to a specific family of physical layer implementations. The 1000BASE-X family of physical layer standards has been adapted from the ANSI X3.230-1994 (Fibre Channel) FC-0 and FC-1 physical layer specifications and the associated 8B/10B data coding method. The 1000BASE-X family of physical layer implementations is composed of 1000BASE-SX, 1000BASE-LX, and 1000BASE-CX. All 1000BASE-X PHY devices share the use of common PCS, PMA, and autonegotiation specifications. The 1000BASE-T PHY uses four pairs of Category 5 balanced copper cabling. The standard also defines its own PCS, which does not use 8B/10B coding (see Table 11.9). From an enterprise network point of view 100BASE-T is very important, as is 1000BASE-LX. For metro access/metro core applications, 1000BASE-LX is important because people have extended the optics reach to cover a (small) metro area.

Autonegotiation provides a 1000BASE-X device with the capability to detect the abilities (modes of operation) supported by the device at the other end of a link segment, determine common abilities, and configure for joint operation. Autonegotiation is performed upon link startup through the use of a special sequence of reserved link code words. Autonegotiation is used by 1000BASE-T devices to detect the abilities (modes of operation) supported by the device at the other end of a link segment, determine

TABLE 11.9   PHYSICAL LAYERS OF GbE

1000BASE-SX short wavelength optical	Duplex multimode fibers
1000BASE-LX long wavelength optical	Duplex single-mode fibers or Duplex multimode fibers
1000BASE-CX shielded jumper cable	Two pairs of specialized balance cabling
1000BASE-T catagory 5 UTP	Advanced multilevel signaling over four pairs of Category 3 balanced copper cabling

common abilities, and configure for joint operation. Autonegotiation is performed upon link startup through the use of a special sequence of fast link pulses.

Next we describe the PCS and PMA sublayer, type 1000BASE-X.

11.2.3.1 Physical Sublayers

A section of the standard specifies the PCS and the PMA sublayers that are common to a family of 1,000-Mbps physical layer implementations, collectively known as 1000BASE-X. As noted, there are currently four PHYs: 1000BASE-CX, 1000BASE-LX, 1000BASE-SX, and 1000BASE-T. The 1000BASE-CX embodiment specifies operation over a single copper media: two pairs of 150-Ω balanced copper cabling. 1000BASE-LX specifies operation over a pair of optical fibers using long-wavelength optical transmission. 1000BASE-SX specifies operation over a pair of optical fibers using short-wavelength optical transmission. 1000BASE-T specifies operation over four pairs of Category 5 balanced copper cabling. The term *1000BASE-X* is used when referring to issues common to any of three subvariants—1000BASE-CX, 1000BASE-LX, 1000BASE-SX—but *not* to refer to 1000BASE-T.

1000BASE-X is based on the physical layer standards developed by ANSI X3.230-1994 (*Fibre Channel Physical and Signaling Interface*). In particular, the standard uses the same 8B/10B coding as Fibre Channel (FC), a PMA sublayer compatible with speed-enhanced versions of the ANSI 10-bit serializer chip, and similar optical and electrical specifications. 1000BASE-X PCS and PMA sublayers map the interface characteristics of the PMD sublayer (including MDI) to the services expected by the reconciliation sublayer. 1000BASE-X can be extended to support any other full-duplex medium requiring only that the medium be compliant at the PMD level. The following are the objectives of 1000BASE-X:

- To support the CSMA/CD MAC;
- To support the 1,000-Mbps repeater;
- To provide for autonegotiation among like 1,000-Mbps PMDs;
- To provide a 1,000-Mbps data rate at the GMII;
- To support cable plants using 150-Ω balanced copper cabling, or optical fiber compliant with ISO/IEC 11801:1995;
- To allow for a nominal network extent of up to 3 km, including (1) 150-Ω balanced links of 25-m span, (2) one-repeater networks of 50-m span (using all 150-Ω balanced copper cabling), (3) one-repeater networks of 200-m span (using fiber), and (4) DTE/DTE links of 3,000m (using fiber);
- To preserve full-duplex behavior of underlying PMD channels;
- To support a BER objective of 10^{-12}.

11.2.3.2 PCS

The PCS interface is the GMII that provides a uniform interface to the reconciliation sublayer for all 1,000-Mbps PHY implementations (e.g., not only 1000BASE-X but also other possible types of gigabit PHY entities). 1000BASE-X provides services to the GMII in a manner analogous to how 100BASE-X provides services to the 100-Mbps MII. The 1000BASE-X PCS provides all services required by the GMII:

- Encoding (decoding) of GMII data octets to (from) 10-bit code-groups (8B/10B) for communication with the underlying PMA;
- Generating carrier sense and collision detect indications for use by PHY's half-duplex clients;
- Managing the autonegotiation process, and informing the management entity via the GMII when the PHY is ready for use.

11.2.3.3 PMA Sublayer

The PMA provides a medium–independent means for the PCS to support the use of a range of serial-bit-oriented physical media. The 1000BASE-X PMA performs the following functions:

- Mapping of transmit and receive code groups between the PCS and PMA via the PMA service interface;
- Serialization (deserialization) of code groups for transmission (reception) on the underlying serial PMD;
- Recovery of clock from the 8B/10B-coded data supplied by the PMD;
- Mapping of transmit and receive bits between the PMA and PMD via the PMD service interface;
- Data loopback at the PMD service interface.

11.2.3.4 PMD Sublayer

1000BASE-X physical layer signaling for fiber and copper media is adapted from ANSI X3.230-1994 (FC-PH), which defines 1,062.5-Mbps, full-duplex signaling systems that accommodate single-mode optical fiber, multimode optical fiber, and 150-Ω balanced copper cabling. 1000BASE-X adapts these basic physical layer specifications for use with the PMD sublayer and mediums. The MDI, logically subsumed within each PMD, is the actual medium attachment, including connectors, for the various supported media.

11.2.3.5 Intersublayer Interfaces

A number of interfaces are employed by 1000BASE-X. Some (such as the PMA service interface) use an abstract service model to define the operation of the interface. An optional physical instantiation of the PCS interface has been defined. It is called the GMII. Another optional physical instantiation of the PMA service interface has also been defined, being adapted from ANSI Technical Report TR/X3.18-1997 (Fibre Channel 10-bit Interface).

11.2.3.6   PCS

PCS Interface (GMII)

The PCS service interface allows the 1000BASE-X PCS to transfer information to and from a PCS client. PCS clients include the MAC (via the reconciliation sublayer) and repeater.

Functions Within the PCS

The PCS comprises the PCS transmit, carrier sense, synchronization, PCS receive, and autonegotiation processes for 1000BASE-X. The PCS shields the reconciliation sublayer (and MAC) from the specific nature of the underlying channel. When communicating with the GMII, the PCS uses an octet-wide, synchronous data path, with packet delimiting being provided by separate transmit control signals (TX_EN and TX_ER) and receive control signals (RX_DV and RX_ER). When communicating with the PMA, the PCS uses a 10-bit-wide, synchronous data path, which conveys 10-bit code groups. At the PMA service interface, code group alignment and MAC packet delimiting are made possible by embedding special nondata code groups in the transmitted code group stream. The PCS provides the functions necessary to map packets between the GMII format and the PMA service interface format.

Use of Code Groups

The PCS maps GMII signals into 10-bit code groups, and vice versa, using an 8B/10B lock coding scheme. Implicit in the definition of a code group is an establishment of code group boundaries by a PMA code group alignment function.

8B/10B Transmission Code

The PCS uses a transmission code to improve the transmission characteristics of information to be transferred across the link. The encodings defined by the transmission code ensure that sufficient transitions are present in the PHY bit stream to make clock recovery possible at the receiver. Such encoding also greatly increases the likelihood of detecting any single or multiple bit errors that may occur during transmission and reception of information. Also, some of the special code groups of the transmission code contain a distinct and easily recognizable bit pattern that assists a receiver in achieving code group alignment on the incoming PHY bit stream.

In 8B/10B encoding, each octet is given a code group name according to the bit arrangement. Each octet is broken down into two groups; the first group contains the three most significant bits (y) and the second group contains the remaining five bits (x). Each data code group is named /Dx.y/, where the value of x represents the decimal value of the five least significant bits and y represents the value of the three most significant bits [2]. For example:

```
/D0.0/  = 000 00000
/D6.2/ = 010 00110
/D30.6/= 110 11101
```

There are also 12 special octets that may be encoded into 10 bits. The PCS differentiates between special and data code words via a signal passed to it from the GMII. Special code words follow the same naming convention as data code words, except they are named /Kx.y/ rather than /Dx.y/.

One of the motivations behind the use of 8B/10B encoding lies with the ability to control the characteristics of the code words such as the number of ones and zeros and the consecutive number of ones or zeros. Another motivation behind the use of 8B/10B encoding is the ability to use special code words, which would be impossible if no encoding was performed. Features of the code are as follows [2]:

- Every 10-bit code group must fit into one of the following three possibilities:

  1. Five ones and five zeros;

  2. Four ones and six zeros;

  3. Six ones and four zeros.

  This helps limit the number of consecutive ones and zeros between any two code groups.

- A special sequence of seven bits, called a comma, is used by the PMA in aligning the incoming serial stream. The comma is also used by the PCS in acquiring and maintaining synchronization. The following bit patterns represent the comma:

```
0011111 (comma+)
1100000 (comma-)
```

  The comma is contained within only the /K28.1/, /K28.5/ and /K28.7/ special code groups. The comma cannot be transmitted across the boundaries of any two adjacent code groups unless an error has occurred [3]. This characteristic makes it very useful to the PMA in determining code group boundaries.

- dc balancing is achieved through the use of a running disparity calculation. Running disparity is designed to keep the number of ones transmitted by a station equal to the number of zeros transmitted by that station. This should keep the dc level balanced halfway between the "one" voltage level and the "zero" voltage level. Running disparity can take on one of two values: positive or negative. In the absence of errors, the running disparity value is positive if more ones have been transmitted than zeros and the running disparity value is negative if more zeros have been transmitted than ones since power-on or reset.

- The 8B/10B encoding scheme was designed to provide a high transition density which makes synchronization of the incoming bit stream easier.

11.2.4   PMD Sublayer and Baseband Medium, Type 1000BASE-LX (Long-Wavelength Laser) and 1000BASE-SX (Short-Wavelength Laser)

Because of the fact that the GbE metro rings now being deployed uses 1000BASE-LX principles, we look at this PMD in some detail. This section of the standard specifies the 1000BASE-SX PMD and the 1000BASE-LX PMD (including MDI) and baseband medium for multimode and single-mode fiber.

11.2.4.1   PMD Sublayer Service Interface

The following specifies the services provided by 1000BASE-SX and 1000BASE-LX PMD. The PMD service interface supports the exchange of encoded 8B/10B characters between PMA entities. The PMD translates the encoded 8B/10B characters to and from signals suitable for the specified medium. The following primitives are defined:

- PMD_UNITDATA.request;
- PMD_UNITDATA.indicate;
- PMD_SIGNAL.indicate.

The PMD transmit function conveys the bits requested by the PMD service interface message PMD_UNITDATA.request(tx_bit) to the MDI. The higher optical power level corresponds to tx_bit = ONE. The PMD receive function conveys the bits received from the MDI to the PMD service interface using the message PMD_UNITDATA.indicate(rx_bit). The higher optical power level corresponds to rx_ bit = ONE. The PMD signal detect function reports to the PMD service interface, using the message PMD_SIGNAL.indicate(SIGNAL_DETECT), which is signaled continuously. PMD_SIGNAL.indicate is intended to be an indicator of optical signal presence. The value of the SIGNAL_DETECT parameter are generated according to the conditions defined in Table 11.10.

11.2.4.2   PMD to MDI Optical Specifications for 1000BASE-SX

The operating range for 1000BASE-SX is defined in Table 11.11. A 1000BASE-SX-compliant transceiver supports both multimode fiber

TABLE 11.10   SIGNAL_DETECT VALUES

RECEIVE CONDITIONS	SIGNAL DETECT VALUE
Input_optical_power ≤ −30 dBm	FAIL
Input_optical_power ≥ receive sensibility AND compliant 1000BASE-X signal input	OK
All other conditions	Unspecified

TABLE 11.11   1000BASE-SX

FIBER TYPE	MODAL BANDWIDTH @ 850 NM (MIN. OVERFILLED LAUNCH) (MHz · KM)	MINIMUM RANGE (M)
62.5-μm MMF	160	2–200
62.5-μm MMF	200	2–275
50-μm MMF	400	2–500
50-μm MMF	500	2–550
10-μm MMF	N/A	Not supported

media types listed in Table 11.11 (i.e., both 50- and 62.5-μm multimode fiber). A transceiver that exceeds the operational range requirement while meeting all other optical specifications is considered compliant (e.g., a 50-μm solution operating at 600m meets the minimum range requirement of 2–550m).

The 1000BASE-SX transmitter needs to meet the specifications defined in Table 11.12. The 1000BASE-SX receiver needs to meet the specifications defined in Table 11.13.

The worst-case power budget and link penalties for a 1000BASE-SX channel are shown in Table 11.14.

11.2.4.3 PMD to MDI Optical Specifications for 1000BASE-LX

The operating range for 1000BASE-LX is defined in Table 11.15. A transceiver that exceeds the operational range requirement while meeting all other optical specifications is considered compliant (e.g., a single-mode solution operating at 5500m meets the minimum range requirement of 2–5,000m). The 1000BASE-LX transmitter is required to meet the specifications defined in Table 11.16. *Conditioned launch* (CL) produces sufficient mode volume so that individual *multimode fiber* (MMF) modes do not dominate fiber performance. This reduces the effect of peak-to-peak *differential*

TABLE 11.12   1000BASE-SX TRANSMIT CHARACTERISTICS

DESCRIPTION	62.5-μM MMF 50-μM MMF	UNIT
Transmitter type	Shortwave laser	
Signaling speed (range)	1.25 $\pm$ 100 ppm	GBd
Wavelength (λ range)	770–860	nm
T_{rise}/T_{fall} (max: 20%–80% λ > 830 nm)	0.26	ms
T_{rise}/T_{fall} (max: 20%–80% λ $\leq$ 830 nm)	0.21	ms
RMS spectral width (max)	0.85	nm
Average launch power (max)	See standard	dBm
Average launch power (min)	−9.5	dBm
Average launch power txf OFF transmitter (max)	−30	dBm
Extinction ratio (min)	9	dB
RIN (max)	−117	dB/Hz
Coupled power ratio (min)	9, CPR	dB

TABLE 11.13   1000BASE-SX RECEIVE CHARACTERISTICS

DESCRIPTION	62.5-μM MMF	50-μM MMF	UNIT
Signaling speed (range)	1.25 $\pm$ 100 ppm	1.25 $\pm$ 100 ppm	GBd
Wavelength (range)	700–860	700–860	nm
Average receive power (max)	0	0	dBm
Receive sensitivity	−17	−17	dBm
Return loss (min)	12	12	dB
Stressed receive sensitivity	−12.5	−13.5	dBm
Vertical eye closure penalty	2.60	2.20	dB
Receive electrical 3–dB upper cutoff frequency (max)	1,500	1,500	MHz

mode delay (DMD) between the launched mode groups and diminishes the resulting pulse-splitting-induced nulls in the frequency response. A CL is produced by using a single-mode fiber offset-launch mode-conditioning patch cord, inserted at both ends of a full-duplex link, between the optical PMD MDI and the remainder of the link segment. The single-mode fiber offset-launch mode-conditioning patch cord contains a fiber of the same type as the cable (i.e., 62.5- or 50-μm fiber) connected to the optical PMD receiver input MDI and a specialized fiber/connector assembly connected to the optical PMD transmitter output.

TABLE 11.14   WORST-CASE POWER BUDGET

PARAMETER	62.5-μM MMF		50-μM MMF		UNIT
Modal bandwidth as measued at 850 nm (minimum overfilled launch)	160	200	400	500	MHz · km
Link power budget	7.5	7.5	7.5	7.5	dB
Operting distance	220	275	500	550	m
Channel insertion loss	2.38	2.60	3.37	3.56	dB
Link power penalties	4.27	4.29	4.07	3.57	dB
Unallocated margin in link power budget	0.84	0.60	0.05	0.37	dB

TABLE 11.15   1000BASE-LX OPERATING RANGE

FIBER TYPE	MODAL BANDWIDTH @ 1,300 NM (MIN. OVERFILLED LAUNCH) (MHz · KM)	MINIMUM RANGE (M)
62.5-μm MMF	500	2–550
50-μm MMF	400	2–550
50-μm MMF	500	2–550
10-μm MMF	N/A	2–5,000

TABLE 11.16   1000BASE-LX TRANSMIT CHARACTERISTICS

DESCRIPTION	62.5-μM MMF	50-μM MMF	10-μM MMF	UNIT
Transmitter type	Longwave laser	Longwave laser	Longwave laser	
Signaling speed (range)	1.25 ± 100 ppm	1.25 ± 100 ppm	1.25 ± 100 ppm	GBd
Wavelength (range)	1,270–1,355	1,270–1,355	1,270–1,355	nm
T_{rise}/T_{fall} (max: 20%–80% λ > 830 nm)	0.26	0.26	0.26	ns
RMS spectral width (max)	4	4	4	nm
Average launch power (max)	−3	−3	−3	dBm
Average launch power (min)	−11.5	−11.5	−11.0	dBm
Average launch power of OFF transmitter (max)	−30	−30	−30	dBm
Extinction ratio (min)	9	9	9	dB
RIN (max)	−120	−120	−120	dB/Hz
Coupled power ratio (CPR)	28 < CPR < 40	12 < CPR < 20	N/A	dB

The 1000BASE-LX receiver needs to meet the specifications defined in Table 11.17. The worst-case power budget and link penalties for a 1000BASE-LX channel are shown in Table 11.18. (Both the 1000BASE-SX and 1000BASE-LX fiber optic cabling need to meet the specifications defined in Table 11.18.)

The fiber optic cabling consists of one or more sections of fiber optic cable and any required connections along the way. It also includes a connector plug at each end to connect to the MDI. The fiber optic cable requirements are satisfied by the fibers specified in IEC 60793-2:1992. Types A1a (50/125-μm multimode), A1 (62.5/125-μm multimode), and B1 (10/125-μm single-mode) fibers are supported with the exceptions noted in Table 11.19.

TABLE 11.17   1000BASE-LX RECEIVE CHARACTERISTICS

DESCRIPTION	VALUE	UNIT
Signaling speed (range)	1.25 ± 100 ppm	GBd
Wavelength (range)	1,270–1,355	nm
Average receive power (max)	−3	dBm
Receive sensitivity	−19	dBm
Return loss (min)	12	dB
Stressed receive sensitivity	−14.4	dBm
Vertical eye closure penalty	2.60	dB
Receive electrical 3-dB upper cutoff frequency (max)	1,500	MHz

TABLE 11.18   WORST-CASE 1000BASE-LX POWER BUDGET

PARAMETER	62.5-μM MMF	50-μM MMF	10-μM MMF	UNIT
Modal bandwidth as measured at 1,300 nm (minimum overfilled launch)	500	400	500	N/A	MHz · km
Link power budget	7.5	7.5	7.5	8.0	dB
Operting distance	550	550	550	5,000	m
Channel insertion loss	2.35	2.35	2.35	4.57	dB
Link power penalties	3.48	5.08	3.96	3.27	dB
Unallocated margin in link power budget	1.67	0.07	1.19	0.16	dB

TABLE 11.19   OPTICAL CABLE AND FIBER CHARACTERISTICS

DESCRIPTION	62.5-μM MMF		50-μM MMF		10-μM MMF	UNIT
Nominal Fiber Specification Wavelength	850	1,300	850	1,300	1,300	nm
Fiber cable attenuation	3.75	1.5	3.5	1.5	0.5	dB/km
Modal bandwidth (min. overfilled launch)	160	500	400	400	N/A	MHz · km
	200	500	500	500	N/A	MHz · km
Zero dispersion wavelength (λ_0)	$1,320 \leq \lambda_0 \leq 1,365$		$1,295 \leq \lambda_0 \leq 1,320$		$1,300 \leq \lambda_0 \leq 1,324$	nm
Dispersion slope (max) (S_0)	0.11 for $1,320 \leq \lambda_0 \leq 1,348$ and 0.001 $(1,458 - \lambda_0)$ for $1,320 \leq \lambda_0 \leq 1,365$		0.11 for $1,300 \leq \lambda_0 \leq 1,320$ and 0.001 $(\lambda_0 - 1,190)$ for $1,295 \leq \lambda_0 \leq 1,300$		0.093	ps/nm^2 · km

11.2.4.4   MDI

The 1000BASE-SX and 1000BASE-LX PMD is coupled to the fiber optic cabling through a connector plug into the MDI optical receptacle. The 1000BASE-SX and 1000BASE-LX MDI optical receptacles are the duplex SC, meeting the following requirements:

- Meet the dimension and interface specifications of IEC 61754-4.
- Meet the performance specifications as specified in ISO/IEC 11801.
- Ensure that polarity is maintained.
- The receive side of the receptacle is located on the left when viewed looking into the transceiver optical ports with the keys on the bottom surface.

11.2.5   PMD Sublayer and Baseband Medium, Type 1000BASE-CX (Short-Haul Copper)

GbE specifies the 1000BASE-CX PMD (including MDI) and baseband medium for short-haul copper. 1000BASE-CX has a minimum operating range of 0.1 to 25m. Jumper cables are used to interconnect 1000BASE-CX PMDs. These cables, however, cannot be concatenated to achieve longer distances. A 1000BASE-CX jumper cable assembly consists of a continuous shielded balanced copper cable terminated at each end with a polarized shielded plug. The 1000BASE-CX PMD performs three functions: transmit, receive, and signal status. Typical applications include interconnection of equipment within data center racks, although MMF solutions are also often used and/or preferred.

11.2.6   PCS, PMA Sublayer, and Baseband Medium, Type 1000BASE-T

The 1000BASE-T PHY is one of the GbE family of high-speed CSMA/CD network specifications. The 1000BASE-T PCS, PMA, and baseband medium specifications are intended for users who want 1,000-Mbps performance over Category 5 balanced twisted-pair cabling systems. 1000BASE-T signaling requires four pairs of Category 5 balanced cabling, as specified in ISO/IEC 11801:1995 and ANSI/EIA/TIA-568-A (1995).

The IEEE 802.3ab clause defines the type 1000BASE-T PCS, type 1000BASE-T PMA sublayer, and type 1000BASE-T MDI. Together, the PCS and the PMA sublayer comprise a 1000BASE-T PHY. Section 40 of the IEEE 802.3 document describes the functional, electrical, and mechanical specifications for the type 1000BASE-T PCS, PMA, and MDI. This clause also specifies the baseband medium used with 1000BASE-T. The following are the objectives of 1000BASE-T:

- Support the CSMA/CD MAC.
- Comply with the specifications for the GMII.
- Support the 1,000-Mbps repeater.
- Provide line transmission that supports full and half-duplex operation.
- Meet or exceed FCC Class A/CISPR or better operation.
- Support operation over 100m of Category 5 balanced cabling.
- Provide a BER of less than or equal to 10^{-10}.
- Support autonegotiation.

The 1000BASE-T PHY employs full-duplex base and transmission over four pairs of Category 5 balanced cabling. The aggregate data rate of 1,000 Mbps is achieved by transmission at a data rate of 250 Mbps over each wire pair. The use of hybrids and cancelers enables full-duplex transmission by allowing symbols to be transmitted and received on the same wire pairs at the same time. Base and signaling with a modulation rate of 125 Mbaud is used on each of the wire pairs. The transmitted symbols are selected from a four-dimensional five-level symbol constellation. Each four-dimensional symbol can be viewed as a 4-tuple (A_n, B_n, C_n, D_n) of one-dimensional quinary symbols taken from the set $\{2, 1, 0, -1, -2\}$. 1000BASE-T uses a continuous signaling system; in the absence of data, idle symbols are transmitted. Idle mode is a subset of code groups in that each symbol is restricted to the set $\{2, 0, -2\}$ to improve synchronization. Five-level PAM5 is employed for transmission over each wire pair. The modulation rate of 125 MBaud matches the GMII clock rate of 125 MHz and results in a symbol period of 8 ns.

A 1000BASE-T PHY can be configured either as a MASTER PHY or as a SLAVE PHY. The MASTER–SLAVE relationship between two stations sharing a link segment is established during autonegotiation. The

MASTER PHY uses a local clock to determine the timing of transmitter operations. The SLAVE PHY recovers the clock from the received signal and uses it to determine the timing of transmitter operations; that is, it performs loop timing. In a multiport to single-port connection, the multiport device is typically set to be MASTER and the single-port device is set to be SLAVE.

The 1000BASE-T PCS couples a GMII to a PMA sublayer. The functions performed by the PCS comprise the generation of continuous code groups to be transmitted over four channels and the processing of code groups received from the remote PHY. The process of converting data bits to code groups is called *4D-PAM5,* which refers to the four-dimensional five-level PAM coding technique used. Through this coding scheme, 8 bits are converted to one transmission of four quinary symbols.

During the beginning of a frame's transmission, when TX_EN is asserted from the GMII, two code groups representing the start-of-stream delimiter are transmitted followed by code groups representing the octets coming from the GMII. Immediately following the data octets, the GMII sets TX_EN = FALSE, upon which the end of a frame is transmitted. The end of a frame consists of two convolutional state reset symbol periods and two end-of-stream delimiter symbol periods. This is followed by an optional series of carrier extend symbol periods and, possibly, the start of a new frame during frame bursting. Otherwise, the end of a frame is followed by a series of symbols encoded in the idle mode.

Between frames, a special subset of code groups using only the symbols {2, 0, −2} is transmitted. This is called *idle mode.* Idle mode encoding takes into account the information of whether the local PHY is operating reliably or not and allows this information to be conveyed to the remote station. During normal operation, idle mode is followed by a data mode that begins with a start-of-stream delimiter. Further patterns are used for signaling a transmit error and other control functions during transmission of a data stream.

The PCS receive function processes code groups provided by the PMA. The PCS receive detects the beginning and the end of frames of data and, during the reception of data, descrambles and decodes the received code groups into octets that are passed to the GMII. The conversion of code groups to octets uses an 8B1Q4 data decoding technique. PCS receive function also detects errors in the received sequences and signals them to the GMII. Furthermore, the PCS contains a PCS carrier sense function, a PCS collision presence function, and a management interface.

The PMA couples messages from the PMA service interface onto the balanced cabling physical medium and provides the link management and PHY control functions. The PMA provides full-duplex communications at 125 MBaud over four pairs of balanced cabling up to 100m in length.

The PMA transmit function comprises four independent transmitters to generate five-level PAM signals on each of the four pairs BI_DA, BI_DB, BI_DC, and BI_DD.

The PMA receive function comprises four independent receivers for five-level PAM signals on each of the four pairs BI_DA, BI_DB, BI_DC, and BI_DD. This signal encoding technique is referred to as 4D-PAM5. The receivers are responsible for acquiring clock and providing code groups to the PCS as defined by the PMA_UNITDATA.indicate message. The PMA also contains functions for link monitoring.

The PMA PHY control function generates signals that control the PCS and PMA sublayer operations. PHY control begins following the completion of autonegotiation and provides the start-up functions required for successful 1000BASE-T operation. It determines whether the PHY operates in a normal state, enabling data transmission over the link segment, or whether the PHY sends special code groups that represent the idle mode. The latter occurs when either one or both of the PHYs that share a link segment are not operating reliably.

11.2.6.1 Signaling

1000BASE-T signaling is performed by the PCS generating continuous code group sequences that the PMA transmits over each wire pair. The signaling scheme achieves a number of objectives including the following:

- FEC coded symbol mapping for data;

- Algorithmic mapping and inverse mapping from octet data to a quartet of quinary symbols and back;

- Uncorrelated symbols in the transmitted symbol stream;

- No correlation between symbol streams traveling both directions on any pair combination;

- No correlation between symbol streams on pairs BI_DA, BI_DB, BI_DC, and BI_DD;

- Idle mode uses a subset of code groups in that each symbol is restricted to the set $\{2, 0, -2\}$ to ease synchronization, start-up, and retraining;

- Ability to rapidly or immediately determine if a symbol stream represents data or idle or carrier extension;

- Robust delimiters for *start-of-stream delimiter* (SSD), *end-of-stream delimiter* (ESD), and other control signals;

- Ability to signal the status of the local receiver to the remote PHY to indicate that the local receiver is not operating reliably and requires retraining;

- Ability to automatically detect and correct for pair swapping and un-expected crossover connections;

- Ability to automatically detect and correct for incorrect polarity in the connections;

- Ability to automatically correct for differential delay variations across the wire pairs.

The PHY operates in two basic modes, normal mode or training mode. In normal mode, PCS generates code groups that represent data, control, or idles for transmission by the PMA. In training mode, the PCS is directed to generate only idle code groups for transmission by the PMA, which enables the receiver at the other end to train until it is ready to operate in normal mode.

11.2.7 Network Diameter

Table 11.20 depicts the maximum segment length for various media.

11.3 10GbE

This section [4] looks at 10–Gigabit Ethernet. Efforts were basically complete at press time to extend the existing IEEE 802.3 standards to support 10-Gbps data rates and to enhance Ethernet to include more direct support for WAN links. Standards for 10GbE were produced as an extension to the existing IEEE standards, with the basic changes being at the PHY (consisting of the PCS, PMA, and PMD sublayers). Specifically, the goal of the P802.3ae standard is to define 802.3 MAC parameters and minimal augmentation of its operating characteristics, physical layer characteristics, and management parameters for transfer of LLC and Ethernet format frames at

TABLE 11.20   MAXIMUM SEGMENT LENGTH FOR VARIOUS MEDIA

MEDIA TYPE	MAXIMUM SEGMENT LENGTH (M)	MAXIMUM MEDIUM ROUND-TRIP DELAY PER SEGMENT (BT) BT = BIT TIMES
Category 5 UTP link segment (1000BASE–T)	100	1,112
Sheilded jumper cable link segment (1000BASE-CX)	25	253
Optical fiber link segment (1000BASE-SX, 1000BASE-LX	316★	3,192

★May be limited by the maximum transmission distance of the link.

10 Gbps using full-duplex operation. 10GbE has two implications for the metro access/metro core networks:

1. As end users deploy 10GbE LAN systems, the optimized WAN connectivity is in the 1–Gbps range. It follows that 10GbE-based end systems would preferably require a *transparent LAN service* (TLS) WAN service perhaps based on this very same technology (or, as an alternative, TLS could also be based on SONET/new generation SONET). Viewed from this angle, 10GbE is a driver for broadband metro core/metro access systems.

2. 10GbE affords the use of a MAC framing mechanism in the WAN, obviating the need to do either translational bridging at layer 2 (e.g., remapping a local MAC frame to an AAL-5 frame) or remapping at layer 3 (e.g., remapping a MAC/IP frame to a PPP/SONET frame, e.g., as shown in Figure 11.15). Typical traditional layer 2 protocols used in the WAN include LAPD for frame relay, PPP and the newer GFP (see Chapter 7), and AAL-5.

We believe that the second application could have *significant* relevance to new generation networks and architectures. The WAN PHY is *compatible* with SONET/SDH, but cannot directly interconnect with a SONET interface unless a gateway interworking function is used either at the CPE level or at the network edge level, as we discuss later. This section also

FIGURE 11.15
Ethernet over
SONET LTE.

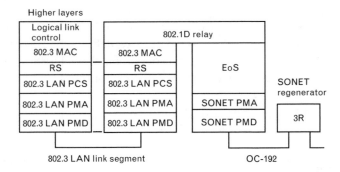

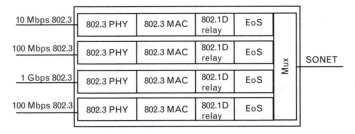

looks at the *10-Gbps Fibre Channel standard* (10GFC). While this technology is more focused on data-center SANs, it represents a potential user of 10GbE-WAN TLS or other forms of TLS in the metro access/metro core.

The 802.3ae standard was scheduled for completion at the time of this writing. Figure 11.16 depicts the timeline of the standardization work. The standard was at Draft 3.2 at press time, and it is about 500 pages long. This section is based on the draft version (3.2) of the standard. Very likely, the functionality described in this chapter will be part of the standard; however, some aspects could still change. The material is included for pedagogical purposes and to stimulate interest in the topic. Readers and developers should refer to the IEEE/standard for the "final word" on the issue.

11.3.1 Overview

A desire exists on the part of developers to bring to the market a "higher speed" (specifically 10-Gbps) LAN technology that can also be used over the WAN. Prior experience in scaling IEEE 802.3 across the range of 1 to 1,000 Mbps, as covered in the earlier part of the chapter, indicates that the cost balance between adapters, switches, and the infrastructure remains roughly constant. 10GbE is expected to continue this trend at least for building/campus applications. In addition to the traditional LAN space, the standard adds parameters and mechanisms that enable deployment of Ethernet over the WAN operating at a data rate that is compatible with OC-192c and SDH VC-4-64c payload rate. The new features expand the Ethernet application space to include WAN links, in order to provide a significant increase in bandwidth while maintaining maximum compatibility with the installed base of IEEE 802.3 interfaces, previous investment in research and development, and principles of network operation and management. The approach has been to define two families of PHYs: (1) a LAN PHY, operating at a data rate of 10.000 Gbps; and (2) a WAN PHY, operating at a data rate compatible with the payload rate of OC-192c/SDH VC-4-64c. Note, however, that the WAN PHY *does not render the PHY compliant with either SONET or SDH* at any rate or format and the WAN interface *is not intended to interoperate directly with interfaces that comply with SONET or SDH standards.* 10GbE is defined for full-duplex mode of operation only. Figure 11.17 depicts a simplified protocol model. 10GbE will improve the performance of LAN backbone and server and gateway

FIGURE 11.16
10GbE timeline for standard development.

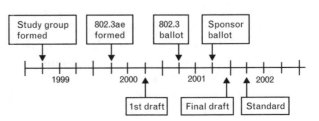

FIGURE 11.17
*Sublayer protocol
model.*

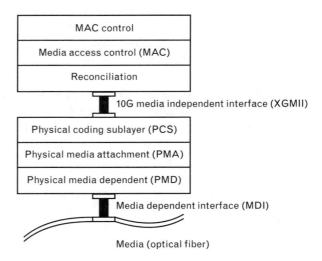

connectivity, switch aggregation, and MAN, WAN, *regional area network* (RAN), and SANs.

11.3.1.1   Goals

The principle of scaling the 802.3 MAC to higher speeds has been well established by previous work within 802.3 committees. The 10-Gbps work built on this experience. The design objectives of the *Higher Speed Study Group* (HSSG) were as follows [5]:

- Preserve the 802.3/Ethernet frame format at the MAC client service interface.
- Meet 802 FR, with the possible exception of Hamming distance.
- Preserve minimum and maximum FrameSize of current IEEE 802.3 standard.
- Support full-duplex operation (only).
- Support star-wired LANs using point-to-point links and structured cabling topologies.
- Specify an optional MII.
- Support proposed standard P802.3ad (on link aggregation).
- Support a speed of 10.000 Gbps at the MAC/PLS service interface.
- Define two families of PHYs (also see Figure 11.18): (1) a LAN PHY operating at a data rate of 10.000 Gbps, and (2) a WAN PHY, operating at a data rate compatible with the payload rate of OC–192c/ SDH VC-4-64c. (These streams clocked at 9.953 Gbps; payload is approximately 9.29 Gbps.) (PCSs known as 10GBASE-X, 10GBASE-R and 10GBASE-W [6].)

FIGURE 11.18
Key parameters for various PHYs.

Stack	10 GbE LAN PHY		10 GE WAN PHY
	Serial	CWDM	Serial
PCS	64B/66B	8B/10B	64B/66B SONET framing
PMA interface	XSBI	XAUI	XSBI
PMD	1550 nm DFB 1310 FP 850 nm VCSEL	1310 CWDM	1550 nm DFB 1310 FP 850 nm VCSEL
Line rate	10.3 Gbps	4x3.125 Gbps	9.953 Gbps

- Define a mechanism to adapt the MAC/PLS data rate to the data rate of the WAN PHY.

- For the LAN PHYs, develop physical layer specifications that support link distances of (1) at least 65m over MMF (the solution is t optimized for this distance), (2) at least 300m over installed MMF ("installed" implies all MMF specified in 802.3z; 62.5-m 160/500 MHz · km FDDI grade is the worst case), (3) at least 2 km over SMF, (4) at least 10 km over SMF, and (5) at least 40 km over SMF.

- Support fiber media selected from the second edition of ISO/IEC 11801.

- The physical layers specified include 10GBASE-S, an 850-nm wavelength serial transceiver that uses two multimode fibers; 10GBASE-L4, a 1,310-nm WWDM transceiver that uses two multimode or single-mode fibers; 10GBASE-L, a 1,310-nm wavelength serial transceiver that uses two single-mode fibers; and 10GBASE-E, a 1,550-nm wavelength serial transceiver that uses two single-mode fibers.

There is broad market potential for this technology and a broad set of applications; multiple vendors and multiple users benefit from the technology. More than 55 companies attended the original IEEE 10 Gigabit Call for Interest conference; attendance and interest have increased steadily since that time. Growth of network and Internet traffic has placed high demand on the existing infrastructure, motivating the development of higher performance links. The 10-Gbps 802.3 solution extends Ethernet capabilities, providing higher bandwidth for multimedia, distributed processing, imaging, medical, CAD/CAM, and other existing applications. New opportunities are added when one defines a mechanism to adapt the MAC/PLS data rate to the data rate and framing of the WAN PHY such as SONET/SDH.

Compatibility with IEEE Standard 802.3
As usual, one wants conformance with CSMA/CD MAC, PLS, 802.2 LLC, and 802 FR. By adapting the existing 802.3 MAC protocol for use at

10 Gbps, this standard maintains maximum compatibility with the installed base of more than 600 million Ethernet nodes. The standard conforms to the full-duplex operating mode of the 802.3 MAC, appropriately adapted for 10-Gbps operation. Half-duplex operation and CSMA/CD itself will not be supported at 10 Gbps. As was the case in previous 802.3 standards, new physical layers have been defined for 10-Gbps operation. The standard also defines a set of systems management objects that are compatible with SNMP system management standards. The MIB for 10-Gbps 802.3 will be extended in a manner consistent with the 802.3 MIB for 10-, 100-, and 1,000-Mbps operation. Therefore, network managers, installers, and administrators will see a consistent management model across all operating speeds.

The 10GbE products are technically feasible and can be developed in a cost-effective manner. For example, bridging equipment that performs rate adaptation between 802.3 networks operating at different speeds has been demonstrated over the years by the broad set of product offerings that bridge between the 10-, 100-, and 1,000-Mbps speeds. Vendors of optical components and systems are developing reliable products that operate at 10 Gbps and also meet worldwide regulatory and operational requirements. Component vendors are confident of the feasibility of physical layer signaling at a rate of 10 Gbps on fiber optic media using a wide variety of innovative low-cost technologies. A target cost increase of three-fold of 1000BASE-X with a ten-fold increase in available bandwidth in the full-duplex operating mode (for the LAN PHY) will result in an improvement in the cost–performance ratio by a factor of 3. This cost model has been validated during both the 100- and 1,000-Mbps Ethernet deployments. (Cost factors are extrapolated from the OC-192c component supplier base and technology curves.)

In 10GbE, the bit rate is faster and the bit times are shorter—both in proportion to the change in bandwidth. The minimum packet transmission time has been reduced by a factor of 10. A rate control mode is added to the MAC to adapt the average MAC data rate to the SONET/SDH data rate for WAN-compatible applications of the standard. Achievable topologies for 10-Gbps operation are comparable to those found in 1000BASE-X full-duplex mode and equivalent to those found in WAN applications. There is significant additional supporting material in the standard for a *10-gigabit media independent interface* (XGMII), a *10-gigabit attachment unit interface* (XAUI), a *10-gigabit sixteen-bit interface* (XSBI), and management.

11.3.1.2 Media

LAN/MAN Media

A number of alternative media types have been developed for LAN/MAN environments. Since 10GbE is not expected to connect directly to users' desktops (at least, not yet), the standards are initially based on optical fiber. Optical PHYs include MMF and SMF using serial and parallel links (see Figure 11.19).

FIGURE 11.19
PHYs in 10GbE.

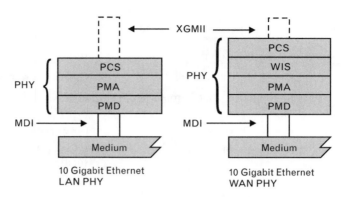

MDI = Medium dependent interface
PCS = Physical coding sublayer
PHY = Physical layer entity
PMA = Physical medium attachment
PMD = Physical medium dependent
WIS = WAN interface sublayer

WAN Media

Support for Ethernet-based communications over long distances is unique to 10GbE and would not be feasible using the original CSMA/CD protocol, because the throughput goes to zero. The full-duplex operation allows 10GbE to operate over long link spans, repeaters, and other transport layers (see Figure 11.19).

The 10GbE WAN physical layer is also defined to be SONET-like. It can be said that 10GbE will be "SONET friendly," even though it is not fully compliant with all of the SONET standards. Some SONET features will be avoided: TDM support, performance requirements, and management requirements (the most costly aspects of SONET).

11.3.1.3   Comparison

Table 11.21 provides a brief comparison between the 1- and 10-Gbps versions of Ethernet. Note once again that the traditional CSMA/CD protocol will not be used and that the fiber will be used as the physical media.

11.3.2   Snapshot of Technical Details

11.3.2.1   10GbE Physical Layer

The 10GbE PHY consists of the following sublayers (see Figure 11.20):

1.   Reconciliation sublayer (RS);

2.   Physical coding sublayer (PCS);

3.   Physical media attachment (PMA);

4.   Physical media dependent (PMD);

5.   An optional *WAN interface sublayer* (WIS) is also defined.

TABLE 11.21   COMPARISON BETWEEN THE 1- AND 10-GBPS ETHERNET

CHARACTERISTIC	1-GBPS ETHERNET	10-GBPS ETHERNET
Physical media	Optical and copper media	Optical media only
Distance	LANs up to 5 km; nointrinsic WAN support	LANs up to 40 km; direct attachment toSONET/SDH equipment for WANs
PMD leverages	Fibre Channel PMDs	Developed new optical PMDs
PCS	Reuses 8B/10B coding	Established new coding schemes (64B/66B) for 10GBASE-W and 10GBASE-R; uses 8B/10B for 10GBASE-X
MAC	Protocol half-duplex (CSMA/CD) as well as full duplex	Full-duplex only

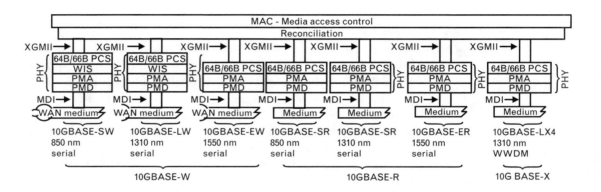

MDI   = Medium dependent interface
PCS   = Physical coding sublayer
PHY   = Physical layer device
PMA   = Physical medium attachment
PMD   = Physical medium dependent
WIS   = WAN interface sublayer
XGMII = 10 Gigabit media independent interface

FIGURE 11.20   *PHYs and supportive sublayers.*

As noted, P802.3ae/D3.2 defines the sublayers that implement two families of PHYs: a *LAN PHY* operating at a data rate of 10.000 Gbps and a *WAN PHY* operating at a data rate and format compatible with SONET STS-192c and SDH VC-4-64c. Seven variants are specified, as seen in Figure 11.21.

The extensive standard specifies a family of physical layer implementations. Table 11.22 specifies the correlation between technologies and required sublayers. The term *10GBASE-X* refers to a specific family of

FIGURE 11.21
The 10GbE layer model.

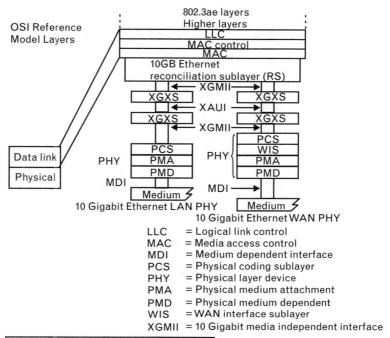

LLC	= Logical link control
MAC	= Media access control
MDI	= Medium dependent interface
PCS	= Physical coding sublayer
PHY	= Physical layer device
PMA	= Physical medium attachment
PMD	= Physical medium dependent
WIS	= WAN interface sublayer
XGMII	= 10 Gigabit media independent interface

10 Gigabit media independent interface (XGMII) The XGMII is designed to connect to a 10-Gbps capable MAC to a 10-Gbps PHY. While conformance with implementation of this interface is not strictly necessary to ensure communication, it is highly recommended, since it allows maximum flexibility in intermixing PHYs and DTEs at 10 Gbps speeds. The XGMII is intended for use as a chip-to-chip interface. No mechanical connector is specified.

10 gigabit attachment unit interface (XAUI) The XAUI is designed to extend the connection between a 10-Gbps capable MAC and a 10-Gbps PHY. While conformance with implementation of allows maximum flexibility in intermixing PHYs and DTE at 10-Gbps speed. The XAUI is intended for use as a chip-to-chip interface. No mechanical connector is specified.

10 gigabit sixteen-bit interface (XSBI) The XSBI is provided as a physical instantiation of the PMA service interface for 10GBASE-R and 10GBASE-W PHYs. While conformance with implementation of this interface is not strictly necessary to ensure communication, it is highly recommended, since it provides a convenient partition between the high-frequency circuitry associated with the PMA sublayer and the logic functions associated with the PCS and MAC sublayers. The XSBI is intended for use as a chip-to-chip interface. No mechanical connector is specified.

TABLE 11.22   CORRELATION BETWEEN TECHNOLOGIES AND REQUIRED SUBLAYERS

NOMENCLATURE	8B/10B PCS & PMA	64B/66B PCS	WIS	SERIAL PMA	850-NM SERIAL PMD	1,310-NM SERIAL PMD	1,550-NM SERIAL PMD	1,310-NM WWDM PMD	10GBASE-X	10GBASE-R	10GBASE-W
10GBASE-SR		Req.		Req.	Req.					X	
10GBASE-SW		Req.	Req.	Req.	Req.						X
10GBASE-LX4	Req.							Req.	X		
10GLRBASE-LR		Req.		Req.		Req.				X	
10GERBASE-LW		Req.	Req.	Req.		Req.					X
10GBASE-ER		Req.		Req.			Req.			X	
10GBASE-EW		Req.	Req.	Req.			Req.				X

physical layer implementations based on the 8B/10B data coding method; the 10GBASE-X family of physical layer implementations is composed of 10GBASE-LX4. The term *10GBASE-R* refers to a specific family of physical layer implementations based on the 64B/66B data coding method; the 10GBASE-R family of physical layer implementations is composed of 10GBASE-SR, 10GBASE-LR, and 10GBASE-ER. The term *10GBASE-W* refers to a specific family of physical layer implementations based on STS-192c/SDH VC-4-64c *encapsulation* of 64B/66B encoded data; the 10GBASE-W family of physical layer standards has been adapted from the ANSI T1.416-1999 (SONET STS-192c/SDH VC-4-64c) physical layer specifications. The 10GBASE-W family of physical layer implementations is composed of 10GBASE-SW, 10GBASE-LW, and 10GBASE-EW. All 10GBASE-R and 10GBASE-W PHY devices share a common PCS specification. The 10GBASE-W PHY devices also require the use of the WIS. A 10GBASE-PHY needs to "talk" to a 10GBASE-PHY protocol peer; it cannot "talk" directly to a SONET protocol peer. The following are the objectives of 10GBASE-R:

- Support the full-duplex Ethernet MAC.

- Provide 10-Gbps data rate at the XGMII.

- Support LAN PMDs operating at 10 Gbps and WAN PMDs operating at SONET STS-192c/SDH VC-4-64c rate.

- Support cable plants using optical fiber compliant with ISO/IEC 11801:1995.

- Allow for a nominal network extent of up to 40 km.

- Support a BER objective of 10^{-12}.

The generic term *10 Gigabit Ethernet* refers to any use of the 10-Gbps IEEE 802.3 MAC (the 10GbE MAC) coupled with any IEEE 802.3 10GBASE physical layer implementation. Interfaces of interest are as follows:

- *XGMII.* The 10-gigabit media independent interface provides an interconnection between the MAC sublayer (specifically, the RS) and PHY. This XGMII supports 10-Gbps operation through its 32-bit-wide transmit and receive paths. The RS provides a mapping between the signals provided at the XGMII and the MAC/PLS service definition. Although the XGMII is an optional interface, it is used extensively in the standard as a basis for functional specification and provides a common service interface.

- *XGMII extender Sublayer (XGXS) and XAUI (pronounced "Zowie").* The 10-gigabit attachment unit interface provides an interconnection between two XGMII extender sublayers to increase the reach of the XGMII. This XAUI supports 10-Gbps operation through its four-lane [7], differential-pair transmit and receive paths [8, 9]. The XGXS provides a mapping between the signals provided at the XGMII and the XAUI. (This interface is also optional.)

Ethernet (including 10GbE) defines a logical MAC–PLS interface between the RS (PHY) and the MAC, as shown in Figure 11.22. The interface operates at a constant 10-Gbps data rate. The interface is defined as a set of service primitives as follows: (1) PLS_DATA.request, (2) PLS_DATA.indicate, and (3) PLS_DATA_VALID.indicate. (To support WANs, the MAC includes "open loop rate control"; this is a MAC mechanism defined by 802.3ae to adapt the 10-Gbps MAC–PLS signaling rate to the 9.29-Gbps effective data rate at the PHY.)

11.3.2.2 RS

P802.3ae/D3.2 defines a reconciliation sublayer and optional XGMII [10]. The purpose is to convert the logical MAC–PLS service primitives to electrical signals; it provides a simple, inexpensive, and easy-to-implement interconnection between the MAC sublayer and the PHY.

11.3.2.3 PCS

The PCS provides for packet delineation and scrambling for the LAN PHY. The transmission code used by the PCS, referred to as 8B/10B, is identical to that specified in Clause 36 of the IEEE 802.3 standard as well as ANSI X3.230-1994 (FC-PH), Clause 11. The PCS maps XGMII characters into 10-bit code groups, and vice versa, using the 8B/10B block coding scheme. Implicit in the definition of a code group is an establishment of code group boundaries by a PCS synchronization process. The 8B/10B transmission code, as well as the rules by which the PCS ENCODE and DECODE functions generate, manipulate, and interpret code groups, is specified in the standard. As seen in Figure 11.20, 10GBASE-LX4 is the only PHY utilizing this encoding. The standard also specifies the PCS that is common to the family of 10-Gbps physical layer implementations known as 10GBASE-R. This PCS can connect directly to one of the

FIGURE 11.22
MAC–PLS interface.

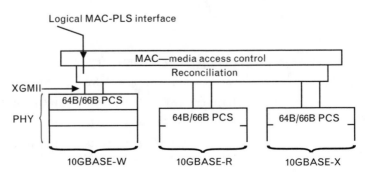

10GBASE-R physical layers: 10GBASE-SR, 10GBASE-LR, and 10GBASE-ER. Alternatively, this PCS can connect to a WIS, which will produce the 10GBASE-W encoding (10GBASE-R encoded data stream encapsulated into frames compatible with SONET and SDH networks) for transport by the 10GBASE-W physical layers, namely, 10GBASE-SW, 10GBASE-LW, and 10GBASE-EW. The term *10GBASE-R* is used when referring generally to physical layers using the PCS defined here. The 10GBASE-R is based on a 64B/66B code. The 64B/66B code supports data and control characters while maintaining robust error detection. 10GBASE-R PCS maps the interface characteristics of the WIS when present, and the PMA sublayer to the services expected by the RS and XGXS. 10GBASE-R can be extended to support any full-duplex medium requiring only that the medium be compliant at the PMA level. 10GBASE-R PCS may be attached through the PMA sublayer to a LAN PMD sublayer supporting a data rate of 10 Gbps or it may be attached to a WAN PMD through the WIS and PMA sublayers. When attached to a WAN sublayer, this PCS adapts the data stream to the WAN data rate.

11.3.2.4 PMA Sublayer

The PMA provides for the serialization and deserialization of the data being transmitted. It is responsible for supporting multiple encoding schemes, since each PMD will use an encoding that is suited to the specific media it supports.

11.3.2.5 PMD Sublayer

A summary of the seven PHY variants is shown in Table 11.23. The data rates of the PMA sublayer, type serial, is as follows:

- 10GBASE-R nominal baud rate1   ˙   0.3125 Gbps;
- 10GBASE-W nominal baud rate   9.95328 Gbps.

TABLE 11.23   SUMMARY OF THE SEVEN PHY VARIANTS

		10GbE Designation	
Description	Reach/Fiber	LAN PHY	WAN PHY
850–nm serial	~85m/MMF	10GBASE-SR	10GBASE-SW
1,310–nm serial	10 km/SMF	10GBASE-LR	10GBASE-LW
1,550–nm serial	40 km/SMF	10GBASE-ER	10GBASE-EW
1,310–nm WDM	10 km/SMF ~300m/MMF	10GBASE-LX4	

11.3.3 Applications

As the number of 100–Mbps Ethernet links at the edge of customers' networks increases, so will the need for 10GbE systems to aggregate 1–Gbps links in data centers. The deployment of a 10GbE–based backbone for a campus application (where the bandwidth is used for a single customer) is probably overkill for the majority of users for the near future. However, a 10GbE–based system that supports an aggregation metro access or metro core ring infrastructure would be in range with evolving requirements. For example, if the typical building could generate 73 Mbps of bandwidth in the near future, then 32 buildings on a ring (a typical design) would need about 2.3 Gbps, assuming we desire no oversubscription. Although this could be supported over next generation SONET, some intrinsic advantages could accrue from using 10GbE technology in the metro.

Figures 11.23 (a)–(d) depict four metro access/metro core applications of the 10GbE technology. Figure 11.23(a) depicts WAN applications driving fiber with the LAN PHY; this is not highly desirable. Figure 11.23(b) shows a WAN application driving fiber with the WAN PHY over a direct SONET infrastructure; this requires an adaptor in the CPE gateway device to adapt the stream to SONET—this could get expensive. Figure 11.23(c) illustrates a WAN application driving fiber with the WAN PHY over a SONET infrastructure supported by 10GbE–ready edge devices; here the network adapts the signal for carriage over SONET/SDH; this architecture is similar to today's TLS-providing networks. Figure 11.23 (d) shows a WAN application driving fiber with the WAN PHY over a WAN-PHY-based infrastructure; although this is less expensive, it requires a "dedicated" fiber just carrying the TLS service; also this implementation has fewer OAM&P features. As noted elsewhere, full OAM&P support is critical to providers that want to deliver resilient carrier-class services. Application such as Figures 11.23(a) and (d) would rely on a STP/RSTP for recovery, while the others would utilize the SONET mechanisms. The architecture of Figures 11.23(a) and (d) are similar to today's GbE-based MANs except that the latter one would use a WAN-PHY-based (SONET framing) infrastructure. Alternatives for building a full-function Ethernet-based metro network to support IP transport are discussed next [11].

11.3.3.1 Packet over Ethernet over SONET-Based Optical (PES)

IP can be transported directly over a full SONET infrastructure using PPP (or GFP) encapsulation, or over an Ethernet segment built from SONET links. SONET would be preferred for long distances and could be also be used at the MAN level. This method, however, has a degree of overhead from the various SONET management functions. PES is based on proven technologies that are widely installed and well understood. Ethernet is scalable, allowing the bandwidth offered to be more closely matched to the

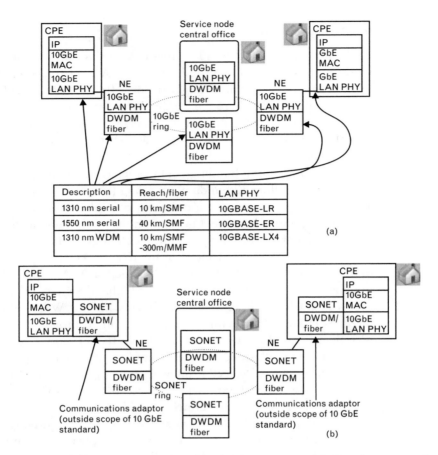

Description	Reach/fiber	LAN PHY
1310 nm serial	10 km/SMF	10GBASE-LR
1550 nm serial	40 km/SMF	10GBASE-ER
1310 nm WDM	10 km/SMF -300m/MMF	10GBASE-LX4

FIGURE 11.23 *(a) WAN applications driving fiber with the LAN PHY (less desirable); (b) WAN applications driving fiber with the WAN PHY over a direct SONET infrastructure;*

anticipated demand. Network management systems can be integrated across the different network speeds and configuration.

Because IP directly over SONET is expensive and difficult to manage when the number of links becomes large, Ethernet over SONET could be used to add a switching capability that can reduce the number of point-to-point links (i.e., a full mesh may not be required). Ethernet allows LAN, MAN, and WAN networks to be combined to form end-to-end connections, thereby reducing the need for format and protocol conversions within the network.

11.3.3.2 Packet over Ethernet over WDM-Based Optical (PEW)

A variation on the previous system is to transport Ethernet over a WDM-based physical layer, with or without a "thin SONET" interface. A small subset of the SONET header and SONET scrambling is used, but the

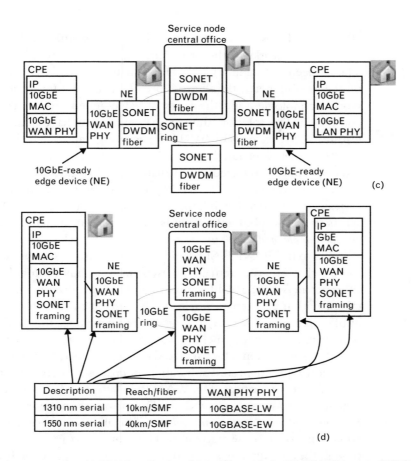

FIGURE 11.23 (CONTINUED)   *(c) WAN applications driving fiber with the WAN PHY over a SONET infrastructure supported by 10GbE-ready edge devices; and (d) WAN applications driving fiber with the WAN-PHY-based infrastructure (less desirable).*

majority of the overhead is eliminated. This solution avoids the complexities of SONET TDM functions, the stringent SONET physical layer specifications, and the need for a separate SONET element management system. A key advantage of PEW is that the overheads of ATM and SONET can be eliminated. Both 1GbE and 10GbE are more affordable, practical, and simpler than ATM, the major alternative for high-speed WANs.

11.3.3.3   Packet over Ethernet over Fiber (PEF)

In the metro area, enterprise customers can use 10GbE over dark fiber to support requirements such as remote hosting, off-site storage or backup, and disaster recovery. Metro service providers can build 10GbE backbones with less complex and expensive POPs from an equipment point of view.

10GbE will be used for aggregating slower access links, will be used in the backbone networks of the providers, and can also provide WAN access. The choice of "building blocks" for a metro service provider will depend on the existing network infrastructure, the need to interconnect older network technologies, and the types of value-added services that are being considered. This solution, though, may or may not have all of the OAM&P features and relies on STP/RSTP for reconfiguration, which is slow.

Note that to enable low-cost WAN PHY implementations, the task force specifically rejected conformance to SONET/SDH jitter, stratum clock, and certain SONET/SDH optical specifications. The WAN PHY is basically a cost-effective link that uses common Ethernet PMDs to provide access to the SONET infrastructure, thus enabling attachment, *when using an appropriate additional interworking device*, of packet-based IP/Ethernet switches to the SONET/SDH and TDM infrastructure. It is also important to note that Ethernet remains an asynchronous link protocol. As in every Ethernet network, 10GbE's timing and synchronization must be maintained within each character in the bit stream of data, but the receiving hub, switch, or router may retime and resynchronize the data. In contrast, synchronous protocols, including SONET/SDH, require that each device share the same system clock to avoid timing drift between transmission and reception equipment and subsequent increases in network errors where timed delivery is critical. The WAN PHY attaches data equipment such as switches or routers to a SONET/SDH or optical network, *when using an appropriate additional interworking device*. This allows simple extension of Ethernet links over those networks. Therefore, two routers will behave as though they are directly attached to each other over a single Ethernet link. Because no bridges or store-and-forward buffer devices are required between them (i.e., no speed conversions are needed), all of the IP traffic management systems for differentiated service operate over the extended 10GbE link connecting the two routers. However, to connect to a SONET/SDH network *an appropriate additional interworking device is needed*. To simplify management of extended 10GbE links, the WAN PHY provides most of the SONET/SDH management information, allowing the network manager to view the Ethernet WAN PHY links as though they are SONET/SDH links. It is then possible to do performance monitoring and fault isolation on the entire network, including the 10GbE WAN link, from the SONET/SDH management station. The SONET/SDH management information is provided by the WAN interface, WIS, which also includes the SONET/SDH framer. The WIS operates between the 64B/66B PCS and serial PMD layers common to the LAN PHY.

11.3.4   WIS, Type 10GBASE-W

The WIS, defined in Clause 50 of the standard, is an optional PHY sublayer that may be used to create a 10GBASE-W PHY that is data rate and format compatible with the SONET STS-192c transmission format defined by ANSI, as well as SDH's VC-4-64c container specified by ITU. The purpose of the WIS is to allow 10GBASE-W equipment to generate Ethernet data streams that may be mapped directly to STS-192c or VC-4-64c streams at the PHY level, without requiring MAC or higher layer processing. These streams are then carried over a SONET/SDH infrastructure via the use of an interworking unit.

The WIS, therefore, specifies a *subset* of the logical frame formats in the SONET and SDH standards. In addition, the WIS constrains the effective data throughput at its service interface to the payload capacity of STS-192c/VC-4-64c, that is, 9.58464 Gbps. Multiplexed SONET/SDH formats are not supported.

The WIS does not render a 10GBASE-W PHY compliant with either SONET or SDH at any rate or format. A 10GBASE-W interface is not intended to interoperate directly with interfaces that comply with SONET or SDH standards, or other synchronous networks. Operation over electrically multiplexed payloads of a transmission network is outside the scope of the standard; such interoperation would require full conformance to the optical, electrical, and logical requirements specified by SONET or SDH, and is outside the scope and intent of the standard.

The achievable topologies with the use of a WIS as part of a 10GBASE-W PHY are identical to those implementable without it. From the perspective of the 10-Gbps MAC layer, a 10GBASE-W PHY with a WIS does not appear different (in either the functions or service interface) from a PHY without a WIS, with the exception of sustained data rate. However, a 10GBASE-W interface implementing a WIS may interoperate only with another 10GBASE-W interface that also implements a WIS.

11.3.4.1 Scope

The WIS clause in the 10GbE standard specifies the functions, features, services, and protocol of the WIS. The WIS may be used with any of the PCS, PMA, and PMD sublayers that are defined for 10GBASE-W; as shown earlier in Figure 11.20, it is placed between the PCS and PMA sublayers within the 10GBASE-W PHY. The WIS is common to all members of the family of 10GBASE-W WAN-compatible PHY implementations, specifically, 10GBASE-SW, 10GBASE-LW, and 10GBASE-EW, as seen earlier in Table 11.22.

The definition of the WIS is based on the subset of signaling rates and data formats standardized by ANSI T1.416-1999 (*Network to Customer Installation Interfaces—Synchronous Optical Network Physical Layer Specification: Common Criteria*), which is in turn based on ANSI T1.105-1995 (*Telecommunications—Synchronous Optical Network—Basic Description Including Multiplex Structure, Rates and Formats*).

The WIS maps the encoded Ethernet data received (transmitted) from (to) the PCS into a frame structure that has the same format as that defined by T1.416–1999, implementing a *minimal* number of the standard SONET overhead fields and functions. WIS does not adhere to the electrical and optical aspects of SONET specified by T1.416–1999, because it is intended to be used with PHYs that conform to the corresponding parameters defined by the 10GBASE-W standard; otherwise, a gateway mechanism is needed to use an existing SONET/SDH infrastructure, either at the CPE level or at the network edge device. The WIS meets all requirements of ANSI T1.416–1999 except the following:

1. Section 5 (jitter);
2. Section 6 (synchronization);
3. Section 7.2.2 (VT1.5 rate–electrical interface);
4. Section 7.4.2 (VT1.5 rate);
5. Section 7.6 (performance and failure alarm monitoring);
6. Section 7.7 (Performance monitoring functions);
7. Annex A (SONET VT1.5 line interface common criteria);
8. Annex B (SONET maintenance signals for the NI);
9. Annex C (receiver jitter tolerance and transfer).

11.3.4.2 Summary of WIS Functions

The following provides a summary of the principal functions implemented by the WIS. In the transmit direction (i.e., when transferring data from the PCS to the PMA), the WIS performs the following functions:

- Mapping of data units received from the PCS via the WIS service interface to the *payload capacity of the* SPE defined for STS-192c;
- Addition of path overhead and fixed stuff octets to generate the actual SPE;
- Creation of frames consisting of line overhead and section overhead octets plus the SPE, and the generation and insertion of section, line, and path BIP (refer to Chapter 6 for a definition of these terms);
- Scrambling of the generated WIS frames;
- Transmission of these frames to the PMA sublayer via the PMA service interface.

In the receive direction, the functions performed by the WIS include these:

- Reception of data from the PMA sublayer via the PMA service interface;

- Delineation of octet boundaries as well as STS-192c frame boundaries within the unaligned data stream from the PMA;

- Descrambling of the payload and overhead fields within the incoming frames;

- Processing of the pointer field within the line overhead, and delineation of the boundaries of the SPE within the received WIS frames;

- Generation and checking of BIP within the section, line, and path overheads;

- Removal of line, section and path overhead columns, as well as fixed stuff columns, in order to extract the actual payload field;

- Handling of errors and exception conditions detected within the incoming WIS frame stream, and reporting these errors to layer management;

- Mapping of octets extracted from the payload capacity of the incoming SPE to data units that are passed to the PCS via the WIS service interface.

11.3.4.3 Sublayer Interfaces

A WIS service interface is provided to allow the WIS to transfer information to and from the 10GBASE-R PCS, which is the sole WIS client. An abstract service model is used to define the operation of this interface. In addition, the WIS utilizes the service interface provided by the PMA sublayer to transfer information to and from the PMA. The 10GbE specification defines these interfaces in terms of bits, octets, data units, and signals; however, implementers may choose other data path widths and other control mechanisms for implementation convenience, provided that the logical models of the service interfaces are adhered to.

The PMA service interface may be optionally instantiated as an actual physical interface, referred to as the XSBI; in this case, the WIS must also implement the client portion of the XSBI.

The WIS service interface supports the exchange of data units between PCS entities on either side of a 10GBASE-W link using request and indicate primitives. Data units are mapped into WIS frames by the WIS and passed to the PMA, and vice versa.

11.3.4.4 Functions Within the WIS

The WIS comprises the WIS transmit and WIS receive processes for 10GBASE-W, together with a synchronization process and a link management function. The WIS transmit process accepts fixed-width tx_data-units from the PCS via the WIS service interface and maps them into the payload capacity of the transmitted WIS frame stream. Fixed stuff octets are added, together with a set of path overhead octets, to create a SPE. Line and section overhead octets are combined with the SPE and then scrambled using a frame-synchronous scrambler to produce the final transmitted WIS frame. The WIS continuously generates one SONET-compatible WIS frame, comprising overhead fields, fixed stuff and payload, every 125 μs. No gaps are present between WIS frames. The data produced by the transmit process are passed to the PMA via the PMA service interface.

The WIS synchronization process accepts data from the PMA (via the PMA Service interface) and performs an alignment operation to delineate both octet and frame boundaries within the received data stream. Aligned and framed data are passed to the WIS receive process where section and line overhead octets are extracted from the WIS frames and processed after descrambling the frame data. The payload pointer within the line overhead is used to delineate the start and end of the received SPE, and the path overhead is extracted from the SPE and processed. Finally, the fixed stuff is removed from the SPE as well, and the resulting data stream is conveyed to the PCS via the WIS service interface. Severe errors detected by the synchronization and receive processes (e.g., loss of WIS frame synchronization) cause a WIS_SIGNAL.indicate primitive to be sent to the PCS with a parameter of FAIL.

11.3.4.5 Payload Mapping

The WIS maps the tx_data–unit:0 and rx_data–unit:0 parameters that are transferred via its service interface to/from the payload capacity of a standard STS–192c SPE structure. This structure is represented as a two-dimensional array with 9 rows and 16,704 columns, each row consisting of one octet of path overhead, 63 octets of fixed stuff, and 16,640 octets of actual payload capacity. The total payload capacity of the SPE comprises 149,760 octets (per WIS frame). The transmission order is from left to right, that is, lower numbered octets are transmitted before higher numbered octets.

11.3.4.6 WIS Frame Generation

As part of the transmit process, the WIS encapsulates the payload generated by the payload mapping function within a series of WIS frames. The receive process performs the reverse operation, extracting payload from the incoming WIS frame stream and submitting it to the payload mapping function. SONET compatibility in the WIS follows ANSI T1.416-1999, with the exceptions noted earlier.

Transmit Path Overhead Insertion
The WIS transmit process inserts path overhead fields as defined in Section 4.2 of ANSI T1.416-1999, and specified in Table 11.24. For the fields where the "Coding" column of Table 11.24 contains "Per T1.416," the field is inserted according to the specifications of ANSI T1.416-1999.

TABLE 11.24   STS PATH OVERHEAD

Overhead Octet	Function	Usage	Coding (Bits 1–8)
B3	STS path error monitoring (path BIP-8)	Supported	Per T1.416
C2	STS path signal label	Specific value	00011010
F2	Path user channel	Unsupported	00000000
G1	Path status	Supported	Per T1.416
H4	Multiframe indicate	Unsupported	00000000
J1	STS path trace	Specific value	See text
N1	Tandem connection maintainence/path data channel	Unsupported	00000000
Z3–Z4	Reserved for path growth	Unsupported	00000000

Note: SONET/SDH and 802.3 differ in bit coding conventions. The values in this table follow SONET bit ordering, in which bit index values range from 1 to 8, from left to right, and bit 8 is the least significant bit.

For the fields where the "Coding" column contains a specific value or "See text," the 10GbE document supersedes the corresponding values in Table 1, SONET Overhead at NIs, in the ANSI document (also see Chapter 6).

Transmit Line Overhead Insertion

The WIS transmit process inserts line overhead fields as defined in Section 4.2 of ANSI T1.416-1999, and specified in Table 11.25. For the fields where the "Coding" column of Table 11.25 contains "Per T1.416," the field is inserted according to the specifications of ANSI T1.416-1999. For the fields where the "Coding" column contains a specific value or "See text," the 10GbE document supersedes the corresponding values in Table 1, SONET Overhead at NIs, in the ANSI document. In addition, line overhead octets not listed in the latter table are set to 00000000 by the transmit process.

Transmit Section Overhead Insertion

The WIS transmit process inserts section overhead fields as defined in Section 4.2 of ANSI T1.416-1999, and specified in Table 11.26. For the fields where the "Coding" column of Table 11.26 contains "per T1.416," the field is inserted according to the specifications of ANSI T1.416-1999. For the fields where the "Coding" column contains a specific value or "See text," the 10GbE document supersedes the corresponding values in Table 1, SONET Overhead at NIs, in the ANSI document. In addition,

TABLE 11.25   LINE OVERHEAD

Overhead Octet	Function	Usage	Coding (Bits 1–8)
B2	Line error monitoring (line BIP-1536)	Supported	Per T1.416
D4–D12	Line data communication channel (DCC)	Unsupported	00000000
E2	Orderwire	Unsupported	00000000
H1–H2	Pointer	Specific value	See text
H3	Pointer action	Specific value	00000000
K1, K2	Automatic protection switch (APS) channel and line remote defect identifier (RDI-L)	Specific value	See text
M0	STS-1 line remote error indication (REI)	Unsupported	00000000
M1	STS-N line remote error indication (REI)	Supported	Per T1.416
S1	Synchronization messaging	Unsupported	00001111
Z1	Reserved for line growth	Unsupported	00000000
Z2	Reserved for line growth	Unsupported	00000000

TABLE 11.26   SECTION OVERHEAD

Overhead Octet	Function	Usage	Coding (Bits 1–8)
A1	Frame alignment	Supported	Per T1.416
A2	Frame alignment	Supported	Per T1.416
B1	Section error monitoring (section BIP-8)	Supported	Per T1.416
D1–D3	Section data communications channel (DCC)	Unsupported	00000000
E1	Orderwire	Unsupported	00000000
F1	Section user channel	Unsupported	00000000
J0	Section trace	Supported	Per T1.416
Z0	Reserved for section growth	Unsupported	11001100

section overhead octets not listed in the latter table are set to 00000000 by the transmit process.

Receive Path, Line, and Section Overhead Extraction

The WIS receive process extracts path, line and section overhead fields as defined in Section 4.2 of ANSI T1.416-1999 and shown in Tables 11.24, 11.25, and 11.26. For the fields where the "Coding" columns of Tables 11.24, 11.25, and 11.26 contain "Per T1.416," the fields are extracted according to the specifications of ANSI T1.416-1999. For the fields where the "Coding" columns contain a specific value or "See text," the 10GbE document supersedes the corresponding values in Table 1, SONET Overhead at NIs, in the ANSI document. Overhead octets marked as unsupported in Tables 11.24, 11.25, and 11.26 are ignored by the receive process. In addition, overhead octets not listed in Table 1, SONET Overhead at NIs, in ANSI T1.416-1999 is also be ignored by the receive process.

Fault Processing

Defects and anomalies detected by the receive process are classified as defined in Section 7.1 of ANSI T1.416-1999. Section, line, and path defects and anomalies are detected and processed as defined by Sections 7.3, 7.4.1, and 7.5 of ANSI T1.416-1999, except that only those defects and anomalies listed in Table 11.27 are processed, with the remainder being ignored.

11.3.4.7 Issues Related to Digital Wrapper

Observers believe that it will be complex, expensive, and time consuming to use the current 10GbE WAN interface in existing transmission networks. These interworking problems will severely impact the use of 10GbE WAN when customers discover it is not compatible with OC-192c interfaces. The problem of mapping 100-ppm 10GbE WAN in OTN G.709 frames was discussed during an IEEE plenary meeting in March 2001 [12].

The current 100-ppm specification will impact interworking with other types of equipment that are already deployed in transmission networks. 10GbE specifies a WAN application with an OC-192c frame in order to provide interconnection between distant equipment such as routers. The existing transmission network is now based on two layers,

TABLE 11.27 WIS-Supported Near-End Events and Far-End Reports

	Physical Media	Section		Line		Path	
	Defect	Anomaly	Defect	Anomaly	Defect	Anomaly	Defect
Near end	LOS	BIP-N(S)	SEF/LOF	BIP-N(L)	AIS-L	BIP-N(P)	LOP-P AIS-P
Far end	N/A	N/A	N/A	RELL	RDLL	RELP	ERDLP

SONET/SDH and DWDM. In the near future, OTN and all optical networks will be introduced. The impossibility of mapping 100-ppm 10GbE WAN inside the new G.709 frame has been demonstrated. The SONET/SDH physical layer was specified more than 10 years ago with an accuracy of 20 ppm. All existing SONET/SDH interfaces have been specified to accept signals within 20 ppm but not 100 ppm. This impacts the following equipment [12]:

- SONET/SDH equipment;

- DWDM with their SONET/SDH interfaces;

- OTN G.709 equipment.

The issue is that 10GbE LAN PHYs and WAN PHY operate in an *independent plesio-isosynchronous* mode [13]. 10GbE is a data-centric optical communications protocol technology in which the transmitters are clock independent of the receivers. Each transmitting interface/port in a data switch can operate with a different and independent clock. The receivers synchronize only with the data stream sent by the upstream transmitter at the data character/frame level. The transmitters operate within a "loose" timing specification because the information in the communications stream is multiplexed at the data frame level (OSI layer 2), not the TDM payload level (OSI layer 1).

ITU-T SG15 OTN (discussed in Chapter 3), on the other hand, is isosynchronous. In general terms, isosynchronous is a transmission mode that uses start and stop bits to identify characters. Both senders an receiver are synchronized. The OTN uses the *digital wrapper* (DW) to multiplex data streams from various sources into common telephony based payloads. DW bandwidths are at 2.5, 10, and 40 Gbps. Multiple data streams from different sources are mapped into the same DW bandwidth at TDM payloads. This is the same functionality that exists in the TDM telephony networks today. All communications steams are treated as "clients" of the network [13]. Normal operation of the network is isosynchronous with each transmitter locked to a common clock source. Alternatively, the transmitter is locked to the receiver, which is locked to an upstream transmitter, which is locked to the common clock. The transmitters operate within a tight timing specification because the information in the communications stream is multiplexed at the TDM payload level (OSI layer 1), not the data frame level (OSI layer 2).

Packet over SONET (POS) can be a guide. The experience with POS indicates that a client system of isosynchronous networks needs to support full isosynchronous clock timing. POS is an optical data communications protocol that was developed and began deployment in 1996. POS began as an independent plesio-isosynchronous protocol with ± 20-ppm clock tolerance to provide for receiver lock at the transmission synchronous client interface, operating in "maintenance mode." This caused synchronization alarms within the transmission network and much higher data errors than occurs over DS-1/DS-3 circuits. (It is hypothesized to be because of the clock pointer offsets that were occurring as the POS interface clock drifted between tolerance boundaries.) When POS interfaces were converted to full support for "loop timing" and/or *primary reference source* (PRS) support, the synchronization alarms and the higher data errors were eliminated [13].

Technical difficulties will be experienced when mapping 10GbE onto OTN. OTN is an isosynchronous network infrastructure. If 10GBASE-W is interfaced to OTN as an OC-192c, it will be a "client" on that isosynchronous network. Experience with POS indicates that 110GBASE-W only has a ± 20-ppm clock tolerance to support functioning as an isosynchronous network "client," and it will have higher than normal data errors and cause "synchronization alarms" within the OTN infrastructure. If the 10GbE standard attempts to support operation within OTN is will need to support full isosynchronous clock timing. This means "loop timing" or PRS support (T1X1.416-1999, Section 6).

802.3 Ethernet standards normally provide for reliable functionality with a wide range of operating conditions [13]. Clock tolerance and compensation provide for (1) different component manufacturers with differing quality tolerances, (2) different system manufacturers with differing quality tolerances, (3) multiple systems in series with the data stream, and (4) different ages of systems within a data network over a period of time. IPG idle bytes are used to provide a wide tolerance for transmitter clock drift accumulation over multiple systems of different age and manufacture quality. For 10-Gbps implementations, the spacing between two packets, from the last bit of the FCS field of the first packet to the first bit of the preamble of the second packet, can have a minimum value of 40 BT, as measured at the XGMII receive signals at the DTE. This IFG shrinkage may be caused by variable network delays and clock tolerances.

G.gfp (now G.7041/Y.1303) in the digital wrapper under OTN cannot support the 10GbE LAN PHY. To support the P802.3ae LAN PHY, OTN G.gfp has to compensate the payload rate. The OTN DW payload rate is 9953.208 Mbps. The OTN G.gfp replaces the IPG, preamble, and SFD with a *header error check* (HEC) frame and extension. G.gfp in the DW wants to use the ability to drop IPG idle octets to compensate for the difference between the LAN PHY and DW payload. A total of 5,765,000 octets per second must be available for deletion or data frame errors can occur; 9,727,620 octets are available under ideal operating conditions. At the extreme allowed operation point of 40 BT of IPG with full load of maximum current MTU frames, only 4,071,660 octets per second are available for deletion.

Fueling this discussion, other experts believe that without OTN the 10GbE WAN PHY will be too expensive to deploy, in terms of the 10GbE WAN PHY requiring entirely new infrastructure. Without OTN, support for 10GbE WAN PHY will not have high-speed service protection and will not be reliable like SONET/SDH.

Proponents for retaining the 100-ppm specification, make the pitch that what is actually required is full isosynchronous support in the 10GbE WAN PHY for proper operation. While OTN can support the 10GbE LAN PHY by deleting octets from the IPG, this works well under ideal operating conditions, but will fail and lose data under allowable reduced IPG operating conditions. The current linear deployment of *high count dense wavelength division multiplexing* (HC DWDM) systems (up to 160 wavelengths) will support the 10GbE WAN PHY without need for modification at ± 20-ppm clock tolerance. A ± 100-ppm clock allowance is an upgrade to the existing HC DWDM systems [13].

Proponents for retaining the 100-ppm spec, also make the pitch that protection switching of wavelength services is done at the optical level, not the electrical level as is the case for SONET/SDH. The optical control plane controls the optical service provisioning and service protection switching, not the optical channel service overheads. Optical DWDM service support for 10GbE WAN PHY does not "require" a transmission clock tolerance of ± 20 ppm. (As noted, for OTN DW, just having a ± 20-ppm clock is inadequate, so OTN DW G.gfp cannot properly support the 10GbE LAN PHY.) But if a lambda is used, deployment of a new infrastructure is not required to support 10GbE WAN; customer and service provider support for 10GbE does not obligatorily require a new infrastructure and does not require ITU-T OTN.

Some look at the problem from a more general perspective. Two solutions are available:

1. Adapt 10GbE WAN to SONET/SDH.
2. Adapt SONET/SDH to 10GbE WAN.

Before we get into a discussion of these solutions, note that some proponents suggest the use of *Ethernet line termination equipment* (ELTE) (e.g., see Figure 11.24 [14]). Others say that ELTE, if they exist at the OC-192c level, cannot be considered as a good solution since this would require insertion of new equipment into the network, which raises questions about management, spacing, and redundancy issues. And since this equipment would have a 20-ppm oscillator, why not put it directly in the OC-192 source? Early SONET/SDH deployments were linear point–to–point systems using linear point–to–point support technology, not the ring systems with the ring-based technology of today. Early linear SONET/ SDH systems used systems for regeneration called the *line regenerating element* (LRE) and the *section regenerating element* (SRE). LREs and SREs provided OAM overhead support and performance monitoring. LREs do full line and section level overhead termination and reinsertion. SREs only do section

FIGURE 11.24
ELTEs: (a) Protocol model, and (b) Ethernet on ELTE.

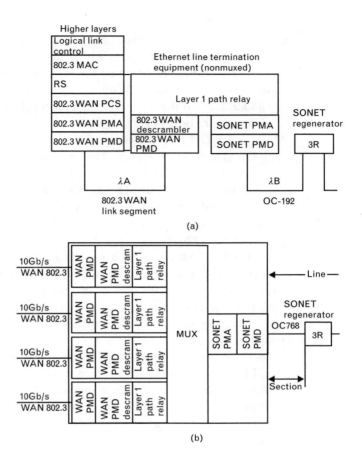

(a)

(b)

level overhead termination and reinsertion. SREs are still used today in extended distance SONET/SDH rings. Except for the additional support of the *data communications channel* (DCC) and *order wire* (OW), the LRE functions are the same as the ELTE functions. Hence, the functional description of the ELTE is the same as the LRE.

Solution 1: Adapt the 10GbE WAN. Insert a 20-ppm option in the WAN application of 10GbE. The results are that logical, new equipment takes care of existing ones, and it is easy and inexpensive to implement: The oscillator generating the 10GbE WAN frame is the only required modification; all other functions impacted by the frequency accuracy, clock recovery, FIFO, and so on, work within 20 ppm since it is specified to work within 100 ppm [12].

Solution 2: Adapt SONET interface to 100 ppm. The clock recovery of input OC-192 interfaces needs to be redesigned to accept 100 ppm (this means replacement of equipment already in the field). The input signal must be adapted to SONET/SDH frequency. The SDH frame allows one pointer event each four frames. Pointer processors could accept 100-ppm deviation, but this has never been requested nor tested. Test equipment is not even qualified for that. Once passed through a pointer processor, the new frame is within 4.6 ppm with 100 pointer rate and can be transported through SONET/SDH. Only the equipment extracting the payload from the 10GbE may present an unexpected rate of pointers and a high level of wander, but it is the function of the terminal equipment to take care of the client requirement [12].

Adapt DWDM equipment. OC-192c clock recovery should also be modified to accept 100 ppm. This implies redesign and update in the field.

Adapt G.709 OTN. OC-192c clock recovery should also be modified to accept 100 ppm. The SDH layer is not accessed in OTN equipment; it is only mapped in the OTN frame. Adaptation refers to terminating (sink) an OC-192c, processing the SONET/SDH pointer, and terminating (source) a new OC-192c frame.

Implementation of such modification is estimated to be about 400K gates. The total complexity of the adaptation is redesign new clock recovery, design a new 400K gate ASIC, and manage a new layer in OTN NEs, adding complexity to the network manager [12].

Committee T1X1 [15] is aware that the IEEE 802.3ae Task Force is standardizing a 10GbE WAN interface, referred to as 10GBASE-W. They have been fully support this initiative. The carrier community represented in T1X1 sees a significant business opportunity in the transport of 10GbE in their metropolitan and long-haul networks. To make this service possible in existing and future transport network infrastructures, T1X1 recently requested that IEEE Working Group 802.3 reconsider the following points:

- The network impact of the ± 100-ppm clock tolerance specified for 10GBASE-W;

- The impact of ELTE on network management for 10GbE transported over the OTN, as specified in ITU-T Rec. G.709.

T1X1 believes that the cost impact of using a 20-ppm oscillator (relative to a target cost for 10GbE equipment) is less than 1% over the cost for a 100-ppm implementation. The graceful accommodation of a 20-ppm 10GBASE-W signal into the extensive SONET/SDH network infrastructure and the next generation OTN core network will facilitate the widespread availability of 10GbE service over great distances.

A ± 20-ppm tolerance will also allow the transport of 10GbE over the OTN without the use of an ELTE. This will allow for end-to-end network management, which is otherwise precluded by the termination of the 10GBASE-W signal at OTN boundaries. The cost of accommodating a ±100-ppm tolerance with stand-alone ELTE in existing SONET/SDH networks will be reflected in the higher cost of the 10GbE service to the end user. T1X1 therefore has requested that the line rate tolerance for the 10GbE be changed to ± 20 ppm.

It is impossible to map 10GbE WAN into G.709 OTN equipment. As noted, there are two solutions:

1. Adapt 10GbE WAN to SONET/SDH; this is the easiest, just change an oscillator.

2. Adapt SONET/SDH to 10 GbE WAN; this is more complex. Clock interfaces on OC-192c need to be redesigned, implying replacement of equipment that is already in the field. This option was discussed in more detail; however, the bottom line is that this option is highly unlikely to happen. A current estimate of 400K gates to change the design exists.

The request has been made again to change the 100-ppm clock to 20 ppm to minimize the impact on existing networks that is sure to happen when adapting the two different systems. These issues were not fully resolved as of press time.

11.3.5 PMD Sublayers

In this section we briefly look at the PMD sublayer and baseband medium, type 10GBASE-S (short-wavelength serial), 10GBASE-L (long-wavelength serial), and 10GBASE-E (extra-long-wavelength serial).

11.3.5.1 PMD to MDI Optical Specifications for 10GBASE-S

The operating range for 10GBASE-S is defined in Table 11.28. A PMD that exceeds the operational range requirement while meeting all other optical specifications is considered compliant (e.g., a 400 MHz·km 50-μm solution operating at 80m meets the operating range requirement of 2–66m). The 10GBASE-S transmitter meets the specifications defined in Table 11.29. Table 11.30 depicts the link power budgets.

TABLE 11.28 10GBASE-S OPERATING RANGE FOR EACH OPTICAL FIBER TYPE

Fiber Type	Modal Bandwidth @ 850 nm (min) (MHz km)	Operating Range (m)
62.5-μm MMF	160	2–26
	200	2–33
50-μm MMF	400	–66
	500	–82
	2,000	2–300

TABLE 11.29   10GBASE-S TRANSMIT CHARACTERISTICS

DESCRIPTION	10GBASE-SW	10GBASE-SR
Signaling speed (nominal)	9.95328 GBd	10.3125 GBd
Clock tolerance (max)	± 20 ppm	± 100 ppm
Wavelength (range)	840–860 nm	
T_{rise}/T_{fall} (max: 20%–80%)	35 ps	
RMS spectral width (max)	See specification	
Average launch power (max)	See specification	
Launch power (min) in OMA	See specification	
Average launch power of OFF transmitter (max)	−30 dBm	
Extinction ratio	3 dB	
RIN_{12}OMA (max)	−125 dB/Hz	
Encircled flux	See specification	

TABLE 11.30   10GBASE-S LINK POWER BUDGETS

PARAMETER	62.5-μM MMF		50-μM MMF			UNIT
Modal bandwidth as measured at 850 nm	160	200	400	500	2,000	MHz·km
Link power budget	7.5	7.5	7.5	7.5	7.5	dB
Operating distance	26	33	66	82	300	m
Channel insertion loss	1.60	1.63	1.75	1.81	2.59	dB
Allocation for penalties	5.90	5.87	5.75	5.69	4.91	dB
Aditional insertion loss allowed	0.84	0.81	0.63	0.57	0.0	dB

Note: Link penalties are used for link budget calculations. They are not requirements and are not meant to be tested.

Tables 11.31, 11.32, and 11.33 depict key characteristics of the 10GBASE-L PMD, whereas Tables 11.34, 11.35, and 11.36 depict key characteristics of the 10GBASE-E PMD.

11.3.5.2   PMD Sublayer and Baseband Medium for WWDM PHY, Type 10GBASE-LX4

This section briefly describes the 10GBASE-LX4 PMD and the baseband medium for both multimode and single-mode optical fiber. Operating this PMD requires the following capabilities: RS, XGMMI (optional), XGXS (optional), XAUI (optional), and 10GBASE-X PCS/PMA.

TABLE 11.31   10GBASE-L OPERATING RANGE

PMD TYPE	NOMINAL WAVELENGTH (NM)	MINIMUM RANGE
10GBASE-L	1,310	2m to 10 km

TABLE 11.32   10GBASE-L TRANSMIT CHARACTERISTICS

DESCRIPTION	10GBASE-LW	10GBASE-LR
Signaling speed (nominal)	9.95328 GBd	10.3125 GBd
Clock tolerance (max)	± 20 ppm	± 100 ppm
Wavelength (range)	1,260 to 1,355 nm	1,260 to 1,355 nm
RMS spectral width (max)	See specification	See specification
Side mode suppression ratio (min)	30 dB	30 dB
Average launch power (max)	0.5 dBm	0.5 dBm
Launch power (min) in OMA	See specification	See specification
Average launch power of OFF transmitter (max)	−30 dBm	−30 dBm
Extinction ratio (min)	4 dB	4 dB
$RIN_{12}OMA$ (max)	−125 dB/Hz	−125 dB/Hz
Return loss (max)	12 dB	12 dB

TABLE 11.33   10GBASE-L LINK POWER BUDGETS

PARAMETER	10GBASE-L
Link power budget	9.4 dB
Operating distance	10 km
Channel insertion loss	7.17 dB
Allocation for penalties	2.96 dB

TABLE 11.34   10GBASE-E OPERATING RANGE

PMD TYPE	NOMINAL WAVELENGTH (NM)	MINIMUM RANGE
10GBASE-E	1,550	2m to 40 km

TABLE 11.35 10GBASE-E TRANSMIT CHARACTERISTICS

DESCRIPTION	10GBASE-EW	10GBASE-ER
Signaling speed (nominal)	9.95328 GBd	10.3125 GBd
Clock tolerance (max)	± 20 ppm	± 100 ppm
Wavelength (range)	1,530 to 1,565 nm	1,530 to 1,565 nm
Side mode suppression ratio (min)	30 dB	30 dB
Average launch power (max)	4.0 dBm	4.0 dBm
Launch power (min) in OMA	−1.39 dBm	−1.39 dBm
Average launch power of OFF transmitter (max)	−30 dBm	−30 dBm
Transmitter and dispersion penalty (max)	3 dB	3 dB
Extinction ratio (min)	3 dB	3 dB
$RIN_{21}OMA$ (max)	−125 dB/Hz	−125 dB/Hz

TABLE 11.36 10GBASE-E LINK POWER BUDGETS

PARAMETER	10GBASE-E
Link power budget	18.0 dB
Operating distance	40 km
Channel insertion loss	13.0 dB
Return loss for any device in the optical link	26 dB
Allocation for penalties	5.00 dB

The operating ranges for 10GBASE-LX4 PMD are defined in Table 11.37. A 10GBASE-LX4-compliant transceiver supports all media types listed in Table 11.37 (i.e., 50- and 62.5-μm multimode fiber, and 10-μm single-mode fiber). A transceiver that exceeds the operational range requirement while meeting all other optical specifications is considered compliant (e.g., a single-mode solution operating at 10,500m meets the minimum range requirement of 2–10,000m). Table 11.38 depicts the transmit characteristics for the 10GBASE-LX4 over each optical fiber type. At press time multiple system vendors are currently designing blades with 10GBASE-SX4, and several have been demonstrated at public forums.

11.3.6 Other Work, Copper PMD

Work was under way at the time of this writing to develop an inexpensive copper PMD that operates at 10m. The goal is to design a link that is significantly cheaper than optical at maximum distance. The link would include two transceivers (monolithic CMOS) and a jumper cable. Designers want to leverage the 1000BASE-CX PMD specification. In one

TABLE 11.37   OPERATING RANGE FOR 10GBASE-LX4 PMD OVER EACH OPTICAL FIBER TYPE

FIBER TYPE	MODAL BANDWIDTH @ 1,300 NM (MIN. OVERFILLED LAUNCH) (MHz·KM)	MINIMUM RANGE (M)
65.5-μm MMF	500	2 to 300
50-μm MMF	400	2 to 240
50-μm MMF	500	2 to 300
10-μm MMF	N/A	2 to 10,000

TABLE 11.38   TRANSMIT CHARACTERISTICS FOR THE 10GBASE-LX4 OVER EACH OPTICAL FIBER TYPE

DESCRIPTION	62.5-μM MMF, 50-μM MMF, 10-μM MMF	
Transmitter type	Longwave laser	
Signaling speed per lane (nominal)	3.125 $\pm$ 100 ppm GBd	
Lane wavelengths (range)	1,269.0–1,282.4 nm 1,293.5–1,306.9 1,318.0–1,331.4 1,342.5–1,355.9	
T_{rise}/T_{fall} (max. 50% to 80% response time)	100	ps
Side mode suppression ratio (min)	0.0	dB
RMS spectral width (max)	0.62	nm
Average launch power, four lanes (max)	5.5	dBm
Average launch power, per lane (max)	−0.5	dBm
Optical modulation amplitude, per lane (max)	750 (−1.25)	mW (dBm)
Optical modulation amplitude, per lane (min)	237 (−6.25)	mW (dBm)
Average launch power of OFF transmitter, per lane (max)	−30	dBm
RIN_{12} (OMA)	−120	dB/Hz

proposal, the PMD would use PAM-5 signaling to reduce the signaling rate to 5 Gbps. Also, the effort looks to leverage the same PHY proposed for SX, LX, and EX for 10GbE. The jumper cable assembly consists of a continuous shielded balanced cable (twinax) terminated at each end with a polarized shielded plug. Early vendor information shows that FC and GbE

CX cables and connectors can support signaling at 2.5 GHz. Today, 2-Gbps FC CX cables are available [16].

11.3.7   ANSI 10GFC

The 10-Gigabit Fibre Channel standard describes in detail extensions to Fibre Channel signaling and physical layer services introduced in ANSI X3.230, FC-PH, to support data transport at a rate in excess of 10 Gbps. This standard was developed by Task Group T11 of Accredited Standards Committee NCITS during 2000–2001. This ensemble of standards is utilized to deploy SANs networks and/or support mainframe-to-peripheral (see next section) connections.

10GFC is a member of the Fibre Channel family of standards. This family includes ANSI X3.230, FC-PH, which specifies the physical and signaling interface. ANSI X3.297, FC-PH-2, and ANSI X3.303, FC-PH-3, specify enhanced functions added to FC-PH. ANSI X3.272, FC-AL, specifies the arbitrated loop topology.

10GFC describes signaling and physical services that may be utilized by the FC-2 level to transport data at a rate in excess of 10 Gbps. The Fibre Channel signaling and physical services include the following:

- Link architecture including retiming;
- Data and line rate specifications;
- Transmission coding;
- FC-1 data path interface;
- Signaling interfaces;
- Physical layer specifications;
- Equalization;
- Connector performance specifications (still being debated);
- Migration services (still being debated);
- Transceiver module form factor and interface(still being debated);
- Management interface and register set (still being debated);
- Link and cable plant management specifications (still being debated);
- Jitter measurement methods and specifications.

These are the original FC standards:

1. ANSI X3.230:1994, *Information Technology—Fibre Channel Physical and Signaling Interface (FC-PH)*;

2. ANSI X3.297:1997, *Information Technology—Fibre Channel— Physical and Signalling Interface-2 (FC-PH-2)*;

3. ANSI X3.272:1996, *Information Technology—Fibre Channel—Arbitrated Loop (FC-AL)*;

4. ANSI X3.303:1998, *Fibre Channel—Physical and Signalling Interface-3 (FC-PH-3)*;

5. ANSI NCITS 332-1999, *Fibre Channel—Arbitrated Loop (FC-AL-2)*.

The Fibre Channel is logically a bidirectional point-to-point serial data channel, structured for high performance information transport. Physically, Fibre Channel is an interconnection of one or more point-to-point links. Each link end terminates in a port. Ports are fully specified in FC-FS, FC-PI and FC-AL-2. *Fiber* is a general term used to cover all physical media supported by Fibre Channel including optical fiber, twisted pair, and coaxial cable.

Fibre Channel is structured as a set of hierarchical and related functions, FC-0 through FC-3. Each of these functions is described as a level. Fibre Channel does not restrict implementations to specific interfaces between these levels.

The physical interface (FC-0), specified in FC-PI, consists of transmission media, transmitters, receivers, and their interfaces. The physical interface specifies a variety of media and associated drivers and receivers capable of operating at various speeds.

The transmission protocol (FC-1), signaling protocol (FC-2), and common services (FC-3) are fully specified in FC-FS and FC-AL-2. Fibre Channel levels FC-1 through FC-3 specify the rules and provide mechanisms needed to transfer blocks of information end to end, traversing one or more links. An Upper Level Protocol mapping to FC-FS constitutes an FC-4, which is the highest level in the Fibre Channel structure. FC-2 defines a suite of functions and facilities available for use by an FC-4.

10GFC describes the signaling and physical interface services that may be utilized by an extended version of the FC-2 level to transport data at a rate in excess of 10 Gbps over a family of FC-0 physical variants. 10GFC additionally introduces port management functions at the FC-3 level.

A Fibre Channel node may support one or more N_Ports and one or more FC-4s. Each N_Port contains FC-0, FC-1, and FC-2 functions. FC-3 optionally provides the common services to multiple N_Ports and FC-4s.

11.3.7.1 FC-3 General Description

The FC-3 level of 10GFC extends the FC-3 levels of FC-FS and FC-AL-2 by adding a port management interface and register set and low-level signaling protocol. The port management interface and register set provides an interconnection between manageable devices within a port and port management entities. The *link signaling sublayer* (LSS) is used to signal low-level link and cable plant management information during the idle stream.

The WIS is an optional sublayer that can be used to create a physical layer that is data rate and format compatible with the SONET STS-192c transmission format defined by ANSI, as well as the SDH VC-4-64c container specified by ITU. The purpose of the WIS is to support 10GFC data streams that may be mapped directly to STS-192c or VC-4-64c streams at the PHY level, without requiring higher layer processing. The WIS specifies a subset of the logical frame formats in the SONET and SDH standards. In addition, the WIS constrains the effective data throughput at its service interface to the payload capacity of STS-192c/VC-4-64c, that is, 9.58464 Gbps. Multiplexed SONET/SDH formats are not supported.

11.3.7.2 FC-2 General Description

The FC-2 level of 10GFC extends the FC-2 levels of FC-FS and FC-AL-2 to transport data at a rate of 10.2 Gbps over a family of FC-0 physical variants. 10GFC provides the specification of optional physical interfaces applicable to the implementation of 10GFC Ports. These interfaces include the XGMII and the XAUI. One or both of these interfaces may typically be present within a 10GFC port.

XGMII

The 10-gigabit media independent interface provides a physical instantiation of a 10.2-Gbps parallel data and control transport within FC-2. Its implementation is typically an internal chip interconnect or chip-to-chip interconnect. The XGMII supports 10.2-Gbps data transport through its 32-bit-wide data and 4-bit-wide control transmit and receive paths.

XAUI

The 10-gigabit attachment unit interface provides a physical instantiation of a 10.2-Gbps four-lane serial data and control transport within FC-2 or between FC-2 and lower levels including FC-1 and FC-0. The XAUI is defined as an XGMII extender. Its implementation is typically a chip-to-chip interconnect including chips within transceiver modules. The XAUI supports 10.2-Gbps data transport through its four 8B/10B based serial transmit and receive paths.

11.3.7.3 FC-1 General Description

The FC-1 level of 10GFC provides the ability to transport data at a rate of 10.2 Gbps over a family of FC-0 physical variants. 10GFC provides the following FC-1 functions and interfaces:

- Direct mapping of FC-1 signals to 10GFC ordered sets;
- 8B/10B transmission code that divides FC-2 data and ordered sets among four serial lanes;
- 64B/66B transmission code that supports FC-2 data and ordered sets over a single serial lane;

- An optional physical interface for use by single-lane serial FC-0 variants, known as XSBI.

FC-1 signals convey FC-2 data as well as frame delimiters and control information to be encoded by FC-1 transmission code. The same conveyance exists in the reverse direction.

8B/10B transmission code is the same as that specified in FC-FS. It is intended for 10.2-Gbps data transport across printed circuit boards, through connectors, and over four separate transmitters and receivers. These four transmitters and receivers may be either optically multiplexed to and from a single fiber optic cable or directly conveyed over four individual fibers.

64B/66B transmission code is intended for 10.2-Gbps data transport across a single fiber optic cable. The primary reason for the development of this code is to provide minimal overhead above the 10.2-Gbps serial data rate to allow the use of optoelectronic components developed for other high-volume 10-Gbps communications applications such as SONET OC-192.

The XSBI provides a physical instantiation of a 16-bit-wide data path that conveys 64B/66B encoded data to and from FC-0. The XSBI is intended to support serial FC-0 variants.

11.3.7.4 FC-0 General Description

The FC-0 level of 10GFC describes the Fibre Channel link. The FC-0 level covers a variety of media and associated transmitters and receivers capable of transporting FC-1 data. The FC-0 level is designed for maximum flexibility and allows the use of a large number of technologies to meet the broadest range of Fibre Channel system cost and performance requirements.

The link distance capabilities specified in 10GFC are based on ensuring interoperability across multiple vendors supplying the technologies (both transceivers and cable plants) under the tolerance limits specified in 10GFC. Greater link distances may be obtained by specifically engineering a link based on knowledge of the technology characteristics and the conditions under which the link is installed and operated. However, such link distance extensions are outside the scope of 10GFC.

Optical Variants

Multiple optical serial physical full-duplex variants are specified to support the transport of encoded FC-1 data transport over fiber-optic medium. The variants include the following:

- Four serial lanes over individual fibers;

- Four serial lanes optically multiplexed over a single fiber;

- One lane serial over one fiber.

Copper Physical Variant

A four-lane electrical serial full-duplex physical variant is specified to support the transport of encoded FC-1 data transport over copper medium. The standards describe these interfaces and the accompanying connector and cable specifications.

Ports, Links, and Paths

Each fiber set is attached to a transmitter of a port at one link end and a receiver of another port at the other link end. When a fabric is present in the configuration, multiple links may be utilized to attach more than one N_Port to more than one F_Port. Patch panels or portions of the active fabric may function as repeaters, concentrators, or fiber converters. A path between two N_Ports may be made up of links of different technologies. For example, the path may have single-fiber multimode fiber links or parallel copper or fiber multimode links attached to end ports but may have a single-fiber single-mode fiber link in between.

Functional Characteristics

FC-PI describes the physical link, the lowest level, in the Fibre Channel system. It is designed for flexibility and allows the use of several physical interconnect technologies to meet a wide variety of system application requirements.

The FC-FS protocol is defined to operate across connections having a BER detected at the receiving node of less than 10^{-12}. It is the combined responsibility of the component suppliers and the system integrator to ensure that this level of service is provided at every node in a given Fibre Channel installation.

FC-PI has the following general characteristics. In the physical media signals a logical "1" is represented by the following properties:

1. *Optical:* the state with the higher optical power;

2. *Unbalanced copper:* the state where the ungrounded conductor is more positive than the grounded conductor;

3. *Balanced copper:* the state where the conductor identified as "+" is more positive than the conductor identified as ".–"

Optical Interface Specification

The standard defines the optical signal characteristics at the interface connector receptacle. Each conforming optical FC attachment shall be compatible with this optical interface to allow interoperability within an FC environment. Fibre Channel links shall not exceed the BER objective (10^{-12}) under any conditions. The parameters specified in this clause support meeting that requirement under all conditions including the minimum input power level. Clause 7 specifies the corresponding interface receptacle. The following physical variants are included:

- 850–nm parallel (four–lane) optics. Specified in this standard;
- 850–nm serial. Fully specified in IEEE P802.3ae Clause 52;
- 850–nm WDM (four–wavelength). Specified in this standard;
- 1,310–nm serial. Fully specified in IEEE P802.3ae Clause 52;
- 1,310–nm WDM (four–wavelength). Fully specified in IEEE P802.3ae Clause 53.

The 850–nm parallel, four–lane variant supports MM short–wavelength data links. The laser links operates at the 3.1875 gigabaud (GBd) rate. The specifications are intended to allow compliance to Class 1 laser safety. Reflection effects on the transmitter are assumed to be small but need to be bounded. A specification of maximum *relative intensity noise* (RIN) under worst case reflection conditions is included to ensure that reflections do not impact system performance. The receiver shall operate within a BER of 10^{-12} over the link's lifetime and temperature range. The spectral specifications for the 850–nm WDM four wavelengths are as follows:

1. 771.5–784.5 nm;
2. 793.5–806.5 nm;
3. 818.5–831.5 nm;
4. 843.5–856.5 nm.

11.3.8   ESCON/FICON

A decade ago, IBM introduced ESCON® (Enterprise Systems CONnection), a proprietary channel architecture to link its large S/390® mainframe computers to controllers and storage throughout an enterprise. Running on optical fiber, ESCON was designed to transfer data at rates of up to 17 MBps (136 Mbps)—an astonishing improvement over what bulky copper "bus and tag" cables could do. Today, even ESCON is hard pressed to keep up with the requirements of mission–critical applications. World-class enterprise server technology and higher capacity storage call for connectivity solutions that support faster devices and link speeds, offer more addressability for greater flexibility, extend topology over longer distances, and increase control unit intelligence.

11.3.8.1   FIber CONnector Technology (FICON)

Introduced by IBM in 1998, FICON delivers higher throughput capability and better utilization than existing ESCON channels. Based on industry–standard high-speed Fibre Channel technology, FICON meets the I/O bandwidth and connectivity needs of e-business workloads and brings significant benefits to traditional business applications as well.

The first storage products to arrive with FICON interfaces were tape drives and printers. Disk drives were added shortly thereafter. Additional FICON products are doubtless on their way. Note these characteristics of FICON:

- FICON is 10 times better for long-distance backup and recovery applications. ESCON data "droop" starts at only 9 km.
- FICON's data droop is negligible up to 100 km.
- FICON's per-channel bandwidth is almost six times ESCON's (100 MBps versus 17 MBps).
- FICON supports up to 480% more I/O operations per second (up to 1,000 for ESCON, compared with FICON's 4,800 maximum on IBM's new zSeries 900 eServer).
- FICON supports 16,000 unit addresses per channel, ESCON only 1,000. Supporting 16 times as many devices, FICON permits more storage granularity.
- FICON utilizes fiber twice as efficiently (full-duplex transfer of data versus half duplex for ESCON).
- FICON allows greater configuration flexibility. ESCON is circuit switched, so there are restrictions on large and small block intermixing. FICON is packet switched and without restrictions.
- FICON relieves "channel-challenged" systems. FICON channel's performance equates to up to 360 ESCON channels for IBM's G5 mainframe, ranging all the way to 928 for the zSeries 900.

FICON is the new high-performance highway for mainframe I/O operations and for channels. Channel Director systems switch up to 256 ESCON channels while maintaining top performance. Recently, we have seen the introduction of new technology that can be installed in the CD/9000 to convert up to 128 ESCON ports into Fibre Channel and FICON ports. Also, for those who require a Fibre Channel director, vendors have brought out 64-port systems supporting all Fibre Channel interfaces, including FICON.

11.3.9   10-Gigabit Ethernet Alliance

The 10–Gigabit Ethernet Alliance was organized to facilitate and accelerate the introduction of 10GbE into the networking market. It was founded by networking industry leaders: 3Com, Cisco Systems, Extreme Networks, Intel, Nortel Networks, Sun Microsystems, and World Wide Packets. Additionally, the alliance will support the activities of the IEEE 802.3

Ethernet committee, foster the development of the 802.3ae (10GbE) standard, and promote interoperability among 10GbE products. The alliance now has more than 100 members.

A diverse collection of 10GbE products from 18 industry-leading network vendors were seen at trade shows in late 2001. GEA member companies whose products were on display in the booths included Agilent, AMCC, Avaya, Inc., Broadcom Corp., CDT Corp., Cisco Systems, Extreme Networks, Inc., Foundry Networks, Intel Corp., Ixia, Lucent Technologies, MindSpeed, Nortel Networks, PMC-Sierra, Inc., Picolight, Spirent Communications, Tyco Electronics, and Velio Communications. The 10GEA stated: "This collection of products and technologies is the result of years of industry collaboration. This demonstration moves us to the next phase of 10 Gigabit Ethernet technology—deploying products in multivendor networks around the world" [17].

ENDNOTES

[1] Overview and Guide to the IEEE 802 LMSC," retrieved from http://grouper.ieee.org/groups/802/overview_07_12_2002.pdf, July 2002.

[2] http://www.iol.unh.edu/training/ge/8B10BEncoding.html.

[3] The exception to this rule is if /K28.7/ is followed by any of the following special or data code groups: /K28.x/, /D3.x/, /D11.x/, /D12.x/, /D19.x/, or /D28.x/, where x is a value in the range 0 to 7, inclusive.

[4] Some of this material was previously included in Minoli, D., et al., *Ethernet-Based Metro Area Networks*, New York: McGraw-Hill, 2002.

[5] Thatcher, J., July 2001 IEEE presentations, grouper.ieee.org/groups/802/3/ae/public/jul01/intro_0701.pdf.

[6] This concept is discussed later, but the notations are as follow: R = 64B/66B encoded without WIS; W = 64B/66B encoded without WIS.

[7] A *lane* is a bundle of signals that constitutes a logical subset of a point-to-point interconnect. A lane contains enough signals to communicate a quantum of data and/or control information between the two endpoints.

[8] The "AUI" portion is borrowed from the Ethernet attachment unit interface. The "X" represents the Roman numeral for 10 and implies 10 Gbps. The XAUI is designed as an interface extender, and the interface, which it extends, is the XGMII. The XGMII is a 64-signal-wide interface (32-bit data paths for each of transmit and receive) that may be used to attach the Ethernet MAC to its PHY. The XAUI may be used in place of, or to extend, the XGMII in chip-to-chip applications typical of most Ethernet MAC to PHY interconnects. The XAUI is a low pin count, self-clocked serial bus that is directly evolved from the Gigabit Ethernet 1000BASE-X PHY. The XAUI interface speed is 2.5 times that of 1000BASE-X. By arranging four serial lanes, the 4-bit XAUI interface supports the 10-times data throughput required by 10GbE.

[9] 10GbE Alliance tutorial materials., http://www.10gea.org.

[10] XGMII is optional, but used as the basis for specifications.

[11] The next three sections are based on the following references: Someshwar, V., "Metropolitan Area Networks Ethernet's Next Gig?," TM 611, Stevens Institute of Technology Class Project, Summer 2001; 10GbE Alliance materials;

Cunningham, D., and B. Lane, "Gigabit Ethernet Networking"; http://www.10gea.org; http://www.10gigabit-ethernet.com; http://www.optical-networks.com; http://www.atrica.com; http://www.cisco.com; http://www.nwfusion.com; http: //www.juniper.net; and http://www.foundry.com.

[12]   Ferrant, J. L., "Interworking Issues Between 10gbe WAN and Existing Transmission Network," IEEE G.802.3ae meeting, St. Louis, MO, May 23–25, 2001.

[13]   Bynum, R., "OTN Digital Wrapper," *The Fine Print,* IEEE P802.3ae Plenary, July 2001.

[14]   Booth, B., et al., "WAN PHY Definitions," retrieved from http://grouper.ieee.org/groups/802/3/10G_study/public/jan00/law_1_0100.pdf.

[15]   White, A., "T1X1 Pitch," IEEE P802.3ae Plenary, May 2001.

[16]   Taborek, R., "10 GbE CX—Short Haul Copper," IEEE 802.3 HSSG Contribution, July 1999.

[17]   Press Release, 10GEA, September 11, 2001.

CHAPTER 12

Transparent LAN Services

This chapter covers the concept of *transparent LAN service* (TLS), also now called *virtual private LAN services* (VPLS) by the MPLS community (we use the two terms interchangeably). The idea of TLS is to place edge NEs on customer premises, so that the handoff to the user is a LAN interface (e.g., 10, 100, 1,000, or 10,000 Mbps), rather than an OC-x interface. The advantage to the end user is that the user can avoid using expensive SONET plug-ins in their routers to run a stack such as POS, in order to connect to the network; instead, they can utilize a low-cost Ethernet router plug-in and an Ethernet cable to the carrier's NE. In effect, the network takes the responsibility of carrying the signal end to end in a transparent fashion, so that the user need not convert its signal into a format suitable to the network; rather, the network does what it has to, to accommodate the user's signal. In turn, an appropriate networking infrastructure (e.g., and ATM cloud, a SONET ring, a next generation SONET ring, an OTN/ASON infrastructure, a GbE or 10GbE metro mechanism, an MPLS/IP network, and so on) is used—Figure 12.1 shows a layer 1 or layer 2 implementation, and Figure 12.2 shows a router-based implementation [1]. One of the useful aspects of TLS is that it can interconnect multiple sites (not just point-to-point connections), each with a single Ethernet drop from the carrier; the carrier networks performs the Ethernet bridging function among all of the sites on the transparent LAN.

Put differently, a service provider offering layer 2 connectivity to multiple customer sites in a manner that is transparent to the *customer edge* (CE) switches or routers is said to be offering TLS. The provider takes care of transporting customer layer 2 frames and switching them across the provider network from one customer site to the other(s) [2]. Versions of this service were introduced in the late 1980s that used point-to-point multiplexers at either end, typically with a DS-3 facility between them. In the early 1990s, some carriers tried to extend *fiber distributed data interface* (FDDI) systems into the metro area to support TLS; unfortunately these systems suffered from lack of security, lack of QoS, lack of traffic management, inability to be restored quickly, and so on. Hence, these architectures soon disappeared. In the mid-1990s TLS started to gain importance; at this time it was delivered over an ATM platform, and carriers such as Teleport Communications Group established a relatively large customer base [3]. In the late 1990s, TLS was "rediscovered" and more than a billion

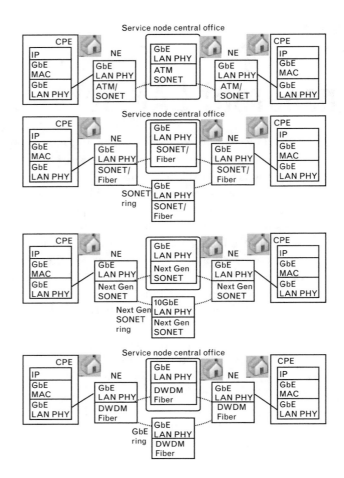

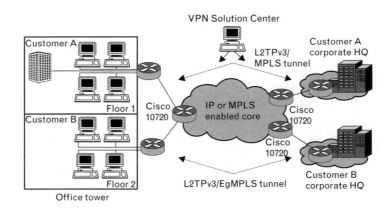

dollars of venture capital money was poured into developing GbE–based metropolitan networks supporting TLS. Unfortunately, the economics of the undertaking, particularly the cost to introduce the fiber into buildings (and at the same time limiting oneself to 1 Gbps of bandwidth when 10 Gbps could easily be achieved over the fiber), as well as the service limitations of that platform, were completely missed by the venture capitalists and by the advocates. With the providers of the time, such as Yipes and virtually all others, either gone or absorbed into other companies, it is clear now that a model that does not properly account for all costs is flawed. But the limitations of the GbE architecture were identical to the discarded FDDI rings in the metro approach: lack of security, lack of QoS, lack of traffic management, and inability to be quickly restored. We hope venture capitalists will spare us from a replay, by funding the introduction of 10GbE rings in the metro, but, of course, known at some future point by some other marketing name on colorful viewgraphs, obfuscating the traditional nature of the concept.

These failures do not take away from the concept of TLS. They merely point out that it has to be delivered over an appropriate platform, such as SONET, next generation SONET, ATM, and now MPLS. Before anachronistic cries of "bandwidth inefficiencies" are voiced, we should take note of the bandwidth glut that has been created by the development of DWDM; therefore, the focus needs to be on providing services, not on rearchitecting a few megabits per second of overhead.

12.1 Architectures

As noted in Figure 12.1, TLS can be delivered over a layer 2 platform such as ATM or MPLS (after all, LAN technologies are layer 2 technologies), or over layer 1 technologies, such as SONET, next generation SONET, or OTN/ASON. (It can also be delivered over a LAN architecture-for-the-metro, but we do not discuss this further at this juncture.) Of late, much discussion has been related to MPLS support. We here touch briefly on a related SONET issue, then focus the discussion on MPLS.

12.1.1 TLS Delivery over SONET-Based Private Lines

When using a layer 1 mechanism such as SONET to deliver TLS, the issue remains that carrying data traffic across a SONET link is not completely bandwidth efficient. For example, a 10-Mbps Ethernet link is typically transported across an STS-1 (51.84-Mbps) link. We believe that this is not a major issue. Anecdotally, when ISDN was first developed, very complex rate adaption mechanisms were developed to pack multiple 2.4-, 4.8-, 9.6-, or 19.2-Kbps DDS links onto a B-channel; now DS-0s are a pure commodity, so no one would use a complex mechanism to pack multiple low-speed channels onto a DS-0. We believe that this is the same now at the STS-1 level. However, the industry has developed complex virtual concatenation mechanisms to pack several Ethernet channels onto an

STS-1. The issue is, however, that soon customers will be looking for 100-Mbps TLS and, hence, some of this work will be of limited value.

Virtual concatenation enables channels of the same group to be combined to enable better efficiencies. Ethernet, for example, could be carried across a VT1.5-6v or a 10.368-Mbps link instead of an STS-1 link (note, by the way, that a 100BASE-T Ethernet might be transported on a single STS-1 in cases where the traffic requirements are matched to that transport speed.) Similarly, a 100-Mbps Ethernet link could be carried across an STS-2 (103.68-Mbps) link instead of an STS-3 (155.520-Mbps) link. The *link capacity adjustment scheme* (LCAS) goes a step further and increases or decreases the capacity of these links without interrupting the traffic flow. Finally, the ANSI's T1X1 committee is taking two steps to improve SONET's performance. The ratified OC-768 specification paves the way for vendors to implement 40-Gbps interfaces on their equipment. At the same time, a new proposal hopes to improve SONET scalability by enabling it to span multiple wavelengths [4].

The basic principle of virtual concatenation [5, 6] is simple [7]. A number of smaller containers are concatenated and assembled to create a bigger container that carries more data. Virtual concatenation is possible for all container sizes from *Virtual Container Level 11/Virtual Tributary Level 1.5* (VC-11/VT-1.5) up to and including *Virtual Container Level 4/Synchronous Transport Signal Level 3c* (VC-4/STS-3c). Smaller containers allow for finer granularity but reduced maximum channel sizes, and they also require the network to be able to switch down to that level. Table 12.1 shows the containers for which virtual concatenation is defined and their bandwidth ranges.

Normally, the use of virtual concatenation provides four key advantages: scalability, efficiency, compatibility, and resiliency, as discussed next.

Scalability

SONET point-to-point links, which are used to connect two remote NEs supporting TLS, can be sized to match the desired data rate to optimize bandwidth efficiency (although this author remains of the opinion that this may or may not be as important an issue in the post-telecom-glut era described in the opening paragraphs of this book as it was in the pre-telecom-glut era—just as PC developers long ago gave up the concept of "super-maximum memory efficiency" in favor of user-friendly interfaces and new applications, telecom people have to move on from what can only be called a "bandwidth efficiency fixation," which probably originated in the late 1950s, 1960s, and early 1970s when bandwidth was a scarce resource, and focus on new applications, not bandwidth efficiency; since "applications" is the third key pillar of telecom in the 2000s as we discussed in Chapter 1, this is where the emphasis must be.). While traditional contiguous concatenation comes in coarse steps, virtual concatenation allows the bandwidth to be tuned in small increments on demand. For example, if

TABLE 12.1   VIRTUAL CONCATENATION BASE CONTAINERS
AND APPROXIMATE BANDWIDTHS

SONET NAME	SDH NAME	MINIMUM SIZE (MBPS)	MAXIMUM SIZE (MBPS)	GRANULARITY (MBPS)
VT1.5	VC–11	1.6	102	1.6
VT2	VC–12	2.2	139	2.2
VT3	–	3.4	217	3.4
VT6	VC–2	6.8	43	6.8
STS-1 SPE	VC–3	48	412,000	48
STS-3c SPE	VC–4	150	38,000	150

a contiguous concatenated STS-12c (599-Mbps payload) was not quite big enough for a specific link, a full STS-48c (2,400-Mbps payload) needs to be allocated. However, with virtual concatenation, an STS-12v (599 Mbps) can be slightly upsized to an STS-13v (649 Mbps) to meet the needs and thereby leave the additional bandwidth for other links. Figure 12.3 shows a typical case where GbE data is transported over a SONET network. With legacy contiguous concatenation, the utilization is poor; with virtual concatenation, an OC–48 link can actually carry two full GbE links and still have 6 STS-1s available to carry other traffic. The question, though, is this: Is it less expensive for the carrier, which may have deployed large DWDM systems, to "waste" 1 Gbps, or is it cheaper to research, buy, engineer, deploy, OSS-integrate, provision, and monitor some new kind of network equipment that handles tighter multiplexing?

FIGURE 12.3
Scalability by using virtual concatenation.

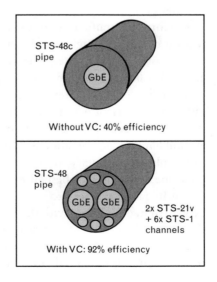

STS-48c
pipe

GbE

Without VC: 40% efficiency

STS-48
pipe

GbE   GbE

2x STS-21v
+ 6x STS-1
channels

With VC: 92% efficiency

Efficiency

Virtually concatenated channels are more easily routed through a network and can also eliminate stranded bandwidth. The orthodox conventional wisdom is that bandwidth is a scarce resource and its efficiency must be maximized. In the author's opinion there are only three places where this remains the case: radio-based links, metro access (but not metro core), and links to remote international locations. Virtual concatenation allows for more efficient usage of an existing network's available bandwidth. The real issue is that the telecom industry has tended to hold back the availability of bandwidth to the user by charging fees way in excess of the cost of bandwidth. This strategy, which the carriers considered its "salvation," backfired and the telecom crash of the early 2000s happened: One cannot anachronistically continue to label a resource scarce when it is not. Economics 101 and the law of supply and demand (e.g., consider any commodity such as grain, potatoes, or porkbellies) say that when the production is high, the price of the commodity goes down—it is very difficult to keep the price of a commodity up when the availability is high. Telecom executives need to rethink their strategies if their companies (and their jobs) are to be saved.

Compatibility

Virtual concatenation works across legacy networks. Only the end nodes of the network are aware of the containers being virtually concatenated, because this is transparent to the network. Hence, with virtual concatenation, large data channels can be routed over older networks that do not support large contiguous channels. Typically, implementing virtual concatenation means purchasing VC-enabled line cards to use with existing SONET equipment.

Resiliency

Best effort type traffic is often carried over unprotected links or in the protection channel of high-priority traffic. In the event of a link failure, the high priority reclaims the protection bandwidth and the best effort traffic across that link is halted. For a contiguous channel, this means that the best effort traffic service is interrupted completely, that is, all data are lost. Individual members of a virtually concatenated channel should be routed as diversely as possible across a network. So if one link goes down, the others are still likely to be operational. The virtually concatenated channel thereby loses only one tributary in the event of a link failure, and the link can still continue to provide the best effort service, albeit with a reduced bandwidth.

While it is beneficial to use virtual concatenation alone, advantages can be gained in coupling it with the LCAS [8]. Virtual concatenation provides the means for creating right-sized channels. But in many applications the size of a right-sized channel changes with time. When a virtual channel is resized, traffic is disrupted and lost. Strict SLAs often limit the number of acceptable traffic disruptions and thereby effectively limit channel resizing. LCAS is the solution to this problem. With LCAS, channels can be resized at any time without disturbing the traffic on the link. Also, connectivity checks are continuously performed and failed links automatically removed and added back as the link is repaired, without intervention of the (slow) network management system. With this capability, carriers can dynamically change the bandwidth allocated to a connection. As an example, bandwidth demand may increase during the nights and at weekends. A customer may subscribe to a 100-Mbps connection that increases to 1,000 Mbps between 2:00 a.m. and 3:00 a.m. every night as their computer system creates data backups. During the daytime this bandwidth is not needed and can be reallocated to other customers.

Virtual concatenation allows for any size of bandwidth. LCAS is a protocol to synchronize the resizing of a pipe in use, so it can be changed without corrupting packets in the process. It also provides automatic recovery of a link after tributary failures. Virtual concatenation can be used without LCAS, but LCAS requires virtual concatenation. LCAS is resident in the H4 POH byte of the SONET overhead, the same byte as virtual concatenation. The H4 bytes from a 16-frame sequence make up a message for both virtual concatenation and LCAS. Virtual concatenation uses 4 of the 16 bytes for its MFI and SQ numbers. LCAS uses 7 other bytes for its purposes, leaving 5 reserved for future development.

While virtual concatenation is a simple labeling of individual STS-1s within a channel, LCAS is a two-way handshake protocol. Status messages are continuously exchanged and consequent actions taken. Each STS-1 carries one of six LCAS control commands:

- Fixed—LCAS not supported on this STS-1.
- Add—Request to add this STS-1 to a channel, thereby increasing the bandwidth of an existing channel or creating a new channel.
- Norm—This STS-1 is in use.
- EOS—This STS-1 is in use and is the last STS-1 of this channel, that is, the STS-1 with the highest SQ number.
- Idle—This STS-1 is not part of a channel. Commands and returns an OK in the link status for this STS-1.
- Do not use—This STS-1 is supposed to be part of a channel, but is removed due to a broken link reported by the destination.

A typical sequence when upsizing a link is as follows:

1. The network management system adds a new trace through the network between the source and destination node.

2. The network management system orders the source to add this new link to the existing channel.

3. The source node starts sending "Add" control commands in this STS-1.

4. The destination notices the add command and returns an OK in the link status for this STS-1.

5. The source sees the OK, assigns this STS-1 an SQ number one higher than currently in use by this channel.

6. At a frame boundary, the source includes this STS-1 in the byte interleaving and sets the control command to "EOS," indicating that this STS-1 is in use and the last in the sequence.

7. The STS-1 that previously was "EOS" now becomes "Norm" because it is no longer the one with the highest SQ number.

Multiple STS-1s can be added to or removed from a link concurrently to allow for fast resizing.

12.1.2   TLS Delivery with 10GbE

As noted, and entry-level manner to deliver TLS is to "grow" the in-building GbE or 10GbE to the metro space by using long-reach optics. However, the issues that come into play are traffic management, QoS (assured bandwidth, delay, jitter), security, restorability, availability of fiber in buildings, and in-building distribution of signals (risers). Some have even advanced a two-standards approach: GbE at the last mile, and the 10GbE standard at the metro core. Unfortunately, this is a difficult-to-manage and difficult-to-scale architecture when looking at the issue from the challenge of delivering carrier-grade services perspective: Users will not place an entire gigabit-per-second stream of highly aggregated company traffic if the availability/robustness is not 99.9999%—and to achieve this level of resiliency this a heavy-duty architecture is required.

A slightly different approach than to simply use long-reach optics is, at least in the 10GbE context, the WAN PHY. As noted in Chapter 11, within the IEEE 802.3 committee, the WAN PHY was developed such that it could integrate more effectively with existing SONET networks. However, the WAN PHY has limitations—the problem relates to clocking. SONET's clock is accurate to $\pm$ 10 picoseconds. The clock originally proposed by the 10GbE group called for an accuracy of $\pm$ 100 ppm, or 100 μs of inaccuracy. The 10GbE group has corrected the problem to some extent, using a clock that's accurate to 20 ppm. However, even this may pose problems. If the carrier has a point-to-point connection, then the transport of native Ethernet frames on the WAN will work, but no SONET muxing or grooming can be in between the end nodes [4].

12.1.3   TLS Delivery via MPLS

MPLS supports the ability to provide virtual private networks over an IP infrastructure [9]. MPLS utilizes a labeling mechanism (see Figure 12.4)

FIGURE 12.4
MPLS label stack.

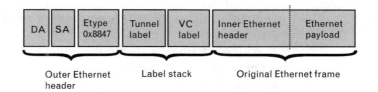

that was described in detail in Chapter 8. One of the applications of this capability is TLS. Table 12.2 depicts a set of technical documents generated at press time to develop the TLS concept further [10]. The IETF has a working group, the MPLS working group, categorized as a sub-IP group, that is responsible for the development of standards that define the core technology; other IETF working groups, such as the Provider Provisioned Virtual Private Networks Working Group, also develop standards that make use of the MPLS technology. The list of current Internet drafts can be accessed at http://www.ietf.org/ietf/1id-abstracts.txt. The list of Internet draft shadow directories can be accessed at http://www.ietf.org/shadow.html. The technical details of TLS support are discussed in the next section. The rest of the chapter is based on a number of the IETF MPLS drafts; these activities are "works in progress." These mechanisms may change in the future, but because they relate to what we consider

TABLE 12.2   MPLS-RELATED TECHNICAL DOCUMENTS PRODUCED BY THE IETF TO SUPPORT TLS

July 2002	Virtual Private LAN Service (VPLS) Management Information Base Using SMIv2
July 2002	Virtual Private LAN Service (VPLS) Solution Using GRE Based IP Tunnels
June 2002	Discovering Nodes and Services in a VPLS Network
June 2002	DNS/L2TP Based VPLS
May 2002	Auto-Discovery of VPLS Membership and Configuration Using BGP-MP
March 2002	Requirements for Virtual Private LAN Services (VPLS)
March 2002	VPLS/LPE L2VPNs: Virtual Private LAN Services Using Logical PE Architecture
March 2002	Virtual Private LAN Services over MPLS (in progress)
February 2002	Requirements for Virtual Private LAN Services
February 2002	Decoupled/Hierarchical VPLS: Commonalities and Differences
February 2002	VPLS Architectures
January 2002	DNS/LDP Based VPLS
November 2001	Decoupled Virtual Private Transparent LAN Services
November 2001	Hierarchical Virtual Private LAN Service
November 2001	VPLS/LPE L2VPNs: Virtual Private LAN Services Using Logical PE Architecture
November 2001	Bandwidth Management in VPLS Networks
November 2001	Virtual Private LAN Service
November 2001	Architecture and Model for Virtual Private LAN Services (VPLS)
November 2001	Decoupled Virtual Private LAN Services
November 2001	Virtual Private LAN Service
November 2001	Transparent VLAN Services over MPLS
November 2001	VPLS/LPE L2VPNs: Virtual Private LAN Services Using Logical PE Architecture

important work, we take the opportunity to publicize these concepts at this juncture.

As noted earlier in the book, MPLS is a framework that allows the introduction of label switching to any combination of layer 3 and layer 2 protocols. MPLS is not a protocol per se; it is a framework of functions. The framework incorporates concepts, mechanisms, and protocols to achieve functions that enhance the current layer 3 and layer 2 technologies. In an MPLS domain, a packet is examined at the ingress point, its headers are parsed, a routing decision is made, and a label is attached to it. The packet is then forwarded to the next router and the label tells the router what to do with the packet. Hence, the switching decision is made based on the label only—not on the layer 3 headers. The router discards the label and attaches a new label to be used by the next router. The process continues until the packet emerges at the egress point. MPLS evolved from technologies that were primarily developed in the mid-1990s. However, it is worth noting that the concept of using label switching for a layer 3 connectionless protocol could be traced back to the mid-1980s. At that time, the high-speed networking community was faced with the challenge of increasing IP datagram forwarding rates several orders of magnitude, to the rates of OC-3 and OC-12. Routing based on IP headers used to be performed in software, consuming many CPU cycles and, hence, performance was limited by the processing power of the CPU. Several network researchers investigated the possibility of using label switching as a means of increasing the forwarding performance in an IP network. Label switching was a much simpler function and could be implemented in hardware, which made it a very promising approach. In the mid-1990s the concept of label switching started drawing attention again, and several technologies were developed based on it, typically in the context of IP and ATM [9]. These were the most notable developments:

- Toshiba's cell switching router, 1995;
- Ipsilon's IP switch, 1996;
- Cisco's tag switching, 1996;
- IBM's *aggregate route-based IP switching* (ARIS), 1996.

In 1997, the MPLS working group was formed with the goal of developing a standard approach for label switching.

Because MPLS provides tunneling of packets from an ingress point to an egress point, it is an attractive technology for VPN applications. Several flavors of VPNs—defined in IETF drafts and RFCs—can be implemented over MPLS. These VPNs can be broadly categorized as either MPLS layer 3 VPNs or MPLS layer 2 VPNs. *Virtual leased line* (VLL) is a form of MPLS layer 2 VPN offered as a service by a service provider. The service connects two customer edge devices at two different locations, as if they had a traditional leased line in between. That is, whatever layer 2 frames the CE device on one end sends are transparently transported to the CE device at the other end of the VLL. MPLS layer 2 VPNs could be used to implement TLS. In the context of MPLS and provider provisioned IP-based VPNs, the preferred term for this service is VPLS, which is mentioned in RFC 2764, *A Framework for IP Based Virtual Private Networks*. A service provider offering layer 2 connectivity to multiple customer sites in a manner that is transparent to the CE devices is said to be offering a VPLS. This is a new term that was chosen because the service resembles the process of connecting CE devices via a switch, that is, all in the same broadcast domain/LAN segment. The provider takes care of transporting customer layer 2 frames and switching them across the provider network from one customer site to the other(s). MPLS layer 2 VPNs could be used to implement this service. In the context of classical layer 2 service provider networks or ATM networks, the term *TLS* is used. In the context of MPLS and provider provisioned IP based VPNs, the terms *TLS* and *VPLS* are often used interchangeably; however, the preferred term for this service is *VPLS,* which is mentioned in RFC 2764, *A Framework for IP Based Virtual Private Networks* [2]. Table 12.3 provides key concepts in VPLS [11].

12.2   MPLS-Based Transparent LAN Services

A lot of work has been done of late by Marc Lasserre and team on describing mechanisms to deliver TLS via MPLS. This section highlights the concepts and then includes the technical details of the key IETF drafts on this topic.

12.2.1   Conceptual Highlights

Up to the present, metropolitan services have been based on TDM technologies, such as SONET. With data traffic becoming more prominent, in

TABLE 12.3   KEY CONCEPTS IN VPLS

VPLS	VIRTUAL PRIVATE LAN SERVICE
VPLS system	A collection of communication equipment, related protocols, and configuration elements that implements VPLS services.
VLAN	A customer VLAN identification using some scheme such as IEEE 802.1Q tags, port configuration, or any other means. A VPLS service can be extended to recognize customer VLANs.
VPLS virtual switch	The virtual switch is a logical switch that has logical ports (e.g., virtual circuits) as its interfaces. Therefore, it has the ability to do regular bridge/switch functionality such as MAC address learning/aging, flooding, forwarding (unicasting, multicasting/broadcasting), and running STP (if needed) per broadcast domain but based on its logical ports.
VPLS virtual port	The logical port of a virtual switch. It is connected to virtual circuit.
VPN	Virtual private network.

addition to these traditional approaches, carriers need to offer data services that are more directly based on Ethernet and IP technologies. The drive to use Ethernet as a communications technology comes from the economic benefits it offers the user in reducing its interworking costs and the flexibility that Ethernet offers. (It is tautological that transport costs are the same regardless of the line protocol one employs, once the protocol overhead is factored in and once the multiplexing density is taken into account. Stating that Ethernet-based NEs are cheaper that SONET NEs is not a well-defined statement if one ignores all the redundancy and restoration capabilities that come with SONET. For example, which company these days can afford multihour-long outages to their operation?)

It is crucial to realize that no store in the world would survive by offering *only* one product. Imagine Starbucks Coffee offering just one beverage. In the late 1990s one saw the emergence of *metro service providers* (MSPs), also known as Ethernet LEC or IP CLECs, that offered *only* TLS, not even DS-1 or DS-3 services. This industry segment has now nearly disappeared, because no store can survive by offering a single product, and because the ELECs totally misunderstood where the costs are for a carrier, specifically entrance, OAM&P, and SGA costs, and what it takes to run an operation. Statements such as "Ethernet access can offer customers much more bandwidth for much less money" are completely unfounded in the following sense: These MSPs are not sustainable *as a business* if they do not charge at least $6,000 MRC for a 10-Mbps metro service, $11,000 MRC for a 100-Mbps service, and $20,000 for a 1,000-Mbps service. One cannot use venture capitalist funds to subsidize a service that is not self-sustaining and keep the company afloat indefinitely. Of course anybody, including AlphaBeta-Gamma, Inc., of Crego, Faz. Premia, VB, Italy, would move thousands of Lexus 430s if they gave them away for $1,000 each, but how long would that business proposition last?

Ignoring all other issues, and focusing on the service issue, MSPs need to offer business customers DS-1/DS-3 private-line services (for voice applications and so on), services already available on ATM or frame relay networks, and TLS; the more services they can offer the better. Stated very simply they need a multiple-architecture, multiple-technology, multiple-service, multiple-QoS capability. Their architecture must be such that it is not limiting or excluding any service. A GbE-in-the-MAN architecture suffers from precisely the same limitations of the FDDI-in-the-MAN architecture that was tried but quickly abandoned in the early 1990s.

Today [12], MSPs rely on VLAN technology and IP networks to offer VLL and TLS. Unfortunately, this is clearly a short-term solution. VLANs were never designed for this usage. The IEEE 802.1Q specification allows a maximum of 4,096 unique VLANs; as soon as more than 4,096 customers need to be supported, MSPs will need new technologies. Furthermore, IP tunnels do not offer the kind of QoS guarantees available with ATM

VCs, nor the level of protection that SONET offers. To truly compete with TDM technologies, new mechanisms are required.

MPLS-based TLS and VLL offer an alternative. Using MPLS-TLS and VLL allows carriers to offer security, traffic engineering, and QoS services to customers across the metro network and into the core network. TLS allows carriers to create a VPN tunnel for every customer through the network. Each VPN tunnel can be provisioned to a customer-specified bandwidth and delay. VLL services, in turn, allow carriers to compete with traditional LECs by offering a bandwidth provisioning point-to-point circuit within the metro. Instead of connecting metro buildings with traditional T1 circuits from a LEC, customers can obtain an Ethernet VLL from a carrier to perform the same service, at a lower cost. Several new technologies address some or all of these problems:

- Stackable VLAN (SVLAN);
- Ethernet in IP or GRE;
- MPLS.

SVLANs solve the 4,096 VLAN limitation by allowing a stack of two 802.1Q headers to be carried in an Ethernet frame, effectively extending the number of VLANs to more than 16 million (4,096 × 4,096). At the same time multiple VLANs can now be multiplexed within a single core VLAN (the top .1Q tag). The *Generic Attribute Registration Protocol* (GARP) and the *GARP VLAN Registration Protocol* (GVRP) can be used to automatically provision VLANs across the backbone. Spanning tree extensions allow fast convergence (in the order of 1 sec). However, SVLANs provide only a partial solution. If customers are given the option to define their own VLAN ID spaces, the core of the network remains limited to 4,096 VLANs. Also, without extended tunneling options such as Ethernet-in-Ethernet tunneling or without the use of a CPE router, the number of MAC addresses handled by the core will not be manageable.

Ethernet in IP or GRE offers the strength and scalability of a routed backbone, while allowing each customer site to define multiple private VLANs that can be tunneled within a very large number of IP tunnels. Provisioning of IP tunnels is not automatic, and the number of IP tunnel address pairs to manage is a major issue. For protection, new protocols are being devised. IP routing protocols take several seconds at best to converge when a failure occurs. The Link Management Protocol, being defined within the IETF, will monitor the link state of any underlying technology and provide fast failure detection. For scalability purposes, the number of tunnels in the core could be minimized by defining hierarchical IP VPNs. But this leads to bandwidth inefficiency as the original Ethernet frame needs to be encapsulated into two IP headers, where the inner IP tunnel is used for intra-POP connectivity and the outer IP tunnel for inter-POP connectivity.

MPLS offers the strength and scalability of IP tunnels while providing means to dynamically provision MPLS tunnels (*label-switch paths*—LSPs). These LSPs can be used for traffic engineering, to create differentiated services, and to offer unique protection schemes. The Martini Internet draft specifies how to transport Ethernet, ATM and FR PDUs, and TDM over MPLS. This Internet draft focuses on point-to-point connectivity. Extensions to this model are needed in order to provide multipoint-to-multipoint support, that is, broadcasting and multicasting support. Note that all of these schemes can be combined. For instance, SVLANs can be used within the POP while IP or MPLS tunneling is used in the core. A possible scenario would be to start deploying MPLS in the core while the edge continues to use more mature technologies such as VLANs or stackable VLANs.

In MPLS, a customer's Ethernet frame is either switched or routed by a CPE device to a PE router known as an MPLS *label edge router* (LER). The PE router determines which VLAN the frame belongs to, either by looking at the 802.1Q header or by determining the VLAN associated with the incoming port. Filters can be applied to the frame so that undesired frames get dropped. For instance, if a CPE router is used, the PE device can check that the source MAC address corresponds to the CPE MAC address. Once the frame is deemed valid, the packet is mapped to a user-defined FEC that defines how specific packets get forwarded. The FEC lookup yields the outgoing port and two labels. The first label at the top of the stack is the tunnel label and is used to carry the frame across the provider backbone. The second label at the bottom of the stack is the VC label and is used by the egress switch to determine how to process the frame. After adding the two MPLS headers, one for each label, the frame is encapsulated into the proper format corresponding to the outgoing interface.

The backbone LSRs only look at the top label to switch the labeled frame across the MPLS domain. It is possible that additional labels get pushed along the way. The top tunnel label is typically removed by the penultimate hop, that is, the hop prior to the egress LER. The egress LER infers from the VC label how to process the frame and then forwards it to the appropriate outgoing port. VC LSPs are usually set up statically or dynamically via the LDP. LDP allows best effort LSPs to be established. When traffic engineering LSPs are required, the *constraint-based LDP* (CR-LDP) and RSVP-TE signaling protocols are used instead. Because resources in metro area networks are usually plentiful, traffic engineering is not necessary, making LDP a good choice for setting up VC LSPs. In the core backbone, because resources are not as easily available, traffic engineering is often required. For this reason, tunnel LSPs are usually established via RSVP-TE. One VC LSP, or multiple differentiated VC LSPs, is established between each customer site belonging to the same VLAN. A single-tunnel LSP carries all the traffic from multiple customers between two locations. By nesting LSPs, that is, by building a forwarding hierarchy, and by limiting the number of core LSPs to the number of locations to interconnect, MPLS offers a very scalable solution.

In Figure 12.5, two different customers are provided with TLS services. Customer A has three different sites, one in San Francisco, one in Chicago and one in New York. Customer B has facilities in San Francisco and New York. The carrier backbone consists of a full mesh of three LSPs (three pairs as discussed later). An end-to-end LSP, established between each location for each customer, is tunneled through a core LSP. For customer A, two VC LSPs are established at each POP. From the San Francisco POP, one VC LSP carries traffic to Chicago and another LSP carries traffic to New York. Similarly, two VC LSPs in Chicago and New York are set up exclusively for customer A. This full mesh of LSPs forms a unique broadcast domain, VLAN A, for customer A. For customer B, only one VC LSP is needed in San Francisco and New York. Customers A and B share the same tunnel LSP between San Francisco and New York LSRs.

Note that since LSPs are unidirectional, a pair of LSPs is actually needed to create a bidirectional pipe. Extensions to the LDP and RSVP-TE signaling protocols are being proposed in order to automatically set up either the reverse path LSP when the first simplex LSP is established or to set up bidirectional LSPs.

The ability to treat pairs of LSPs as virtual interfaces that can be added to a VLAN allows transparent bridging to operate. When a broadcast frame or a frame with an unknown destination needs to be sent, the frame is flooded on all the LSPs that are part of the VLAN. The LER performs the packet replication across the LSPs as the frame enters the MPLS domain. Once MAC addresses have been learned, frames are only sent on the

FIGURE 12.5
TLS across MANs.

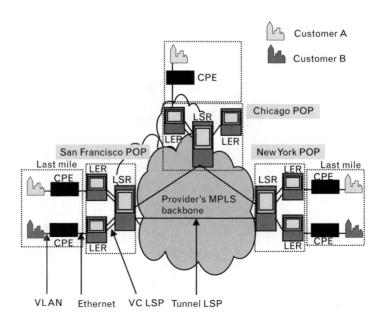

proper LSP. When a new MAC address is learned on an inbound LSP, it needs to be associated with the outbound LSP that is part of the same pair.

Because of the MPLS tunneling hierarchy (see Figure 12.6), the VC label is not visible until the frame reaches the egress LER. The egress LER infers from the VC label the type of traffic being carried, such as ATM, FR, or Ethernet, and how to handle the corresponding frame. For ATM AAL-5 traffic, the frame needs to be carried across the fabric to the proper output port and VPI/VCI. For Ethernet traffic, the VC label can be used to determine which VLAN the frame belongs to and the outgoing port or to perform an extended L2 lookup. The VC LSP creates a per-customer tunnel that isolates traffic from other customers and offers the same level of security as a FR or ATM virtual circuit.

As a frame crosses an MPLS domain, several headers get added and several fields get changed. Figure 12.7 shows how an Ethernet frame originated from a customer site is transformed into a labeled frame and sent across other Ethernet links. The first hop, the CPE device, is an Ethernet switch in our example that will not change any field. As the frame enters the service provider network, the LER adds a two-label MPLS header to the original frame. The LER then adds another Ethernet header since the outgoing interface is also an Ethernet link. This outer Ethernet header contains the source MAC address of the LER and the destination MAC address of the next MPLS hop, and the MPLS Ethernet type (0x8847 for unicast traffic and 0x8848 for multicast). The original Ethernet is obviously untouched and carries the MAC addresses of the original sender and the actual recipient. The tunnel label is swapped by each transit LSR as the labeled frame crosses the MPLS cloud. At the same time, outer source and destination MAC addresses also get changed for the current hop and next hop MAC addresses, exactly like a traditional router. When the frame reaches the penultimate hop, the tunnel label is popped off and the labeled

FIGURE 12.6
MPLS tunneling hierarchy.

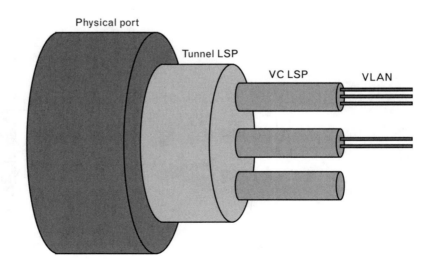

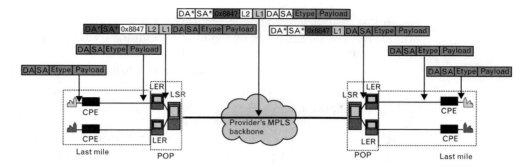

FIGURE 12.7   *MPLS encapsulation.*

frame is sent to the egress LER. The LER uses the VC label to infer the output port, pops off the last label, removes the outer Ethernet header, and transmits the original Ethernet frame toward the recipient.

12.2.2   Technical Details

Marc Lasserre, Vach Kompella, and others have prepared "Virtual Private LAN Services over MPLS," an Internet draft document that is a "work in progress" that describes a VPLS solution over MPLS, also known as TLS. VPLS simulates an Ethernet virtual 802.1D bridge [13, 14] for a given set of users. It delivers a layer 2 broadcast domain that is fully capable of learning and forwarding on Ethernet MAC addresses that is closed to a given set of users. Many VLS services can be supported from a single PE node. Existing Internet drafts specify how to provide point-to-point Ethernet L2 VPN services over MPLS. The draft that we provide here defines how multipoint Ethernet services can be supported.

12.2.2.1   Overview

Ethernet [15] has become the predominant technology for LAN connectivity and is gaining acceptance as an access technology, specifically in MANs and WANs, respectively. An Ethernet port is used to connect a customer to the PE router acting as an LER. Customer traffic is subsequently mapped to a specific MPLS L2 VPN by configuring L2 FECs based on the input port ID and/or VLAN index depending on the VPLS service.

Broadcast and multicast services are available over traditional LANs. MPLS does not support such services currently. Sites that belong to the same broadcast domain and that are connected via an MPLS network expect broadcast, multicast, and unicast traffic to be forwarded to the proper location(s). This requires MAC address learning/aging on a per-LSP basis and packet replication across LSPs for multicast/broadcast traffic and for flooding of unknown unicast destination traffic.

The primary motivation behind VPLS is to provide connectivity between geographically dispersed customer sites across MAN/WAN network(s), as if they were connected using a LAN. The intended application for the end user can be divided into the following two categories:

1. Connectivity between customer routers: LAN routing application;

2. Connectivity between customer Ethernet switches: LAN switching application.

Reference [16] defines how to carry L2 PDUs over point-to-point MPLS LSPs, called VC LSPs. Such VC LSPs can be carried across MPLS or GRE tunnels. This section describes extensions to [16] for transporting Ethernet/802.3 and VLAN [17] traffic across multiple sites that belong to the same L2 broadcast domain or VPLS. Note that the same model can be applied to other 802.1 technologies. It describes a simple and scalable way to offer virtual LAN services, including the appropriate flooding of broadcast, multicast, and unknown unicast destination traffic over MPLS, without the need for address resolution servers or other external servers, as discussed in [18].

The following discussion applies to devices that serve as LERs on an MPLS network that is VPLS capable. The behavior of transit LSRs that are considered a part of MPLS network is not discussed. The MPLS network provides a number of LSPs that form the basis for connections between LERs attached to the same MPLS network. The resulting set of interconnected LERs forms a private MPLS VPN in which each LSP is uniquely identified at each MPLS interface by a label.

12.2.2.2   Bridging Model for MPLS

An MPLS interface acting as a bridge must be able to flood, forward, and filter bridged frames (see Figure 12.8). The set of PE devices interconnected via transport tunnels appears as a single 802.1D bridge/switch to customer C1. Each PE device will learn remote MAC addresses to VC LSP associations and learns directly attached MAC addresses on customer facing ports.

FIGURE 12.8
Bridging model for MPLS.

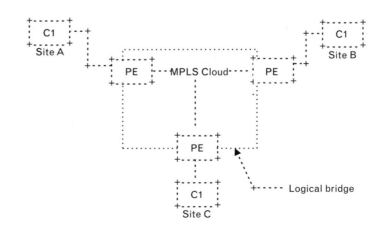

While [15] shows specific examples using MPLS transport tunnels, other tunnels that can be used by pseudowires, for example, GRE, L2TP, and IPSEC, can also be used, as long as the originating PE can be identified, since this is used in the MAC learning process.

The scope of the VPLS lies within the PEs in the service provider network, highlighting the fact that apart from customer service delineation, the form of access to a customer site is not relevant to the VPLS [18]. The PE device is typically an edge router capable of running a signaling protocol and/or routing protocols to exchange VC label information. In addition, it is capable of setting up transport tunnels to other PEs to deliver VC LSP traffic.

Flooding and Forwarding

One of the attributes of an Ethernet service is that all broadcast and destination unknown MAC addresses are flooded to all ports. To achieve flooding within the service provider network, all address unknown unicast, broadcast, and multicast frames are flooded over the corresponding pseudowires[1] to all relevant PE nodes participating in the VPLS. In the MPLS environment this means sending the PDU through each relevant VC LSP.

Note that multicast frames are a special case and do not necessarily have to be sent to all VPN members. For simplicity, the default approach of broadcasting multicast frames can be used. Extensions explaining how to interact with 802.1 GMRP protocol, IGMP snooping, and static MAC multicast filters will be discussed in a future revision if needed.

To forward a frame, a PE must be able to associate a destination MAC address with a VC LSP. It is unreasonable and perhaps impossible to require PEs to statically configure an association of every possible destination MAC address with a VC LSP. Therefore, VPLS-capable PEs must have the capability to dynamically learn MAC addresses on both physical ports and virtual circuits and to forward and replicate packets across both physical ports and virtual circuits.

Address Learning

Unlike BGP VPNs [19], reachability information does not need to be advertised and distributed via a control plane. Reachability is obtained by standard learning bridge functions in the data plane.

As discussed previously, a pseudowire consists of a pair of unidirectional VC LSPs. When a new MAC address is learned on an inbound VC LSP, it needs to be associated with the outbound VC LSP that is part of the same pair. The state of this logical link can be considered as up as soon as

---

1.   Pseudowire is a concept defined by the IETF PWE3 (Psuedo Edge to Edge Emulation) WG. It emulates the canonical attributes of a (typical layer 2) service, such as Frame Relay, PPP, T1, Ethernet, ATM, and so forth over a packet-switched network. The packet switched network could be an IP network or an MPLS network. It "sort of" is like a wire, but not exactly.

both incoming and outgoing LSPs are established. Similarly, it can be considered as down as soon as one of these two LSPs is torn down. Standard learning, filtering, and forwarding actions, as defined in [13, 14, 17], are required when a logical link state changes.

LSP Topology

PE routers typically run an IGP between them, and are assumed to have the capability to establish MPLS tunnels. Tunnel LSPs are set up between PEs to aggregate traffic. VC LSPs are signaled to demultiplex the L2 encapsulated packets that traverse the tunnel LSPs. In an Ethernet L2VPN, it becomes the responsibility of the service provider to create the loop free topology. For the sake of simplicity, we assume that the topology of a VPLS is a full mesh of tunnel and pseudowires.

Loop Free L2 VPN

For simplicity, a full mesh of pseudowires is established between PEs. Ethernet bridges, unlike FR or ATM where the termination point becomes the CE node, have to examine the layer 2 fields of the packets to make a switching decision. If the frame is a destination unknown, broadcast, or multicast frame, the frame must be flooded. Therefore, if the topology is not a full mesh, the PE devices may need to forward these frames to other PEs. However, this would require the use of an STP to form a loop free topology, which may have characteristics that are undesirable to the provider. The use of a full mesh and split-horizon forwarding obviates the need for an STP.

Each PE must create a rooted tree to every other PE router that serves the same L2 VPN. Each PE must support a split-horizon scheme in order to prevent loops, that is, a PE must *not* forward traffic from one pseudowire to another in the same VPN (since each PE has direct connectivity to all other PEs in the same VPN).

Note that customers are allowed to run STP, such as when a customer has "back door" links used to provide redundancy in the case of a failure within the VPLS. In such a case, STP BPDUs are simply tunneled through the MPLS cloud.

LDP-Based Signaling

To establish a full mesh of pseudowires, all PEs in a VPLS must have a full mesh of LDP sessions. Once an LDP session has been formed between two PEs, all pseudowires are signaled over this session (see Figure 12.9). In [20], the L2 VPN information is carried in a label mapping message sent in downstream unsolicited mode, which contains the following VC FEC TLV. The VC, C, VC Info Length, Group ID, and Interface parameters are as defined in [20].

Reference [15] defines a new VC type value in addition to the values already defined in [20] and shown in Table 12.4. VC types 0x0004 and 0x0005 identify VC LSPs that carry VLAN tagged and untagged Ethernet traffic, respectively, for point-to-point connectivity.

FIGURE 12.9
LDP-based signaling.

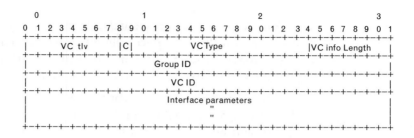

TABLE 12.4 VC TYPE VALUES

VC TYPE	DESCRIPTION
0x0001	FR DLCI
0x0002	ATM AAL-5 VCC transport
0x0003	ATM transparent cell transport
0x0004	Ethernet VLAN
0x0005	Ethernet
0x0006	HDLC
0x0007	PPP
0x8008	CEM
0x0009	ATM VCC cell transport
0x000A	ATM VPC cell transport
0x000B	Ethernet VPLS

Reference [15] defines a new VC type, Ethernet VPLS, with code point 0x000B to identify VC LSPs that carry Ethernet traffic for multipoint connectivity. The Ethernet VC type is described later. For VC types 0x0001 to 0x000A, the VC ID identifies a particular VC. For the VPLS VC type, the VC ID is a VPN identifier globally unique within a service provider domain.

Note that the VCID as specified in [20] is a service identifier, identifying a service emulating a point–to–point virtual circuit. In a VPLS, the VCID is a single service identifier, identifying an emulated LAN segment.

The use of the VCID as the VPN-ID creates some challenges for inter-provider VPLS service and this issue will be addressed in the future revision.

Ethernet VPLS VC Type

VPLS Encapsulation Actions   In a VPLS, a customer Ethernet packet without preamble is encapsulated with a header as defined in [16]. A customer Ethernet packet is defined as follows:

• If the packet, as it arrives at the PE, has an encapsulation that is used by the local PE as a service delimiter, then that encapsulation is stripped before the packet is sent into the VPLS. As the packet exits the VPLS, the packet may have a service-delimiting encapsulation inserted.

• If the packet, as it arrives at the PE, has an encapsulation that is not service delimiting, then it is a customer packet whose encapsulation should not be modified by the VPLS. This covers, for example, a packet that carries customer-specific VLAN-IDs that the service provider neither knows about nor wants to modify.

By following the preceding rules, the Ethernet packet that traverses a VPLS is always a customer Ethernet packet. Note that the two actions, at ingress and egress, of dealing with service delimiters are local actions that neither PE has to signal to the other. They allow, for example, a mix-and-match of VLAN tagged and untagged services at either end, and they do not carry across a VPLS a VLAN tag that may have only local significance. The service delimiter may be a VC label also, whereby an Ethernet VC given by [16] can serve as the access side connection into a PE. An RFC 1483 PVC encapsulation could be another service delimiter. By limiting the scope of locally significant encapsulations to the edge, hierarchical VPLS models can be developed that provide the capability to network-engineer VPLS deployments, as described later.

VPLS Learning Actions   Learning is done based on the customer Ethernet packet, as defined earlier. The *forwarding information base* (FIB) keeps track of the mapping of customer Ethernet packet addressing and the appropriate pseudowire to use. We define two modes of learning: qualified and unqualified learning.

In *unqualified learning,* all of the customer's VLANs are handled by a single VPLS, which means that they all share a single broadcast domain and a single MAC address space. This means that MAC addresses need to be unique and nonoverlapping among customer VLANs or else they cannot be differentiated within the VPLS instance, which can result in loss of customer frames. An application of unqualified learning is port-based VPLS service for a given customer (e.g., a customer with a nonmultiplexed UNI interface for which the entire traffic is mapped to a single VPLS instance).

In *qualified learning,* each customer VLAN is assigned to its own VPLS instance, which means each customer VLAN has its own broadcast domain and MAC address space. Therefore, in qualified learning, MAC addresses among customer VLANs may overlap with each other, but they will be handled correctly since each customer VLAN has its own FIB; that is, each customer VLAN has its own MAC address space. Because VPLS broadcasts multicast frames, qualified learning offers the advantage of limiting the broadcast scope to a given customer VLAN.

12.2.2.3 MAC Address Withdrawal

It *may* be desirable to remove or relearn MAC addresses that have been dynamically learned for faster convergence. In [15], an optional MAC TLV is introduced that is used to specify a list of MAC addresses that can be removed or relearned using the address withdraw message. The address withdraw message with MAC TLVs *may* be supported in order to expedite learning of MAC addresses as the result of a topology change (e.g., failure of the primary link for a dual-homed MTU). If a notification message is sent on the backup link (blocked link), which has transitioned into an active state (e.g., similar to the Topology Change Notification message of 802.1w RSTP), with a list of MAC entries to be relearned, the PE will update the MAC entries in its FIB for that VPLS instance and send the message to other PEs over the corresponding directed LDP sessions.

If the notification message contains an empty list, this tells the receiving PE to remove all of the MAC addresses learned for the specified VPLS instance except the ones it learned from the sending PE (MAC address removal is required for all VPLS instances that are affected). Note that the definition of such a notification message is outside the scope of the document, unless it happens to come from an MTU connected to the PE as a spoke. In such a scenario, the message will be just an address withdraw message as noted earlier.

MAC TLV

MAC addresses to be relearned can be signaled using an LDP address withdraw message that contains a new TLV, the MAC TLV. Its format is described next. The encoding of a MAC TLV address is the 6-byte MAC address specified by IEEE 802 documents [14] (see Figure 12.10).

- U bit—unknown bit. This bit must be set to 0. If the MAC address format is not understood, then the TLV is not understood and must be ignored.

- F bit—forward bit. This bit must be set to 0. Since the LDP mechanism used here is targeted, the TLV must *not* be forwarded.

FIGURE 12.10
MAC TLV.

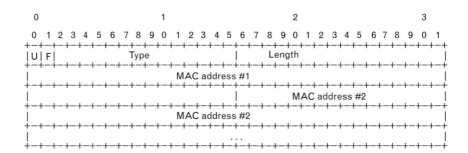

- Type—type field. This field must be set to 0x0404 (subject to IANA approval). This identifies the TLV type as MAC TLV.
- Length—length field. This field specifies the total length of the TLV, including the Type and Length fields.
- MAC address—This is the MAC address being removed.

The LDP address withdraw message contains an FEC TLV (to identify the VPLS in consideration), a MAC address TLV and optional parameters. No optional parameters have been defined for the MAC address withdraw signaling.

Address Withdraw Message Containing MAC TLV

When MAC addresses are being removed or relearned explicitly, for example, the primary link of a dual–homed MTU–s has failed, an address withdraw message can be sent with the list of MAC addresses to be relearned. The processing for MAC TLVs received in an address withdraw message is as follows. For each MAC address in the TLV:

- Relearn the association between the MAC address and the interface/pseudowire over which this message is received.
- Send the same message to all other PEs over the corresponding directed LDP sessions.

For an address withdraw message with an empty list:

- Remove all the MAC addresses associated with the VPLS instance (specified by the FEC TLV) except the MAC addresses learned over this link (over the pseudowire associated with the signaling link over which the message is received).
- Send the same message to all other PEs over the corresponding directed LDP sessions.

The scope of a MAC TLV is the VPLS specified in the FEC TLV in the address withdraw message. The number of MAC addresses can be deduced from the length field in the TLV.

Further descriptions of how to deal with failures expeditiously with different configurations will be described in other documents, such as [21].

12.2.2.4   Operation of a VPLS

This section shows an example of how a VPLS works. The following discussion uses Figure 12.11, which shows a VPLS that has been set up between PE1, PE2 and PE3. Initially, the VPLS is set up so that PE1, PE2, and PE3 have a full mesh of tunnels between them for carrying tunneled traffic. The VPLS instance is assigned a VCID (a 32-bit quantity that is unique across the provider network across all VPLSs). (Allocation of

FIGURE 12.11
Operation of a
VPLS.

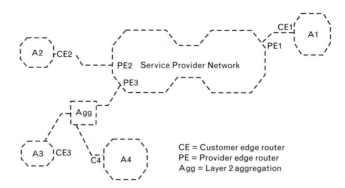

CE = Customer edge router
PE = Provider edge router
Agg = Layer 2 aggregation

domainwide unique VCIDs is outside the scope of the draft.) For our example, say PE1 signals VC label 102 to PE2 and 103 to PE3, and PE2 signals VC label 201 to PE1 and 203 to PE3.

Assume a packet from A1 is bound for A2. When it leaves CE1, say it has a source MAC address of M1 and a destination MAC of M2. If PE1 does not know where M2 is, it will multicast the packet to PE2 and PE3. When PE2 receives the packet, it will have an inner label of 201. PE2 can conclude that the source MAC address M1 is behind PE1, since it distributed the label 201 to PE1. It can therefore associate MAC address M1 with VC label 102.

MAC Address Aging

PEs that learn remote MAC addresses need to have an aging mechanism to remove unused entries associated with a VC label. This is important both for conservation of memory as well as for administrative purposes. For example, if a customer site A is shut down, eventually, the other PEs should unlearn A's MAC address. As packets arrive, MAC addresses are remembered. The aging timer for MAC address M should be reset when a packet is received with source MAC address M.

12.2.2.5   Hierarchical VPLS Model

The solution just described requires a full mesh of tunnel LSPs between all of the PE routers that participate in the VPLS service. For each VPLS service, $n \times (n - 1)$ VCs must be set up between the PE routers. Although this creates signaling overhead, the real detriment to large-scale deployment is the packet replication requirements for each provisioned VC on a PE router. Hierarchical connectivity, described in [15], reduces signaling and replication overhead to allow large-scale deployment.

In many cases, service providers place smaller edge devices in multiple-tenant buildings and aggregate them into a PE device in a large CO facility. In some instances, standard IEEE 802.1Q tagging techniques may be used to facilitate mapping CE interfaces to PE VPLS access points (referred to as Q-in-Q).

It is often beneficial to extend the VPLS service tunneling techniques into the MTU domain. This can be accomplished by treating the MTU device as a PE device and provisioning VCs between it and every other edge, as an basic VPLS. An alternative is to utilize [16] VCs or Q-in-Q VCs between the MTU and selected VPLS-enabled PE routers. Q-in-Q encapsulation is another form of L2 tunneling technique, which can be used in conjunction with MPLS signaling, as described later. This section focuses on this alternative approach. The VPLS mesh core tier VCs (hub) are augmented with access tier VCs (spoke) to form a two-tier hierarchical VPLS (H-VPLS).

Spoke VCs may include any L2 tunneling mechanism, expanding the scope of the first tier to include nonbridging VPLS PE routers. The non-bridging PE router would extend a spoke VC from a layer 2 switch that connects to it, through the service core network, to a bridging VPLS PE router supporting hub VCs. We also describe how VPLS-challenged nodes and low-end CEs without MPLS capabilities may participate in a hierarchical VPLS.

Hierarchical Connectivity

This section describes the hub-and-spoke connectivity model and describes the requirements of the bridging capable and nonbridging MTU devices for supporting the spoke connections. For the remainder of this discussion, we refer to a bridging-capable MTU device as MTU-s and a nonbridging capable PE device as PE-r. A routing and bridging capable device will be referred to as PE-rs.

Spoke Connectivity for Bridging-Capable Devices   As is displayed in Figure 12.12, consider the case where an MTU-s device has a single connection to the PE-rs device placed in the CO. The PE-rs devices are connected in a basic VPLS full mesh. To participate in the VPLS service, MTU-s device creates a single point-to-point tunnel LSP to the PE-rs device in the CO. We call this the *spoke connection*. For each VPLS service, a single spoke pseudowire is set up between the MTU-s and the PE-rs based on [20] or its extension [22]. Unlike traditional pseudowires that terminate on a physical (or a VLAN-tagged logical) port at each end, the spoke VC terminates on a virtual bridge instance on the MTU-s and the PE-rs devices.

The MTU-s device and the PE-rs device treat each spoke connection like an access port of the VPLS service. On access ports, the combination of the physical port and/or the VLAN tag is used to associate the traffic to a VPLS instance while the pseudowire tag (e.g., VC label) is used to associate the traffic from the virtual spoke port with a VPLS instance, followed by a standard L2 lookup to identify which customer port the frame needs to be sent to.

The signaling and association of the spoke connection to the VPLS service may be done by introducing extensions to the LDP signaling as specified in [20].

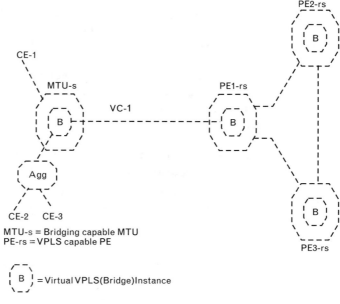

FIGURE 12.12
*MTU-s device has a
single connection to the
PE-rs device.*

MTU-s Operation   MTU–s device is defined as a device that supports layer 2
switching functionality and does all of the normal bridging functions of
learning and replication on all of its ports, including the virtual spoke port.
Packets to unknown destination are replicated to all ports in the service
including the virtual spoke port. Once the MAC address is learned, traffic
between CE1 and CE2 will be switched locally by the MTU–s device saving
the link capacity of the connection to the PE–rs. Similarly traffic between
CE1 or CE2 and any remote destination is switched directly onto the spoke
connection and sent to the PE–rs over the point–to–point pseudowire.

Because the MTU–s is bridging capable, only a single pseudowire is
required per VPLS instance for any number of access connections in the
same VPLS service. This further reduces the signaling overhead between
the MTU–s and PE–rs.

If the MTU–s is directly connected to the PE–rs, other encapsulation
techniques such as Q–in–Q can be used for the spoke connection pseudowire. However, to maintain a uniform end–to–end control plane based
on MPLS signaling, [20] can be used for distribution of pseudowire tags
(e.g., Q–in–Q tags or VC labels) between MTU–s and PE–rs.

PE-rs Operation   The PE–rs device supports all of the bridging functions for
VPLS service and supports the routing and MPLS encapsulation, that is, it
supports all the functions described. The operation on the PE–rs node is
identical to that described previously, with one addition. A point–to–point
VC associated with the VPLS is regarded as a virtual port (see earlier discussion on service delimiting). The operation on the virtual spoke port is

identical to the operation on an access port as described in the earlier section. As shown in Figure 12.12, each PE-rs device switches traffic between aggregated access VCs that look like virtual ports and the network side VPLS VCs.

Advantages of Spoke Connectivity   Spoke connectivity offers several scaling and operational advantages for creating large-scale VPLS implementations, while retaining the ability to offer all of the functionality of the VPLS service:

- Eliminates the need for a full mesh of tunnels and full mesh of VCs per service between all devices participating in the VPLS service.
- Minimizes signaling overhead because fewer VC-LSPs are required for the VPLS service.
- Segments VPLS nodal discovery. MTU-s needs to be aware of only the PE-rs node although it is participating in the VPLS service that spans multiple devices. On the other hand, every VPLS PE-rs must be aware of every other VPLS PE-rs device and all of its locally connected MTU-s and PE-r.
- Addition of other sites requires configuration of the new MTU-s device but does not require any provisioning of the existing MTU-s devices on that service.
- Hierarchical connections can be used to create VPLS service that spans multiple service provider domains. This is explained in a later section.

Spoke Connectivity for Nonbridging Devices   In some cases, a bridging PE-rs device may not be deployed in a CO or a multiple-tenant building while a PE-r might already be deployed. If there is a need to provide VPLS service from the CO where the PE-rs device is not available, the service provider may prefer to use the PE-r device in the interim. In this section, we explain how a PE-r device that does not support any of the VPLS bridging functionality can participate in the VPLS service. As shown in Figure 12.13, the PE-r device creates a point-to-point tunnel LSP to a PE-rs device. Then for every access port that needs to participate in a VPLS service, the PE-r device creates a point-to-point [16] VC that terminates on the physical port at the PE-r and terminates on the virtual bridge instance of the VPLS service at the PE-rs.

PE-r Operation   The PE-r device is defined as a device that supports routing but does not support any bridging functions. However, it is capable of setting up VCs between itself and the PE-rs [16]. For every port that is supported in the VPLS service, a VC is set up from the PE-r to the PE-rs [16]. Once the VCs are set up, there is no learning or replication function required on part of the PE-r. All traffic received on any of the access ports is transmitted on the VC. Similarly all traffic received on a VC is transmitted

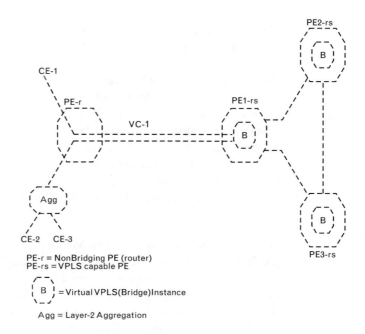

PE-r = NonBridging PE (router)
PE-rs = VPLS capable PE

B = Virtual VPLS (Bridge) Instance

Agg = Layer-2 Aggregation

to the access port where the VC terminates. Thus traffic from CE1 destined for CE2 is switched at PE-rs and not at PE-r.

This approach adds more overhead than the bridging capable (MTU-s) spoke approach since a VC is required for every access port that participates in the service versus a single VC required per service (regardless of access ports) when a MTU-s type device is used. However, this approach offers the advantage of offering a VPLS service in conjunction with a routed Internet service without requiring the addition of new MTU device.

PE-rs Operation   The operation of PE-rs is independent of the type of device at the other end of the spoke connection. Whether there is a bridging-capable device (MTU-s) at the other end of the spoke connection or a nonbridging device (PE-r) at the other end of the spoke connection, the operation of PE-rs is exactly the same. Thus, the spoke connection from the PE-r is treated as a virtual port and the PE-rs device switches traffic between the virtual port, access ports, and the network side VPLS VCs once it has learned the MAC addresses.

Redundant Spoke Connections   An obvious weakness of the hub and spoke approach described thus far is that the MTU device has a single connection to the PE-rs device. In case of failure of the connection or the PE-rs device, the MTU device suffers total loss of connectivity.

In this section we describe how the redundant connections can be provided to avoid total loss of connectivity from the MTU device. The mechanism described is identical for both MTU-s and PE-r types of devices.

Dual-Homed MTU Device   To protect from connection failure of the VC or the failure of the PE-rs device, the MTU-s device or the PE-r is dual-homed into two PE-rs devices, as shown in Figure 12.14. The PE-rs devices must be part of the same VPLS service instance. An MTU-s device will set up two VCs (one each to PE- rs1 and PE-rs2) for each VPLS instance [16]. One of the two VCs is designated as *primary* and is the one that is actively used under normal conditions, while the second VC is designated as *secondary* and is held in a standby state. The MTU device negotiates the VC labels for both the primary and secondary VC, but does not use the secondary VC unless the primary VC fails. Because only one link is active at any given time, a loop does not exist and, hence, an 802.1D spanning tree is not required.

Failure Detection and Recovery   The MTU-s device controls the usage of the VC links to the PE-rs nodes. Because LDP signaling is used to negotiate the VC labels, the hello messages used for the LDP session can be used to detect failure of the primary VC. On failure of the primary VC, the MTU-s device immediately switches to the secondary VC. At this point the PE3-rs device that terminates the secondary VC starts learning MAC addresses on the spoke VC. All other PE-rs nodes in the network think that CE-1 and CE-2 are behind PE1-rs and may continue to send traffic to PE1-rs until they learn that the devices are now behind PE3-rs. The relearning process can take a long time and may adversely affect the connectivity of higher level protocols from CE1 and CE2. To enable faster convergence, the PE3-rs device where the secondary VC got activated may send out a flush message, using the MAC TLV as defined in

FIGURE 12.14
Two VCs.

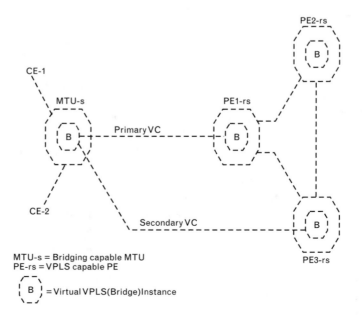

Section 12.2.2.6, to all other PE-rs devices participating in the VPLS service. Upon receiving the message, all PE-rs flush the MAC addresses associated with that VPLS instance.

Multidomain VPLS Service   Hierarchy can also be used to create a large-scale VPLS service within a single domain or a service that spans multiple domains without requiring full-mesh connectivity between all VPLS capable devices. Two fully meshed VPLS networks are connected using a single LSP tunnel between the VPLS gateway devices. A single VC is setup per VPLS service to connect the two domains. The VPLS gateway device joins two VPLS services to form a single multidomain VPLS service. The requirements and functionality required from a VPLS gateway device remained to be specified at the time of this writing.

12.2.2.6   Hierarchical VPLS Model Using Ethernet Access Network

In the previous section, a two-tier hierarchical model that consists of hub-and-spoke topology between MTU-s devices and PE-rs devices and a full-mesh topology among PE-rs devices was discussed. In this section the two-tier hierarchical model is expanded to include an Ethernet access network. This model retains the hierarchical architecture discussed previously in that it includes MTU-s devices and PE-rs devices and also utilizes a full-mesh topology among PE-rs devices. The motivation for an Ethernet access network is that Ethernet-based networks are currently deployed by some service providers to offer VPLS services to their customers. Therefore, it is important to provide a mechanism that allows these networks to integrate with an IP or MPLS core to provide scalable VPLS services. One can categorize Ethernet access networks into the following three groups:

1.   Based on the existing 802.1Q standard (this is comparable to the situation where the customer comes in on a VLAN-tagged port);
2.   Based on an extension to the IEEE 802.1Q standard that tunnels 802.1Q VLANs using a service provider 802.1Q tag (referred to as Q-in-Q);
3.   Based on Q-in-Q tunneling with ability to distribute .1Q tags using MPLS control plane.

For the first category, the MTU-s and all the other nodes in the access network (excluding PE-rs devices) have standard 802.1Q Ethernet switch capability. However, the PE-rs device is a VPLS-capable router as described previously.

For the second category, in addition to the functionality described in the preceding category, the MTU-s and the PE-rs are required to support Q-in-Q tunneling capabilities, and the Ethernet nodes in between are required to handle larger data frames (to accommodate the additional encapsulation).

The third category requires the MTU-s and the PE-rs to support LDP signaling for the distribution of Q-in-Q tags (i.e., MTU-s and PE-rs to have MPLS signaling capability, without MPLS encapsulation) in addition to the functionality described in categories 1 and 2 described above. Single-sided signaling [23] can be used to distribute Q-in-Q tags.

Note that since the Ethernet access network can have any arbitrary topology, the standard 802.1D STP may be required for loop detection and prevention. However, the use of a spanning tree is topology dependent and may or may not be required.

The connectivity within the service provider core is unchanged. Thus, a PE-rs may need to run STP on its Ethernet access interfaces and split-horizon on its MPLS/IP network interfaces. In a topology where MTU-s devices are directly connected to PE-rs devices, STP is not required on the access network.

Figure 12.15 shows a VPLS network and several possible ways of connecting customer CE devices to the network. As can be seen, CEs can be either connected directly to PE-rs (or PE-r) or they can be first aggregated by an MTU-s and then connected to PE-rs or they can be connected via an Ethernet Access network to the PE-rs.

Port-Based Versus VLAN-Based VPLS Operation

Where a customer uses a port-based VPLS service, all customer traffic received from that port, regardless of whether it has VLAN tags or not, is directed to a single VPLS instance. In the MTU-s, this is done by assigning a *service provider VLAN* (SP-VLAN) tag to that customer port. If the customer traffic is already tagged, the SP-VLAN serves as the outer tag. The SP-VLAN tag serves as a VPLS identifier that identifies a FIB associated with that particular VPLS. MAC address learning is done using unqualified learning. The customer packets are then forwarded through the FIB to the

FIGURE 12.15
Several possible ways of connecting customer CE devices to the network.

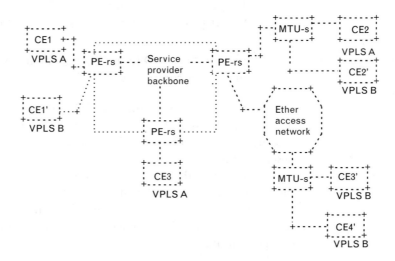

appropriate pseudowire based on the destination MAC address. In the reverse direction, the pseudowire tag identifies the VPLS instance and thus the FIB. The packet is forwarded to the proper Ethernet port based on the destination MAC address. When the packet leaves the PE-rs toward the MTU-s, it will be appended with the SP-VLAN tag associated with that VPLS.

Where a customer uses a VLAN-based VPLS service, if the traffic within each customer VLAN is to be isolated from each other and each one has its own broadcast domain, then each customer VLAN is mapped to a single VPLS instance. If the service provider assigns the customer VLAN tags, then the service provider can ensure the uniqueness of these VLAN tags among different customers, and a given customer VLAN tag can be used as a VPLS identifier. However, if customers assign their own VLAN tags independently, then the MTU-s must map each customer VLAN into a unique SP-VLAN. Subsequently, the SP-VLAN will be used as a VPLS identifier to index the proper FIB and to forward traffic based on destination MAC address. The operation of PE-rs in this case remains same as before. Figure 12.16 shows the Q-in-Q tunneling encapsulation that is applied when an Ethernet data plane is used between MTU-s and PE-rs.

12.2.3 VPLS Solution Using GRE-Based IP Tunnels

In this section we present a VPLS solution that uses *generic routing encapsulation* (GRE) and IP tunnels [23]. It is anticipated that use of IP tunnels will simplify the back-end tunneling plane of VPLS solutions, especially when compared to MPLS-based solutions. The charter of the PPVPN WG specifies VPLS and layer 2 VPN. Reference [23] presents a IP tunnel-based solution for VPLS. Presently published VPLS solutions are based on MPLS. According to proponents, the GRE-based solution is much simpler and scalable. GRE tunnels provide a designated pathway across the shared WAN and encapsulate traffic with new packet headers, which ensures

FIGURE 12.16
*Q-in-Q tunneling
encapsulation.*

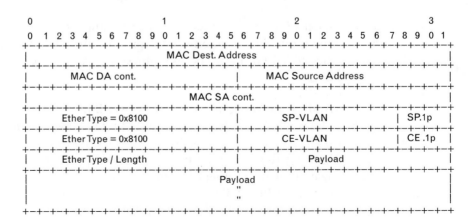

delivery to specific destinations. The network is private because traffic can enter a tunnel only at an endpoint.

VPLS is used in extending private LANs across public networks. It is anticipated that VPLS will gain wider deployment with the growth of MANs. Currently, several VPLS solutions have been proposed. Almost all of these solutions use MPLS as the service provider tunneling protocol. Use of MPLS essentially adds one more protocol that must be maintained. Most of service provider backbones are based on IP. Hence, the ability to carry VPLS traffic using the same native protocol simplifies the operations of the service provider, thus reducing the cost of providing VPLS services.

Reference [23] provides a VPLS solution based on IP tunnels. Reference [24] specifies encapsulation of Ethernet frames over IP using GRE [25]. GRE is used to encapsulate VPLS packets over IP tunnels. The draft proposes to extend methods specified in [24] to provide VPLS services.

12.2.3.1   Overview of the Solution

Reference [26] provides a detailed set of VPLS requirements and related reference models. Any VPLS solution must at a minimum support the following requirements:

- Representation of VPLS users;
- Traffic separation (that is, methods to demultiplex traffic belonging to different users);
- Address learning and unicast forwarding;
- Broadcast/multicast/unknown traffic forwarding;
- Membership and configuration discovery.

Representation of Users
Different users of VPLS services are treated as a separate VPN and represented by a unique VPLS-ID.

Traffic Separation
Each PE device is required to identify the VPLS instance before making any forwarding decisions. In the proposed solution, it is suggested that the embedded GRE key be used for this purpose. See Section 12.2.3.4 for details.

Address Learning and Unicast Forwarding
Unicast MAC addresses need only be forwarded to the destinations that own that MAC address. Traditional layer 2 devices accomplish this by learning the addresses. In the proposed solution, addresses are learned against the originating PE device (tunnel). A given PE or tunnel can carry multiple VPLS traffic streams that belong to different VPLS-ID. Hence we propose for each PE device to distribute a unique key for each VPLS-ID.

Thus, the receiving PE can uniquely identify the VPLS of the packet using the embedded GRE key; and the packet originator PE by the source IP address of the IP header. In other words, MAC addresses are learned against a VC. See Section 12.2.3.2 for VC representation.

Broadcast/Multicast/Unknown Traffic Forwarding

Broadcast/multicast/unknown traffic is required to be replicated to all members of that VPLS-ID. The originating PE accomplishes this by replicating the packet over each VC that is part of the *virtual port* (VP) of that VPLS-id. See section 12.2.3.2 for details.

Membership and Configuration Discovery

Each PE device that is a member of a given VPLS must acquire information about other member PE devices of the VPLS. Information includes the IP address of the PE, associated GRE key (VC ID), and other optional information. Such information can either be manually configured or automatically discovered. Automatic VPLS discovery solutions are presented in [27–29].

12.2.3.2 VCs and VPs

VCs

A separate VC is used for each participating PE device. The sending device must replicate broadcast, unknown, and multicast packets over each VC. The VC-ID is locally significant only. The VC-ID (GRE key) is locally allocated and distributed (manually or automatically) to all other devices. Remote sending devices must encode the local receiving devices VC-ID (GRE key) in the encapsulation header. Hence, the following mappings are needed when address learning:

```
VC-ID (GRE Key in the Header) —> VPLS-ID(local)
Source IP Address, VLPS-ID -> VC-ID (GRE Key of the
remote)
```

All of the addresses are learned against the VC-ID remote. Consider the following example:

```
MAC A—> PE-A           —————————————————> VC-AB== GREKey_AB

VC-BA== GREKey_BA  <————————————— PE-B
```

MAC address is serviced by PE-A. Now FIB at PE-B would looks like

```
MAC Address        Destination
A,VPLS_ID          Encap=GRE, DstIp=IPPE-A, VC= GREKey_BA
```

VP

The VP concept is used to avoid loops. A VP is defined as set of VCs that belong to a given VPLS. A given VC can be a member of one and only one

VP. A VP is defined in the scope of a VPLS. For simplicity and scalability reasons, it is always assumed that there is a set of fully meshed tunnels between member PE devices. Hence, to avoid loops, packets received on a VC must not be sent back on another VC that is part of the same VP. A typical broadcast FIB entry may appear as follows:

```
MAC address           Destination
  **,VPLS-ID          Encap=GRE, VP= {( srcIp=IPPE-A, VC=
                      GREKey_BA)....}
```

12.2.3.3 Discussion

The IP tunnel-based VPLS solution presented has several advantages compared to MPLS-based VPLS solutions:

- *Simplicity*. There is no additional control plane; it uses the native IP plane for tunnels;
- *Scalability*. Scalability of MPLS is mainly governed by software scalability. Due to the simplicity of the GRE key management, the proposed solution is expected to scale to a much higher degree than MPLS-based solutions.

However, the proposed solution requires a larger encapsulation header compared to MPLS-based solutions. For smaller packets, especially, this overhead can be significant.

12.2.4 Architectural Alternatives

This section defines a reference architecture for a VPLS system. It is based on [30]. The Internet draft describes possible VPLS architectures and the merits of each. Each VPLS architecture is described in terms of its logical components and their relationship to each other, as well as the mapping of these logical components to physical network elements. By understanding these logical components, which are fundamental building blocks for any VPLS system, one can easily compare different VPLS architectures and understand the pros and cons of each. A VPLS system may support one or more of these architectures simultaneously. Having described each VPLS architecture, the draft explores the different logical components of the reference architecture, and describes how they relate to a service provider backbone (whether MPLS enabled or not). Finally, it examines the operation of each VPLS architecture.

12.2.4.1 Introduction

The primary motivation behind VPLS is to provide connectivity between geographically dispersed customer sites across MANs and WANs, as if they were connected through a LAN. Furthermore, the intended application for the end user can be divided into the following two categories:

• Connectivity between site routers—LAN routing application;

• Connectivity between site Ethernet switches—LAN switching application.

The LAN routing application is intended for the interconnection of routers via a MAN or WAN, and for providing a broadcast domain among these routers. Traditionally, the interconnection of VPN sites over a MAN/WAN has required a mesh of overlay virtual circuits between routers. Adding a new router to the interconnection mesh potentially requires significant reconfiguration. The VPLS service is intended to mitigate this interconnection problem by leveraging native broadcast capabilities for router discovery, along with a simple point-to-cloud service model. This service model simplifies the task of configuration and provisioning for the end customer routers by providing a virtual LAN/bridge system. Therefore, to the routers it looks as if they are connected via a local LAN switch even though they are located in different sites.

The LAN switching application, as its name implies, provides a virtual LAN switching service to its end users, and it is used for interconnecting customer switches across a MAN/WAN network. Therefore, if several sites within a VPN subscribe to this service, it looks as if these sites are colocated within a campus and are connected via a local Ethernet switch providing bridging services for them. This service is not anticipated to be widely deployed in large-scale networks because of limited scalability and lower network efficiency. However, for small/medium business customers, this service can be very viable to extend the bridging domain between VPN sites across a MAN/WAN network. By extending the bridging domain, the customer only needs to use low-cost Ethernet switches as CPEs.

The underlying VPLS service for supporting these two types of applications is fundamentally the same. However, each application has different requirements in terms of control frame processing, MAC address usage, BPDUs handling, and so on.

Figure 12.17 shows a VPLS system with two VPLS instances—one for each customer, referred to as *VPLS A* and *VPLS B*. This figure shows that each customer CPE device (whether router or switch) connects to a PE device within the service provider POP. Note that when the phrase "a VPLS" is used by itself, it refers to a VPLS instance and not a VPLS system.

One of the main scenarios for the VPLS service is the delivery of an Ethernet service to multiple customers colocated in one or more MxUs over the metro area. MxUs can be apartments or *multiple-dwelling units* (MDUs), office buildings or *multiple-tenant units* (MTUs), and hotels or *multiple-hospitality units* (MHUs). In such cases, it would be much more cost effective to place an aggregation device at the colocation site, rather than connecting each CPE directly to a PE within the service provider POP. This device would be used to aggregate all customers' traffic and carry it via an uplink to the service provider POP as opposed to having a separate

FIGURE 12.17
*A VPLS system with
two VPLS instances.*

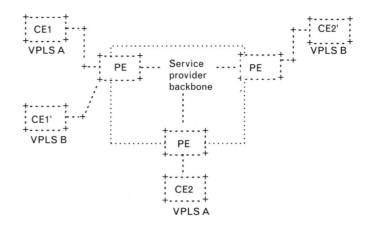

uplink for each customer. Because of a large number of colocation sites and thus the large number of aggregation devices in a metro area, it is important for the aggregation device to be truly low cost. In such cases, the VPLS functionality is distributed between the aggregation device in the colocation site and the PE sitting in the service provider POP. Apart from the obvious cost reductions, a VPLS system based on a distributed-PE model is also able to provide access bandwidth efficiency and system scalability. These additional benefits will be described in more detail in a later section.

Figure 12.18 extends the VPLS system architecture to include distributed-PE functionality. As can be seen in the figure, the VPLS system can have a mix of distributed and nondistributed (single) PEs. In the case of a nondistributed PE, we refer to it simply as PE and it is typically located in the service provider's central office or POP. In the case of a distributed PE, we refer to the PE located at the customer site as PE-CLE and the PE

FIGURE 12.18
DIFFERENT WAYS
FOR CEs AND
PE-CLEs TO
CONNECT.

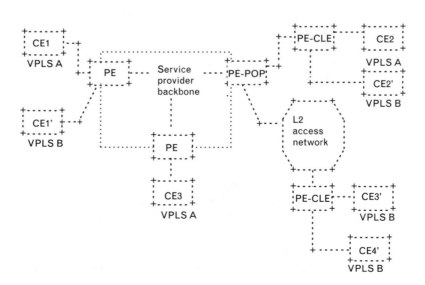

located in the service provider's POP as PE-POP. The service provider's backbone network connecting PEs/PE-POPs can be based on either MPLS or IP.

As shown in Figure 12.18, there are several different ways for CEs and PE-CLEs to connect to the service provider's network. CEs and PE-CLEs can be either connected directly to PE-POPs via a given transport mechanism such as SONET, fiber, CWDM/DWDM, and so on, or they can be connected via an Ethernet access network. In the later case, care must be taken to avoid loops in the access network using standard mechanisms such as STP. Furthermore, there may be cases (although not typical) where several customer sites are connected to the same PE-CLE. This may imply certain restrictions on the customer's network based on the capability of the PE-CLE as will be elaborated on in a later section.

Reference Architecture

The reference architecture described in this section is based on the model defined in [31] with several extensions. The model defined in [31] is primarily intended for non-VPLS systems with nondistributed PEs (e.g., a single PE is used to connect a CE device to the service provider's network). A non-VPLS L2VPN can be described in terms of its components as defined in [31]:

- Attachment VCs;
- Emulated VCs;
- Emulated tunnels;
- Autodiscovery;
- Autoconfiguration.

In [30], three additional components are introduced as follows for a VPLS system that will be described in subsequent sections of this draft. The first component is mandatory for a VPLS system (along with the first three non-VPLS components described earlier) and the other two components are needed for a distributed-PE model (and more specifically for a particular distributed-PE model).

- *Virtual switch instances* (VSIs);
- Attachment tunnels;
- Extension VCs.

The only components that are dependent on network type (MPLS vs. IP) are emulated VCs and emulated tunnels. In [31] an attachment VC is defined as the virtual circuit between a CE and its associated PE, and an emulated VC is defined as the virtual circuit between two PEs. Therefore, an end-to-end virtual circuit between two CEs can be defined as an ordered triple <Attachment VC, emulated VC, Attachment VC>. The mapping between attachment VC and its corresponding emulated VC is one-to-one mapping and is performed on each side by its associated PE. Furthermore, the mapping between attachment VC and its corresponding emulated VC is fixed and is established at the time of end-to-end circuit establishment. The attachment and emulated VCs are always terminated in the PEs and thus L2VPN processing is only limited to PE nodes and transparent to P nodes.

In a VPLS system, the attachment VCs can be considered as connected to a virtual switch (referred to as a VSI) corresponding to that VPLS instance at each PE. Furthermore, the emulated circuits can be considered as the virtual circuits connecting these virtual switches located at each PE.

A VPLS for a given customer can be defined as a set of virtual switches (one in each PE that is part of the VPLS) connected to each other via emulated VCs and to CEs via attachment VCs. This means that for a VPLS system, the relationship between the attachment and emulated VCs is no longer fixed. The dynamic mapping of these entities (which is sometimes one-to-one for unicast and sometimes one-to-many for multicast or broadcast) is provided by each virtual switch through the use of the destination MAC address and is performed on a frame-by-frame basis.

Given the preceding description of a VPLS system, a customer VPLS instance can be defined in terms of its attachment VCs, its emulated VCs, and its set of virtual switches as follows: < Ingress Attachment VCs, Ingress VSIs, Emulated VCs, Egress VSIs, Egress Attachment VCs>. Figure 12.19 depicts the attachment and the emulated VCs and their relationship for a given L2VPN with three sites via (1) non-VPLS network and (2) VPLS network.

What differentiates the VSI from a typical Ethernet switch is its ability to do MAC address learning, flooding, forwarding, and so on, based on virtual circuits (e.g., based on emulated VCs) per broadcast domain. For example, if a physical port of a PE carries several emulated VCs, then the VSI can replicate packets (for flooding or broadcasting) across these emulated VCs over the same physical port. This is in contrast to a typical switch that may only perform packet replication across physical ports for a given broadcast domain.

A VPLS-capable PE will have one VSI per VPLS instance and it can either be transparent to a customer's VLANs or it can recognize the customer's VLANs. If the PE is transparent to the customer's VLANs, then the associated VSI will handle the customer's VLANs transparently; however, if the PE needs to recognize the customer's VLANs and to provide isolation among them, then separate VSIs for each customer VLAN should be maintained. Each VSI corresponds to a separate broadcast domain and may be shared between customer sites for the creation of extranet or Internet connectivity. The requirement for maintaining a separate broadcast domain per VPLS instance is identified in [32].

FIGURE 12.19
*Attachment and
emulated VCs.*

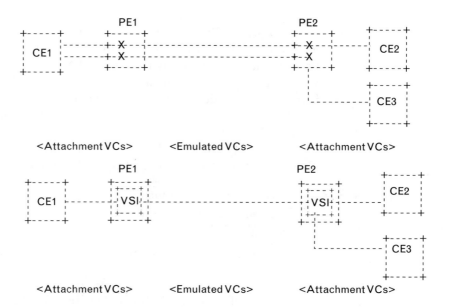

Because customers' VLANs can be overlapping, the PE should have the capability of associating customers' VLANs with a unique identifier for identification of the VSIs. Therefore, the ability to (1) map customers VLANs into unique identifiers (for VLAN-aware PE) and (2) to perform VSI functionality with emulated VCs constitutes major data plane requirements for a VPLS-capable PE.

For a PE that needs to recognize customers' VLANs and to maintain isolation among them, if each customer assigns their VLANs independently, then overlapping VLANs can occur within that PE. Therefore, the PE is required to provide additional capability so as to map customers' VLANs into unique VSI identifiers. However, if the service provider assigns the VLANs to its customers, then assignment can be done such that customers' VLANs are unique within a PE and thus the VLAN-ID itself can be used as the VSI identifier, thus simplifying the requirement for the PE or the PE-CLE (in the case of the distributed-PE model). However, this simplification at the PE comes at the expense of the customer losing the flexibility of assigning his or her own VLANs (which may be acceptable for certain deployment models).

The next few sections describe different VPLS architectures. A VPLS system may support one or more of these architectures simultaneously.

Single-PE Model

In the single-PE model, each PE is capable of performing the full set of VPLS functions and it can be either located in a customer premise or at the service provider POP. If the PE is located in the customer premise, then the link between the CE and the PE is local and many CEs can be aggregated via the upstream service provider link. However, it may cost too much to place a PE in the customer premise or the number of customer located PEs may present some system scalability issues. If the PE is located in the service provider POP, then one MAN link is required per customer, which may in turn result in an increase in the cost of facilities and thus an increase in operational costs. The next section shows how the use of a distributed-PE can address these issues effectively when a single-PE environment is not desirable.

In a single-PE model, all VPLS functionality is performed within a single PE. In terms of the data plane operation, these functions constitute:

- Attachment VC termination;
- Emulated VC termination;
- Tunneling of emulated VCs;
- Maintaining a VSI per VPLS instance (for bridging among attachment and emulated VCs).

In terms of the control plane operation, these functions constitute:

- Signaling of emulated tunnels (if needed);
- Signaling of emulated VCs;
- Autodiscovery of peer PEs per VPLS instance;
- Autoconfiguration of a VPLS instance.

As mentioned previously the PE needs to perform VSI functionality (such as MAC address learning, flooding, and unicasting/broadcasting on a per-emulated VC basis). In addition to this, if the PE is required to recognize customer's VLANs and to provide isolation among these different VLANs (by using a separate broadcast domain per VLAN), then it also needs to have the capability of mapping customers' VLANs into unique VSI identifiers if each customer assigns his or her VLANs independently.

Distributed-PE Model

Some of the main motivations behind a distributed-PE model include the following:

- The ability to deploy a low-cost edge device at the customer premise;
- Improving access bandwidth efficiency by multiplexing different customer traffic (in contrast to a single-PE model located at the service provider's POP);
- Improving access bandwidth efficiency by avoiding packet replication on the access uplink for multicast/broadcast traffic;
- Improving system scalability versus single-PE functionality within the customer premise in terms of (1) number of BGP peers, (2) number of IGP peers, (3) number of routes in the edge device, (4)

number of MPLS labels in the edge device, and (5) number of MPLS or IP tunnels among PEs (only a subset of routes and MPLS labels needs to installed in PE–CLEs; whereas in a single-PE model all routes and MPLS labels are installed in PEs—whether they are located at customer sites or service provider POPs).

As mentioned previously, the most important data plane functionality of a VPLS-capable PE is the ability to perform VSI functionality. In the case of the distribute-PE model, two cases should be considered: Either the PE-POP is capable of performing such functionality or it is not. If the PE-POP is capable of such functionality, then a true low-cost Ethernet switch–based PE-CLE can be used.

If the PE-POP does not have the ability to perform VSI functionality (e.g., it is a router), then this functionality needs to be performed by the PE-CLE. Note that if a PE-CLE needs to perform such functionality, then it must have similar capability as MPLS/IP switches so as to support multiple virtual circuits per physical interface, and to treat these VCs as logical ports of a VSI, so that it can do flooding and broadcasting across these VCs per broadcast domain. As mentioned before, existing Ethernet switches may not perform these functions on a per-VC basis. If their data planes need to be modified for such functionality, then one can easily extend them to use MPLS labels or IP information as well.

Based on the preceding discussion, one can define two types of PE-CLE. The first type is an Ethernet–based PE-CLE that requires the PE-POP to perform VSI functionality. The second type is an MPLS/IP-based PE-CLE, which does VSI functionality and thus the PE-POP can be just an MPLS or an IP router.

Based on which type of PE-CLE is used in a distributed-PE model, different sets of data plane and control plane functions are required at the PE-CLE and PE-POP. Another way of categorizing a distributed-PE model is based on control plane functionality—more specifically in terms of emulated signaling and autodiscovery. Based on these two functions, there can be four categories of distributed-PE models as follows:

1. Emulated signaling and autodiscovery are both performed at the PE-POP;

2. Emulated signaling at PE-CLE and autodiscovery at PE-POP;

3. Emulated signaling at PE-POP and autodiscovery at PE-CLE;

4. Emulated signaling and autodiscovery are both performed at the PE-CLE.

The third category does not offer any real benefits in terms of system scalability and thus is not considered further. The fourth option can be considered as a degenerate case for a single-PE model. Since the PE-CLE is capable of terminating and signaling emulated VCs and performing autodiscovery, it can basically perform all other functions required for a VPLS service. Therefore, only the first two options are examined further.

As one might have guessed, a distributed-PE model based on Ethernet-based PE-CLE uses the first control plane category (e.g., both emulated signaling and autodiscovery are performed at the PE-POP). In contrast, a distributed model based on an MPLS/IP-based PE-CLE can use either the first or second control plane category. As will be shown later, for an MPLS/IP-based PE-CLE, the second control plane category offers additional benefits over the first one. An example of a distributed-PE model using an MPLS-based PE-CLE and the first control plane category (e.g., emulated signaling and autodiscovery both performed at the PE-POP) is described in [33].

In the following two sections, reference architectures for Ethernet-based and MPLS/IP-based PE-CLE are described and their relationship in terms of signaling/autodiscovery categorization is further examined.

Ethernet-Based PE-CLE   A distributed-PE model with Ethernet-based PE-CLE provides a true low-cost edge device and at the same time it provides all of the benefits of a distributed model in terms of motivational factors considered earlier. This model provides a natural migration in offering VPLS services for carriers with an existing Ethernet access network. An Ethernet-based PE-CLE not only provides access bandwidth efficiency by multiplexing different customers traffic but also it provides bandwidth efficiency by not replicating frames over its access uplink connected to the PE-POP (which is not the case for MPLS/IP-based PE-CLEs). Also, since in this model the PE-POP performs both emulated VC signaling and autodiscovery, all the signaling and routing scalability issues are addressed as well.

Now let's examine the reference architecture for this model with respect to its components. Because Ethernet-based PE-CLEs do not support VSI functionality (e.g., MAC address learning, flooding, forwarding, and so on) based on virtual connections per broadcast domain, emulated VCs must be terminated in the PE-POP. Also, since emulated VCs signaling is performed at the PE-POP, the autodiscovery should also be performed there as well (autodiscovery is used for endpoint identification in setting up emulated VCs). This means that the Ethernet-based PE-CLE model fits very naturally into the first control plane categorization of emulated signaling and autodiscovery being performed at the PE-POP.

Now that the emulated VCs are terminated at the PE-POP and the VSIs are also located at the PE-POP, the question is how to get to the attachment VCs that come into the PE-CLE (it should be noted that emulated VCs are always terminated at the node that has the VSI functionality). There are two possible options:

1.   To tunnel the attachment VCs from the PE-CLE to the PE-POP and do further processing of the attachment VCs at the PE-POP;

2. To extend the attachment VCs from the PE-CLE into the PE-POP using extension VCs (e.g., by mapping attachments VCs into extension VCs at the PE-CLE).

The tunneling of the attachment VCs can be performed by placing an outer .1Q tag over a given customer traffic. This is also referred to as .1Q-in-.1Q encapsulation or, for short, Q-in-Q. The provision of an Ethernet switch for this functionality is rather simple and a number of vendors currently provide such functionality in their Ethernet switches. The main advantage of this approach is that it requires no modifications to Ethernet switches that can support Q-in-Q encapsulation, so these switches can be readily used as PE-CLEs. If the service provider is required to recognize customers' VLANs, then the PE-POP requires the capability of creating a unique VSI identifier per customer's VLAN by looking at both the outer and the inner .1Q tags, which is referred to as *double-tag lookup*. However, if the service provider is not required to recognize customers' VLANs, then no double-tag lookup functionality is needed at the PE-POP and the PE-POP can simply use the outer .1Q tag as the VSI identifier since the outer tag has been assigned to ensure uniqueness at each PE-POP.

The second option of extending the attachment VCs into the PE-POP requires additional functionality in the PE-CLE that may not be currently available in Ethernet-based switches. This functionality allows the mapping of a customer VLAN (e.g., attachment VC) into an internal VLAN (extension VC), which is unique within the PE-CLE. This mapping is needed when separate VPLS instances need to be maintained per customer VLANs in order to provide isolation among customer's VLANs (e.g., customers' VLANs can overlap with each other and the service provider does not impose any restriction in customers' VLAN assignment per [32]). Note that this option is not needed if the service provider doesn't need to recognize customers' VLANs and thus the first option without double-tag lookup at the PE-POP is sufficient.

In summary, in a distributed-PE model based on Ethernet-based PE-CLE, the VPLS functions are distributed between the PE-CLE and the PE-POP, with the following functions being performed at the PE-CLE:

- Tunneling of attachment VCs to the PE-POP *OR*;
- Attachment VC termination and origination of extension VCs toward the PE-POP.

The following VPLS functions are performed in the PE-POP:

- Attachment tunnels termination (if used);
- Attachment VC termination *OR* extension VC termination;
- Emulated VCs—signaling and termination;
- Emulated tunnels—signaling and termination;

- Maintaining a VSI per VPLS instance;
- Autodiscovery;
- Autoconfiguration.

MPLS/IP-Based PE-CLE   In cases where the PE-POP is an MPLS or an IP router without switching capability, the task of performing VSI functionality resides in the PE-CLE. The emulated VCs can be setup at the PE-CLE using either MPLS or IP encapsulation—thus the name MPLS/IP-based PE-CLE.

In this distributed-PE model, the emulated VCs are terminated at the PE-CLE and the bridging functionality between a customer's attachment VCs and the corresponding emulated VCs are provided using a VSI per VPLS at the PE-CLE. Unlike the Ethernet-based PE-CLE, there is no need for attachment tunnels or extension VCs since the VSIs are located in the PE-CLEs.

As mentioned previously, there are two types of control plane options for this distributed-PE model: (1) to have both emulated signaling and autodiscovery performed at the PE-POP and (2) to have emulated signaling at the PE-CLE and autodiscovery at the PE-POP. Note that in both options, the termination of the emulated VCs is performed at the PE-CLE (emulated VCs are terminated in the same node that has the VSI functionality). However, in the first option, the signaling of emulated VCs is done among the PE-POPs and then the relevant information is conveyed to the PE-CLEs; whereas, in the second option, the signaling of emulated VCs is done directly between the PE-CLEs (e.g., using directed LDP).

The advantage of performing the signaling for emulated VCs at the PE-CLE is that the emulated VC gets set up as a single segment VC between the two PE-CLEs and, furthermore, the emulated VC is transparent to the PE-POP (e.g., only emulated tunnels would be visible to the PE-POPs and, therefore, there is less overhead in terms of configuration, state information, and processing for the PE-POP). However, if the signaling of emulated VCs is performed at the PE-POP, then the emulated VC will have multiple segments and multiple sets of labels (e.g., one set for each segment) need to be assigned as suggested by [33]—one set of labels for the customer-facing segment at each PE and another set for the WAN/MAN segment.

In addition to the configuration burden of maintaining multiple sets of labels, there is additional overhead for coordinating and installing routes among these different label sets. Because of the drawbacks of signaling of emulated VCs at the PE-POP, this option is not considered any further and instead, in a later section, the operational scenario for a single-segment emulated VC (e.g., signaling of emulated VCs at the PE-CLE) will be elaborated on further.

Based on the single-segment emulated VC model (e.g., signaling of emulated VCs to be done at the PE-CLE), the VPLS functions can be partitioned as follows with the following functions residing at the PE-CLE:

• Attachment VCs termination;

• Emulated VCs—termination and signaling;

• Emulated tunnels—signaling and termination;

• Maintaining a VSI per VPLS instance.

The following functions reside at the PE-POP: autodiscovery and autoconfiguration.

12.2.4.2 Emulated VCs

For a VPLS system, a set of emulated VCs is used to connect different VSIs belonging to the same VPLS instance. Therefore, emulated VCs are terminated at the nodes with VSI functionality.

For a VPLS system over an MPLS network, the encapsulation of emulated VCs are performed based on [34] and the corresponding signaling is performed based on [35].

For a VPLS system over an IP network, the encapsulation and the signaling of the emulated VCs can be performed based on [36].

We define two additional types of emulated VCs for a VPLS system that are the counterparts of the Ethernet VCs defined in [35] for non–VPLS L2VPN. Their VC type descriptions are as follows:

• 0x000C VLAN VPLS;

• 0x000D Ethernet VPLS.

The reason for the definition of these two types of VCs is that the system operation with respect to transport of VLAN tagged and untagged frames, as well as handling of control packets and BPDUs, differs for each of these VC types.

An emulated VC, which is of type VLAN VPLS, corresponds to a single customer's VLAN. Therefore, for scenarios where the service provider needs to recognize customers' VLANs and provide a separate broadcast domain for different VLANs, then this emulated VC type is used. Furthermore, only the frames with the specified tag are allowed to be transported over this emulated VC. The emulated VC of type VLAN VPLS corresponds to customers' UNI ports that are multiplexed (e.g., each UNI ports carries multiple attachment VCs). A customer UNI port is a physical port between the customer's CE and the service provider's PE.

However, an emulated VC of type Ethernet VPLS corresponds to all the traffic that comes through a customer UNI port—both VLAN tagged and untagged traffic, including BPDUs. In other words, the emulated VC acts as a pseudowire corresponding to the customer's Ethernet port. This type of emulated VC is used in applications where the service provider does not need to recognize customers' VLANs and thus it only provides a single broadcast domain (e.g., single VPLS) per customer or set of customer's ports. This type of emulated VC can be also viewed as corresponding to customers' UNI ports that are not multiplexed (e.g., each UNI ports corresponds to a single attachment VC).

In multiplex UNI scenarios for LAN bridging applications, not only is isolation of customers' VLANs needed (and different VLANs can go to different switches), but BPDU packets also need to be transported among these switches. For these applications, besides having a separate set of emulated VCs of type VLAN VPLS for each customer's VLAN, another set of emulated VCs of the same type is used for untagged frames, including BPDUs. Therefore, in multiplex UNI applications, the untagged frames are transported over the emulated VCs of type VLAN VPLS (e.g., untagged frames can be considered as frames with special tags).

It is important that these two types of emulated VCs be differentiated and processed accordingly by the PEs involved in setting up a VPLS instance, since during autoconfiguration it provides a compatibility check between attachment VCs and their corresponding emulated VCs.

12.2.4.3 Emulated Tunnels

An emulated tunnel is used to carry emulated VCs between two PEs and to make them transparent to the P nodes within the network. The general requirements for these tunnels are given in [31]. The emulated tunnels are set up based on one of the existing methods specified for L3VPNs [19] and are independent from L2VPN applications (e.g., if MPLS is used then tunnels either can be MPLS-TE using RSVP signaling or can be regular MPLS LSP using LDP signaling).

In the reference architectures that we have defined for the single-PE and both of the distributed-PE models, the emulated tunnels are terminated at the same node where the emulated VCs themselves are terminated. This means that in the case of the distributed-PE model with Ethernet-based PE-CLE, the tunnels are terminated at the PE-POP; and in the case of the distributed-PE model with MPLS/IP-based PE-CLE, the tunnels are terminated at the PE-CLE.

12.2.4.4 Attachment Tunnels

The attachment tunnels are used only for the distributed-PE model and more specifically with Ethernet-based PE-CLEs. Attachment tunnels can be between PE-CLEs and PE-POPs that are either directly connected or via a .1Q access network. In the later case, these tunnels are transparent to the .1Q access network and are treated as regular VLANs with slightly larger payloads.

Attachment tunnels are formed by placing an outer .1Q tag in front of the frames (either tagged or untagged) entering a PE-CLE via a customer UNI port. This outer tag is referred to as a *provide edge VLAN* (PE-VLAN). In the case of a nonmultiplexed UNI, the PE-VLAN is the same for all ports of a customer belonging to the same VPLS instance, and in the case of a multiplexed UNI, the PE-VLAN is the same for all ports of a customer that shares one or more VPLS instances.

The uniqueness of the PE-VLAN needs to be guaranteed within a PE-POP domain since there are multiple PE-CLEs per PE-POP, and PE-VLANs are used to differentiate between different customers. In other words, PE-VLAN is a local parameter that gets shared between PE-POP and its associated PE-CLEs and it is not transported over the emulated VCs—instead it can be derived from the emulated VCs associated with that VPLS instance.

In the case of a multiplexed UNI, an attachment tunnel is used to transport all of the customer's VLANs (e.g., attachment VCs) to the PE-POP, and at the PE-POP both the PE-VLAN and CE-VLAN are used together as a VSI identifier (e.g., the double-tag lookup mechanism described previously). However, for a nonmultiplex UNI, the attachment tunnel degenerates into carrying only a single attachment VC associated with that UNI, and at the PE-POP only the PE-VLAN is used as a VSI identifier.

Encapsulation

Figure 12.20 shows the encapsulation format for customer tagged frames as they are carried between PE-CLE and PE-POP. The encapsulation format for customer untagged frames is the same as with 802.1Q.

Signaling

To configure an attachment tunnel between the PE-CLE and the PE-POP, the minimum pieces of information that need to be passed to the PE-CLE are (1) attachment tunnel ID or PE-VLAN and (2) customer-ID/VPN-ID associated with the attachment tunnel ID or PE-VLAN. The PE-CLE, on receiving the message that contains the association between PE-VLAN and customer-ID/VPN-ID, looks for which

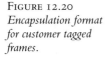

FIGURE 12.20
Encapsulation format for customer tagged frames.

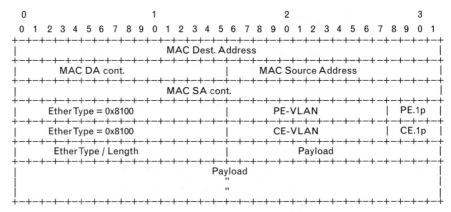

of its ports are configured for that customer-ID/VPN-ID and then applies the PE-VLAN ID to the respective port(s).

If in addition to the attachment tunnel configuration, the PE-CLE's port(s) configuration is also done through the PE-POP, then the PE-POP can resolve the association between PE-VLAN and the associated port and thus send the PE-VLAN + port-ID information along with port configuration information to the PE-CLE (e.g., PE-CLE does not need to receive customer-ID/VPN-ID info). Note that if the PE-CLE's port configuration is done via the PE-POP, then this shall include all relevant configuration information including QoS parameters and thus the signaling protocol shall support it as well.

Signaling of the attachment tunnels can be accomplished through extensions to an existing protocol (e.g., VLAN trunking protocol) and these extensions would only require software changes to the Ethernet-based PE-CLE (no hardware/ASIC changes are required). This signaling protocol along with its extensions will be specified in the future.

12.2.4.5   Extension VCs

Extension VCs are used in conjunction with Ethernet-based PE-CLEs in the distributed-PE model for multiplexed UNI ports and they are used in lieu of the attachment tunnels. Because extension VCs are only used for multiplexed UNI ports, the PE-CLE is required to have VLAN-mapping capability, which may not be currently available in standard Ethernet switches. This capability is needed in order to map overlapping VLANs of different customers into unique internal VLANs (PE-VLANs) so that at the PE-POP, these PE-VLANs can be used as VSI identifiers.

Encapsulation

The encapsulation format for extension VCs is exactly the same as those for 802.1Q since only customers' VLANs are mapped to PE-VLANs and everything else remains the same.

Signaling

Similar to the attachment tunnels, a signaling protocol can be used to configure the extension VCs between the PE-CLE and the PE-POP. At a minimum, the signaling protocol needs to convey the PE-VLAN and the associated VPN-ID to the PE-CLE. The PE-CLE uses the VPN-ID to identify which VLAN under which port needs to be mapped to a given PE-VLAN.

Again similar to the attachment tunnels, if, besides configuration of the extension VCs, the PE-CLE's ports and VLANs also need to be configured via the PE-POP, then only PE-VLAN information needs to be passed to PE-CLE along with the customer port and VLAN configuration. (For example, there is no need to pass VPN-ID information to the PE-CLE since the PE-POP can already resolve the association between PE-VLAN and customer port–ID + VLAN.)

Signaling of the extension VCs can also be accomplished through extensions to an existing protocol (e.g., VLAN trunking protocol) and these extensions would only require software changes to the Ethernet-based PE-CLE. This signaling protocol along with its extensions will be specified in the future.

12.2.4.6   Virtual Switch Instance

As mentioned previously, for a customer's VPLS, there exists a VSI in each PE associated with that VPLS. The VSI, as its name implies, is a virtual switch that can have logical ports (e.g., virtual circuits) as its interfaces. Therefore, it has the ability to do regular bridge/switch functionality such as MAC address learning/aging, flooding, forwarding (unicasting, multicasting/broadcasting), and running STP (if needed) per broadcast domain but, based on its logical ports (regular Ethernet switches may only perform these functions on their physical ports).

In the case of distributed-PE models, VSIs can be located either in PE-CLEs or PE-POPs based on their capabilities as described previously. VSIs of a VPLS are connected to each other through emulated VCs, which act as pseudowires. If all VSIs are connected directly with each other through a full mesh of emulated VCs, then one can avoid running STP in the PEs by disabling packet forwarding between any two emulated VCs in a VSI (e.g., packets arriving from emulated VCs have to go out on attachment VCs). This is referred to as *split-horizon* in [37]. However, if the VSIs are not directly connected with each other, then loops can exist and other means for loop prevention can be devised. The use of STP in such scenarios requires further study. Note that if STP were used for loop prevention between the PEs, then it would be different from the customers' STPs (e.g., customers' BPDUs are tunneled transparently through the network).

Although customers' BPDUs are passed transparently through the provider's network, certain scenarios would require monitoring of the customers' BPDUs and taking the appropriate action. For example, if a customer is running 802.1w and a change occurs in its network topology, then the corresponding VSI, on receiving a *topology change notification* (TCN), will flush the MAC addresses learned from all other ports except the one on which it received the TCN. This processing is needed in order to achieve a convergence time of the order of milliseconds as provided by 802.1w. Also, in some other scenarios where 802.1w is not used, monitoring of customers' BPDUs may be needed for accelerating aging time-outs of all ports' MAC addresses upon receiving a TCN.

12.2.4.7   Autodiscovery

Autodiscovery is used to discover all of the PEs associated with a VPLS instance for setting up emulated VCs among the corresponding VSIs.

There are two basic mechanisms for autodiscovery. One is based on directory services and it is described in [38] and the other is based on BGP, which was first described as part of [19], but has been extended and generalized in [39].

In directory-based autodiscovery, IP addresses of all participating PEs for a VPLS are added as records under the VPN-ID (e.g., domain name) in the directory. When a new attachment VC is configured, it is configured with a VPN-ID, and the PE queries the directory for the IP addresses of all other PEs for that VPLS and then tries to establish an emulated VC (if one does not already exist) between its VSI and other VSIs within the VPLS system.

In BGP-based autodiscovery, similar to the directory-based scheme, the VPN-ID is associated with attachment VCs when they are configured; however, in contrast with the directory-based approach, the IP addresses of the member PEs need not to be entered in another server. Instead, each PE running the BGP protocol distributes its VPN-ID information along with its IP address to all other PEs through MP-BGP *network layer reachability information* (NLRI).

In the case of the distributed-PE model (either Ethernet-based or MPLS/IP-based), the directory query or BGP protocol is performed by the PE-POP. Furthermore, if the distributed-PE model is MPLS/IP based, then the discovered information (e.g., IP addresses of the peer PEs) for a VPLS needs to be conveyed to the PE-CLE since emulated VCs are set up from there (e.g., a single-segment emulated VC between two PE-CLEs).

12.2.4.8 Autoconfiguration

Autoconfiguration is used to tie all of the pieces of a VPLS together. Sometimes autodiscovery and autoconfiguration are used loosely and in lieu of each other; however, we make a clear distinction between them in order to clearly define the functionality of each. Autodiscovery is only limited to member discovery of a VPLS instance (e.g., IP addresses of all the PEs participating in a VPLS). However, autoconfiguration uses some components of a VPLS (including autodiscovery) as a triggering mechanism for configuring other components of that VPLS.

As mentioned previously, a VPLS can be defined by a quintuple set as follows: <ingress Attachment VCs, ingress VSIs, Emulated VCs, egress VSIs, and egress Attachment VCs>. When an attachment VC is configured and becomes active, autoconfiguration associates the attachment VC to a VSI for that VPLS, and uses it to trigger autodiscovery for finding member PEs. Upon discovery of the member PEs, the autoconfiguration uses the information to trigger configuration of emulated VCs if they are not already set up by the previous triggers.

Because an emulated VC is a full-duplex circuit and it consists of two unidirectional circuits, when configuring an emulated VC, both circuits need to be configured. We use the single-sided signaling mechanism for

L2VPNs as described in [40] for this purpose, which basically means the setup of the return leg is triggered by the setup of the forward leg. The signaling of the emulated VC also triggers the activation of the egress attachment VC if it is not already activated. Autodiscovery in conjunction with single-sided signaling helps to confine the configuration changes for adding a new customer site to a single PE. Therefore, on configuring the attachment VC associated with the new customer site on the corresponding PE, the configuration of all other pieces is done automatically as described earlier. Note that in the case of directory-based autodiscovery, the directory server might also need to be configured (for adding the IP address of the PE associated with the new site as a new record under the domain name if it does not exist).

In the case of a distributed-PE model where the configuration of attachment VCs is performed at the PE-POP, then autoconfiguration causes the configuration of an attachment VC to also trigger the signaling to the PE-CLE. In the case of an Ethernet-based PE-CLE, the signaling is for configuring attachment tunnels or extension VCs at the PE-CLE as well as configuring the attachment VCs at the PE-CLE. In the case of an MPLS/IP-based PE-CLE, the signaling is for conveying the VPLS membership information (e.g., IP addresses of associated PEs) to the PE-CLE as well as configuring the attachment VCs at the PE-CLE.

12.2.4.9   System Operation

In the following sections, we describe the system operation for different VPLS system architectures in terms of control and data plane operation.

Single-PE

In setting up a VPLS for a customer, first different components of the VPLS (attachment VCs, emulated VCs, and VSIs) need to be configured and hooked up to each other. Then forwarding information needs to be installed in the VSIs. The former task is done within the domain of the control plane by use of different signaling protocols and autoconfiguration. The later task of installing forwarding information is done solely in the data plane by use of MAC address learning and flooding (this is in contrast to L3VPN in which the control plane is used for installing prefix information). Thus, a VPLS can be considered a collection of virtual switches connected with each other and to customers' devices via logical ports, and the control plane is used for setting up these virtual switches and connecting them with each other and the customers' devices.

Let's look at the control plane operation for a VPLS in more details. Upon configuring an attachment circuit at a PE (for connection to a new customer's site), the autoconfiguration mechanism on that PE triggers autodiscovery (either using BGP or DNS) to discover other PEs participating in this VPLS. Then the PE tries to set up an emulated VC between its VSI and every other VSI of the discovered PEs if one already doesn't exist.

Once the setup of the emulated VCs is completed, the new attachment VC is connected to the VPLS and the new CE can participate in data plane functionality.

The operation of the data plane is the same as for regular Ethernet switches except that the VSIs have the ability to perform Ethernet functionality (MAC address learning, flooding, forwarding, and so on) based on logical ports as mentioned previously. Once the VSIs learn the MAC addresses of the new CE and associate them with the logical ports over which they come from, then they perform layer 2 forwarding based on these MAC addresses.

Distributed-PE with Ethernet-Based PE-CLE

A VPLS in this model can be considered to be a collection of virtual switches connected with each other via emulated VCs and connected to customers' devices via attachment VCs and either attachment tunnels or extension VCs. Therefore, the network side of the VSI in this case is the same as the one for the single-PE model, and configuration and operation of emulated VCs are the same as the single-PE model and are not considered further.

The customer-facing side of the VSI is different and it is connected to customers' devices via attachment VCs in conjunction with extension VCs or attachment tunnels. As described previously, when configuring attachment VCs via the PE-POP, the IDs for attachment tunnels or extension VCs are allocated by the PE-POP and conveyed to the PE-CLEs via the appropriate signaling along with the attachment VCs' configuration information. The PE-CLE in turn uses this information to set up its end of the attachment tunnels or extension VCs as well as to set up the attachment VCs to the CEs. Once all of the pieces are set up, the VPLS is ready for data plane operation.

In case of the attachment tunnels for nonmultiplexed UNI ports, all of the customer's traffic on a port (either tagged or untagged) gets sent through the attachment tunnel to the PE-POP, and at the PE-POP based on the tunnel ID, the corresponding VSI associated with that VPLS is selected and the traffic is switched through. Note that in the case of a nonmultiplexed UNI port, the VPLS and in turn the associated VSIs are transparent to customers' VLANs. However, for the multiplexed UNI ports, the VPLS is associated with a given customer's VLAN and thus it is important to examine the customers' VLANs (e.g., attachment VCs) at the PE-POP. Because customers' VLANs can overlap with each other, in order to uniquely identify them, both the attachment tunnel ID and attachment VC ID need to be looked at together (e.g., double-tag lookup) to identify the corresponding VSI. After VSI identification, the packets are forwarded to their destinations based on their destination MAC address. Note that when attachment tunnels are used for multiplexed ports, the PE-CLE is oblivious to the attachment VCs carried inside these tunnels. Therefore, if more than one customer site is connected to the same

PE-CLE and these sites use overlapping MAC addresses across different VLANs, then these VLANs are not isolated from one another at the PE-CLE. Although such situations are rare, it is worth mentioning that there are solutions for this scenario such as the use of sticky MAC addresses; however, because of the rarity of this scenario, they are not discussed here.

If extension VCs are used in lieu of attachment tunnels for multiplexed UNI ports, then the extension VC ID can be used directly to identify the corresponding VSI. However, as mentioned previously, in order to use the extension VCs, the PE-CLE must have the ability to map between the attachment VCs and the extension VCs at the ingress port. This mapping is needed in order to map overlapping customers' VLAN IDs into unique extension VCs for proper identification of the VSIs. Once a VSI is selected, then packet forwarding is performed as before.

Distributed-PE with MPLS/IP-Based PE-CLE

In this model virtual switches can be considered connected with each other via longer logical ports (emulated VCs) that span across multiple service provider devices (PE-CLEs and PE-POPs). In the previous cases, an emulated VC was either between two PEs or between two PE-POPs. However, in this case an emulated VC is between two PE-CLEs and it passes through the PE-POPs transparently. The difference between this distributed-PE model and the single-PE model is primarily in the control plane and once the emulated VCs are set up among the VSIs, then the data plane operation is identical to the one for the single-PE model, whereas the same thing cannot be said for the distributed-PE model with Ethernet-based PE-CLE.

Because the setup of emulated VCs originates from the PE-CLE, the tunneling of these VCs also needs to originate from the PE-CLE. As discussed before, a signaling protocol is used to pass the IP addresses of other PE-CLEs discovered via the autodiscovery mechanism from the PE-POP to the PE-CLE. On receiving the IP addresses, the PE-CLE installs them in its routing table and then uses them as /32 prefixes to establish the tunnels (for emulated VCs). Tunnel establishment is accomplished through one of the existing methods (e.g., downstream unsolicited LDP or RSVP-TE). When tunnels are established, a directed LDP session is set up for emulated VC signaling between this PE-CLE and the other discovered PE-CLEs for that VPLS. Note also that if the configuration of the PE-CLE is done through the PE-POP, then the PE-POP can also be configured with the IP address of the PE-CLE and convey this IP address to the PE-CLE via the signaling protocol; however, if the attachment VCs in the PE-CLE are not configured through the PE-POP, then a simple IGP protocol such as RIP can be used to pass the IP address of the PE-CLE to the PE-POP.

Once the attachment VCs are configured and the emulated VCs are set up at the PE-CLE to connect the VSI to other VSIs, then the VPLS is ready for its data plane operation and furthermore its data plane operates exactly like the one for the single-PE model.

12.2.5 Commonalities and Differences Between Decoupled and Hierarchical Architectures

This section is based on [41], which describes commonalities and differences between decoupled and hierarchical architectures to support scalable VPLS. The need to maintain a full mesh of control connections (LDP, BGP, and so on) and transport paths between all PEs that are service aware may impose a scalability limit on the non–decoupled VPLS architecture. On the other hand, a VPLS-based solution is required to meet the scaling requirement where the PEs facing customer devices and attached to a core network need to perform MAC learning for all the VPLS services.

Proposed decoupled and hierarchical architectures [33, 42, 43] are aimed at improving customer MAC address management on the provider core network and reducing the number of control and transport connections needed between service aware devices by introducing levels in the network. The draft is an attempt to describe commonalities and differences between these architectures.

As discussed later, the differences among the three architectures are not critical and the draft could facilitate discussion toward merging and enhancing the current decoupled/hierarchical VPLS proposals. The draft discussed here is viewed as a step toward a unified decoupled/hierarchical VPLS solution. Figure 12.21 illustrates a general decoupled/hierarchical VPLS architecture.

This section first describes the functions required to support VPLS. In Section 12.2.5.2, a description of the commonalities and differences between the architectures in the control plane is given. In Section 12.2.5.3, a description of the commonalities and differences between the architectures in the forwarding plane is given.

12.2.5.1 Functions to Support VPLS

Following are some of the functions needed to create VPLS. Reference [42] describes most of these functions in more detail. Control Plane Functions are as follows:

- VPLS membership autodiscovery;
- Transport tunnel signaling;
- Service label signaling;
- VPLS configurations.

Forwarding plane functions are as follows:

FIGURE 12.21 *General decoupled/ hierarchical VPLS architecture.*

- MAC learning;
- Customer VLAN processing;
- Customer traffic prioritizing, policing, and shaping;
- Customer packet encapsulation;
- Packet replication/flooding;
- Service label demultiplexing.

12.2.5.2 Control Plane Architecture

The VPN membership autodiscovery and service label signaling functions described in [33, 42, 43] contain both point-to-point and point-to-multipoint signaling. Point-to-point signaling connectivity implies that every device has a signaling connection to all of its peer devices. An example of point-to-point signaling is targeted LDP. Point-to-multipoint signaling refers to a situation in which there exists some mechanism where a device has only a few signaling connections to some entity. The mechanism then relays the information to all the peers of the device. An example of point-to-multipoint signaling is BGP with RR.

In both decoupled and hierarchical architectures, there is a need to distinguish between the control plane functions between the "Core" devices and control plane functions between "Edge" and "Core" devices. Creating levels in control plane connections to enhance control plane scalability is the common theme in all proposed decoupled and hierarchical VPLS architectures. Core devices serve as the signaling gateways between their subtending Edge devices and the rest of the provider network.

Architecture Between Core Devices

Proposals [33, 42, 43] all describe full-mesh transport tunnel signaling connections between Core devices. Reference [43] combines its autodiscovery signaling with service label signaling and uses point-to-point signaling connections. Targeted LDP is the mechanism described in [43]. The HVPLS architecture creates a full mesh of autodiscovery/service label signaling connections between Core devices.

Reference [33] also combines its autodiscovery signaling with service label signaling. However, it can use point-to-point or point-to-multipoint signaling connections to convey autodiscovery information to its peer Core devices. BGP is the mechanism described in [33].

Reference [42] describes more of an architecture framework. The document does not describe a particular autodiscovery and service label signaling scheme.

Architecture Between Core and Edge Devices

References [33, 43] describe Edge devices connected to Core devices via point-to-point links. However, these point-to-point links can be either physical or virtual (FR, ATM, and so on). With physical links, transport tunnel signaling is not required between Core and Edge devices. However, sets of point-to-point transport tunnel signaling connections between Core and Edge devices are required for creating and maintaining point-to-point virtual links.

Reference [42] uses switched Ethernet transport as the mechanism to move traffic between Core and Edge devices. This scheme is similar to MPLS-in-IP [44], where transport tunnels do not exist, thus there is no need for a transport tunnel signaling scheme.

Reference [43] combines its autodiscovery signaling with service label signaling and uses point-to-point signaling connections. There is a single autodiscovery/service label signaling connection between the Core and every Edge devices. The mechanism described again is targeted LDP.

Reference [42] does not specifically describe a particular autodiscovery and service label signaling scheme between Core and Edge devices. However, [45] describes a point-to-multipoint mechanism utilizing the broadcast capability of the switched Ethernet transport. All Core and Edge devices in the same switched Ethernet transport domain can receive autodiscovery and service label information from other service aware devices within the same switched Ethernet transport.

Reference [33] does not specifically describe a particular autodiscovery and service label signaling scheme between Core and Edge devices.

Configurations of Decoupled VPLS

Proposal [43] requires Edge devices to create a single point-to-point spoke VC, for each VPLS service supported on this Edge device, to the Core device using [20]. The Edge device negotiates the VC labels with the Core device. Addition of other Edge devices to add new customer sites requires configuration of the new Edge device but no configuration changes to existing Edge devices.

Proposal [33] describes both mechanisms where configurations can be partly carried out on Edge and Core devices and only on Core devices. However, the focus is on creating a protocol between Edge and Core devices such that configuration information can be carried from Core device to the Edge devices. With over provisioning, addition of new Edge devices need not require additional configuration. But when not overprovisioned, configuration needs to be done on Core devices and/or new Edge devices.

Reference [42] proposes VPLS configurations to be carried out at Edge devices or at Core devices. The objective is to configure customer-facing ports on Edge devices and associate the port/endpoints with the VPLS service. It supports use of signaling between Edge devices and Core devices to distribute configuration information or use of autodiscovery mechanism. Reference [45] is an example of a signaling and autodiscovery mechanism for Edge devices to distribute information within the LPE.

Reference [42] also supports dual VPN membership schemes within the Edge and Core devices and among the Core devices. In such cases, VPN membership mapping function is configured at the Core devices.

12.2.5.3   Forwarding Plane Architecture

In [33], all *virtual bridge* (VB) instances only exist in Edge devices. The DTLS scheme basically creates full-mesh tunnels through service labels between all VBs that have membership in the same VPLS.

Like [33], all VB instances in [42] exist only in the Edge devices. Reference [42] also creates full-mesh tunnels through service labels between all VBs that have membership to the same VPLS. However, unlike [33], traffic between two Edge devices in the same switched Ethernet transport domain does not need to go through the Core device. If one reduces the switched Ethernet transport to a point-to-point Ethernet connection, then [33, 42] have similar forwarding plane architectures.

In [43], VB instances exist in both Edge and Core devices. Reference [43] creates full-mesh tunnels through service labels between all VBs in Core devices that have membership to the same VPLS. Each VB in the Core also has a hub-and-spoke structure to its Edge devices. The hub is the Core VB and the spokes are the Edge VBs. Reference [43] basically creates a tree structure in the forwarding plane where the base of the tree composes of a full mesh of Core VBs. Each Core VB then has branches to its related Edge VBs.

MAC learning:

- Reference [42]: Edge devices only;
- Reference [33]: Edge devices only;
- Reference [43]: Both Edge and Core devices.

Customer VLAN processing:

- Reference [42]: Edge devices only;
- Reference [33]: Edge devices only;
- Reference [43]: Both Edge and Core devices.

Customer traffic prioritizing, policing, and shaping:

- Reference [42]: Edge devices only;
- Reference [33]: Edge devices only;
- Reference [43]: Both Edge and Core devices.

Customer packet encapsulation:

- Reference [42]: Edge devices only;
- Reference [33]: Edge devices only;
- Reference [43]: Both Edge and Core devices.

Packet replication/flooding:

- Reference [42]: Edge devices, switched Ethernet transport, and Core devices;
- Reference [33]: Edge devices only;
- Reference [43]: Both Edge and Core devices.

Service label demultiplexing:

- Reference [42]: Edge devices only;
- Reference [33]: Edge devices only;
- Reference [43]: Both Edge and Core devices.

ENDNOTES

[1]   Retrieved from http://www.cisco.com/warp/public/cc/pd/rt/10700/10720rt/prodlit/vbov_ov.htm.

[2]   Retrieved from http://www.foundrynet.com/technologies/mpls/faqs.html.

[3]   MFS Datanet offered a similar service in the same time frame.

[4]   Greenfield, D., "Optical Standards: A Blueprint for the Future," *Network Magazine,* October 5, 2001.

[5]   *Synchronous Optical Networks (SONET),* ANSI T1X1.5, 2001-062, January 2001.

[6]   *G.707 Network Node Interface for the Synchronous Digital Hierarchy (SDH),* ITU, April 2000.

[7]   Portions of the remainder of this section are based on Olsson, F., J. Shupenis, and A. Gunn, "Virtual Concatenation + LCAS Providing Scalable SONET/SDH Bandwidth," Agilent Technologies White Paper, 2002; note that all editorializations are ours.

[8]   *G.LCAS Link Capacity Adjustment Scheme (LCAS) for Virtual Concatenation,* ITU, October 2001.

[9]   Although MPLS usually carries IP, some MPLS-only switches exist for which the use of IP is incidental and limited to control plane functions.

[10]   Retrieved from http://www.mplsworld.com/archi_drafts/vpn/archi_vpn.htm#020304d.

[11]   Elangovan, P., "Virtual Private LAN Service (VPLS) Management Information Base Using SMIv2," IETF Draft, draft-paari-ppvpn-vpls-mib-00.txt, July 2002.

[12]   Portions of this section are based on Lasserre, M., "MPLS Based Transparent LAN Services," Riverstone Networks White Paper, retrieved from http://www.riverstonenet.com/technology/tls.shtml.

[13]   *MAC Bridges,* Original 802.1D—ISO/IEC 10038, ANSI/IEEE Std 802.1D-1993.

[14]   *Information Technology—Telecommunications and Information Exchange Between Systems—Local and Metropolitan Area Networks—Common Specifications—Part 3: Media*

Access Control (MAC) Bridges: Revision, 802.1D. This is a revision of ISO/IEC 10038: 1993, 802.1j-1992 and 802.6k-1992. It incorporates P802.11c, P802.1p and P802.12e. ISO/IEC 15802-3:1998.

[15] Lasserre, M., et al., "Virtual Private LAN Services over MPLS," Internet draft document draft-lasserre-vkompella-ppvpn-vpls-02.txt, June 2002. Copyright 2001 The Internet Society (2001). All Rights Reserved. This document and translations of it may be copied and furnished to others, and derivative works that comment on or otherwise explain it or assist in its implementation may be prepared, copied, published and distributed, in whole or in part, without restriction of any kind, provided that the above copyright notice and this paragraph are included on all such copies and derivative works. It is inappropriate to use Internet drafts as reference material or to cite them other than as "work in progress."

[16] *Encapsulation Methods for Transport of Ethernet Frames Over IP and MPLS,* draft-martini-ethernet-encap-mpls-00.txt, work in progress, April 2002.

[17] *IEEE Standards for Local and Metropolitan Area Networks: Virtual Bridged Local Area Networks,* 802.1Q, ANSI/IEEE Draft Standard P802.1Q/D11, July 1998.

[18] *Requirements for Virtual Private LAN Services (VPLS),* draft-ietf-ppvpn-vpls-requirements-00.txt, work in progress.

[19] Rosen, E., and Y. Rekhtery, *BGP/MPLS VPNs,* RFC 2547, March 1999.

[20] Martini, L., et al., *Transport of layer 2 Frames over MPLS,* draft-martini-l2circuit-trans-mpls-09.txt, work in progress, April 2002.

[21] *Bridging and VPLS,* draft-finn-ppvpn-bridging-vpls- 00.txt, work in Progress, June 2002.

[22] *Single-Sided Signaling for L2VPN,* draft-rosen-ppvpn-l2-signaling-01.txt, work in progress, February 2002.

[23] This section is based on Senevirathne, T., "Virtual Private LAN Service (VPLS) Solution Using GRE Based IP Tunnels," IETF draft-tsenevir-gre-vpls-01.txt, February 2002. This is a "work in progress." Copyright 2002 The Internet Society. All Rights Reserved. This document and translations of it may be copied and furnished to others, and derivative works that comment on or otherwise explain it or assist in its implementation may be prepared, copied, published and distributed, in whole or in part, without restriction of any kind, provided that the above copyright notice and this paragraph are included on all such copies and derivative works.

[24] Senevirathne, T., *Ethernet over IP—A Layer 2 VPM Solution Using Generic Routing Encapsulation (GRE),* work in progress, July 2001.

[25] Hanks, S., *Generic Routing Encapsulation (GRE),* RFC 1701, October 1994.

[26] Augustyn, W., *Requirements for Virtual Private LAN service (VPLS),* work in progress, August 2002.

[27] Senevirathne, T., *Distribution of 802.1Q VLAN information using Opaque LSA,* work in progress, January 2001.

[28] Senevirathne, T., *Use of BGP-MP for Layer 2 VPN Membership Discovery,* work in progress, July 2001.

[29] Ould-Brahim, H., et al., *Using BGP as an Auto-Discovery Mechanism for Network Based VPNs,* January 2002.

[30] Sajassi, A., D. Lee, and J. Guichard, "VPLS Architectures," Internet draft-sajassi-vpls-architectures-00.txt, February 20, 2002. Copyright 2001 The Internet Society. All Rights Reserved. This document and translations of it may be copied and furnished to others, and derivative works that comment on or otherwise explain it or assist in its implementation may be prepared, copied, published and distributed, in

whole or in part, without restriction of any kind, provided that the above copyright notice and this paragraph are included on all such copies and derivative works.

[31] Rosen, E., *An Architecture for L2VPNs,* draft-rosen-ppvpn-l2vpn-00.txt, May 2001.

[32] Augustyn, W., *Requirements for Virtual Private LAN Services (VPLS),* draft-augustyn-vpls-requirements-00.txt, April 2002.

[33] Kompella, K., *Decoupled Virtual Private LAN Services,* draft-kompella-ppvpn-dtls-01.txt, May 2002.

[34] Martini, L., et al., *Encapsulation Methods for Transport of Layer 2 Frames over MPLS,* draft-martini-l2circuit-encap-mpls-01.txt, February 2001.

[35] Martini, L., et al., *Transport of Layer 2 Frames over MPLS,* draft-martini-l2circuit-trans-mpls-05.txt, February 2001.

[36] *Layer Two Tunneling Protocol, L2TP,* draft-ietf-l2tpext-l2tp-base-01.txt.

[37] Lasserre, M., et al., *Transparent LAN Service over MPLS,* draft-lasserre-vkompella-ppvpn-tls-00.txt, November 2001.

[38] Heinanen, J., *DNS/LDP Based VPLS,* draft-heinanen-dns-ldp-vpls-00.txt, work in progress, January 2002.

[39] Ould-Brahim, H., et al., *Using BGP as an Auto-Discovery Mechanism for Network-Based VPNs,* draft-ietf-ppvpn-bgpvpn-auto-02.txt, work in progress, January 2002.

[40] Rosen, E., *Single-Sided Signaling for L2VPNs,* draft-rosen-ppvpn-l2-signaling-00.txt, March 2001.

[41] Chen, M., et al., *Decoupled/Hierarchical VPLS: Commonalities and Differences,* Internet draft-chen-ppvpn-dvpls-compare-00.txt, March 2002. Copyright 2001 The Internet Society. All Rights Reserved. This document and translations of it may be copied and furnished to others, and derivative works that comment on or otherwise explain it or assist in its implementation may be prepared, copied, published and distributed, in whole or in part, without restriction of any kind, provided that the above copyright notice and this paragraph are included on all such copies and derivative works. It is inappropriate to use Internet drafts as reference material or to cite them other than as "work in progress."

[42] Ould-Brahim, H., et al., *VPLS/LPE: Virtual Private LAN Service Using the Logical PE Architecture,* draft-ouldbrahim-l2vpn-lpe-00.txt, work in progress, November 2001.

[43] Khandekar, S., et al., *Hierarchical Virtual Private LAN Service,* draft-khandekar-ppvpn-hvpls-mpls-00.txt, work in progress, November 2001.

[44] Worster, T., et al., *MPLS Label Stack Encapsulation in IP,* draft-worster-mpls-in-ip-05.txt, work in progress, July 2001.

[45] Knight, P., et al., *Logical PE Auto-Discovery Mechanism,* draft-knight-l2vpn-lpe-ad-00.txt, work in progress, November 2001.

CHAPTER 13

Digital Video and Multimedia Technologies and Applications

This chapter looks at a set of digital video and multimedia applications that can run on the broadband infrastructure that has been discussed in the body of this book. As noted in this book, applications are the key to a reinvigorated telecom industry; we see the need for a sustained effort in this arena, spanning the entire telecom space. Here, however, we focus only on video/multimedia applications. To date, digital video has not represented a significant revenue stream for carriers, although new cable TV systems as well as satellite-based video distribution systems, particularly *direct broadcast satellite* (DBS), utilize these digital video techniques. The opportunity is to identify services and applications that can increase that revenue stream.

Real-time high-quality video is an isochronous application that requires both end-to-end QoS as well as end-to-end broadband transmission channels. Both QoS and broadband connectivity need to be realized over the LANs, MANs, WANs, and GANs (namely, the Internet) that comprise the end-to-end connection. It should be noted, however, that although real-time broadcast video is completely isochronous, nonreal-time video delivery applications do exist that do not require synchronized timing between the sender and the receiver; for example, distribution of video content to ReplayTV and Tivo-type devices. Even broadcast video need not be completely isochronous given "large enough" playout buffers for digitized streaming media: If one allows some reasonable amount of delay jitter, the demands on the network characteristics could be reduced. Video or multimedia delivery comes in four different types: (1) real-time interactive presentations (videoconferencing for example), (2) one-way real-time ("broadcast") presentations of near-live content; (3) one-way presentations of archived content, and (4) off-line delivery of content. Each of these has different performance characteristics. One can trade off delay-jitter, available bandwidth, and reliability to optimize for any of these four alternatives. For example, DBS gets away with somewhat limited link margins because of the FEC it uses. On data networks, with nonreal-time, one-way delivery, one can use erasure codes to compensate for lost portions of "video file," tolerating a higher error or drop rate at the expense of either more time or bandwidth. Our presentation explores these various themes, but focuses on high-quality digital video.

This chapter explores issues related to digital video in the area of trends, standards, applications, and challenges. The following topics are addressed:

1. The emergence of video-based applications, including *digital TV* (DTV) broadcast;

2. The ITU-T H.320/H.323 standards;

3. The ISO MPEG family of standards, including MPEG-1, MPEG-2, MPEG-4, MPEG-7, and MPGEG-21.

Portions of the discussion focus on support of video over the intranet and Internet, and the intrinsic need for QoS in these networks.

13.1 Introduction

Interest in high-quality video-based digital communications for business as well as for entertainment applications goes back several years [1]. At one end, digital methods are now planned to be deployed for the support of a new generation of television known as *advanced TV* (ATV) or DTV, although there are several approaches to the matter (terrestrial broadcast, satellite broadcast, and cable TV distribution). At the other end, there is interest in delivering desktop business video applications via the IP suite of protocols, and over the corporate intranet, the Internet, or extranets. In addition to traditional room-based videoconferencing, vertically integrated applications, such as distance learning and telemedicine, are also receiving some attention of late. PC-based multimedia is becoming ever-more popular with the availability of inexpensive cameras and reasonable desktop throughput. Furthermore, as mobile communications become a basic part of modern life, the next step is to support mobile multimedia applications; in fact, standards and equipment are already evolving to support handheld or laptop-based video telephony.

During the recent past, a major change in the area of TV distribution has occurred. While in some countries wireless video is equivalent to the reception of terrestrial analog TV signals via a rooftop antenna, in other countries TV programs are received via satellite, cable, or even *multichannel microwave distribution system* (MMDS) links. In the future TV programs may also be distributed over the Internet. Already, companies such as CNN and MSNBC offer nonreal-time TV-program video clips. The reader should also keep in mind that while DSL is now being deployed in the Internet access environment, it was initially positioned to deliver video-on-demand ("video dialtone") by the Bell Operating Companies. More generally, the advent of DTV transforms the traditional TV channel into a data transmission medium that supports large data rates (e.g., upward of 20 Mbps) at very low bit error rates ($<10^{-11}$). In this new environment, the term *television* loses part of its original meaning: Digital television is no longer restricted to transmitting sound and images; instead, it becomes a data broadcasting mechanism that is fully transparent to the content [2].

A number of driving forces are encouraging the development of video-based digital services. First and foremost there is the demand for higher quality in the video and audio spectrum. Second, one finds an array of technological developments that are, in fact, spurring innovation and deployment. No longer will barely discernible "talking heads" with unsynchronized lips and video, plagued by annoying encoding delays and jumping or exploding heads be tolerated by the business user community for videoconferencing applications. Twenty years of this is enough. Fortunately, new technologies are entering the field at the commercial level that have the potential to address problems that have been holding back the deployment of video services for years. These technologies span both high-performance *digital signal processing* (DSP), as well as broadband transmission. Until now, transmission of business video had to be accomplished using the ISDN Basic Rate Service at 128 Kbps or, at most, at 384 Kbps (here employing inverse multiplexers to merge several ISDN channels into one aggregate channel). Now broadband services are becoming available, both at the data link layer (e.g., TLS and ATM), as well as at the network layer (e.g., via high-performance IP routers and MPLS switches). In general, emerging digital video applications can be classified in three categories:

1. Corporate/institutional video transmission (classical systems, as well as the newer H.323-based systems);

2. Stored digital video for multimedia applications;

3. Entertainment video (video broadcast, including postproduction distribution).

Because of the significant amount of information contained in a full-motion video stream, for example, up to 200 Mbps for traditional video and even higher for *high-definition TV* (HDTV), video compression is an absolute must.

In corporate/institutional video transmission, the news in video relates to the development of QoS-based networks that support video transmission, along with the introduction of IP/LAN-based (specifically, H.323) coders for desktop applications. Standardization and the use of ubiquitously available IP networks will go a long way toward fostering the proliferation of these video services. These developments drive the general need for QoS-based communications. At this time, most discussions about QoS have an enterprise network point of reference; however, because many corporate networks use the Internet in some fashion, this translates ultimately into a requirement on the Internet itself. This is particularly the case when enterprise users wish to employ VPNs. In reference to stored digital video, the news relates to the general availability of DVDs, along with the supporting digital encoding methods. The news in entertainment video is the plan now under way to broadcast DTV commercially, particularly in terrestrial networks, but also in cable and satellite environments. There is a desire to move up from the rather limited resolution baseline of the present. However, one of the factors possibly holding back the deployment is the high price of the TV sets, which remains in the "several thousand dollars" range.

To facilitate ubiquitous communications, standards are required. The video space is no different. The *Moving Picture Expert Group* (MPEG), a working group of ISO/IEC in charge of the development of standards for coded representation of digital audio and video, was established in 1988. High-quality digital encoding schemes such as MPEG-1 and MPEG-2 are now fairly entrenched. MPEG-4 and MPEG-7 are poised to see deployment. The first such standard in the family, the ISO MPEG-1 (ISO 11172-2, 1993), is a "high-quality" audiovisual coding standard supporting the storage and retrieval of multimedia information on a CD-ROM up to about 1.5 Mbps. ISO MPEG-2 (ISO 13818-2) followed, in 1995; this standard addresses broadcast TV applications. MPEG-4 is a new flexible low-bandwidth standard under recent development. The advanced capabilities of this standard enable it to support a number of evolving applications, ranging from wireless videophones to Internet multimedia presentations, broadcast TV, and DVD. MPEG-4 enables robust video transmission over noisy communication channels (such as wireless video links). Naturally, a standard that supports these diverse functionalities and associated applications turns out by necessity to be fairly complex. MPEG-4 became an official international standard in January 1999 [3]. In 1996 MPEG started work on a new standard known as MPEG-7. The new member of the MPEG family, called the *multimedia content description interface*, extends the limited capabilities of proprietary solutions in identifying content that exists today, notably by including more data types. MPEG-7 specifies a standard set of descriptors that can be used to describe various types of multimedia information. MPEG-7 will also standardize ways to define other descriptors as well as structures for the descriptors and their relationships. This description (i.e., the combination of descriptors and description schemes) will be associated with the content itself, to allow fast and efficient searching for material of interest to a user. MPEG-7 will also standardize a language to specify description schemes, that is, a *description definition language*. Audiovisual material that has MPEG-7 data associated with it can be indexed and searched for. This "material" may include still pictures, graphics, 3D models, audio, speech, video, and information about how these elements are combined in a multimedia presentation (scenarios, composition information). Work on the new standard MPEG-21, *Multimedia Framework*, started in 2000.

In summary, the MPEG group has produced MPEG-1, the standard on which such products as video CD and MP3 are based; MPEG-2, the standard on which such products as DTV settop boxes and DVD are based; MPEG-4, the standard for multimedia for the fixed and mobile Web; and MPEG-7, the standard for description and search of audio and visual content. Note that originally another standard, MPEG-3, was designed to support HDTV with sampling dimensions up to $1,920 \times 1,080$ at 30 frames per second (fps). This standard has now been abandoned since, with a little tweaking, MPEG-2 works well for HDTV at rate of 20 and 40 Mbps. Hence, HDTV is now part of the MPEG-2 high-1440 level specification (this topic is discussed in a later section).

Wrapping up this preliminary discussion of standards, Figures 13.1 through 13.3 identify key video standards, and Figures 13.4 through 13.7 provide more detailed listings of pertinent standards. (Table 13.1 amplifies some elements of Figure 13.7.)

FIGURE 13.1
Standards/MPEG Committee.

Standards/MPEG Committee

• The Moving Picture Experts Group is a working group of ISO/IEC in charge of the development of international standards for compression, decompression, processing, and coded representation of moving pictures, audio, and their combination

• MPEG has produced:
  - MPEG-1, the standard for storage and retrieval of moving pictures and audio on storage media (approved Nov. 1992)
  - MPEG-2, the standard for digital television (approved Nov. 1994)
  - MPEG-4, version 1, the standard for multimedia applications (approved Oct. 1998). Version 2 (approved Dec. 1999)
  - MPEG-7, content representation standard for information search (Dec. 1999)

• MPEG usually holds three meetings a year. These comprise plenary meetings and subgroup meetings on Requirements, Delivery, Systems, Video, Audio, SNHC, Test, Implementation, DSM, and Liaison. MPEG meetings are attended by some 300 experts from 20 countries

FIGURE 13.2
Standards/MPEG 7.

• MPEG-7 is concerned with the standardization of a "Multimedia Content Description Interface" which will allow the description, identification, and access of audiovisual information
  - Preliminary work for the MPEG-7 International Standard completed 2000

• Audiovisual information is becoming available in digital form

• Finding the desired information is increasingly difficult
  - Many text-based search engines are available on the World Wide Web
  - Searching information is, however, not possible for audiovisual content, as no generally recognized or standardized descriptions of these material exist

• MPEG-7 to provide a solution...
  - Will extend today's search capabilities to include more information types
  - MPEG-7 will specify a standardized description of various types of multimedia information

FIGURE 13.3
Key DTV standards.

- ATSC A/49 ghost cancellation signal for NTSC
- ATSC A/52 digital audio compression for ATSC HDTV
- ATSC A/53, A/54 ATSC HDTV standard
- ATSC A/57 program, episode, and version ID for ATSC HDTV
- ATSC A/63 method for handling 25/50 Hz video for ATSC HDTV
- ATSC A/65 program and system information protocol for ATSC HDTV
- Dolby Digital
- EIA-770 analog RGB video interfaces for SDTV and HDTV
- IEEE 1394 Trade Organization

Digital Television Standards

ATSC Formats and Typical Compression Rates

ATSC Format Number	Image Format	Scanning Rate (Hz)	# Pixels/ Sec (Mpels)	Active Bit Rate Assumes 4:2:2[+] 10-bit input	Compression Ratio for 1 Program per ATSC channel	Bit Rate (Mbps) Normalized to 55:1	# of Local Channel Capacity
15,33	640 x 480 4 x 3	P24,P23.98	7.4	147	8:1	3 Mbps	17%
16,34		P30,P29.97	9.2	184	10:1	3 Mbps	17%
17,35		I30,I29.97	9.2	184	10:1	3 Mbps	17%
18,36		P60,P59.94	18.4	369	20:1	7 Mbps	22%
11,29	720 x 486 4 x 3	P24,P23.98	8.4	167	9:1	3 Mbps	17%
12,30		P30,P29.97	10.5	210	12:1	4 Mbps	22%
13,31		I30,I29.97	10.5	210	12:1	4 Mbps	22%
14,32		P60,P59.94	20.2	404	23:1	7 Mbps	39%
7,25	720 x 483 4 x 3	P24,P23.98	8.1	162	9:1	3 Mbps	17%
8,26		P30,P29.97	10.1	202	11:1	4 Mbps	22%
9,27		I30,I29.97	10.1	202	11:1	4 Mbps	22%
10,28		P60,P59.94	20.9	418	23:1	8 Mbps	44%
4,22	1280 x 720 16 x 9	P24,P23.98	22.1	442	24:1	8 Mbps	64%
5,23		P30,P29.97	27.6	552	31:1	10 Mbps	55%
6,24		P60,P59.94	55.3	1106	61:1	18 Mbps	100%
1,19	1920 x 1080 16 x 9	P24,P23.98	49.8	996	55:1	18 Mbps	100%
2,20		P30,P29.97	62.2	1,244	69:1	18 Mbps	100%
3,21		I30,I29.97	62.2	1,244	69:1	18 Mbps	100%

Notes:
+ For HD Formats 4:2:2
refers to ratio of T samples
to Pb and Pr samples rather
than relationshiop
to base sampling frequency

FIGURE 13.4   *Digital television standards: ATSC formats and typical compression rates. (Courtesy Miranda Technologies, Montreal, Canada.)*

System Nomenclature	Display Aspect Ratio	Samples per Total Line (S/TL)	Total Lines per Frame	Samples per Active Line (S/AL)	Active Lines per Frame	Scanning Format	Frame Rate (Hz)	Line Rate (kHz)	Interface Sampling Frequency fs (MHz)†	Active Bit Rate (Mbps)	Total Bit Rate (Mbps)	SMPTE Scan Format Standard	SMPTE Serial Interface Standard	ATSC Table 3
SDTV 4fsc 525	4:3	910	525	768	486	2:1 interlace	29.97	15.73	14.32	111.9	143.2	244M	259M-A	
4fsc 625	4:3	1135	625	948	576	2:1 interlace	25	15.63	17.73	136.5	177.3	EBU	259M-B	
601 525	4:3/16:9	858	525	720	486	2:1 interlace	29.97	15.73	13.50	209.7	270.0	125M/267M	259M-C	27,31
601 625	4:3/16:9	864	625	720	576	2:1 interlace	25	15.63	13.50	207.4	270.0	EBU	259M-C	
601-18MHz 525	16:9	1144	525	960	486	2:1 interlace	29.97	15.73	18.00	279.7	360.0	267M	259M-D	
720×480/59.94/1:1	16:9	858	525	720	483	1:1 progressive	59.94	31.47	27.00	416.9	540.0* **	293M	294M	28
720×480/30/1:1	16:9	?	525	720	?	1:1 progressive	30	15.75					259M-C	8
720×480/29.97/1:1	16:9	?	525	720	?	1:1 progressive	29.97	15.73					259M-C	26
HDTV 1280×720/60/1:1	16:9	1650	750	1280	720	1:1 progressive	60	45.00	74.25	1105.9	1485.0	296M	292M	6
1280×720/59.94/1:1	16:9	1650	750	1280	720	1:1 progressive	59.94	44.96	74.18	1104.8	1483.5	276M	292M	24
1280×720/50/1:1	16:9	1980	750	1280	720	1:1 progressive	50	37.50	74.25	921.6	1485.0	296M*	292M	
1280×720/30/1:1	16:9	3300	750	1280	720	1:1 progressive	30	22.50	74.25	553.0	1485.0	296M*	292M	5
1280×720/29.97/1:1	16:9	3300	750	1280	720	1:1 progressive	29.97	22.48	74.18	552.4	1483.5	296M*	292M	23
1280×720/25/1:1	16:9	3960	750	1280	720	1:1 progressive	25	18.75	74.25	460.8	1485.0	296M *	292M	
1280×720/24/1:1	16:9	4125	750	1280	720	1:1 progressive	24	18.00	74.25	442.4	1485.0	296M*	292M	4
1280×720/23.92/1:1	16:9	4125	750	1280	720	1:1 progressive	23.98	17.99	74.19	442.0	1483.8	296M*	292M	22
1920×1035/60/2:1	16:9	2200	1125	1920	1035	2:1 interlace	30	33.75	74.25	1192.3	1485.0	260M	292M	
1920×1035/59.94/2:1	16:9	2200	1125	1920	1035	2:1 interlace	29.97	33.72	74.18	1191.1	1483.5	260M	292M	
1920×1080/60/1:1	16:9	2200	1125	1920	1080	1:1 progressive	60	67.50	148.50	2488.3	2970.0	274M		
920x1080/59.94/1:1	16:9	2200	1125	1920	1080	1:1 progressive	59.94	67.43	148.35	2485.8	2967.0	274M		
1920x1080/50/1:1	16:9	2640	1125	1920	1080	1:1 progressive	50	56.25	148.50	2073.6	2970.0	274M		
1920x1080/50/1:1 (1250)	16:9	2376	1250	1920	1080	1:1 progressive	50	62.50	148.50	2073.6	2970.0	295M		
1920x1080/60/2:1	16:9	2200	1125	1920	1080	2:1 interlace	30	33.75	74.25	1244.2	1485.0	274M	292M	1
1920x1080/59.94/2:1	16:9	2200	1125	1920	1080	2:1 interlace	29.97	33.72	74.18	1242.9	1483.5	274M	292M	21
1920x1080/50/2:1	16:9	2640	1125	1920	1080	2:1 interlace	25	28.13	74.25	1036.8	1485.0	274M	292M	
1920x1080/50/2:1 (1250)	16:9	2376	1250	1920	1080	2:1 interlace	25	31.25	74.25	1036.8	1485.0	295M	292M	
1920x1080/30/1:1	16:9	2200	1125	1920	1080	1:1 progressive	30	33.75	74.25	1244.2	1485.0	274M	292M	2
1920x1080/29.97/1:1	16:9	2200	1125	1920	1080	1:1 progressive	29.97	33.72	74.18	1242.9	1483.5	274M	292M	20
1920x1080/25/1:1	16:9	2640	1125	1920	1080	1:1 progressive	25	28.13	74.25	1036.8	1485.0	274M	292M	
1920x1080/24/1:1	16:9	2750	1125	1920	1080	1:1 progressive	24	27.00	74.25	995.3	1485.0	274M	292M	1
1920x1080/23.98/1:1	16:9	2750	1125	1920	1080	1:1 progressive	23.98	26.98	74.19	994.5	1483.8	274M	292M	19
1920x1080/30 (sF)	16:9	2200	1125	1920	1080	1:1 Segmented	30	33.75	74.25	1244.2	1485.0	274M **	292M	2

FIGURE 13.5 *Digital television standards: Digital production formats. (Courtesy Miranda Technologies, Montreal, Canada.)*

System Nomenclature	Display Aspect Ratio	Samples Per Total Line (S/TL)	Total Lines Per Frame	Samples Per Active Line (S/AL)	Active Lines Per Frame	Scanning Format	Frame Rate (Hz)	Line Rate (kHz)	Interface Sampling Frequency fs (MHz)†	Active Bit Rate (Mbps)	Total Bit Rate (Mbps)	SMPTE Scan Format Standard	SMPTE Serial Interface Standard	ATSC Table 3
1920x1080 /29.97(sF)	16:9	2200	1125	1920	1080	1:1 Segmented	29.97	33.72	74.18	1242.9	1483.5	274M **	292M	20
1920x1080 /25(sF)	16:9	2640	1125	1920	1080	1:1 Segmented	25	28.13	74.25	1036.8	1485.0	274M **	292M	
1920x1080 /24(sF)	16:9	2750	1125	1920	1080	1:1 Segmented	24	27.00	74.25	995.3	1485.0	274M **	292M	1
1920x1080 /23.98(sF)	16:9	2750	1125	1920	1080	1:1 Segmented	23.98	26.98	74.19	994.5	1483.8	274M **	292M	19

Notes:

* These formats are proposed to be added to SMPTE 296M

** These formats are proposed to be added to SMPTE 274M

*** Although the total data rate of 720x480/1:1 is 540 Mbps, the actual interface may be any one of:
Single Link At 1.485 Gbps (proposed)
Single Link At 540 Mbps (proposed)
Dual Link At 270 Mbps
Single 4:2:0 Link at 360 Mbps

† Refers to Y sampling frequency for component formats and overall sampling frequency for composite formats

FIGURE 13.5 (CONTINUED)

Format	Interface Standard	Tape Size	Sampling Format	Video Sample Resolution (bits)	Essential Video Bit Rate (Mbps)*	Recorded Data Bit Rate (Mbps)	Recorded Video Rate (Mbps)	Compression Type	Approximate Compression Ratio	# Audio Tracks	Audio Sample Resolution (bits)	Audio Sampling Frequency (kHz)
D1	SMPTE 259M-C	19mm	4:2:2	8	172	225	172	—	—	4	16-20	48
D2	SMPTE 259M-A	19mm	4Fsc	8	94	127	94	—	—	4	16-20	48
D3	SMPTE 259M-A	1/2"	4Fsc	8	94	125	94	—	—	4	16-20	48
D5	SMPTE259M-C	1/2"	4:2:2	10	220	300	220	—	—	4	16-20	48
DCT	SMPTE 259M-C	19mm	4:2:2	8	174	125	88	DCT	2:1	4	16-20	48
DV	IEEE-1394	6mm	4:1:1 (525/60) 4:2:0 (625/50)	8	125	35.5	25	DV	5:1	2	16	48
D7 - DVCAM	IEEE-1394	6mm	4:1:1 (525/60) 4:2:0 (625/50)	8	125	35.5	25	DV	5:1	2	16	48
DVC-Pro	IEEE I-1394, SMPTE 259M-C SMPTE 305M(SDTI)	6mm	4:1:1 (525/60) 4:2:0 (625/50)	8	125	41.8	25	DV	5:1	2	16	48
DVC-Pro 50	SMPTE 259M-C, 305M(SDTI)	6mm	4:2:2	8	168	100	50	DV	3.3:1	4	16	48
DVC-Pro100	SMPTE 259M-C, 305M(SDTI)	6mm	4:2:2	8	168	165	100	DV	1.7:1	4	16	48
D9 - Digital-S	SMPTE 259M-C, 305M(SDTI)	1/2"	4:2:2	8	168	83.6	50	DV	3.3:1	2	16	48
Betacam SX	SMPTE 259M-C, 305M(SDTI)	1/2"	4:2:2	8	176	40	18	MPEG-2	10:1	4	16	48
Digital Betacam	259M-C	1/2"	4:2:2	10	219	128	95	DCT	2.3:1	4	16-20	48
HD-CAM	SMPTE 259M-C, 305M(SDTI)	1/2"	3:1:1	8	994	185	140	DCT	7.1:1	4	20	48
HD-D5	259M-C, 292M	1/2"	4:2:2	10	1244	300	276	DCT	4.5:1	4	20	48
D6	259M-C, 292M	19mm	4:2:2	8	995	1180	995	—	—	10/12	20-24	48

Notes:* refers to the data presented to the compression encoder after 8 versus 10 bit sampling and resolution subsampling processes, if any.

FIGURE 13.6   *Digital television standards: Digital tape formats. (Courtesy Miranda Technologies, Montreal, Canada.)*

FIGURE 13.7
*Digital television
standards: MPEG-2
compression levels and
profile chart.
(Courtesy Miranda
Technologies,
Montreal, Canada.)*

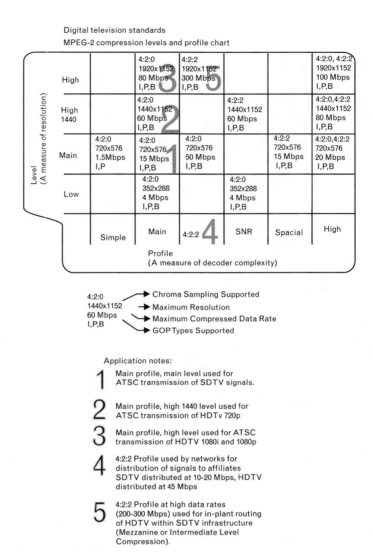

Digital television standards
MPEG-2 compression levels and profile chart

Application notes:

1   Main profile, main level used for
ATSC transmission of SDTV signals.

2   Main profile, high 1440 level used for
ATSC transmission of HDTv 720p

3   Main profile, high level used for ATSC
transmission of HDTV 1080i and 1080p

4   4:2:2 Profile used by networks for
distribution of signals to affiliates
SDTV distributed at 10-20 Mbps, HDTV
distributed at 45 Mbps

5   4:2:2 Profile at high data rates
(200-300 Mbps) used for in-plant routing
of HDTV within SDTV infrastructure
(Mezzanine or Intermediate Level
Compression).

For corporate video applications, a set of coding (compression) standards has been used that goes back a number of years, in the form of traditional ITU-T H.261/H.262 recommendations (under the H.320 umbrella). Recently, two new recommendations have emerged for the support of IP-based multimedia (H.323) and for mobile IP-based telephony (H.324). Applications include stand-alone videophones, PC-based multimedia applications, inexpensive voice/data modems, World Wide Web browsers with live video, video-only security cameras, and others [4]. H.323 also supports the VoIP mechanism discussed in Chapter 10. Figures 13.8, 13.9, and 13.10 depict various aspects of coding systems.

On the entertainment side, in 1996 the FCC adopted a DTV transmission standard for terrestrial broadcast in the United States. Since the

TABLE 13.1   SOME ADDITIONAL DETAILS ON KEY DIGITAL TAPE STANDARDS

D1   A format for digital videotape recording working to the ITU-R 601, 4:2:2 standard using 8-bit sampling. The tape is 19 mm wide and allows up to 94 min. to be recorded on a cassette. Being a component recording system it is ideal for studio or postproduction work with its high chrominance bandwidth allowing excellent chroma keying. Also multiple generations are possible with very little degradation and D1 equipment can integrate without transcoding to most digital effects systems, telecines, graphics devices, disk recorders, and so on. Because it is a recording system, there are no color framing requirements. Despite the advantages, D1 equipment is not extensively used in general areas of TV production, at least partly due to its high cost.

D2   The VTR standard for digital composite (coded) PAL or NTSC signals. It uses 19-mm tape and records up to 208 min on a single cassette. Neither cassettes nor recording formats are compatible with D1. D2 has often been used as a direct replacement for 1-in. analog VTRs. Although offering good stunt modes and multiple generations with low losses, because it is a coded system, coded characteristics are present. The user must be aware of cross-color, transcoding footprints, low chrominance bandwidths, and color framing sequences. Employing an 8-bit format to sample the whole coded signal results in reduced amplitude resolution making D2 more susceptible to contouring artifacts.

D3   A VTR standard using 1/2-in. tape cassettes for recording digitized composite (coded) PAL or NTSC signals sampled at 8 bits. Cassettes are available for 50 to 245 min. Because this uses a composite signal, the characteristics are generally as for D2 except that the 1/2-in. cassette size has allowed a full family of VTR equipment to be realized in one format, including a camcorder.

D4   There is no D4. Most DVTR formats hail from Japan where 4 is regarded as an unlucky number.

D5   A VTR format that uses the same cassette as D3 but recording component signals sampled to ITU-R 601 recommendations at 10-bit resolution. With internal decoding, D5 VTRs can play back D3 tapes and provide component outputs. Being a noncompressed component digital video recorder means that D5 enjoys all the performance benefits of D1, making it suitable for high-end postproduction as well as more general studio use. Besides servicing the current 625- and 525-line TV standards, the format also has provision for HDTV recording by use of about 4:1 compression (HD D5).

D6   A digital tape format that uses a 19-mm helical-scan cassette tape to record uncompressed HDTV material at 1.88 Gbps. D6 is currently the only high-definition recording format defined by a recognized standard. D6 accepts both the European 1250/50 interlaced format and the Japanese 260M version of the 1125/60 interlaced format, which uses 1,035 active lines. It does not accept the ITU format of 1,080 active lines. ANSI/SMPTE 277M and 278M are D6 standards.

D7   This has been assigned to DVCRPO.

D16   A recording format for digital film images that makes use of standard D1 recorders. The scheme was developed specifically to handle Quantel's Domino (*Digital Opticals for Movies*) pictures and record them over the space that sixteen 625-line digital pictures would occupy. In this way, three film frames can be recorded or played every 2 sec. Playing the recorder allows the film images to be viewed on a standard monitor; running at 16 speed shows full motion direct from the tape.

mid–1980s, many people have closely followed the intense efforts directed at developing a North American over-the-air DTV standard. These efforts have included standards for digital program origination [known as *production standards,* and typically developed within the *Society of Motion Picture and Televisions Engineers* (SMPTE)] and the recently publicized DTV terrestrial broadcasting standard. (The DTV transmission standard was developed by a private sector consortium known as the Grand Alliance, under the supervision of the FCC's Advisory Committee on Advanced Television Services [5].) As a consequence of this interindustry cooperation (spanning

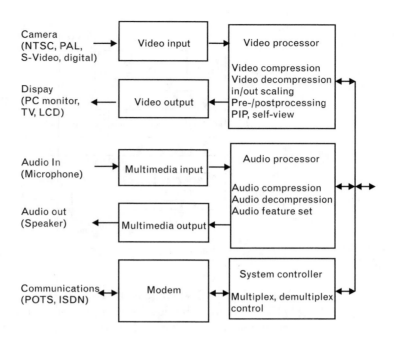

FIGURE 13.8
Typical coder.

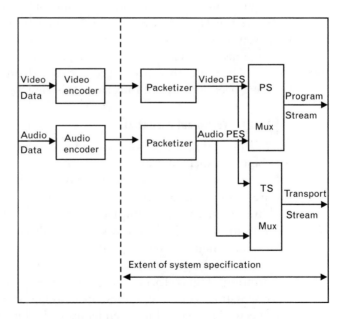

FIGURE 13.9
*Elements of an
MPEG encoder.*

broadcasters, manufacturers of both professional and consumer electronics, computing, motion picture film, and telecommunications industries), the standards that have emerged are cost effective and flexible [6] (the TV sets, however, are not). The baseline DTV terrestrial broadcasting standard encompasses both digital SDTV (*standard definition TV,* i.e., 525–line–based

FIGURE 13.10
MPEG-2 encoder.

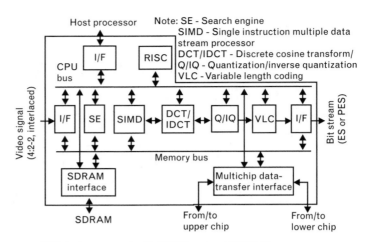

NTT device uses embedded RISC engine, 18 DSP-like processors

digital television) and digital HDTV. Although it took a lot of effort for the DTV standard to be developed and approved, in the view of many, that actually now seems the easier part of the job in light of the challenges that lie ahead for the entire industry. After all, a single analog video format having just one resolution and one frame rate is planned at this juncture to be retired in favor of a multiplicity of digital video formats varying in both resolution and frame rate. The FCC has mandated deadlines at specific time instances in the future for deployment of digital services [7].

The rest of this chapter provides details on the digital video themes that have been introduced thus far in this view of the industry.

13.1.1 Technical Essentials

In this section we provide some basic technical essentials. We expand on these essentials later on in the chapter.

Color spaces employed in the discussion of both analog and digital TV are mathematical representations of color. They can be represented as three-dimensional vector spaces; namely, every color can be represented as a vector that is described by supplying the scaling factors for an x-, y-, and z-component. Three common color spaces of interest in TV/video are RGB, YUV, and YCbCr (other color spaces exist, e.g., CMYK used in printing). Figure 13.11 depicts the three-dimensional nature of color representation. All three of these spaces support *component* video; that is, each "vector" is retained independently such that each pixel is represented, for example, $P_x = (R_x, G_x, B_x)$. RGB is typically employed in studio applications. YUV is combined in *National Television Standards Committee* (NTSC) to provide a *composite* signal used in transmission. YCbCr is used in uncompressed digital video. Figure 13.12 illustrates the two approaches to video, namely, component video and composite video, whereas Figure 13.13 focuses on composite video.

FIGURE 13.11
Three-dimensional nature of color representation.

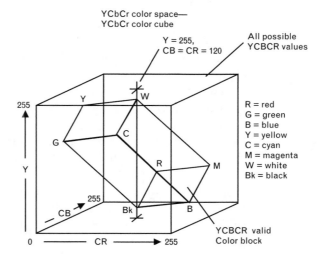

FIGURE 13.12
Two approaches to video.

- Composite (video)
  Luminance and chrominance are combined along with
  the timing reference 'sync' information using one of the
  coding standards - NTSC, PAL or SECAM - to make composite video

- Component (video)
  Video signal where luminance and chrominance remain as separate
  components, e.g. analogue components in Beatcam VTRs,
  digital components Y, Cr, Cb in ITU-R 601

- RGB is also a component signal
- Component video signals retain maximum luminance and chrominance bandwidth

FIGURE 13.13
Composite video.

- Composite (video)
  - Process, which is an analogue form of video
    compression, restricts the bandwidths (image detail) of components
  - In composite process, color is literally added to the monochrome
    (luminance) information using a visually acceptable technique

    · Since human eyes have far more luminance resolving power than for color,
      color sharpness (bandwidth) of coded single is reduced to far below
      that of the luminance
    · Provides a good solution for transmission but it becomes difficult,
      if not impossible, to accurately reverse the process (decode) into
      pure luminance and chrominance which limits its use in postproduction

- Composite signal:
  - RGB components from the camera
    are generally translated to a set of
    color difference components (such
    as Y, R-Y, B-Y) before being
    encoded to NTSC or PAL for
    transmission (In modern equipment
    all of these operations may take
    place in the camera)
  - Composite signal must be decoded
    in the receiver to a color difference
    format, then translated to RGB for
    display

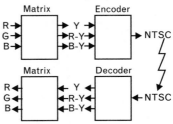

The video signal originating from a TV studio camera is considered a baseband component. The three-color components (red, green, and blue) are distinct signals. Because early NTSC systems needed to maintain compatibility with black-and-white TV sets, the R, G, B color space in video signals was converted to the Y, U, V color space. Y is the luminous information (lightness). U is defined as $U = B - Y$, and V is defined as $V = R - Y$.

Less bandwidth needs to be assigned to color difference signals (U and V) than to the luminance signal. If RGB-to-YUV color space conversion is done maintaining the full bandwidth chroma (hue plus saturation), such conversion is called a *4:4:4 sampling*. If the conversion is carried out on chroma samples for every other pixel, then it is termed a *4:2:2 sampling* scheme. The 4:2:2 sampling halves the chroma resolution horizontally, resulting in a 33% saving in bandwidth compared to a 4:4:4 sampling, yet with little perceptible loss in video quality. A 4:2:0 sampling reduces the chroma bandwidth even more, halving the overall bandwidth [7]. More on this is discussed in Section 13.3. Figure 13.14 summarizes some of these concepts.

Consumer video applications almost invariably make use of large, low-resolution displays. Typically the displays are viewed in a dim environment and from a distance. Interlaced video with 50 or 60 fields per second is the norm (see Figure 13.15). The sources of the video are analog (e.g., VCRs, settop boxes, camcorders—these are single/fixed resolution sources with interlaced video); the sources can also be digital (e.g., DVD, digital settop boxes, digital VCRs—these also are single/fixed resolution sources, for instance, 720 × 480 pixels for NTSC or 720 × 576 pixels for PAL, with interlaced video).

PC video uses small, high-resolution displays. Typically the displays are viewed in a bright environment, closed up. The video display is of the noninterlaced kind (see Figure 13.16), but the video source is generally

FIGURE 13.14
*Signals and color
spaces.*

Video signal originating from a TV studio camera is considered a baseband component

Three color components (red, green, and blue) are distinct signals

Early NTSC systems need to maintain compatibility with black-and-white TV sets => R, G, B color space in video signals was converted to Y, U, V color space

Y is luminous information (lightness), U=B–Y and V=R–Y

Less bandwidth needs to be assigned to color difference signals (U and V) than to luminance signal

If RGB-to-YUV color space conversion is done maintaining full bandwidth chroma (hue plus saturation), it is called to 4:4:4 sampling

If conversion is carried out on chroma samples every other pixel, it is termed a 4:2:2 sampling scheme

- 4:2:2 sampling halves chroma resolution horizontally, resulting in a 33 percent saving in bandwidth compared to a 4:4:4 sampling
- A 4:2:0 sampling reduces chroma bandwidth even more, halving overall bandwidth

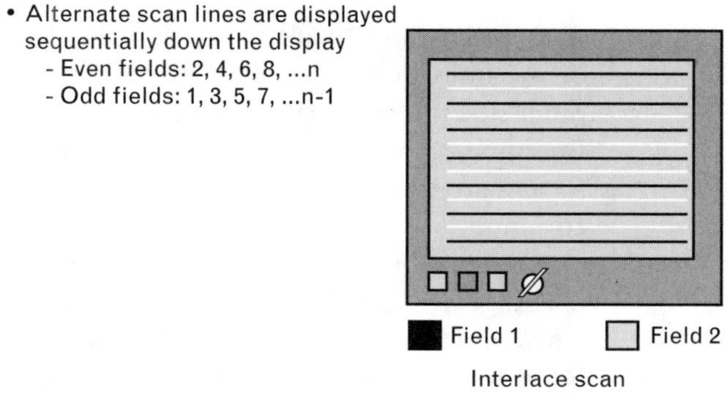

FIGURE 13.15
*Consumers versus PC
video—Interlaced
video.*

- Alternate scan lines are displayed
  sequentially down the display
    - Even fields: 2, 4, 6, 8, ...n
    - Odd fields: 1, 3, 5, 7, ...n-1

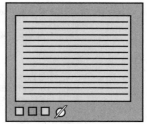

■ Field 1   ☐ Field 2

Interlace scan

FIGURE 13.16
*Consumers versus PC
video—Noninterlaced
video.*

- Scan lines are displayed sequentially
  down the display
    - 1, 2, 3, 4, ...n
- Also called progressive scanning

Progressive scan

interlaced. The video is displayed in a window; real-time scaling of the video window may be required. Analog video inputs can be from a VCR, a settop box, or a camcorder; digital video inputs can be from a DVD, digital settop boxes, CD-ROMs, and so on. Digital PC video entails the use of compression such as MPEG-1, MPEG-2, MPEG-4, motion JPEG, wavelets, or fractals.

PCs use square pixels (namely, horizontal and vertical sampling intervals are spatially the same). Consumer video uses rectangular pixels (horizontal and vertical sampling intervals are spatially different). NTSC's square pixel resolution is 640 × 480 (active portion) with a 12.2727-MHz sample rate. PAL's square pixel resolution is 768 × 576 (active portion) with a 14.75-MHz sample rate. NTSC's rectangular pixel resolution is 720 × 480 (active portion) with a 13.5-MHz sample rate. PAL's rectangular pixel resolution is 720 × 576 (active portion), with a 13.5-MHz sample rate.

Interlacing is a technique the camera uses to take two snapshots of a scene within a frame time. During the first scan, it creates one field of video, containing even-numbered lines, and during the second, it creates another, containing the odd-numbered lines. This technique, which is used in NTSC video, is used to reduce flicker and to provide higher brightness on the TV receiver for a specified frame rate (and bandwidth). Computer-generated video is generally scanned in a progressive manner, in which one frame of video contains all of the lines in their proper order. Equipment providers from the computer side prefer the progressive mode, but would like a lower frame rate. Equipment providers from the TV set side call for the inclusion of multiple formats, because the use of interlaced formats will initially be more common.

The YCbCr color space is defined by ITU-R BT.601. Y has nominal range 16–235; Cb and Cr have nominal ranges of 16–240, with zero corresponding to 128. Figure 13.17 shows how one can convert from the RGB to the YcbCr space. Y and CbCr components have different bandwidth and dynamic ranges. Here are some typical sampling ratios for Y:Cb:Cr:

- 4:4:4 (see Figure 13.18);
- 4:2:2 (see Figure 13.19);

FIGURE 13.17
YCbCr color space.

```
BT.601 RGB/YCbCr Equations
Y = (77/256)R' + (150/256)G' + (29/256)B'
Cb = (-44/256)R' - (87/256)G' + (131/256)B' + 128
Cr = (131/256)R' - (110/256)G' - (21/256)B' + 128
R' = Y + 1.371 (Cr - 128)
G' = Y - 0.336 (Cb -128) - 0.698 (Cr - 128)
B' = Y + 1.732 (Cb -128)

Equations assume R'G'B' nominal range of 16-235 (Black = 16, White = 235)
 - Many ICs/PCs overlook this, resulting in minor color errors
 - Can modify equations for R'G'B' range of 0-255 to properly support PCs
```

FIGURE 13.18
4:4:4 sampling.

Each sample has Y, Cb, and Cr data
- 13.5 or 18-MHz Sample rate
- 720 or 960 active samples per line

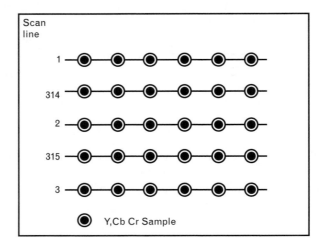

• 4:1:1 (see Figure 13.20);

• 4:2:0 (see Figure 13.21).

Figure 13.22 depicts some examples of implementations.

FIGURE 13.19
4:2:2 sampling.

4:2:2 YCbCr

• Consumer and PC applications
• Each sample has Y data
  - 13.5 or 18-MHz sample rate
  - 720 or 960 active samples per line
• Every other sample (horizontally)
  has Cb and Cr data
  - 6.75 or 9 MHz sample rate
  - 360 or 480 active samples per line

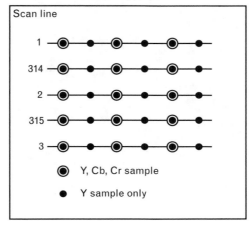

FIGURE 13.20
4:1:1 sampling.

4:1:1 YCbCr
• Consumer applications (DVC and TV)

• Each sample has Y data
  - 13.5 or 18-MHz sample rate
  - 720 or 960 active samples per line

• ConEvery fourth sample (horizontally)
  has Cb and Cr data
    - 3.375 or 4.5-MHz sample rate
    - 180 or 240 active samples per line

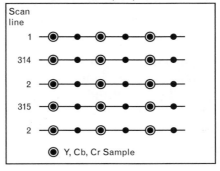

FIGURE 13.21
4:2:0 sampling.

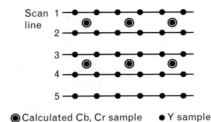

4:2:0 YCbCr

Two Types of 4:2:0 Notation
- Version used for internal processing by MPEG 1, H.261, and H.263: Noninterlaced video only
- Version used for internal processing by MPEG 2: Noninterlaced or interlaced video

For MPEG 1, H.261, and H.263

⊚ Calculated Cb, Cr sample   ● Y sample

For MPEG 2, (Noninterlaced)

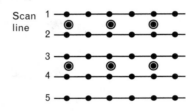

⊚ Calculated Cb, Cr sample   ● Y sample

FIGURE 13.22
Common YCbCr formats for the YCbCr color space.

Professional video	Consumer/ PC video
32-bit 4:4:4:4	16-bit 4:2:2
24-bit 4:4:4	8-bit multiplexed 4:2:2
16-bit 4:2:2	12-bit 4:1:1
10-bit data common	

13.1.2   Corporate Applications and QoS Support

13.1.2.1   Background on Corporate/Institutional Video and Multimedia Applications

Significant gains in automation and productivity have been seen in the corporate landscape during the past 20 years. These gains have been made possible by the plethora of application software that has emerged. The near–ubiquitous penetration of "standardized" PC operating systems, such as Microsoft's Windows on the client side and Windows NT and Unix on the server side, friendly *graphical user interfaces* (GUIs) along with network GUIs (specifically, WWW browsers), and the TCP/IP apparatus have made this software and productivity revolution possible. The 1990s saw widespread introduction of client/server and then Web-based intranet/ Internet systems in the corporate environment.

As we move forward, a whole gamut of *new applications* need to be deployed in order to reach the next plateau in business support tools, and in order to rein in ensuing additional productivity gains. Multimedia, desktop videoconferencing, *computer video integration* (CVI), voice over data networks, and *computer telephony integration* (CTI), to list a few, are expected to play an increasingly important role in the corporate landscape during the next decade, from both intranet and Internet perspectives. Business video takes the form of PC-based conferencing, multimedia, video–server–based computer-based training, reception of digitized broadcast video on a PC window, and imaging-based document management/workflow systems [8, 9].

Multimedia, as a general field, is a technology that is based on the multisensory nature of people and the ability of computers to store, manipulate, and display information such as video, graphics, audio, and text. Multimedia has been enabled by the synergistic confluence of the PC, the television, and the optical file server. Broadband communication networks are another key technical driver. There are now many practical multimedia business applications, including presentation development, kiosks, computer-based training, preparation of business presentations, on-line magazines, *computer-based training* (CBT), and desktop videoconferencing [10].

Industry observers agree that multimedia is one of the key technologies influencing how people will use computers over the coming years. Basically all new PCs and laptops now support multimedia. Corporations are examining the possibility of putting multimedia to work for them, to support their transition to a competitive business posture in the context of the global economy. Multimedia is not a single technology, but a class of technologies and applications that span two (voice and data) or more (voice, data, video, and graphics) media. Multimedia can operate delivering from as little as 56 Kbps of information to a user, to as much as 100, 155, 622, 1,000 Mbps, or even higher. The 1.5- to 6.0-Mbps data rate per user is a basic range that is being designed to, by multimedia technology developers, for example, for instructional purposes. MPEG-1 products operating at the 1-Mbps rate have seen good penetration.

Initial multimedia applications were confined to the desktop, where all required information resides in a PC-attached CD-ROM videodisk. Recent enterprise networking history has shown that *stand-alone islands* are untenable over time, even in the case of traditional business applications. Companywide connectivity is expected in the next few years with regard to multimedia and digital video. Desktop videoconferencing, access to remote libraries of video or multimedia material, access to archived multimedia corporate records, and downloading of server-based multimedia instruction are just some examples of desirable networked multimedia applications, requiring both local and wide area connectivity. Beyond the desktop-based platforms, multimedia requires high-capacity digital networks to provide real-time services such as retrieval, messaging, conversation, and distribution; here too, QoS-enabled networks are needed. In fact, what has held back broadband applications in general, and multimedia in particular, has been the lack of adequate bandwidth and QoS, not only at the WAN level, but also at the local level for both private and public networks. Communication technology of adequate capacity, quality of service, and flexibility is a critical factor that will be required to enable multimedia to migrate from dedicated desktop systems to more efficient distributed systems, making more, preferably most, employees of a corporation actual users of multimedia applications. As noted, the 1990s are burgeoning with new high-quality, high-speed digital services at the local and wide area level.

During the 1990s, a battery of new standards was approved by the ITU-T for LAN/intranet/Internet-based multimedia. These standards promise to enable interworking among products and lower prices. The key new umbrella standard is H.323, which is discussed in more detail in a later section. Once standards are widely supported, a technology may see rapid introduction [11].

In addition to networked multimedia per se, a variety of vendors and suppliers are pursuing the objective of bringing real-time two-way video to the corporate desktop. For example, multimedia conferencing enables designers in remote locations to review and/or work cooperatively on the same project using PCs that incorporate text, graphics, audio, visual, and tactile (touch-screen) capabilities. An increasing number of companies are now utilizing videoconferencing as part of their normal business practices even though most of the applications are still in the conference room-to-conference room arrangement, rather than being desktop based. Such desktop conferencing systems, however, are now beginning to appear.

Proponents make the case that videoconferencing on the desktop is only a matter of time. They quote well-known "benefits" that are direct carryovers from traditional videoconferencing: (1) reduced travel expenses, (2) more effective use of time, and (3) the ability to connect dispersed work groups. Conference room systems cost from $10,000 to $50,000, whereas desktop systems can be as inexpensive as $200 [12]. "Mid-range" systems cost between $2,000 and $5,000.

Video and multimedia have two important characteristics that impact the type of transmission technology that can be employed at the LAN or WAN level for its delivery [10, 13–25], as discussed next.

The first characteristic is that the packetized video must be delivered with low and predictable end-to-end delay. Additionally, the delay variation must be small. Usually one cannot use store-and-forward methods except for nonreal-time video, unless (1) the network is absolutely well-tuned or (2) the data rate is very small (e.g., 2 × 64 Kbps in ITU-T H.261) or (3) the network supports QoS. Traditional protocols, not to say routers, may or may not have the ultimate performance capabilities for supporting hundreds of simultaneous corporate users using the enterprise backbone or LAN for videoconferencing applications [26].

The second characteristic is that simple digitization of a video signal can yield from 200 Mbps for traditional full-motion NTSC video, to 1 Gbps for HDTV. Recently developed digital compression algorithms and supporting hardware reduce these values by about 100-fold or better. Compression is an economical method for storage and transmission of video in limited-bandwidth environments, including an organization's enterprise network. Compression methods include ISO MPEG schemes and ITU-T H.261/H.263, in addition to some vendor proprietary methods [27]. However, the data rate is still fairly high as follows:

- Desktop videoconferencing applications using H.320/H.261 ITU-T standards produce from 128 to 768 Kbps per user; the newer H.263 provides video support at 28.8 Kbps or higher.

- Entertainment video and distance learning applications using MPEG-1/MPEG-2 standards produce 1.5 or 6 Mbps, respectively, per user.

- Multimedia applications, using MPEG-1 or proprietary methods (for example, Intel's Indeo, digital video interactive, or wavelets), produce about 1.5 Mbps per user.

Note that until the mid-1990s a number of vendors used Motion Joint Photographic Expert Group (JPEG), which is an adaptation from still-photography—this method typically requires from 10 to 30 Mbps; other broadcast-level digitization schemes used in the recent past required 45 Mbps.

This brief discussion on data rates should make it immediately clear that traditional *shared* Ethernet LANs are marginally adequate for the support of video in the corporation. Studies indicate that, unless the video quality on PCs approaches the quality to which users are accustomed to on their TV sets or VCRs, the deployment of video in the organization will be negatively impacted. This baseline requirement leads to the realization that one needs signal bandwidths of 1.5 to 6 Mbps *per user* for some business applications (lower resolution will be acceptable for other applications).

An increasing number of vendors and users have taken the approach of using 10BASE-T Ethernet switches to achieve the desired networkwide connectivity (in addition to routers to interconnect IP subnetworks). Although these switches are fairly inexpensive (e.g., $400 per port or less) and are readily available from dozens of vendors, an enterprisewide video architecture based on such an approach can be somewhat limiting. For example, 100-Mbps Ethernet and GbE may be better suited: 100 Mbps could support from 15 to 60 video users at 6 or 1.5 Mbps, respectively, unless one used switched Ethernet. GbE has been deployed as a backbone technology and 10GbE systems are also appearing. A high-speed switch would still be required to support segment-to-segment or segment-to-server connectivity. By contrast, ATM *dedicates* appropriate bandwidth resources to *each* user as a matter of course, up to the maximum line card speed or the speed of the switch's backplane [28]. The issue, however, is more obvious in the WAN: A 100-Mbps Ethernet infrastructure in two buildings with an interbuilding router backbone operating at T1 rates will be totally limited by the bottleneck in bandwidth. It is imperative that bandwidth and QoS issues be addressed in LAN/WAN networks, while also taking into account bandwidth demands on existing legacy applications before video-based technology can emerge.

For WAN applications, ATM technology is nominally the underpinning technology for high-end networked multimedia applications; services such as ISDN and frame relay can support midrange applications. However, one may not want to run native ATM because of the changes required, and because these changes would be required throughout many elements of the corporate network. Preferably, one wants to be able to run IP over ATM, so that only a few backbone routers to ATM need to be upgraded. Several approaches have evolved to facilitate this hybrid network architecture. Some companies can utilize POS, but this assumes that the user needs at least 155 Mbps or more between endpoints in question, which may or may not be the case.

13.1.2.2   Broadcast Applications

On February 2, 1997, Tim Russert closed NBC's broadcast of *Meet the Press* with a historical statement that owed much to recent developments in both digital broadcast technology and network politics: "...*Meet the Press* made television history. You didn't see or hear it because your set isn't equipped yet, but this was the first network program to be broadcast in digital high definition, or HDTV. It brings you crisp, movie-quality images and CD-quality sound. HDTV sets will be available in late 1998, and eventually this technology will be the industry standard." This test broadcast marked the implementation of television's most significant technological development since its invention. In a medium that has seen little fundamental innovation in the way its message reaches the populace, the historic *Meet the Press* broadcast suggests not only that a new technology is entering the scene, but also that broadcasters grasp its significance and its potential for improving the quality and flexibility of television services. The HDTV revolution has begun and proponents claim it will have far-reaching consequences for broadcasters, manufacturers, and consumers [6].

With its 16:9 aspect ratio, six-channel audio, and screen sizes in the 36- to 100-in. range, HDTV will offer consumers a "cinematic" entertainment experience in their home. It is projected that, as terrestrial cable and satellite services vie for share in a dynamic and competitive marketplace, receiver prices will drop, fostering the introduction of this technology. (After about 7 years in the making, however, the receiver costs are still too high.) The industry has waited many years for DTV's emergence as a technologically and economically feasible method of television production and broadcast. The U.S. broadcast industry has a large archive of high-definition 35-mm motion picture films; these film archives have constituted the source material for about 70% of traditional television prime-time programming. This readily available programming source can be retransferred on high-definition systems that are already appearing on the market (it is expected that such retransferred material will provide a great deal of the programming content for HDTV's first few years). This is similar to the issuing of AAD audio discs long before DDD discs became more of the norm.

Packetized digital video (e.g., digitized video carried in ATM cells) is entering not only the corporate world, but the entire broadcasting business, both over the air and over cable [27]. These advances have been called "broadcasting's third revolution; the first two were radio, then television" [29]. Realizing the impact that the first two events had, one can appreciate the implied impact of the third. DTV units not only can carry a mix of video, audio, and data services today, but they also provide the mechanism for adding new services in the future without making receivers already in the field obsolete. For example, during a commercial, a browsable brochure can also be downloaded, in the "data channel," which can be consulted via a PC to obtain information such as product specifications, available options, dealers, and so on.

Digitization entered the TV studio even before HDTV made its presence felt. Advances in VLSI ICs have made possible a new class of studio equipment: digital equipment for NTSC video. New production switchers, routers, and tape machines all support digital component 4:2:2 video. Standards such as ITU-R 601 support high-quality production: The ITU-R 601 standard for broadcast video has an active resolution of 710 pixels (picture elements) by 485 lines. Also, there are standards for parallel and serial video data interchange between equipment, standards such as SMPTE-259D, a 360-Mbps interface between, say, digital tape machines and video routers. Although many TV studios still use analog NTSC equipment, many others have started making the transition to full digital facilities for producing NTSC programs [7].

Hence, even prior to actual over-the-air digital transmission for end-to-end digital delivery, the broadcast industry is seeing the decline of the analog videotape as the medium for storing video. Videographic designers have been using disks for some time as their primary medium, with videotape reserved for output and occasional input; the trend is now expected to permeate the industry. Vendors of video servers are emphasizing the advantages of tapeless production and tapeless distribution, including "digital ad insertion." Disk-based storage of video is convenient not only because it is digital and is random accessible, but because the information can be transmitted over an (internal) backbone data network, for studio workflow purposes. Digital technology is making it possible to include more functions into video editing systems now reaching the market; these editing systems are available at prices that were unthought of a few years ago [30].

Uncompressed digital editing has been used during the past few years, but with the introduction of compressed video, the need has arisen to also deal with these newer formats. The 6-Mbps data rate of MPEG-2 makes it easier to store, download, process, distribute, and archive this material. Video compression is entering the teleproduction industry. Applications include disk-based nonlinear editors, video file servers, and on-air automated playback systems. Soon, similar teleproduction techniques will also enter the mainstream corporation, whether in the public relations department, the training department, the library, the production floor, or the departmentalized server. Improvements in compression technology create a dynamic environment of innovation as to how a company employs video technology. MPEG encoders can now be found for desktop applications, ranging from as little as $500 to $10,000 (high-end real-time systems); many were around $3,000 at press time. We are in a transition era where disk-based compression systems are replacing videotape; the use of nonlinear editors to produce video products continues to increase [31]. The data rates for these applications range from 50 to 3 Mbps, which relate to compression rates ranging from 4.5:1 to 72:1. (Note that a 2-hour film feature requires about 5 GB.)

At this juncture, the move to DTV will be the catalyst for a total digitization of the broadcast operation. The bottom line is that video studios will have to be redesigned. In the equipment room, a routing switcher (or router), which can have hundreds of video input and output ports, handles all of the video in a TV station. Currently this is an analog matrix. At the push of a button, the routing switcher allows easy connectivity among many video cameras, tape machines, and other studio equipment. The production switcher is the main piece of equipment in the control room and is used to handle special effects, like video fades and wipes, and to insert commercials. In a conventional television studio, NTSC signals are routed on coaxial cables from one piece of equipment to another. With the advent of DTV, the switcher will likely have to be a bona fide multigigabit digital ATM switch or a GbE/10GbE switch. Fibre Channel SAN could also be used.

Digital HDTV and digital SDTV are now supported by industry-developed standards, as was underscored by some of the figures presented earlier in this chapter. The standard that is at the base of the end-to-end digitization of TV was developed by the *Advanced Television Systems Committee* (ATSC) [32] and "adopted" by the FCC. However, when the FCC ruled on the specifics of the digital transmission parameters for DTV, they left the choice of video "payload" (mix of channels, as well as resolution) up to the broadcasters. The FCC has aimed at leaving choices but no ambiguities in the definition of the digital terrestrial "channel" to be used by broadcasters [6]. Furthermore, SMPTE has delineated for broadcasters, producers, and manufacturers all of the specific parameters for HDTV and SDTV equipment. The NTSC standard defines a video frame as containing a total of 525 interlaced lines, such that all the odd lines are scanned before all of the even lines at about a 30-Hz frame rate. Television studio equipment relies on this fixed frame structure for timing and synchronization. The new ATSC standard mandates compressing the video and audio signals as well as using packetized transport for video, audio, and data packets. Clearly, the transition from analog NTSC to the compressed digital ATSC high-definition standard will completely transform how a television studio routes, stores, processes, and transmits the new television signal [7].

The new ATSC standard [33] defines four digital television formats. This standard encompass both digital HDTV and SDTV. These formats are described by the number of pixels per line, the number of lines per video frame, the frame repetition rate, the frame structure (interlaced or progressive), and the aspect (width-to-length) ratio, as follows:

- 1,920 × 1,080 pixels, 60 interlaced frames per second (also 30 or 24 progressive), 16:9 aspect;

- 1,280 × 720 pixels, 60 or 30 or 24 progressive frames per second; 16:9 aspect;

- 704 × 480 pixels, 60 or 30 or 24 progressive frames per second (also 60 interlaced); 16:9 aspect (also 4:3);

- 640 × 480 pixels, 60 or 30 or 24 progressive frames per second (also 60 interlaced); 4:3 aspect.

The ATSC DTV transmission standard (ATSC Doc. A/53) specifies the parameters for digital HDTV and SDTV video and audio formats. The ATSC DTV standard defines a transmission system flexible enough to transmit HDTV and/or multiplexed SDTV at different times during the programming day. At the highest resolution level (necessary for true HDTV), the ATSC/SMPTE standards requires a resolution of 1,920 (horizontal) × 1,080 (vertical, or number of scanning lines) for digital sampling structure. This format can be transmitted digitally at 60 pictures/second interlaced (normal "live" television for sports, and so on) or, alternatively, at 24 or 30 frames per second progressive-scan (for 24-fps film-originated material) [6]. The same DTV transmission system can support a number of digital 525-line multiplexed SDTV channels (in the standard 4:3 aspect ratio or the new 16:9 wide-screen aspect ratio). Within the 19.4-Mbps total video "payload" allowed by the DTV standard, broadcasters are free to determine the number of SDTV channels they will offer, by selecting their own individual balance between picture quality and channel data rate.

In fact, as noted earlier, the standard includes the two HDTV formats: the 1,920-pixel-by-1,080-line interlaced, and the 1,280-pixel-by-720-line progressive-scan format. Most of the HDTV broadcast equipment now available or emerging is designed for the 1,920 × 1,080 interlaced video format only (progressive-scanning HDTV cameras and monitors are not yet available); 24-fps film translated to video has a progressive format. The progressively scanned 704-pixel-by-480-line video with a 16:9 aspect ratio will probably see some penetration. Compared to the NTSC standard, it has a wider picture and eliminates artifacts affecting interlaced video (such as the line crawl that affects some scenes containing slow vertical motion). The ATSC standard also supports SDTV formats (interlaced and 704-pixel-by-480-line or 640-pixel-by-480-line formats). Given the fact that most of the existing studios support one or the other of these two formats, most local production will probably be of this kind.

The bit stream being transmitted may change according to the nature of the program, for example, news, followed by an archive segment, followed by a commercial. The ATSC standard recommends that the receiver "seamlessly," and without loss of video, continue to display all of these formats in the native format of the television receiver.

The ATSC standard specifies MPEG-2 as the video compression standard. In an end-to-end context, note that the ATSC standard allows controlled coding/decoding because if the video is coded (compressed) and decoded more than a handful of times, the picture quality rapidly degrades, unless a hierarchical (tiered) compression approach is utilized. For example, for video production, the compression can be 4:1 and the coding format can be I-frames (intraframes) only [34]. For contribution video, the compression can be 10:1 and use a IPPP coding structure (where P represents the predicted frames). For storage, the compression can be 25:1 with IPIP coding structure. For transmission, the compression can be 50:1 with IPBBBBP coding structure (where B is a bidirectionally interpolated frame).

While the advent of DTV is a given, it is important at this time that broadcasters and manufacturers work together and take the steps necessary to ensure that the full promise of digital HDTV is realized. An important first step for broadcasters to undertake will be the dual transition from composite analog to direct-digital component program origination, and from the standard 4:3 aspect ratio to the new 16:9 wide-screen format. At this time, direct-digital 4:2:2 SDTV origination offers good video quality with high SNRs. The quality of these programs will be fully appreciated when the signals are digitally manipulated (compressed) for transmission over the DTV channels in the near future; their quality will be appreciated even more when the upconversions to the digital HDTV are required for the next decade or so.

Manufacturers have been steadily developing sequential generations of HDTV program origination products, despite the small market of digital broadcasters and program producers. For example, in digital SDTV, Sony has continued its support of the international digital 4:2:2 component video standard that is emerging as the preferred basis for digital 525-line program creation and the associated MPEG-2 digital 4:2:0 transmission format, which is now formally a part of the ATSC DTV transmission standard. A 4:2:2 digital component wide-screen camcorder has emerged as an early entrant system for field acquisition of digital high-end SDTV program material. Sony's subsequent development of an entire digital broadcast product line based on the recently standardized MPEG 4:2:2 Profile at Main Level brought in a secondary digital "layer" that can constitute the "workhorse" heart of an entire digital broadcasting operation. For example, Betacam SX is the 4:2:2 recording format associated with that system, and it ushers in a new era of high-performance and highly cost-effective camcorders for digital 4:2:2 applications. The 4:2:2 platform is now supported in anticipation of SDTV's crucial role within the overall DTV dynamic [6]. A new generation of HDTV studio cameras, an all-digital HDTV camcorder, digital HDTV switchers and multiple-effects system, and studio monitors has been unveiled recently.

If the bit rate used within the studio is in the range of 200 to 270 Mbps, then the compressed data can be stored in and routed by uncompressed standard definition equipment that complies with SMPTE 259. Both Sony and Panasonic have proposed systems allowing D1 and D5 tape equipment currently in wide use to be used to store compressed data (see Table 13.1). The 270-Mbps version of SMPTE 259 can be used to store and route up to 200-Mbps data, and the 360-Mbps version can be used for bit rates up to 270 Mbps. But as more equipment supporting HDTV pictures becomes available, digital SDTV will give way to HDTV programming [7].

The new TV studio must support HDTV as well as SDTV/NTSC equipment. It also must allow compressed operations (like storage and splicing). Effects such as instantaneous cutaway, edits, and so on need to be supported. With current production equipment these studio effects can readily be achieved. Production equipment for compressed video also needs to support these capabilities. An intracoded system can do this, since all intracoded frames are independent I-frames, so that altering (or deleting) one frame does not affect others frames in the bit stream. Things are more complex for intercoded systems. A high-speed digital network will route the compressed bit stream and other data around the studio. There will also be transcoders to convert one compressed format into another. In the early stages, video production will be preformed on uncompressed video, but as compressed technology advances, more production will be done on compressed video (splicing, editing, and effects are more complicated in compressed-video environments).

Broadcasters are seizing the initiative at this time, in preparing for DTV. CBS station WRAL-TV in North Carolina (officially the first station in the United States to transmit digitally) and five other stations have been licensed by the FCC to begin experimental digital HDTV broadcasting. Many more have been licensed by the FCC to begin experimental digital HDTV broadcasting, and many more have license applications under review. An industry initiative, the HDTV Model Station (formally licensed as WHD-TV, and housed in the host station WRC-TV in Washington, D.C.), has created a working laboratory for HDTV equipment. The HDTV Model Station is serving as a proving ground where manufacturers and broadcasters can address technical issues associated with implementation of a simulcast digital DTV/analog NTSC operation. Also, five member stations of the Public Broadcasting Service (PBS) have formed the Broadcasting Digital Alliance to cooperatively address the technical and programming issues of DTV. To advance the cause of DTV broadcasting, PBS has also recently arranged for satellite feeds of the compressed ATSC DTV signal format to allow stations and laboratories (involved in DTV receiver development) to receive this signal. There will also be new operator requirements. For example, there is the matter of reeducating camerapersons, directors, set designers, video operators, and postproduction specialists in the ways of wide-screen and wide-angle origination production. As with any technological innovation, there will be a considerable learning curve [6].

The FCC's approval of the DTV standard means that the timetable for broadcasters to begin the transition to digital television has been initialized. The rollout of DTV will begin in the larger U.S. markets. The FCC has assigned additional channels to broadcasters for digital transmission and has mandated a rapid build-out plan. The four top network affiliates in each of the 30 top markets were required to be on the air by November 1998, with all the commercial stations on the air by 2002. The FCC also called for the return of a second (NTSC) channel by the year 2006. These early services will be a combination of SDTV and HDTV.

13.1.2.3   QoS Support

Many Fortune 500 companies now use packet network services (e.g., FR, ATM, IP, and MPLS) for corporate data, intranet, and e-commerce applications. In the enterprise network, new LAN technologies, such as switched Ethernet and/or FE or GbE, have been deployed. At the WAN level, networks have shifted in two directions, compared with, for example, a decade ago:

1. The movement away from TDM transmission facilities of fixed bandwidth, particularly when focusing at the whole (nationwide) network, rather than just the access tail, which does continue to be based on traditional telephony facilities, such as DS-1, DS-3 lines, or OC-3;

2. The movement away from dedicated point-to-point lines, which become impractical as the number of interconnected sites increases, and toward the use of packet technology (whether at the

network layer in the form of IP–routed systems or at the data link layer, such as frame relay and, to a lesser extent, ATM).

It follows that there is interest in addressing the question of services and media integration. Integration has found reasonable effectiveness in the frame relay context for the support of *small office/home office* (SoHo) locations. In the end, an integrated network has a great deal of appeal for transmission efficiencies and technology/network management reasons. The two trends just noted have both benefits and drawbacks. In general, the benefits have weighted on the side of economics and scalability; the drawbacks have weighted on the side of performance and QoS. The goal, therefore, is to achieve integration and at the same time support QoS.

Video streams have different requirements compared with traditional data flows that supported applications such as e-mail, word processing, and financial analysis. Interactive video/multimedia applications, as well as more traditional but mission–critical applications, require enterprise networks that support QoS. (As noted in the introduction to this chapter, nonreal–time delivery of video content is not as dependent on QoS.) Appropriate local area, campus, wide area, and international communication infrastructures will be needed to support this move to interactive digital video and other broadband applications, including QoS support. (These topics were discussed in earlier chapters.) The need for QoS is driven by three factors, at this time:

1. Support of voice over *packet networks* (intranets and Internet) with ensuing statistical gains and lower costs. There has been interest, going back to at least the mid–1970s if not earlier, in utilizing integrated networks that support all of the organization's media, because of the efficiencies in transport and management that would result [35–38].

2. Support of desktop video over the enterprise network and the Internet. Standards such as ITU-T H.323 are seeing penetration in corporate LANs. H.320-based systems are also in place at this time. The H.320/H.323 standards make video on LANs possible at the technology and interoperability level; however, as soon as more than a few users fire up conferences (particularly multipoint conferences/multicasts), the network may be severely impacted and the quality of the conference impacted. IETF's *Session Initiation Protocol* (SIP) (RFC 3261) is also expected to play a role in the future.

3. Support of priority-based data applications over the intranet and Internet.

QoS includes guaranteed bandwidth-on-demand (minimum, average, and peak), as well as predictable (small) end-to-end delay, delay variation (jitter), and unit (block, cell, frame) loss. QoS-enabled networks are needed not only for time-sensitive applications, such as voice and video, but also to support data applications, which, as networks become more congested and more integrated (both at the corporate level and at the Internet level), need to receive a guaranteed level of performance (see Table 13.2).

The QoS issue has to be addressed at an end-to-end level, that is, on a LAN-WAN (Internet)-LAN basis, as well as at multiple layers of the communication protocol suite (specifically, the data link layer and network layer). From an end-to-end perspective, one has to look at the various LAN technologies and understand their QoS support; then one has to do the same for the WAN technologies. On the protocol layer view, one could look at the problem as being addressed end to end, regardless of the underlying subnetworks, if the QoS reservation protocol is end to end (end system to end system) and if every element at the layer under discussion (e.g., routers if one is looking at the IP/*intserv*/*diffserv* level) supports the QoS fulfillment mechanism. When considering the discrete subnetworks, LAN technologies tend to have the issue of QoS driven by protocol considerations (e.g., contention issues) rather than by bandwidth (and, hence, cost) issues. For WANs this is exactly the opposite: The QoS support in protocols may exist, but the bandwidth is expensive, so the QoS fulfillment has to be done in a cost-effective manner.

New significant industry work is now under way to address QoS in data networks. Successful as well as unsuccessful attempts have been made to address the integration/QoS issue over the years (see Table 13.3). Efforts in the 1980s and 1990s in ISDN and ATM have also been aimed at voice support in general and multimedia in particular. Efforts on the data side have included support of voice in LANs (such as IEEE 802.9 and FDDI II) and enhancements to routers, IP (e.g., IPv6 and RSVP), and network-layer handling of packets (such as MPLS).

Key mechanisms showing promise include *intserv*, *diffserv*, network layer switching/MPLS, RTP, and ATM UNI. At the same time, existing services such as frame relay, may or may not be upgraded to support QoS, which could turn out to be a problem.

TABLE 13.2 APPLICATIONS AND QoS

	QoS REQUIRED	APPLICATIONS
Nonreal-time data	Little or none	Data file transfer, imaging, simulation, and modeling
Nonreal-time multimedia	Little or none	Exchange text e-mail, exchange audio/video e-mail, Internet browsing with voice and video, intranet browsing with voice and video
Real-time one-way	Various QoS levels	Multimedia playback from server, broadcast video, distance learning, surveillance video, animation playback
Real-time interactive	Various medium or high QoS levels	Videoconferencing, audioconferencing, process control

TABLE 13.3   SUPPORT OF QoS ON EXISTING, EVOLVING, OR OBSOLETE TECHNOLOGIES

TECHNOLOGY	KIND	INTEGRATION SUPPORT	QoS SUPPORT	SUCCESS
100–Mbps Ethernet (shared/switched)	STDM, LAN	Yes	Some, based on reduced congestion	Prevalent
ATM LAN	STDM, LAN	Yes	Yes	Very limited deployment
ATM WAN	STDM, WAN	Yes	Yes	Medium deployment now
Ethernet (shared)	STDM, LAN	Yes with some effort	Minimal	Ubiquitous
Ethernet (switched)	STDM, LAN	Yes	Some, based on reduced congestion	Ubiquitous
FDDI II	TDM-like in circuit channel, LAN	Yes	Yes, but fixed channels	No deployment
Frame relay	STDM, WAN	Possible now	Limited: via engineering, no intrinsic support (FRF.13 just begins to address the issue)	Wide-scale deployment
Frame relay on ISDN	STDM, WAN	Possible	Limited: via engineering, no intrinsic support	No deployment
IEEE 802.1Q and .1P	STDM LAN	Possible	Yes	No deployment as of yet
IEEE 802.9	TDM-like in circuit channel, LAN	Yes	Yes, but fixed 64–Kbps channels	No deployment
IP	STDM, LAN/WAN	Yes	Limited; perhaps "gigarouters"★ have more punch and can lower delay and delay variation compared to existing routers	Ubiquitous
IPv6	STDM, LAN/WAN	Yes	Possible to some degree if vendors implement priority features QoS support is identical for both IPv4, IPv6: the priority field needs appropriate support (procedurally, from a router engine power, and from a link capacity point of view) to achieve QoS goals; IPv6 could possibly be worse because the packets are longer due to the longer addresses	Not yet deployed
ISDN 2B+D/23B+D/H0 /H11	TDM, WAN	Yes	Yes, but fixed bandwidth	Limited deployment for integration purposes; mostly for Internet access
ISDN packet on B	TDM/STDM, WAN	Yes	Unlikely, depends on packet switch	No deployment
ISDN packet on D	TDM/STDM, WAN	Yes	Unlikely, depends on packet switch	No deployment

TABLE 13.3   SUPPORT OF QOS ON EXISTING, EVOLVING, OR OBSOLETE TECHNOLOGIES (CONTINUED)

TECHNOLOGY	KIND	INTEGRATION SUPPORT	QOS SUPPORT	SUCCESS
Network layer switching (flavors)	STDM, LAN/WAN	Yes	Some, based on the fact that a more efficient treatment of IP is possible	Limited deployment
Packet	STDM, WAN	Yes	Unlikely, depends on packet switch	Limited deployment now
RSVP with IP	STDM, LAN/WAN	Yes	Yes, but it is only a reservation mechanism; IP and (perhaps) ATM needed	Only now being deployed
RTP	STDM, LAN/WAN	Yes	Yes	Limited deployment

* While "gigarouters" are not strictly required to lower delay and delay variation (shorter queues being required for that purpose), these routers operate at line speed and have the computing power to manage the queues appropriately and keep them at a minimum (assuming, of course, that the outgoing link has adequate capacity to service all the incoming traffic).

Fortunately, as of press time, router vendors were starting to build products to support QoS, in order to facilitate integrated voice/data/video networking at the corporate branch level, as well as over the intranet and Internet. Leading-edge vendors are including *intserv* and/or *diffserv* capabilities in their routers. They are also modifying their routers to accommodate better buffer management to deliver predictable QoS in a standard IP environment.

Some IP "purists" feel that the practical use of videoconferencing in the workplace has demonstrated that QoS, *diffserv*, and the other alphabet soup of acronyms are not necessary or, probably, even cost effective for most organizations wanting to use public networks. These purists feel that with no special QoS, and running inside a CPE-based IP VPN, performance is "reasonably" adequate, and the IT department could get major pushback if they were to come to the table with a proposal for premium QoS or *diffserv*-based network service. Naturally, there is always a trade-off between quality and price. Considering the general availability of bandwidth that was (supposedly) created by the telecom boom, there are no reasons to compromise on quality, keeping in mind that an uncongested network that is resource rich will often result in a high QoS. (Naturally, appropriate queue management will always be required, otherwise even large pools of bandwidth will result in poor QoS for some applications.)

In summary, to carry video on enterprise networks, QoS issues, both at the protocol and at the technology level, need to be addressed.

13.2   ITU-T H.320/H.323

As noted earlier, in order to achieve ubiquity, there is a need for standardization. Today one can call anyone in the world, dial into any modem, or use any Ethernet network interface card and achieve the desired connectivity. The same is needed with regard to videoconferencing. Fortunately, standards have evolved over the years; unfortunately, these standards have been adopted only slowly. This section addresses the technologies that are evolving to support desktop business video. Codecs receive considerable attention, because they are the basis for the entire video application. Codecs must be compatible (i.e., conforming to industry standards, e.g., ITU-T H.320), be relatively inexpensive (e.g., a $100 chipset), and support reasonable quality.

13.2.1   H.320

ITU-T H.320 is a key group of standards (see Figure 13.23) that was developed for traditional videoconferencing applications. ITU-T H.320 is the basis for the majority of standardized "low-bit-rate" videoconferencing systems in use today. H.320 enables a variety of applications such as telehealth, distance learning, and travel-free conferencing. This standard supports data rates from 64 Kbps to 2.048 Mbps. It is intended for traditional CBR communications channels such as DS-1/T1, E1, ISDN Primary Rate Interface, and single or multiple Basic Rate Interface.

H.320 utilizes the H.261 video encoding specification, which in turn uses the p × 64 codec. Adoption of standards ensures compatibility. Some vendors have introduced proprietary systems that claim to be better, but in fact do not represent a significant improvement. H.320 operates up to 1.544 Mbps (for U.S. implementations) per user. (For comparison, motion JPEG can go as high as 20–45 Mbps per user and offer significant quality improvements.) Field introduction of H.320, however, has been slower than initially expected. Industry groups such as the Personal Conferencing Workgroup (PCWG) and International Multimedia Teleconferencing Consortium (IMTC) have addressed the interworking details of H.320 and T.120 (whiteboards). A specification for LAN communication support of H.320, specifically H.323, has emerged, addressing the interoperability question. (H.323 is discussed later.)

FIGURE 13.23
H.320 standards.

(Narrowband switched digital ISDN)

ITU-T H.261: video codec
ITU-T G.711: 64-kbps audio codec
ITU-T G.722: 48, 56 or 64-kbps audio codec
ITU-T G.728: 16-kbps audio codec
ITU-T H.221: multiplexing
ITU-T H.230: control
ITU-T H.231: multipoint
ITU-T H.233: encryption
ITU-T H.234: encryption key handling
ITU-T H.242: control
ITU-T H.243: control
ITU-T T.120: data

A number of vendors have looked at the use of H.320 over emulated channels delivered via ATM. ATM offers circuit emulation service, where if the user employs ATM Adaptation Layer 1, the network provides a high–quality cell transfer (low loss and low jitter) to emulate a DS-1/T1 channel (in whole or slotted in 64-Kbps channels). H.321 adapts H.320 to ATM, but it retains all other capabilities of H.320.

13.2.2   H.323

The relatively new H.323 standard provides a foundation for audio, video, and data communications across LAN and IP networks, including the Internet. Specifically, H.323 describes terminals, equipment, and services for multimedia communication over LAN and IP networks that do not provide a guaranteed quality of service. H.323 terminals and equipment may carry real-time voice, data, and video, or any combination, including videotelephony. The LAN over which H.323 terminals communicate may be a single segment or it may be multiple segments with complex intranet topologies. Note that operation of H.323 terminals over the multiple LAN segments (including the Internet) may actually result in poor performance since the possible means by which QoS might be assured on such types of LANs/internetworks is beyond the scope of the recommendation. H.323 is also utilized in the context of VoIP. Figures 13.24, 13.25, and 13.26 highlight some aspects of H.323.

H.323 terminals may be integrated into personal computers or implemented in standalone devices such as videotelephones. Support for voice is mandatory in the standard, while data and video are optional, but if

FIGURE 13.24
H.323 highlights.

> H.323 described terminals, equipment, and services for multimedia communication over LAN and IP network that do not provide a guaranteed quality of service
>
> H.323 terminals and equipment may carry real-time voice, data and video, or any combination, including videotelephony
>
> LAN over which H.323 terminals communicate may be a single segment or multiple segments with complex intranet topologies
>
> Industry migration to H.323 is not going to happen overnight, there is a need for internetworking between two systems

FIGURE 13.25
H.323 environment.

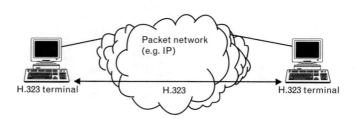

H.323 terminal   H.323   H.323 terminal

FIGURE 13.26
H.323 model.

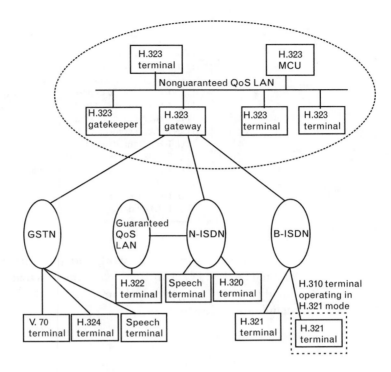

supported, the ability to use a specified common mode of operation is required, so that all terminals supporting that media type can interwork. Other components in the H.323 series include H.225.0 packet and synchronization, H.245 control, H.261 and H.263 video codecs, G.711, G.722, G.728, G.729, and G.723 audio/speech codecs, and the T.120 series of multimedia communications protocols (see Figure 13.27).

H.323 terminals may be used in multipoint configurations and may interwork with H.310 terminals on B-ISDN, H.320 terminals on N-ISDN, H.321 terminals on B-ISDN, H.322 terminals on guaranteed QoS LANs (e.g., IEEE 802.9; see Figure 13.28), H.324 terminals on the PSTN and wireless networks (see next section), and V.70 terminals on PSTN.

FIGURE 13.27
H.323 components.

• Components include

  - H.225.0 packet and synchronization
  - H.245 control
  - H.261 and H.263 video codecs
  - G.711, G.722, G.728, G.729, and G.723 audio/speech
  - T.120-series of multimedia communications protocols

• H.263 video codec has been developed as an improvement over H.261 (today's standard video codec for videoconferencing on the ISDN)

• H.263 can operate at a range of bit rates

  - applicable at rates below 32 kbps
  - reasonable quality for video containing limited motion

FIGURE 13.28
*H.322
videoconferencing.*

> (Guaranteed bandwidth packet-switched networks)
>
> ITU-T H.261: video codec
> ITU-T H.262: video codec
> ITU-T G.711: 64-Kbps audio codec
> ITU-T G.722: 48, 56 or 64-Kbps audio codec
> ITU-T G.728: 16-Kbps audio codec
> ITU-T H.221: multiplexing
> ITU-T H.230: control
> ITU-T H.231: multipoint
> ITU-T H.233: encryption
> ITU-T H.234: encryption key handling
> ITU-T H.242: control
> ITU-T H.243: control
> ITU-T T.120: data

As a brief history, H.323 work started in May 1995. The standard was adopted by the ITU–T in 1996. New versions that support additional functionality have been published since then and current H.323 components include the following:

· Terminals;

· Gatekeepers;

· Gateways (H.323 to H.320/H.324/POTS);

· MCUs: multipoint controllers (MCs) and multipoint processors (MPs).

The H.323 terminal (see Figures 13.29, 13.30, and 13.31) encompasses two versions: (1) corporate network (high quality) and (2) Internet (optimized for low bandwidth 28.8/33.6 Kbps; G.723.1 and H.263).

FIGURE 13.29
H.323 terminal.

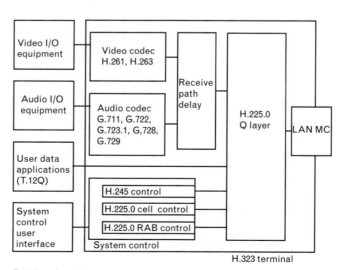

Peripherals and
peripheral functions

FIGURE 13.30
*Terminal's protocol
stack.*

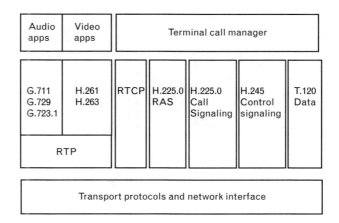

FIGURE 13.31
H.323 umbrella.

(Nonguaranteed bandwidth packet switched networks, i.e, Ethernet)

ITU-T H.261: video codec
ITU-T H.262: video codec
ITU-T G.711: 64-Kbps audio codec
ITU-T G.722: 48, 56, or 64-Kbps audio codec
ITU-T H.223: audio codec
ITU-T G.728: 16-Kbps audio codec
ITU-T H.729: audio codec
ITU-T H.225.0: multiplexing
ITU-T H.323: multipoint
ITU-T H.233: encryption
ITU-T H.234: encryption key handling
ITU-T H.245: control
ITU-T T.120: data

The terminal may have a built-in multipoint capability for ad hoc conferences. Multicast capabilities (specifically, multiple-unicast) allow a few people into a call without centralized mixing or switching. The H.323 *Gatekeeper* (see Figure 13.32) supports the following functions:

- Address translation:
  - H.323 alias to transport (IP) address based on terminal registration;
  - "E-mail-like" names can be supported;
  - "Phone-number-like" names can be supported.
- Admission control:
  - Permission to complete call;
  - Can apply bandwidth limits;
  - Method to control LAN traffic.
- Management of gateway:
  - H.320, H.324, POTS, and so on.

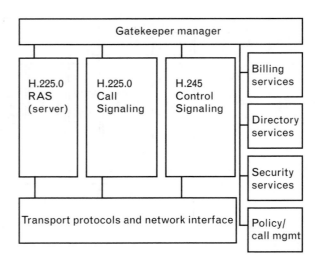

• Call signaling:

   • Is able to route calls in order to provide supplementary services or
     to provide MC functionality;

• Call management/reporting/logging.

   An industry migration to H.323 is not going to happen overnight;
hence, there is a need for interworking between the two systems. H.323
gateways provide global (world wide) connectivity and interoperability from
LAN. For example, interoperability between H.323 and H.320, as well as
H.323 to H.324, may be supported. Also, it maps call setup signaling (Q.931
to H.225.0), control (H.242/H.243 to H.245), and media (FEC, multiplex,
rate matching, audio transcoding, T.123 translation). Figures 13.33 through
13.36 depict various aspects of the gateway functionality.

   Related to multipoint functionality, one has the MC portion of a traditional MCU, which manages common modes and capabilities, and the
MP portion of a traditional MCU, which mixes or switches audio. It is not

• Primary goal in the development of H.323 was interoperability with other
  terminal types

• Interoperability is achieved through the use of common recommendations,
  procedures, and messages

• A gateway unit is specified to perform any network or signaling
  translation required for interoperability

• Interoperability of H.323 with other terminal types

   - H.320 Narrowband integrated services digital network(N-ISDN)
   - H.321 Broadband integrated services digital network (B-ISDN)
   - H.322 IsoEthernet
   - H.324 General switched telephone network (GSTN)
   - H.310 Asynchronous transfer mode (ATM)

FIGURE 13.34
Gateways.

- Gateway provides appropriate translation between transmission formats (for example H.225.0 to/from H.221) and between communications procedures (for example H.245 to/from H.242)
- Gateway also performs call setup and clearing on both the LAN side and the switched circuit networks (SCN) side
- Translation between video, audio, and data formats may also be performed in the Gateway
- In general, the purpose of the Gateway (when not operating as a multipoint control unit) is to reflect the characteristics of a LAN endpoint to an SCN endpoint, and the reverse, in a transparent fashion
- Gateway has the characteristics of an H.323 terminal or MCU on the LAN, and of SCN terminal or MCU on the SCN

FIGURE 13.35
Gatekeeper functions.

1. Translation from LAN aliases for terminals and gateways to IP or IPX addresses

   - Calls originating outside the LAN and received by a gateway may use an E.164 telephone number (e.g., 555-555-5555) to address the destination terminal
   - Similarly a call origination on the LAN may use an alias to address the destination terminal
   - Gatekeeper translates this telephone number or alias into the network address (e.g. 204.222.32.156) for destination terminal
   - Source endpoint can't then address the destination endpoint on the LAN

2. Provides a mechanism for network administrators to control the amount of video telephony on the network

   - Done through admission control: terminals must get permission from the gatekeeper to place or accept a call
   - Permission includes a limit in the amount of bandwidth the terminal may use on the network
   - Mechanism is provided for requesting changes to the bandwidth during a call; however, the gatekeeper has the final say on how much bandwidth can be used for a given call
     · Effect is to limit the total conferencing bandwidth to some fraction of the total available; the remaining capacity is left for e-mail, file transfers, and other LAN protocols

FIGURE 13.36
Gateway protocol stack.

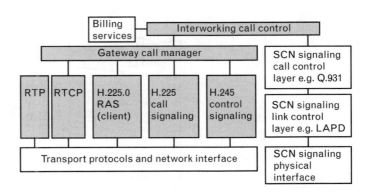

necessarily coresident with MC (e.g., MC running multicast conference with each terminal mixing audio). A traditional MCU supports the following functions (which are not necessarily unique to H.323) [39]:

- Media distribution:

  - *Unicast:* Send media to one terminal (centralized in MP; traditional model).

  - *Multicast:* Send to each receiver directly.

  - *Hybrid:* Does some of each.

- Manage ad hoc multipoint calls:

  - Join, invite, control conference modes.

- Traditional MCU applications.

- Multiprotocol through utilization of gateways.

Figures 13.37 through 13.43 depict the call establishment process in H.323.

FIGURE 13.37
*H. 323 call
establishment.*

H.323 Call establishment

- T1 sends the RAS ARQ messge on the RAS channel to the gatekeeper for registration. T1 requests the use of direct call signaling.
- The gatekeeper confirms the admission of T1 by sending ACF to T1. the gatekeeper indicates in ACF that T1 can use direct call signaling.
- T1 sends an H.225 call signaling "setup" message to T2 requesting a connection.
- T2 responds with an H.225 "call proceeding" message to T1.
- Now T2 has to register with the gatekeeper. It sends an RAS ARQ message to the gatekeeper on the RAS channel.
- The gatekeeper confirms the registration by sending an RAS ACF message to T2
- T2 alerts T1 of the connection establishment by sending an H.225 "alerting" message.
- Then T2 confirms the connection establishment by sending an H.225 "connect" message to T1, and the call is established.

FIGURE 13.38
*H. 323 call
establishment flow.*

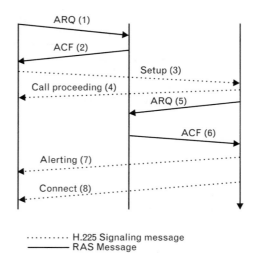

FIGURE 13.39
*H. 323 control
signaling flows.*

H.245 control channel is established between T1 and T2. T1 sends an H.245 "TerminalCapabilitySet" message to T2 to exchange its capabilities.

T2 acknowledges T1's capabilities by sending an H.245 "TerminalCapabilitySetAck" message.

T2 exchanges its capabilities with T1 by sending an H.245 "TerminalCapabilitySet" message.

T1 acknowledges T2's capabilities by sending an H.245 "TerminalCapabilitySetAck" message.

T1 opens a media channel with T2 by sending an H.245 "openLogicalChannel" message. The transport address of the RTCP channel is included in the message.

T2 acknowledges the establishment of the unidirectional logical channel from T1 to T2 by sending an H.245 "openLogicalChannel" message. Included in the acknowledging message are the RTP transport address allocated by T2 to be used by the T1 for sending the RTP media stream and the RTCP address received from T1 earlier.

Then, T2 opens a media channel with T1 by sending an H.245 "openLogicalChannel" message.
The transport address of the RTCP channel is included in the message.

T1 acknowledges the establishment of the unidirectional logical channel from T2 to T1 by sending an H.245 "openLogicalChannelAck" message. Included in the acknowledging message are the RTP transport address allocated by T1 to be used by the T2 for sending the RTP media stream and the RTCP address received from T2 earlier. Now the bidirectional media stream communication is established.

FIGURE 13.40
Signaling flows.

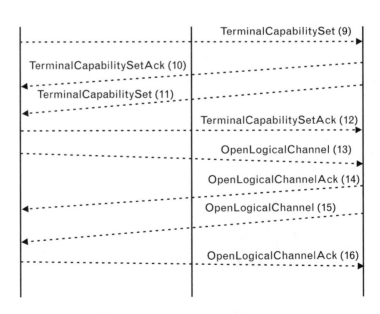

The H.263 [40] video codec (*Version 2, Video Coding for Low Bit Rate Communication,* ITU-T H.263, January 1998) that is used in H.323 has been developed as an improvement over H.261, which is today's standard

FIGURE 13.41
*H. 323 media stream
and media control
flows.*

- T1 sends the RTP encapsulated media stream to T2
- T2 sends the RTP encapsulated media stream to T1
- T1 sends the RTCP message to T2
- T2 sends the RTCP message to T1

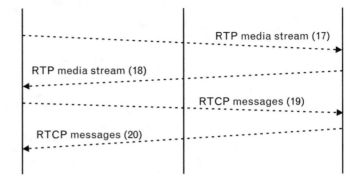

RTP media stream (17)

RTP media stream (18)

RTCP messages (19)

RTCP messages (20)

- - - - - - - - - - RTP media stream and RTCP message

FIGURE 13.42
H. 323 call release.

T2 initiates the call release. It sends an H.245 "EndSesssionCommand" message to T1.

T1 releases the call endpoint and confirms the release by sending an H.245 "EndSessionCommand" message to T2.

T2 completes the call release by sending an H.225 "release complete" message to T1.

T1 and T2 disengage with the gatekeeper by sending a RASDRQ message to the gatekeeper.

Gatekeeper disengages T1 and T2 and confirms by sending DCF messages to T1 and T2.

FIGURE 13.43
*H.323 call release
sequence.*

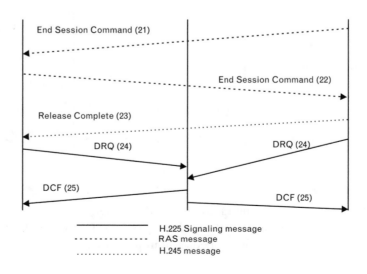

End Session Command (21)

End Session Command (22)

Release Complete (23)

DRQ (24)   DRQ (24)

DCF (25)   DCF (25)

————————— H.225 Signaling message
- - - - - - - - - - RAS message
· · · · · · · · · · · · · H.245 message

video codec for videoconferencing on the ISDN. H.263 (see Figures 13.44 through 13.47) can operate at a range of bit rates, but is particularly applicable at rates below 32 Kbps where reasonable quality is still possible for video containing limited motion. *Quarter common intermediate format*

FIGURE 13.44
H. 263.

H.263 video codec has been developed as an improvement on H.261

H.263 can operate at a range of bit rates, applicable at rates below 32 kbps

QCIF resolution is the most common format for these rates

Video input is a sequence of digitized pictures (frames) at a rate of 30 frames/s

At QCIF resolution, each picture is divided into 11 x 9 macroblocks (MB)

MBs consist of 16 x 16 pixels, which are further subdivided into four 8 x 8 blocks

FIGURE 13.45
Video source coding algorithm.

Main elements are:

Prediction, block transformation, and quantization

A hybrid of interpicture prediction to utilized temporal redundancy and transform coding of the remaining signal to reduce spatial redundancy is adopted

Decoder has motion compensation capability, allowing optional incorporation of this technique in the coder

Half pixel precision is used for the motion compensation, as opposed to Recommendation H.261, where full pixel precision and a loopfilter are used

Variable length coding is used for the symbols to be transmitted

FIGURE 13.46
H.263: Video coding for low-bit-rate communication.

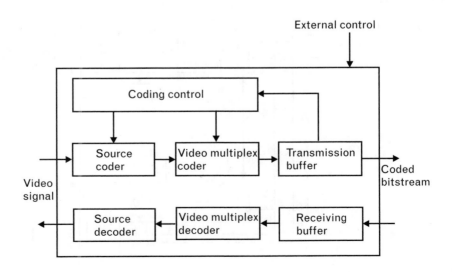

FIGURE 13.47
Number of pixels per line for each of the standardized H.263 picture formats.

Picture format	Number of Pixels for Luminance (dx)	Number of Lines for Luminance (dy)	Number of Pixels for Chrominance (dx/2)	Number of Lines for Chrominance (dy/2)
sub-Q CIF	128	96	64	48
Q CIF	176	144	88	72
CIF	352	288	176	144
4 CIF	704	576	352	288
16 CIF	1408	1152	704	576

(QCIF) resolution (176 × 144 pixels) is the most common format for these rates. The video input is a sequence of digitized pictures (frames) at a rate of 30 fps. At QCIF resolution, each picture is divided into 11 × 9 macroblocks (MBs). The MBs consist of 16 × 16 pixels, which are further subdivided into four 8 × 8 blocks. In general, when there is significant activity in one part of the image, the MBs corresponding to this area generate more bits than other parts of the image.

Two modes can be selected for each MB, implemented by a switch S: INTRA and INTER (the selection of the path is beyond the scope of the recommendation and is a local implementation issue); see Figures 13.48 and 13.49. In the INTRA mode, no temporal dependency from the previous frames is postulated. The picture is directly coded using a *discrete cosine transform* (DCT; the working of DTC is discussed in a later section). In the INTER mode, the MB is predicted from the previously reconstructed frame using compensation. The output from the frame memory is identical to the decoded frames at the decoder for error-free transmission. Hence, the

FIGURE 13.48
Coder supporting INTRA and INTER coding.

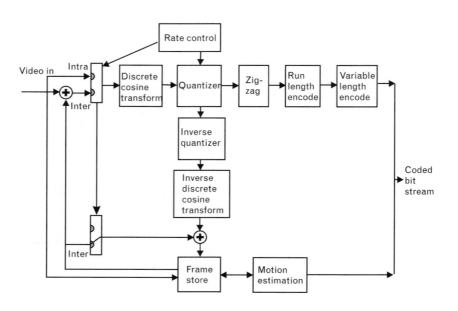

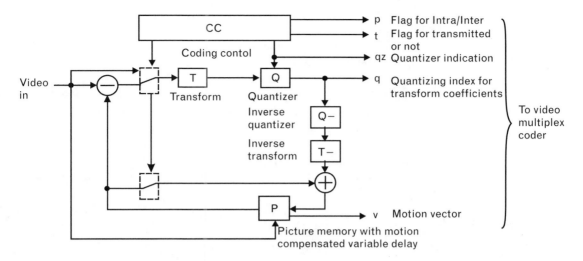

p   Flag for Intra/Inter
t   Flag for transmitted
     or not
qz   Quantizer indication
q   Quantizing index for
     transform coefficients

To video
multiplex
coder

FIGURE 13.49   *Functional view of coder.*

same prediction can be formed at the encoder and decoder. Motion compensation implies that the current MB is predicted by a 16×16 block in the previous frame that is spatially shifted in accordance with a motion vector. In the ideal case, the remaining prediction error is negligible, and no more information need be transmitted. In general, however, the remaining prediction error is encoded using a DCT for each 8×8 block. The transform coefficients are quantized and encoded as a series of zero runs and quantizer levels. Then, run-level pairs, motion vectors, and mode information are entropy coded along with other side information, resulting in variable-length code words that are multiplexed to the video bit stream [4]. Naturally, the INTER mode is preferred if possible, because of its efficiency.

The H.223 multiplex protocol interleaves video, audio, data, and control streams into a combined bit stream, supporting dynamic allocation of bandwidth to the individual channels. Video typically requires the largest portion of the total bit rate. The protocol consists of a lower multiplex layer, which actually mixes the different media streams, and a set of adaptation layers, which perform logical framing, sequence numbering, error detection, and error correction by retransmission as appropriate to each media type [4]. Hence, H.223 is a connection-oriented multiplexer that is designed to mix any number of channels on a circuit-switched network (for low bit rates, however, generally no more than one video, audio, and data channel are used in each direction). H.223 is byte oriented and provides very low multiplex delay.

New capabilities are being added to this protocol. As it existed originally, H.223 was not robust in dealing with transmission errors since it was designed to work with V.34 modems on channels providing low error rates; hence, H.223 was not suitable for wireless applications without some extensions. These extensions have now been included in annexes to the original recommendation. The additions allow a hierarchical, multilevel multiplexing mechanism that allows engineering trade-offs between overhead, complexity, and robustness.

The original, default, and simplest level is level 0. In this V.34-based version, packets are variable length and delimited by an 8-bit header containing a multiplex code (MC) identifying the content of the packet. What follows is the payload, which in general can consist of a mix of bits from various sources. The end boundary of the packet is determined by the next appearance of the 8-bit synchronization flag. (Bit stuffing is performed on all data between synchronization flags is to avoid flag misinterpretation.) The major vulnerabilities of H.223 level 0 arise from the bit stuffing, and from the short and consequently vulnerable synchronization flags and headers. In level 1 bit stuffing is not performed, and a longer synchronization flag is used. The flag can be simulated by the data, but such simulations are considered not to be problematic. In level 2, additional robustness is provided by lengthening and adding error protection to the header that describes the contents of the packets. There is also a level 3; it uses FEC. The packet marker (PM) is a 1-bit field that identifies the length of the payload in bytes. This provides additional redundancy by informing the receiver where the next sync flag can be expected.

A terminal has to support many control messages. Those messages are defined in the H.245 multimedia system control protocol. The control messages are transmitted over a reliable data link layer with error recovery. Recovery via retransmission is accomplished using either the V.42 *link access procedure for modems* (LAPM) or *Simplified Retransmission Protocol* (SRP).

As noted in the introduction of this section, there is interest in supporting wireless multimedia. Indeed, with relatively minor extensions, specifically an error-tracking approach adopted by the ITU-T for H.263 Version 2 and with minor extensions of H.245 to include an additional control message video, one now has a multimedia capability for wireless environments.

In 1995 the ITU-T standardized the *Algebraic Code Excited Linear Prediction* (ACELP) voice algorithms for the coding of speech signals in wide area networks; ACELP is used for compression rates at or below 16 Kbps. ITU-T G.729 (*Conjugate Structure ACELP*, CS-ACELP) is now an international standard that compresses 64-Kbps PCM streams as used in typical voice transmission to as low as 8 Kbps. ITU-T G.728 (*Low Delay CELP*, LD-CELP) is an international standard used to compress voice to 16 Kbps. ITU-T G.723.1 compresses voice to rates as low as 5.3 Kbps (it also operates at 6.3 Kbps). This standard is used in the H.323 recommendation for conferencing over LANs. G.723.1 is considered a good first step and is best suited for intranets and controlled point-to-point IP-based connections. G.729A, a simplified version of G.729, operates at 8 Kbps and, therefore, is slightly better in quality than G.723.1.

13.2.3 H.324

ITU-T H.324 is a recommendation for terminals for low–bit–rate multi-media communication that may consist of real–time voice, data, and video, or any combination, including videotelephony [4] (see Figures 13.50 and 13.51). The standard makes the assumption that the transmission is based on V.34 modems operating over the widely available GSTN (also known as the PSTN in North America). Developers hold the position that H.324 terminals are likely to play a major role in future multimedia applications. In fact, a gamut of H.324-enabled products have already been developed and are being sold in rapidly increasing numbers.

In the model, only the modem (V.34), multiplex (H.223 [41]), and control protocol (H.245) are mandatory [42]; the data, video, and audio streams are optional. By design, even the most basic H.324 terminals can interwork with more advanced terminals. During the setup of the session, the terminals negotiate a common set of capabilities used for the connection.

There is interest in using a coding scheme for mobile/wireless applications that has the characteristic of being based on some existing method (to reduce both the standardization interval and the product-development interval) and of requiring a low bit rate. H.324 is a good candidate. As noted, H.324 is a complete multimedia terminal operating with V.34/V.8 modems at 28.8 Kbps. The system can operate as low as 10 Kbps and as high as a few hundred kilobits per second. Hence, it is a good candidate for mobile (wireless) applications because wireless applications are at the low end of the bit rate range. However, a number of extensions are needed to deal with the increased error rate of wireless/mobile channels. In 1994 the ITU–T started

FIGURE 13.50
H.324
videoconferencing.

```
(PSTN or POTS, the analog phone system)

ITU-T H.261: video codec
ITU-T H.263: video codec
ITU-T G.723: audio codec
ITU-T H.223: multiplexing
ITU-T H.233: encryption
ITU-T H.234: encryption key handling
ITU-T H.245: control
ITU-T V.34: modem
ITU-T T.120: data
```

FIGURE 13.51
H.324/C
videoconferencing.

```
(Mobile radio)

ITU-T H.261: video codec
ITU-T H.263: video codec
ITU-T G.723: audio codec
ITU-T H.223A: multiplexing
ITU-T H.233: encryption
ITU-T H.234: encryption key handling
ITU-T H.245: control
ITU-T T.120: data
```

an *ad hoc group* (AHG) to investigate the use of H.324 in mobile environments. This group, which is now part of ITU-T Study Group 16, Question 11, also handles work on error-resilience video coding. (The error issue is now embodied in ITU-T Study Group 16, Question 15.)

H.324 employs the V.34 modem to transmit and receive the bit stream generated by the H.223 multiplex. No data compression or retransmission capabilities are supported. This approach simplifies replacement of the V.34 modem with other connection-oriented transmission systems; in particular, a "wireless interface" with a range of bit rates can be used in place of V.34. This, however, requires certain changes in the setup procedures (for example, for the establishment of a connection). The changes necessary to replace the modem with a wireless interface are identified in Annex C of the H.324 recommendation.

13.2.4   SIP

SIP (RFC 2543, March 17, 1999) is a signaling protocol for Internet conferencing, telephony, presence, events notification, and instant messaging. SIP has evolved as an alternative to H.323 for both VoIP and videoconferencing applications. SIP is independent of the packet layer and can use unreliable datagram service, because it provides its own reliability mechanism. Although SIP is typically used over UDP or TCP, it can be run over IPX, FR, ATM AAL-5, or X.25. SIP was developed within the IETF MMUSIC (*Multiparty Multimedia Session Control*) working group, with work proceeding since September 1999 in the IETF SIP working group [43]. A number of standardization organizations and groups are using or considering SIP: (1) the IETF PINT working group; (2) 3GPP; (3) Softswitch Consortium; (4) IMTC and ETSI Tiphon, which are working on SIP-H.323 interworking; (5) the PacketCable DCS specification; and (6) SpeechLinks, for moving between speech-enabled sites.

SIP provides the protocol mechanisms so that end systems and proxy servers can provide services such as the following [43]: call forwarding; callee and calling "number" delivery, where numbers can be any (preferably unique) naming scheme; *personal mobility*, that is, the ability to reach a called party under a single, location-independent address even when the user changes terminals; terminal-type negotiation and selection, in which a caller can be given a choice about how to reach the party, for example, via Internet telephony, mobile phone, an answering service, and so on; terminal capability negotiation; caller and callee authentication; blind and supervised call transfer; and invitations to multicast conferences. Obviously it has applications in corporate videoconferencing, as an alternative to H.323.

Extensions of SIP to allow third-party signaling are available, for example, for click-to-dial services, fully meshed conferences and connections to MCUs, as well as mixed modes and the transition between those. SIP addresses users by an e-mail-like address and reuses some of the infrastructure of e-mail delivery such as DNS MX records or using SMTP EXPN for address expansion. SIP addresses (URLs) can also be embedded in Web pages. SIP is addressing neutral, with addresses expressed as URLs of various types such as SIP, H.323, or telephone (E.164). Table 13.4 highlights the history of the protocol and key related events [43].

13.2.5 H.310

H.310 is an MPEG-2 audio and video service (see Figure 13.52). H.310 defines broadband audiovisual (MPEG-2) terminals as follows:

- Send-only terminals supporting AAL-1, AAL-5 and AAL-1 and -5 (known respectively as SOT-1, SOT-5, and SOT-1&5);

- Receive-only terminals supporting AAL-1, AAL-5 and AAL-1 and 5 (known respectively as ROT-1, ROT-5, and ROT-1&5);

- Receive-and-send terminals supporting AAL-1, AAL-5 and AAL-1 and 5 (known respectively as RAST-1, RAST-5, and RAST-1&5).

For example, ROT-5 terminals can operate in the native communications mode that consists of H.222.1 with 11172-3 layer II, H.262, H.245 as the audio, video, and end-to-end control protocols, respectively. RAST-5 terminals can operate in the native communications mode that consists of H.222.1 with G.711, H.262, H.245 as the audio, video, and end-to-end control protocols, respectively. H.310 requires RAST terminals to interoperate with H.320/321 terminals (see Figure 13.53). However, RAST-5 needs a gateway for AAL-1 adaptation, H.221 multiplex, and H.242 control message generation in the user's network in order to interwork with these H.320/321 terminals.

To support the conferencing application, the service uses a full MPEG-2 encoder/decoder pair, including stereo audio. The service is optimized for interactive applications where excessive delay is a consideration. Delay is reduced by encoding I-frames only, or I- and P-frames (discussed elsewhere). As a result of reduced coding complexity, the operating bandwidth will be higher than encoding I-, B-, and P-frames. The user may be able to have some control over the bandwidth/delay trade-off for various applications and/or bandwidth.

The VoD service defined in ATM Forum af-saa-0049.000 uses a full MPEG-2 decoder including stereo audio, working from a video server. This MPEG-2 decoder needs to interwork with the MPEG-2 encoder functions, up to the full bandwidth using the IBP encoding mechanism. The service is targeted at applications that are not fully interactive and can accept coding delays. The objective is to achieve maximum possible video quality at a given bit rate at the expense of delay.

TABLE 13.4   HISTORY OF SIP AND KEY RELATED EVENTS

January 23, 2003   RFC 3313, *Private Session Initiation Protocol (SIP) Extensions for Media Authorization*, published.

January 13, 2003   *Requirements for Resource Priority Mechanisms for the Session Initiation Protocol* approved as an informational RFC.

December 19, 2002   RFC 3388, *Grouping of Media Lines in Session Description Protocol (SDP)*, published.

December 17, 2002   *Mapping of Media Streams to Resource Reservation Flows* approved as proposed standard.

December 16, 2002   *The Refer Method* approved as proposed standard.

December 16, 2002   *A Session Initiation Protocol (SIP) Event Package for Registrations* approved as proposed standard.

December 13, 2002   RFC 3327, *Session Initiation Protocol (SIP) Extension Header Field for Registering Non-Adjacent Contacts*, published.

December 6, 2002   RFC 3398, *Integrated Services Digital Network (ISDN) User Part (ISUP) to Session Initiation Protocol (SIP) Mapping*, published.

December 6, 2002   RFC 3326, *The Reason Header Field for the Session Initiation Protocol (SIP)*, published.

December 6, 2002   RFC 3428, *Session Initiation Protocol (SIP) Extension for Instant Messaging*, published.

December 4, 2002   RFC 3323, *A Privacy Mechanism for the Session Initiation Protocol (SIP)*, published.

December 4, 2002   RFC 3325, *Private Extensions to the Session Initiation Protocol (SIP) for Asserted Identity within Trusted Networks*, published.

November 26, 2002   The *Session Initiation Protocol* (SIP) and *Session Description Protocol* (SDP) *Static Dictionary for Signaling Compression (SigComp)* and *Compressing the Session Initiation Protocol* approved as proposed standards.

November 7, 2002   *DHCPv6 Options for SIP Servers* approved as proposed standard.

November 5, 2002   PDF versions of the core SIP specifications made available.

November 1, 2002   CPL schema definition.

September 23, 2002   *Internet Media Types message/sipfrag* approved as proposed standard.

September 20, 2002   *Session Initiation Protocol Extension for Instant Messaging* approved as proposed standard.

September 2002   RFC 3311, *The Session Initiation Protocol (SIP) UPDATE Method*, published.

September 2002   RFC 3372, *Session Initiation Protocol for Telephones (SIP-T): Context and Architectures*, published.

September 1, 2002   IETF liaison statement to 3GPP.

August 28, 2002   *SDP Simple Capability Declaration* approved as proposed standard.

August 27, 2002   RFC 3361, *Dynamic Host Configuration Protocol (DHCP-for-IPv4) Option for Session Initiation Protocol (SIP) Servers*, published.

August 26, 2002   *Grouping of m Lines in SDP* approved as proposed standard.

August 15, 2002   RFC 3351, *User Requirements for the Session Initiation Protocol (SIP) in Support of Deaf, Hard of Hearing and Speech-Impaired Individuals*, published.

July 22, 2002   List of differences between draft-ietf-sip-rfc2543bis-09 and RFC 3261.

July 9, 2002   *SIP for Telephones (SIP-T): Context and Architectures* (draft-ietf-sipping-sipt-04.txt) approved as a BCP.

July 3, 2002   RFC 3261, *SIP: Session Initiation Protocol*; RFC 3262, *Reliability of Provisional Responses in Session Initiation Protocol (SIP)*; RFC 3263, *Session Initiation Protocol (SIP): Locating SIP Servers*; RFC 3264, *An Offer/Answer Model with Session Description Protocol (SDP)*; RFC 3265, *Session Initiation Protocol (SIP)-Specific Event Notification*; and RFC 3266, *Support for IPv6 in Session Description Protocol (SDP)*, published.

June 24, 2002   *Signaling Compression* approved as a proposed standard; *SigComp—Extended Operations* and *Signaling Compression Requirements & Assumptions* approved for publication as informational RFCs.

June 24, 2002   *Extensions to the Session Initiation Protocol (SIP) for Asserted Identity* and *Short Term Requirements for Network Asserted Identity* approved for publication as informational RFCs.

June 24, 2002   *Reason Header Field for the Session Initiation Protocol* and *SIP Extension for Registering Non-Adjacent Contacts* approved as proposed standards.

May 17, 2002   SIP UPDATE and resource management approved as proposed standards.

May 6/7, 2002   SIP/SIPPING interim meeting in Las Vegas.

TABLE 13.4   HISTORY OF SIP AND KEY RELATED EVENTS (CONTINUED)

April 6, 2002   SIP DHCP approved as proposed standard.

March 7, 2002   SIP "bis" and related documents approved as proposed standards.

February 22, 2002   The revised SIP specification, along with *Reliability of Provisional Responses in SIP, SIP: Locating SIP Servers, SIP-Specific Event Notification*, and *An Offer/Answer Model with SDP* submitted for IETF last call.

February 5, 2002   CPL was approved as a proposed standard.

January 18, 2002   RFC 3219, *Telephony Routing over IP (TRIP)*, published.

The tenth SIP interoperability test event held April 22–26, 2002 in Cannes, France; hosted by ETSI.

December 2001   The ninth SIP interoperability test event took place in San Diego; hosted by Nuera.

October 26, 2001   A completely editorially revised version of the SIP specification released for comment.

The eighth SIP interoperability test event took place August 13–17, 2001, in Cardiff, United Kingdom.

July 25, 2001   AOL submits statement on use of SIMPLE to FCC.

June 29, 2001   Picture gallery of some SIP products and icons for SIP servers.

May 29, 2001   RFC2543bis (-03) draft.

May 4, 2001   Information about the 8th SIP Interoperability Test Event made available.

April 14, 2001   Search feature added.

April 11, 2001   SIP interoperability test event has a new logo, courtesy of Ubiquity.

April 10, 2001   RFC 3087, *Control of Service Context using SIP Request-URI*, published.

Feb. 1, 2001   RFC 3050, *Common Gateway Interface for SIP* (sip-cgi), published.

November 30, 2000   Caller preferences draft in WG last call until December 24, 2000.

November 29, 2000   Guidelines for authors of SIP extensions draft in WG last call until December 24, 2000.

November 24, 2000   RFC2543bis (-02) draft.

November 17, 2000   CPL in IESG last call.

RFC 2976, *The SIP INFO Method*, published.

The sixth SIP interoperability test event took place December 5–8, 2000, at Sylantro and Sun in Silicon Valley, California.

June 20, 2000   SIP Forum founded. "SIP Forum is a nonprofit association whose mission is to promote awareness and provide information about the benefits and capabilities that are enabled by SIP."

The fifth SIP interoperability test event took place August 8–10, 2000, at pulver.com in Melville, New York.

June 15, 2000   RFC 2848, *The PINT Service Protocol: Extensions to SIP and SDP for IP Access to Telephone Call Services*, published.

Added SIP internship and job listing.

The fourth SIP interoperability test event took place April 17–19, 2000, in Rolling Meadows (near Chicago), Illinois, hosted by 3Com.

February 28, 2000   *The SIP INFO Method* draft is in IETF last call for proposed standard.

September 1999   A new IETF working group on SIP was created.

The third SIP interoperability test event took place December 6–8, 1999, in Richardson, Texas; hosted by Ericsson.

The second SIP interoperability test event took place August 5–6, 1999, at pulver.com in Melville, New York.

The first SIP interoperability test event (known as "the bake-off") took place April 8–9, 1999, at Columbia University, New York.

SIP is a proposed standard (February 2, 1999) published as RFC 2543 (March 17, 1999).

New list of public SIP servers.

Courtesy of http://www.cs.columbia.edu/sip.

FIGURE 13.52
H.310.

```
(Broadband ISDN or ATM, LAN)

ITU-T H.261: video codec
ITU-T  H.262 (MPEG 2): video codec
ITU-T G.711: 64-Kbps audio codec
ITU-T G.722: 48, 56, or 64-Kbps audio codec
ITU-T G.728: 16-Kbps audio codec
MPEG 2 audio
ITU-T H.222.0: multiplexing
ITU-T H.222.1 (MPEG): multiplexing
ITU-T H.233: encryption
ITU-T H.245: encryption key handling
ITU-T H.245: control
ITU-T T.120: data
```

FIGURE 13.53
*H.321
videoconferencing.*

```
(Broadband ISDN or ATM, LAN)

ITU-T H.261: video codec
ITU-T  H.263: video codec
ITU-T G.711: 64-Kbps audio codec
ITU-T G.722: 48, 56, or 64-Kbps audio codec
ITU-T G.728: 16-Kbps audio codec
ITU-T H.221: multiplexing
ITU-T H.231: multipoint
ITU-T H.233: encryption
ITU-T H.234: encryption key handling
ITU-T H.242: control
ITU-T T.120: data
```

13.3 Basic Compression Concepts and the MPEG Family

13.3.1 Digital Video Compression Overview

Compression algorithms are critical to the viability of digital video, digital video distribution, VoD, multimedia, and other video services. This section provides an overview of digital video compression as well as a synopsis of some MPEG-1, MPEG-2, and MPEG-4 principles. MPEG-7 is also briefly introduced. The standards discussions is based directly on the ISO standards; such material is included here in order to promulgate the use of open standards; however, developers and other parties working on products of commercial value should refer directly to the original documentation, especially since only some key highlights are included.

Dozens of video standards/formats are available worldwide for TV video (without even counting variant broadcasting schemes beyond the basic NTSC, PAL, and SECAM methods). Table 13.5 summarizes the video standards discussed or alluded to in this section. This section focuses on the higher end standards. (The previous section focused on the lower end business conferencing standards, but many of the underlying principles and technologies are similar.) In order for digital television to successfully

TABLE 13.5   PLETHORA OF VIDEO FORMATS AND CODING (IN ALPHABETICAL ORDER)

CD-I	Digital consumer-electronic format
D1/CCIR 601	Digital production standard
D2	Digital production standard
DS-3 based	Digital U.S. commercial methods (nonstandard)
DVD-video	Digital consumer-electronic format
DVI/Indeo	Early de facto digital standards for multimedia
H.261 et al	Digital videoconferencing format
H.263	Digital videoconferencing format for LAN/IP networks
HDTV	High-definition digital scheme discussed earlier in chapter
JPEG/motion JPEG	Digital compression format (principally for still video, but also for some video)
MPEG-1	Digital compression full-motion video (low-end entertainment video and multimedia)
MPEG-2	Digital compression full-motion video (high-resolution)
MPEG-4	Robust low-data-rate compression full-motion video
NTSC	Analog U.S. and Japanese format
PAL	Analog European format
SECAM	Analog French and Eastern Europe format
Vendor-based	Digital vendor-specific videoconferencing formats

enter the market, it is important for the agreed-on audio and video compression technique to be used on an industrywide basis. It is also important for the transport approach to be standardized; standards such as MPEG-2 and ATM do exactly that.

13.3.1.1   Compression Methods

Video can be considered a sequence of frames, in which each frame is an array of pixels. The goal of a video coding algorithm is to remove redundant information and greatly reduce the data rate. Two types of redundancies exist in video: redundancy within a single frame and redundancy between adjacent frames. There are two classes of compression algorithms: *lossless* algorithms and *lossy* algorithms. Another way of classifying compression algorithms is as entropy coding and source coding.

In lossless compression, all of the information contained in the uncompressed message can be faithfully recovered by the decompressor. For example, instead of sending a 100-bit message 0111111111...111111 one could compress it as x0y1, where x and y are octets that take values 0 (base 10) to 255 (base 10). In this case, one would send (00000001)0 (01100011)1, which is only 18 bits long, and yet the receiver would still be able to recover the message exactly. Lossless compression algorithms are symmetrical; namely, either the sender or receiver can perform the compression and decompression with the same level of computational complexity and without loss of data integrity. Compression of *data* material, either for transmission or for storage, clearly requires lossless methods. Many hardware and software products implement lossless compression. They typically double or quadruple the storage capacity on a disk (i.e., have a compression ratio of about 2:1 or 4:1), or double the apparent speed of a communication line. These algorithms can also be applied to files that represent voice or image information. Because the redundancy is higher, the compression ratios can be as high as 10:1. However, this is both (1) less effective than the compression obtained with specialized lossy techniques and (2) less than the information bandwidth reduction that is sought (typically 100:1 or even 200:1). Lossy compression algorithms do not aim at retaining all of the information; instead they want to retain just enough to be adequate for the task at hand. Lossy algorithms result in slightly degraded pictures. The advantage of these algorithms is that they can achieve 100:1 or 200:1 compression ratios.

As stated, there in another way of looking at compression, namely, source coding versus entropy coding. Source coding deals with *features* of the source material and encompasses lossy algorithms. Source coding can be further classified as intraframe and interframe coding. Intraframe coding is used for the first picture of a sequence and for downstream pictures after some major change of scenery. Intraframe coding is used for sequences of similar pictures (even for those including moving objects). Intraframe coding removes only the spatial redundancy within a picture; interframe coding also removes the temporal redundancy between pictures. Entropy coding, on the other hand, achieves compression by using statistical properties of the coded signal and is, in theory, lossless.

Video can be compressed using lossless or lossy methods. For lossless compression, video can be digitized according to the ITU-R 601 standard; the bit rate is approximately 165 Mbps. Although useful in a number of high-end commercial applications, this data rate is simply too high for user-level applications. Lossy methods, such as MPEG-1 and MPEG-2, are more appropriate for VoD and digital video distribution.

Two other methods are on the horizon: fractal and wavelets. The fractal transform uses Mandelbrot's approach of using simple equations to generate natural-looking images in a high level of detail. It is believed by experts to be a good compression scheme, particularly for still images of nature. Based on equations, it can be expanded to sizes even larger than the original, leading to claims of greater compression compared to other schemes. Images are segmented into domains that can be described as squeezed-down, distorted versions of larger parts or "ranges" of the same image [44]. Artifacts include softness and substitution of details by other details typically undetectable in natural scenes. Packages ranging in cost from $500 to $10,000 are available for desktop applications. The wavelet method is also based on mathematical techniques. A wavelet codec transforms a picture into a set of different spatial representations, some of which contain insignificant high frequencies, and one of which contains all the important low-frequency information. This scheme can also compress audio. Artifacts include softness, small random noise, and edge halos. Products based on this approach are also appearing.

13.3.1.2 Traditional Digital Video of Broadcast Quality

As covered briefly earlier, an image is composed of three elements: a luminance (brightness) element and two chrominance (color) elements. These elements come into play in the digitization process. There are two nearly lossless methods of digitizing television signals: digital component video and digital composite video.

As discussed earlier, *digital component video* is a time-multiplexed digital stream of three video signals: luminance Y, C_r (R − Y) and C_b (B − Y). The 4:2:2 refers to the ratio of sampling rates for each component. This format is also often called D1, referring to the tape format associated with the digital component recording. CCIR/ITU-R Recommendation 601 was adopted in 1982, after 8 years of study and compromise among European, Japanese, and North American approaches. The standard accommodates equally well NTSC, PAL, and SECAM formats. Typical digital component video encoding systems have utilized the following parameters:

- *NTSC:* Luminance sampling frequency 13.5 MHz. Sampling frequency for color differences: 6.75 MHz. Pixels: 858 × 525 (about 720 × 484 for the active image area);
- *PAL:* Luminance sampling frequency of 17.734475 MHz. Sampling frequency for color differences: 8.867236. Pixels: 910 × 525 (about 768 × 484 for the active image area).

At the final stage, the word length for digital image delivery is usually between 8 and 10 bits, but to maintain precision more may be utilized, particularly in the early stages of off-line processing (e.g., 16 bits). Because there is no one single sampling rate to obtain digital video, conversion between digital formats requires transcoding, not only of the formats, but also of the sampling frequencies.

The CCIR/ITU-R 601 standard support both the 525-line, 60 fields/second format and the 625-line 50 fields/second format (MPEG-2 covered later aims at providing similar quality but at much smaller data rate) [45].

The other encoding method is *digital composite video*, known as $4f_{sc}$. This format, applicable to NTSC, PAL, and SECAM, also consists of three components: Y, I, and Q. However, I and Q are not multiplexed but are instead quadrature modulated and summed to the Y component. The result is a single information stream sampled at four times the color subcarrier rate. The term $4f_{sc}$ refers to "4 the frequency of the subcarrier." This format is often called D2, referring to the associated tape format. Typical digital composite video encoding systems have utilized the following parameters:

- *NTSC:* Luminance sampling frequency of 14.31818 MHz (four times the frequency of the NTSC subcarrier). Sampling frequency for color differences: 7.15909. Pixels: 910 × 525 (about 768 × 484 for the active image area).

The transmission in digital form of a composite television signal (whether NTSC, PAL, or SECAM, particularly for contribution networks) requires, according to the sampling rates used and the number of bits employed to represent the signal, a rate of 100 to 150 Mbps. In component video (e.g., 4:2:2), the bit rate can reach 216 to 270 Mbps; uncompressed HDTV requires about 1 Gbps [46]. Compression is unquestionably required.

In the United States, DS-3 transmission facilities support about 45 Mbps; in Europe, E3 facilities support about 34 Mbps; also note that DS-1 supports 1.544 Mbps and E1 supports 2.048 Mbps. These rates have defined, at the pragmatic level, the boundaries for the commercially available video encoding algorithms during the past 15 years. For example, vendor-proprietary methods have been used in the past in the United States to encode TV signals at 45 Mbps for remote delivery (e.g., between TV studios in two cities), utilizing commonly available telecommunication carrier services. Several suppliers have developed broadcast-quality DS-3 video coders using "relatively mild compression" on 525/60-Hz video. Uncompressed pulse-code modulation of NTSC signals would require a minimum sampling rate of 8.4 MHz for a 4.2-MHz bandwidth, and 8 or 9 bits per sample, resulting in an uncompressed rate of 67.2 to 75.6 Mbps (in reality higher data rates are needed, because, as noted later, the sampling rate is higher). In Europe, the encoding of interest is 34 Mbps. These network rates clarify why the CCIR published Recommendation 723, which supports standardized video coding at 32 to 45 Mbps for full-resolution video/TV signals (720 pixels per line, 483/576 lines per frame, 59.94/50 2:1-interlaced frames per second). However, there is an interest in bringing the video data rate down further, as discussed in the next subsections.

13.3.1.3 Compression Algorithms in Common Use

The video compression requirements vary between various applications, digital storage media, transport method (e.g., terrestrial and DBS), and video programming type (e.g., talk shows versus sporting events). Nonetheless, it is important that easy interworking and movement between such media be accomplished. Many factors come into play, such as timing, program stream reconstruction, synchronization, demultiplexing/remutiplexing, packetizing/repacketizing, and encryption [47]. It has been the objective of the standards study groups to limit the extent of the specifications to a minimum and to define only what it takes to accomplish meaningful interoperability.

The mid-1980s saw the emergence of ITU-T's Recommendation H.261, supporting video compression and coding at p×64 Kbps (p = 1, 2, ..., 30). These standards (compared earlier in Section 13.2) are suited for videotelephony and videoconferencing, but are not deemed appropriate for entertainment-quality VoD programming (they could, however, be used in conjunction with digital video in support of these other services, perhaps in support of telecommuting [48]).

The 1980s also saw the emergence of a standard originated by ISO, which was formally adopted at the end of 1992. This standard, ISO/IEC 11172 (also known as MPEG-1), provides video coding for digital storage media with a rate of 2 Mbps or less. See Figure 13.54. H.261 and MPEG-1 standards provide picture quality similar to that obtained with a VCR. Both of these standards are characterized by low-bit-rate coding and low spatial resolution. H.261 supports 352 pixels per line, 288 lines per frame, and 29.97 noninterlaced frames per second (lower resolution is also supported). MPEG-1 typically supports 352 pixels per line, 240/288 lines per frame, and 29.97/25 noninterlaced frames per second. Many useful video and multimedia applications require higher resolution than this, in order to provide an acceptable level of quality to the user. Real-time encoding/decoding also introduces delays that increase with decreasing data rates. For example, current-technology encoding at 128 Kbps may result in unacceptable delays for a quality videoconference.

MPEG-1 was developed as a video compression standard to be used with CD-ROMs. The compression ratio is about 100:1; however, the quality of the picture is marginal for generic broadcast and cable applications of high-action movies and sporting events. MPEG-1 employs a source input format for motion video and associated audio with a data rate up to 1.5 Mbps. Figure 13.55 depicts the MPEG decoder functionality.

The MPEG standard embodies the concepts of (1) groups of frames and (2) interpolated frames (the presence of interpolated frames is optional). Each group of frames contains a frame that is intraframe coded only, to facilitate random access. There will also be predicted frames. Interpolated frames are formed from adjacent (past and future) *keyframes* (both the stand-alone and predicted frames can be used for interpolation). A group of frames consists of a single stand-alone frame, several predicted

FIGURE 13.54
*MPEG-1
specifications.*

ISO/IEC 11172-1: system
ISO/IEC 11172-2: video
ISO/IEC 11172-3: audio
ISO/IEC 11172-4: conformance testing
ISO/IEC 11172-5: simulation software

FIGURE 13.55
MPEG decoder.

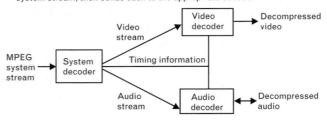

- System decoder extracts the timing information from the MPEG system stream and sends it to the other system components
- System decoder also demultiplexes the video and audio streams from the system stream; then sends each to the appropriate decoder

frames, and one or more interpolated frames positioned between key-frames. The bandwidth allocated to each type of frame typically conforms to the ratio of 5:3:1 (intraframe coded, predicted, and interpolated, respectively). "Future" frames may be employed to predict intermediary frames. (This can be done not only for stored material, but also for real-time material by buffering a few frames for analysis.) Figures 13.56 through 13.67 depict some of the concepts described in this paragraph.

FIGURE 13.56
Frames.

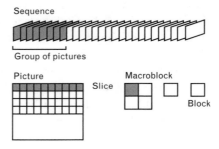

FIGURE 13.57
Frame concepts.

Video sequence
- Begins with a sequence header (may contain additional sequence headers), includes one or more groups of pictures, and ends with an end-of-sequence code

Group of pictures (GOP)
- A header and a series of one or more pictures intended to allow random access into the sequence

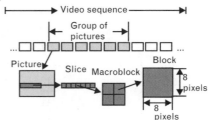

Picture
- The primary coding unit of a video sequence. A picture consists of three rectangular matrices representing luminance (Y) and two chrominance (Cb and Cr) values. The Y matrix has an even number of rows and columns. The Cb and Cr matrices are (usually) one-half the size of the Y matrix in each direction (horizontal and vertical)

FIGURE 13.58
More terms.

> Slice
>
> • One or more "contiguous" macroblocks...order of the macroblocks within a slice is from left-to-right and top-to-bottom.
>   -Slices are important in the handling of errors. If the bitstream contains an error, the decoder can skip to the start of the next slice.
>   Having more slices in the bitstream allows better error concealment, but uses bits that could otherwise be used to improve picture quality.

FIGURE 13.59
Macroblock.

• Macroblock
  - A 16-pixel by 16-line section of luminance components and the corresponding 8-pixel by 8-line section for the two chrominance components
  - Figure shows spatial location of luminance and chrominance components

Y   Cb   Cr

1	2
3	4

5   6

  - A macroblock contians four Y blocks, one Cb block, and one Cr block
    · Numbers correspond to the ordering of the blocks in the data stream, with block 1 first

FIGURE 13.60
Arrangement of blocks in a macroblock.

> Each picture is divided either into gruops of blocks (GOBs) or into slices
>
> A GOB comprises up to k*16 lines, where k depends on the number of lines in the picture
>
> Number of GOBs per picture is 6 for sub-QCIF, 9 for QCIF, and 18 for CIF, 4 CIF, and 16 CIF
>
> GOB numbering is done by sue of vertical scan of the GOBs, starting with the upper GOB (number 0) and ending with the bottom-most GOB
>
> Each GOB is divided into macroblocks
>
> Cr = R - Y
> Cb = B - Y

FIGURE 13.61
Cb and Cr.

• Diagram shows the relative x-y locations of the luminance and chrominance components
  - Note that for every four luminance values, there are two associated chrominance values: one Cb value and one Cr value
    · (location of the Cb and Cr values is the same, so only one circle is shown in the figure)

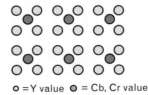

o = Y value   ◐ = Cb, Cr value

FIGURE 13.62
Picture types.

Predicted pictures
  - Predicted pictures, or P-pictures, are coded with
    respect to the nearest previous I- or P-picture
    · Technique is called forward prediction

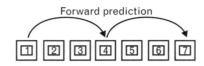

  - Like I-pictures, P-pictures serve as a prediction reference for B-pictures
    (bi-directionally interpolated pictures) and future P-pictures
    · However, P-pictures use motion compensation to provide more compression than
      is possible with I-pictures. Unlike I-pictures, P-pictures can propagate coding errors
      because P-pictures are predicted from previous reference (I- or P-) pictures

FIGURE 13.63
*Picture types: Intra
pictures.*

Intra pictures
- Intra pictures, or I-pictures, are coded using only information present in picture itself
- I-pictures provide potential random access points into the compressed video data
- I-pictures use only transform coding and provide moderate compression
- I-pictures typically use about two bits per coded pixel

FIGURE 13.64
*Picture types:
Bidirectional.*

Bidirectional pictures
- Bidirectional pictures, or B-pictures, are pictures that use
  both a past and a future picture as a reference
  · Technique is called bidirectional prediction

bidirectional prediction

· B-pictures provide the most compression and do not propagate errors because
  they are never used as a reference
· Bidirectional prediction also decreases the effect of noise by averaging two pictures

FIGURE 13.65
*Video stream
composition.*

MPEG algorithm allows encoder to choose the frequency and location
of I-pictures
- Choice is based on application's need for random accessibility and
  location of scene cuts in video sequence. In applications where random
  access is important, I-pictures are typically used two times a second.
  Encoder also chooses the number fo B-pictures between any pair of
  reference (I- or P- pictures)
- Choice is based on factors such as the amount of memory in the encoder and
  the characteristics of the material being coded. For example, a large class
  of scenes have two bidirectional pictures separating successive
  reference pictures.

FIGURE 13.66
A typical arrangement of I-, P-, and B-pictures in the order in which they are displayed.

• A typical arrangement of I-, P-, and B-pictures, in the order in which they are displayed

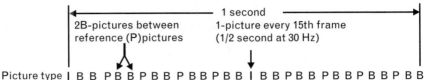

|← 1 second →|
2B-pictures between reference (P)pictures   1-picture every 15th frame (1/2 second at 30 Hz)

Picture type I B B P B B P B B P B B P B B I B B P B B P B B P B B P B B
Display order 1 2 3 4 5 6 7 8 9 10 11 12 13 14 15 16 17 18 19 20 21 22 23 24 25 26 27 28 29 30

FIGURE 13.67
Video stream.

• MPEG encoder reorders pictures in video stream to present pictures to decoder in most efficient sequence
  - In particular, the reference pictures needed to reconstruct B-pictures are sent before the associated B-pictures

Display order

| I | B | B | P | B | B | P |
| 1 | 2 | 3 | 4 | 5 | 6 | 7 |

Video stream order

| I | P | B | B | P | B | B |
| 1 | 4 | 2 | 3 | 7 | 5 | 6 |

During the mid-1990s, both ISO and ITU-T worked on new high-quality video coding standards: full-motion, reasonable-resolution digital video is sought in the 4- to 20-Mbps range. MPEG started its study on the second phase work (known as MPEG-2) in 1990, with completion in 1995. MPEG-2 (ISO/IEC 13818) supports both full CCIR/ITU-R 601 resolution as well as HDTV. The rates of interest range from 2 to 20 Mbps. As noted, work on MPEG-4 has been undertaken at the time of this writing (see Figure 13.68).

In a hybrid coder, a type of coder commonly used, an estimate of the next frame to be processed is created from the current frame. The difference between this estimate and the actual value of the variable(s) of interest contained in the next frame, when it arrives at the coder, is encoded by an appropriate mechanism. One of the more common examples of this type of

FIGURE 13.68
MPEG-2 specifications.

ISO/IEC 13818-1: systems
ISO/IEC 13818-2: video
ISO/IEC 13818-3: audio
ISO/IEC 13818-4: conformance testing
ISO/IEC 13818-5: simulation software
ISO/IEC 13818-6: command and control
ISO/IEC 13818-7:nonbackwards compatible audio formats
ISO/IEC 13818-9: real-time interface

coder is the motion-compensated DCT coder used in MPEG-1 and MPEG-2. Motion compensation capitalizes on the correlation between successive frames of a video sequence. A motion-compensation algorithm assigns a velocity (that is, a speed and direction to a moving object). Constant velocity makes predicting the next frame of video fairly straightforward. Misprediction, however, can cause the loss of two or three frames of video. Even when true motion is not at play in a scene, motion compensation algorithms improve the data rate by seeking a block of identical (or nearly identical) values in the previous frame and spatially close to the block to be encoded; this block is then used to formulate a prediction. Only the information used to find the prediction block is sent to the decoder [46].

The DCT (see Figure 13.69) concentrates the remaining signal into a small number (64 to be precise) of coefficients that can be quantized and efficiently represented. This coder is three to four times as efficient as one that uses no prediction, but it is sensitive to transmission errors and does not permit random access. This coder is utilized (with modifications to facilitate random access) in the MPEG-1 standard; the basic DCT method (see later discussion) is also used in the JPEG still-image standard. (Some developers have also used JPEG to encode video, but this is now generally on the decline.) As discussed earlier, encoding/decoding falls into two main types [49]:

1. *Interframe.* These encoders/decoders use combinations of key, motion-predicted and interpolated frames to achieve high compression ratios with low data rates. Examples of these types of algorithms include the various MPEG algorithms.

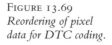

FIGURE 13.69
Reordering of pixel data for DTC coding.

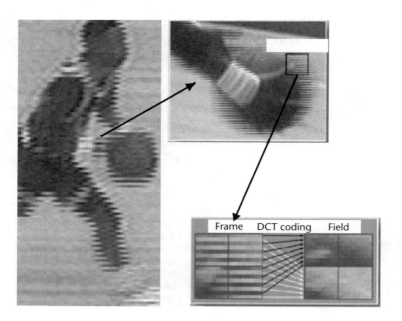

2. *Intraframe*. These systems compress every frame (and sometimes every field) of video individually. These algorithms (e.g., motion JPEG, but also MPEG with I-frames) offer the advantage of frame-accurate editability; however, they produce from 2 to 10 times more data than the interframe algorithms.

Coding standards such as H.261 and H.263 logically partition the images to be encoded into rows of macroblocks known as *groups of blocks* (GOBs). From the encoder's point of view, each of the current frames of the video sequence is partitioned into rectangular regions of 16 × 16 pixels called *macroblocks* (see Figures 13.70, 13.71, and 13.72). For each macroblock, the motion estimation stage of the video encoder computes the motion vectors that best represent the location in the previous frame where

FIGURE 13.70
Macroblocks.

• Each macroblock relates to 16 pixels by 16 lines of Y and
  the spatially corresponding 8 pixels by 8 lines of CB and CR

• That is, a macroblock consists of four luminance blocks
  and two spatially corresponding color difference blocks

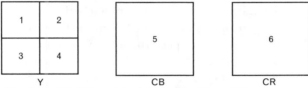

1	2
3	4

Y   CB   CR

 - Each luminance block thus relates to 8 pixels by 8 lines of Y
 - Each chrominance block thus relates to 8 pixels by 8 lines of CB or CR, but these last are in
   4:2:2 mode, resulting in 4 pointes for them in each 8x8 Y block (4:2:2 -> "8:4:4," namely: 8x8:4x4:4x4)

• A GOB comprises one macroblock row for sub-QCIF, QCIF and CIF, two macroblock
  rows for 4CIF and four macroblock rows for 16CIF

FIGURE 13.71
A macroblock.

• 16x16 Y pixels
• 8x8 Cr
• (not shown) 8x8 Cb

-[also remember that 16x16
  is four 8x8; that is, four 8x8 Y
  and four 4x4 Cr]

[or instead of four 4x4
  you can say one 8x8 for Cr/Cb]

FIGURE 13.72
Pixel coding.

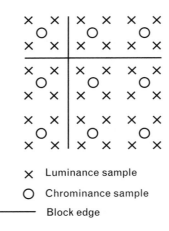

× Luminance sample

○ Chrominance sample

—— Block edge

image pixels of similar intensity values occur. The motion compensation stage applies these motion vectors to the corresponding macroblocks in the previous frame and computes a motion-compensated frame, and the difference with the current frame is then computed. The frame representing these differences is called the *residual frame*. The residual frame represents the information in the current frame that cannot be predicted from the previous image. This residual frame is then coded by the DCT stage. The DCT is typically performed on 8 × 8 pixel blocks. The quantized DCT coefficients are then coded using VLC schemes such as Huffman coding. Hence, for every frame of the video sequence, the video encoder transmits motion vector information, DCT information, and some overhead header information. To achieve further coding efficiency, the header and motion vector information are also coded using VLC techniques. These motion-compensated frames of the video sequence are known as *predictive frames* (P-frames) or *interpolated frames* (B-frames, being bidirectionally interpolated). Some frames of the video sequence are coded completely with respect to themselves with no motion compensation, and these are known as *intraframes* (I-frames). Intraframes do not have any motion vector information associated with them [50].

Until the recent development of chips able to support real-time encoding, the issue of compression algorithm asymmetry was important. (Powerful DSP chips are being developed by a number of vendors to support video processing.) With symmetric algorithms, the compression process requires the same amount of clock time as decompression (playback); asymmetric compression requires considerably more clock time than decompression [49]. In theory this makes the decompression possible on cheap low-end equipment. Recent advances in chip power make this a less important issue, particularly since the suppliers are coming out with a single chip that can be programmed to be an MPEG-1 encoder, MPEG-1 decoder, MPEG-2 encoder, MPEG-2 decoder, a JPEG encoder, a JPEG decoder, or an H.261 encoder/decoder [51].

A typical MPEG-2 decoder available at press time supports the following functions:

- MPEG-2 13818 standard for audio, video, and multiplexing;
- MPEG audio layer I and II;
- I-only, IP, IBP, and IBBP coding;
- 0.8- to 15-Mbps bit;
- 4:2:0 sampling resolution, 8 bits, or 4:2:2 sampling resolution, 10 bits;
- Two audio channels;
- CCIR 601 and SIF formats;
- 5.1 audio channels and support for AC-3;
- Multiple input/output ports (A/V switch).

High-end MPEG-2 systems for telemedicine, distance learning, and business video applications went for $10,000 to $25,000 at press time.

13.3.1.4 Brief Description of DCT

This section provides a short description of some key DCT features, by discussing DCT in the context of still-picture encoding. The DCT-based operation for the case of a single-component [52] (i.e., monochrome) includes a *forward DCT* (FDCT) function and an *inverse DCT* (IDCT) function.

One can think of DCT-based compression as compression of a stream of 8 × 8 blocks of gray-scale image samples. Each 8 × 8 block (represented by 64-point values known as $f(x,y)$, $0 \le x \le 7$, $0 \le y \le 7$) makes its way through each processing stage, yielding output in compressed form. For progressive-mode codecs, an image buffer is placed between the quantizer and the entropy coding module; this enables the image to be stored and then sent out in multiple scans with follow-up information aimed at successively improving the quality of the received image.

Each 8 × 8 block of source image (frame) samples can be viewed as a 64-point discrete signal that is a function of the two spatial dimensions x and y. At the input to the encoder, these 64 source image samples are cranked through the following equation [53]: For $0 \le u \le 7$, $0 \le v \le 7$ calculate the following 64 values:

$$F(u,v) = 0.25 \times C(u) \times C(v) \times \left\{ \sum_{x=0}^{7} \sum_{y=0}^{7} f(x,y) \times \cos\left[(2x+1) \times u\pi / 16\right] \right.$$
$$\left. \times \cos\left[(2y+1) \times v\pi / 16\right] \right\}$$

where $C(u) = C(v) = 1/\mathrm{sqrt}(2)$ for u, $v = 0$ and $C(u) = C(v) = 1$ otherwise. Mathematically, the FDCT takes the input signal and decomposes it into

64 orthogonal basis vector signals. The output of the FDCT is a set of 64 basis signal amplitudes, which are known as *DCT coefficients*. The coefficient for the vector (0,0) is called the *DC coefficient*; all other coefficients are called *AC coefficients*. The DC coefficient generally contains a significant fraction of the total image energy. Because sample values typically vary slowly from point to point across an image, the FDCT processing achieves data compression by concentrating most of the signal in the lower values of the (u,v) space. For a typical 8 × 8 sample block from a typical source image, many, if not most, of the (u,v) pairs have zero or near-zero coefficients and therefore need not be encoded. At the decoder, the IDCT reverses this processing step. One can use 8-bit or 12-bit source image samples; 12-bit samples, however, require fairly large computational resources for FDCT or IDCT calculations.

In principle, the DCT introduces no loss to the source image samples; it just transforms them to a domain where they can be more efficiently encoded. This means that if the FDCT and IDCT could be computed with perfect accuracy and if the DCT coefficients were not quantized, the original 8 × 8 block could be recovered exactly. But, as just seen, the FDCT and the IDCT equations contain transcendental functions (i.e., cosines). Consequently, no finite-time implementation can compute them with perfect accuracy. In fact, a *number* of algorithms have been proposed to compute these values approximately. No single algorithm is found to be optimal for all implementations: An algorithm that runs optimally in software usually does not operate optimally in firmware (say, for a programmable DSP) or in hardware.

Given the finite precision of the DCT inputs and outputs, an interworking challenge arises: Coefficients calculated by two different algorithms (say, one the sender and one in the receiver) or even by independently designed implementations of the same FDCT or IDCT algorithm (which differ only minutely in the precision of the cosine terms or intermediate results) will result in slightly different outputs from identical inputs.

Each of the 64 DCT coefficients obtained at the output of the FDCT is then uniformly quantized by utilizing a 64-element quantization table, which must be specified by the application (or user). Each element can take an integer value from 1 to 255 (or 1,023), that specifies the step size of the quantizer for its corresponding DCT coefficient. The purpose of quantization is to achieve further compression by discarding information that is not visually significant. Quantization is a lossy process and is the principal source of lossiness in DCT–based encoders.

When the aim is to compress the image or frame as much as possible but without visible artifacts, each step size is chosen to be the perceptual threshold of human vision. These thresholds are functions of the source image characteristics, display characteristics, and viewing distance.

After the quantization process, the DC coefficient, representing a sort of average of the value of the 64 image samples, is handled separately.

Because a high correlation usually exists between the DC coefficients of adjacent 8 × 8 blocks, the quantized DC coefficient is encoded differentially, namely, as the difference between the current value and the previous value. To facilitate entropy coding, the quantized AC coefficients are ordered into the "zigzag" sequence. This ordering helps the entropy coding process by placing low-coordinate coefficients (which are more likely to be nonzero) before high-coordinate coefficients (see Figures 13.73 and 13.74).

The last step for DCT-based encoding is entropy coding itself. This step achieves additional lossless compression by encoding the quantized DCT coefficients more compactly, based on their statistical characteristics. Entropy coding can be viewed as a two-step process. The first step converts the sequence of quantized coefficients (ordered as discussed earlier) into an intermediate sequence. The second step converts the symbols to a stream in which the symbols no longer have externally identifiable boundaries. Two entropy coding methods are in common use: Huffman coding and arithmetic coding. The sequential codec uses *Huffman coding,* but codecs with both methods are specified for all modes of operation. *Arithmetic coding* produces about 10% better compression than Huffman; however, it is more complex. Huffman coding requires that one or more sets of code tables be specified. The same tables used to compress an image are needed to decompress it. Huffman tables may be predefined and used within an

FIGURE 13.73
DCT operation.

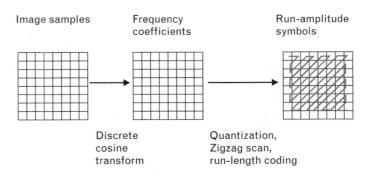

FIGURE 13.74
Zigzag.

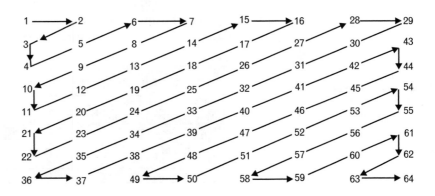

application as defaults, or developed specifically for a given image in an initial statistics-gathering pass prior to actual compression. Figure 13.75 depicts the number of operations required to handle compression.

13.3.2 JPEG and Motion JPEG

The JPEG standard has been developed jointly by both ISO and ITU-T (hence, the nomenclature "joint") for compression of still images. It can compress typical images from 1/10 to 1/50 of their uncompressed bit size without visibly affecting image quality. JPEG is the first international digital image compression standard for multilevel continuous-tone still images (both gray scale and color). Some applications to which JPEG addresses itself include color facsimile, quality newspaper wirephoto transmission, desktop publishing, graphic arts, and medical imaging. Some have used JPEG to support video transmission (particularly in the medical industry where one wants to be able to adjust the resolution of the transmission in real time, e.g., when viewing the display of a set of X-rays).

JPEG utilizes a methodology based on the DCT, discussed in the previous section. It is a symmetrical process, with the same complexity for coding and for decoding. JPEG will be an important compression standard for a number of applications since it works relatively well and is already available in the marketplace, as evidenced by vendor support. Many digital video cameras, fax machines, copiers, and scanners now include JPEG chips.

Some vendors also use JPEG methods for encoding of full-motion video, NTSC TV signals in particular. This is known as *motion JPEG*. JPEG plays a limited role in digital video at this time. Some Internet video applications utilize JPEG because of its relative simplicity. While JPEG was not designed for full-motion video, it can accommodate it with some restrictions. For example, audio is not supported in an integrated fashion. One of the limiting factors of this use of the algorithm is that it works independently from frame to frame; hence, it cannot reduce the redundancies that exist between frames. Some view the fact that JPEG performs only intraframe compression as a benefit in the sense that it offers "fast" random access to any frame of the video material. Other full-motion video compression techniques performing interframe compression rely on periodic transmission of a reference frame—if the reference frame is sent every 20

FIGURE 13.75
Example of number of operations required to handle compression.

Frames per second	15
Macroblocks per frame	396
Macroblocks per second	5,940
Motion vectors per macroblock	1,024
Motion vectors per second	6,082,560
SOADs per motion vector	256
Operations per SOAD	3
Operations per second	4,671,406,080

frames one may have to wait for as long as 19 frames before the reference frame is received; this would equate to a wait of 20/60 or 0.33 second. With JPEG, one needs to wait only for the time required to decompress one frame, that is, 0.04 second.

Network-based applications using JPEG for full-motion video are not likely to be widely implemented because JPEG is bandwidth intensive. For video material displayable at a PC monitor at medium resolution, that is, 640 × 480 pixels with 24 bits for color representation, JPEG is required to compress about 1 MB per frame, or 30 MBps (240 Mbps) to a lower value. The downloading, displaying, and manipulation of full-screen video in digital form is a daunting task, even if one were to achieve a 50:1 compression. This is why standards such as MPEG-1 and MPEG-2 are needed for any digital video transmission except perhaps for desktop multimedia applications.

13.3.3   MPEG-1

As noted, the ISO/IEC/JTC1/SC29/WG11 MPEG working group has produced a specification for coding of combined video and audio information. It is directed at video display, in contrast to still-image display which JPEG addresses. MPEG specifies a decoder and data representation for retrieval of full-motion video information from digital storage media in the 1.5- to 2-Mbps range. Hence, it can (nominally) be used in an ADSL context. Its principal goal, however, was to support storage of digital video. The specification is composed of three parts: Part 1, *Systems*; Part 2, *Video*; and Part 3, *Audio*. The system part specifies a system coding layer for combining coded video and audio, and it provides the ability to also combine private data streams and streams that may be defined at a later date. The specification describes the syntax and semantic rules of the coded data stream.

As mentioned earlier, MPEG-1 uses three types of frames: intra (I) picture frames, predicted (P) frames, and bidirectional interpolated (B) frames. I-frames are compressed using only the information in that frame using the DTC algorithm. An incoming video signal of 1 sec will contain at least two I-frames. P-frames are derived from the preceding I-frames (or from other P-frames) by predicting motion forward in time; P-frames are compressed to approximately 60:1. Bidirectional B interpolated frames are derived from the I-and P-frames, based on previous and frame referencing; B-frames are required to achieve the low average data rate [49].

Because MPEG allows coding comparisons across multiple frames, it can yield compression ratios of 50:1 to 200:1. The MPEG algorithm is asymmetrical. Namely, it requires more computational complexity (hardware) to compress full-motion video than to decompress it. This is useful for applications where the signal is produced at one source but it is distributed to many. MPEG chips on the market at press time (a handful) provide 200:1 compression to yield VHS quality at 1.2 to 1.5 Mbps; they also can provide 50:1 compression for broadcast quality at 6 Mbps.

MPEG's system coding layer specifies a multiplex of elementary streams such as audio and video, with a syntax that includes data fields directly supporting synchronization of the elementary streams. The system data fields also assist in the following tasks [54, 55]:

1. Parsing the multiplexed stream after a random access;
2. Managing coded information buffers in the decoders;
3. Identifying the absolute time of the coded information.

The system semantic rules impose some requirements on the decoders; however, the encoding process is not specified in the ISO document and can be implemented in a variety of ways, as long as the resulting data stream meets the system requirements.

The video encoder receives uncoded digitized pictures called *video presentation units* (VPUs) at discrete time intervals; similarly, at discrete time intervals, the audio digitizer receives uncoded digitized blocks of audio samples called *audio presentation units* (APUs). Note that the times of arrival of the VPUs are not necessarily aligned with the times of arrival of the APUs.

The video and audio encoders, respectively, encode digital video and audio as described in the MPEG specification Parts 2 and Part 3, producing coded pictures called *video access units* (VAUs) and coded audio called *audio access units* (AAUs). These outputs are referred to generically as elementary streams. The system encoder and multiplexer produce a multiplex stream [referred to as M(i)] containing the elementary streams as well as system layer coding (described later). Audio is supported from 32 to 384 Kbps and can consist of a single channel or two stereo channels.

The MPEG system specification includes a syntax with three coding layers above the layer implicitly defined by the elementary streams: (1) the ISO 11172 stream layer, (2) the pack layer, and (3) the packet layer, as discussed next:

- The ISO 11172 *stream layer* includes a sequence of packs followed by an end code.

- The *pack layer* includes the SCR field, the *Mux rate* field, an optional system header packet, and the packet layer (the Mux rate field bounds the rate of bytes per second as measured by the current and succeeding SCR values and the number of coded bytes intervening).

- The *packet layer* is comprised of packets containing information from individual elementary streams (there is information from exactly one elementary stream in each packet). The packet contents and all system layer coding are octet aligned, although the individual coding elements within elementary streams may not necessarily be octet aligned. Each packet consists of a packet start code followed by the packet length (in octets, ranging up to $2^{16} - 1$). Sixty-nine different values of packet start codes are currently defined: 16 are for video, 32 are for audio, 2 are private, 1 is for padding; the remaining 18 codes are reserved for future use (of which the Multimedia and Hypermedia Experts Group may take 16). Private data is unrestricted, other than by the syntax and STD model that applies to the entire stream. Three optional fields may be included in the packet: STD buffer size, PTS, and DTS. Last, one finds the packet information from the elementary stream. The amount of packet information is limited only by the total available packet length (decremented by the data in the packet header itself) and by constraints imposed by the decoder.

The method of multiplexing the elementary streams (VAUs and AAUs) is not directly specified in MPEG. However, some constraints must be followed by an encoder and multiplexer in order to produce a valid MPEG data stream. For example, it is required that the individual stream buffers must not overflow or underflow. The sizes of the individual stream buffers impose limits on the behavior of the multiplexer. Decoding of the MPEG data stream starting at the beginning of the stream is straightforward because there are no ambiguous bit patterns. Starting the decoding operation at random points requires locating pack or packet start codes within the data stream.

An important aspect of MPEG is clock synchronization. Synchronization is a fundamental aspect of communication. The principle may be known to the reader in the context of synchronizing various nodes on a network to enable character, block, or message recovery, because these data entities are transferred from the sender to the receiver. In a multimedia context, synchronization is even more "intimate" in the sense that the various signal objects comprising the combined signal must be stored, retrieved, and transmitted with precise timing relationships. The basic mechanism to achieve the desired synchronization is, to a large extent, the same in both contexts.

The *system time clock* (STC) is a reference time operating at 90 kHz. The STC is not necessarily phase-locked to the audio or the video sample clocks. The STC produces 33 bit time values (binary values from 0 to $2^{33} - 1$) incremented at 90 kHz. For some (but not necessarily all) VPUs and APUs arriving at the encoder, the value of the STC is determined and stored with the presentation units through the coding, transmission, and decoding processes. These values are called *presentation time stamps* (PTS). The time stamps, using a reference frequency of 90 kHz and unsigned binary values from 0 to $2^{33} - 1$, allow unique identification of operating time within the information stream over an interval exceeding 24 hours [54, 55].

In addition to the PTS, two other time stamps are associated with the coded information stream itself that are used to ensure synchronization between the various decoders and the information stream source in a decoding system:

1. *System clock references* (SCR). These are generated as samples of the STC such that the value of the SCR equals the value of the STC at the time the last octet of the SCR exits the system encoder.

2. *Decoding time stamps* (DTS). These are similar to PSTs, except that the permutation ordering of pictures in video coding process is reflected in the DTS values.

To achieve synchronization in multimedia systems that decode multiple video and audio signals originating from a storage or transmission medium, the decoding system must have a *time master*. MPEG does not specify which entity is the time master. The time master can be (1) any of the decoders, (2) the information stream source, or (3) an external time base; all other entities in the system (decoders and information sources) must slave their timing to the master. If a decoder is taken as the time master, the time when it presents a presentation unit is considered to be the correct time for the use of the other entities. Decoders can implement phase-locked loops or other timing means to ensure proper slaving of their operation to the time master. If an information stream source is considered the time master, the SCR values indicate the correct time at the moment that these values are received. The decoders then use this information of what the correct time is to pace their decoding and presentation timing. If the time base is an external entity, all of the decoders and the information sources must slave the timing to the external timing source.

13.3.4   MPEG-2

The compression schemes discussed so far do not produce adequate quality for full-motion video (although JPEG supports high quality, its data rate is too high). MPEG-2 was developed by ISO/IEC/JTC1/SC29/WG11, and is known as ISO/IEC 13118. This is probably going to be the most important video compression standard for digital video and VoD applications. The committee (WG11) consists largely of U.S. and European companies involved in or interested in video and audio compression [45]. As noted earlier, its goal is to provide CCIR/ITU-R quality for NTSC, PAL, and SECAM and also to support HDTV quality (this being a relatively newer requirement). MPEG-2 work is now driven by the desire to accomplish global unification of digital TV program generation, editing, storage, retrieval, transport, and display [47]. The standard provides a set of agreed-on methodologies for audio and video compression, and transport of complex multiplexes of associated data and related data services.

MPEG-2 has been very successful commercially, with significant acceptance in the market, not only for broadcast TV but also for other applications (e.g., distance learning). MPEG-2 has been used in DBS, DVD, and HDTV. There were plans to develop another compression standard for HDTV (this was referred to as MPEG-3, as noted in

Section 13.1); however, planners found that MPEG-2 was suitable (hence, it was decided to include HDTV as a separate profile of MPEG-2).

The objectives of newer video coding standards for full-motion video that have been developed recently, specifically MPEG-2, are as follows:

1. Picture quality should be higher than that of the current NTSC/PAL/SECAM broadcast systems.

2. Compression to bit rates in the range of 4 to 6 Mbps for NTSC/PAL/SECAM material and 6 to 10 Mbps for television signals conforming to CCIR Rec. 601. These target rates are already achievable in experimental systems.

3. The standard(s) need to be flexible enough to allow both high performance/high-complexity and low-performance/low-complexity (e.g., intraframe mode only operation) codec systems.

4. The standard(s) should take into account existing standards. Compatibility consideration enables smooth migration of new standards while maintaining interoperability among equipment conforming to the old and new generation standards.

5. Compatibility should be maintained to the extent possible. There are two types of compatibility: upward/downward compatibility (addressing different picture format sizes) and forward/backward compatibility (addressing different generation standards). A system is upward compatible if a higher resolution receiver is able to decode material from the signal transmitted by a lower resolution encoder. A system is downward compatible if a lower resolution receiver is able to decode material from the signal or part of the signal transmitted by a higher resolution encoder. A system is forward compatible if the new standard decoder is able to decode material from a signal or part of the signal of an existing standard encoder. A system is backward compatible if an existing standard decoder is able to decode material from the signal or part of the signal of a new standard encoder.

The following four layers come into play in the discussion of MPEG-2 coding:

- *Block:* The smallest coding unit in the MPEG algorithm. It consists of 8×8 pixels and can be one of three types: luminance (Y), red chrominance (C_r), and blue chrominance (C_b). The block is the basic unit in intraframe coding.

- *Macroblock:* The basic coding unit in the MPEG algorithm. It is a 16×16 pixel segment in a frame. Because each chrominance component has one-half the vertical and horizontal resolution of the luminance component, a macroblock consists of four Y, one C_r, and one C_b block.

- *Slice:* A horizontal strip within a frame. It is the basic processing unit in the MPEG coding scheme. Coding operations on blocks and macroblocks can only be performed when all pixels for a slice are available. A slice is an autonomous unit since coding a slice is done independently of its neighbors. Typically each frame contains 30 slices of 512 × 16 pixels.

- *Picture:* The basic unit of display. It corresponds to a single frame in a video sequence. The spatial dimensions of a frame are variable and are determined by the requirements of an application. Typically these dimensions are 512 × 480 pixels, which is similar to NTSC broadcast quality.

MPEG-2 initially consisted of three *profiles* (simple, main and next), each further divided into four *levels* (high level type 1, high level type 2, main level, and low level)—the simple profile does not support bidirectional B frames [56], eliminating the need to store them (see Table 13.6). The levels refer primarily to the resolution of the video produced; for example, the low level refers to a standard image format video with a resolution of 352 × 240 (also known as *source input format*—SIF). The main level conforms with the CCIR/ITU-R 601 quality. The high level is for HDTV.

The main level/main profile was standardized in 1993 (with chips available in 1994). Toward the end of 1993, the structure of profiles and levels was changed to reflect likely applications of the standard. The levels were modified to high level, high-1440 level, main level, low level; the

TABLE 13.6   INITIAL MPEG-2 PROFILES AND LEVELS

	SIMPLE PROFILE	MAIN PROFILE	NEXT PROFILE
High level type 1: Supports 1,152 lines per frame (lpf), 1,920 pixels/line (ppl), and 60 fps. This equates with 62.7 M pixels per second (pps) or 60 Mbps.		HDTV (U.S.)	HDTV (European)
High level type 2: Supports 1,152 lps, 1,440 ppl, and 60 fps. This gives 47 Mpps or 60 Mbps.		HDTV (U.S.)	HDTV (European)
Main level: Supports 576 lpf, 720 ppl, and 30 fps. This gives 10.4 Mpps or 15 Mbps.	Cable TV industry	Video broadcasting industry	
Low level: Supports 288 lpf, 352 ppl, and 30 fps. This gives 2.5 Mpps or 4 Mbps.			

profiles were modified to simple profile, main profile, SNR [57] scalable profile, spatially scalable profile, and high profile [58]. A previous profile known as *main + profile* was split into two scalable profiles, anticipating situations of network congestion, whereby the scalable profiles can drop information if it cannot be transported. The *next profile* was renamed *high profile* to reflect the fact that it applies to high-resolution video applications (see Table 13.7).

The MPEG-2 main profile is the one of greatest current interest [45]. It can handle images from the lowest level with MPEG-1 quality, up through broadcast quality at the main level, to HDTV at the high level.

The expectation is that MPEG-2 will play a critical role in most industrial and consumer applications in the foreseeable future. Although MPEG-1 was developed for computer applications and, hence, supports progressive scanning only, MPEG-2 is to be used in the TV world, and, hence, supports interlaced scanning.

The input to the MPEG-2 encoder is digital component video. The standard covers audio compression, video compression, and transport. In the transport area it defines [47]:

1. *Program streams:* a grouping of audio, video and data elemental components having a common time relationship, and being generally "associated" for delivery, storage, and playback;

2. *Transport streams:* a collection of program streams or elementary streams (video, audio, data) that have been multiplexed in a non-specific relationship for the purpose of transmission.

These efforts at the system layer are aimed at providing a basic data structure; a data structure in this context is viewed as the "semantics and syntax" of a data stream that can serve as a common format for local usage (e.g., storage, edit, and so on) and for broadcast. A number of basic structural elements have been defined and are expected to become part of the

TABLE 13.7 MPEG-2 CONFORMANCE POINTS

	SIMPLE PROFILE	MAIN PROFILE	SNR SCALABLE PROFILE	SPATIALLY SCALABLE PROFILE	HIGH PROFILE
High level		x★			x
High-1140 level		x		x	x
Main level	x	x	x		x
Low level		x	x		

★ x = point of concordance (likely combinations of levels and profiles in actual applications).

system layer syntax. Pivotal to this structure is the fact that the transport stream is based on packet principles. The packets (in the 130- to 192-octet range) contain digital information from a single elementary stream or data type [59]. Each packet has a header of up to 4 octets that provides information such as packet ID, clear/scramble indication, key (even/odd), and continuity counter.

13.3.5 MPEG-4

MPEG-4 is the next audiovisual coding standard that ISO developed after MPEG-1 and MPEG-2. Unlike the previous two standards, which, as covered earlier, had a clear application in mind (e.g., storage or transmission), MPEG-4 is a broad umbrella standard with a number of different technologies aimed at different applications. Initially MPEG-4 was focused at low-bit-rate communications, but later on the scope was expanded to include multimedia coding. Interestingly, MPEG-4 is efficient across a range of bit rates, from a few kilobits per second to tens of megabits per second. Figures 13.76 and 13.77 provide some highlights of the standard.

MPEG-4 is the first real multimedia representation standard, allowing interactivity and a combination of natural and synthetic material, coded in the form of objects. (It models audiovisual data as a composition of these objects.) MPEG-4 provides the standardized technological elements enabling the integration of the production, distribution, and content access paradigms of the fields of interactive multimedia, mobile multimedia, interactive graphics, and enhanced digital television.

MPEG-4 developers are looking at frame reconstruction from a narrow bandwidth command string, speech and video synthesis, and

FIGURE 13.76
MPEG-4 approach.

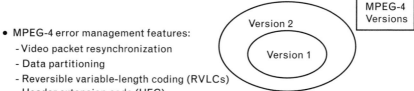

- MPEG-4 Committee has taken an approach of versioning to standard formation process
- Version 1... Version 2...

- MPEG-4 Version 1 includes a number of useful tools, and Version 2 is expected to include some others being developed by a standards body
- Expect that MPEG-4 Version 2 will be backward-compatible with MPEG-4 Version 1

- MPEG-4 error management features:
  - Video packet resynchronization
  - Data partitioning
  - Reversible variable-length coding (RVLCs)
  - Header extension code (HEC)
- When compressed video data is transmitted over noisy channels, bit and burst errors corrupt bit-stream
- A video decoder that is decoding this compromised bit-stream will lose synchronization with the encoder

FIGURE 13.77
MPEG-4 highlights.

MPEG-4 Video standard will support the decoding
of conventional rectangular images and video as well
as the decoding of images and video of arbitrary
shape

MPEG-4 VL BV Core Coder

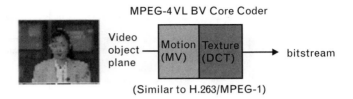

(Similar to H.263/MPEG-1)

Generic MPEG-4 coder

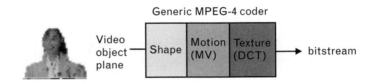

the use of fractal geometry, computer visualization, and artificial intelligence to build accurate data from minimal frames. Some of the highlights of the standard include the following [50] (also see Figures 13.78 through 13.83):

- The ability to efficiently encode mixed media data such as video, graphics, text, images, audio, and speech, called *audiovisual objects* (AVOs);

- The ability to create a compelling multimedia presentation by composing these mixed media objects by a compositing script;

- Error resilience to enable robust transmission of compressed data over noisy communication channels;

- The ability to encode arbitrarily shaped video objects;

FIGURE 13.78
MPEG-4 coded representation of media objects.

MPEG-4 Coded representation of media objects

- A media object in its coded form consists of descriptive elements that allow it to handle the object in an audiovisual scene as well as of associate streaming data, if needed. It is important to note that in its coded form, each media object can be represented independent of its surroundings or background.
- Coded representation of media objects is as efficient as possible while taking into account the desired functionalities. Examples of such functionalities are error robustness, easy extraction and editing of an object, or having an object available in a scalable form.

FIGURE 13.79
MPEG-4 and
VRML.

MPEG-4 provides a standardized way to describe a scene, allowing,
for example, to:

- Place media objects anywhere in a given coordinate system;
- Apply transforms to change the geometrical or acoustical appearance of
  a media object;

- Group primitive media objects in order to form compound media objects;

- Apply streamed data to media objects, in order to modify their attributes
  (e.g., moving texture belonging to an object; animating parameters animating
  a moving head);

- Change, interactively, the user's viewing and listening points anywhere
  in the scene.

Scene description builds on several concepts from VRML in terms of both its
structure and the functionality of object composition nodes and extends it.

FIGURE 13.80
MPEG-4 scenes.

Figure gives an example that highlights the way in which an audiovisual scene
in MPEG-4 is described as composed of individual objects:

- Figure contains compound media objects that group primitive media
  objects together;

- Primitive media objects correspond to leaves in the descriptive tree while
  compound media objects encompass entire subtrees;

- As an example: visual object corresponding to the talking person and the
  corresponding voice are tied together to form a new compound media
  object, containing both the aural and visual components of a talking person.

Such grouping allows authors to construct complex scenes, and enables
consumers to manipulate meaningful (sets of) objects.

FIGURE 13.81
MPEG-4
environment.

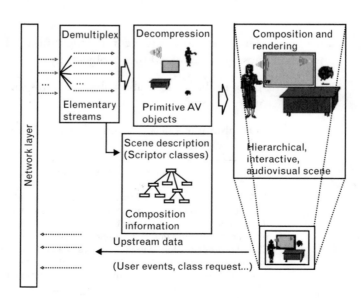

FIGURE 13.82
MPEG-4 example.

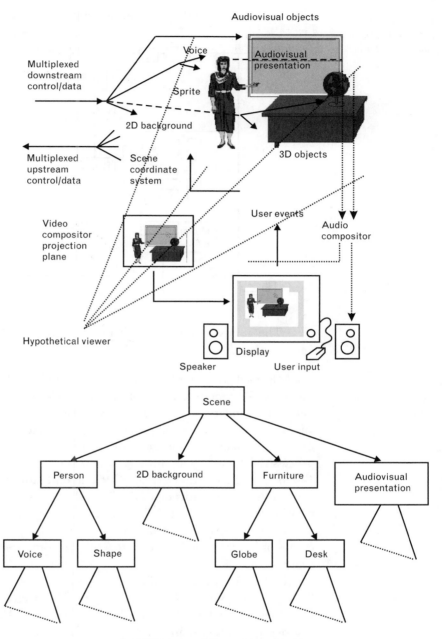

An MPEG-4 scene follows a hierarchical
structure which can be represented as a
directed acyclic graph...Each node of the
graph is a media object

- Multiplexing and synchronization of the data associated with these objects so that they can be transported over network channels providing a QoS appropriate to the nature of the specific objects;

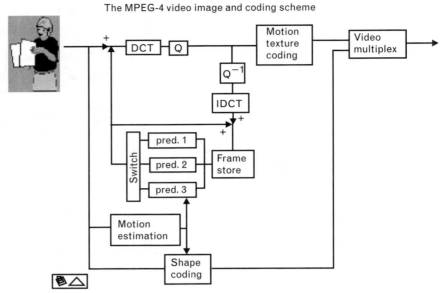

FIGURE 13.83
*MPEG-4 video image
and coding scheme.*

The MPEG-4 video image and coding scheme

• The ability to interact with the audiovisual scene generated at the receiver.

MPEG-4 is a multimedia coding standard; it standardizes tools not only for video coding but also for audio, graphics, and text coding. The standard also includes a systems part that describes how the audio, video, text, and graphics are synchronized. The MPEG-4 committee has taken the approach of versioning to the standard formation process. MPEG-4, with formal as its ISO/IEC designation "ISO/IEC 14496," was finalized in October 1998 and became an international standard in the first months of 1999. The fully backward compatiable extensions under the title of MPEG-4 Version 2 were frozen at the end of 1999, to aquire the formal international standard status early in 2000. Several extensions were added since, and work on some specific items is still in progress.

Table 13.8 depicts a timetable for MPEG-4 as well as MPEG-2, MPEG-7, and MPEG-21 extensions. It was expected that MPEG-4 Version 2 will be backward-compatible with MPEG-4 Version 1.

Because of the desire for usage in noisy channels, MPEG-4 incorporates several error management features. These mechanisms provide important capabilities such as resynchronization, error detection, data recovery, and error concealment. The mechanisms are as follows:

• Video packet resynchronization;

• Data partitioning;

• Reversible variable-length coding (RVLCs);

STD	PT	EDIT.	PRO.	DESCRIPTION	CFP	WD	CD	FCD	FDIS
2	1	2000	Amd.3	Carriage of AVC Content			02/05	02/12	03/07
2	7	1997	Amd.1	Embedding of bandwidth extension			02/10	03/03	03/07
4	1	2003	Amd.7	Use of AVC in MPEG-4 Systems			02/07	03/03	03/07
4	1	2003	Amd.8	Use of MPEG-7 in MPEG-4 Systems		02/07	03/03	03/07	03/12
4	2	2003	Amd.4	Error resilience in scalable enhancement layer		02/07	02/12	03/03	03/07
4	3	2001	Amd.2	Extension 2 – Parametric Audio extension		01/12	02/12	03/07	03/12
4	4	2002	Amd.3	Visual New Level and Tools			02/07	02/12	03/07
4	4	2003	Amd.4	IPMP Extension, MUW and AFX Conformance		02/07	02/12	03/07	03/12
4	4	2003	Amd.5	Error Resilience Scalable Profile Conformance			03/03	03/07	03/12
4	4	2004	Amd.6	AVC			03/07	03/12	04/03
4	5	2003	Amd.4	IPMP Extension, MUW and AFX Reference SW		02/10	02/12	03/07	03/12
4	5	2003	Amd.5	Error Resilience Scalable Profile Reference Software			03/03	03/07	03/12
4	5	2004	Amd.6	AVC		03/03	03/07	03/12	04/03
4	9	200x	Amd.1	Reference Hardware Description	02/07	03/07	04/03		04/03
4	11	2002	Amd.2	Advanced Text and 2D Graphics		02/07	03/03	03/07	03/12
4	15	2003	1st Ed.	AVC File Format			02/07	02/12	03/07
7	1	2001	Amd.1	Systems Extensions	02/03	02/07	02/12	03/07	03/12
7	3	2001	Amd.1	Visual Descriptors Extensions		02/05	03/03	03/07	03/12
7	4	2001	Amd.1	Audio Descriptors Extensions		01/12	02/05	02/10	03/07
7	5	2001	Amd.1	Multimedia Description Schemes Extensions		01/12	02/05	02/10	03/07
7	6	2001	Amd.1	Reference software extensions		01/12	02/05	03/03	03/07
7	7	2004	Amd.1	Conformance extensions		03/03	03/07	03/10	04/03
21	4	200x	1st Ed.	IPMP Framework		03/07	03/10	04/03	04/07
21	5	200x	1st Ed.	Rights Expression Language		01/12	02/07	02/12	03/07
21	6	200x	1st Ed.	Rights Data Dictionary		01/12	02/07	02/12	03/07
21	7	200x	1st Ed.	Digital Item Adaptation	02/03	02/05	02/12	03/07	03/12
21	8	200x	1st Ed.	Reference software		03/03	03/07	03/10	04/03
21	9	200x	1st Ed.	File Format		02/07	03/03	03/07	03/12
21	10	200x	1st Ed.	Digital Item Processing	02/12	03/03	03/12	04/03	04/07
21	11	200x	1st Ed.	Evaluation Tools for Persistent Association		02/12	03/10		04/03
21	12	200x	1st Ed.	Test Bed for MPEG-21 Resource Delivery		02/12	03/12		04/07

Source: http://mpeg.telecomitalialab.com/workplan.htm

• Header extension code (HEC).

When the compressed video data is transmitted over noisy channels, bit and burst errors corrupt the bit stream. A video decoder that is decoding this compromised bit stream will lose synchronization with the encoder (i.e., the decoder is unable to identify the precise location in the image where the current data belongs). If rectifying measures are not in place, the quality of the decoded video degrades rapidly and significantly. One way to deal with this problem is for the encoder to insert resynchronization markers in the bit stream. When the decoder detects an error it can then search for the resynchronization marker and regain resynchronization.

A horizontal row of macroblocks for QCIF images comprises a GOB (in the case of CIF images, each row of macroblocks consists of two GOBs). To provide synchronization resilience, the H.263 encoder has the option of inserting resynchronization markers at the beginning of each GOB. The smallest section to which the error can be isolated and concealed is, therefore, one row of macroblocks. Furthermore, when the resynchronization markers are restricted to be at the beginning of the

FIGURE 13.84
Description and synchronization of streaming data for media objects.

> • Media objects may rely on streaming data that is conveyed in one ore more elementary streams
>   - All streams associated to one media object are identified by an object descriptor
> • Allows handling hierarchically encoded data as well as the association of metainformation about content (object content information) and the intellectual property rights associated with it

FIGURE 13.85
Delivery of data streaming.

> • Synchronized delivery of streaming information from source to destination, exploiting different QoS as available from the network, is specified in terms of the aforementioned synchronization layer and a delivery layer containing a two-layer multiplexer
> • First multiplexing layer is managed according to the DMIF specification, part 6 of the MPEG-4 standard
> - Multiplex may be embodied by the MPEG-defined ElexMux too, which allows grouping of Elementary Streams (ESs) with a low multiplexing overhead
> - Multiplexing at this layer may be used, for example, to group ES with similar QoS requirements, reduce the number of network connections or the end-to-end delay
> • "TransMux"(Transport Multiplexing) layer models the layer that offers transport services matching the requested QoS
> - Only the interface to this layer is specified by MPEG-4 while the concrete mapping of data packets and control signaling must be done in collaboration with the bodies that have jurisdiction voer the respective transport protocol
> - Any suitable existing transport protocol stack such as (RTP)/UDP/IP, (AAL5)/ATM, or MPEG-2's Transport Stream over a suitable link layer may become a specific TransMux instance. Choice is left to end users/service providers, and allows MPEG-4 to be used in a wide variety of operation environments

GOBs, it is only possible for the decoder to isolate the errors to a fixed GOB independent of the image content.

MPEG-2 has a similar (optional) slice resynchronization methodology (see Figures 13.84 and 13.85). MPEG-4 provides an improved method, because the MPEG-4 encoder is not restricted to inserting the resynchronization markers only at the beginning of each row of macroblocks. (H.263 Version 2 adopted a resynchronization scheme similar to MPEG-4 in an additional annex.) The encoder has the option of dividing the image into video packets, each made up of an integer number of consecutive macroblocks. These macroblocks can span several rows of macroblocks in the image and can even include partial rows of macroblocks [50]. One simple approach is for the MPEG-4 encoder to insert a resynchronization marker every 1,000 bits [60]. Note that when the MPEG-4 encoder inserts the resynchronization markers at uniformly spaced bit intervals, the macroblocks interval between the resynchronization markers is shorter in the high-activity areas and longer in low-activity areas. This implies that in the presence of a short burst of errors, the decoder can quickly localize the error to within a few macroblocks for high-activity areas of the images and retain the image quality in these areas (also see Figure 13.86).

After detecting an error in the bit stream and resynchronizing to the next marker, the decoder has isolated the data in error to be in the

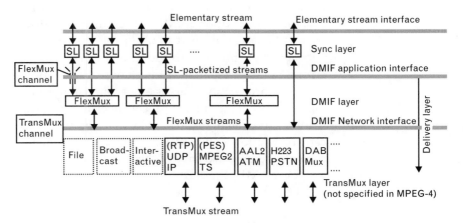

FIGURE 13.86
*Streaming data
protocol stack.*

- Protocol makes it possible to:
    - Identify access units, transport timestamps and clock reference information and identify data loss.
    - Optionally interleave data from different elementary streams into FlexMux streams
    - Convey control information to:
        - · Indicate the required QoS for each elementary stream and FlexMux stream;
        - · Translate such Qos requirements into actual network resources;
        - · Associate elementary streams to media objects
        - · Convey the mapping of elementary streams to FlexMux and TransMux channels

macroblocks between the two resynchronization markers. The video decoder discards all of these macroblocks and replaces the luminance and chrominance of these macroblocks with the luminance and chrominance from the corresponding macroblocks in the previous frame to conceal the errors.

Between two markers, the motion and DCT data for each macroblock are coded together. Consequently, when the decoder detects an error, whether the error occurred in the motion or DCT part, all of the data in the stream have to be discarded: Because of the uncertainty of the exact location where the error occurred, the decoder cannot determine that the motion or DCT data on any macroblocks in the packet are not erroneous.

The data partitioning mode in MPEG-4 partitions the data within a video packet into motion part and a texture part separated by a unique *motion boundary marker* (MBM). The MBM identifies the end of the motion data and the beginning of the DCT data. When an error is detected in the motion section, the decoder flags an error and replaces all the macroblocks in the current packet with skipped blocks until the next resynchronization marker (resynchronization occurs at the next successfully read resynchronization marker). If any subsequent video packets are lost before resynchronization, those packets are replaced by skipped macroblocks as well. When an error is detected in the texture section (and no errors are detected in the motion section) the NMB motion vectors are used to perform motion compensation (NMB is the number of macroblocks in the video packet). The texture part of all of the macroblocks is discarded and the decoder resynchronizes to the next resynchronization marker [50]. If an error is not detected in the motion or texture section of the bit stream, and the resynchronization marker is not found at the end of decoding all of the macroblocks of the current packet, an error is flagged. In this case, only the texture part of all of the macroblocks in the current packet needs to be discarded. Motion compensation can still be applied for the NMB macroblocks because we have higher confidence in the motion vectors since we got the MBM [50].

The other two methods (reversible variable-length codes and header extension codes) also provide additional error management capabilities for MPEG-4. Figures 13.87 through 13.90 depict other aspects of MPEG-4.

13.3.6 MPEG-7

The MPEG-7 standard [61], Version 6.0, is dated December 2001. MPEG-7, formally named *Multimedia Content Description Interface,* is a standard for describing the multimedia content data that support some degree of interpretation of the information's meaning, which can be passed onto, or accessed by, a device or a computer code. MPEG-7 is not aimed at any

FIGURE 13.87
*Delivery multimedia
integration framework
(DMIF).*

• DMIF (Delivery Multimedia Integration Framework) is a session protocol
  for the management of multimedia streaming over generic delivery
  technologies.

  - In principle, it is similar to FTP... the only (but essential) difference is that FTP
    returns data. DMIF returns pointers to where to get the (streamed) data.

  - When FTP is run, the very first action it performs is the setup of a session with
    the remote side...later, files are selected and FTP sends a request to
    download them, the FTP peer will return the files in a separate connection.

  - Similarly, when DMIF is run, the very first action it performs is the setup of a
    session with the remote side. Later, streams are selected and DMIF sends
    a request to stream them, the DMIF peer will return the pointers to the
    connections where the streams will be streamed (and establishes the
    connection themselves).

• Conceptually, a "real" remote application accessed through a network,
  (e.g., IP or ATM-based), is no different than an emulated remote producer
  application getting content from a broadcast source or from a disk.

  - However, in the former case, the messages exchanged between the two entities
    have to be normatively defined to ensure interoperability (these are the DMIF
    Signaling messages), while in the latter case the interfaces between the two
    DMIF peers and the emulated Remote Application are internal to a single
    implementation and need not be considered in the specification.

FIGURE 13.88
*An additional view of
DMIF.*

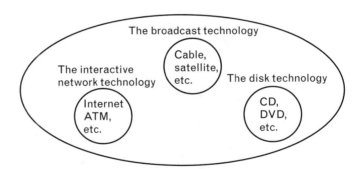

one application in particular; rather, the elements that MPEG–7 standardizes support as broad a range of applications as possible.

Accessing audio and video used to be a simple matter—simple
because of the simplicity of the access mechanisms and because of the
poverty of the sources. A large amount of audiovisual information is
becoming available in digital form, in digital archives, on the World Wide
Web, in broadcast data streams, and in personal and professional databases,
and this amount is only continuing to grow. The value of information
often depends on how easy it can be found, retrieved, accessed, filtered,
and managed.

FIGURE 13.89
DMIF communication architecture.

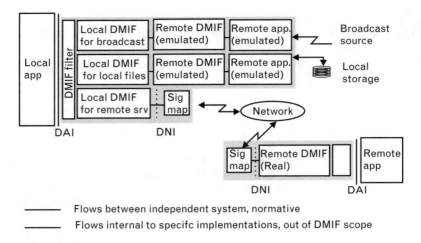

——— Flows between independent system, normative

——— Flows internal to specifc implementations, out of DMIF scope

FIGURE 13.90
Intellectual property.

- MPEG-4 provides mechanisms for protection of intellectual property rights (IPR)

- Achieved by supplementing the coded media objects with an optional Intellectual Property Identification (IPI) data set, carrying information about the contents, type of content, and (pointers to) rights holders

- The data set, if present, is part of an elementary stream descriptor that describes the streaming data associated with a media object

- The number of data sets to be associated with each media object is flexible; different media objects can share the same data sets or have separate data sets

- The provision of the data sets allows the implementation of mechanisms for audit trail, monitoring, billing, and copy protection

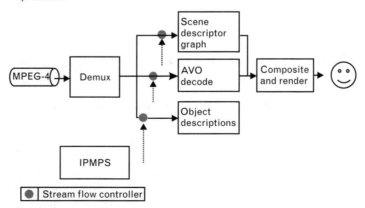

The transition between the second and third millennium abounds with new ways to produce, offer, filter, search, and manage digitized multimedia information. Broadband is being offered with increasing audio and video quality and speed of access. The trend is clear: In the next few years, users will be confronted with such a large amount of content provided by multiple sources that efficient and accurate access to this almost infinite amount of content seems unimaginable today. In spite of the fact that users have increasing access to these resources, identifying and managing them efficiently is becoming more difficult, because of the sheer volume. This applies to professional as well as end users. The question of identifying and managing content is not just restricted to database retrieval applications such as digital libraries, but extends to areas such as broadcast channel selection, multimedia editing, and multimedia directory services. This challenging situation demands a timely solution to the problem. MPEG-7 is the answer to this need.

The MPEG-7 standard provides a rich set of standardized tools to describe multimedia content. Both human users and automatic systems that process audiovisual information are within the scope of MPEG-7. MPEG-7 offers a comprehensive set of audiovisual description tools (the metadata elements and their structure and relationships, that are defined by the standard in the form of descriptors and description schemes) to create descriptions (i.e., a set of instantiated description schemes and their corresponding descriptors at the users will) that will form the basis for applications enabling the needed effective and efficient access (search, filtering, and browsing) to multimedia content. This is a challenging task given the broad spectrum of requirements and targeted multimedia applications, and the broad number of audiovisual features of importance in such context.

MPEG-7 has been developed by experts representing broadcasters, electronics manufacturers, content creators and managers, publishers and intellectual property rights managers, telecommunication service providers, and academia. More information about MPEG-7 can be found at the MPEG-7 Web site (http://mpeg.cselt.it) and the MPEG-7 Alliance Web site (http://www.mpeg-industry.com). These Web pages contain links to a wealth of information about MPEG, including many publicly available documents, several lists of frequently asked questions, and links to other MPEG-7 Web pages.

13.3.6.1 Context of MPEG-7

More and more audiovisual information is available from many sources around the world. The information may be represented in various forms of media, such as still pictures, graphics, 3D models, audio, speech, or video. Audiovisual information plays an important role in our society, regardless of whether it is recorded in such media as film or magnetic tape or originates in real time from some audio or visual sensors and regardless of whether it is analog or, increasingly, digital. While audio and visual information used to be consumed directly by the human being, in an increasing number of cases, the audiovisual information is created, exchanged, retrieved, and reused by computational systems. This may be the case for such scenarios as image understanding (surveillance, intelligent vision, smart cameras, and so on) and media conversion (speech to text, picture to speech, speech to picture, and so on). Other scenarios are information retrieval (quickly and efficiently searching for various types of multimedia documents of interest to the user) and filtering in a stream of audiovisual content description (to receive only those multimedia data items that satisfy the user's preferences). For example, a code in a television program triggers a suitably programmed personal video recorder to record that program, or an image sensor triggers an alarm when a certain visual event happens. Automatic transcoding may be performed from a string of characters to audible information or a search may be performed in a stream of audio or video data. In all of these examples, the audiovisual information has been suitably "encoded" to enable a device or a computer code to take some action.

Audiovisual sources will play an increasingly pervasive role in our lives, and there will be a growing need to have these sources processed further. This makes it necessary to develop forms of audiovisual information representation that go beyond the simple waveform or sample-based, compression-based (such as MPEG-1 and MPEG-2) or even object-based (such as MPEG-4) representations. Forms of representation that allow some degree of interpretation of the information's meaning are necessary. These forms can be passed onto, or accessed by, a device or a computer code. In the examples given earlier, an image sensor may produce visual data not in the form of PCM samples (pixels values) but in the form of objects with associated physical measures and time information. These could then be stored and processed to verify whether certain programmed conditions are met. A PVR could receive descriptions of the audiovisual information associated with a program that would enable it to record, for example, only news with sports being excluded. Products from a company could be described in such a way that a machine could respond to unstructured queries from customers making inquiries.

MPEG-7 is a standard for describing the multimedia content data that will support these operational requirements. The requirements apply, in principle, to both real-time and nonreal-time as well as push and pull applications. MPEG-7 does not standardize or evaluate applications. In the development of the MPEG-7 standard, applications have been used for understanding the requirements and evaluation of technology. It must be made clear that the requirements are derived from analyzing a wide range of potential applications that could use MPEG-7 descriptions. MPEG-7 is not aimed at any one application in particular; rather, the elements that MPEG-7 standardizes support as broad a range of applications as possible.

13.3.6.2 MPEG-7 Objectives

In October 1996, MPEG started a new work item to provide a solution to the questions posed. The new member of the MPEG family, MPEG-7, provides standardized core technologies allowing description of audiovisual data content in multimedia environments. It extends the limited capabilities of proprietary solutions in identifying content that exists today, notably by including more data types.

As mentioned, audiovisual data content that has MPEG-7 data associated with it may include still pictures, graphics, 3D models, audio, speech, video, and composition information about how these elements are combined in a multimedia presentation (scenarios). A special case of these general data types is facial characteristics. MPEG-7 description tools do not, however, depend on the ways the described content is coded or stored. It is possible to create an MPEG-7 description of an analog movie or of a picture that is printed on paper in the same way as is done with digitized content.

MPEG-7, like the other members of the MPEG family, is a standard representation of audiovisual information satisfying particular requirements. The MPEG-7 standard builds on other (standard) representations such as analog, PCM, and MPEG-1, -2, and -4. One functionality of the MPEG-7 standard is to provide references to suitable portions of them. For example, perhaps a shape descriptor used in MPEG-4 is useful in an MPEG-7 context as well, and the same may apply to motion vector fields used in MPEG-1 and MPEG-2.

MPEG-7 allows different granularity in its descriptions, offering the possibility to have different levels of discrimination. Even though the MPEG-7 description does not depend on the (coded) representation of the material, MPEG-7 can exploit the advantages provided by MPEG-4 coded content. If the material is encoded using MPEG-4, which provides the means to encode audiovisual material as objects having certain relations in time (synchronization) and space (on the screen for video or in the room for audio), it will be possible to attach descriptions to elements (objects) within the scene, such as audio and visual objects.

Because the descriptive features must be meaningful in the context of the application, they will be different for different user domains and different applications. This implies that the same material can be described using different types of features, tuned to the area of application. To consider the example of visual material, a lower abstraction level would be a description of, for example, shape, size, texture, color, movement (trajectory), and position ("Where in the scene can the object be found?"); and for audio:

key, mood, tempo, tempo changes, and position in sound space. The highest level would give semantic information: "This is a scene with a barking brown dog on the left and a blue ball that falls down on the right, with the sound of passing cars in the background." Intermediate levels of abstraction may also exist.

The level of abstraction is related to the way the features can be extracted: Many low-level features can be extracted in fully automatic ways, whereas high-level features need much more human interaction. Next to having a description of what is depicted in the content, other types of information about the multimedia data must also be included:

The form. An example of the form is the coding format used (e.g., JPEG, MPEG-2) or the overall data size. This information helps in determining whether the material can be read by the user's terminal.

Conditions for accessing the material. This includes links to a registry with intellectual property rights information and price.

Classification. This includes parental rating and classification of content into a number of predefined categories.

Links to other relevant material. This information may help the user speed the search.

The context. In the case of recorded nonfiction content, it is very important to know the occasion of the recording (e.g., Olympic Games, 1996, the 200m men's hurdle final).

In many cases, it is desirable to use textual information for the descriptions. Care was taken, however, to ensure that the usefulness of the descriptions is as independent from the language area as possible. A very clear example where text comes in handy is in giving names of authors, titles, places, and so on. Therefore, MPEG-7 description tools allow us to create descriptions (i.e., a set of instantiated description schemes and their corresponding descriptors at the user's will) of content that may include the following:

- Information describing the creation and production processes of the content (director, title, short feature movie);

- Information related to the usage of the content (copyright pointers, usage history, broadcast schedule);

- Information of the storage features of the content (storage format, encoding);

- Structural information on spatial, temporal or spatiotemporal components of the content (scene cuts, segmentation in regions, region motion tracking);

- Information about low-level features in the content (colors, textures, sound timbres, melody description);

- Conceptual information of the reality captured by the content (objects and events, interactions among objects);

- Information about how to browse the content in an efficient way (summaries, variations, spatial and frequency subbands, and so on);

- Information about collections of objects;

- Information about the interaction of the user with the content (user preferences, usage history).

All of these descriptions are of course coded in an efficient way for searching, filtering, and so on. To accommodate this variety of complementary content descriptions, MPEG-7 approaches the description of content from several viewpoints. The sets of description tools developed on those viewpoints are presented here as separate entities. However, they are interrelated and can be combined in many ways. Depending on the application, some will be present and others can be absent or only partly present.

A description generated using MPEG-7 description tools will be associated with the content itself, to allow fast and efficient searching for, and filtering of, material that is of interest to the user.

MPEG-7 data may be physically located with the associated AV material, in the same data stream or on the same storage system, but the descriptions could also live somewhere else on the globe. When the content and its descriptions are not colocated, mechanisms that link the multimedia material and their MPEG-7 descriptions are needed; these links will have to work in both directions.

MPEG-7 addresses many different applications in many different environments, which means that it needs to provide a flexible and extensible framework for describing audiovisual data. Therefore, MPEG-7 does not define a monolithic system for content description but rather a set of methods and tools for the different viewpoints of the description of audiovisual content. Having this in mind, MPEG-7 is designed to take into account all of the viewpoints under consideration by other leading standards such as TV Anytime, Dublin Core, SMPTE Metadata Dictionary, and EBU P/Meta. These standardization activities are focused on more specific applications or application domains, whereas MPEG-7 has been designed to be as generic as possible. MPEG-7 also uses XML as the language of choice for the textual representation of content description, as XML schema have been the basis for the *Description Definition Language* (DDL) that is used for the syntactic definition of MPEG-7 description tools and for allowing extensibility of description tools (either new MPEG-7 ones or application-specific ones). Considering the popularity of XML, usage of it will facilitate interoperability with other metadata standards in the future. The main elements of MPEG-7's standard are as follows:

- *Description tools,* that is, descriptors (D) that define the syntax and the semantics of each feature (metadata element), and description schemes (DS) that specify the structure and semantics of the relationships between their components, which may be both descriptors and description schemes;

- A *DDL* to define the syntax of the MPEG-7 description tools and to allow the creation of new description schemes and, possibly, descriptors and to allow the extension and modification of existing description schemes;

- *System tools* to support binary coded representation for efficient storage and transmission, transmission mechanisms (both for textual and binary formats), multiplexing of descriptions, synchronization of descriptions with content, management and protection of intellectual property in MPEG-7 descriptions, and so on.

13.3.6.3   Scope of the Standard

MPEG-7 addresses applications that can be stored (on-line or off-line) or streamed (e.g., broadcast, push models on the Internet) and can operate in both real-time and nonreal-time environments. A "real-time" environment in this context means that the description is generated while the content is being captured.

Figure 13.91 shows an abstract block diagram of a possible MPEG-7 processing chain, included here to explain the scope of the MPEG-7 standard. This chain includes feature extraction (analysis), the description itself, and the search engine (application). To fully exploit the possibilities of MPEG-7 descriptions, automatic extraction of features (or descriptors) will be extremely useful. It is also clear that automatic extraction is not always possible, however. As noted earlier, the higher the level of abstraction, the more difficult automatic extraction is, and interactive extraction tools will be put to good use. However, as useful as they are, neither automatic nor semiautomatic feature extraction algorithms are inside the scope of the standard. The main reason is that their standardization is not required to allow interoperability, which leaves space for industry competition. Another reason not to standardize analysis is to allow for the ability to make expected improvements in these technical areas. The search engines, filter agents, and any other program that can make use of the description are also

FIGURE 13.91
Scope of MPEG-7.

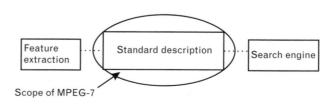

Scope of MPEG-7

not specified within the scope of MPEG-7; again this is not necessary, and here too, competition will produce the best results.

To provide a better understanding of the terminology introduced above (i.e., descriptor, description scheme, and DDL), please refer to Figures 13.92 and 13.93. Figure 13.92 shows the relationship among the different MPEG-7 elements introduced here. The DDL allows the definition of the MPEG-7 description tools, both descriptors and description schemes, providing the means for structuring the Ds into DSs. The DDL also allows for the extension to specific applications of particular DSs. The

FIGURE 13.92
MPEG-7 main elements.

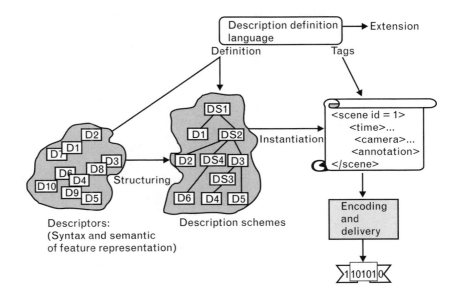

FIGURE 13.93
Abstract representation of possible applications using MPEG-7.

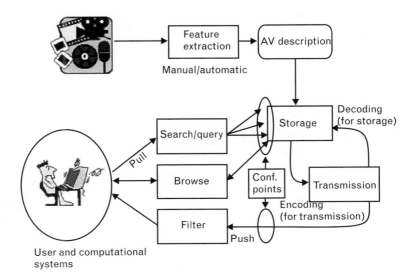

description tools are instantiated as descriptions in textual format (XML) thanks to the DDL (which is based on XML schema). A binary format of descriptions is obtained by means of the BInary format for MPEG-7 data (BiM) defined in the systems part.

Figure 13.93 explains a hypothetical MPEG-7 chain in practice [62]. From the multimedia content an audiovisual description is obtained via manual or semiautomatic extraction. The AV description may be stored (as depicted in the figure) or streamed directly. If we consider a pull scenario, client applications will submit queries to the descriptions repository and will receive a set of descriptions matching the query for browsing (just for inspecting the description, for manipulating it, for retrieving the described content, and so on). In a push scenario, a filter (e.g., an intelligent agent) will select descriptions from the available ones and perform the programmed actions afterward (e.g., switching a broadcast channel or recording the described stream). In both scenarios, all modules may handle descriptions coded in MPEG-7 formats (either textual or binary), but only at the indicated conformance points is it required to be MPEG-7 conformant (because they show the interfaces between an application acting as information server and information consumer).

The emphasis of MPEG-7 is on the provision of novel solutions for audiovisual content description. Thus, addressing text-only documents was not among the goals of MPEG-7. However, audiovisual content may include or refer to text in addition to its audiovisual information. MPEG-7 therefore has standardized different description tools for textual annotation and controlled vocabularies (see later discussion), taking into account existing standards and practices.

Besides the descriptors themselves, the database structure plays a crucial role in the final retrieval's performance. To allow the desired fast judgment about whether the material is of interest, the indexing information will have to be structured, for example, in a hierarchical or associative way.

13.3.6.4   MPEG-7 Application's Areas

The elements that MPEG-7 standardizes provide support to a broad range of applications (for example, multimedia digital libraries, broadcast media selection, multimedia editing, home entertainment devices, and so on). MPEG-7 will also make the Web as searchable for multimedia content as it is searchable for text today. This would apply especially to large content archives, which are being made accessible to the public, as well as to multimedia catalogs enabling people to identify content for purchase. The information used for content retrieval may also be used by *agents*, for the selection and filtering of broadcasted "push" material, or for personalized advertising. Additionally, MPEG-7 descriptions will allow fast and cost-effective use of the underlying data, by enabling semiautomatic multi-media presentation and editing.

All applications' domains making use of multimedia will benefit from MPEG-7. Considering that at the present day it is hard to find someone who is not using multimedia, please extend the list of the examples below using your imagination:

- Architecture, real estate, and interior design (e.g., searching for ideas);
- Broadcast media selection (e.g., radio channel, TV channel);
- Cultural services (history museums, art galleries, and so on);
- Digital libraries (e.g., image catalog, musical dictionary, biomedical imaging catalogs, and film, video and radio archives);
- E-commerce (e.g., personalized advertising, on-line catalogs, directories of e-shops);
- Education (e.g., repositories of multimedia courses, multimedia search for support material);
- Home entertainment (e.g., systems for the management of personal multimedia collections, including manipulation of content; home video editing; searching a game; karaoke);
- Investigation services (e.g., human characteristics recognition, forensics);
- Journalism (e.g., searching speeches of a certain politician using his name, his voice, or his face);
- Multimedia directory services (e.g., Yellow Pages, tourist information, geographical information systems);
- Multimedia editing (e.g., personalized electronic news service, media authoring);
- Remote sensing (e.g., cartography, ecology, natural resources management);
- Shopping (e.g., searching for clothes that you like);
- Social (e.g., dating services);
- Surveillance (e.g., traffic control, surface transportation, nondestructive testing in hostile environments).

The way MPEG-7 data will be used to answer user queries or filtering operations is outside the scope of the standard. The type of content and the query do not have to be the same; for example, visual material may be queried and filtered using visual content, music, speech, and so on. It is the responsibility of the search engine and filter agent to match the query data to the MPEG-7 description. Consider these few query examples:

- Play a few notes on a keyboard and retrieve a list of musical pieces similar to the required tune, or images matching the notes in a certain way, for example, in terms of emotions.

- Draw a few lines on a screen and find a set of images containing similar graphics, logos, ideograms, and so on.

- Define objects, including color patches or textures, and retrieve examples among which you select the interesting objects to compose your design.

- On a given set of multimedia objects, describe movements and relations between objects and then search for animations fulfilling the described temporal and spatial relations.

  - Describe actions and get a list of scenarios containing such actions.

- Using an excerpt of Pavarotti's voice, obtain a list of Pavarotti's records, video clips where Pavarotti is singing, and photographic material portraying Pavarotti.

13.3.6.5 Method of Work and Development Schedule

The method of development has been comparable to that of the previous MPEG standards. MPEG work is usually carried out in three stages: definition, competition, and collaboration. In the definition phase, the scope, objectives, and requirements for MPEG-7 were defined. In the competitive stage, participants worked on their technology by themselves. The end of this stage was marked by the MPEG-7 evaluation following an open *call for proposals* (CfP). The CfP asked for relevant technology fitting the requirements. In answer to the CfP, all interested parties, regardless of whether they participate or have participated in MPEG, were invited to submit their technology to MPEG. Some 60 parties submitted, in total, almost 400 proposals, after which MPEG made a fair expert comparison between these submissions.

Selected elements of different proposals were incorporated into a common model (the eXperimentation Model, or XM) during the collaborative phase of the standard with the goal of building the best possible model, which was in essence a draft of the standard itself. During the collaborative phase, the XM was updated and improved in an iterative fashion, until MPEG-7 reached the committee draft stage in October 2000, after having advanced through several versions of working drafts. Improvements to the XM were made through *core experiments* (CEs). CEs were defined to test the existing tools against new contributions and proposals, within the framework of the XM, according to well-defined test conditions and criteria. Finally, those parts of the XM (or of the working draft) that corresponded to the normative elements of MPEG-7 were standardized. The MPEG-7 development schedule has been as shown next:

| Call for Proposals | October 1998 |

Call for Proposals   October 1998

Evaluation   February 1999

First version of working draft   December 1999

Committee draft   October 2000

Final committee draft   February 2001

Final draft international standard   July 2001

International standard   September 2001

13.3.6.6   MPEG-7 Parts

The MPEG-7 standard consists of the following parts:

MPEG-7 Systems: the binary format for encoding MPEG-7 descriptions and the terminal architecture;

MPEG-7 Description Definition Language: the language for defining the syntax of the MPEG-7 description tools and for defining new description schemes;

MPEG-7 Visual: the description tools dealing with (only) visual descriptions;

MPEG-7 Audio: the description tools dealing with (only) audio descriptions;

MPEG-7 Multimedia Description Schemes: the description tools dealing with generic features and multimedia descriptions;

MPEG-7 Reference Software: a software implementation of relevant parts of the MPEG-7 standard with normative status;

MPEG-7 Conformance: guidelines and procedures for testing conformance of MPEG-7 implementations (under development at press time);

MPEG-7 Extraction and use of descriptions: informative material (in the form of a technical report) about the extraction and use of some of the description tools (under development at press time).

13.3.7   MPEG-21

13.3.7.1   Introduction [63]

Today, many elements exist to build an infrastructure for the delivery and consumption of multimedia content. There is, however, no "big picture" to describe how these elements, either in existence or under development, relate to each other. The aim for MPEG-21 is to describe how these various elements fit together. Where gaps exist, MPEG-21 will recommend which new standards are required. ISO/IEC JTC 1/SC 29/WG 11 (MPEG) will then develop new standards as appropriate while other relevant standards may be developed by other bodies. These specifications will be integrated into the multimedia framework through collaboration between MPEG and these bodies.

The result is an open framework for multimedia delivery and consumption, with both the content creator and content consumer as focal points. This open framework provides content creators and service providers with equal opportunities in the MPEG-21-enabled open market. This will also benefit the content consumer by providing them with access to a large variety of content in an interoperable manner.

The vision for MPEG-21 is to define a multimedia framework *that can enable transparent and augmented use of multimedia resources across a wide range of networks and devices* used by different communities. For a detailed examination and description of the requirements for the MPEG-21 multimedia framework, readers are advised to refer to the MPEG-21 technical report, "Vision, Technologies and Strategy" [64], the current version of which can be found on the MPEG home page (http://mpeg.telecomitalialab.com).

13.3.7.2   MPEG-21 Multimedia Framework

Currently, multimedia technology provides the different players in the multimedia value and delivery chain (from content creators to end users) with an excess of information and services. Access to information and services from almost anywhere at anytime can be provided with ubiquitous terminals and networks. However, no complete solutions exist that allow different communities, each with their own models, rules, procedures, interests, and content formats, to interact efficiently using this complex infrastructure. Examples of these communities are the content, financial, communication, computer, and consumer electronics sectors and their customers. Developing a common multimedia framework will facilitate cooperation between these sectors and support a more efficient implementation and integration of the different models, rules, procedures, interests, and content formats. This will enable an enhanced user experience.

The multimedia content delivery chain encompasses content creation, production, delivery, and consumption. To support this, the content has to be identified, described, managed, and protected. The transport and delivery of content will occur over a heterogeneous set of terminals and networks within which events will occur and require reporting. Such reporting will include reliable delivery, the management of personal data and preferences taking user privacy into account and the management of financial transactions.

A multimedia framework is required to support this new type of multi-media usage. Such a framework requires that a shared vision, or road map, be understood by its architects, to ensure that the systems that deliver multimedia content are *interoperable* and that transactions are simplified and, if possible, *automated*. This should apply to the infrastructure requirements for content delivery, content security, rights management, secure payment, and the technologies enabling them—and this list is not exhaustive.

The MPEG-21 multimedia framework will identify and define the key elements needed to support the multimedia delivery chain as described earlier and the relationships between them and operations supported by them. Within the parts of MPEG-21, MPEG will elaborate the elements by defining the syntax and semantics of their characteristics, such as interfaces to the elements. MPEG-21 will also address the necessary framework functionality, such as the protocols associated with the interfaces, and mechanisms to provide a repository, composition, conformance, and so on. The following seven key elements are defined in MPEG-21:

1. *Digital item declaration:* a uniform and flexible abstraction and interoperable schema for declaring digital items;

2. *Digital item identification and description:* a framework for identification and description of any entity regardless of its nature, type, or granularity;

3. *Content handling and usage:* provide interfaces and protocols that enable creation, manipulation, search, access, storage, delivery, and (re)use of content across the content distribution and consumption value chain;

4. *Intellectual property management and protection:* the means to enable content to be persistently and reliably managed and protected across a wide range of networks and devices;

5. *Terminals and networks:* the ability to provide interoperable and transparent access to content across networks and terminals;

6. *Content representation:* how the media resources are represented;

7. *Event reporting:* the metrics and interfaces that enable users to understand precisely the performance of all reportable events within the framework.

MPEG-21 recommendations will be determined by interoperability requirements, and their level of detail may vary for each framework element. The actual instantiation and implementation of the framework elements below the abstraction level required to achieve interoperability will not be specified.

13.3.7.3 MPEG-21 Scope

The scope of MPEG-21 could be described as the integration of the critical technologies enabling transparent and augmented use of multimedia resources across a wide range of networks and devices to support functions such as content creation, content production, content distribution, content consumption and usage, content packaging, intellectual property management and protection, content identification and description, financial management, user privacy, terminals and network resource abstraction, content representation, and event reporting.

From its background in key technology and information management standards related to the management, delivery, and representation of multimedia content, MPEG is well positioned to initiate such an activity. However, it is recognized that the integration of such disparate technologies can only be achieved by working in collaboration with other bodies.

13.3.7.4 User Model

One needs to set out the user requirements in the multimedia framework. A user is any entity that interacts in the MPEG-21 environment or makes use of a digital item. Such users include individuals, consumers, communities, organizations, corporations, consortia, governments, and other standards bodies and initiatives around the world. Users are identified specifically by their relationship to another user for a certain interaction. From a purely technical perspective, MPEG-21 makes no distinction between a "content provider" and a "consumer"—both are users. A single entity may use content in many ways (publish, deliver, consume, and so on), and so all parties interacting within MPEG-21 are categorized as users equally. However, a user may assume specific or even unique rights and responsibilities according to its interaction with other users within MPEG-21.

At its most basic level, MPEG-21 provides a framework in which one user interacts with another user and the object of that interaction is a digital item commonly called *content*. Some such interactions are creating content, providing content, archiving content, rating content, enhancing and delivering content, aggregating content, delivering content, syndicating content, retail selling of content, consuming content, subscribing to content, regulating content, and facilitating and regulating transactions that occur from any of the above. Any of these are "uses" of MPEG-21, and the parties involved are users. The seven MPEG-21 key elements support these transactions, as shown in Figure 13.94.

13.3.7.5 Overview of Digital Items

Within any system (such as MPEG-21) that proposes to facilitate a wide range of actions involving digital items, there is a strong need for a very concrete description that defines exactly what constitutes such an "item." Clearly there are many kinds of content, and probably just as many possible

FIGURE 13.94
*Event reporting, by
creating metrics and
interfaces, further
describes specific
interactions.*

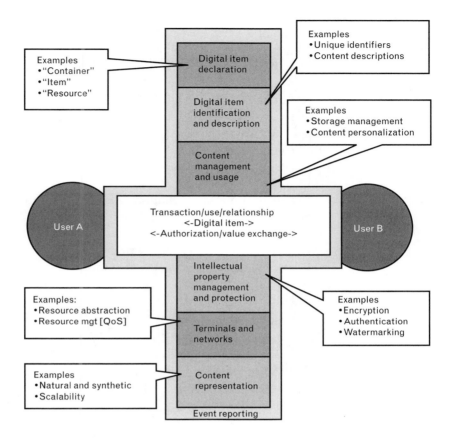

ways of describing it. This presents a strong challenge to lay out a powerful and flexible model for digital items that can accommodate the myriad forms that content can take (and the new forms it will assume in the future). Such a model is truly useful only if it yields a format that can be used to represent any digital items defined within the model unambiguously and communicate them, and information about them, successfully. Consider the example of a simple "Web page" as a digital item. A Web page typically consists of an HTML document with embedded "links" (or dependencies on) to various image files (e.g., JPEGs and GIFs) and possibly some layout information (e.g., style sheets). In this simple case, it is a straightforward exercise to inspect the HTML document and deduce that this digital item consists of the HTML document itself, plus all of the other resources on which it depends.

Now let's modify the example to assume that the "Web page" contains some custom scripted logic (e.g., JavaScript) to determine the preferred language of the viewer (among some predefined set of choices) and to either build/display the page in that language or to revert to a default choice if the preferred translation is not available. The key point in this modified example is that the presence of the language logic clouds the question of exactly what constitutes this digital item now and how this can be unambiguously determined.

The first problem is one of actually determining all of the dependencies. The addition of the scripting code changes the declarative "links" of the simple Web page into links that can be (in the general case) determined only by running the embedded script on a specific platform. This could still work as a method of deducing the structure of the digital item, *assuming* that the author intended each translated "version" of the Web page to be a separate and distinct digital item.

This assumption highlights the second problem: It is ambiguous whether the author actually intends for each translation of the page to be a stand-alone digital item, or whether the intention is for the digital item to consist of the page with the language choice left unresolved. If the latter is the case, it makes it impossible to deduce the *exact* set of resources that this digital item consists of, which leads back to the first problem.

The problem stated above is addressed by the digital item declaration. A *digital item declaration* (DID) is a document that specifies the makeup, structure, and organization of a digital item. Part 2 of MPEG-21 contains the DID specification.

13.3.7.6   MPEG-21 Work Plan

MPEG-21 has established a work plan for standardization. Three parts of standardization within the multimedia framework had already been started at press time (Note that the technical report is Part 1 of the MPEG-21 standard):

Digital Item Declaration (DID—Part 2): This work item was expected to become an international standard in 2002.

Digital Item Identification and Description (DII&D—Part 3): This work item was also expected to become an international standard in 2002.

Intellectual Property Management and Protection (IPMP—Part 4): This work was expected to become an international standard in 2002.

Rights Expression Language (REL—Part 5): This work has now commenced.

Rights Data Dictionary (RDD—Part 6): This work has now commenced.

Digital Item Usage Environment Description (DIUED—Part 7): An initial CfP was launched in 2001.

Table 13.8 provides additional details on timetables.

ISO/IEC TR 21000-1: MPEG-21 Multimedia Framework, Part 1: Vision, Technologies, and Strategy

A technical report has been written to describe the multimedia framework and its architectural elements together with the functional requirements for their specification that was formally approved in September 2001. The title, *Vision, Technologies and Strategy*, has been chosen to reflect the fundamental purpose of the technical report, which is to:

- Define a *vision* for a multimedia framework to enable transparent and augmented use of multimedia resources across a wide range of networks and devices to meet the needs of all users;

- Achieve the integration of components and standards to facilitate harmonization of *technologies* for the creation, management, transport, manipulation, distribution, and consumption of digital items;

- Define a *strategy* for achieving a multimedia framework by the development of specifications and standards based on well-defined functional requirements through collaboration with other bodies.

MPEG-21 Part 2: DID

The purpose of the DID specification is to describe a set of abstract terms and concepts to form a useful model for defining digital items. Within this model, a digital item is the digital representation of "a work" and, as such, it is the thing that is acted on (managed, described, exchanged, collected, and so on) within the model. The goal of this model is to be as flexible and general as possible, while providing for the "hooks" that enable higher level functionality. This, in turn, will allow the model to serve as a key foundation in the building of higher level models in other MPEG-21 elements (such as IPMP). This model specifically does not define a language in and of itself. Instead, the model helps to provide a common set of abstract concepts and terms that can be used to define such a scheme or to perform mappings between existing schemes capable of DID, for comparison purposes. The DID technology is described in three normative sections:

1. *Model*. The DID model describes a set of abstract terms and concepts for defining digital items. A digital item is the digital representation of "a work"; it is the thing that is acted on (managed, described, exchanged, collected, and so on) within the model.

2. *Representation*. Normative description of the syntax and semantics of each of the DID elements, as represented in XML. This section also contains some non-normative examples for illustrative purposes.

3. *Schema*. Normative XML schema comprising the entire grammar of the DID representation in XML.

The following sections describe the semantic "meaning" of the principle elements of the DID model. Please note that in the following descriptions, the defined elements in *italics* are intended to be unambiguous terms within this model.

Container   A *container* is a structure that allows *items* and/or *containers* to be grouped. These groupings of *items* and/or *containers* can be used to form logical *packages* (for transport or exchange) or logical *shelves* (for organization). *Descriptors* allow for the "labeling" of *containers* with information that is appropriate for the purpose of the grouping (e.g., delivery instructions for a *package*, or category information for a *shelf*). Note that a *container* itself is not an *item*; *containers* are groupings of *items* and/or *containers*.

Item   An *item* is a grouping of sub*items* and/or *components* that are bound to relevant *descriptors*. *Descriptors* contain information about the *item*, as a representation of a work. *Items* may contain *choices*, which allow them to be customized or configured. *Items* may be conditional (on *predicates* asserted by *selections* defined in the *choices*). An *item* that contains no subitems can be considered an entity—a logically indivisible work. An *item* that does contain *subitems* can be considered a compilation—a work composed of potentially independent subparts. *Items* may also contain *annotations* to their subparts.

The relationship between *items* and digital items (as defined in ISO/IEC 21000-1:2001, *MPEG-21 Vision, Technologies and Strategy*) could be stated as follows: *items* are declarative representations of digital items.

Component   A *component* is the binding of a *resource* to all of its relevant *descriptors*. These *descriptors* are information related to all or part of the specific *resource* instance. Such *descriptors* will typically contain control or structural information about the *resource* (such as bit rate, character set, start points, or encryption information) but not information describing the "content" within. Note that a *component* itself is not an *item*; *components* are building blocks of *items*.

Anchor   An *anchor* binds *descriptors* to a *fragment,* which corresponds to a specific location or range within a *resource.*

Descriptor   A *descriptor* associates information with the enclosing element. This information may be a *component* (such as a thumbnail of an image, or a text *component*) or a textual *statement*.

Condition   A *condition* describes the enclosing element as being optional, and links it to the *selection*(s) that affect its inclusion. Multiple *predicates* within a *condition* are combined as a conjunction (an AND relationship).

Any *predicate* can be negated within a *condition*. Multiple *conditions* associated with a given element are combined as a disjunction (an OR relationship) when determining whether to include the element.

Choice   A *choice* describes a set of related *selections* that can affect the configuration of an *item*. The *selections* within a *choice* are either exclusive (choose exactly one) or inclusive (choose any number, including all or none).

Selection   A *selection* describes a specific decision that will affect one or more *conditions* somewhere within an *item*. If the *selection* is chosen, its predicate becomes true; if it is not chosen, its *predicate* becomes false; if it is left unresolved, its *predicate* is undecided.

Annotation   An *annotation* describes a set of information about another identified element of the model without altering or adding to that element. The information can take the form of *assertions*, *descriptors*, and *anchors*.

Assertion   An *assertion* defines a full or partially configured state of a *choice* by asserting true, false or undecided values for some number of *predicates* associated with the *selections* for that *choice*.

Resource   A *resource* is an individually identifiable asset such as a video or audio clip, an image, or a textual asset. A *resource* may also potentially be a physical object. All *resources* must be locatable via an unambiguous address.

Fragment   A *fragment* unambiguously designates a specific point or range within a *resource*. *Fragment* may be *resource*-type specific.

Statement   A *statement* is a literal textual value that contains information, but not an asset. Examples of likely *statements* include descriptive, control, revision tracking, or identifying information.

Predicate   A *predicate* is an unambiguously identifiable declaration that can be true, false, or undecided. Figure 13.95 is an example showing the most important elements within this model, how they are related, and, as such, the hierarchical structure of the DID model.

MPEG-21 Part 3: Digital Item Identification and Description
The scope of the digital item identification and description (DII&D) specification includes the following:

- Determines how to identify uniquely and describe digital items (and parts thereof) and other entities.

Figure 13.95
*Relationship of the
principle elements
within the DID
model.*

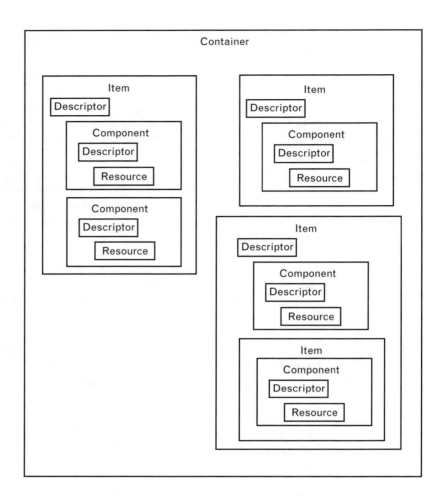

- The relationship between digital items (and parts thereof) and existing identification systems contains a list of relevant identification systems. This is not an exhaustive list and is subject to change over time.

- The relationship between digital items (and parts thereof) and relevant description schemes contains a list of relevant description schemes. This is not an exhaustive list and is subject to change over time.

The DII&D specification does not specify the following:

- New identification systems for the content elements for which identification and description schemes already exist and are in use (e.g., ISO/IEC 21000-3 does not attempt to replace the ISRC, as defined in ISO 3901, for sound recordings);

- Normative description schemes for describing content (apart from the identification DS).

Descriptive metadata and identifiers covered by this specification can be associated with digital items by including them in a specific place in the DID. This place is the statement element. Examples of likely statements include descriptive, control, revision tracking, and/or identifying information. Figure 13.96 shows this relationship. The shaded boxes are the subject of the DII&D specification, whereas the other boxes are defined in the DID specification.

Digital items and their parts within the MPEG-21 framework are identified by encapsulating uniform resource identifiers into the identification DS. A *uniform resource identifier* (URI) is a compact string of characters for identifying an abstract or physical resource, where a resource is defined as "anything that has identity."

The requirement that an MPEG-21 digital item identifier be a URI is also consistent with the statement that the MPEG-21 identifier may be a *uniform resource locator* (URL). The acronym URL refers to a specific subset of URI that is in use today as pointers to information on the Internet; it

FIGURE 13.96
Relationship between DID and DII&D.

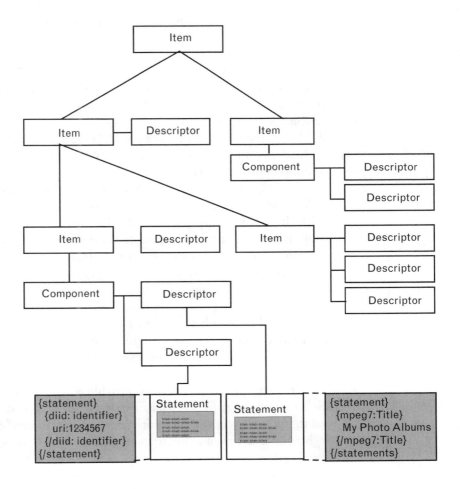

allows for long-term to short-term persistence depending on the business case.

In some cases, it may be necessary to use an automated resolution system to retrieve the digital item (or parts thereof) or information related to a digital item from a server (e.g., in the case of an interactive on-line content delivery system).

MPEG-21 Part 4: Intellectual Property Management and Protection

The fourth part of MPEG-21 defines an interoperable framework for *intellectual property management and protection* (IPMP). Fairly soon after MPEG-4, with its IPMP hooks, became an international standard, concerns were voiced within MPEG that many similar devices and players might be built by different manufacturers, all MPEG-4, but many of them not interworking. This is why MPEG decided to start a new project on more interoperable IPMP. The project includes standardized ways of retrieving IPMP tools from remote locations and exchanging messages between IPMP tools and between these tools and the terminal. It also addresses authentication of IPMP tools, and has provisions for integrating rights expressions according to the rights data dictionary and the rights expression language.

MPEG-21 Part 5: Rights Expression Language

Following an extensive requirements gathering process that started in January 2001, MPEG issued a CfP during its July meeting in Sydney for a *rights data dictionary* (RDD) and a *rights expression language* (REL). Responses to this CfP were processed during the December meeting in Pattaya and the evaluation process established an approach for going forward with the development of a specification.

An REL is seen as a machine-readable language that can declare rights and permissions using the terms as defined in the RDD. An REL is intended to provide flexible, interoperable mechanisms to support transparent and augmented use of digital resources in publishing, distributing, and consuming of electronic books, broadcasting, digital movies, digital music, interactive games, computer software, and other creations in digital form, in a way that protects digital content and honors the rights, conditions, and fees specified for digital contents. It is also intended to support specification of access and use controls for digital content in cases where financial exchange is not part of the terms of use, and to support exchange of sensitive or private digital content.

The REL is also intended to provide a flexible interoperable mechanism to ensure that personal data are processed in accordance with individual rights and to meet the requirement for users to be able to express their rights and interests in a way that addresses issues of privacy and use of personal data.

A standard REL should be able to support guaranteed end-to-end interoperability, consistency, and reliability between different systems and services. To do so, it must offer richness and extensibility in declaring rights, conditions, and obligations; ease and persistence in identifying and associating these with digital contents; and flexibility in supporting multiple usage/business models.

MPEG-21 Part 6: RDD

Following the evaluation of submissions in response to a CfP, the specification of a RDD commenced in December 2001. The following points summarize the proposed scope of this specification:

1. RDD provides a set of clear, consistent, structured, and integrated definitions of terms for use in the MPEG-21 REL.

2. Terms in RDD are categorized as *Primitive, Native, Adopted,* and *Mapped*. The definitions of *Primitive and Native* terms are determined by the governance process of the RDD. Definitions of *Adopted* and *Mapped* terms are determined externally.

3. RDD is a semantic network through which the definitions of terms are developed through the medium of its primary data model (the Context Model) supported by two secondary models (the Resource Model, and the Ascriptive Model).

4. RDD terms are drawn from a continually expanding and diverse range of governed descriptive, legal, and commercial metadata systems and schemes, supporting the description of rights and permissions in digital items, physical objects, and abstract entities, incorporated within MPEG standards as well as those defined and governed elsewhere.

5. Terms will be added to the RDD or modified in accordance with its declared governance process.

6. RDD supports interoperability, so that metadata necessary for the management of rights and permissions can cross in and out of domains in an automated or partially automated way with the minimum ambiguity or loss of semantic integrity.

7. *Primitive, Native,* and *Adopted* terms within RDD do not define intellectual property rights or other legal entities. RDD *Primitive, Native,* and *Adopted* terminology implies no assumptions about the nature or extent of specific legal rights, the commerce (or other) models through which rights may be exploited or protected, or the legal frameworks within which they operate.

8. RDD includes the terms from all metadata schemes and systems that have been mapped to it.

Digital Item Adaptation

The goal of the terminals and networks key element is to achieve interoperable transparent access to distributed advanced multimedia content by shielding users from network and terminal installation, management, and implementation issues. This will enable the provision of network and terminal resources on demand to form user communities where multimedia content can be created and shared, always with the agreed/contracted quality, reliability, and flexibility, allowing the multimedia applications to connect diverse sets of users, such that the quality of the user experience will be guaranteed.

Toward this goal, the adaptation of digital items is required. This concept is illustrated in Figure 13.97. As shown in this conceptual architecture, a digital item is subject to a resource adaptation engine, as well as a descriptor adaptation engine, which together produce the adapted digital item.

It is important to emphasize that the adaptation engines themselves are non-normative tools of digital item adaptation. However, descriptions and format-independent mechanisms that provide support for digital item adaptation in terms of resource adaptation, descriptor adaptation, and/or QoS management are within the scope of the requirements.

A preliminary CfP was issued in 2001. The work is anticipated to result in one or more new parts for the MPEG-21 standard.

13.3.7.7   Proposals and Recommendations for Further Work

The following recommendations for WG11 standardization activities with respect to the MPEG-21 multimedia framework are proposed.

FIGURE 13.97
Concept of digital item adaptation.

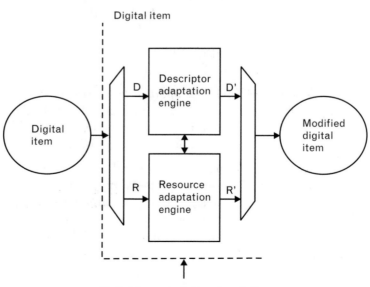

Digital item adaptation descriptions

Persistent Association of Information with Digital Items

As a logical extension to the ongoing specification of the DID and DII&D, MPEG intends to consider the requirements for the persistent association of information with content. MPEG experts are currently defining the functional requirements for the persistent association of information with content and how this interacts with the rest of the MPEG-21 technical architecture.

The term *persistent association* is used to categorize all of the techniques for managing identifiers with content. This will include the carriage of identifiers within the context of different content file formats, including file headers, and embedded into content as a watermark. It also encompasses the ability for identifiers associated with content to be protected against their unauthorized removal and modification.

Content Handling and Usage

MPEG-21 needs to define interfaces and protocols for search, storage, and management of digital items and descriptions that enable and support:

1. User(s) to express their preferences;
2. User(s) to locate relevant content in the network, device, and so on;
3. The integration and interoperability of different asset management systems;
4. Content lifetime management and associated configurable policies;
5. Tracking of changes to digital items and descriptions;
6. User(s) to identify where all copies of content they own are located with associated usage restrictions;
7. Definition of interfaces and protocols for user profile management and metrics that enable and support;
8. Creating, modifying, and managing user profiles;
9. Creating, tracking, and packaging of content usage metrics information;
10. Interchange formats for user profiles with other systems;
11. Definition of interfaces and protocols to bring the benefits of intelligent agents within the framework;
12. To operate, intelligent agents need a representation of the user's self (user profile), a knowledge about the specific domain (an ontology), and a standard language that allows the nonhuman entities to entertain a dialogue with other nonhuman entities (which will again possess knowledge about the humans they represent and a shared ontology) to achieve the goal that has been set;
13. To allow that a single language (Agent Communication Language) be defined. Therefore, a standardized representation of user information will be needed. Ontologies for the different domains

will also need to be referenced, when available, and their develop-ment stimulated when not available. In some specific cases, WG11 may need to develop specific ontologies itself.

Terminals and Networks

To achieve interoperable transparent access to distributed digital items by shielding users from network and terminal installation, management, and implementation issues, MPEG-21 should standardize the following:

- APIs and associated protocols (behavior) for terminal QoS management;

- NPIs and associated protocols (behavior) for network QoS management;

- APIs and associated protocols (behavior) for joint terminal and net-work QoS management;

- Rules for QoS contract negotiation and implementation;

- APIs enabling QoS agent technologies.

Content Representation

The goal of the content representation item is to provide, adopt, or inte-grate CR technologies able to efficiently represent MPEG-21 content in a scalable and error-resilient way. The content representation of the media resources will need to be synchronizable and multiplexed and allow interaction.

Event Reporting

MPEG-21 event reporting should standardize metrics and interfaces for performance of all reportable events in MPEG-21 and provide a means of capturing and containing these metrics and interfaces that refer to identi-fied digital items, environments, processes, transactions, and users.

Such metrics and interfaces will enable users to understand precisely the performance of all reportable events within the framework. Event reporting must provide users with a means of acting on specific interac-tions, as well as enabling a vast set of out-of-scope processes, frameworks, and models to interoperate with MPEG-21.

ENDNOTES

[1]   Portions of this material were synthesized from the numerous articles, books, and pa-pers the author has published in the recent past on the topic of digital video.

[2]   Reimers, U., "Digital Video Broadcasting," *IEEE Communications Magazine,* June 1998, pp. 104ff.

[3]   A wealth of information is available on MPEG at the official MPEG Web site, http://www.cselt.it/mpeg.

[4] Farber, N., and J. Villasenor, "Extensions of ITU-T Recommendations for Error-Resilient Video Transmission," *IEEE Communications Magazine,* June 1998, pp. 120ff.

[5] The standards are formally documented by the Advanced TV Standards Committee.

[6] Thorpe, L. J., "Digital Television Is Coming," *TV Broadcast,* March 1997, pp. 26ff.

[7] Bhatt, B., D. Birks, and D. Hermreck, "Digital Television: Making It Work," *IEEE Spectrum,* October 1997, pp. 19ff.

[8] Minoli, D., "Supporting Multimedia and Other Evolving Applications Using Broadband," in *Annual Review of Communications,* Vol. 50, Chicago: International Engineering Consortium, 1997, pp. 729ff.

[9] *Desktop Video Communications,* July/August 1997, p. 45.

[10] Minoli, D., and B. Keinath, *Distributed Multimedia Through Broadband Communication Services,* Norwood, MA: Artech House, 1994.

[11] Borthic, S., "Turning Up the Heat at DVC West," *Desktop Video Communications,* July/August 1997, pp. 7ff.

[12] Davis, A., "Cameras for Desktop Videoconferencing," *Desktop Video Communications,* March/April 1997, pp. 18ff.

[13] Minoli, D., "Digital Video," in *The Telecommunications Handbook*, K. Terplan and P. Morreale, (eds.), New York: IEEE Press, 2000.

[14] Minoli, D., "Digital Video Compression: Getting Images Across a Net," *Network Computing,* July 1993, pp. 146ff.

[15] Minoli, D., "Distributed Multimedia, Bringing the Infrastructure Up to the Challenge," *WAN Connections/Communications Week,* August 1993, pp. 60ff.

[16] Minoli, D., *Multimedia: Opportunities for Carriers and Service Providers,* Market Report, Probe Research Corporation, June 1993.

[17] Minoli, D., *Distance Learning Applications,* Broadband Networking, DataPro Report 1015BBN, November 1993.

[18] Minoli, D., "Concocting a Recipe for the Right Multimedia Mix," *Network World,* September 12, 1994, pp. L10.

[19] Minoli, D., *Imaging Communications,* Broadband Networking, DataPro, Report, June 1994.

[20] Minoli, D., *Videoconferencing,* Broadband Networking, DataPro Report, April 1994.

[21] Minoli, D., "Designing Scalable Networks," *Network World Collaboration,* January 10, 1994, pp. 17ff; Minoli, D., *Communications-Based Imaging,* Market Report, Probe Research Corporation, September 1994.

[22] Minoli, D., *An Assessment of Digital Video and Video Dialtone Technology, Regulation, Services, and Competitive Markets,* Market Report on Convergence Strategies & Technologies, DataPro, April 1995.

[23] Minoli, D., "1995: The Year of Video in Enterprise Nets," *Network World,* December 5, 1994, p. 21.

[24] Minoli, D., *Video Dialtone (VDT): Overview,* Market Report 1090CNS, DataPro, May 1995.

[25] Minoli, D., *Video Compression Schemes,* Market Report on Convergence Strategies & Technologies, DataPro, May 1995.

[26] For "near-real-time" broadcast-type applications, it is reasonable to have playout buffering on the order of hundreds of milliseconds if there is no concern about the turn-around time. For corporate communications that are predominately one way

("addresses to the masses"), this works just fine, and practically it may not require the added expense of premium network transport

[27]   Minoli, D., *Video Dialtone Technology, Approaches, and Services: Digital Video over ADSL, HFC, FTTC, and ATM,* New York: McGraw-Hill, 1995.

[28]   These observations relate to the campus/building network. This discussion does not imply that ATM to the desktop is superior to other technologies, such as 100-Mbps Ethernet, only that bandwidth and QoS (e.g., frame delay, frame delay variation, frame loss) must be accounted for.

[29]   Pensinger, G., "Sarnoff Assumes Role in Broadcasting's 'Third Revolution,'" *TV Broadcast,* June 1995, pp. 19ff.

[30]   Eggers, R., "On-Line, Off-Line Editing Systems Converging," *TV Broadcast,* June 1995, pp. 19ff.

[31]   Van Pelt, J., "Objective Testing for Video Compression," *TV Broadcast,* June 1995, p. 86.

[32]   This is a cooperative committee of manufacturers and broadcasters that promotes the DTV standard and certifies the new equipment.

[33]   See http://www.atsc.org for ATSC Standard Documents A/53 for video and A/52 for audio.

[34]   These terms are discussed later in the chapter.

[35]   Minoli, D., "Packetized Speech Networks, Part 1: Overview," *Australian Electronics Engineer,* April 1979, pp. 38–52.

[36]   Minoli, D., "Packetized Speech Networks, Part 2: Queuing Model," *Australian Electronics Engineer,* July 1979, pp. 68–76.

[37]   Minoli, D., "Packetized Speech Network, Part 3: Delay Behavior and Performance Characteristics," *Australian Electronics Engineer,* August 1979, pp. 59–68.

[38]   Minoli, D., and E. Minoli, *Voice over Packet Networks,* New York: Wiley, 1998.

[39]   Picturetel promotional literature.

[40]   Shortly after the original version 1 of H.263 was formally adopted by the ITU-T in March 1996, the ITU-T launched a new effort to further improve H.263 without changing the basic concept of block-based motion compensation. This effort resulted in version 2 of the H.263 video coding standard, which is informally known as H.263+ and was adopted by the ITU-T in February 1998. It contains a number of optional feature enhancements that have been added to the already existing options in an upward-compatible way.

[41]   H.223, *Multiplexing Protocol for Low Bit Rate Multimedia Communication,* was adopted in 1996. H.245, *Control Protocol for Multimedia Communication,* was also adopted in 1996.

[42]   Although the recommendation specifies the use of a V.34 modem, extension to other transmission systems is doable.

[43]   Retrieved from http://www.cs.columbia.edu/sip.

[44]   Doyle, B., "Crunch Time for Digital Video," *Newmedia,* March 1994, pp. 47ff.

[45]   Lockwood, L. W., "MPEG-2: A Wide Ranging Standard," *Communications Technology,* October 1993, pp. 16ff.

[46]   Barezzani, M., et al,, "Compression Codecs for Contribution Applications," *Electrical Communication,* 3rd Quarter, 1993, pp. 220ff.

[47]   Wechselberger, T., "Conditional Access and Encryption Options for Digital Systems," *Communications Technology,* November 1993, pp. 20ff.

[48]   Eldib, O., and D., Minoli, *Telecommuting,* Norwood, MA: Artech House, 1995.

[49]   Walker, P. E., "Squeezing the Picture: Video Compression," *Broadcast Engineer,* February 1994, pp. 54ff.

[50]   Talluri, R., "Error-Resilient Video Coding in the ISO MPEG-4 Standard," *IEEE Communications Magazine,* June 1998, pp. 112ff.

[51]   *Multimedia Week,* March 14, 1994, p. 8.

[52]   A source image/frame contains three image components (also called colors, spectral bands, or channels), for example, RGB. Each component consists of an array of samples, to which the DCT can be applied in turn. A sample is expressed as an integer with precision P bits with values in the range $[0, 2^P - 1]$; all samples of all components within the same source image/frame must have the same precision (P can be 8 or 12), but image components may be sampled at different rates compared to each others.

[53]   Wallace, G. K., "The JPEG Still Picture Compression Standard," *Communications of the ACM,* Vol. 34, No. 4, April 1991, pp. 30ff.

[54]   MacInnis, A. G., "The MPEG Systems Coding Specification," *Signal Processing: Image Communication,* Vol. 4, 1992, pp. 153–159.

[55]   ISO/IEC JTC1/SC29/WG11 CD1-11172, Coding of Moving Pictures and Associated Audio for Digital Storage Media at up to 1.5 Mbits/s, Part 1 (Systems), November 1991.

[56]   B-frames are storage areas where the incoming frame is compared to the preceding frame and used to predict the next one. This, however, requires that the decoder have more than two frames of video memory, thus more than doubling the memory requirement from 1 to 2 MB.

[57]   Editorial note: The acronym SNR is not defined even in the standard, which says: "SNR scalability: A type of scalability where the enhancement layer(s) contain only coded refinement data for the DTC coefficients of the base layer."

[58]   *Video Technology News,* December 20, 1993, p. 8.

[59]   One advantage of this packetized approach, besides efficient transport (e.g., over ATM), is the fact that the packets can be encrypted in support of conditional access.

[60]   The spacing of the resynchronization markers in MPEG-4 is recommended based on the bit rates: For 24 Kbps it is recommended that they be inserted at intervals of 480 bits; for bit rates between 25 and 48 Kbps, every 736 bits.

[61]   Section 13.3.6 is based directly on the ISO MPEG-7 standard.

[62]   There can be other streams from content to user; these are not depicted here. Furthermore, it is understood that the MPEG-7 coded description may be textual or binary, because there might be cases where a binary efficient representation of the description is not needed, and a textual representation would suffice.

[63]   This section is based directly on MPEG-21 ISO documentation.

[64]   ISO/IEC TR 21000-1:2001(E), Part 1: Vision, Technologies and Strategy.

About the Author

At the time of this writing, Mr. Minoli was chief executive officer and founder of Leading Edge Networks Incorporated (LENI). LENI is a high-tech incubator/management company whose goal is to facilitate the deployment and operation of advanced leading-edge secure networking and IT technologies in the areas of wireless, metro Ethernet, and next generation ACDs and IP PBXs.

Mr. Minoli has many years of corporate experience with marquee end-user organizations and carriers. He has expertise in the total delivery of technology, systems, and applications related to secure, cost-effective corporate enterprise data networking, voice-over-IP, Internet-based systems, call center systems, and secure wireless LANs and WANs. On the carrier side, he has extensive experience in broadband communications, next generation optics, IP/ATM/FR/MPLS, QoS, metro Ethernet, voice-over-IP/ATM/FR/MPLS, Internet architectures and design, and multimedia applications.

Most recently, through the LENI incubator, Mr. Minoli was founder and/or cofounder of two high-tech companies: Global Wireless Services, a Hi-Fi hotspot IEEE 802.11b provider of mobile Internet/data services in high-end marinas, and InfoPort Communications Group, an optical and Gigabit Ethernet metro carrier. At Global Wireless Services, efforts entailed establishing the technical direction of the company; designing, deploying, and turning up access points, bridges, routers, and AAA servers; establishing a networkwide IP numbering plan; supporting VoIP services and Web portal services; undertaking site surveys and producing engineering packages; deploying encryption and LEAP-based security; configuring end users, placing them in service, and supporting them; supporting interactions with Internet providers as well with other vendors; and establishing company engineering, turnup, and provisioning processes. At InfoPort Communications, efforts entailed conceiving, launching, staffing, and funding the company; rapidly deploying a next generation metro-optical network in New York City; and establishing the technical direction of the firm as a multiple-architecture (RPR, RSTP, VLAN, GbE, 10GbE, CWDM, next generation SONET), multiple-technology, multiple-service (Ethernet private lines, VoIP, access, and so forth), multiple-QoS Gigabit Ethernet metro-level provider.

In the 1990s Mr. Minoli was a senior executive (vice president of packet services) at TCG/AT&T, where he deployed several large data networks with cumulative P&L $250 million capex, $75 million opex, $125 million direct revenue, and $4 billion of revenue impacted. Mr. Minoli started the broadband/IP services operation at Teleport Communications Group in late 1994; he was "Broadband Data Employee No. 2" and built the operation into a $25 million per year business. Mr. Minoli's team deployed 2,000 backbone/concentration/access routers and 100 ATM/FR switches in 20 cities in 5 years; the team turned up 1,500 active broadband ports and secured 400 broadband customers. Prior to AT&T/TCG, Mr. Minoli worked at DVI Communications (principal consultant), where he managed the deployment of three major corporate ATM networks, including one for the Federal Reserve Bank. From 1985 to 1994, Mr. Minoli worked at Bellcore/Telcordia (technology manager) where he thoroughly researched all aspects of broadband data networking, from service creation to standards writing and from architecture design to marketing and so on. In the mid-1980s he worked at Prudential Securities (assistant vice president), deploying dozens of data networks, including a 300-node VSAT branch data network. At ITT Worldcom (1980–1984), he deployed a fax-over-packet network and automated nearly all carrier operations by developing over a dozen OSSs, including Customer Care/CRM. At Bell Labs, he worked on internal and public packet networks, and at Network Analysis Corporation, he undertook ARPA work on voice-over-packet, wireless networks, and integrated communications.

Mr. Minoli has been a columnist for trade periodicals (*ComputerWorld*, *NetworkWorld*, and *Network Computing*) for several years. He has made presentations at 75 trade conferences and has written technology reports for *Gartner* over a 17-year period for CIOs and network planners. He has written five full-fledged market reports for *Probe Research Corporation*. He has documented his hands-on practitioner's technical expertise in 37 well-received industry-led books and more than 300 articles. Mr. Minoli has also taught graduate optical networking/datacom, voice-over-IP, multimedia, and e-commerce at Stevens Institute (1994–2003), New York University (1984–2000), Rutgers University (1992–1994), Carnegie Mellon University (1992–1993), and Monmouth University (1997) as an adjunct professor for 18 years. He is often sought after for advice by companies, patent attorneys, and venture capitalists, and has been involved in mezzanine investments, providing in-depth reviews of technology and market baseline for high-tech companies in the nanofabrication, multimedia, digital video, CTI/Java, VSAT, TMN mediation, and telemedicine arenas (for high-tech investments exceeding $150 million).

Index

For further information on these and other Artech House titles, including previously considered out-of-print books now available through our In-Print-Forever® (IPF®) program, contact:

Artech House
685 Canton Street
Norwood, MA 02062
Phone: 781-769-9750
Fax: 781-769-6334
e-mail: artech@artechhouse.com

Artech House
46 Gillingham Street
London SW1V 1AH UK
Phone: +44 (0)20 7596-8750
Fax: +44 (0)20 7630-0166
e-mail: artech-uk@artechhouse.com

Find us on the World Wide Web at:
www.artechhouse.com